# 3rd International Conference on Building Materials and Construction (ICBMC 2018)



IOP Conference Series: Materials Science and Engineering
Volume 371

Nha Trang, Vietnam
23 – 25 February 2018

ISBN: 978-1-5108-6449-8
ISSN: 1757-8981

**Printed from e-media with permission by:**

Curran Associates, Inc.
57 Morehouse Lane
Red Hook, NY 12571

**Some format issues inherent in the e-media version may also appear in this print version.**



Printed by Curran Associates, Inc. (2018)

For permission requests, please contact the Institute of Physics
at the address below.

Institute of Physics
Dirac House, Temple Back
Bristol BS1 6BE UK

Phone: 44 1 17 929 7481
Fax: 44 1 17 920 0979

techtracking@iop.org

**Additional copies of this publication are available from:**

Curran Associates, Inc.
57 Morehouse Lane
Red Hook, NY 12571 USA
Phone: 845-758-0400
Fax: 845-758-2633
Email: curran@proceedings.com
Web: www.proceedings.com

# Table of contents

## Volume 371

**2018 3rd International Conference on Building Materials and Construction**

**(ICBMC 2018)23–25 February 2018, Nha Trang, Vietnam**

**Accepted papers received: 22 May 2018**
**Published online: 19 June 2018**

### Preface



### Papers

**Building Materials**

**Civil and Urban Engineering**

**Construction Engineering and Management**

**Material Science and Engineering**

# PREFACE

The 3rd International Conference on Building Materials and Construction (ICBMC) held in Nha Trang, Vietnam from February 23-25, 2018 was well supported. ICBMC 2018 was dedicated to issues related to building materials and construction.

ICBMC 2018 presented a golden opportunity for everyone present to exchange new ideas and experiences, to establish business or research relations, and to identify global as well as regional partners for future collaboration. Distinguished professors were also invited to deliver keynote speeches relating to the latest developments and trends in their respective fields. The keynote speakers included Professor Low Sui Pheng from the National University of Singapore, Professor Zongjin Li from the University of Macau and Professor Jin-Keun Kim from Korea Advanced Institute of Science and Technology.

The conference has four sessions; namely with topics covering Building Materials and Structures; Material Science and Civil Engineering; Construction and Urban Engineering; as well as Concrete Materials and Technology. Participants have benefitted greatly from these four sessions that were presented successfully.

This set of conference proceedings presents a selection of papers submitted to the conference from various universities, research institutes and industries. All papers were subjected to peer-review by conference committee members and a panel of international reviewers. The papers were selected based on the quality and relevance to the conference. This volume presents recent advances in the field of civil and structural engineering, materials science and engineering, materials properties, measurement methods and applications etc.

We would like to thank all the authors who have contributed to this volume and also to the organizing committee, reviewers, speakers, co-chairpersons, sponsors and all the conference participants for their strong support to ICBMC 2018 to make this a successful event. We look forward to meeting you in ICBMC 2019!

Prof. Low Sui Pheng
National University of Singapore
March 5, 2018

IOP Conf. Series: Materials Science and Engineering **371** (2018) 011002 doi:10.1088/1757-899X/371/1/011002

# Organizing Committee

**Conference Co-Chairs**

Prof. Low Sui Pheng, National University of Singapore, Singapore

Prof Zongjin Li, University of Macau (UM), China

Prof. Jin-Keun Kim, Korea Advanced Institute of Science and Technology, South Korea

**Program Chairs**

Assoc. Prof. Wong Wah Sang, University of Hong Kong, Hong Kong

Assoc. Prof. Mohammad Arif Kamal, Aligarh Muslim University, India

**Organizing Chair**

Ms. Yang Dan, Hong Kong Chemical, Biological & Environmental Engineering Society, Hong Kong

**International Technical Committee**

Prof. Guillaume Habert, Swiss Federal Institute of Technology (ETH Zurich), Switzerland

Prof. Mahen Mahendran, Queensland University, Australia

Prof. Manicka Dhanasekar, Queensland University, Australia

Prof. Stephen Kajewski, Queensland University of Technology, Australia

Prof. Robin Drogemuller, Queensland University, Australia

Prof. Christos G. Papakonstantinou, University of Thessaly, Greece

Prof. Walid Larbi, Conservatoire National des Arts et Métiers, France

Prof. Kim Van Tittelboom, Ghent University, Belguim

Prof. Zbyšek Pavlík, Czech Technical University, Czech Republic

Prof. Milena Pavlíková, Czech Technical University in Prague, Czech Republic

Prof. Tan Kiang Hwee, National University of Singapore, Singapore

Prof. Murat Dicleli, Middle East Technical University, Turkey

Prof. Kefei Li, Tsinghua University, China

Prof. Peng Pan, Tsinghua University, China

Prof. Cheng Sun, Harbin Institute of Technology, China

Prof. Chien-Yuan Chen, National Chiayi University, Taiwan

Prof. Ahsan Habib, The University of Tokyo, Japan

Prof. Ippei Maruyama, Nagoya University, Japan

Prof. Myunghoun Jang, Jeju National University, South Korea

IOP Conf. Series: Materials Science and Engineering **371** (2018) 011003 doi:10.1088/1757-899X/371/1/011003

# Peer review statement

All papers published in this volume of *IOP Conference Series: Materials Science and Engineering* have been peer reviewed through processes administered by the proceedings Editors. Reviews were conducted by expert referees to the professional and scientific standards expected of a proceedings journal published by IOP Publishing.

IOP Conf. Series: Materials Science and Engineering **371** (2018) 012001 doi:10.1088/1757-899X/371/1/012001

# Synergy Action of Glass Powder and Foaming Additive in Production of Lightweight Cement-Based Materials

**J Pokorný[1], M Pavlíková[1], M Záleská[1], J. Studnička[1] and Z Pavlík[1]**

1 Faculty of Civil Engineering, Czech Technical University in Prague, Thákurova 7, 166 29 Prague 6, Czech Republic
E-mail: pavlikz@fsv.cvut.cz

**Abstract.** Waste glass in the form of finely grounded powder can be, due to its assumed pozzolanic behaviour, used in cement and concrete industry as a Supplementary Cementing Material (SCM) helping to reduce the consumption of natural resources and emission of greenhouse gases. In this sense, the presented paper is focused on the use of waste borosilicate glass in production of cement-based lightweight composites. For both Portland cement and glass admixture, chemical composition, pozzolanic activity and basic physical characteristics were accessed. In composite mixes, cement binder was partially replaced with glass powder in the amount of 15 and 20 mass %. In order to produce lightweight materials, foaming agent in the amount of 0.58 ml/kg of blended binder was used. For the developed materials, bulk density, matrix density, total open porosity, mechanical resistance, water transport parameters and thermal properties were tested The obtained experimental results revealed pozzolanic activity of micro scale milled waste glass that led to the decrease in porosity and thus improved mechanical resistance of the investigated composites. High porosity of foamed concretes significantly improves their thermal-insulation function and helps to save raw materials for their production.

## 1. Introduction

Concrete as well as other cement-based materials belong among worldwide spread man-made building materials used not only in 20$^{th}$ century but also to these days. About 800 million tons of concrete was consumed in the U.S. in 2007, and the world consumption was estimated at 11 billion tons. It approximately corresponds to 1.7 ton for every living human being [1]. Excessive cement and concrete production is closely tied with high nature resources consumption as well as energy claims. In addition, cement production as highly polluting process contributes about 5-8 % to global $CO_2$ emissions [2].

The application of supplementary cementitious materials in cement and concrete production introduce a promising method for mitigating environmental impacts and in many cases providing economic benefits. In general, concrete practice utilizes a wide variety of SCMs as in the form of active mineral admixtures as well as filler materials. Active powders, so called pozzolans, are defined according to the standard EN 197-1 (2011) [3], compartmentalized into nature and technogenic (artificial) material sort. Among very often incorporated admixtures is considered fly ash, silica fume, diatomite powder, blast furnace slag, etc. which are consumed in huge amount [4], [5], [6].

Glass waste represents urgent environmental problem all over the world from the point of its nonbiodegradable origin. This unused commodity is left on spread landfills releasing pollutants influencing water and soil [7]. In 2007, the world total production of glass was estimated about 89.4 million tons. From this amount European Union contributes by 38.3 million tons, corresponding approximately to 30 % of the total word's output. About 83 % of glass production is attributed to container and flat glass [8]. From a sustainable point of view recycling procedure provides a suitable solution in waste glass reduction, nevertheless, this process is limited for different types of glass and their sorted combinations. Glass contains significant amount of silica and as a finely milled waste material can be very interesting alternative to common SCMs. However, its properties, such as chemical composition and particle size, affect its pozzolanic character. In this respect some authors, such as Khmiri et al. [9] and Shi et al. [10], recommend to mill the glass material with the help of ball mill to micro size range with particle size below 100 μm.

Glass powder incorporation into cement based materials positively helps to improve workability of fresh mix [1], however, recommended dosage supporting compressive strength is usually between 10 up to 20 mass %. Nassar and Soroushian [11] examined a different percentage substitution of cement by glass powder in portion of 15; 20 and 23 mass %. The compressive strength results revealed higher strengths for concretes containing 15 and 20 mass % of waste glass compared with control samples.

Nevertheless, ordinary produced concretes cannot always meet the specific requirements of civil engineers. In recent years, the trend of application of lightweight concretes in construction has rapidly increased. This building material excels in a low content of input substances for its manufacturing as well as due to its advanced properties, such as low weight and very good sound and thermal insulation function. Together with fire resistance are lightweight composites predetermined as sound and thermal insulation layers, material for production of lightweight blocks and an infill material for lightweight composite panels [12], [13].

This paper is aimed at application of waste borosilicate glass in the form of fine powder in production of lightweight concretes. Developed materials were characterized by their physical properties, mechanical and liquid water transport parameters as well as their thermal properties. Experimental investigation carried out on both ordinary fine-grained and foamed concretes revealed positive effect of glass powder that acted as a suitable cement replacing material and foaming technology enabled produce an environmentally sustainable material with thermal-insulation function.

## 2. Experimental

### *2.1. Used materials*

The composites preparation was done in two separate steps. At first, the basic reference mix containing Portland cement and fine silica aggregate was enriched by glass powder. Subsequently, the foaming additive was admixed into both reference and modified concrete mix. As a main binder substance, Portland cement 42.5 R produced by Heidelberg Cement Group Radotin, Czech Republic, was used. Part of cement was substituted by an artificially prepared borosilicate glass powder (GP) in the dosage of 15 and 20 mass %. Collected discarded glass vessels were washed in an ultrasonic bath, dried, manually crushed and milled in a disc mill. After milling, newly made waste glass powder disposed of an average particle size of 48.3 μm.

Chemical composition of applied materials is presented in Table 1.

**Table 1.** Chemical composition of used powders.

| Substance | CEM 42.5 R | GP |
|---|---|---|
| | Content (mass %) | |
| $SiO_2$ | 19.00 | 78.5 |
| $Al_2O_3$ | 4.31 | 2.94 |
| $Fe_2O_3$ | 2.40 | 0.19 |
| CaO | 62.90 | - |
| MgO | 1.80 | - |
| $K_2O$ | 0.82 | 1.09 |
| $Na_2O$ | 0.14 | 4.55 |
| $TiO_2$ | 0.28 | - |
| $SO_3$ | 3.24 | 10.50 |
| $P_2O_5$ | 0.16 | 0.16 |

IOP Conf. Series: Materials Science and Engineering **371** (2018) 012001 doi:10.1088/1757-899X/371/1/012001

The binders' composition was accessed on the basis of classical chemical analysis. The particular materials were melted together with $LiBO_2$, and then dissolved into the solution. The solution composition was analysed with ICP spectrometer.

Basic material characteristics determined for both used binder materials are given in Table 2. The values of specific surfaces were measured according the standard EN 196-6 (2010) [14] with the help of Blaine apparatus. Obtained data revealed approximately 1.4 higher specific surface of glass powder compared to Portland cement. Pozzolanic activity of milled glass was determined by the modified Chapelle test (NF P 18-513:2012 [15]) in which reacts 1g of tested powder material with 2g CaO in water. In general, added mineral admixture is considered pozzolana active when 1 g of sample absorbs more than 650 mg of $Ca(OH)_2$.

**Table 2.** Basic material characteristics determined for Portland cement and glass powder.

| Material | Specific surface ($m^2kg^{-1}$) | Density ($kgm^{-3}$) | Loss on ignition at 1000 °C (%) | Pozzolanic activity ($mgCa(OH)_2 1g^{-1}$) |
|---|---|---|---|---|
| CEM I | 360 | 3 129 | 0.40 | - |
| GP | 517 | 2 345 | 0.02 | 719 |

*2.2. Developed mixtures and sampling*

All prepared mixes contained three fractions of silica sand 0.0/0.5; 0.5/1.0 and 1.0/2.0 (Filtrační písky, Ltd.) which were combined together in mas ratio 1/1/1. The amount of batch water was kept constant with water/binder ratio 0.5. In case of foamed lightweight composites, foaming additive (FA) MasterCell 285 produced by BASF Stavební hmoty, Ltd., Czech Republic was added to batch water in the amount of 0.58 ml/kg of binder. Chemical additive contained 30 mass % of a dry matter and had density of 1.12 $kgm^{-3}$. Residual water of foaming agent was deducted from the total amount of batch water. Composition of all developed mixes is given in Table 3.

**Table 3.** Composition of developed mixes.

| Substance | Content in $kg/m^3$ | | | | | |
|---|---|---|---|---|---|---|
| | Dense composites | | | Foamed composites | | |
| | R-C | GP 15-C | GP 20-C | R-F | GP 15-F | GP 20-F |
| CEM I 42.5 | 496.2 | 421.8 | 397.0 | 320.8 | 272.7 | 256.6 |
| GP | - | 74.4 | 99.2 | - | 48.1 | 64.2 |
| Sand 0.0/0.5 | 496.2 | 496.2 | 496.2 | 320.8 | 320.8 | 320.8 |
| Sand 0.5/1.0 | 496.2 | 496.2 | 496.2 | 320.8 | 320.8 | 320.8 |
| Sand 1.0/2.0 | 496.2 | 496.2 | 496.2 | 320.8 | 320.8 | 320.8 |
| FA | - | - | - | 0.22 | 0.22 | 0.22 |
| Water | 248.1 | 248.1 | 248.1 | 160.2 | 160.2 | 160.2 |

From Table 3, there are evident savings in dosage of input raw materials for production of lightweight concretes. It was due to the use of foaming additive that allows increase of material volume compared to the reference concrete mix.

Mixing procedure included dosage of dry raw materials in a mixer with vertical axis Spar D 200-Band and their mixing at speed 96 rpm (1st speed regime) for 1 min. After that, batch water with or without liquid foaming agent was added and mixing procedure continued another 1 minute. Then, fresh mix was manually mixed in order to remove the scum from the bottom of mixing vessel. Materials without foaming were after that mixed in 2nd speed regime (196 rpm) for 2 min and their mixing was finished in 1st speed regime that was used for 1 min. Preparation of foamed mixes was finished in 2nd speed regime applied for 4 min.

From fresh mixes were casted prisms having dimensions of 40 × 40 × 160 mm and cubes with side of 70 mm. Samples prepared from dense mixes were casted in two layers, whereas each layer was compacted on vibrating table for 30 s.

In case of foamed concrete specimens, every layer was compacted by 3 hits on a table. Casted specimens were left in laboratory conditions at temperature of 23 ± 1 °C and 45 ± 5 % of relative humidity. In order to prevent excessive evaporation of batch water and autogenous shrinkage, the specimens were covered with PE foil for 1 day. Then these were unmoulded and stored in water (dense concrete samples) or in laboratory conditions covered with PE foil and water sprayed ones per day (foamed concrete samples) for overall time of 28 days.

### *2.3. Measured properties*

*2.3.1. Particle size distribution.* Except specific surface, cement and GP were also characterized by their particle size distribution. For the measurement, red and green laser equipped analyser Analysette 22 – Micro Tec Plus (Fritch) was applied. This apparatus is able to detect grains with diameter from 2 nm up to 2 000 μm. In the measurement, spherical grains' shape is assumed.

*2.3.2. Penetration time.* The influence of glass admixture as well as consequent effect of chemical additive on rheology of fresh concretes were characterized by the measurement of penetration time in accordance with the technical standard EN 1015-9 (1999) [16], where the value of compressive stress caused by penetration cylinder is supposed to reach 0.5 MPa. This stress is considered as a border value for suitable concrete processing. Experiments were performed at 23 ± 1 °C for minimum two samples of each studied concrete mix.

*2.3.3. Basic physical properties.* For dried hardened specimens cured for 28 days, basic physical parameters, such as bulk density, matrix density and total open porosity, were accessed. The bulk density measurements were performed with utilization of gravimetric method according to the standard EN 1015-10 (1999) [17]. The relative expanded uncertainty of the density test was 2.3 %. For determination of the matrix density, device Pycnomatic ATC (Thermo Scientific) working on a helium pycnometry principle, was used. The total open porosity was calculated on the basis of measured matrix and bulk densities. The relative expanded uncertainty of the total open porosity test was 3.6 %.

*2.3.4. Strength parameters and dynamic moduli measurements.* Mechanical resistance of the examined materials was studied by flexural and compressive strength tests that were performed according to the standard EN 1015 – 11 (1999) [18]. Three point bending test was realised on three prismatic samples with dimensions of 40 × 40 × 160 mm. On the rest of broken prisms, the compressive strength was measured. The expanded measuring uncertainty of the both strength tests was 2 % [19]. Among mechanical parameters testing, dynamic moduli measurement according to the standard EN 12504-4 was done. For this purpose, device Dio 562 NLF (Starmans Electronics) was applied [20]. The relative expanded uncertainty of the dynamic modulus measurement was 2 %.

*2.3.5. Liquid water transport properties.* One dimensional free water uptake test was carried out according the procedure described in the standard EN 1015-18 (2002) [21]. Hardened cubic samples were previously water and vapour proof insulated by epoxy resin on all lateral sides. The face side of each sample was immersed 1-3 mm into water on the top of a water reservoir, allowing samples to suck water from free water surface. The sample mass was measured in the chosen times interval. The water absorption coefficient $A$ ($kgm^{-2}s^{-1/2}$) was calculated from the linear part of the dependence of the increasing sample mass on the square root of suction time. From the known weight of dry and fully saturated sample, the saturated moisture content value $w_{sat}$ ($kgm^{-3}$) was calculated. The apparent moisture diffusivity $\kappa_{app}$ ($m^2s^{-1}$) was calculated based on $w_{sat}$ and $A$ data using procedure reported by Roels at al. [22].

*2.3.6. Thermo-physical properties.* Among the thermal properties, thermal conductivity $\lambda$ ($Wm^{-1}K^{-1}$), thermal diffusivity $a$ ($m^2s^{-1}$) and volumetric heat capacity $c$ ($Jm^{-3}K^{-1}$) were measured with the help of commercially produced device ISOMET 2114 (Applied Precision) working on a dynamic measurement principle [23]. The measuring range of the thermal conductivity is from 0.015 to 6 $Wm^{-1}K^{-1}$. The accuracy is 5 % of reading + 0.001 $Wm^{-1}K^{-1}$. The measurement reproducibility is 3 % of reading in operation temperature range from 0 to 40°C. The measuring range of the volumetric heat capacity is $4 \times 10^4 - 4 \times 10^6$ $Jm^{-3}K^{-1}$ with accuracy 15 % of reading + $1 \times 10^3$ $Jm^{-3}K^{-1}$. The reproducibility of the volumetric heat capacity measurement is 3 % of reading + $1 \times 10^3$ $Jm^{-3}K^{-1}$.

## 3. Results and discussion

Particle size distribution curves of cement and glass powder are graphed in Figure 1. The both measured distribution curves exhibit bimodal character with two local maxima. In case of Portland cement, curve peaks correspond to the particle diameter 8 μm and 30.2 μm respectively. Curve characterising glass admixture is more spread with local maxima at 5 μm and 80.4 μm. In comparison with cement binder, used mineral admixture contains a considerable amount of coarser particles that may also bring positive filler effect. On the other hand, higher volume of coarser grains is generally linked with a lower pozzolanic activity of SCMs [24].

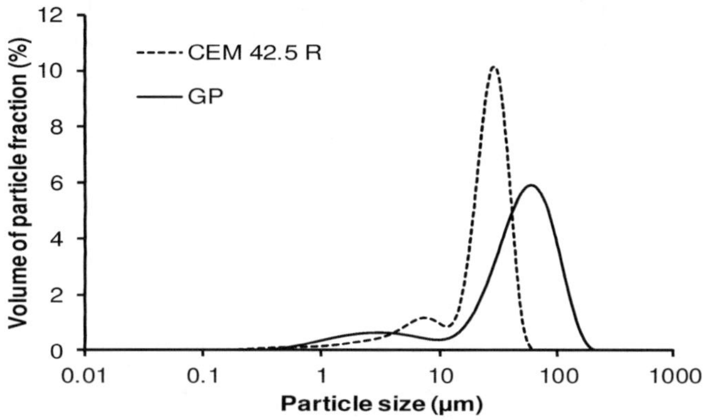


**Figure 1.** Particle size distribution of cement and glass powder.

Basic physical properties measured on hardened samples cured for 28 days in a laboratory or in water are summarized in Table 4. Portland cement substitution by milled glass brought reduction in the total open porosity especially in case of dense fine-grained concretes. The observed differences in porosity were 8.1 % for GP 15-C and 10.6 % for GP 20-C related to the reference material (R-C). This behaviour is attributed to the pozzolanic activity of glass and additionally to the partial filler effect due to the presence of coarser GP grains. On the other hand, combined action of mineral admixture and foaming agent used in production of lightweight samples caused only slight decrease in porosity that varied in the relative expanded uncertainty of the applied testing method. In comparison with hardened concretes, foamed materials reached approximately three times higher porosity values. This data points to the significant weight decrease of foamed materials compared to the reference concrete. Based on that, better thermal-insulation function of lightweight concretes can be anticipated.

**Table 4.** Basic physical properties measured on hardened composites.

| Material | Bulk density ($kgm^{-3}$) | Matrix density ($kgm^{-3}$) | Total open porosity (%) |
|---|---|---|---|
| R-C | 2 107 | 2 511 | 16.1 |
| GP 15-C | 2 105 | 2 471 | 14.8 |
| GP 20-C | 2 110 | 2 465 | 14.4 |
| R-F | 1 336 | 2 495 | 46.5 |
| GP 15-F | 1 337 | 2 491 | 46.3 |
| GP 20-F | 1 331 | 2 485 | 46.4 |

IOP Conf. Series: Materials Science and Engineering **371** (2018) 012001 doi:10.1088/1757-899X/371/1/012001

Table 5 shows mechanical parameters and penetration time data measured for examined concrete samples. Obtained mechanical parameters, represented by the flexural strength, compressive strength and dynamic moduli exhibited distinctly increasing character for dense concrete samples modified with GP. Similar improvement in mechanical resistance we observed also for lightweight concretes, however, the differences compared to the reference foamed concrete were not such remarkable. In general, the measured mechanical parameters correspond with the total open porosity data. From the quantitative point of view the lightweight materials reached low strength, nevertheless, they are worth because of their specific advanced properties.
Incorporation of GP into fresh concrete mixes has no negative impact on workability when the values of spreading oscillated between 160/160 mm and 180/180 mm. However, significant changes in workability were recorded from the point of view of measured penetration times. One can see a gradual shortening of penetration time with the increasing dosage of GP in the dense concrete mix. On the other hand, penetration times measured for foamed mixes were distinctively higher compared to the dense concrete mixes. The prolongation of penetration time was a consequence of encased air bubbles network uniformly dispersed in a fresh mix. Certain effect may be also attributed to foaming agent, which as a chemical additive might possibly affect hydration process of Portland cement.

**Table 5.** Strength characteristics and penetration times accessed after 28 days.

| Material | Flexural strength (MPa) | Compressive strength (MPa) | Dynamic moduli (GPa) | Penetration time (min) |
|---|---|---|---|---|
| R-C | 9.1 | 54.1 | 32.2 | 145 |
| GP 15-C | 11.2 | 59.2 | 34.3 | 130 |
| GP 20-C | 11.5 | 61.2 | 35.4 | 115 |
| R-F | 1.7 | 4.7 | 4.6 | 355 |
| GP 15-F | 2.0 | 4.9 | 4.8 | 480 |
| GP 20-F | 1.8 | 4.7 | 4.6 | 460 |

Material parameters characterizing the liquid water transport are expressed in Table 6. Results obtained for dense concretes with GP showed a slight decrease in the water absorption coefficient and thus sorptivity. This was accompanied by the drop in the capillary moisture content when the lowest values were recorded for concrete with 20 mass % of GP. On the contrary, foamed concretes exhibited practically unchanged water transport characteristics regarding to their initial state. As evident from different structural nature, lightweight samples, as more porous materials, are able to easily transport liquid water which clarifies their increased apparent moisture diffusivity.

**Table 6.** Liquid water transport properties.

| Material | $A$ ($kgm^{-2}s^{-1/2}$) | $w_{cap}$ ($kgm^{-3}$) | $\kappa_{app}$ ($m^2s^{-1}$) | $S$ ($ms^{-1/2}$) |
|---|---|---|---|---|
| R-C | 0.018 | 174.4 | $1.09 \times 10^{-8}$ | $1.82 \times 10^{-5}$ |
| GP 15-C | 0.017 | 172.2 | $9.74 \times 10^{-9}$ | $1.70 \times 10^{-5}$ |
| GP 20 C | 0.016 | 169.9 | $8.92 \times 10^{-9}$ | $1.61 \times 10^{-5}$ |
| R-F | 0.025 | 455.2 | $3.04 \times 10^{-9}$ | $2.52 \times 10^{-5}$ |
| GP 15-F | 0.024 | 450.0 | $2.87 \times 10^{-9}$ | $2.41 \times 10^{-5}$ |
| GP 20-F | 0.025 | 453.0 | $3.14 \times 10^{-9}$ | $2.55 \times 10^{-5}$ |

Thermal properties measured on dried cubic samples are summarized in Table 7. Compared to the reference material R-C, a gradual increase in both the thermal conductivity and diffusivity with increasing amount of GP in dense concretes was observed. This is in agreement with results presented above and clearly corresponds with porosity data. Moreover, the explanation of heat transport parameters may be provided by comparison of the thermal conductivities measured for all dry input substances. The particular raw materials were placed into glass cylinder and vibrated for 10 sec.

Obtained data revealed that the average value of the thermal conductivity for borosilicate glass powder was 0.84 $Wm^{-1}K^{-1}$; 0.10 $Wm^{-1}K^{-1}$ for cement, and 0.56 $Wm^{-1}K^{-1}$ in case of silica sand mix.
As was expected on the basis of measured open porosities, the thermal conductivity values accessed for foamed samples were three times lower contrary to the dense concretes. On this account, lightweight composites are able to provide much better thermal-insulation function than ordinary concretes.
For all analysed concrete samples, the volumetric heat capacity corresponded to the porosity data and other investigated material characteristics.

**Table 7.** Thermo-physical properties of tested composites.

| Material | $\lambda$ ($Wm^{-1}K^{-1}$) | $a$ ($\times 10^{-6}\ m^2s^{-1}$) | $c$ ($\times 10^6\ Jm^{-3}K^{-1}$) |
|---|---|---|---|
| R-C | 2.10 | 1.14 | 1.84 |
| GP 15-C | 2.15 | 1.16 | 1.85 |
| GP 20-C | 2.5 | 1.40 | 1.86 |
| R-F | 0.76 | 0.48 | 1.59 |
| GP 15-F | 0.81 | 0.50 | 1.60 |
| GP 20-F | 0.76 | 0.47 | 1.59 |

## 4. Conclusions

A combined effect of waste borosilicate glass powder and foaming additive on properties of both fresh and hardened concretes was examined. Tests performed on fresh non-foamed mixtures showed a gradual shortening of penetration times with increasing portion of glass admixture attributed to its pozzolanic activity. On the other hand, foamed mixes reached desired penetration pressure much later due to presence of encased bubbles network. Observed decreasing porosity values corresponding with the better mechanical resistance determined for dense concrete samples indicated positive filler effect and pozzolanic activity of glass powder. Three times higher porosity of foamed concretes led to the significant reduction in strength characteristics, however, on the other hand to the enhanced thermal-insulation function. Lightweight foamed composites disposing with advanced properties may provide a possible material sort applicable for specific construction requests.
Our future work will be aimed at the possible use of different kinds of industrial by-products as mineral pozzolanic admixtures for production of foamed lightweight composites and research on the synergic performance of pozzolanic materials and foaming agents of different origin.

## 5. **References**

[1] Aliabdo A A, Elmoaty A E M A, Aboshama A Y 2016 *Constr. Build. Mater.* **124** 866
[2] Pade C and Guimaraes M 2007 *Cem. Concr. Res.* **37** 1348
[3] EN 197-1 Cement – Part 1: Composition, specifications and conformity criteria for common cements, CEN 2011
[4] Islam G M S, Rahman M H and Kazi N 2017 *Int. J. Sustain. Built. Environ.* **6** 37
[5] Kastis D, Kakali G, Tsivilis S and Dtamatakis M G 2006 *Cem. Concr. Res.* **36** 1821
[6] Pokorny J, Pavlikova M, Zaleska M, Rovnanikova P, Pavlik Z Coagulated Silica - a-SiO2 admixture in cement paste *AIP Conf. Proceed.* **1752** 040023
[7] Jani Y and Hogland W 2014 *J. Environ. Chem. Eng.* **2** 1767
[8] Schmitz A, Kaminski J, Scalet B M and Soria A 2011 *Energy Policy* **39** 142
[9] Khmiri A, Samet B and Chaabouni M *Constr. Build. Mater.* **28** 680
[10] Shi C, Wu Y, Riefler Ch and Wang H 2005 *Cem. Concr. Res.* **35** 987
[11] Nassar R and Soroushian P 2012 *Constr. Build. Mater.* **29** 368
[12] Nguyen T T, Bui H H, Ngo T D and Hguyen G D 2015 *Mater. Des.* **130** 103
[13] Flores-Johnson E A and Li Q M 2012 *Compos. Struct.* **94** 1555
[14] EN 196-6 Methods of testing cement – Part 6: Determination of fineness, CEN 2010
[15] NFP 18 513:2012 Pozzolanic Addition For Concrete - Metakaolin - Definitions, Specifications And Conformity Criteria, AFN, 2012

[16] EN 1015-9 Methods of test for mortar for masonry - Part 9: Determination of workable life and correction time of fresh mortar, CEN 1999
[17] EN 1015-10 Methods of test for mortar for masonry - Part 10: Determination of dry bulk density of hardened mortar, CEN 1999
[18] EN 1015 – 11 Methods of test for mortar for masonry - Part 11: Determination of flexural and compressive strength of hardened mortar, CEN 1999
[19] Pokorny J, Pavlikova M, Medved I, Pavlik Z, Zahalkova J and Rovnanikova P 2016 Influence of various amount of diatomaceous earth used as cement substitute on mechanical properties of cement past *AIP Conf. Proceed.* **1738** 280007
[20] Pavlik Z, Keppert M, Pavlikova M, Zumar J, Fort J and Cerny R 2014 *Cement Wapno Beton* **19** 67
[21] EN 1015-18 Methods of test for mortar for masonry - Part 18: Determination of water absorption coefficient due to capillarity action of hardened mortar, CEN 2002
[22] Roels S., Carmeliet J., Hens H., Adan O., Brocken H., Cerny R., Pavlik Z., Hall Ch., Kumaran K., Pel L. and Plagge, R. 2004. *J. Therm. Env. Build. Sci.* **27** 307
[23] Pavlik Z, Trnik A, Ondruska J, Keppert M, Pavlikova M, Volfova P, Kaulich V and Cerny R 2013 Int. J. Thermophys. **34** 851
[24] Afshinnia K and Rangaraju P R 2015 *Constr. Build. Mater.* **81** 257

**Acknowledgments**
The authors gratefully acknowledge the financial support received from the Czech Science Foundation, under project No 17-04215S and by the CTU in Prague under project No SGS17/166/OHK/3T/11.

IOP Conf. Series: Materials Science and Engineering **371** (2018) 012002 doi:10.1088/1757-899X/371/1/012002

# Effect of Waste Expanded Polypropylene-Based Aggregate on Mechanical and Thermal Properties of Lightweight Concrete

**M Záleská[1], M Pavlíková[1], J. Studnička[1] and Z Pavlík[1]**

1 Department of Materials Engineering and Chemistry, Faculty of Civil Engineering, Czech Technical University in Prague, Thákurova 7, 166 29 Prague 6, Czech Republic
E-mail: pavlikz@fsv.cvut.cz

**Abstract.** Resource efficiency and building energy consumption is in recent years an open question for construction industry. Recycled plastic aggregate can be used as an alternative aggregate to produce lightweight concrete with low environmental impact of both concrete and waste plastic and moreover, the obtained lightweight concrete is characterized by improved thermal insulation properties. In the present study, waste expanded polypropylene (EPP) aggregate is used for partial replacement of natural aggregate to produce lightweight cement-based composites containing synthetic coagulated amorphous $SiO_2$ as supplementary cementitious material. For the applied EPP, basic physical properties were measured and specific attention was paid to thermal transport and storage properties that were studied in dependence on compacting time. The effect of EPP content on the mechanical characteristics and on the thermal properties of developed material was investigated for 28 and 90 days water cured samples. Composite with incorporated EPP aggregates exhibits enhanced thermal insulation properties with sufficient mechanical resistance and can be considered as promising material for building subsoil or floor structures.

## 1. Introduction

The building industry has consumed a vast amount of natural resources and also been responsible for a significant energy usage. This is expected to increase further with economic development where more people will need quality housing. Therefore, any building material that minimizes the usage of natural resources or use waste materials to a certain extent could have promising future [1].

Since the 20th century, plastics have been used increasingly in a large range of products because of their favourable properties, including low density, high strength-to-weight ratio, high durability, ease of design and manufacture, and low cost [2]. The top three markets for plastics are packaging, building and construction and automotive. However, large amounts of plastic waste give much pressure on the environment due to very low biodegradability of plastics [3]. The global production of plastics in 2014 was around 311 Mt. In 2014 in Europe, 25.8 Mt of post-consumer plastic waste ended up in the waste upstream. Plastic waste recycling in Europe have increased to almost 30 %, however landfilling and energy recovery still constitutes a large part of plastic waste disposal [4].

Reuse of waste and recycled plastic materials in concrete mix is able to combine both economic growth and environmental protection. A large number of studies undertaken on concrete containing waste plastic particles have demonstrated that plastic aggregate can reduce the cost and significantly improve unit weight and the thermal and acoustic insulation parameters of developed composites [5], [6], [7]. On the other hand, many studies reported on worsening of mechanical properties of concrete with incorporated plastic waste aggregate [2], [6], [8].

IOP Conf. Series: Materials Science and Engineering **371** (2018) 012002 doi:10.1088/1757-899X/371/1/012002

Since most of the published literature is focused on the use of polyethylene terephthalate (PET) and high density polyethylene (HDPE) aggregates, and because the recycle ratio for polypropylene (PP) is only 0.6 %, which makes it one of the least recycled post-consumer plastics [9], this paper is concerned to study the effect of expanded PP particles as partial silica sand replacement for the development of lightweight concrete with improved thermal insulation properties.

## 2. Experimental

Expanded polypropylene used in this study was obtained as shredded waste from aircraft model production. Basic physical and thermal properties of EPP aggregate were measured at first. Lightweight concrete was then prepared using Portland cement CEM I 42.5 R (Czech-Moravian cement, Ltd., HeidelbergCement group, Mokrá plant), silica sand (Filtrační písky, Ltd., affiliate Chlum u Doks), synthetic coagulated amorphous $SiO_2$ (Silica VP4, AV EKO-COLOR, Ltd.), water and EPP waste, which was used as partial replacement of silica sand of 60 % by volume. Dynamon SX 45 (Mapei Ltd.) superplasticizer was used to achieve good workability. As ultra-lightweight aggregate tends to float and can result in a poor mix distribution and segregation [10], Silica VP4 as admixture for improvement of dispersion of EPP particles in the cement matrix was used. The silica sand was mixed from three size of sand with granulometry PG1 (0.06 3/0.5 mm), PG2 (0.08/1 mm) and PG3 (1/2 mm) respectively in weight ratio 1:1:1. The water/cement ratio was 0.5 for all investigated mixtures. For comparative reasons, the reference sample without EPP was prepared as well. Table 1 gives chemical characteristics of cement and silica VP4. In Table 2, composition of tested mixtures is shown.

**Table 1.** Chemical composition of cement and silica VP4.

| Chemical composition (mass %) | Cement CEM I 42.5 R | Silica VP4 |
|---|---|---|
| $SiO_2$ | 19.5 | 91.1 |
| $Al_2O_3$ | 4.7 | 0.1 |
| $Fe_2O_3$ | 3.2 | 0.03 |
| CaO | 64.2 | 0.00 |
| MgO | 1.3 | 0.00 |
| $Na_2O$ | 0.09 | 0.78 |
| $K_2O$ | 0.78 | < 0.01 |
| $SO_3$ | 3.2 | 0.00 |
| Loss on ignition (LOI) | 3.3 | 4.75 |

**Table 2.** Mix design of studied lightweight concretes.

| | CEM (g) | VP4 (g) | Sand (ml) | EPP (ml) | Water (g) | Dynamon (g) | w/c |
|---|---|---|---|---|---|---|---|
| **Ref** | 1387.5 | 112.5 | 3300 | 0 | 750 | 30 | 0.5 |
| **EPP** | 1387.5 | 112.5 | 1300 | 2000 | 750 | 30 | 0.5 |

The casted prismatic samples 40 × 40 × 160 mm and cubic samples with side dimension of 70 mm were left for 1 day at laboratory conditions and then they were unmoulded and cured for 28 days (Ref28, EPP28) resp. 90 days (Ref90, EPP90) in water.

### *2.1. EPP aggregates*

For the basic physical characterization of EPP grains, their powder density, matrix density and grain-size distribution were measured. The matrix density was determined using automatic helium pycnometer Pycnomatic ATC (Thermo Scientific) with fully integrated temperature control with precision of ± 0.01 °C and real multi volume density analyser. The matrix density achieved value of 105 $kgm^{-3}$, which ranks EPP aggregates among artificial ultra-lightweight (density of less than 300 $kg/m^{-3}$) nonabsorbent aggregate as classified by Babu and Babu [10]. The grain-size analysis was realized using standard sieve method with sieves of following mesh dimensions: 0.063; 0.125; 0.25; 0.5; 1.0; 2.0; 4.0; 8.0; 16.0; 31.5 and 63.0 mm. Obtained results shows that size of all EPP particles is below 8 mm, what is favourable for their use as partial replacement of natural aggregate.

To determine thermal insulation properties of EPP aggregates, powder density and thermal properties namely thermal conductivity $\lambda$ ($Wm^{-1}K^{-1}$), volumetric heat capacity $C_v$ ($Jm^{-3}K^{-1}$) and thermal diffusivity $a$ ($m^2s^{-1}$) of EPP particles were measured in dependence on compacting time. During the measurement, EPP aggregate was put into graduated cylinder and compacted using the vibration exciter. From the known mass of the sample in cylinder and its volume, the powder density was evaluated. To determine thermal parameters, the commercially produced device ISOMET 2114 (Applied Precision, Ltd.) was used as a representative of transient impulse methods [11], [12]. The measurement is based on an analysis of the temperature response of the analysed material to the heat flow impulses. For the measurement of EPP aggregate, needle probe was used. The measurement accuracy was 5 % of reading + 0.001 W/mK for the thermal conductivity in the range 0.015-0.70 $Wm^{-1}K^{-1}$ and 10 % of reading for the thermal conductivity ranging from 0.70 $Wm^{-1}K^{-1}$ to 6.0 $Wm^{-1}K^{-1}$. The accuracy of the volumetric heat capacity was 15 % of reading + $1 \times 10^3$ $Jm^{-3}K^{-1}$.

### *2.2. Basic physical characteristics*

As fundamental physical characteristics of prepared composites, bulk density, matrix density and total open porosity were measured. Samples were first dried in a vacuum drier at 60 °C. The matrix density was measured using automatic helium pycnometer as described above. The bulk density was determined using the measurement of sample sizes and dry mass of the sample according to EN 12390-7 [13]. The total open porosity was determined on the basis of bulk density and matrix density measurement. The relative expanded uncertainty of porosity test was 5 %.

### *2.3. Mechanical properties*

The mechanical properties of casted materials were evaluated on 40 × 40 × 160 mm prismatic samples, which were cured 28 resp. 90 days in water. Mechanical resistance of researched concretes was characterized by compressive strength, flexural strength and Young´s modulus. The flexural strength was tested according the standard EN 12390-5 [14]. The compressive strength was determined by following the standard EN 12390-3 [15] on the broken halves of samples from the flexural strength tests. The loading area was 40 × 40 mm. The relative expanded uncertainty of the both testing methods was 1.4 %. The Young´s modulus was obtained on a dynamic principle using the pulse ultrasonic method. This method is based on the measurement of travel time of ultrasonic wave launched from the device and passing through the material. The measurement was done on 40 × 40 × 160 mm prismatic samples in a longitudinal direction using device DIO 562 (Starman's Electronics) working on the frequency of 50 KHz [16]. Before the ultrasonic measurement, the samples were dried in a vacuum drier at 60 °C. The relative expanded uncertainty of the dynamic Young's modulus measurement was 2 %.

### *2.4. Thermal properties*

In order to quantify the influence of the plastic aggregates on improvement of heat transport and storage properties of prepared composites, their thermal conductivity, volumetric heat capacity and thermal diffusivity were obtained using device ISOMET 2114 (see above). Before the particular tests, the samples were dried at 60 °C in a vacuum drier until their constant mass was achieved.

## 3. Results and discussion

The data on matrix density, bulk density and porosity of tested materials is given in Table 3.

**Table 3.** Basic physical properties of tested materials.

| Material | Matrix density ($kgm^{-3}$) | Bulk density ($kgm^{-3}$) | Porosity (%) |
|---|---|---|---|
| Ref28 | 2306 | 1984 | 14.0 |
| EPP28 | 1798 | 1437 | 20.1 |
| Ref90 | 2320 | 1998 | 13.9 |
| EPP90 | 1800 | 1443 | 19.8 |

The bulk density of samples containing EPP aggregates reached value of 1800 $kgm^{-3}$; this composite can be therefore classified as lightweight concretes (bulk density < 2000 $kgm^{-3}$) in D2.0 class according to the standard EN 206-1 [17]. As expected, the matrix and bulk density values were for samples after 90 days of curing slightly greater than for specimens after 28 days of curing and conversely, porosity decreased with age of samples. In accordance with the literature [8], the use of plastic aggregate resulted in the increased porosity compared to the reference mixture due to the addition of materials with an extremely heterogeneous morphology, which modified workability and amount of entrained air in the mixture.

The mechanical properties of the tested composites are introduced in Table 4.

**Table 4.** Material properties of tested materials.

| Material | Compressive strength (MPa) | Flexural strength (MPa) | Young´s modulus (GPa) |
|---|---|---|---|
| Ref28 | 53.6 | 9.7 | 25.7 |
| EPP28 | 20.8 | 5.1 | 8.1 |
| Ref90 | 54.3 | 10.7 | 25.8 |
| EPP90 | 21.4 | 5.9 | 10.9 |

Babu and Babu [10] stated that the strength of concrete is mainly influenced by the strength of the aggregate. This finding together with the higher porosity and with the low bond strength between the surface of EPP particles and the cement paste resulting from the hydrophobic nature of EPP led to the decrease in mechanical resistance of the developed EPP lightweight concrete in comparison with concrete without plastic aggregate. These results are in accordance with results of other studies, where plastics were used as aggregate for production of concretes/mortars (e.g., [2], [5], [6], [7], [18]). As expected, all mechanical characteristics increased with curing time for both reference samples and samples with incorporated EPP, due to the prolonged hydration time and development of denser concrete structure.

Thermal parameters of plastic waste aggregate together with corresponding powder density are presented in Table 5. With increasing powder density, the thermal conductivity as well as the volumetric heat capacity increased. This is due to the reduction of air gaps during compaction that can be anticipated during preparation of light-weight concrete with incorporated EPP aggregate. Generally, the EPP aggregate had significantly better thermal properties than silica aggregates commonly used in the concrete industry [19].

**Table 5.** EPP aggregate properties in dependence on compacting time.

| Compaction time (s) | Powder density ($gcm^{-3}$) | $\lambda$ ($Wm^{-1}K^{-1}$) | $C_v$ ($\times 10^6$ $Jm^{-3}K^{-1}$) | $a$ ($\times 10^{-6}$ $m^2s^{-1}$) |
|---|---|---|---|---|
| 0 | 0.0189 | 0.042 | 0.049 | 0.96 |
| 10 | 0.0224 | 0.043 | 0.055 | 0.80 |
| 30 | 0.0233 | 0.044 | 0.058 | 0.75 |
| 60 | 0.0239 | 0.044 | 0.061 | 0.72 |
| 180 | 0.0243 | 0.047 | 0.062 | 0.69 |

Table 6 gives the thermal properties of the investigated composites in the dry state.

**Table 6.** Thermal properties of tested materials.

| Material | $\lambda$ ($Wm^{-1}K^{-1}$) | $C_v$ ($\times 10^6$ $Jm^{-3}K^{-1}$) | $a$ ($\times 10^{-6}$ $m^2s^{-1}$) |
|---|---|---|---|
| Ref28 | 1.86 | 1.81 | 1.09 |
| EPP28 | 0.69 | 1.58 | 0.47 |
| Ref90 | 1.88 | 1.82 | 1.10 |
| EPP90 | 0.71 | 1.66 | 0.50 |

The presence of EPP particles with advantageous thermal properties caused substantial reduction in the thermal conductivity of the developed lightweight concrete. The replacement of silica sand by EPP aggregates of 60 % in volume is resulting in a 63 % reduction in the thermal conductivity compared to the reference material. Similarly, Herrero et al. [5] stated that 20 % in mass addition of PE waste aggregate leads to the thermal conductivity below one half of its original value. The data obtained show the positive effect of EPP aggregates content on the moderation of heat transport in the developed material.

## 4. Conclusions

The paper has reported the results of experimental study on the mechanical and thermal properties of lightweight concrete with incorporated waste EPP-based aggregate. These results indicate that use of plastic aggregates decreases the mechanical resistance of the developed materials. However, for intended application in multi-layered floor structures or as the core of sandwich panels the mechanical parameters of EPP modified material are quite sufficient. Lightweight concrete with EPP exhibits enhanced thermal insulation properties, which indicates promising potential for future development in terms of building energy efficiency as well as plastic waste disposal.

## 5. References

[1] Dissanayake D M K W, Jayasinghe C, Jayasinghe M T R 2017 *Energy Build.* **135** 85
[2] Ozbakkaloglu T and Gu L 2016 *Waste Manag.* **51** 19
[3] Yang S, Yue X, Liu X and Tong Y 2014 *Const. Build. Mater.* **84** 444
[4] PlasticsEurope, Plastics – the Facts 2015, www.plasticseurope.org, consulted January 2017
[5] Ruiz-Herrero J L, Nieto D V, López-Gil A, Arranz A, Fernández A, Lorenzana A, Merino S, De Saja J A and Rodríguez-Pérez M A 2016 *Constr. Build. Mater.* **104** 298
[6] Sharma R and Bansal P P 2016 *J. Clean. Prod.* **112** 473
[7] Saikia N and de Brito J 2012 *Constr. Build. Mater.* **34** 385

IOP Conf. Series: Materials Science and Engineering **371** (2018) 012002 doi:10.1088/1757-899X/371/1/012002

[8] Colangelo F, Cioffi R, Liguori B and Iucolano F 2016 *Composites Part* B **106** 234
[9] Ozbakkaloglu T, Gu L and Gholampour A 2017 *J. Mater. Civ. Eng.* **29** 1
[10] Babu K G and Babu D S 2003 *Cem. Concr. Res.* **33** 755
[11] Pavlík Z, Záleská M, Pavlíková M and Černý R 2014 The thermal and mechanical performance of cement-based composites with enhanced thermal insulation properties *Heat Transfer XIII: Simulation and Experiments in Heat and Mass Transfer* ed B Sundén and C A Brebia (WIT Press, Southampton) pp 251-260
[12] Pavlik Z, Trnik A, Ondruska J, Keppert M, Pavlikova M, Volfova P, Kaulich V and Cerny R 2013 *Int. J. Thermophys.* **34** 851
[13] EN 12390-7 Testing hardened concrete – Part 7: Density of hardened concrete CEN 2009
[14] EN 12390-5 Testing hardened concrete – Part 5: Flexural strength of test specimens CEN 200
[15] EN 12390-3 Testing hardened concrete – Part 3: Compressive strength of test specimens CEN 2009
[16] Keppert M, Reiterman P, Pavlik Z, Pavlikova M, Jerman M and Cerny R 2010 *Cement Wapno Beton* **15** 187
[17] EN 206-1 Concrete – Part 1: Specification, performance, production and conformity CEN 2014
[18] Alqahtani F K, Ghataora G, Khan M I and Dirar S 2017 *Constr. Build. Mater.* **148** 386
[19] Kiyohashi H, Sasaki S and Masuda H 2003/2004 *High Temp.-High Press.* **35/36** 179

**Acknowledgments**
The authors gratefully acknowledge the financial support received from the Czech Science Foundation, under project No 17-04215S - Thermal insulation composites containing waste plastic based fillers.

IOP Conf. Series: Materials Science and Engineering **371** (2018) 012003 doi:10.1088/1757-899X/371/1/012003

# Thermal Stability of Lightweight Concrete with Incorporated Regranulated Polypropylene Aggregate

**L Scheinherrová[1], M Záleská[1], M Pavlíková[1] and Z Pavlík[1]**

1 Department of Materials Engineering and Chemistry, Faculty of Civil Engineering, Czech Technical University in Prague, Thákurova 7, 166 29 Prague 6, Czech Republic
E-mail: pavlikz@fsv.cvut.cz

**Abstract.** Plastic waste and its low recycling rate make a significant contribution towards the pollution of the environment. Therefore, it is essential that plastic waste is utilised in different applications, such as aggregates in concrete. Since the coefficient of thermal expansion of polypropylene (PP) is significantly higher than for the ordinary concrete, the resulting mismatch in thermal expansion behaviour of the lightweight concrete can cause many problems, especially when the external temperatures reach the higher values. In this paper, an investigation of a manufactured plastic aggregate as a partial replacement for natural silica aggregate in concrete is presented. For concrete manufacturing, regranulated polypropylene (R-PP) waste coming from PP pipes production was used for the partial replacement of silica sand in concrete mix composition by 10, 20 and 30 mass % respectively. The studied materials were heated up to 120 °C. The basic physical properties before and after thermal treatment were studied. To analyse the thermal strain and the coefficient of thermal expansion of the studied materials, the horizontal thermodilatometry was applied.

## 1. Introduction

Plastic materials have become a very important part of our everyday life. Their properties, such a low density, strength, user-friendly design, long life and especially low cost make them easily accessible what leads to their growing consumption. Since their commercial development in the 1930s and 1940s they have been widely used in packaging, buildings and constructions, medical delivery systems, communication materials, automotive and industrial applications, or for example in security systems [1, 2]. Global plastic resin production reached about 311 Mt in 2014 which is a 670% increase since 1975 [3]. While plastics production around the world grows steadily, the situation in Europe has been stable within the last several years. It reaches the values around 55–59 Mt per year [3]. The most popular plastic material is polypropylene (19.2% of the total plastics utilization), mainly used for folders, food packaging or car bumpers [3]. The worldwide utilization of plastics brings many problems with their following recycling since there are still many countries where the landfilling is still the first option how to treat plastic waste. Even when the Czech Republic belongs to the list of these countries, it has doubled the volume of plastics being recycled in less than a decade. While it reached only 7% rate of recycling at the beginning of the millennium, it recorded about 30% in 2014. This result put the Czech Republic in the second place in recycling PET bottles and in the fourth place in the whole Europe's overall packaging recycling performance [4]. There is also an increasing amount of companies interested in plastics waste recycling, what helps to improve the numbers of recycled plastics.

The utilization of waste and recycled plastic materials as a part of a concrete structure makes it not only economical, but it also helps to reduce the disposal problems of an increasing amount of plastic waste. There have been many research works concerned about the use of recycled plastics in concrete as aggregate. One of the most important concrete properties to study are the basic physical properties, which define its structure and can describe expected behaviour of the fresh/hardened concrete mixture. Al-Manaseer and Dalal [5] reported that bulk density of concrete decreases with an increasing amount of plastic aggregates content and that this decrease is directly related to the plastic aggregates content. This trend is one of the reasons why the plastic waste is mostly used in the design of lightweight concrete [6]. The plastic aggregate can reduce not only the weight, but it also reduces the cost of concrete. Besides this, it improves the thermal and acoustic insulation parameters, as documented by many research papers [7-9]. This is given by the fact that plastic waste aggregate has significantly lower thermal conductivity than the natural aggregates [7]. On the other hand, the incorporation of the plastics into the concrete mixtures leads to a decrease of the mechanical properties [7, 9]. Furthermore, since their coefficient of thermal expansion (CTE) is significantly higher than for the plain concrete, the resulting mismatch in thermal expansion behavior can cause many problems when the external temperatures varies from -30 °C to + 40 °C and more [10]. Some research works offered to combine the plastics of the appropriate sizes (nanoscales) with glass fiber reinforcement to reduce the CTE [10, 11].

In this paper, the utilization of plastic waste coming from the production of plastic pipes for pressure distribution of water in the design of lightweight concrete is studied. The R-PP waste is used as a partial replacement of natural aggregates by 10, 20, and 30 mass% respectively. The intention of this research paper is to study the influence of plastic particles on the basic physical properties and thermal properties of lightweight concrete after its exposure to the temperature of 120 °C.

## 2. Experimental

Regranulated waste polypropylene (R-PP) used in this research paper was obtained from the production of plastic pipes for pressure distribution of water, namely from the company FV Plast (Čelákovice, Czech Republic). The plastic waste (see Figure 1) was being cut into small pieces of the 2–8 mm. Its matrix density was determined using automatic helium pycnometer Pycnomatic ATC (Thermo Scientific), and it was found to be 921 $kgm^{-3}$.

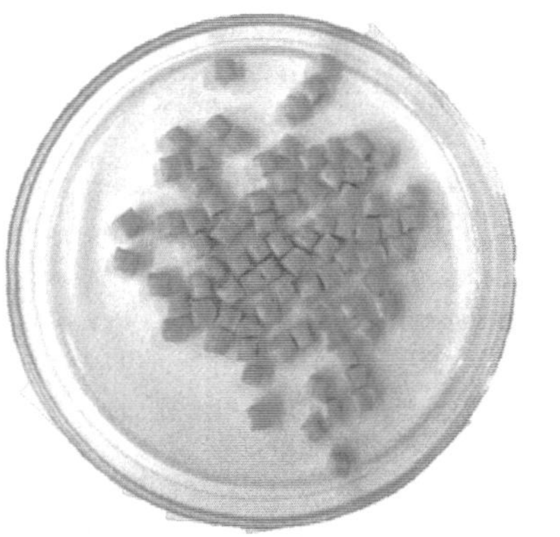

**Figure 1.** RPP waste used for the preparation of the lightweight concrete mixtures.

**Figure 2.** Studied mixtures with different amounts of RPP waste.

Lightweight concrete was prepared using an ordinary Portland cement CEM I 42.5 R (Czech-Moravian cement, Ltd., HeidelbergCement group, Mokrá plant), silica sand of fraction of 0–2 mm (Filtrační písky Ltd., affiliate Chlum u Doks, Czech Republic), batch water, and R-PP waste which was used as a partial replacement of natural aggregates by 10, 20, and 30 mass% respectively. The water to cement ratio was chosen as 0.5 for all studied mixtures. The composition of the studied materials is summarized in Table 1.

The reference sample without any RPP waste was labelled as REF; the samples containing 10, 20 and 30 mass% of R-PP waste were labelled in accordance to the amount of plastics used in the lightweight concrete structure.

**Table 1:** Mix proportions of lightweight concrete with incorporated PP waste.

| Sample | Mass (g) | | | | Water/cement ratio |
|---|---|---|---|---|---|
| | Cement | Natural aggregate | Water | PP waste | |
| REF | 450 | 1350 | 225 | - | 0.5 |
| R-PP-10 | 450 | 1215 | 225 | 135 | 0.5 |
| R-PP-20 | 450 | 1080 | 225 | 270 | 0.5 |
| R-PP-30 | 450 | 945 | 225 | 405 | 0.5 |

From the prepared mixtures specimens with dimensions of 40 × 40 × 160 mm for the determination of the basic physical properties; and 15 × 15 × 160 mm for the horizontal dilatometry analysis were casted (see Figure 2). The samples were kept for one day at laboratory conditions and then they were unmoulded and put in water for 28 days. After reaching this age, the samples were carefully dried in a vacuum at 60 °C to constant mass.

*2.1. Basic physical properties*

Basic physical properties of the studied materials, such as bulk density, matrix density and total open porosity were determined. The bulk density was calculated from the measurement of sample size and its dry mass. The matrix density was measured by automatic helium pycnometer Pycnomatic ATC (Thermo Scientific). The total open porosity was calculated on the basis of the knowledge of these two parameters, as it was already described in [12]. The bulk density tests were done on five samples having size 40 × 40 × 160 mm. Fragments of samples were used for pycnometric measurements. All measurements took place in an air-conditioned laboratory at the temperature of 23 ± 1 °C and 25–30% relative humidity. Each result represents the average value from five measured values. The relative expanded uncertainty of applied testing method was 5%.

The basic physical properties of examined samples were measured as after 28 days of water curing as well as after their thermal treatment. The particular samples were heated from the room temperature of 25 °C to 120 °C with a heating rate of 1 $°Cmin^{-1}$.

*2.2. Thermal properties*

One of the main motivations of this paper was the determination of the influence of R-PP waste on the thermal properties of the designed lightweight concretes. Since the thermal strain is one of the most important parameters in case of sun exposure of materials'surface, the horizontal thermodilatometry was chosen for this analysis. This method is in accordance with EN 13471 [13]. For the measurement, a unique horizontal thermodilatometer designed by Clasic CZ, Ltd. (Řevnice, Czech Republic) was used. The apparatus consists of a furnace for placing the studied material. It is heated using a resistance heating and the temperature distribution is controlled using a thermocouple type S. The temperature is displayed using Orbit Merret OM 351. The Sylvac ms229 digital indicator having resolution 0.001 mm and repeatability 2 μm measures the length changes, and its calibration is provided by Micro Precision Calibration Inc. accredited according ISO 17025 [14]. The calibration was done using an alumina rod. Rectangular prisms having dimensions of 15 × 15 × 160 mm were used for the measurement. The analysis was performed in a horizontal arrangement under a static air atmosphere within the temperature interval from 25 °C to 120 °C with a heating rate of 1 $°Cmin^{-1}$. The length changes (%) of the studied materials were measured as a function of temperature. The results were used for the calculation of the linear thermal expansion coefficient $\alpha$ ($°C^{-1}$), as described in equation (1)

$$\alpha = \frac{d(l_T / l_0)}{dT} = \frac{d\varepsilon}{dT}, \tag{1}$$

where $l_0$ is the original length of the sample (m), $l_T$ is the temperature dependent length (m), $T$ is the temperature (°C or K), and $\varepsilon$ the thermal strain (-) of the studied sample.
Three samples of the each designed material were used for the analysis. The mean value was then determined and presented in this study.

## 3. Results and discussion

### *3.1. Basic physical properties*

The results of the basic physical properties, such as matrix density, bulk density and total open porosity of examined materials before and after thermal treatment are summarized in Table 2 and Table 3, respectively. The matrix density of the reference material was 2508 $kgm^{-3}$. The results of the materials with incorporated R-PP waste indicate that the matrix density tends to decrease by 12%, 19%, and 26% for R-PP-10, R-PP-20, and R*PP-30, respectively, compared to the reference samples. This trend is attributed to the density of the plastic waste which is significantly lower than the sand used which is in accordance with data reported in literature [15]. The matrix density values did not change after the heating up to 120 °C. The bulk density of REF samples reached value of 2064 $kgm^{-3}$. The bulk densities significantly decreased with an increasing amount of R-PP waste in the studied mixtures and reached values in the range between 1533–1837 $kgm^{-3}$. Therefore, in accordance to the standard EN 206-1 [16] these materials can be classified as lightweight concretes having the bulk density values lower than 2000 $kgm^{-3}$. The total open porosity was also calculated and was found to be lower than 18% for all studied mixtures. The materials R-PP-10 and R-PP-20 reached lower porosity values compared to the reference samples; while the samples having 30 mass% of the RPP waste had a slightly higher total open porosity. However, the differences in porosity were in the range of measuring uncertainty. The lower amounts of RPP waste (up to 20 mass%) caused that the nonporous structure of this plastic waste decreased the total open porosity of cement composite, whereas the workability and therefore air absorption in the structure of the fresh mixes were not negatively affected. On the other hand the higher amount of plastic waste slightly increased the porosity due to its bigger particles what led to the general coarsening of composites inner structure [17]. This “formation a new type of porosity” is discussed also in [18].

**Table 2.** Basic physical properties of tested materials before thermal treatment.

| Material | Matrix density ($kgm^{-3}$) | Bulk density ($kgm^{-3}$) | Total open porosity (%) |
|---|---|---|---|
| REF | 2508 | 2064 | 17.7 |
| R-PP-10 | 2202 | 1837 | 16.6 |
| R-PP-20 | 2035 | 1690 | 17.0 |
| R-PP-30 | 1865 | 1533 | 17.8 |

The basic physical properties of the developed composites were studied also after the thermal treatment. Because there was no obvious difference in the results when the samples were cooled down to the room temperature, the bulk density and total open porosity were calculated from the known parameters during heating at the maximal temperature of 120 °C. The mass change in this temperature range was not expected to be changing because the samples were dried before the tests were done. The size of the samples changed mainly in the horizontal direction during the horizontal thermodilatometry measurements. Based on the new maximal lengths of the samples, new characteristics were calculated. The results are summarized in Table 3. As it can be seen from the obtained results, the bulk density of the studied materials slightly decreased with the temperature applied which led to the increase of the total open porosity. Increasing values of the total open porosity then can cause a decrease in the mechanical properties, which has been widely discussed in literature [15, 19-21].

**Table 3.** Basic physical properties of tested materials after thermal treatment.

| Material | Bulk density ($kgm^{-3}$) | Total open porosity (%) |
|---|---|---|
| REF | 2063 | 17.7 |
| R-PP-10 | 1827 | 17.0 |
| R-PP-20 | 1679 | 17.5 |
| R-PP-30 | 1518 | 18.6 |

From the results of the basic physical properties before and after thermal treatment from 25 °C to 120 °C, it can be concluded that this temperature interval does not significantly affect these material characteristics.

*3.2. Thermal properties*

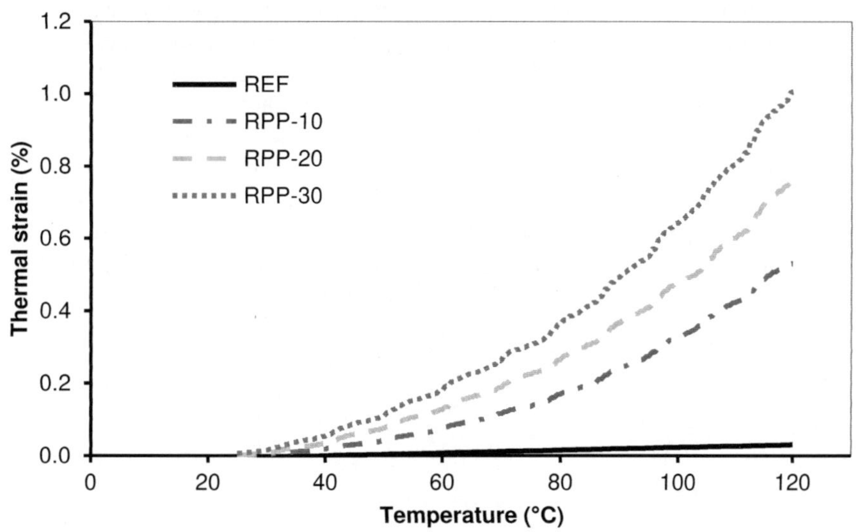


**Figure 3.** Thermal strain curves of the studied lightweight concrete mixtures as functions of temperature.

The results of the thermal properties obtained from the horizontal thermodilatometry are shown in the Figure 3 and 4, respectively.

The results of the thermal strain (calculated in %) of the REF samples were significantly lower than in case of the samples having R-PP waste in the mix. The strain of the designed mixtures is strongly linked to the expansion of the aggregates, as aggregates represent about 75-80% of the volume of concrete [22]. Recently, in the study presented by Koňáková et al. [23], the influence of basalt and silicate aggregates on the thermal expansion of cement-based composites was studied. They confirmed that the basalt aggregates are more stable in the term of thermal expansion in a comparison to the silica aggregates. On the other hand, the thermal strain of both studied aggregates in their study was almost the same in the temperature interval from 25 °C to 120 °C, which is the scope of this study. The thermal strain of the REF samples achieved the highest value of 0.03% of the initial length at the temperature of 120 °C. The trend of the obtained curve was almost linear with no significant increase, which is in a good accordance with the fact that the samples were dried before the thermal treatment, and there was no water evaporation occurred. The effect of the sand aggregates is usually observed at

higher temperatures [24]. Since the other materials presented in our study contained up to 30 mass% of R-PP waste, which is thermally not stable and expands more significantly than the plain concrete under a thermal load, the results of the thermal strain are mainly in accordance to the behaviour of polypropylene [10]. The maximal thermal strains of the samples labelled as RPP-10, RPP-20, and RPP-30 were determined as 0.53%, 0.76%, and 1.00%, respectively. These values are almost ten times higher than for the REF samples what should be taken into account in their further applications. Nevertheless, these results show that the cementitious materials produced from R-PP waste achieved better thermal properties than other PP modified materials, as it can be seen in literature [25].
Since the coefficient of thermal expansion is calculated from the thermal strain values, it also highly depends on the type of aggregates used in the mixture [24, 26]. The CTE of cement paste is usually between $11.0 \times 10^{-6}$ °C$^{-1}$ and $20 \times 10^{-6}$ °C$^{-1}$, while the CTEs of lightweight concretes are around $7 \times 10^{-6}$ °C$^{-1}$ and $13 \times 10^{-6}$ °C$^{-1}$, respectively. These values depend on many other factors besides the type of aggregates; mainly on how the hardened concrete is cured, its relative humidity, the proportion of the coarse versus fine types of aggregates, the mineral admixtures, and some small fluctuations can be observed due to age of samples [24, 26-30]. Thermal strain/CTE can be also affected with varying water to powder ratio (the fine material content, including cement), as observed in [24]. It shows that the thermal behaviour of concrete is in accordance to its air void system, which is the result of the porosity and the water to cement ratio of concrete. The system of voids in concrete can lead to the decrease of the internal pressure of the material resulting in a lower thermal strain. This is probably one of the reasons why the thermal strain of the REF material was found low in our study. Since the thermal strain was found to be almost a straight line in the temperature interval from 25 °C to 120 °C, the CTE was considered as a constant value based on its determination from equation (1). It was $4.23 \times 10^{-6}$ which is in accordance with the results of the self-consolidating normal aggregate concrete designed in [24]. The content of R-PP waste has a significant impact on the results of the CTE of studied materials, as it can be seen in Figure 4. The higher amount of R-PP waste leads to an increase of the CTE. In a contrary to the REF material results, the CTE of the materials with the R-PP waste increases with the temperature in the entire studied temperature range.

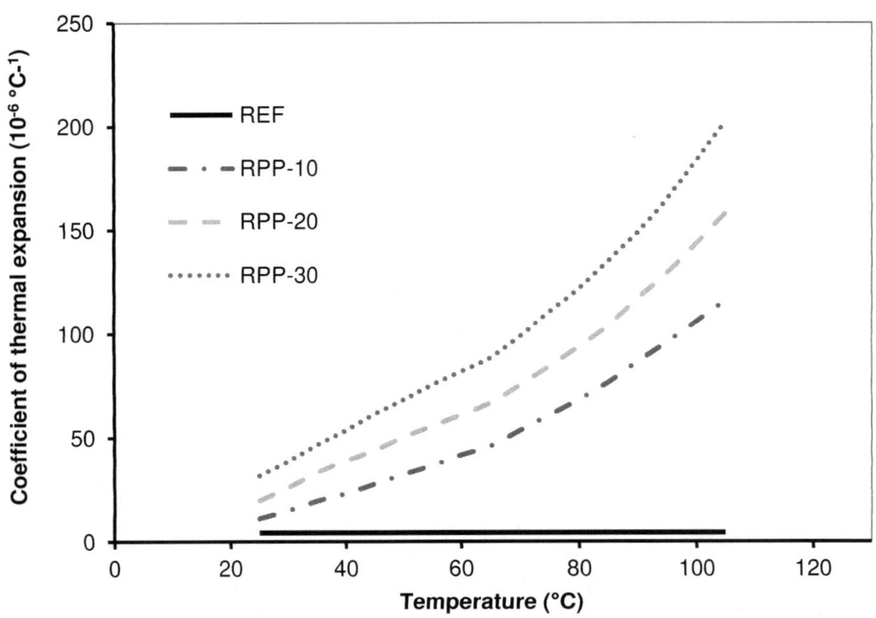


**Figure 4.** CTE curves of the studied lightweight concretes as functions of temperature.

The CTEs of the studied materials at chosen temperatures are summarized in Table 4. The CTEs of R-PP-10, R-PP-20, and R-PP-30 were calculated as $11.13 \times 10^{-6}$ °C$^{-1}$, $19.76 \times 10^{-6}$ °C$^{-1}$ and $31.70 \times 10^{-6}$ °C$^{-1}$, at the temperature of 25 °C, respectively. It is challenging to find a literature dealing with the same plastics waste like in our case to compare our results. To our best knowledge, the closest results were found in [25], where 30 mass% talc-filled PP composites reached values of 50 × $10^{-6}$ °$C^{-1}$.

**Table 4.** CTEs of tested materials at selected temperatures.

| Temperature (°C) | Coefficient of thermal expansion (× $10^{-6}$ °$C^{-1}$) | | | |
|---|---|---|---|---|
| | REF | R-PP-10 | R-PP-20 | R-PP-30 |
| 25 | 4.22 | 11.13 | 19.76 | 31.70 |
| 65 | 4.22 | 46.03 | 66.80 | 88.56 |
| 105 | 4.22 | 116.14 | 158.06 | 202.48 |

The replacement of silica sand by RPP waste aggregates of 30 mass% is resulting in a significant increase of the thermal strain and the CTEs of studied materials which has a negative effect on the thermal stability of the designed materials.

## 4. Conclusions

In this research study, the effect of the use of R-PP waste as a partial replacement of silica sand in concrete mixture composition on chosen properties of lightweight materials was studied. The plastic waste coming from PP pipes production was used for the partial replacement of silica sand in concrete mix composition by 10, 20 and 30 mass% respectively. To observe thermal stability of R-PP modified materials, they were heated up to 120 °C. From the results of the basic physical properties, it can be concluded that the higher R-PP content decreases the bulk density, which is mostly related to the low weight of R-PP. The total open porosity of the studied materials was lower than 18%. These characteristics were almost not affected by the thermal exposure to 120 °C.

Replacing the silica sand with R-PP waste increases both the thermal strain rate and the CTE of the hardened lightweight materials. The extension effect increases with increasing volume of R-PP waste in the mixture. Therefore, an improvement of the thermal stability of the studied materials is necessary to allow their wider use.

## 5. References

[1] Siddique R, Khatib J and Kaur I 2008 *Waste Manag.* **28** 1835
[2] Jambeck J R, Geyer R, Wilcox C, Siegler T R, Perryman M, Andrady A, Narayan R, and Lavender K L 2015 *Science* **347** 768
[3] PlasticsEurope, Plastics - the Facts 2015, www.plasticseurope.org, consulted October 2017
[4] Linnenkoper K, Czech Republic a PET bottle recycling leader., www.recyclinginternational.com, consulted October 2017
[5] Al-Manaseer A and Dalal T 1997 *Concr. Int.* **19** 47
[6] Yang S, Yue X, Liu X and Tong Y 2015 *Const. Build. Mater.* C **84** 444
[7] Saikia N and de Brito J 2012 *Const. Build. Mater.* C **34** 385
[8] Sharma R and Bansal P P 2016 *J. Clean. Prod.* **112** 473
[9] Ruiz-Herrero J L, Nieto D V, López-Gil A, Arranz A, Fernández A, Lorenzana A, Merino S, de Saja J Á and Rodríguez-Pérez M A 2016 *Const. Build. Mater.* C **104** 298
[10] Lee H, Fasulo P D, Rodgers W R and Paul D R 2006 *Int. J. Sci. Tech. Polym.* **47** 3528
[11] Yui H 1992, *Conf. Plastics Age*, Tokyo 65
[12] Antepara I, Pavlík Z, Žumár J, Pavlíková M and Černý R 2016 *Mater. Sci.* **22** 88
[13] EN 13471 Thermal insulating products for building equipment and industrial installations - Determination of the coefficient of thermal expansion CEN 2001
[14] ISO 17025 General requirements for the competence of testing and calibration laboratories ISO 2005
[15] Ismail Z Z and Al-Hashmi E A 2008 *Waste Manag.* **28** 2041
[16] EN 206-1 Concrete – Part 1: Specification, performance, production and conformity CEN 2014
[17] Záleská M, Pavlíková M and Pavlík Z 2016 *AIP Conf. Proc.*
[18] Colangelo F, Cioffi R, Liguori B, Iucolano F 2016 *Composites Part B* C **106** 234

[19] Rahmani E, Dehestani M, Beygi M H A, Allahyari H and Nikbin I M 2013 *Constr. Build. Mater.* **47** 1302
[20] Saikia N and de Brito J 2014 *Const. Build. Mater.* C **52** 236
[21] Panyakapo P and Panyakapo 2008 *Waste Manag.* **28** 1581
[22] Kong D L Y and J G Sanjayan 2010 *Cem. Concr. Res.* **40** 334
[23] Koňáková D, Špedlová V, Čáchová M, Vejmelková E and Černý R 2014 *Adv. Mat. Res.* **1054** 17
[24] Uygunoglu T and Topcu I B 2009 *Const. Build. Mater.* **23** 3063
[25] Wu G, Nishida K, Takagi K, Sano H and Yui H 2004 *Int. J. Sci. Tech. Polym.* **45** 3085
[26] Zhou C, Shu X and Huang B 2014 *Const. Build. Mater.* C **68** 10
[27] Shah P and Ahmad S 1994 *CRC Press* 141-374
[28] Alungbe G D, Tia M and Bloomquist D G 1992 *Transp Res Rec.* **1335** 44
[29] Shin H C and Chung Y 2011 *LTRC* **2**
[30] Shui Z, Zhang R, Chen W and Xuan D 2010 *Const. Build. Mater.* C **24** 1761

**Acknowledgments**

Authors gratefully acknowledge the financial support received from the Czech Science Foundation, under projects No 17-04215S - Thermal insulation composites containing waste plastic based fillers and No SGS17/166/OHK1/3T/11 - Application of secondary raw materials in the development of composite materials for building industry.

IOP Conf. Series: Materials Science and Engineering **371** (2018) 012004 doi:10.1088/1757-899X/371/1/012004

# Performance of Concrete Blended with Inhibitors and Slag Exposed to Chloride Environment

**Anita N Borade[1,3] and B Kondraivendhan[2, 3]**
[1]Research Scholar
[2]Assistant Professor
[3]Applied Mechanics Department, Sardar Vallabhbhai National Institute of Technology, Surat-395007, Gujarat, India
Email: boradeanita@yahoo.com,kondraivendhan78@gmail.com.

**Abstract.** In this paper, the performance of concrete with organic inhibitor and slag on compressive strength, chloride penetration resistance and half-cell potential were studied for different water/binder (w/b) ratios and different curing ages. Organic inhibitor (Amines)was admixed with cement during the manufacturing process itself of the cement named as corrosion resistance cement (CRC). The influence of this cement on compressive strength was studied over different curing ages such as 7, 28, 56, 90, 180 and 360 days. Rapid chloride penetration test (RCPT) was conducted to explore the resistance to chloride penetration. Half-cell potential was observed for the period of one year as a measure of corrosion tendency of reinforced concrete (RC) exposed to 3%, 5% and 7.5% of sodium chloride (NaCl) solutions. The results of the concrete with inhibitor (CRC) were compared with the concrete with Portland slag cement (PSC). Significant improvement was observed in compressive strength with increase in curing ages. The performance of CRC concrete shows remarkable improvement in corrosion potential (delayed corrosion initiation) as compared to PSC concrete.

## 1. Introduction

The premature deterioration in reinforced concrete (RC) structures due to its exposure to chloride laden environment and it reduces the service life of structure considerably. It consumes a huge amount for repairs, maintenance and restoration of such structures. Deterioration of RC structures is more aggressive in chloride laden environment as compared to carbonation [1, 2]. The chloride induced corrosion is a major cause of deterioration of reinforced concrete. Chlorides are introduced either in fresh concrete through contaminated water, aggregate and admixtures containing chlorides or externally in hardened stage through deicing salts and also the structures exposed to marine environment [3, 4].Reinforcement in the concrete is naturally protected by a thin passive layer by the alkaline environment. This passive layer is stable till the pH of the pore solution in surrounding concrete is more than 12.5. Due to the ingress of harmful impurities like chlorides and carbonation through the cover concrete, the alkalinity of concrete get reduced which destroy the passive film. This gives access to chlorides to reach at the reinforcement level. Corrosion initiation of reinforcement takes place, when Chlorides reach to reinforcement level and exceeds the critical level[5-7]. Various researcher suggested different systems to reduce or delay the corrosion initiation periods such as use of pozzolanic, supplementary cementitious materials, admixtures for concrete and epoxy coating for steel etc. [8-10].Among the various efforts taken to reduce the corrosion of RC structures, use of inhibitors was an effective method to protect the steel embedded in concrete exposed to aggressive environment owing to its lower cost and ease in application [11,12]. Corrosion inhibitors can increase service life through prolonging the corrosion initiation time or decreasing the corrosion rate [13].

IOP Conf. Series: Materials Science and Engineering **371** (2018) 012004 doi:10.1088/1757-899X/371/1/012004

Inhibitors are classified depending on the basis of the mechanism of protection, whether it affects the anodic reaction, cathodic reaction or both are called as anodic, cathodic and mixed inhibitor respectively [14]. On the basis of application, inhibitors are either mixed into fresh concrete (mixed-in inhibitors) as admixtures or applied on the surface of hardened concrete (migrating inhibitors, MCI). The chemical nature of inhibitors defines the organic and inorganic. Though, inhibitors had great acceptance in the industries due to excellent anticorrosive proprieties, many showed up as a secondary effect, not ecofriendly. Thus the study is going on searching for environment friendly and safe inhibitors; like the organic inhibitors [15].Amine and ester are organic based admixed inhibitors have dual actions in concrete. The nitrogen (amine) compound part is responsible for its strong film formation on the metal surface by adsorption due to the strong molecular affinity for metal surface whereas the hydrocarbon (carboxylate) ester compound creates at hydrophobic layer helps to repel moisture away from the steel and acts as a pore-blocking agent which blocks the ingress of the chlorides ions ,inhibit both the anodic and cathodic processes[16-18].

Organic inhibitor also called as mixed inhibitors covers the metal surface by adsorption which is depends on type of electrolyte, surface charge of the metal and structures of inhibitors. Mixed inhibitors protect the metal in three possible ways: physical adsorption, chemisorption and film formation. Electrostatic attraction between the metal surface and inhibitor is called as Physical adsorption. These inhibitors consist of both positively charged (cations) and negatively charged ions (anions) [12, 19].

## 2. Experimental Procedure

### *2.1. Materials and mix proportions:*

For the present investigation, Corrosion resistant Cement (CRC) and Portland slag Cement (PSC) confirming to IS 455-1995 was used [20]. The physical test such as blain surface area, specific gravity confirming to IS 2720-Part-3and loss on ignition(LOI) for both the cements were carried out as per Indian Standard (IS) 1727-2004. The oxide elements in both the cements were analyzed by XRF (X-ray fluorescence) as presented in table 1. The Properties of C.A, F.A and concrete mix design are similar to authors' previous work [21].

**Table 1.**Chemical and Physical Properties of Binders.

| Comp-ound | $SiO_2$ | $Al_2O_3$ | $Fe_2O_3$ | CaO | MgO | $SO_3$ | $Na_2O_3$ | $K_2O$ | Loss on Ignition (%) | Specific Gravity | Blain Surface Area ($m^2/kg$) |
|---|---|---|---|---|---|---|---|---|---|---|---|
| CRC | 32.47 | 13.31 | 3.14 | 44.61 | 1.79 | 3.49 | 0.28 | 0.89 | 0.8 | 2.8 | 408.97 |
| PSC | 25.53 | 12.89 | 1.53 | 48.17 | 5.29 | 5.44 | 0.42 | 0.79 | 1.8 | 3.0 | 361.8 |

### *2.2 Methods*

*2.2.1.Compressive strength.* In order to study the influence of inhibitors (CRC) and PSC on compressive strength of concrete, concrete cube specimens were prepared as given in table 2. The specimens were vibrated mechanically for proper compaction. To avoid the evaporation of moisture from the wet concrete, the specimens were covered with plastic sheet. After 24 hours of setting, the specimens were demoulded and cured for the different curing ages in normal tap water. Results were reported for average of triplicate specimens for each variable.

*2.2.2.Rapidchloride ion penetration test (RCPT).* The rapid chloride penetration test was used to assess the permeability of concrete mixed with corrosion inhibitors. For RCPT measurement cylindrical specimen were cast as mentioned in table 2.After 28 days of curing the concrete cylinders were sliced to 100mm diameter and 50mm thick disc specimens. The disc samples were preconditioned confirming as per ASTM C 1202-12 [22]. For measurements, two sides of the disc were sealed in Plexiglas containers with silicon sealant. One side of the container is filled with 3 % sodium chloride (NaCl) solution (cell connected to the negative terminal) and the other side is filled with 0.3N sodium hydroxide (NaOH) solution (cell connected to the positive terminal) as shown in figure 1.

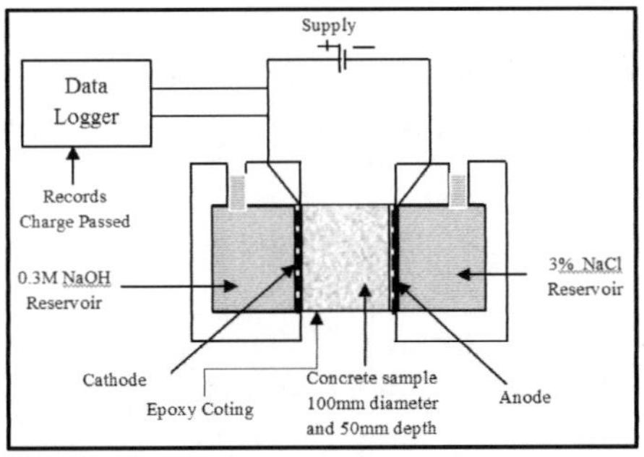


**Figure1.**ASTM C1202-12 line diagram.

**Figure 2.**Typical laboratory test setup for RCPT.

The specimen subjected to RCPT by impressing a 60 V potential and current was measured at every 30 minutes over the period of 6 hours. The total charge passed through the cell at the end of 6 hours was then calculated in order to determine the penetrability. A higher level of charge passed indicates higher permeability of the concrete. The typical laboratory test set up is shown in figure 2.

**Table 2.** Details of Test Specimen.

| Sr. No | Specimen | Size | Test | Quantity (No.s) |
|---|---|---|---|---|
| 1 | Cube | 150 x 150 x 150 mm | Compressive Strength | 180 |
| 2 | Cylinder | Dia=100mm, L=200mm | Rapid chloride Penetration | 24 |
| | (Disc) | Dia=100mm, L=50mm | Test (RCPT) | (72) |
| 3 | Beam | 280 x 115 x 150 mm | Half Cell Potential (HCP) | 54 |

*2.2.3.Half-cell potential (HCP) measurement.* The tendency of any metal to react with an environment is indicated by the potential it develops in contact with the environment. The half-cell potential (HCP) is an electrochemical technique most widely used for the assessment of corrosion risk the reinforced bars in concrete. This technique is based on measuring the electrochemical potential of the steel rebar with respect to a standard reference electrode placed on the surface of the concrete and can provide an indication of the corrosion risk of the steel as shown in figure 3.

**Table 3.** Conditions of Reinforcement Corrosion Related to Half-cell Potential (mV) Measurements to Relative Reference Electrodes [23].

| Sr. No | Hg/HgCl2 (SCE) | Likely Corrosion Condition |
|---|---|---|
| 1 | < -426 mV | Sever corrosion |
| 2 | < -276 mV | High (< 90% risk of corrosion) |
| 3 | -126 to -275 mV | Intermediate corrosion |
| 4 | > -126 mV | Low ( 10% risk of corrosion) |

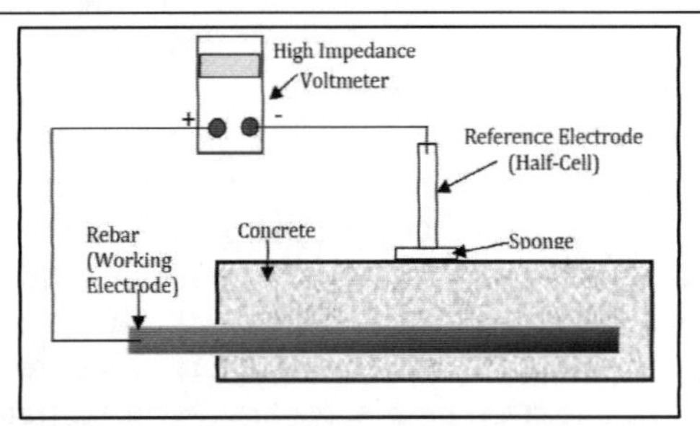


**Figure3.** Schematic representation of Half-cell potential (HCP) measurement.

The measured values are influenced by concrete cover, concrete resistivity (moisture content) and oxygen availability. It should be noted that it only indicate the probability of corrosion and not the actual corrosion rate. The measurement of corrosion potential is the primary step to understand the corrosion tendency. From the measured HCP risk of corrosion can be categorized as per ASTM C876-2007 [23]which is reproduced in table 3.

For corrosion potential measurement ASTM G109 [24]RC beam specimen were prepared for the same concrete mixture with thermo-mechanically treated (TMT)reinforced bar of 12mm diameter confirming to IS 1786 [28]. The composition of steel was same as previously reported work [21].The reinforced bar is preconditioned before casting. The details of prepared steel specimen is explained schematically in figure 4. The preconditioning of bar was carried out as per the procedure mentioned in ASTM G1-03 [25] was also explained in figure 4.

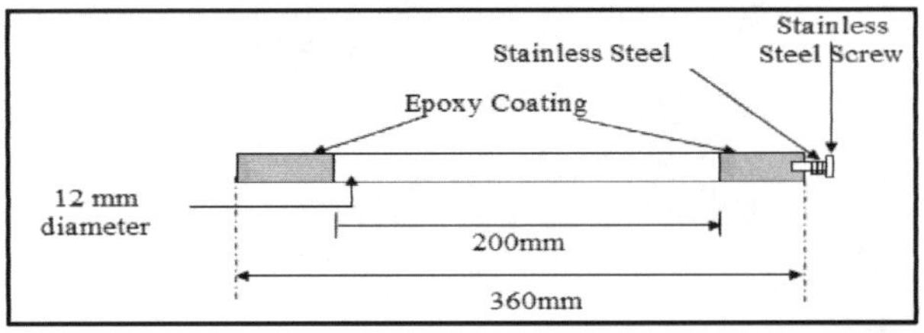


**Figure 4.** Line diagram of preconditioned steel specimen.

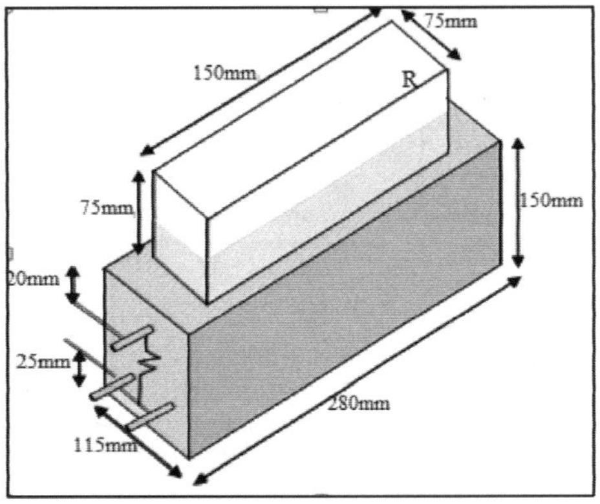


**Figure 5**. Line diagram of ASTM G109-2007 beam specimen.

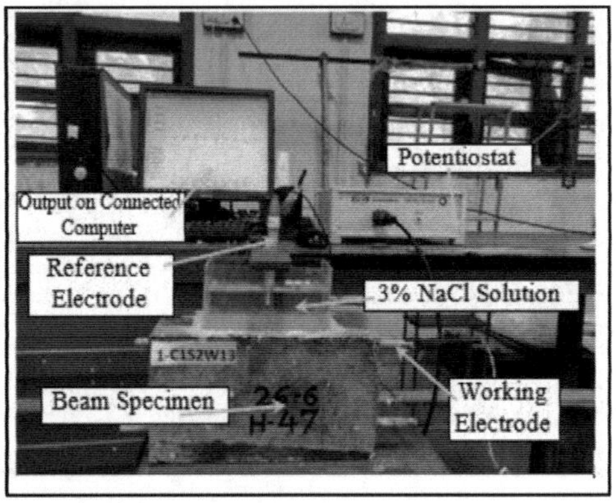


**Figure 6.**Typical laboratory Set of corrosion potential measurement ofASTM–G109 beam specimen.

The beam specimens size and quantity as specified in table 2 [24] were prepared with TMT two steel bars embedded at bottom with a cover of 25mm and one steel bar centrally at the top with the clear cover of 20mm from top. After 28 days moist curing the specimens were removed from tank for the further preparation. A Plastic reservoir of 150mm x 75mm x 75mm was centrally fixed on the top of the beam with silicon sealant to pour the NaCl solution during exposure. Epoxy coating was applied on four vertical faces of beam and on the top the portion outside of reservoir to restrict the entry of contamination from the surrounding throughout the measurements. Before the specimens subjected to chloride ingress, a wire was connected through lugs to bottom bar. Top bar was connected by a 100 Ohm resistor to bottom bars through common terminal of bottom steel bars in each specimen as shown in figure 6.The beam specimens then subjected to alternate wetting and drying i.e. 3 %, 5% and 7.5 % NaCl solution in order to accelerate reinforcement corrosion. One cycle consists of seven days ponding in respective NaCl solution and fourteen days drying in an open atmosphere. This wetting and drying cycle were chosen so that the specimens remain partially saturated and oxygen also available in abundance. Due to this availability of sufficient oxygen, the steel reinforcement was not polarized to immune state (large negative potential) and corrosion can proceed unobstructed when initiated [26].HCP measurements were monitored using ACM Potentiostat, Model No. 1824 [23]. A saturated calomel electrode (SCE) was used as a reference electrode and placed on the surface. The positive terminal of the potentiostat was connected to the working electrode (rebar) and the common terminal was connected to the reference electrode. The typical laboratory set up was done as shown in figure 6

HCP was measured in three different locations very near to the rebar and the average potentials were recorded. The cycle was continued for 260 days. From the results a potential vs. time plot was drawn using the average potentials obtained.

**3. Results and Discussion:**

Compressive strength, RCPT and corrosion potential were monitored over the period of one year. The results of the tests are discussed as below.

*3.1. Compressive strength*

The specimens prepared were exposed to different curing ages such as 7, 28, 56, 90, 180 and 360 days. After the curing period was over, the compressive strength was measured by using compression strength testing machine (CTM) of 3000 ton capacity at loading rate of 140kg/cm$^2$/min conforming to IS 516 [27] as shown in figure 7. It was observed that strength gaining at the age of 7 and 28 days in case of CRC concrete was more as compared to PSC. Reduction in the compressive strength was observed with increase in w/b ratio in the range of 13.07% and 42.5% for CRC whereas for PSC, it was reduced to 17.17% and 28.75% for w/b ratio 0.5 and 0.55 respectively as compared to w/b 0.45 at 180 days. This reduction was 6.39 % and 19.88% for PSC and 11.96% and 40.81% for CRC at 360 days. For CRC concrete w/b ratio affects the compressive strength considerably as compared to PSC. This may be due to hydrophobic nature of inhibitors forms micro bubbles during casting make the concrete porous reduces the strength with increase in w/b ratio.

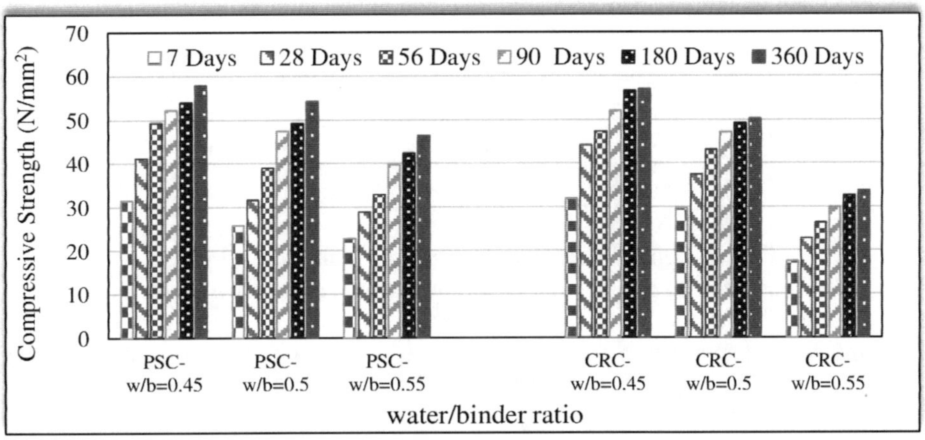


**Figure 7.** Compressive strength of concrete.

For all w/b ratios, compressive strengths were increased with increase in curing ages as expected. The workability of concrete with CRC was increased with increase in w/b ratio which makes concrete porous and reduces the compressive strength experimentally. It was observed that reduction in the w/b ratio is possible in concrete with CRC.The rate of strength gaining at all w/b ratios was fast up to 90 days, afterward becomes slower. In the case of PSC, the pozzolanic reaction is very slow at early age, hence the rate of gain of compressive strength was lower at an early age and it was even less than CRC but after 90 days, there was not much difference in strength between PSC and CRC. It was reported in the literature that when inhibitors mixed in concrete, compressive strength is reduced, while the inhibition effect of corrosion inhibition in chloride exposure is improved [28].

*3.2. RCPT*

It was observed that concrete prepared from CRC was better in chloride penetration (lower coulomb charge passed) than concrete prepared from PSC at early age. With increase in w/b ratio the charge passed was increases but decreased with curing age for all cases as expected. This was because of refinement of pore structure by hydration products owing to pozzolanic action over the age in concrete

IOP Conf. Series: Materials Science and Engineering **371** (2018) 012004 doi:10.1088/1757-899X/371/1/012004

with PSC and pore blocking by inhibitors in CRC. The charge passed in coulomb in all concrete specimens was low.

From the figure 8, it is observed that the concrete mixed with inhibitor systems show lower coulomb values than the PSC Concrete. These differences refer to the difference in the electrical conductivity of concrete which is relatedto many factors such as: mix proportions, use of supplementary cementing materials, chemical admixtures etc. These factors had very significant effects on the concentration of conductive ions in the pore solution.Chloride diffusion studies revealed that the inhibitor mixed concretes exhibit very low permeability [14].

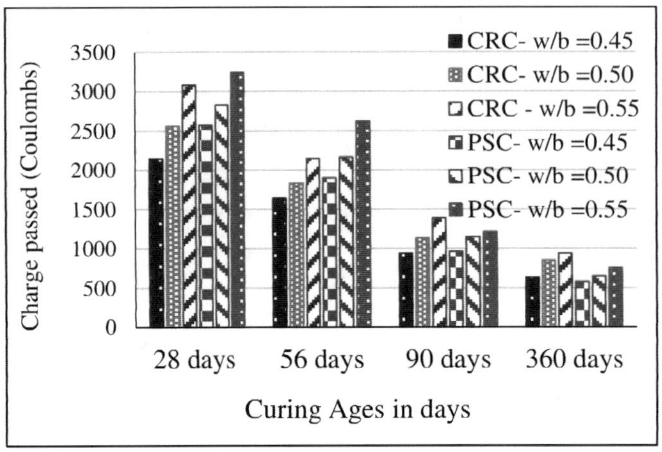


Figure 8. The total charge passing through different concrete specimens with time.

*3.3 Half -cell potential measurement (HCP)*

The half-cell potential of steel reinforcement in concrete measured periodically with potentiostat with saturated calomel reference electrode (SCE) is as shown in fig. 9 and fig. 10 for 3%, 5% and 7.5% chloride exposure respectively.

Corrosion potentials in all specimens were relatively negative after one week of exposure -140 mV to -200 mV versus SCE. This appears to propose that the steel in all specimens is actively corroding (HCP: -126 to -275=Intermediate corrosion) [23], whichcontradicts normal experience. From two weeks of ageonwards, potentials shifted to positive values (about -100 mV). The shift observed was more positive in CRC concrete, indicating that the steel had passivity.

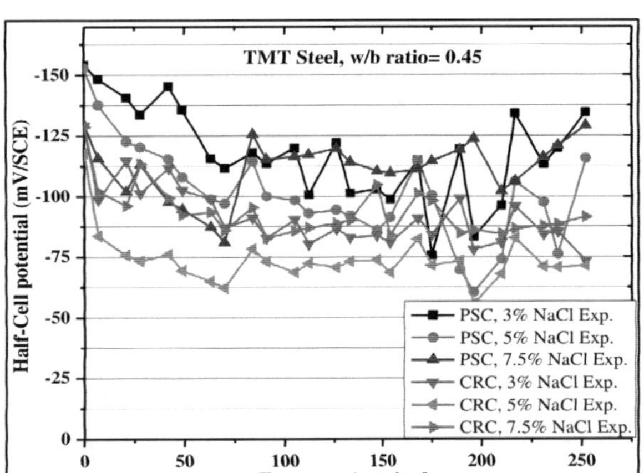


Figure 9. Average HCP of ASTM-G109 beam specimen for different NaCl exposures

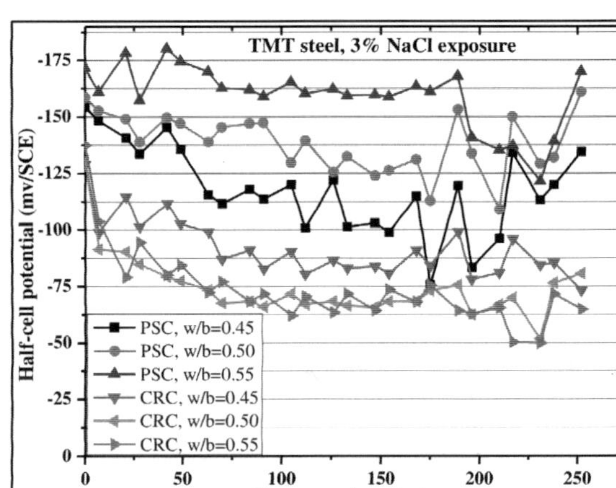


Figure 10. Average HCP of ASTM-G109 beam specimenfor different w/b ratio

IOP Conf. Series: Materials Science and Engineering **371** (2018) 012004 doi:10.1088/1757-899X/371/1/012004

It was observed from the figures 9 and 10 in the concrete with PSC that the increase in w/b ratio increases the negative potential. Potential values were below (-126mV) up to 60 days which indicates corrosion risk is very less [23]. Afterwards, the potential values were increased to the risk of intermediate corrosion (-126 mV to -256 mV). It was also observed that the potential values were moving toward positive side after first few cycles .This may be due to the pozzolanic activity of slag over the time refines the micro-pores, which restrict the ingression of moisture and oxygen up to rebar level. From thc observation of figures 9 and 10, CRC performs better than PSC in corrosion resistance. Potential values were exhibiting in the low limit of corrosion till 200 days of chloride exposure to different chloride concentration solutions. In general, it is observed that the potential values of CRC are negative at lower w/b ratio and positive at higher w/b ratio.

In general, it was observed that the potential values of CRC are less negative at lower w/b ratio compared to higher w/b ratio. Corrosion measurements of steel in concrete normally fluctuate with time. The degree of fluctuation can depend upon the corrosion mechanism occurring [28, 29].The apparent fluctuation in potential values in summer and rainy season was observed. It was noted that the corrosion potential was affected by temperature (more negative) as well as humidity (shift to positive).

The rate of potential change was slow in CRC concrete as compared to PSC. The corrosion potential was observed for the period of 260 days from exposure (about 310 days from casting) between -90mV to -180mV for PSC whereas for CRC concrete the values were below -100mV for all concrete mixes. This shows the improvement in corrosion activity by inhibitor compared to PSC.

Improved performance is exhibited in the case of CRC concrete specimens for all the tests including compressive strength, RCPT and half-cell potential.

## 4. Conclusion

Based on the experimental studies on concrete with CRC and PSC the following conclusions were drawn:

1. At early ages the pozzolanic activity in concrete with PSC is very slow as it gives lower early age strength as compared to CRC.
2. Corrosion-inhibitor has no effect on strength but may influence the delayed onset of corrosion.
3. The HCP of concrete with PSC show more negative potential with the increase in NaCl solution exposure as 3%, 5% and 7.5% and decrease with increase with w/b ratio irrespective ofcuring ages. Whereas, in concrete prepared with CRC, the reverse nature was observed.
4. From half-cell potential measurements, it was observed that the concrete prepared with CRC exhibited more positive potential values as compared to concrete prepared with PSC.
5. The corrosion initiation is delayed due to use of inhibitors in concrete. Improved performance is exhibited in the case of concrete prepared with CRC concrete specimens for all the tests including compressive strength, RCPT and half-cell potential.

## 5. Acknowledgement

The authors are grateful to m/s Krishna Conchem Products Pvt. Ltd., Mumbai for supply of Corrosion Resistance Cement (CRC) and the partial financial support extended to this investigation.

## 6. References

[1]. ElsenerL B 2001 *Corrosion inhibitors for steel in concrete – state of the art report* 1st edition, no.**35** 68 London, EFC Publications.
[2]. Soylev T A and Richardson M G 2008 *Construction and Building Materials*, **22** pp. 609-622.
[3]. Broomfield J P 1997*Corrosion of steel in concrete: Understanding, investigation andrepair*, E & FN Spon, New York.
[4]. Ormellese M, Berra M, BolzoniF and Pastore S 2006 *Cement and Concrete Research***36**, no. 3 pp. 536–547.
[5]. Bertolini B, Elsener B, Pedeferri P and Polder R 2004**392**, Weinheim, Wiley-VCH.
[6]. Liu Y 1996 Ph.D. thesis, Virginia Polytechnic Institute and State University, p 117.
[7]. Andrade C, Alonso C and Molina F J1993 *Materials Structures***26**pp. 453-464.
[8]. Ann K Y and Song H W 2007 *Corr. Sci***49**pp 4113-33.
[9]. Mangat PS, Khatib J M and Molloy BT 1994. *Cem Concr Compos***16(2)** pp 73–81.

[10]. Chindaprasirt P, Chotithanorm C, Cao H T and Sirivivatnanon V*Journal of Construction and Building Materials***21(2)** pp. 356-361.
[11]. Mike B and Otio B 2010 *ACI Material Journal***107** pp. 481-489.
[12]. Tang Y, Zhang G and Zuo Y 2012 *Construction and Building Material***28** pp. 327-332.
[13]. Elsener B 2011 Corrosion inhibitors for steel in concrete – un update on the ongoing discussion Frankfert, pp 30-41
[14]. Song H W and Saraswathy V 2006 *J Hazard Mater.***138** (2) pp 226-33
[15]. Gaidis J M 2004 *Cement Concrete Composite***26** pp. 181-189.
[16]. Gaidis J and Rosenberg A 1987 *Cement Concrete and Aggregates***9** pp. 30-33.
[17]. Jamil HE, Montemorb MF, BoulifcR, ShririaA and Ferreiraad M G 2003, *Electrochimica Acta* **48** pp 3509-3518.
[18]. Angst U M, Buchler M and Schlump J *Mater Struct* (2016) **49**: 2807.
[19]. Ormellese M, Berra M, Bolzoni F and Pastore T 2006 *Cem. and Con. Rese.* **36** pp 536 – 547.
[20]. IS 455-1995, Portland Slag Cement – Specification Bureau of Indian Standards, New Delhi.
[21]. Borade A N and Kondraivendhan B 2017 *IOP Conf. Ser.: Mater. Sci. Eng.***216,** 012022.
[22]. ASTM C1202-12, 2012 Standard test method for electrical indication of concretes ability to resist chloride ion penetration," in Proceedings of the *ASTM International*, West Conshohocken, PA, USA.
[23]. ASTM C876-09, 2009 Standard test method for corrosion potentials of uncoated reinforcing steelin concrete', *ASTM International* West Conshohocken.
[24]. ASTM G109-07 2013 Standard Test Method for Determining Effects of Chemical Admixtures on Corrosion of Embedded Steel Reinforcement in Concrete Exposed to Chloride Environments in Proceedings of the *ASTM International*, West Conshohocken, PA, USA.
[25]. ASTM G1-03 2011 Standard practice for preparing, cleaning and evaluating corrosion test specimens in Proceedings of the *ASTM International,* West Conshohocken, PA, USA.
[26]. PradhanB and Bhattacharjee B 2009 RCJ. Mater. Civ. Eng.**21** pp 543-552.
[27]. IS 516 1959 Indian standards of methods of tests for strength of concrete *Bureau of Indian Standards*, New Delhi.
[28]. Morris W and Vazquez M 2004, *Cement and Concrete Research*, **32** pp 259-267.
[29]. Law D W, Cairns J, Millard S G and Bungey J H 2004 *NDT & E International***37** pp 381–388

# Anticorrosion Effect of Silane Type Surface Penetrants on RC Degraded by Carbonation

**T Kondo[1], Y Kadota[2], Y Kawanishi[3], Y Nakamoto[3], K Yokoi[4] and Y Yamada[5]**

1 Associate Professor, National Institute of Technology, Kochi College, Kochi, JAPAN
2 National Institute of Technology, Kochi College, Kochi, JAPAN
3 DAIDO Corporation, Co., Ltd, Osaka, JAPAN
4 Professor, National Institute of Technology, Kochi College, Kochi, JAPAN
5 Assistant Professor, National Institute of Technology, Kochi College, Kochi, JAPAN
E-mail: tkondou@ce.kochi-ct.ac.jp

**Abstract.** We studied on the anticorrosion effect by applying silane type surface penetrants on RC degraded by carbonation. Two types of carbonation residue and two types of environmental conditions after accelerated carbonation were prepared and investigation was made on the corrosion behaviour of reinforcing bars by applying surface penetrants. As a result, by blocking water penetration by applying surface penetrants, there was a difference in anticorrosive effect in the behaviour of half-cell potential, polarization resistance, and investigation of reinforcing bar corrosion after the end of exposure. From the above, this paper reports on the possibility of realizing anticorrosion effect by suppressing invasion of moisture from concrete surface to carbonated RC.

## 1. Introduction

One of the deterioration of concrete structures is carbonation. When the carbonation residue becomes a certain value or less, the alkali component in the concrete disappears and the anticorrosion performance of the reinforcing bar decreases. For this reason, the corrosion rate of reinforcing bars has been reported to be dependent on concrete moisture content and carbonation residue [1]. Also, the past study reported that the concrete had been peeled off remarkably at the rain-hanging place when the carbonation residue of concrete is lower than a certain value, in the actual structures [2]. In this case, it was reported that when the carbonation residue becomes a certain value or less, the spalling of the cover concrete became remarkable at the rain hanging place. In addition, the STANDARD SPECIFICATIONS FOR CONCRETE STRUCTURES-2013 [Maintenance] [3] shows that the splashed area is involved in the durability of concrete structures.

Therefore, in order to suppress corrosion of the reinforcing bars of the concrete structure and achieve long-term service, it is considered effective to apply the surface penetrants of silane type to prevent invasion of water which is penetrated from the surface of concrete [4]. However, the case of the discussion on the relationship between the corrosion rate of reinforcing bars and the water content by applying the surface penetrants to the RC whose carbonation residue is lower than the corrosion occurrence limit has been reported few [5]. For this reason, there are many unclear points about the anticorrosive effect by using silane type surface penetrants in a state where carbonation occurs in the actual structure, but cracks do not occur on the concrete surface. If the silane type surface penetrants is

IOP Conf. Series: Materials Science and Engineering **371** (2018) 012005 doi:10.1088/1757-899X/371/1/012005

applied in this state and the anticorrosive effect of the reinforcing bars can be confirmed, it is considered to be very advantageous in the management of the structure.
In this study, we examined the anticorrosive effect which is obtained by applying a silane type surface-penetrants to carbonated RC.

## 2. Test methods

### *2.1. Test parameter of this test*

Test parameters and levels were shown in Table 1. In order to investigate the anticorrosive property by applying silane type surface penetrants in the carbonated state, there were two kinds of whether silane type surface penetrants was applied or not. The carbonation residue was classified into two types in order to confirm the effect on corrosion protection performance. In order to confirm the involvement of water in the concrete, two types of drying and water were set for the exposure environment.

**Table 1.** Test parameters and levels.

| Test parameters | Levels |
|---|---|
| Surface Penetrants | Penetrants, Non-penetrants (2 levels) |
| Carbonation Residue | 0mm， -10mm (2 levels) |
| Exposure Condition | Dry, Water (2 levels) |

### *2.2. Test parameter of this test*

Mix promotion of this test was shown in Table 2, and outline of this test was shown in Figure 1. For specimens, reinforcing bars of 13 mm in diameter were placed in prism specimens of 100 mm × 100 mm × 70 mm. For the reinforcing bars embedded in the concrete, corrosion prevention treatment was carried out with tapes of about 10 mm from both ends of the specimens to prevent corrosion at the end portions.
Water cement ratio was set at 70%. Adjustment was made with the admixture so that the target slump was 12 ± 2.5 cm and the air amount was 4.5 ± 1.5%. Regarding the maximum size of the coarse aggregate, it was 10 mm because cover thickness of these specimens was 20 mm.

**Table 2.** Mix proportion of this test.

| W/C | Unit (kg/m$^3$) | | | | AE arrangement |
|---|---|---|---|---|---|
| (%) | W | C | S | G | (cc) |
| 70 | 171 | 244 | 956 | 914 | 33.6 |

### *2.3. Test condition*

Time table of this test was shown in Figure 2. From 14 days after casting, specimens were placed in a carbonation tank under conditions of temperature 20℃ and $CO_2$ concentration 5% in order to carbonate. Accelerated carbonation was continued until predetermined carbonation depth was reached.
After reaching the predetermined carbonation depth, the surface penetration was applied to the surface with small cover depth. For the surface penetration, 200g/m$^2$ of silane type surface penetrants was applied. Subsequently, the specimens were transferred under dry and water environments and exposed.

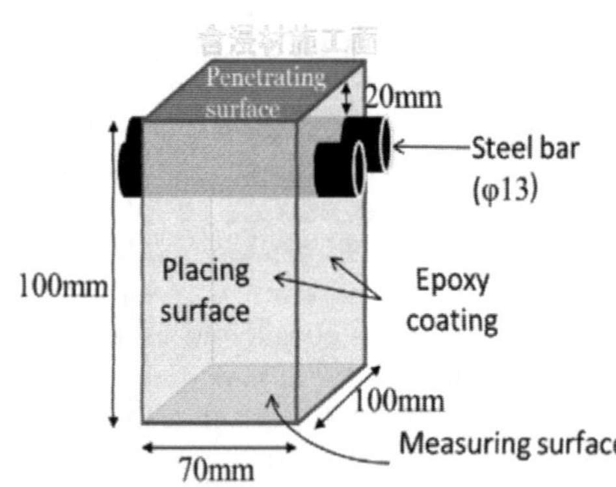


**Figure 1.** Outline specimen.

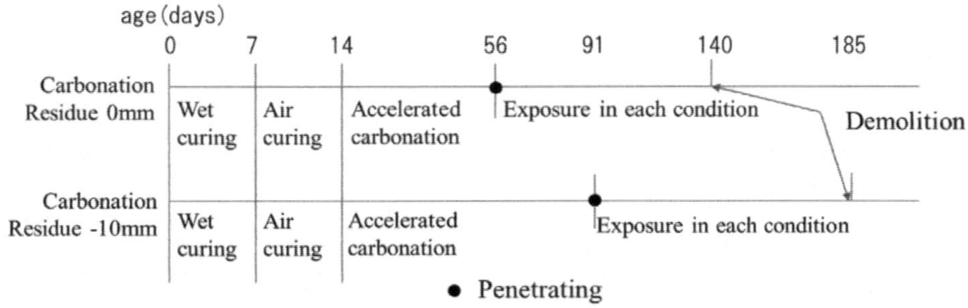


**Figure 2.** Time table of this test.

Specimens under drying conditions were placed under conditions of 20℃ temperature and 60% R.H. As for the water condition, specimens were placed under the conditions of temperature 20℃ and R.H. 90% constant, so that the penetration surface was immersed in water. Two specimens were prepared for one factor, and one specimen was crashed in about three months after the exposure. The other one was continuously placed under exposed environmental conditions, and specimens were left until the age of 260 days.

*2.4. Test method*

After installing the specimens in the carbonation tank, the corrosion condition of the reinforcing bars was monitored using the electrochemical method over time. For this monitoring, we measured the half-cell potential and the polarization resistance. Also, water content of concrete was measured using separately prepared specimens. Using an electric resistance type concrete / mortar moisture meter, measurements were made up to 30 mm in the depth direction by 10 mm. It was conducted every 4 weeks.

Three months after the exposure period, one specimen was taken out and disassembled, and the corrosion condition of the rebar was measured. Measurement items were corrosion area ratio and diameter measurement. The corrosion area was measured by wrapping the taken rebar with tracing paper and tracing the corroded area. In addition, the diameter of the reinforcing bar was measured in two directions orthogonal to one cross section after rust removal according to the JCI-SC 1 method. Measured at intervals of 10 mm, the percentage with sound reinforced bars was taken as the residual diameter ratio.

## 3. Test result

*3.1. Penetration depth*

Penetration depth was shown in Figure 3. There was no difference in penetration depth due to the carbonation residue.

*3.2. Water contents*

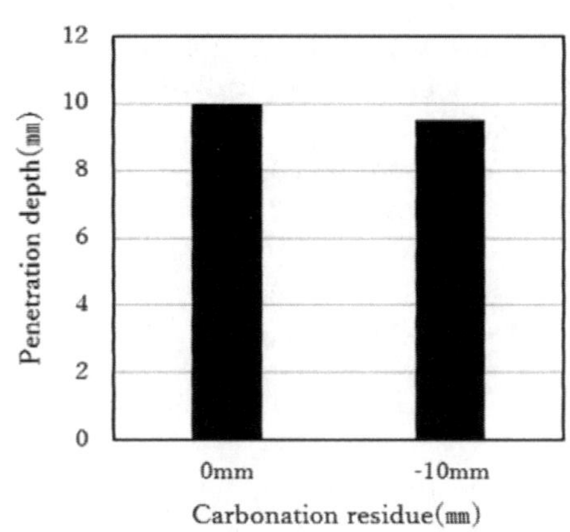


**Figure 3.** Penetration depth.

IOP Conf. Series: Materials Science and Engineering **371** (2018) 012005 doi:10.1088/1757-899X/371/1/012005

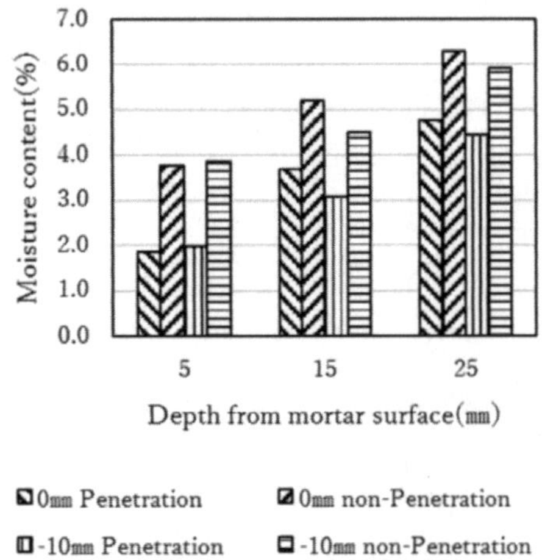


**Figure 4.** Water contents (Under water condition ).

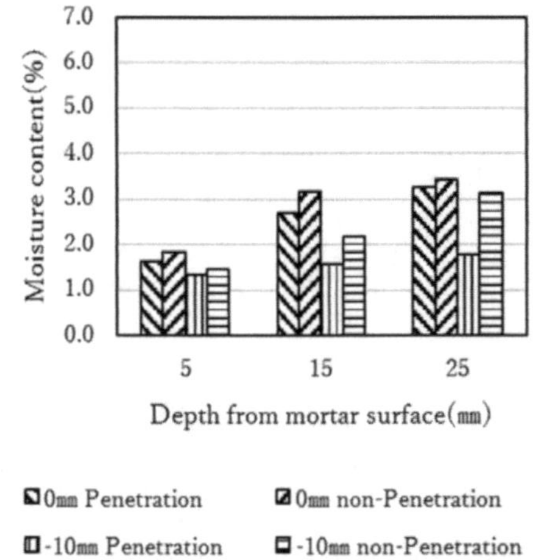


**Figure 5.** Water contents (Under dry condition).

The distribution in the depth direction of the water content from the specimen surface under water condition at age of 200 days was shown in Figure 4, and the water content distribution under the drying condition was shown in Figure 5. Under the dry condition and water condition, the water content on the concrete surface was small and the water content tended to increase toward the inner of the concrete. It was considered that the silane type surface penetrants showed the same tendency both in dry condition and in water condition because it had water shielding and moisture permeability [4]. However, since it was difficult for water vapor permeation to prohibit, even in specimens applied with surface penetration in water condition specimens, it was considered that the water content increased as the depth from the surface increased. Under water conditions, it showed a tendency to suppress the water content of about 2% at any depth by application of surface penetration.

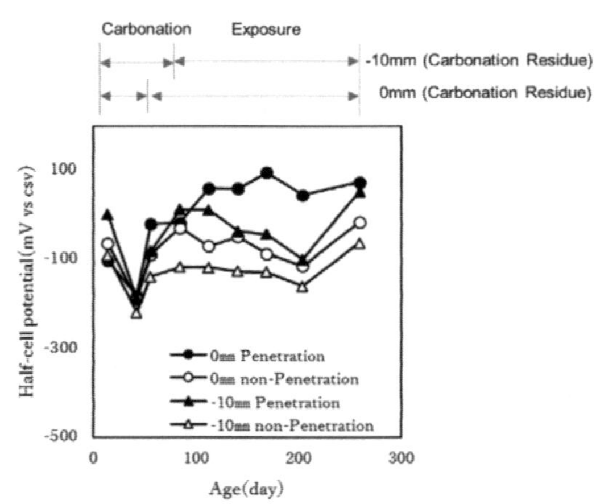


**Figure 6.** Half-cell potential (Under dry condition).

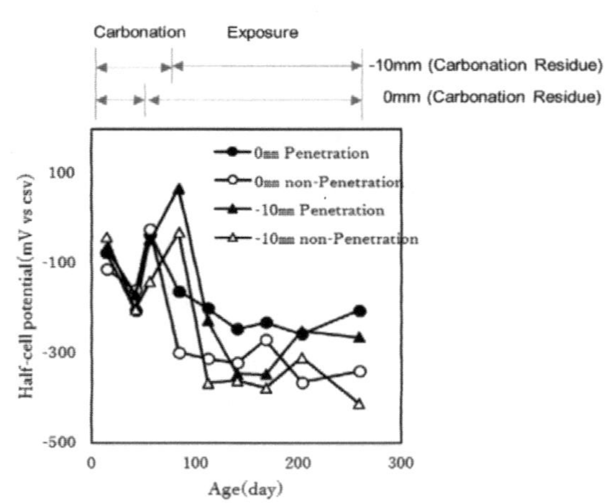


**Figure 7.** Half-cell potential (Under water condition).

### *3.3. Half-cell potential*

Changes with time of half-cell potential under dry conditions were shown in Figure 6. After the accelerated carbonation period, as shown in Figure 6, specimens placed under dry conditions

exhibited nobler potential than -200 (mV vs csv) in any of the carbonation residues. In ASTM C 876, it is judged that "there is no corrosion with a probability of 90% or more". In addition, the presence or absence of application of sur face penetration showed a noble tendency in the applied specimens. This was considered to be due to the blockage of water penetration, and it was considered that the surface penetration was more likely to be in a stable state when applied. In addition, irrespective of applied or non-applied surface penetration, it was showed a low potential on the specimens with small carbonation residue.
Changes with time of half-cell potential under water conditions were shown in Figure 7. Under water conditions, half-cell potential showed a low value in any carbonated residue specimens after changing the placement environment. Half-cell potential of the specimens applied with surface penetration changed from -200 (mV vs csv) to -350 (mV vs csv). In ASTM C 876, corrosion potential is a value judged to be "indeterminate". After setting to the water condition, the non-applied specimens were shifted at a base potential lower than -350 (mV vs csv). In ASTM C 876, it is judged that "there is corrosion with a probability of 90% or more". Therefore, as for the comparison of half-cell potential according to the application state of the surface penetration, it was possible to confirm the situation where the applied specimens were shifted at the noble potential during the exposure period. This was thought to be due to the prevention of invasion of water by applying surface penetration. In addition, similar to the drying condition, it can be confirmed that specimens change to the base side as a whole in the specimens with large carbonation residue.

### *3.4. Polarization resistance*

Changes with time of the polarization resistance under dry conditions were shown in Figure 8, and changes with time of the polarization resistance under water conditions were shown in Figure 9. In Figure 8, there was a tendency that the polarization resistance increases with the passage of days. It is thought that resistance to corrosion increases due to the increase in concrete strength with the passage of age. In addition, the specimen applied with the surface penetration showed a tendency that the polarization resistance was small in the specimen with large polarization resistance and small carbonation residue. This indicated that the water content at the rebar position was small in Figure 5, and Iijima et al. pointed out that the rebar corrosion rate depends on the water content and the carbonation residue, and this time it was considered the same [1].
As shown in Figure 9, the polarization resistance tended to decrease with the passage of days until the age of 200 days after changing the exposure environment. However, according to the comparison based on whether or not the surface penetrants was applied, since the specimen applied with the surface penetration had a large polarization

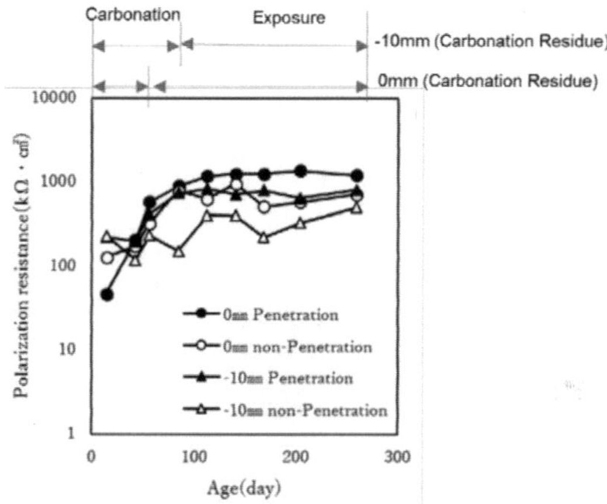


**Figure 8.** Polarization resistance (Under dry condition).

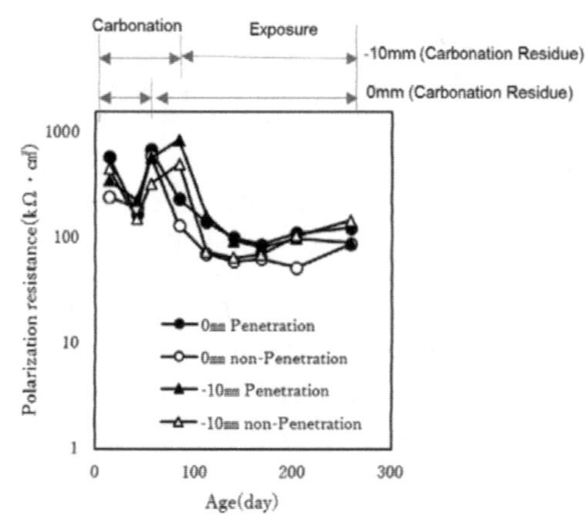


**Figure 9.** Polarization resistance (Under water condition).

resistance, corrosion due to blocking invasion of water from the concrete surface could be suppressed. However, after 200 days of age, no clear difference in polarization resistance due to the presence or absence of surface pen etration application could be confirmed. Because W/C is 70%, it might be affected by oxygen penetration from the surface not applied with surface penetrants. Therefore, although it was necessary to further consider the behaviours of aged more than 200 days, it was considered that corrosion control effect by applying surface penetration was shown for before 200 days.

### *3.5. Corrosion area rate*

The corrosion area ratio under dry condition was shown in Figure 10, and the corrosion area ratio under water condition was shown in Figure 11 among the rebar which was taken out from the specimen 3 months after changing the exposure environment.

From Figure 10, under dry conditions, no corrosion was observed in the carbonation residue 0mm specimen. In the carbonation residue -10mm specimens, corrosion was confirmed irrespective of the presence or absence of the surface penetrants, but both were less than 0.5%. From Figure 11, corrosion was confirmed in all specimens under water conditions. For the applied specimen, no difference was observed due to the carbonation residue, but in the non-applied specimen, the smaller the carbonation residue was, the larger it became. Therefore, it could be confirmed that water is involved in corrosion in specimens that have undergone carbonation.

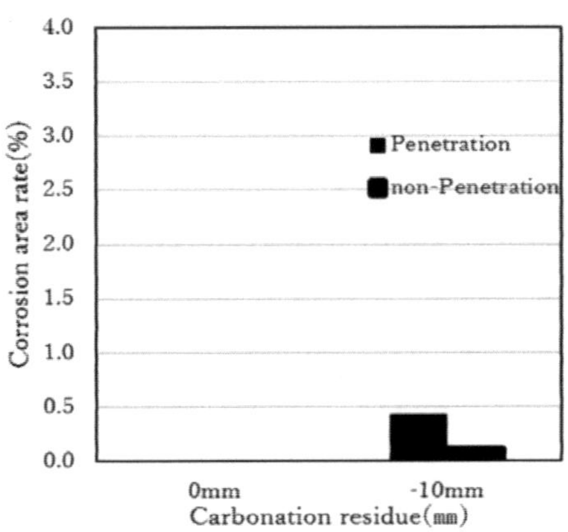


**Figure 10.** Corrosion area rate (Under dry condition).

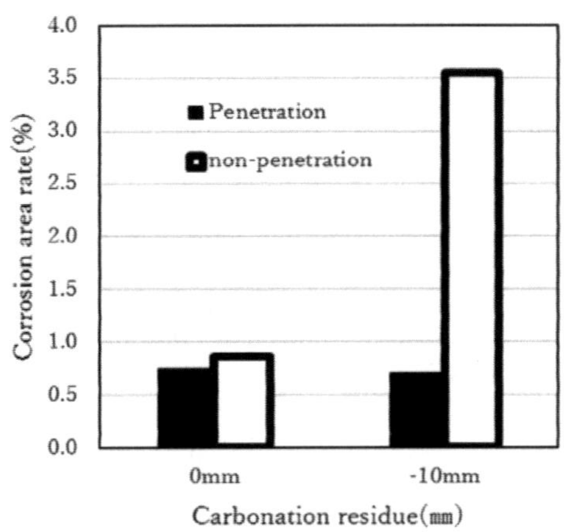


**Figure 11.** Corrosion area rate (Under water condition).

### *3.6. Rebar diameter*

The average residual diameter ratio of the non-applied specimens to the applied specimens after the rust removal for the rebars in the specimens left under the water condition was shown in Fig 12. In all of the carbonation residues, the diameter of the applied specimen was larger than that of the non-applied specimens. As with the other results, it was considered that the penetration of water into the concrete was suppressed by the application of the surface penetration. In addition, since the corrosion main factor of this test was carbonation, local corrosion hardly occurred [6], so it is considered that there was no large difference in residual diameter ratio due to the carbonation residue.

Therefore, in the category of this test, it is considered that the anticorrosion effect can be obtained by applying the silane type surface penetration to RC which has been carbonated. However, as shown in Figure 9, as the time elapses, no difference in polarization resistance due to application of the surface penetration has been confirmed, so it is considered necessary to further study the trend over a longer period.

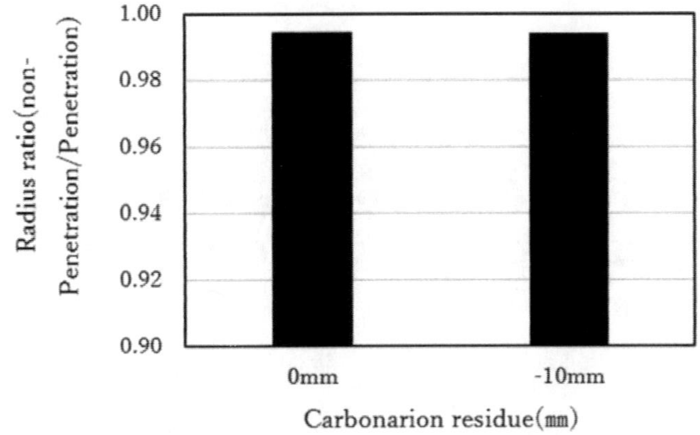


**Figure 12.** Residual radius ratio of rebar after exposure (Under water condition) (Radius ratio = non-penetration specimen / penetration specimen).

## 4. Conclusions

The purpose in this research is confirmation of the usefulness with applying the silane type surface-penetrants for a corrosion prevention to RC which promoted the carbonation. Therefore, we evaluated the corrosion behaviour of the reinforcing bars arranged inside the carbonated RC-specimens whose surface had been coated with the surface penetrants of silane type. The main findings obtained in this research are shown below.

1. When silane type surface penetration was applied to concrete subjected to carbonation, there was a tendency that the half-cell potential showed a noble value in both dry and water conditions. When installed under water conditions, polarization resistance also showed a large initial resistance by application of a surface penetration, but no difference was confirmed with the passage of the age.
2. We measured the ratio of the corrosion area of rebar which had been removed from RC specimen. The RC specimen had been put in water during about 3 months. As the result, the ratio of the corrosion area increased with decreasing the carbonation residue.
3. From the above results, it was shown that the anticorrosion effect can be expected by applying the silane type surface penetration to the carbonated RC.

## 5. References

[1] Iijima T, Kudo T and Tamai Y 2009 Effect of Temperature on Corrosion Rate of Reinforcing Bar in Concrete *RTRI REPORT, Vol.23, No.6,* (Tokyo: Railway Technical Research Institute) p 11-16 (in Japanese)

[2] Ishibasi T, Furuya T, Hamazaki N and Suzuki H 2002 Investigation of Falling on Concrete Fragments from RC Structures *Journal of JSCE, No.711/V-56* (Tokyo: Japan Society of Civil Engineers) p 125-134 (in Japanese)

[3] Japan Society of Civil Engineers 2013 *STANDARD SPECIFICATIONS FOR CONCRETE STRUCTURE Maintenance* (Tokyo: Japan Society of Civil Engineers) (in Japanese)

[4] Japan Society of Civil Engineers 2005 *Recommendation for Concrete Repair and Surface Protection of Concrete Structures* (Tokyo: Japan Society of Civil Engineers) (in Japanese)

[5] Harasawa Y and Iyota T A Study on Application of Surface Penetrants Materials for the Deteriorated Structure 2014 *Proceedings of the Concrete Structure Scenarios, Vol.14,* (Kyoto: Japan Society of Material Science) p547-552 (in Japanese)

[6] Ota K, Tottori S, Kitago Y and Miyagawa T Corrosion Characteristics of Rebars caused by Initial Chloride Ion and Carbonation of Concrete 2000 *Proceedings of the Japan Concrete Institute, Vol.22, No.1* (Tokyo: Japan Concrete Institute) p175-180 (in Japanese)

IOP Conf. Series: Materials Science and Engineering **371** (2018) 012006 doi:10.1088/1757-899X/371/1/012006

# Experimental Study on the Properties of Artificial Lightweight Aggregate Concrete Reinforced with Carpet Waste Fiber

**Y Haryanto[1], G H Sudibyo[1] and N G Wariyatno[1]**

[1]Department of Civil Engineering, Faculty of Engineering, Jenderal Soedirman University, Jl. Mayjend Sungkono Km 5, Purbalingga, Indonesia
E-mail: yanuar_haryanto@yahoo.com

**Abstract**. Lightweight concrete is a low-density concrete that can be made by replacing coarse aggregate with artificial lightweight aggregate such as bloated clay, crushed bricks, or coarse fly ash. This paper presents a study of the influence of carpet waste fiber on the properties of artificial lightweight aggregate concrete. Tests are performed on 30 concrete cylinders of 150 mm in diameter and 300 mm in height with various additions of carpet waste fiber, namely 0%, 0.25%, 0.50%, 0.75%, and 1.00%. Superplasticizer is added at 2% of cement weight for convenience of casting. We find that using artificial lightweight aggregate can reduce concrete density to a maximum of 1.94 t/m$^3$. The compressive strength of artificial lightweight aggregate concrete increases by 17.22% with addition of 0.44% carpet waste fiber, while the split tensile strength of artificial lightweight aggregate concrete increase by 75.18% with addition of 0.85% carpet fiber.

## 1. Introduction

Concrete is an old and extensively used building material, because it has several advantages over other materials, including its ability to support high compressive stress, its relatively low price, and its ability to incorporate materials found in the surrounding environment. In addition, concrete has good resistance against weather, corrosion, and fire. However, drawbacks to concrete include its low tensile strength and its brittleness. Concrete also has a relatively high density, 2.4 t/m$^3$ [1], influencing the amount of load that can be supported by a building structure. Improvements in concrete to overcome weaknesses include making lightweight concrete and addition of fiber into concrete mixtures [2, 3]. This study complements previous studies by characterizing artificial lightweight aggregate concrete reinforced with carpet waste fiber.

Fibers in concrete may bridge cracks in the concrete matrix, providing resistance against propagating and widening before cracks collapse. According to Wang [4, 5, 6], properties of fiber reinforced concrete generally depend on the strength, type, and fiber properties used. There are many types of fibers available in the concrete industry and each type of fiber has advantages and limitations. Fiber selection is primarily based on concrete application. Some types of fiber which may be used in concrete are, steel fiber, glass fiber, natural fiber, synthetic fiber (carbon and polypropylene), and recycled fiber (carpet fiber, fiber from soda cans, and steel fiber from waste tires). Most waste fiber consists of natural materials and synthetic polymer material such as cotton, wool, silk, polyester, nylon, and polypropylene. Waste fiber is consumed and disposed of in large amounts. The carpet industry produces a large amount of waste materials mostly composed of fiber with 50%-70% nylon and 15%-25% polypropylene [7, 8, 9]. Typical carpet fiber is shown in Figure 1.

IOP Conf. Series: Materials Science and Engineering **371** (2018) 012006 doi:10.1088/1757-899X/371/1/012006

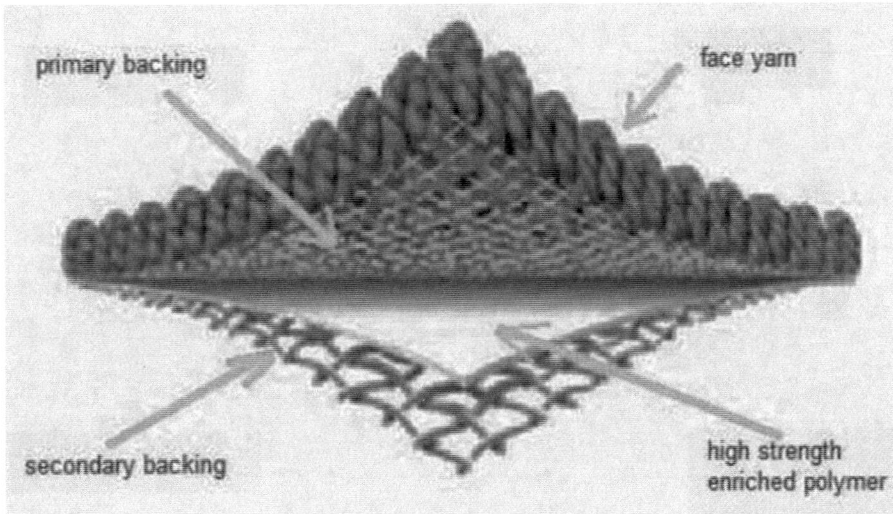


**Figure 1.** Typical carpet fiber [10].

The artificial lightweight aggregate used in this study is manufactured by *Loka Perintisan Bahan Bangunan Lokal Cilacap*. Use of aggregate is intended to reduce dependency on gravel as concrete construction material. Continuous utilization of gravel as a concrete material may affect natural environments, because gravel is a non-renewable resource. Puskim [11] states that shale has abundant potential as an artificial lightweight aggregate material in Indonesia. Haryanto [12] reports the impact resistance of lightweight aluminum fiber concrete with artificial lightweight aggregate, with the most significant increase in impact resistance and absorption energy obtained by adding aluminum fiber 0.75%, resulting in 10 strokes (46.60 J) or an increase of 250% for the first crack, and 15 strokes (69.90 J) or an increase of 300% for total collapse. Addition of 1.00% aluminum fiber decreases the workability level making concrete mixture solidification difficultly resulting in porous concrete with decreased impact resistance and absorption energy.

Kristiyanto [13] tested specific gravity, pore volume, and compressive strength of no-fine lightweight concrete with artificial lightweight aggregate of 15-20 mm, with a water to cement ratio of 0.45, and a planned mixture variation comparing volume of cement:aggregate, including 1:2, 1:4, 1:6, 1:8, and 1:10. The test specimen was a 15 x 15 x 15 cm cube submerged under water and tested on days 7, 14, and 28. The test specimen was found to be lightweight concrete with a maximum specific gravity and pores volume of 1.96 ton/$m^3$ and 34.94%, respectively. The concrete compressive strength was 37.66 MPa on day 28, with a cement content of 571 kg/$m^3$. The concrete compressive strength increased in line with increasing specific gravity and decreasing pore volume.

## 2. Experimental work

The materials used for concrete mixtures include multi-purpose cement, fine aggregate as sand from the Serayu river, Purbalingga, and coarse aggregate as artificial lightweight aggregate (Figure 2) made in Cilacap. The fine aggregate has specific gravity of 2.50 with a fine modulus of 3.55. The artificial lightweight aggregate has a specific gravity of 1.43 with a fine modulus of 6.48. The carpet waste fiber (Figure 3) has a specific gravity of 1.19, is 50 mm x 3 mm x 2 mm, and is added at 0.00%, 0.25%, 0.50%, 0.75%, and 1.00% of the concrete volume. Superplasticizer at 2% of the cement weight is added for convenience of casting. Concrete mixture compositions are calculated as shown in Table 1.

IOP Conf. Series: Materials Science and Engineering **371** (2018) 012006 doi:10.1088/1757-899X/371/1/012006

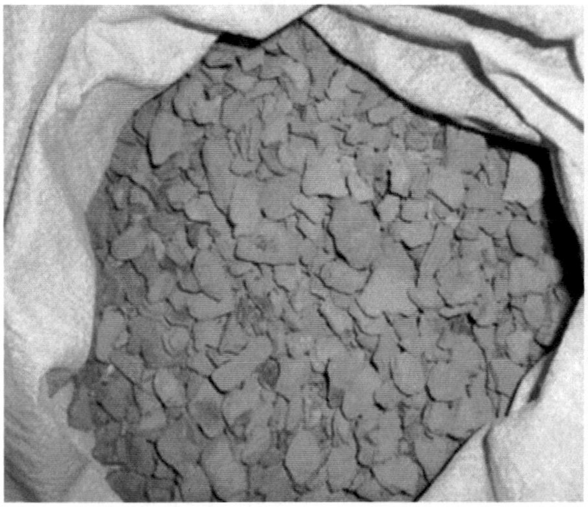

**Figure 2.** Carpet waste fiber used [3, 14].

**Figure 3.** Carpet waste fiber used [14].

**Table 1.** Concrete mixture compositions.

| Fractional Volume (%) | Carpet Fiber (g) | Cement (kg) | Sand (kg) | Artificial Lightweight Aggregate (kg) | Water (kg) |
|---|---|---|---|---|---|
| 0.00 | 0.00 | 375 | 750 | 525 | 81 |
| 0.25 | 2997.50 | 375 | 750 | 525 | 81 |
| 0.50 | 5995.00 | 375 | 750 | 525 | 81 |
| 0.75 | 8992.50 | 375 | 750 | 525 | 81 |
| 1.00 | 11990.00 | 375 | 750 | 525 | 81 |

To verify the concrete mixture workability, a slump test was conducted according to SNI 1972: 2008 [15]. A compressive strength test (Figure 4) and a split cylinder strength test (Figure 5) are conducted on day 28 on 30 cylindrical test specimens with a height of 300 mm and a diameter of 150 mm, in accordance with SNI 1974: 2011 and SNI 03-2491-2002 [16, 17]. The concrete compressive strength was determined using equation (1), while split tensile strength was determined using equation (2).

$$f_c' = \frac{P}{A} \tag{1}$$

$$f_{ct} = \frac{2P}{\pi L d} \tag{2}$$

in which:

$f_c'$ = Compressive strength (MPa)
$P$ = Maximum load (N)
$A$ = Area of specimen ($mm^2$)
$f_{ct}$ = Split tensile strength (MPa)
$L$ = Length of specimen (mm)
$D$ = Diameter of specimen (mm)

**Figure 4.** Compressive strength test [3, 18].

**Figure 5.** Split tensile strength test [3, 18].

## 3. Results and discussion

### *3.1. Workability*

The slump value decreases with addition of carpet waste fiber. Use of fiber in concrete mixture increase the surface area covered by water, decreasing the free water, which decreases concrete workability due to difficulty mixing [19]. The slump value of concrete without carpet waste fiber is 4.5 cm and increased to 19 cm by adding superplasticizer (Figure 6).

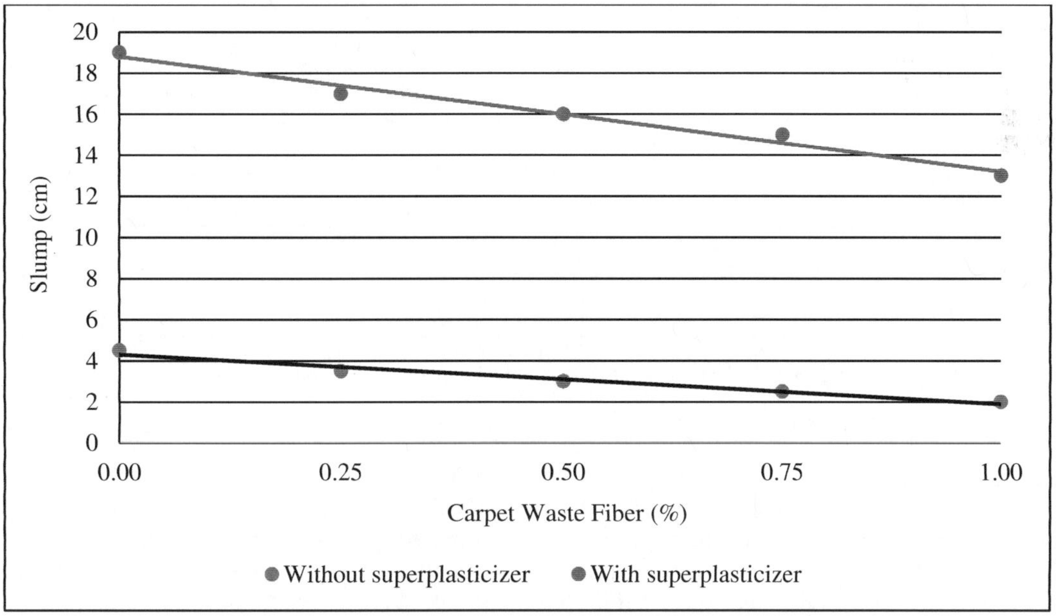


**Figure 6.** Effect of carpet waste fiber on slump value.

### *3.2. Density*

The use of artificial lightweight aggregate results in a maximum concrete density of 1.94 t/m$^3$, classifying it as a lightweight concrete (specific gravity less than 2 t/m$^3$), which can be used for medium structures [20]. Addition of carpet waste fiber increases concrete density to the optimal content of 0.67%. Afterwards, concrete density reduces due to increased pore density. The test results on concrete density are presented in Table 2, while the effect of carpet waste fiber on concrete density is shown in Figure 7.

**Table 2.** Concrete density.

| Fiber Content (%) | Density (t/m$^3$) | | | | | | Avg. (t/m$^3$) |
|---|---|---|---|---|---|---|---|
| | 1 | 2 | 3 | 4 | 5 | 6 | |
| 0.00 | 1.85 | 2.00 | 1.91 | 1.92 | 1.92 | 1.91 | 1.92 |
| 0.25 | 1.91 | 1.96 | 1.94 | 1.92 | 1.92 | 1.91 | 1.93 |
| 0.50 | 1.98 | 1.89 | 1.94 | 1.92 | 1.92 | 1.92 | 1.93 |
| 0.75 | 1.92 | 1.94 | 1.94 | 1.94 | 1.96 | 1.92 | 1.94 |
| 1.00 | 1.96 | 1.92 | 1.91 | 1.94 | 1.92 | 1.92 | 1.93 |

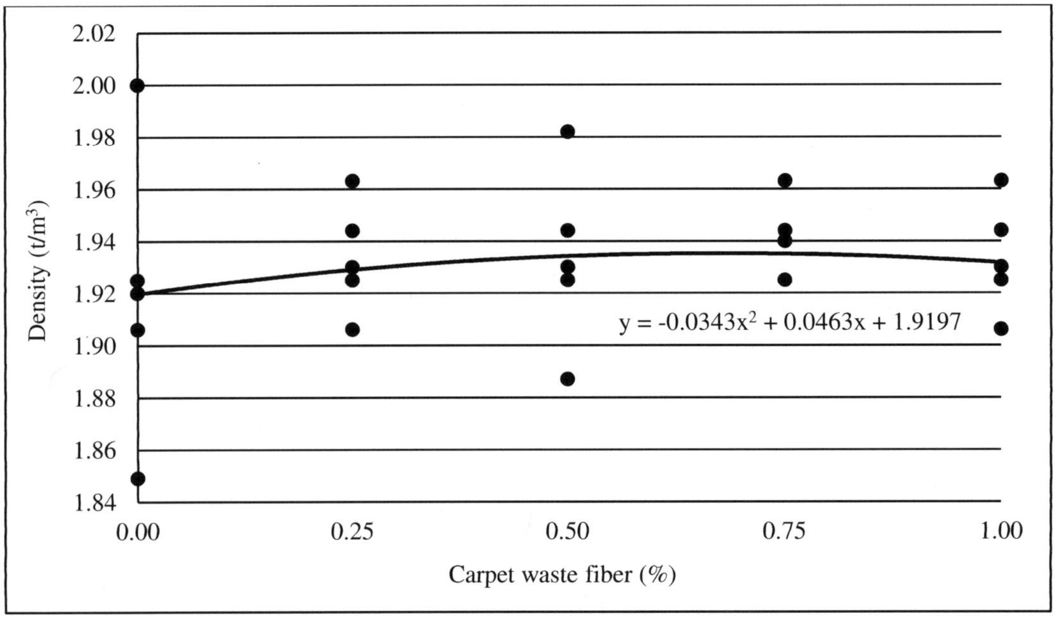


**Figure 7.** Effect of carpet waste fiber on concrete density.

### *3.3. Compressive strength*

Carpet waste fiber increase the compressive strength of artificial lightweight aggregate concrete up to certain content, followed by a decrease. Fiber makes concrete more resistant to cracking and impact. However, difficulty in mixing may result in inhomogeneous concrete mixtures and difficulty in solidifying may result in porous concrete due to trapped air, decreasing compressive strength [1]. The effect of carpet waste fiber on compressive strength (Figure 8) indicates that the compressive strength of artificial lightweight aggregate concrete can increase up to 17.22% by adding 0.44% carpet waste fiber.

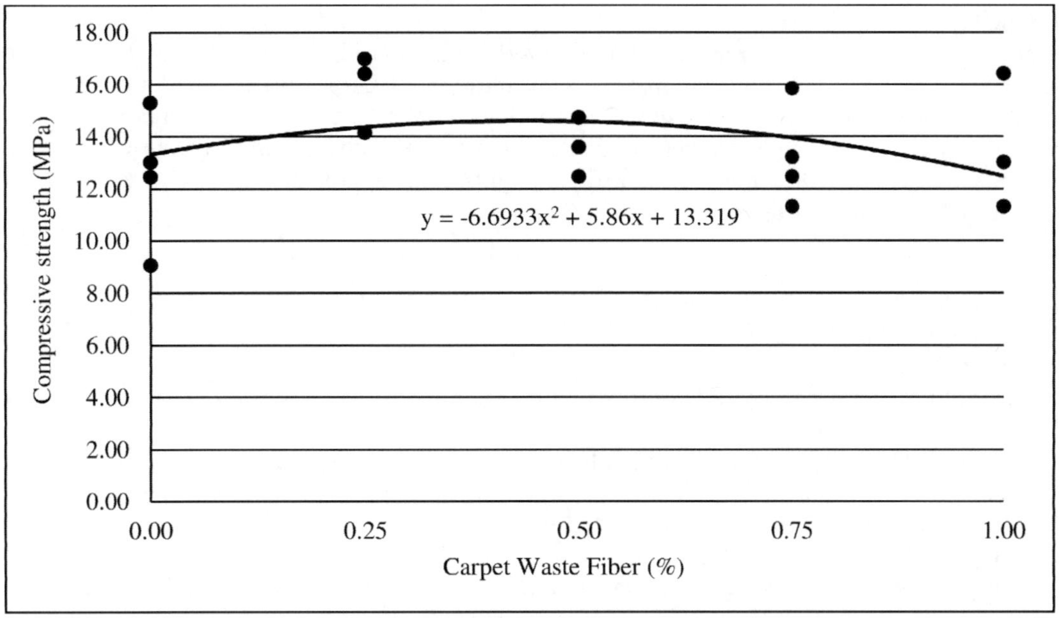


**Figure 8.** Effect of carpet waste fiber on compressive strength.

*3.4. Split tensile strength*

The split tensile strength of artificial lightweight aggregate concrete increases by up to 75.18% with addition of 0.85% carpet waste fiber (Figure 9). Artificial lightweight aggregate concrete without carpet fiber is brittle, with the test specimen split in two along the diameter. Meanwhile, artificial lightweight aggregate concrete with addition of carpet waste fiber is ductile with preliminary small cracks. This is in accordance with the fiber bridging and resisting distribution of cracking, allowing concrete paste to support a load even if it has cracked [12].

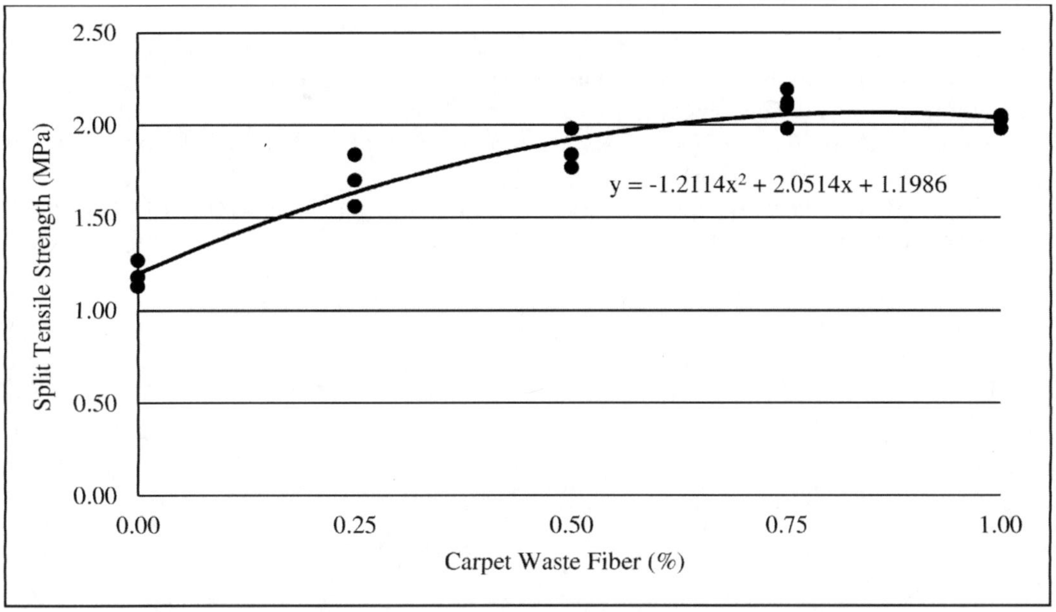


**Figure 9.** Effect of carpet waste fiber on split tensile strength.

IOP Conf. Series: Materials Science and Engineering **371** (2018) 012006 doi:10.1088/1757-899X/371/1/012006

## 4. Conclusions

Addition of superplasticizer can increase concrete workability up to a slump value of 19 cm. Concretes formed from artificial lightweight aggregate with addition of carpet waste fiber are classified as lightweight concrete with a maximum specific gravity value of 1.94 $t/m^3$. The compressive strength of artificial lightweight aggregate concrete increases by 17.22% with addition of 0.44% carpet waste fiber, while the split tensile strength of artificial lightweight aggregate increases by 75.18% with addition of 0.85% carpet waste fiber.

## 5. References


[1] Tjokrodimuljo K 2007 *Teknologi Beton* (Yogyakarta: Nafiri)
[2] Haryanto Y, Hermanto N I S, Pamudji G and Wardana K P 2017 *IOP Conf. Ser. Mater. Sci. Eng.* **267** 012016
[3] Haryanto Y, Widyaningrum A, Sudibyo G H and Maryoto A 2017 *MATEC Web of Conf.* **138** 01021
[4] Wang Y 1997 *J. of Mater. in Civ. Eng.* **8** 103
[5] Wang Y 2006 *Utilization of Recycled Carpet Waste Fibers For Reinforcement Concrete and Soil* (Cambridge: Wood Head Publishing Ltd.)
[6] Wang Y 2010 *Waste Biomass* **1** 135
[7] Herlihy J 1997 *Recycling in The Carpet Industry* (Carpet and Rug Industry)
[8] Schmidt H and Cieslak M 2008 *Waste Management* **28** 1182
[9] Joao M L R 2009 Mater. Research **12** 63
[10] Mohammadhosseini H and Abdul Awal, A S M 2013 Inter. J. of Research in Eng. and Tech. **2** 464
[11] Puskim 2002 *Modul Pengembangan Lempung (Shale) untuk Agregat Ringan Buatan (ALWA)* (Jakarta: Puskim)
[12] Haryanto Y 2006 Din. Rek. **2** 36
[13] Kristiyanto H 2006 J. Tek. Sip. FT UCY **1** 15
[14] Haryanto Y 2016 Techno **17** 73
[15] The National Standardization Agency of Indonesia 2008 *SNI 1972: 2008 Cara Uji Slump Beton*
[16] The National Standardization Agency of Indonesia 2011 *SNI 1974: 2011 Cara Uji Kuat Tekan Beton dengan Benda Uji Silinder*
[17] The National Standardization Agency of Indonesia 2002 *SNI 03-2491-2002 Metode Pengujian Kuat Tarik Belah Beton*
[18] Haryanto Y and Sudibyo G H 2016 *Contribution of Civil Engineering Toward Building Sustainable City* (Surabaya: Post Graduate Program, Civil Engineering Department, Faculty of Civil Engineering and Planning, ITS)
[19] Sudarmoko 2000 *Beton Fiber Lokal Untuk Non-Struktural* (Yogyakarta: Pusat Antar Universitas Ilmu Teknik, UGM)
[20] Neville A M 2011 *Properties of Concrete*, Fifth Edition (London: The English Language Book Society and Pitman Publishing)

IOP Conf. Series: Materials Science and Engineering **371** (2018) 012007 doi:10.1088/1757-899X/371/1/012007

# Effect of High Loss on Ignition-Fly Ash on Properties of Concrete Fully Immersed in Sulfate Solution

**Si-Huy Ngo[1], Trong-Phuoc Huynh[2], Thanh-Tam Thi Le[1], Ngoc-Hang Thi Mai[1]**

[1] Department of Engineering and Technology, Hong Duc University, No. 565, Quang Trung Street, Dong Ve Ward, Thanh Hoa City 440000, Vietnam
[2] Department of Rural Technology, College of Rural Development, Can Tho University, Campus II, 3/2 St., Ninh Kieu Dist., Can Tho City 900000, Vietnam
E-mail: htphuoc@ctu.edu.vn

**Abstract.** This paper presents the experimental results of an investigation on the engineering properties and durability of high loss on ignition-fly ash concrete fully immersed in 5% sodium sulfate solution. The fly ash used in this study is a raw material with the loss on ignition of 15.8%, which is much higher than the upper limitation of 6% as stipulated by ASTM C618. Eight concrete mixtures were designed with different water-to-binder ratios of 0.35 and 0.45. For each water-to-binder ratio, the fly ash was used to replace 0%, 10%, 20%, and 30% amount of cement. Test results show that the workability of fresh concrete increased, while its unit weight decreased with increasing the fly ash replacement level. In addition, the concrete mixtures with 10% and 20% fly ash content showed the higher compressive strength, especially at long-term ages, than that of the fly ash-free concrete mixture. Further, all of the fly ash concrete mixtures exhibited the good quality with ultrasonic pulse velocity values of higher than 4100 m/s.

## 1. Introduction

The use of fly ash in concrete is not only environmental desirable and energy saving, but also improves the mechanical properties and durability of concrete. The 10-30% cement replacement by fly ash resulted in improving the compressive strength of concrete [1-4]. Some other studies have found that fly ash could be used to replace up to 60% cement in concrete with better or comparable compressive strength [5-8]. In addition, the workability of concrete increased with increasing the fly ash content [1, 2, 5]. Moreover, the combination of fly ash and ground granulated blast-furnace slag enhanced both short- and long-term properties of concrete [8].

Previous studies have demonstrated that the presence of fly ash improved the resistance of mortar and concrete to sulfate attack. Sahmaran et al. [9] showed that the utilization of about 13-32% fly ash increased the sulfate resistance of mortar, while Chindaprasirt et al. [10] reported that the use of fly ash incorporating rice husk ash increased the compressive strength and reduced the expansion of mortar. It is noted that mortar samples in both studies were immersed in 5% sodium sulfate solution. Moreover, Sumer [11] investigated the sulfate resistance of concrete containing class-F and class-C fly ash in 10% magnesium sulfate solution. Test results exhibited that, both class-F and class-C fly ash significantly improved the resistance to sulfate attack of concrete with 10-17% fly ash replacing cement. With the use of fly ash to replace 50% cement, Torii et al. [12] indicated that compressive strength of fly ash concrete was significantly greater than that of fly ash-free concrete exposed to the 10% sodium sulfate solution. Under haft-buried in sulfate soil, the 20% replacement of cement by fly ash improved the sulfate resistance of concrete [13]. The utilization of fly ash with other mineral admixtures as slag and silica fume was very effective in against sulfate attack [8, 14].

However, the properties of fly ash concrete have been strongly depended on the quality of fly ash used [15]. It is noted that most of the previous studies used the selected fly ash with the loss on ignition ranged from 0.45 to 3.2%, satisfying the requirement of below 6% as stipulated by ASTM C618 [16]. In order to fill more information to the gap in the literature, the main objective of this study is to investigate the performance of concrete made with a high loss on ignition-fly ash and fully immersed in a sulfate solution.

## 2. Materials and experimental works

### *2.1. Materials and mixture proportions*

The fly ash used in this study was a raw material, which was sourced from Nghi Son thermal power plant in Vietnam, with a loss on ignition of 15.8%. The specific gravity of fly ash and cement were 2.16 and 3.12, respectively. Natural sand with fineness modulus of 2.97, density of 2.62 $T/m^3$, dry rodded weight of 1.50 $T/m^3$, moisture content of 4.35%, and water absorption capacity of 1.08% was used as the fine aggregate. The coarse aggregate used was crushed stone with the nominal maximum size of 12.5 mm, density of 2.69 $T/m^3$, dry rodded weight of 1.39 $T/m^3$, moisture content of 0.25%, and water absorption capacity of 0.08%. The superplasticizer with a specific gravity of 1.15 was used for all concrete mixtures with a dosage of 1% of the total amount of binder materials.

Table 1 shows the mix proportions for all concrete mixtures, which were designed with different water-to-binder ratios of 0.35 and 0.45. For each water-to-binder ratio, four concrete mixtures were designed with 0%, 10%, 20%, and 30% fly ash replacing cement. All of the concrete mixtures shown in Table 1 were designed in accordance with ACI 211.91 [17]. Nomenclature of the mixtures is described as follows: M35 and M45 denote the water-to-binder ratio of 0.35 and 0.45, respectively; the last numbers (0, 10, 20, and 30) are the percentages of fly ash replacing cement.

**Table 1.** Mix proportions for the preparation of concrete samples.

| Sample ID. | w/b | Concrete ingredients proportion (kg/m$^3$) | | | | | |
|---|---|---|---|---|---|---|---|
| | | Cement | FA | Sand | Stone | Water | SP |
| M35-0 | 0.35 | 514.1 | 0.0 | 985.2 | 751.1 | 175.5 | 5.1 |
| M35-10 | | 459.7 | 50.6 | 978.1 | 745.7 | 174.3 | 5.1 |
| M35-20 | | 405.1 | 101.5 | 970.9 | 740.3 | 173.0 | 5.0 |
| M35-30 | | 352.3 | 150.7 | 964.0 | 735.0 | 171.8 | 5.0 |
| M45-0 | 0.45 | 400.4 | 0.0 | 1082.1 | 751.8 | 175.7 | 4.0 |
| M45-10 | | 358.3 | 39.8 | 1076.0 | 747.5 | 174.7 | 4.0 |
| M45-20 | | 316.7 | 79.2 | 1069.9 | 743.3 | 173.7 | 4.0 |
| M45-30 | | 275.6 | 118.1 | 1063.9 | 739.1 | 172.7 | 3.9 |

### *2.2. Samples preparation and test methods*

The concrete ingredients were mixed in the laboratory using a mechanical mixer pan. Right after mixing, fresh concrete properties including slump and unit weight were measured. After that, the cylindrical concrete samples with a diameter of 100 mm and a height of 200 mm were prepared. These samples were de-molded 24 hours after casting and then fully immersed in 5% sodium sulfate solution until the testing days. The compressive strength and ultrasonic pulse velocity (UPV) tests were conducted at 3, 7, 14, 28, 56, and 91 days in accordance with ASTM C39 and ASTM C597, respectively. The reported values herein are the average value of three concrete samples.

## 3. Results and discussion

### *3.1. Characteristics of raw materials*

Chemical compositions of both cement and fly ash used in this study are shown in Table 2. With a low amount of calcium and the sum of silicon dioxide ($SiO_2$), aluminum oxide ($Al_2O_3$), and iron oxide ($Fe_2O_3$) greater than 70%, this fly ash was classified as class-F according to ASTM C618 [16]. It is noted that the loss on ignition of this fly ash was 15.8%, which is significantly higher than that of fly ash used in previous studies as aforementioned in the literature review. On the other hand, Table 3 presents the toxicity characteristics of the fly ash. As shown, a number of heavy metals of this fly ash were significantly lower than the allowable level stipulated by Vietnamese Standard [18].

**Table 2.** Chemical compositions of cement and FA.

| Compositions (wt. %) | $SiO_2$ | $Al_2O_3$ | $Fe_2O_3$ | CaO | MgO | $SO_3$ | $K_2O$ | $Na_2O$ | Others |
|---|---|---|---|---|---|---|---|---|---|
| Cement | 22.4 | 5.3 | 4.0 | 55.9 | 2.8 | 2.1 | 0.8 | 0.3 | 4.5 |
| FA | 48.4 | 20.4 | 4.8 | 2.8 | 1.4 | 0.3 | 1.1 | 0.8 | 4.3 |

**Table 3.** Toxicity characteristics of the fly ash.

| Items | Level of toxicity leached of heavy metals (mg/kg) | | | | | |
|---|---|---|---|---|---|---|
| | As | Cd | Pb | Hg | Zn | $Cr^{6+}$ |
| Fly ash | < 0.8 | < 0.6 | 16.2 | < 1.0 | 35.1 | < 1.0 |
| Regulatory limit | 40 | 10 | 300 | 4 | 5000 | 100 |

*3.2. Properties of fresh concrete*

The slump and unit weight of all concrete mixtures are given in Table 4. The slump values of M35 mixtures were lower than those of the corresponding M45 mixtures. This is because the workability of fresh concrete is closely associated with the water-to-binder ratio. Thus, the higher water-to-binder ratio resulted in higher workability. Moreover, for the same water-to-binder ratio, the slump of fresh concrete increased with increasing the replacement level of cement by fly ash. This is attributable to the spherical shape of fly ash particle and its dispersive ability. This finding is similar to the experimental results from previous studies [1, 2, 5]. It means that the effect of raw fly ash with a high loss on ignition on the workability of fresh concrete is similar to that of the selected fly ash with a low loss on ignition.
On the other hand, Table 4 also shows that the unit weight of fresh concrete decreased with increasing the fly ash content. This is due to the specific gravity of fly ash is much lower than that of cement. Additionally, when water-to-binder ratio increased, the unit weight of concrete decreased. Similarly, it is attributable to the highest specific gravity of cement as compared with other ingredients in concrete. Thus, more amount of cement resulted in the higher unit weight of concrete.

**Table 4.** Properties of fresh concrete mixtures.

| Sample ID. | w/b | SP dosage ($kg/m^3$) | FA content (%) | Slump (mm) | Fresh unit weight ($kg/m^3$) |
|---|---|---|---|---|---|
| M35-0 | 0.35 | 5.1 | 0 | 0 | 2580 |
| M35-10 | | 5.1 | 10 | 6 | 2550 |
| M35-20 | | 5.0 | 20 | 12 | 2510 |
| M35-30 | | 5.0 | 30 | 35 | 2490 |
| M45-0 | 0.45 | 4.0 | 0 | 36 | 2530 |
| M45-10 | | 4.0 | 10 | 45 | 2510 |
| M45-20 | | 4.0 | 20 | 57 | 2480 |
| M45-30 | | 3.9 | 30 | 70 | 2460 |

*3.3. Compressive strength*

Figures 1 and 2 show the compressive strength development of the M35 and M45 concrete mixtures that fully immersed in 5% sodium sulfate solution until 91 days. For the water-to-binder ratio of 0.35, the

IOP Conf. Series: Materials Science and Engineering **371** (2018) 012007 doi:10.1088/1757-899X/371/1/012007

control concrete mixture without fly ash (M35-0) showed the highest compressive strength at early ages. After 14 days age, the concrete mixture with 10% fly ash (M35-10) showed the highest compressive strength. At 91 days age, compressive strengths of the M35-10 and M35-20 mixtures were higher than that of the control mixture (M35-0), while the compressive strength of M35-30 mixture was similar to that of the control mixture.

The performance of concrete against sulfate attack is associated with the permeability performance and the density of concrete structure. At the beginning, cement hydration dominantly happened to form a large amount of C-S-H gel (Equation 1), which provided the high compressive strength for M35-0 mixture at an early state. After several weeks, the pozzolanic reaction significantly happened. The higher resistance of the M35-10 and M35-20 mixtures against sulfate attack than that of the M35-0 mixture is due to the conversion of $Ca(OH)_2$ into a secondary C-S-H gel by pozzolanic reaction (Equation 2). The reduction in $Ca(OH)_2$ led to the high density of concrete structure. In general, the compressive strength of non-fly ash concrete mixture exhibited a lower increasing rate than that of fly ash concrete mixtures. It is attributable to the sulfate attack reaction (Equations 3-6) become more dominant after several weeks. The expansion of gypsum and ettringite formed during the reaction process led to a reduction in compressive strength of the concrete. The lower reduction rate in compressive strength of fly ash concrete is due to the denser structure that prevents the penetration of sulfate ion into concrete. In addition, the lower compressive strength of the concrete mixture with 30% fly ash is due to the presence of un-hydrated/ partially hydrated fly ash particle in concrete. Previous studies [1-8] have proved that the use of fly ash with optimal dosage improved the properties of concrete. However, the optimal dosage of fly ash was varying depended on its properties and concrete proportions. If the content of fly ash was over the optimal level, a part of it acted as fine aggregate instead of acting as a cementitious material.

For the water-to-binder ratio of 0.45, the concrete mixture with 10% fly ash (M45-10) showed the highest compressive strength value, while the concrete mixture with 30% fly ash (M45-30) exhibited the lowest compressive strength value. It is noted that the compressive strengths of concrete mixture with 20% fly ash (M45-20) and control mixture (M45-0) were similar before 56 days age. At 91 days age, however, M45-20 mixture showed a higher compressive strength than the M45-0 mixture, while M45-30 mixture showed a similar compressive strength to M45-0 mixture. Similar to M35 mixtures, the higher compressive strength of M45-10 and M45-20 mixtures than that of M45-0 mixture at long-term ages is also due to the denser structure of fly ash concrete compared with that of the control concrete without fly ash. This is associated with the pozzolanic reaction to form the secondary C-S-H gel in fly ash concrete. This finding proved that the sulfate attack resistance of concrete with a high loss on ignition-fly ash used in this study is similar to that of concrete with a low loss on ignition-fly ash used in previous studies [8-14]. It also means that the high loss on ignition-fly ash can be used in concrete under sulfate attack condition.

Cement hydration:

$$\left(C_3S, C_2S\right) + H_2O \rightarrow \text{C-S-H} + Ca\left(OH\right)_2 \tag{1}$$

Pozzolanic reaction:

$$Ca\left(OH\right)_2 + SiO_2 \rightarrow \text{C-S-H} \tag{2}$$

Sulfate attack reaction:

$$Na_2SO_4.2H_2O + Ca\left(OH\right)_2 \rightarrow CaSO_4.2H_2O + 2NaOH \tag{3}$$

$$Na_2SO_4.2H_2O + \text{C-S-H} \rightarrow CaSO_4.2H_2O + \text{N-S-H} \tag{4}$$

$$C_4AH_{13} + 3CaSO_4.2H_2O + 14H_2O \rightarrow C_3A.3CaSO_4.32H_2O + Ca\left(OH\right)_2 \tag{5}$$

$$C_3A.CaSO_4.12H_2O + 2CaSO_4.2H_2O + 16H_2O \rightarrow C_3A.3CaSO_4.32H_2O \tag{6}$$

IOP Conf. Series: Materials Science and Engineering **371** (2018) 012007 doi:10.1088/1757-899X/371/1/012007

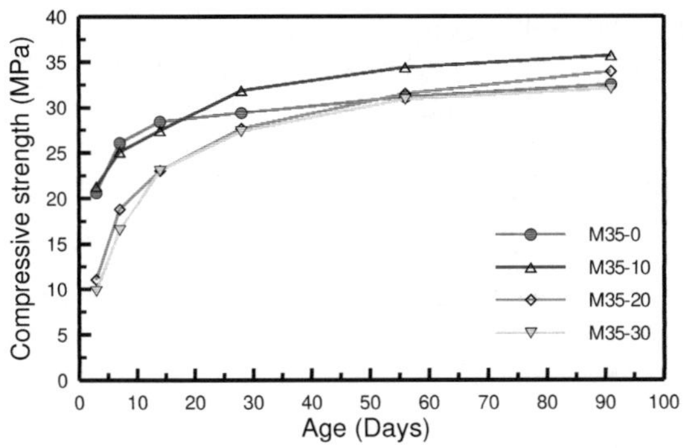


**Figure 1**. Compressive strength development of M35 mixtures.

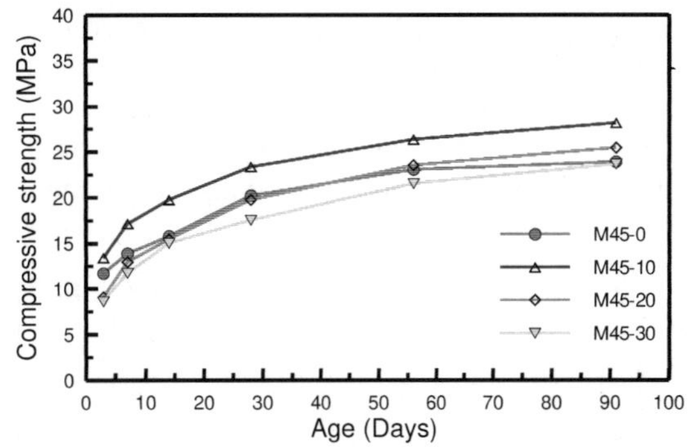


**Figure 2**. Compressive strength development of M45 mixtures.

*3.4. Strength efficiency of cement*

The strength efficiency of cement (SEC) of the M35 and M45 mixtures are presented in Figures 3 and 4, respectively. SEC is a yielded compressive strength provided by per kilogram of cement. It is related to the production cost, energy consumption, and carbon dioxide emission during the production of cement and concrete. In general trend, the concrete mixtures with 30% fly ash showed the highest SEC value, whereas the concrete mixtures without fly ash showed the lowest SEC value. In other words, high fly ash content provided high SEC value. It means that the utilization of fly ash is very effective in terms of production cost and environmental benefit.

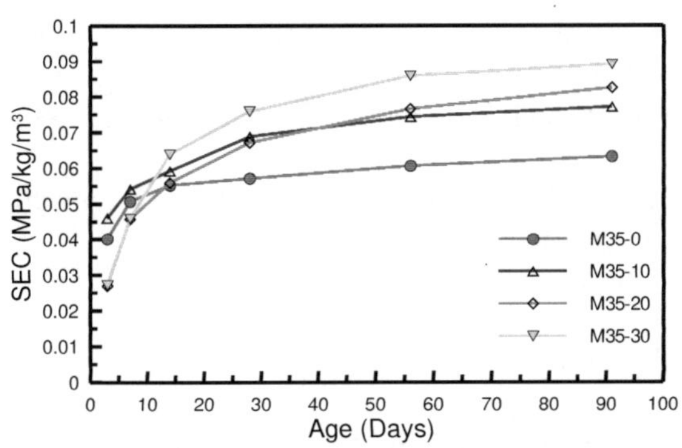


**Figure 3**. Strength efficiency of cement of M35 mixtures.

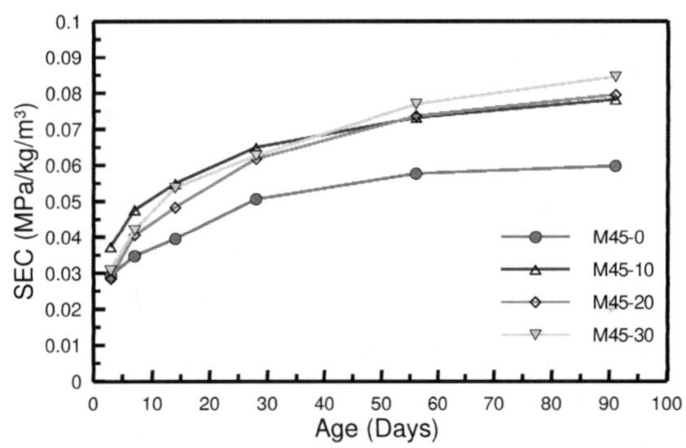


**Figure 4**. Strength efficiency of cement of M45 mixtures.

*3.5. Ultrasonic pulse velocity*

In order to evaluate the durable properties of concrete such as the presence of voids and cracks, porosity, and permeability, the ultrasonic pulse velocity (UPV) test was conducted. The UPV value is often associated with a concrete structure density and is an indirect indicator of concrete quality. Figures 5 and 6 present the UPV values of the M35 and M45 concrete mixtures over the time, respectively. For the water-to-binder ratio of 0.35, the fly ash-free concrete mixture (M35-0) showed the highest UPV value. However, at 91 days, the UPV value of the M35-10 concrete mixture was close to that of the M35-0 concrete mixture. For the water-to-binder ratio of 0.45, UPV values of concrete mixtures with no fly ash (M45-0) and 10% fly ash (M45-10) were similar. While the increasing rate of UPV values of two others with 20% and 30% fly ash content (M45-20 and M45-30) was higher than that of control concrete mixture without fly ash (M45-0). Due to the higher specific gravity of cement in comparison with that of fly ash, the control mixture showed a higher density and a faster cement hydration than the fly ash concrete at an early state. Thus, the UPV value of control concrete mixtures was high at early state. However, after 56 days, the inner structure of fly ash concrete was denser due to the contribution of dominant pozzolanic reaction to convert $Ca(OH)_2$ into a secondary C-S-H gel. This led to the higher increasing rate of UPV value of fly ash concrete than the control concrete mixtures. Moreover, all concrete mixtures had UPV values of higher than 4100 m/s. As suggested by Carcano and Moreno [19], the concrete sample with UPV value of above 4100 m/s can be classified as very good quality. This also means that all of the concrete mixtures prepared with a high loss on ignition-fly ash showed the good quality, which is comparable to that of the concrete made with selected fly ash.

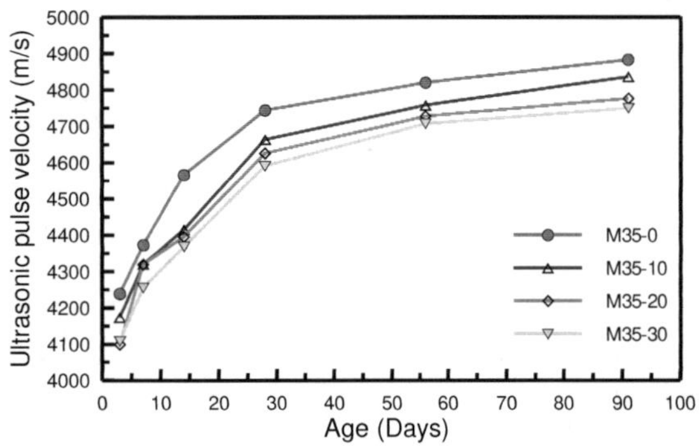


**Figure 5**. Ultrasonic pulse velocity values of M35 mixtures.

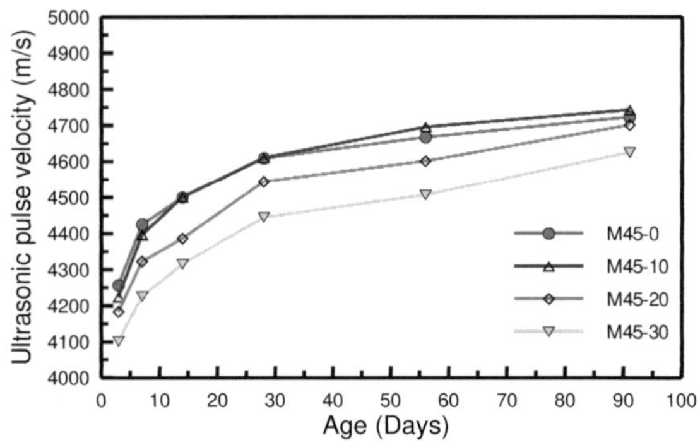


**Figure 6**. Ultrasonic pulse velocity values of M45 mixtures.

## 4. Conclusions

The present study presents the performance of concrete made with a high loss on ignition-fly ash that fully immersed in 5% sodium sulfate solution. Based on the experimental results, the following conclusions may be drawn:

- The workability of fresh concrete mixture increased and its unit weight reduced with increasing the fly ash content.
- Under 5% sodium sulfate solution, the concrete made with 10% and 20% fly ash replacing cement showed an improved compressive strength, while concrete made with 30% fly ash showed a comparable compressive strength value to the control concrete without fly ash in long-term ages.
- The strength efficiency of cement values of all concrete mixtures increased with increasing fly ash replacement level.
- All of the fly ash concrete samples showed the good quality with the ultrasonic pulse velocity values of greater than 4100 m/s.
- The experimental results of the present study demonstrated that the fly ash with a high loss on ignition can be used in concrete under sulfate condition with similar or even better concrete properties.

## 5. References

[1] Wankhede P R and Fulari V A Effect of fly ash on properties of concrete 2014 *Int. J. Emerg. Technol. Adv. Eng.* **4** 284–9

[2] Marthong C and Agrawal T P Effect of fly ash additive on concrete properties 2012 *Int. J. Eng. Res. Appl.* **4** 1986–91

[3] Bansal R, Singh V and Pareek R K Effect on compressive strength with partial replacement of fly ash 2015 *Int. J. Emerg. Technol.* **6** 1–6

[4] Mohamed H A Effect of fly ash and silica fume on compressive strength of self-compacting concrete under different curing conditions 2011 *Ain Shams Eng. J.* **2** 79-86

[5] Nail T R and Ramme B W Effect of high-lime fly ash content on water demand, time of set, and compressive strength of concrete 1990 *ACI Mater. J.* **87** 619–26

[6] Oner A, Akyuz S and Yildiz R An experimental study on strength development of concrete containing fly ash and optimum usage of fly ash in concrete 2004 *Cem. Concr. Res.* **35** 1165–71

[7] Siddique R Performance characteristics of high-volume class F fly ash concrete 2003 *Cem. Concr. Res.* **34** 487–93
[8] Li G and Zhao X Properties of concrete incorporating fly ash and ground granulated blast-furnace slag 2003 *Cem. Concr. Compos.* **25** 293–9
[9] Sahmaran M, Kasap O, Duru K and Yaman I O Effects of mix composition and water-cement ratio on the sulfate resistance of blended cements 2007 *Cem. Concr. Compos.* **29** 159–67
[10] Chindaprasirt P, Kanchanda P, Sathonsaowaphak A and Cao H T Sulfate resistance of blended cements containing fly ash and rice husk ash 2007 *Constr. Build. Mater.* **21** 1356-61.
[11] Sumer M Compressive strength and sulfate resistance properties of concretes containing Class F and Class C fly ashes 2012 *Constr. Build. Mater.* **34** 531-6.
[12] Torii K, Taniguchi K and Kawamura M Sulfate resistance of high fly ash content concrete 1995 *Cem. Concr. Res.* **25** 759–68
[13] Irassar E F, Maio A D, and Batic O R Sulfate attack on concrete with mineral admixtures 1996 *Cem. Concr. Res.* **26** 113–23
[14] Thomas M D A, Shehata M H, Shashiprakash S G, Hopkins D S and Cail K Use of ternary cementitious systems containing silica fume and fly ash in concrete 1999 *Cem. Concr. Res.* **29** 1207–1214
[15] Bilodeau A and Malhotra V M High-Volume Fly Ash System Concrete Solution for Sustainable Development 2000 ACI mater. J. **97** 41-7
[16] ASTM C618 2005 *Standard specification for coal fly ash and raw or calcined natural pozzolan for use in concrete*
[17] ACI Committee 211 1991 *Standard practice for selecting proportions for normal, heavyweight, and mass concrete*
[18] QCVN 07 *National technical regulation on hazardous waste thresholds* (Vietnamese Standard)
[19] Carcano R S and Moreno E I Evaluation of concrete made with crushed limestone aggregate based on ultrasonic pulse velocity 2008 *Constr. Build. Mater.* **22** 1125-31.

IOP Conf. Series: Materials Science and Engineering **371** (2018) 012008 doi:10.1088/1757-899X/371/1/012008

# Properties of Portland Cement Type V Mortar Mixed with Ground Rice Husk Ash and Limestone Powder

**Burachat Chatveera[1] and Permphon Srinourn[2]**

[1] Associate Professor, Department of Civil Engineering, Faculty of Engineering Thammasat University, Rangsit Campus, Pathum Thani, Thailand.
[2] Ph.D. Candidate, Department of Civil Engineering, Faculty of Engineering Thammasat University, Rangsit Campus, Pathum Thani, Thailand.
E-mail: cburacha@engr.tu.ac.th, permphon@hotmail.com

**Abstract.** The objective of this research is to study the Properties of Portland cement Type V mortar mixed with Ground Rice Husk Ash (GRHA) and two types of limestone powder (LS1 and LS2). The physical and chemical properties of Portland cement Type V, ground rice husk ash, and limestone powder were tested. The properties of mortar such as water requirement, compressive strength, compressive strength loss and drying shrinkage were investigated. The solutions used in this study were sodium sulfate ($Na_2SO_4$) and magnesium sulfate ($MgSO_4$). The concentrations of both solutions were 5% by weight in accordance with the ASTM C 1012 standard. GRHA, LS1, and LS2 have particle sizes of 29, 12 and 128 μm. The replacement levels of GRHA, LS1, and LS2 in Portland cement Type V were 0%, 20% and 40% wt. of cementitious materials. The ratio of water to cementitious material was controlled based on the water requirements conforming to flow value at 110 ± 5%. The results showed that the compressive strength of mortar was decreased with increase the percentage replacements of GRHA and LS in Portland cement Type V. At 180 days, the compressive strength of all mortars was lower than that of the control mortar. Mortar containing 20% of GRHA and LS (C80R10LS110) was the highest development in compressive strength and as high as that of the control mortar. From the results of sulfate resistance, it was found that the higher was the percentage replacements of GRHA and LS in Portland cement Type V, the lower was the strength loss and drying shrinkage of mortar. The strength loss of mortars in magnesium sulfate solution was more than those in sodium sulfate solution.

## 1. Introduction

Thailand has coastal areas, the two sides of the eastern seaboard, Gulf of Thailand, the Pacific and the west coast of the Andaman Sea in Indian Ocean. The long coastline is about 2,600 km, covering 23 coastal provinces with a population of over 11 million people live in areas that are critical habitats such as areas for industry, commerce and tourism. At present, many coastal areas have building constructions to accommodate the growth of the economy and society. Building-sites near the sea, like a dam abutment pier lighthouse, etc., must take into account the effects of corrosion due to sulfate and concrete. If you do not take into account the effects of corrosion due to sulfate in the beginning of the construction, the concrete can be damaged in the near future. The need to repair damaged concrete will be very costly. Features of the cement used must be appropriate to the use of various types. Standard

testing according to ASTM C150 is divided into five of Portland cement. The Sulphate-resistant Portland cement is Portland Cement Type V. These include cement seal blue elephant seals, sharks and TPI blue.
Portland cement type 5 can withstand sulfate solution to protect steel from rust compared to other types of cement. [1] Therefore, choosing the right kind of cement and pozzolan materials can help reduce the impact of sulfates and chlorides in seawater to develop and improve some properties of concrete that can be used. Concrete includes inert materials such as limestone, crushed stone, sand and dust profile that looks quite capable of increasing the adhesion of concrete. In addition, the ability to pour reduces oily water and makes surface finishing of concrete easier. However, the use of inert material to replace cement to make concrete were reduced because of the lower volumes produced. A pozzolan material is a material with silica and alumina as the main component. It will not qualify interface by itself but when it reacts with the calcium hydroxide and water temperature, pozzolan materials substances will have capable interface. Pozzolan materials include calcined clay, coal ash or fly ash, silica fume and ground RHA.

## 2. Research Objective

2.1. To study the chemical composition and physical properties of powdered limestone, rice husk ash, Portland cement and its five categories.

2.2. To study the development of calcium hydroxide, cavity volume, distribution of the cavity and the durability of mortar and limestone powder, rice husk ash in cement grinding at cement type 5.

2.3 To determine the optimum proportion of powdered limestone and crushed husk ash in Portland cement type 5.

## 3. Scope of research

3.1. Portland cement type 5 (TPI blue) from TPI Polene Public Company Limited from Saraburi province was use in this studied.

3.2. Limestone powder from TPI Polene Public Company Limited (TPIPL) with the magnitude from 0.0 to 0.1 mm (0-100 microns) using as LS1 and the magnitude from 0.1 to 0.6 mm (100-600 microns) using as LS2.

3.3. RHA from thermal power plant RICE MILL GRANNARY Ltd. (Thailand) [2]

3.4. Water to binder ratio (w/b) based on the flow test roll to spread the flow equals 110 ± 5 mm.[3]

3.5. Physical properties and chemical properties of Portland cement type V, limestone powder and rice husk ash, including specific gravity, bulk density, moisture content, specific surface area in the form of Blaine fineness, particle size distribution, X - Ray Diffraction (XRD) and surface characteristics of the particles by using Scanning Electron Microscope (SEM).

3.6. Air content testing for the fresh properties of mortar and the mechanical properties of mortar was testing the compressive strength at the age of 3, 7, 28, 60, 90, 120 and 180 days. [4]

3.7. The Porosity testing of mortars using the Mercury Intrusion Porosimetry (MIP) technique.

3.8. Test for calcium hydroxide ($Ca(OH)_2$) of the mortar using the technique Thermal Gravimetric Analysis (TGA).

3.9 The durability testing of mortar including loss of compression, expansion, drying shrinkage, and autogenous shrinkage due to the sodium sulfate and magnesium sulphate attack.

## 4. Benefits of research

4.1. Keeping in mind the basic features of the five types of Portland cement, rice husk ash and limestone powder.

4.2. Keeping in mind the capabilities and behavior to get underway to replace Portland cement type V with rice husk ash and limestone powder.

4.3. Keeping in mind the mixture of Portland cement type V husk ash and suitable limestone powder.

## 5. Framework

The concept of bringing powdered limestone and rice husk ash, which is produced by the agricultural industry from grain mills came from power plants that use heat energy from burning the chaff. Brick kiln as inert materials and pozzolan supplies in concrete improves certain properties of the concrete and reduce the cost of it. There is also a byproduct of the domestic materials to make useful and add value to the RHA.

## 6. Results and discussions

### *6.1. Chemical composition and Physical properties*

The test results of the chemical composition of the powders are shown in Table 1. It shows that Portland cement type 5 has CaO as its main component with 60.50% and $SiO_2$ content of 18.40 %. The crushed rice husk ash amount has $SiO_2$ as the main component with 88.75%. The amount of CaO is the main component of limestone powder LS1 (12 μm) and LS2 (128 μm) with almost similar 46.77% and 44.94%, respectively. The amount of magnesium oxide (MaO) of the two kinds of powdered limestone were also almost similar with 2.38% and 2.92%, respectively.

**Table 1**. The Chemical Composition of Portland Cement, Rice Husk Ash and Limestone Powder

| Chemical composition (Percentage by Weight) | Portland cement Type 5 | RHA crushed | Limestone Powdered LS1 | Limestone Powdered LS2 |
|---|---|---|---|---|
| 1. Silicon dioxide ($SiO_2$) | 18.40 | 88.75 | 8.97 | 11.27 |
| 2. Aluminium oxide ($Al_2O_3$) | 4.80 | 0.85 | 1.02 | 0.80 |
| 3. Iron oxide ($Fe_2O_3$) | 3.50 | 1.07 | 0.37 | 0.32 |
| $SiO_2$+$Al_2O_3$+$Fe_2O_3$ | 26.70 | 90.67 | 10.36 | 12.39 |
| 4. calcium oxide (CaO) | 60.50 | 0.44 | 46.77 | 44.64 |
| 5. magnesium oxide (MgO) | 1.00 | 0.07 | 2.38 | 2.92 |
| 6. Potassium oxide ($K_2O$) | 0.40 | 1.74 | 0.13 | 0.14 |
| 7. sodium oxide ($Na_2O$) | 0.30 | 0.05 | 0.02 | 0.02 |
| 8. Sulfur trioxide ($SO_3$) | 2.60 | 0.02 | 0.33 | 0.12 |
| 9. free lime (Free CaO) | 1.30 | 0.06 | - | - |
| 10. Percent weight loss,burn (LOI) | 1.82 | 7.27 | 39.54 | 38.68 |

The physical properties of the materials used in the study are shown in Table 2, the specific surface area by Blaine of cement, rice husk ash, crushed limestone powder LS1 and LS2 is equal to 3400, 4900, 6375 and 409 $cm^2$/ g. The specific gravity of 3.17, 2.12, 2.76 and 2.77, respectively, the demand for water is equal to 100, 109, 101 and 105, respectively, the index was compared with mortar cement, usually when the mortar cement is already 7 days existing. 96, 82, 84 and 79%, respectively, at 28 days is equal to 94, 86, 89 and 81 respectively.

Considering the chemical composition and physical properties of rice husk ash in Table 3, which is a material pozzolan properties through standards ASTM C 618 [5] RHA grinding four hours so it can be used in place of cement. (Purachatra and baby Anay, 2008).

Fine limestone powder is a semi-inert material that reacts chemically and dissipates the powder. The cement reacts with water to hydrate faster. Calcium oxide (CaO) reacts with water to form calcium hydroxide. $Ca(OH)_2$ can be used as a basic substance in the pozzolanic reaction.

**Table 2.** The physical properties of rice husk ash, Portland cement and limestone powder.

| Physical Properties | Portland Cement Type 5 | RHA crashed | Limestone Powdered LS1 | Limestone Powdered LS2 |
|---|---|---|---|---|
| 1.motar cement (percent) | 0.46 | 0.26 | 0.15 | 0.17 |
| 2.With Blaine specific surface area (Sq. Cm per gram) | 3.400 | 4.900 | 6.375 | 409 |
| 3.specific gravity | 3.17 | 2.12 | 2.76 | 2.77 |
| 4.Resolution (parade size, percentage holding) | | | | |
| - 75 micrometer | 0.61 | 13.00 | 11.04 | 88.98 |
| - 45 micrometer | 4.47 | 30.34 | 19.90 | 91.12 |
| - 36 micrometer | 6.89 | 39.00 | 23.40 | 91.68 |
| 5.Saength Index | | | | |
| At the age of 7 days (the equivalent of the control sample) | 100 | 82 | 84 | 79 |
| At the age of 28 days (the equivalent of the control sample) | 100 | 86 | 89 | 81 |
| 6.The demand of water (per cent) | 100 | 109 | 101 | 105 |

**Table 3.** The properties of the rice husk ash compared with ASTM C 618 standard.

| Chemical composition (Percentage by weight) | ASTM C 618 (Class N) | RHA crushed |
|---|---|---|
| 1. Silicon dioxide ($SiO_2$) | - | 88.75 |
| 2. Aluminium oxide ($Al_2O_3$) | - | 0.85 |
| 3. Iron oxide ($Fe_2O_3$) | - | 1.07 |
| $SiO_2+Al_2O_3+Fe_2O_3$ | lowest 70 percent | 90.67 |
| 4. calcium oxide (CaO) | - | 0.44 |
| 5. magnesium oxide (MgO) | - | 0.07 |
| 6. Potassium oxide ($K_2O$) | - | 1.74 |
| 7. sodium oxide ($Na_2O$) | Up 1.5 percent | 0.05 |
| 8. Sulfur trioxide ($SO_3$) | Up 4.0 percent | 0.02 |
| 9. free lime (Free CaO) | - | 0.06 |
| 10. Percent weight loss,burn (LOI) | Up to 10 percent | 7.27 |
| **Physical Properties** | | |
| 1. moisture content (percent) | Up 3.0 percent | 0.26 |
| 2. With Blaine specific area. (Sq.Cm per gram) | | 4,900 |
| 3. Specific gravity | - | 2.12 |
| 4. Resolution (particle size , percentage holding) | - | |
| - 75 micrometer | - | 13.00 |
| - 45 micrometer | | 30.34 |
| - 36 micrometer | | 39.00 |
| 5. Strength Index | | |
| At the age of 7 days (the equivalent of the control sample) At the age of 28 days (the equivalent of the control sample) | lowest 75 percent<br>lowest 75 percent | 82<br>86 |
| 6. The demand of water (per cent) | Up to 115 percent | 109 |

## *6.2. Characteristics of the powder Particles*

*6.2.1. Particle Portland Cement Type 5.* The characterize the particles of Portland cement type 5 with SEM, magnified 7500 times, respectively, in 1500 and found that the particles of Portland Cement Type 5 has edges. Amorphous Surface roughness and particle size mixed together as shown in Figure 1(a) and (b).

IOP Conf. Series: Materials Science and Engineering **371** (2018) 012008 doi:10.1088/1757-899X/371/1/012008

Figure 1 (a) SEM 1500-fold magnification

Figure 1 (b) SEM 7500-fold magnification

*6.2.2. Particle RHA.* The characteristics of rice husk ash particles before grinding with SEM, it was expanded in 1500 and 7500 and found that a relatively coarse particles has a high porosity, smooth and hollow as shown in Figure 2(a) and (b).

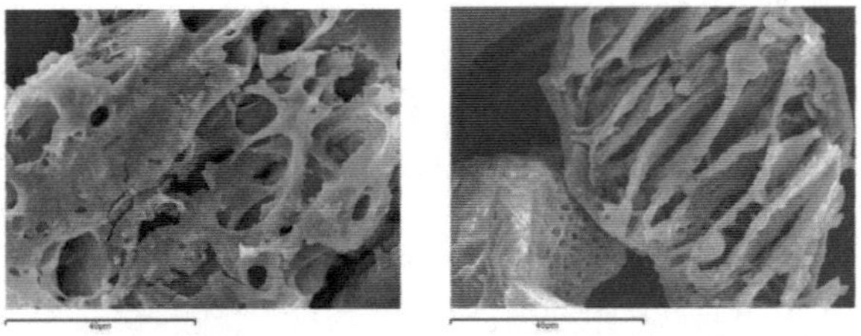

Figure 2 (a). SEM 1500-fold magnification

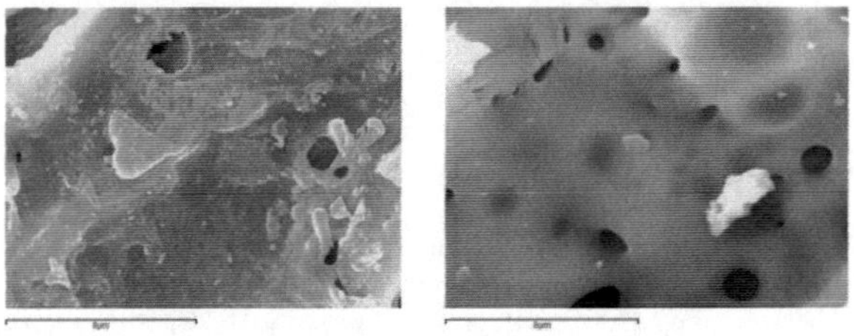

Figure 2 (b). SEM 7500-fold magnification

*6.2.3. RHA grinded for four hours.* The enlarged photos of RHA grinded for four hours were smaller in size. There's Resolution Cavities and the pores are small compared with RHA that are not crushed as shown in Figure 3.

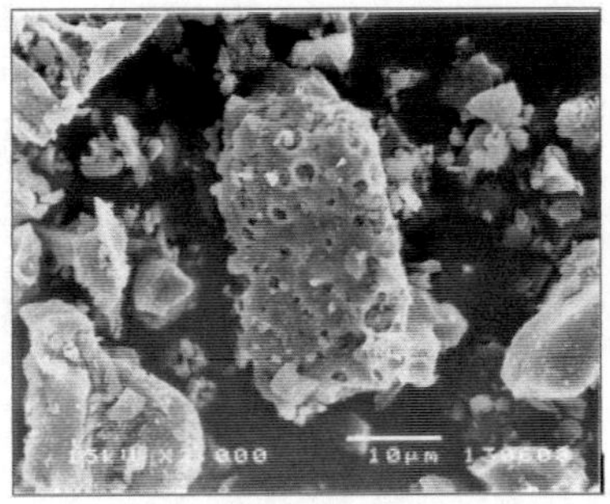

Figure 3. SEM 7500-fold magnification

*6.2.4. limestone powder particles.* The characterized the particles of powdered limestone LS1 that was expanded in 1500 and 7500, respectively, found that powdered limestone particles are smaller and more close to the Portland Cement Type 5 which is a triangle. Overall particle size is not much different as shown in Figure 4(a) and (b).

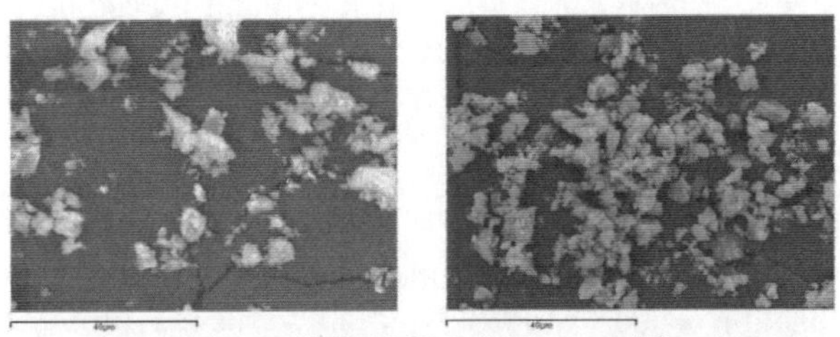

Figure 4 (a). SEM 1500-fold magnification

Figure 4 (b). SEM 7500-fold magnification

The characterized the particles of powdered limestone LS2 that was expanded in 1500 and 7500 respectively as a shaped texture mixed with angles and particle size larger than the LS1 as shown in Figure 5(a) and (b).

Figure 5 (a). SEM 1500-fold magnification

IOP Conf. Series: Materials Science and Engineering **371** (2018) 012008 doi:10.1088/1757-899X/371/1/012008

Figure 5 (b). SEM 7500-fold magnification

### *6.3. Strength of cement mortar*

Test compressive strength of mortar mix, rice husk ash, crushed and powdered limestone to replace Portland cement type 5 percent 0, 20, 30 and 40 shown in Figure 1.1 to 1.3 showed that at the age of 180 days Morse. Guitar mix crushed rice husk ash and limestone powder every test. Compressive strength less mortar control. The test is similar compression control most. Cement is a collection of 80 RHA crushed limestone powder, 10 percent LS1 (12 micron) of 10 (C80R10LS1 # 10) was 96%. Replacing Portland cement type 5 with crushed rice husk ash and limestone powder in a ratio of 20 percent by weight of the series of mortar mix, rice husk ash and limestone powder grinding LS1 (12 microns) with the most strength. Followed by a mortar mixer were crushed husk ash. And testing of mortar mix, rice husk ash and limestone powder grinding LS2 (128 micron) in order to extract the details below.

Ages 3-28 showed that compressive strength test of mortar mixed with crushed rice husk ash and limestone powder LS1 most valuable is a collection tested C80R10LS1 # 10 with a compressive strength at 28 days was 89%. Ages 28-90 day test mortar mix crushed husk ash and limestone powder LS1 most valuable is a collection tested C80R10LS1 # 10 with a compressive strength at 90 days was 94%. The age range 90-180 days with test compression is most C80R10LS1 # 10 sets a compression of 96% in this segment has been observed that the compressive strength of the mortar mix all RHA. 20 has compressive strength better. The compression of 96%, which is close to the test mortar mix husk ash and limestone powder LS1.

Replacing Portland cement type 5 with crushed rice husk ash and limestone powder was found that 30 % of test mortar mix husk ash and limestone powder LS1 has the most strength. Followed by mortar mix husk ash plain. And testing of mortar mix, rice husk ash and limestone powder LS2, respectively, as detailed below.

Ages 3-28 showed that compressive strength test of mortar mixed with crushed rice husk ash and limestone powder LS1 most valuable is a collection tested C70R20LS1 # 10 with a compressive strength at 28 days was 69%. Ages 28-90 day test of mortar mix, rice husk ash and limestone powder LS1 most valuable is a collection tested C70R20LS1 # 10 with a compressive strength at 90 days was 75%. The age range 90-180 days with test compression most C70R20LS1 # 10 is set equal to 82 % and compressive strength of mortar mixed with 30% of all RHA crushing compressive strength better. The compression of 80%, which is close to the test mortar mix husk ash and limestone powder LS1.

Replacing Portland cement type 5 with crushed rice husk ash and limestone powder was found that 40 percent of test mortar mix crushed husk ash and limestone powder LS1 has the most strength. Followed by mortar mix husk ash plain. And testing of mortar mix, rice husk ash and limestone powder LS2 respectively. Ages 3-28 showed that compressive strength test mortar mix husk ash and limestone powder LS1 most valuable is a collection tested C60R30LS1 # 10 with a compressive strength at 28 days was 61%. Ages 28-90 day test of mortar mix, rice husk ash and limestone powder LS1 most valuable is a collection tested C60R30LS1 # 10 with a compressive strength at 90 days was 65%. The age range 90-180 days with test compression most C60R20LS1 # 10 is set equal to 65 % and compressive strength of mortar mixed with 40% of all RHA crushing compressive strength better. The compression of 65%, which is close to the test mortar mix husk ash and limestone powder LS1. Test results can be deduced. LS1 replaced with limestone powder particle size is small and the nearby cement. [6]Can interpolate the distribution of cement, as well as small particles also help fill the gap of the cavity. Make the structure more dense paste. The ability to get a reaction and hydration thus outperformed LS2 limestone powder particles which are larger.

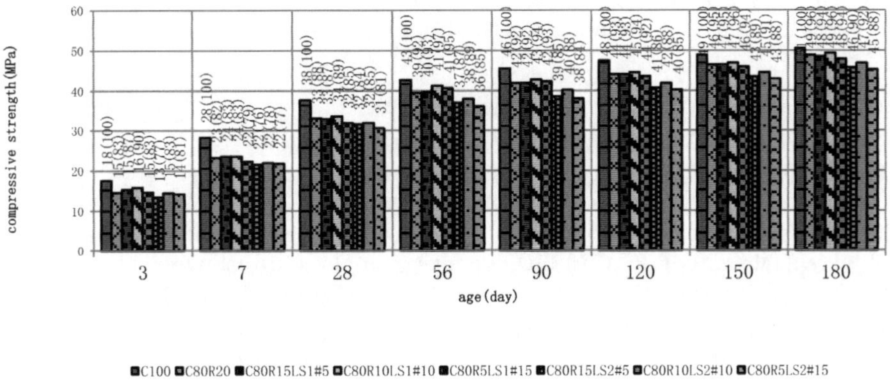


Figure 6. The compressive strength of the mortar mix, rice husk ash, crushed limestone powder and 20 percent. Replace Portland cement in five categories.

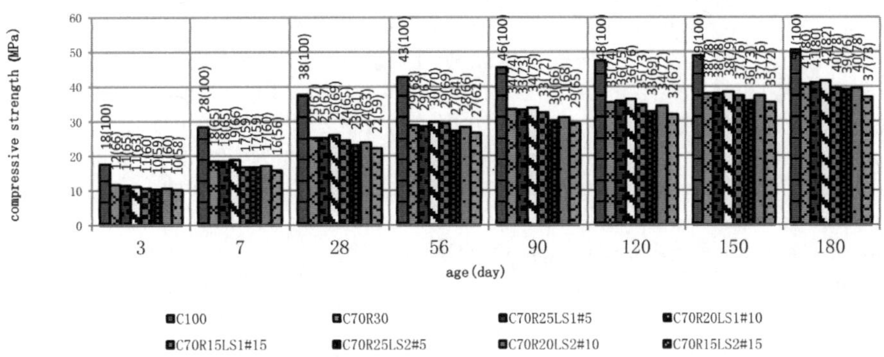


Figure 7. The compressive strength of the mortar mix, rice husk ash, crushed limestone powder and 30 percent. Replace Portland cement in five categories.

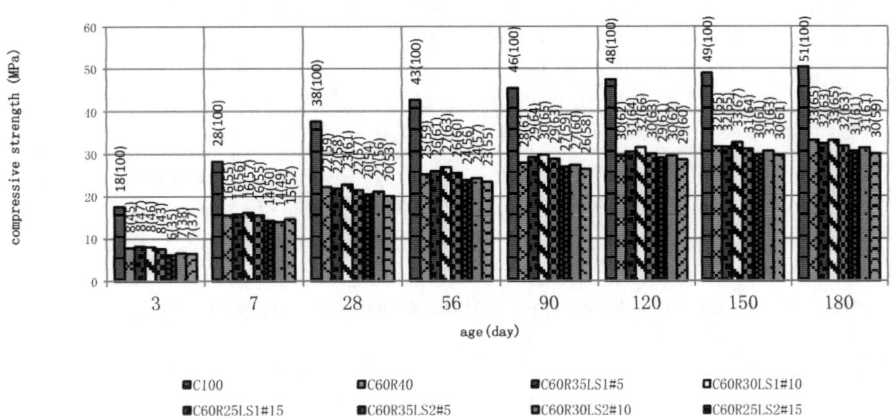


Figure 8. The compressive strength of the mortar mix, rice husk ash, crushed limestone powder and 40 percent. Replace Portland cement in five categories.

### *6.4. Loss of strength due to sulfate solution*

*6.4.1. The loss of strength due to the solution of sodium sulfate ($Na_2SO_4$).* When the tension caused by the expansion force over the surface of the mortar can cause skin cracking mortar. Sodium sulfate solution can penetrate into the expansion pack sample within the clot causing cracking mortar, resulting in decreased strength.

Test loss of strength mortar mix crushed husk ash and limestone powder. Replace the cement sector at five in the ratio of 0, 20, 30 and 40 wt soaked in a solution of sodium sulfate concentration of 5 percent by weight shown in Figure 5.39 (a) 5.45 (a) the mortar control. The loss of strength at 180 days of age was 28.70 percent. Replacing Portland cement by RHA crushed all five categories. The ratio of 0, 20, 30 and 40 was found that at the age of 180 days are likely to lose strength decreased. RHA crushed by replacing all of 30 (C70R30) are losing strength lowest percentage of 15.50.
Replacing Portland cement type 5 with rice husk ash and limestone powder series LS1 (12 microns) were more likely to lose strength increased by test C70R25LS1 # 5 with a loss of compressive strength compared to the age of 180 days. 16.38 percent due to the replacement of limestone powder to reduce the amount of rice husk ash. As a result, the amount of $SiO_2$ reduced the amount of $Ca(HO)_2$ reacts with sulphate ions can be increased. Gypsum was born and tourism guide ring, which is causing the expansion. [7]
Replacing powdered limestone LS1 (12 micron) are losing strength less than replacing powdered limestone LS2 (128 micron) is because the particle size LS1 is smaller than the LS2 ability to interpolate better structured tightening force. Compressive strength is less loss. [8]
The hydrolysis of rice husk ash with Portland cement consisted of hydration of Portland cement with water. The calcium hydroxide from the hydride product reacts with the rice husk as calcium silicate hydrate. The structure of rice husk ash paste is similar to that of Portland cement paste. The amount of calcium silicate hydrate is increased and the calcium hydroxide is reduced. Depending on the amount of rice husk ash Calcium hydroxide is reduced when the amount of rice husk is increased. Due to high silica content and low calcium hydroxide, CSH has a lower C: S ratio than CSH obtained from Portland cement reaction with water. This is the main reason that concrete with rice husk ash has resistance to corrosion of acid and sulfate.

*6.4.2. The loss of strength due to the solution of magnesium sulfate ($MgSO_4$).* Loss of strength due to magnesium sulfate solution is the reaction of magnesium sulfate and calcium hydroxide creating gypsum and magnesium hydroxide. The magnesium hydroxide, thus, lowers the pH of the solution causing an instability of calcium silicate hydrate. Calcium silicate hydrate ruptures so that the pH value increases creating Calcium hydroxide solution and silicon. Magnesium hydroxide can react with the silicon solution creating a magnesium silicate hydrate, which is capable of interfacing.
Test loss of strength mortar mix crushed husk ash and limestone powder. Replace Portland cement in five categories in the ratio of 0, 20, 30 and 40 wt soaked in a solution of magnesium sulfate. The concentration of 5 percent by weight shown in Figure 5.39 (b) of 5.45 (b), the mortar has lost control over the 180-day compressive strength of 34.30 percent.

Replacing Portland cement type 5 with rice husk ash and limestone powder series LS1 (12 microns) were more likely to lose strength decreased by test (C80R5LS1 # 15) are losing strength lowest at age 180. at 30.10 percent due to the replacement of limestone powder to reduce the amount of rice husk ash. Water demand reduction Free water in the cavity left by the reaction is reduced. When the hollow spaces and reduced absorption of sulfate ions into the more difficult to reduce the sulfate corrosion. [9], [10]
Replacing Portland cement type 5 with rice husk ash with powdered limestone LS2 (128 microns) tend to lose strength increases. The LS2 limestone powder particle size. A material fielded poorly. The structure is less tight. The absorption of sulfate ions reacts easier. Loss of strength is increased.

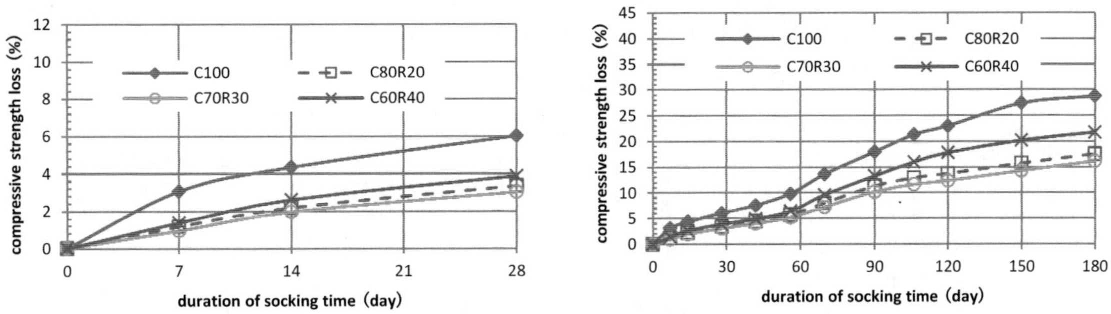


(a) Soak in a solution of sodium

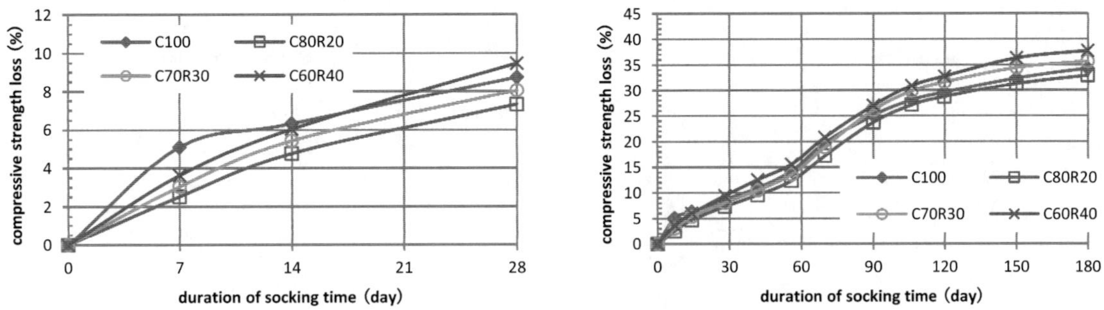


(b) Soak in a solution of magnesium sulfate

Figure 9. Losing strength as sodium sulfate and magnesium sulfate slurry of mortar mix RHA are crushed. Replacement of Portland cement in the ratio of 5 to 20 percent, by weight, 30 and 40.

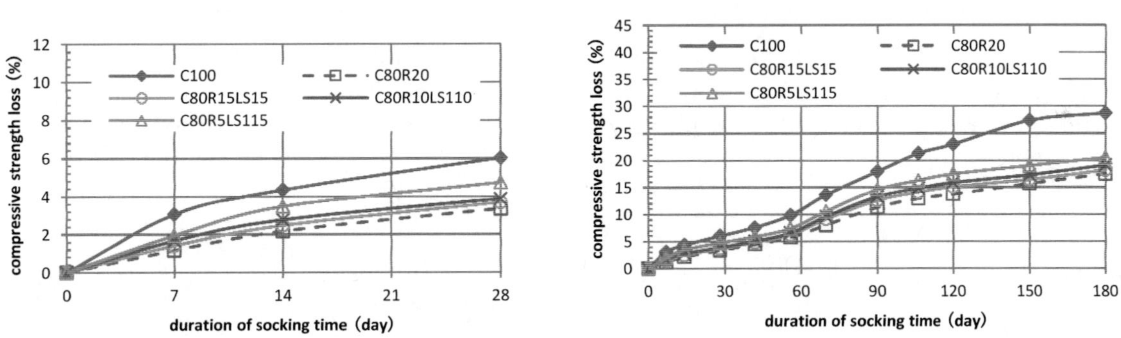


(a) Soak in a solution of sodium

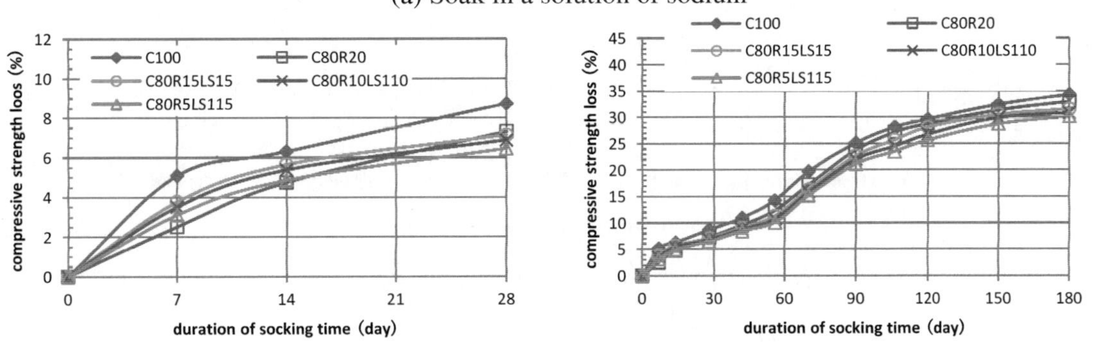


(b) Soak in a solution of magnesium sulfate

IOP Conf. Series: Materials Science and Engineering **371** (2018) 012008 doi:10.1088/1757-899X/371/1/012008

Figure 10. Losing strength as sodium sulfate and magnesium sulfate slurry of mortar mix, rice husk ash and limestone powder grinding LS1.Replacement of Portland cement in the ratio of 5 to 20 percent by weight.

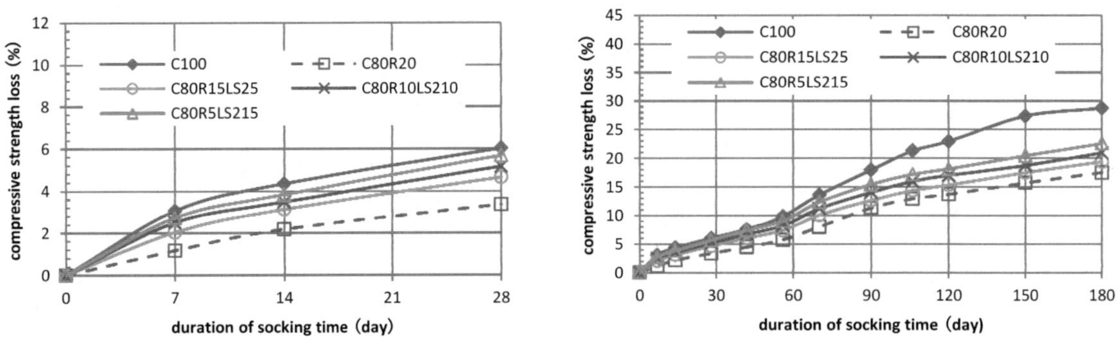


(a) Soak in a solution of sodium

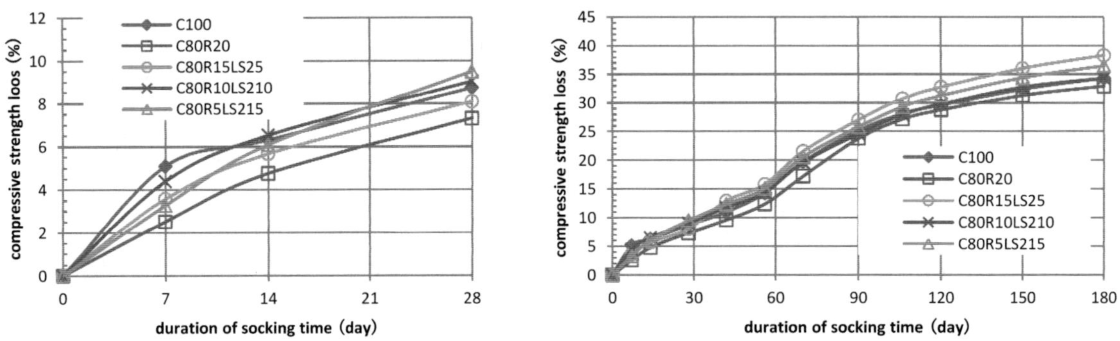


(b) Soak in a solution of magnesium sulfate

Figure 11. losing strength as sodium sulfate and magnesium sulfate slurry of mortar mix, rice husk ash and limestone powder grinding LS2. Replacement of Portland cement in the ratio of 5 to 20 percent by weight.

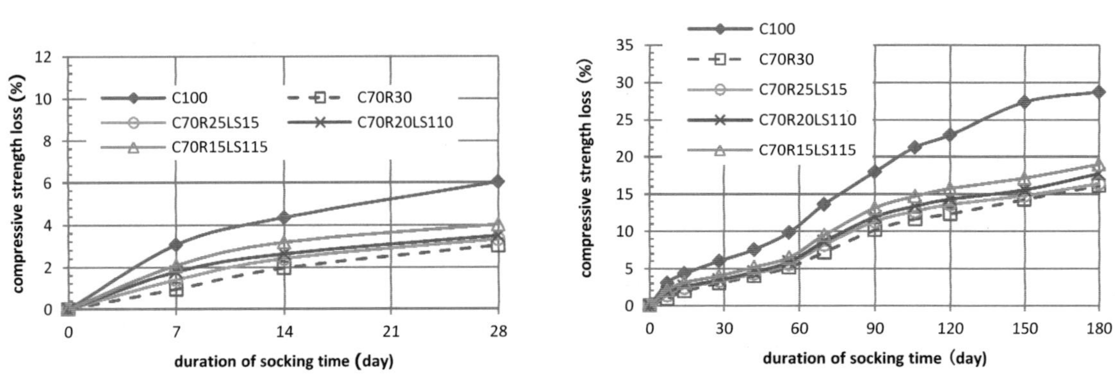


(a) Soak in a solution of sodium

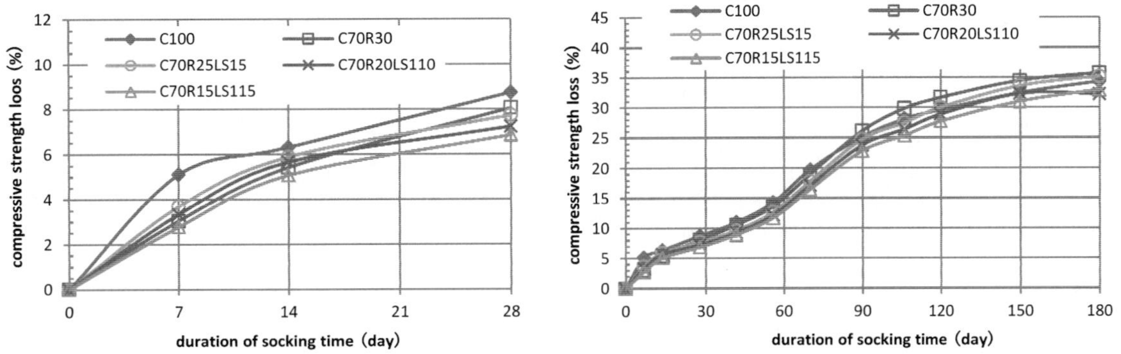


(b) Soak in a solution of magnesium sulfate

Figure 12. losing strength as sodium sulfate and magnesium sulfate slurry of mortar mix, rice husk ash and limestone powder grinding LS1. Replacement of Portland cement in the ratio of 5 to 30 percent by weight.

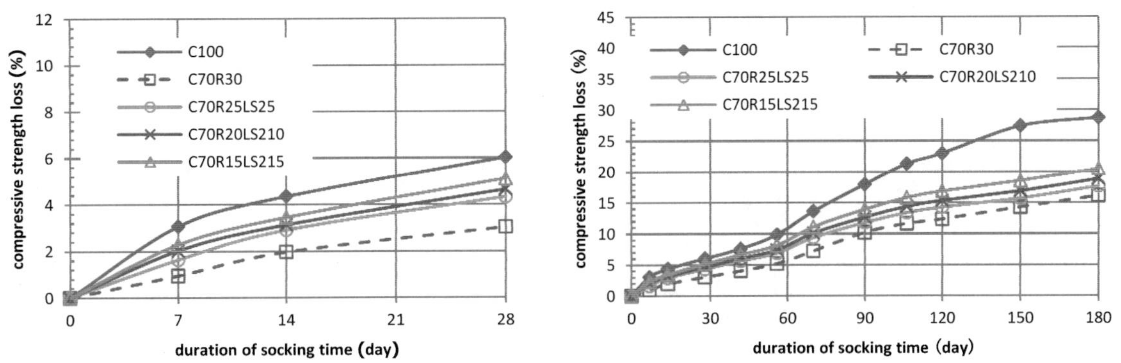


(a) Soak in a solution of sodium

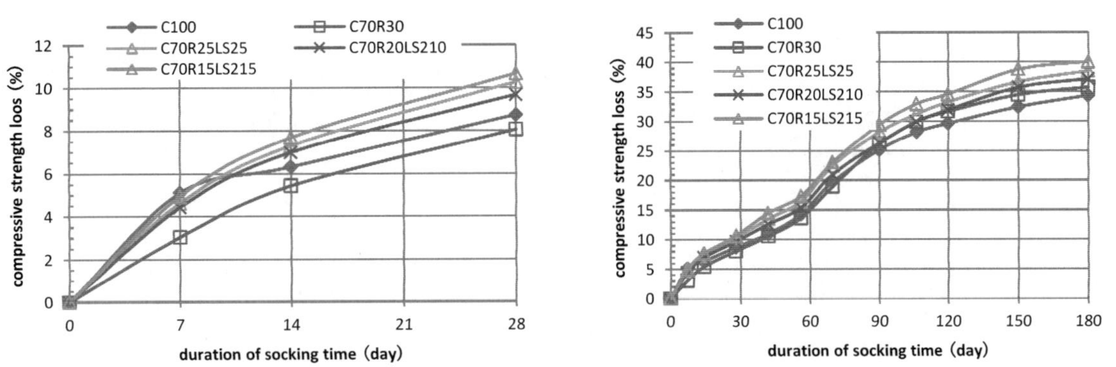


(b) Soak in a solution of magnesium sulfate

Figure 13. losing strength as sodium sulfate and magnesium sulfate slurry of mortar mix, rice husk ash and limestone powder grinding LS2. Replacement of Portland cement in the ratio of 5 to 30 percent by weight.

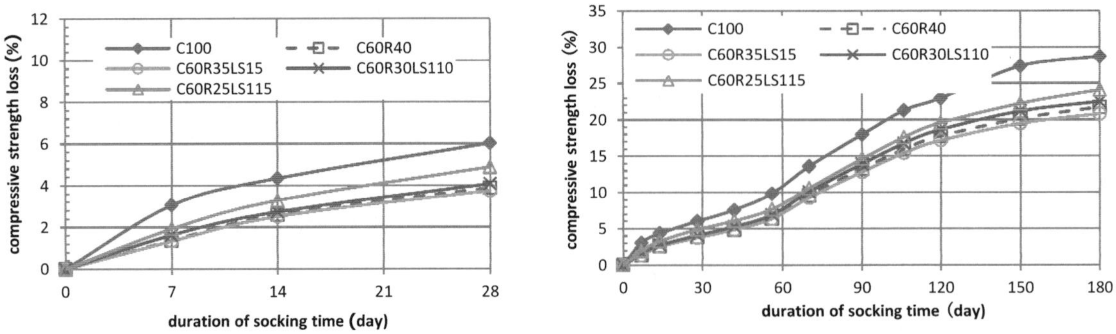


(a) Soak in a solution of sodium

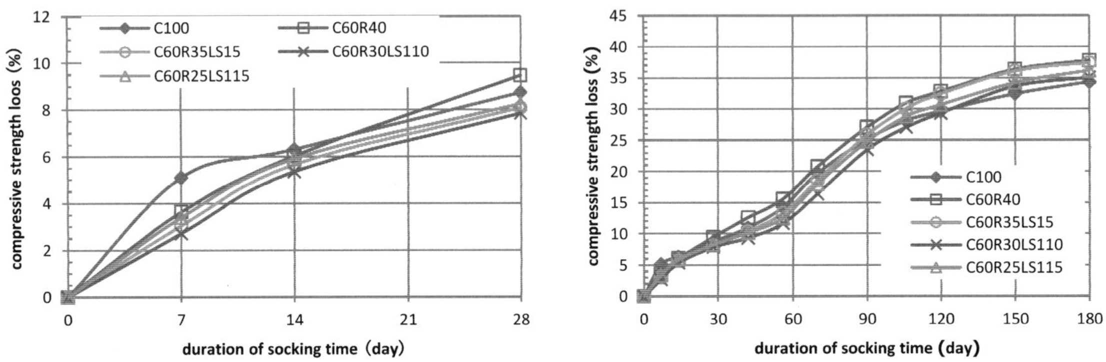


(b) Soak in a solution of magnesium sulfate

Figure 14. losing strength as sodium sulfate and magnesium sulfate slurry of mortar mix, rice husk ash and limestone powder grinding LS1.Replacement of Portland cement in the ratio of 5 to 40 percent by weight.

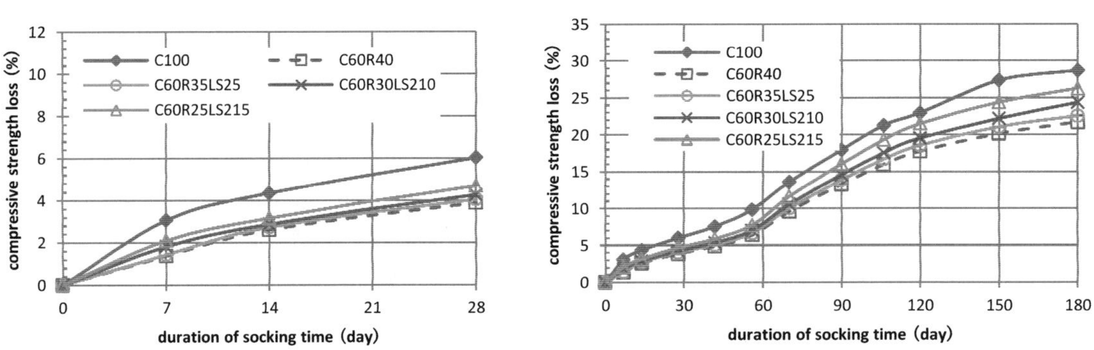


(a) Soak in a solution of sodium

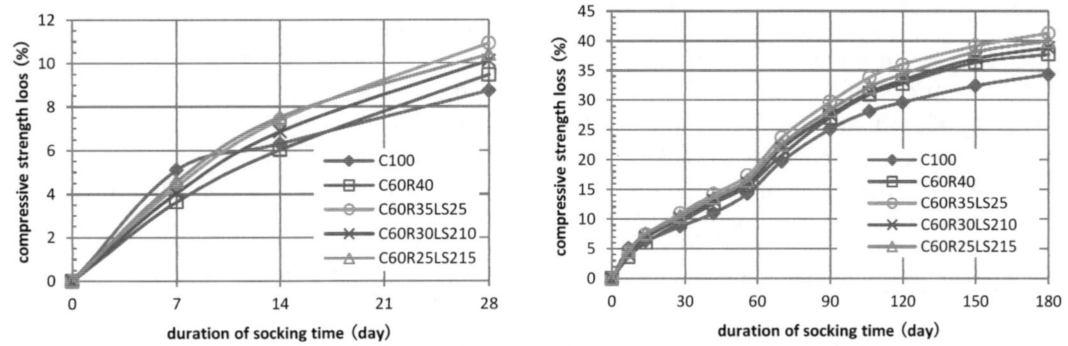


(b) Soak in a solution of magnesium sulfate

Figure 15. losing strength as sodium sulfate and magnesium sulfate slurry of mortar mix, rice husk ash and limestone powder grinding LS2. Replacement of Portland cement in the ratio of 5 to 40 percent by weight.

## 7. Conclusions

7.1. Considering the physical properties of rice husk ash grinding for 4 hours, specific surface area is larger than particles and Portland cement type 5 that led to the replacement of Portland cement type 5. It was found that the demand for water and air, increasing the supply of renewable limestone powder with water and air with lower demand as the rate of limestone powder to the water needs of 12 micrometers.

7.2. Considering the chemical properties of rice husk ash grinding for four hours to compare the pozzolan materials by ASTM C 618 was found to be a pozzolan materials in accordance with ASTM C 618 C but with powdered limestone. Not according to many, it is the only renewable materials in cement.

7.3. The compressive strength of the mortar were replaced by rice husk ash grinding for four hours on at 5 am to the development of early decline. Replaced with limestone powder, 12 microns in size, together with the development are improved. But to replace it with powdered limestone measuring 128 micrometers, the development was reduced to 180 days of age showed that compressive strength of mortar test of every batch is less than the cement mortar. RHA replaced by crushed limestone powder mixed with 10 percent 12 percent 10 micrometers compressive strength similar to control of mostly 96 percent.

7.4. The tests showed that the TGA and MIP replaced RHA crushed limestone powder and 12 micrometers in size with the size and volume of the cavity Capistrano Larry Burrows all fell. RHA and reacting pozzolan Nick calcium silicate hydrate increases. Notices of calcium hydroxide is reduced because the reaction pozzolan Nick RHA.

7.5. Contractions found that replacing the four hours of grinding of rice husk ash and limestone powder in Portland Cement Type 5 reduces shrinkage dry well in the following order. Replaced with crushed rice husk ash and limestone powder, 12 microns were replaced with crushed rice husk ash. And replace it with rice husk ash, crushed limestone and powder measuring 128 microns by shrinking Genius automatically finds replaced by RHA grinding four hours and limestone powder in Portland Cement Type 5 models, reduce shrinkage and replace it with rice husk ash, crushed limestone powder. measuring 128 microns by shrinking Genius that automatically replaced with rice husk ash and limestone powder grinding four hours in Portland cement Type 5 models reduce shrinkage. Auto Genius has the following order. Replaced with crushed rice husk ash and limestone powder 128 micrometers is replaced by RHA are crushed. And replace it with crushed rice husk ash and limestone powder, 12 microns, respectively.

7.6. The solution expansion sulfate cement were soaked in a solution of sodium sulfate has expanded over magnesium sulfate solution RHA replace Portland cement grinding four hours in five categories, reduce the expansion of sodium soluble. Replacing limestone powder with a size of 12 microns to help reduce the growth of sulfate solution a bit more. But instead of powdered limestone measuring 128 micrometers to grow. Evaluation of the endurance of standard ASTM C 1157 showed that replaced RHA grinding for four hours and limestone powder mixture to the expansion of no more than 0.05%, which is an expansion of high durability, age 6 months.

7.7. Loss of strength due to the lime solution soak in a solution of magnesium sulfate, sodium sulfate particles lose strength RHA replace Portland cement grinding four hours in five categories, reduce the loss of strength in a solution of sodium sulfate. But in a solution of magnesium sulfate to change the value of the loss is less dense. Replacing limestone powder with a size of 12 microns, reducing the loss of strength due to the sulfate solution has a bit more. But instead of powdered limestone measuring 128 micrometers loss of strength due to increased sulfate solution.

7.8. The use of rice husk ash and limestone powder for construction work. It is important to thoroughly grind the rice husk ash and fine limestone powder as the cement can penetrate into the cavity of the rice husk ash. Homogeneity and consistency in the water need to be reduced because of the higher compression.

## 8. References

[1] Chindaprasirt P and Jaturapitakkul C 2006 *Cement Pozzolan and Concrete* (Thailand Concrete Association Bangkok) pp 214-244

[2] Columnna V B 1974 *The effect of Rice husk Ash in Cement and Concrete Mixes* (Master's Thesis) No 678

[3] American Society for Testing and Materials 2007 *ASTM C1437 - 07 Standard Test Method for Flow of Hydraulic Cement Mortar* (Annual Book of ASTM Standard) Vol 4.01

[4] American Society for Testing and Materials 2008 *ASTM C109 / C109M - 08 Standard Test Method for Compressive Strength of Hydraulic Cement Mortars (Using 2-in. or [50-mm] Cube Specimens)* (Annual Book of ASTM Standard) Vol 4.01

[5] American Society for Testing and Materials 2008 *ASTM C618 - 08a Standard Specification for Coal Fly Ash and Raw or Calcined Natural Pozzolan for Use in Concrete* (Annual Book of ASTM Standard) Vol 4.02

[6] Isaia G C Gastaldini A L G and Moraes R 2003 *Physical and pozzolanic action of mineral additions on the mechanical strength of high-performance concrete* (Cement and Concrete Composites) Vol 25(1) pp 69-76

[7] Chatveera B Makul N and Rodanan A 2006 *Durability of Cement Mortar Containing Black Rice Husk Ash under Sodium Sulfate and Magnesium Sulfate Attack* (KMUTT Research and Development Journal) Vol 29 No 1 pp 55-71

[8] Tangpagasit J Songmue S Jaturapitakkul C and Kiattikomol K 2005 *Study of Strength Activity Index of Mortar Due to Hydration Reaction, Packing Effect, and Pozzolanic Reaction of Rice Husk-Bark Ash and Palm Oil Fuel Ash* (KMUTT Research and Development Journal) Vol 28 No 4 pp 465-476

[9] Bendsted J and Barnes P 2002 *Structure and Performance of Cement* (London, English)

[10] Cohen MD and Bemter A 1998 *Durability of Portland Cement-Silica Fume Pastes in Magnesium Sulfate and Sodium Sulfate Solution* (ACI Materials Journal) No 85-M18 pp 148

# Compressive Strength of Concrete with Sewage Sludge Ash (SSA)

**Thevaneyan Krishta David[1] and Sivasan Karan Nair[2]**

1Faculty of Civil Engineering, Universiti Teknologi MARA (UiTM), Shah Alam, Malaysia
2Department of Mechanical Engineering, Politeknik Premier Ungku Omar, Ipoh, Malaysia
Email: thevaneyan@salam.uitm.edu.my

**Abstract.** This study was carried out to determine the density and compressive strength of concrete using Sewage Sludge Ash (SSA) as combination of cement and fine aggregate replacement. The SSA suggested replacing the cement in the concrete because it has similar Pozzolanic properties of Ordinary Portland cement (OPC). Besides that, the particle size of the SSA after incineration process is similar to the fine aggregate that used in the concrete. The incineration process of the SSA takes 2 hours at the temperature of 1000°C. The sieving process will take place after incineration process and the SSA will be used as fine aggregate replacement. The remaining SSA will be grounded with ball mill and used as cement replacement. In this study, the compressive strength design by using DOE method to achieved 30 MPa (target mean strength of 46 MPa) at 28 days. The density of the SSA concrete has higher value compare to control sample because the incineration at 1000°C will increase the density of SSA. Previous study shows that 10% of SSA replaces cement in a concrete give same compressive strength value as control sample and the acceptable percentage of fine aggregate is between 10 to 25 percentages. Comparatively, the compressive strength of concrete when SSA used as combination of cement and fine aggregate replacement also give higher value compare to control sample.

## 1. Introduction

Sludge is an inevitable by-product of wastewater treatment [9]. Based on the perspective Sludge Production Factor (SFD), about 5.3 million $m^3$ per year of sewage sludge was produced by national sewage company such as Indah Water Konsortium [11]. The projected sludge generation in the year of 2035 is estimated at 10 million $m^3$ per year with average 2% of solid content [11]. Based on the Indah Water [11], the sludge reuse options in Malaysia are land reclamation, composting, building materials, reforestation, and power generation.

Common methods of disposal are by land filling, ocean dumping, and spreading on reclaimed land. However, these common methods of disposal of sludge are poses complex problem since it can affect the environment such as air, land and water pollution. One of the options to minimize these problems is by applying the Sewage Sludge Ash (SSA) to replace certain percentage of cement and aggregate in the concrete. Therefore, many researchers have carried out the study on utilization of wasted materials as building and construction materials in order to reduce the environmental problem. The study on the possible use of the of sewage sludge ash as building and construction materials have been carried out by various researchers. This study is very useful in order to minimize the environmental problems.

Besides that, the SSA will have a useful value or commercial value when it can be used to minimize the environmental problem and used as a useful material.
Since the methods of the disposal of sludge can affect the environment as well as the materials' cost of the construction project, there is a need for alternatives to solve these problems. Disposal of sludge is an unsatisfactory solution especially when it disposes directly into landfill [13]. Therefore, there are studies carried out about the possible use of the sludge and sludge ash as building and construction materials by various researchers. The reuse of sewage sludge into construction materials not only minimize the disposal problem but also economic, ecological and energy saving advantage [9].
The behaviours of concrete when SSA used to replace cement and its optimum percentage have been studied. Besides that, the acceptable percentage of fine aggregate used in a concrete also have been studied. However, the behaviours of concrete such as its density and compressive strength when the optimum percentage (10%) of SSA used as cement replacement combine acceptable percentage of SSA used as fine aggregate replacement are not clearly study. Apart from minimize the environmental problems, this study will help to reduce the usage of cement and fine aggregate in a concrete as well as reduce the cost of the materials. Therefore, these behaviours of the concrete will be determined in this study.
The strength of concrete is defined as the maximum load or stress that can carry by the concrete [5] [10]. Strength in compression is one of the important behaviour in cement-based materials. Based on the study that carried out by [1], the 10% of the SSA replace cement in concrete conformed to design 28 days compressive strength when cured more than 28 days. [1] conclude that 10% of the SSA replace cement in concrete could apply to unreinforced concrete to reduce the use of cement. [9] also concludes that compressive strength of concrete cubes with 10% of SSA to replace cement was about same strength as control strength. Therefore sludge ash could be recommended for non structural use such as walkways, pavements and also drains [9]. Based on the studied that have been carried out by many researchers, 10% of SSA can be used to replace cement in concrete because the compressive strength shows the little decrease at 28 days. Increasing the SSA percentage as cement replacement in a concrete mix design will decrease the compressive strength of the concrete [12]. [12] concluded that there is potential of domestic waste sludge powder or SSA used as partial cement replacement.
In this study, the SSA was used as combination to replace cement and fine aggregate (sand) in one concrete sample. Therefore, the objectives in this study are to determine the density and compressive strength of the concrete when 10% of SSA replacement of cement and adding 10% of SSA replacement of fine aggregate.

**Figure 1.** Process of production of SSA.

IOP Conf. Series: Materials Science and Engineering **371** (2018) 012009 doi:10.1088/1757-899X/371/1/012009

## 2. Methodology

### *2.1. Production of SSA*

Based on studies done previously by many researchers [2], the temperatures of 1000°C and above are mostly used. The lower temperature induces slow and low strength development because of low of CaO, $SiO_2$ and $Al_2O_3$[8]. In this study, Dried Sewage Sludge (DSS) was collected from Sewage Treatment plant and was incinerated at 1000°C for 2 hours at the furnace to produce sewage sludge ash (SSA) (Figure 1). The process follows with fast cooling in an open air. After which some of SSA was sieved and used as replacement of fine aggregate in the concrete cube. Remaining amount of SSA was then grinded using ball-mills inside the Ball Mill Machine for 12 hours at a speed of 30 rpm. The number of ball-mills used in this study was six (6) of large size and fourteen (14) of small sizes. The grinded SSA was used as 10% replacement of cement and the ungrounded SSA of 10% was used as replacement to fine aggregate.

### *2.2. Preparation of Concrete Cube*

In this study, the concrete mixture was designed (Table 1 & 2) according to [4] British Standard (BS 1881) to achieve the target mean strength of 46 MPa at ages of 28 days using 0.47 water cement ratio. Cube mould of 150mm×150mm×150mm was used to cast the concretes following standard procedures. These concrete samples were left for 24 hours in the laboratory for ambient drying process before placing them in curing tank.

**Table 1:** Concrete Mix Design per $m^3$ and per Trial Mix (water-cement ratio=0.47)

| Quantities | Cement (kg) | Water (kg) | Fine Aggregate (kg) | Course Aggregate (kg), Size of 10 mm |
|---|---|---|---|---|
| Per $m^3$ (nearest 5 kg) | 436 | 205 | 587 | 1192 |
| Per trial mix of 0.00375 $m^3$ | 1.64 | 0.77 | 2.20 | 4.47 |

**Table 2:** Concrete Sample Mix Design (w/c ratio = 0.47)

| Sample | No. of Sample | Cement (kg) | Water (kg) | Fine Aggregate (kg) | Course Aggregate (kg), Size of 10 mm | SSA (kg) | |
|---|---|---|---|---|---|---|---|
| | | | | | | Grounded | Ungrounded |
| S0* | 9 | 14.76 | 6.93 | 19.80 | 40.23 | - | - |
| S1* | 2 | 2.96 | 1.54 | 3.96 | 8.94 | 0.32 | 0.44 |
| **Total** | **11** | **17.72** | **8.47** | **23.76** | **49.17** | **0.32** | **0.44** |

- S0*: Concrete sample contained 0% of SSA
- S1*: Concrete sample contained 10% of grounded SSA (cement replacement) and 10% of ungrounded SSA (fine aggregate replacement)

### *2.3. Concrete Testing*

*2.3.1. Density Test.* After the concrete samples reach their curing time, their density was determined according to [3] BS 1881: Part 114: 1983 (Method for determination of density of hardened concrete) using equation (1). The value of the density of each sample was calculated and recorded. Finally, its average value was determined accordingly.

$$\text{Density} = \frac{\text{Mass of Concrete}}{\text{Volume of Concrete}} \left(\frac{\text{kg}}{\text{m}^3}\right) \quad (1)$$

*2.3.2. Compressive Strength Test.* Compressive strengths of the concrete samples were determined according to BS: Part 116: 1983 (Method for determination of compressive strength of concrete cubes). Compression Machine Test was used to obtain the maximum loads and compressive strength. The loading to the concrete samples, were applied continuously at nominal rate within the range 0.2 $N/mm^2.s$ to 0.4 $N/mm^2.s$ until no greater load can be sustained. By using the compressive Machine Test, the maximum load is obtained and is recorded. By using equation (2), the compressive strength of each sample was determined. The average of the concrete cube samples also was calculated and recorded.

$$\text{Compressive strength} = \frac{\text{Maximum load}}{\text{Cross Sectional Area of Cube Sample}} \left(\frac{\text{kN}}{\text{m}^2}\right) \quad (2)$$

## 3. Result and Discussion

*3.1. Density of Concrete*

The relationship of density of concrete cube and curing time is expressed in Figure 2 below.

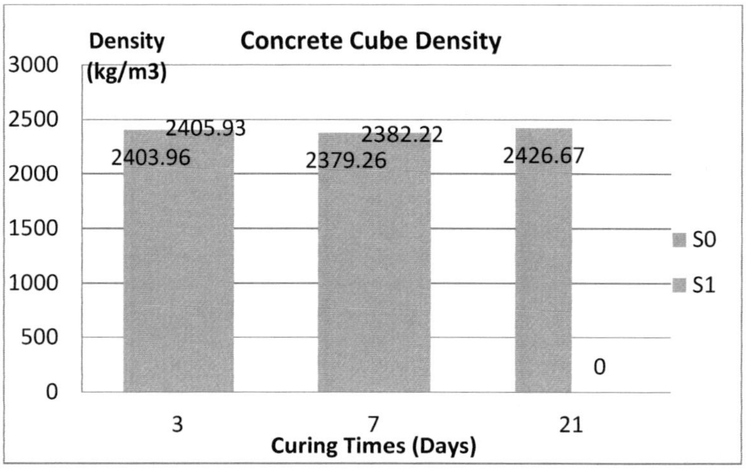


**Figure 2:** Relationship between concrete cube density and curing time.

For the control sample of the cube, the average mass is calculated. The average density of the control sample at curing time of 3, 7 and 21 days is 2 403.96 $kg/m^3$, 2 379.26 $kg/m^3$ and 2 426.67 $kg/m^3$ respectively. The density value for the concrete cube that contain combination of 10% of SSA to replace cement and 10% of SSA to replace fine aggregate at 3 and 7 days is 2 405.93 $kg/m^3$ and 2 382.22 $kg/m^3$ respectively.

The density of control samples fall between 2379 $kg/m^3$ to 2426 $kg/m^3$. The densities of SSA concrete samples fall between 2380 $kg/m^3$ and 2426 $kg/m^3$. Based on [7], the density of SSA concrete may be influenced by incineration temperature. Since the incineration temperature used in this study was 1000°C, the density values of the SSA concrete samples were found to be slightly higher than control samples. Therefore, the result may conclude that the usage of SSA in concrete may increase the density of concrete.

*3.2. Compressive Strength*

The relationship of compressive strength of concrete cube and curing time is expressed in Figure 3 below.

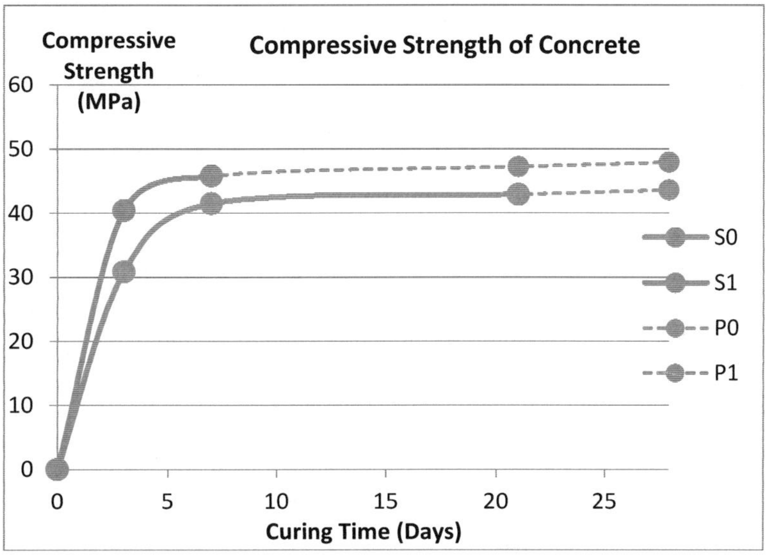


**Figure 3:** Relationship between compressive strength and curing time.

- S0: Concrete sample contained 0% of SSA
- S1: Concrete sample contained 10% of grounded SSA (cement replacement) and 10% of ungrounded SSA (fine aggregate replacement)
- P0: Prediction of concrete sample contained 0% of SSA
- P1: Prediction of concrete sample contained 10% of grounded SSA (cement replacement) and 10% of ungrounded SSA (fine aggregate replacement)

For the control sample, the average of maximum load is calculated. Compressive strength of concrete cubes is obtained using Compressive Strength Formula, where the parameter needed is average of maximum load and cross sectional area. The compressive strength of the control sample at curing time of 3, 7 and 21 days is 30.8 MPa, 41.5 MPa and 42.9 MPa respectively. The compressive strength value for the concrete cube that contain combination of 10% of SSA to replace cement and 10% of SSA to replace fine aggregate at 3 and 7 days is 40.4 MPa and 45.8 MPa respectively.
Based on the result obtained at 3 and 7 days of curing time, the compressive strength of concrete cube that contain SSA as cement replacement and fine aggregate replacement is higher compared to control sample. The compressive strength of concrete cubes with 10% of SSA as cement replacement was about same strength as control strength [9]. The optimum percentage of SSA to replace cement is 10% was used to combine with 10% of SSA to replaced fine aggregate. Referring to [6], the acceptable replacement percentage of SSA to replace fine aggregate in the concrete is up to 25%.
According to the available data, prediction was done for projected strength of SSA concrete (figure 3). The P0 and P1 are referring to the prediction value of compressive strength for concrete sample and SSA concrete respectively. The projected value for concrete sample at 28 days reaches 43.6 MPa. The projected compressive strength value for the SSA sample is 47.2 MPa for 21 days and 47.9 MPa for 28 days respectively. The projected values of the SSA sample found to be attaining the target mean strength at 21 days.
Therefore, it can concluded that the combination of 10% SSA used as cement and fine aggregate replacement in the concrete may give significantly higher compressive strength value compared to control sample. Besides that, the combination of 10% SSA used as cement and fine aggregate replacement in the concrete also were found to be attaining higher projected value of strength compared to control sample at 28 days of curing.

## 4. Conclusion and Recommendations

### *4.1. Conclusion*

IOP Conf. Series: Materials Science and Engineering **371** (2018) 012009 doi:10.1088/1757-899X/371/1/012009

From the study carried out, it may be concluded that:

1. Based on the result obtained, the density value of concrete cube that contained combination of SSA has insignificant differences or almost of the same as control sample. It shows the incineration temperature at 1000°C for 2 hours have insignificant effect on the concrete cube density.
2. Besides that, the compressive strength of combination 10% of cement and 10% of fine aggregate replacement is significantly higher than control sample. The concrete cube sample that contained SSA reached target mean strength value at 21 days. This shows that SSA may have potential to be used in a concrete with incineration temperature of 1000°C for 2 hours. The usage of SSA in a concrete will be beneficial since it can replace two materials in a concrete which is cement and fine aggregate.

As a conclusion, the density and the compressive strength value of the concrete have been determined and analyzed. Therefore, the objective of this study has been achieved.

### *4.2. Recommendation*

Based on the results obtained in this study, the following recommendations are suggested in order to improve and further enhance the outcome of studies related to SSA concrete;

1. Studies can be conducted by increasing the number of sample for each curing time in order to obtained more reliably accurate result.
2. Compressive strength test to be carried out for samples of 28 days curing time in order to verify the projected value.
3. Studies to be conducted to determine the optimum percentage of SSA replacement for fine aggregate when it combines with 10% of SSA as replacement in the concrete.
4. Further studies can be conducted using different incineration temperature and the duration in order to obtain less dense concrete.

## 5. References

[1]. Chang, F. C., Lin, J. D., Tsai, C. C., & Wang, K. S. (2010). Study on Cement Mortar and Concrete made with Sewage Sludge Ash. *Water Science and Technology* .

[2]. Cheeseman, C. R., & Virdi, G. S. (2005). Properties and Microstructure of Lightweight Aggregate Produced from Sintered Sewage Sludge Ash. *Resources, Conservation and Recycling 45* , 18-30.

[3]. British Standard- BS 1881: Testing Concrete (1983). *Part 114: Method of Determination of Density of Hardened Concrete.* London: British Standard Institution.

[4]. British Standard- BS 1881: Testing Concrete. *Part 116: Method for Determination of Compressive Strength.* London: British Standard Institution.

[5]. Jackson, N., & Dhir, R. K. (1996). *Civil Engineering Materials* (Vol. V). New York: Palgrave.

[6]. Kazberuk, M. K. (2011). Application of SSA as Partial Replacement of Aggregate in Concrete. *Polish J. of Environment Stud. Volume 20* , 365-370.

[7]. Lynn, C. J., Dhir, R. K., Ghataora, G. S., & West, R. P. (2015). Sewage Sludge Ash Characteristics and Potential for Use in Concrete. *Construction and Building Materials 98* , 767-779.

[8]. Naamane, S., Rais, Z., & Taleb, M. (2016). The Effectiveness of the Incineration of Sewage Sludge on the Evolution of Physicochemical and Mechanical Properties of Portland Cemeny. *Construction and Building Materials* , 783-789.

[9]. Tay, J. H., & Show, K. Y. (1992). Utilization of Municipal Wastewater Sludge as Building and Construction Materials. *Resources, Conservation and Recycling* , 191-204.

[10]. Teychenne, D. C., Franklin, R. E., & Erntroy, H. C. (1997). *Design of Normal Concrete Mixes.* London: Bulding Research Establishment.

[11]. Zaini, U., & Salmiati. (2011). *Management of Sludge in Malaysia.* Institute of Environmental and Water Resource Management.

[12]. Kartini, K., Dahlia Lema, A. M., Dyg., S. Q., A., A., Anthony, A. D., Nuraini, et al. (2015).Incinerated Domestic Waste Water Sludge Powder as Sustainable Replacement Material for Concrete.
[13]. Smol, M., Kulczycka, J., Henclik, A., Gorazda, K., & Wzorek, Z. (2015). The Possible Use of Sewage Sludge Ash (SSA) in the Construction Industry as a Way Towards a Circular Economy. *Journal of Cleaner Production* , 44-45.

**Acknowledgement**
The completion of this study would have been not possible if not dependent on the support of my undergraduate students Sumba anak Ensanum and Alister from the Faculty of Civil Engineering, Universiti Teknologi MARA, Shah Alam. They hence paid equal contribution to the study for which I always feel profound gratitude in my heart.

IOP Conf. Series: Materials Science and Engineering **371** (2018) 012010 doi:10.1088/1757-899X/371/1/012010

# Flexural Strength of Sewage Sludge Ash (SSA) Concrete

**Thevaneyan Krishta David[1] and Sivasan Karan Nair[2]**

[1]Faculty of Civil Engineering, UniversitiTeknologi MARA (UiTM), Shah Alam, Malaysia
[2]Department of Mechanical Engineering, Politeknik Premier Ungku Omar, Ipoh, Malaysia
E-mail:thevaneyan@salam.uitm.edu.my

**Abstract.**Studies about the sewage sludge becoming popular since the sewage sludge can be treated to produce sewage sludge ash (SSA) which has the potential to replace a percentage of cement.This study focuses on the flexural strength of the sewage sludge ash (SSA) concrete for 3, 7 and 21 days of curing relatively to its density.In this study 10 percentages of SSA was used as a replacement to cement and fine aggregates.SSA had known to exhibits pozzolanic properties similar to Ordinary Portland Cement (OPC). SSA was also used to replace the fine aggregates sinceby products of incineration as the particle size distribution has similar sizes to fine aggregates normally used in concrete mix. Besides, many studies also mention that SSA used in concrete may produce lightweight concrete. The SSA was incinerated at 1000°C for 2 hours followed by cooling in open air. Some of the SSA was sieved to be used as fine aggregates and some was grounded to be used as cement. Number of sample that was prepared for this study was 14 and it was design to achieve compression strength of 30MPa at 28 days based on DOE method. The flexural strength of concrete for cement replaced with SSA was lower than the control sample but it has an early flexural strength. The flexural strength of concrete for fine aggregate replaced with SSA was higher compared to the control sample.

## 1. Introduction

Environmental problem has become a global issue and lots of studies have being carried out as to identify possible and reliable ways in reducing the problem and improving ones way of life. Disposal of sewage sludge to the environment is a serious concern both in developing and developed countries. Studies about the sewage sludge became popular in recent years since the sewage sludge can be treated to produce sewage sludge ash (SSA) which has the potential to replace certain percentages of cement, which may reduce the environmental impact of disposing the sewage sludge. Besides, this may also contributes towards reducing cost of construction using Ordinary Portland Cement (OPC).

The SSA is produce by incineration of the dried sewage sludge (DSS) and in past practice, it is disposed in a controlled landfills. Disposing SSA could result in increased amount of land required to dispose as well as ground contamination since it contains heavy metals. Therefore, as said earlier applying SSA to replace concrete to a certain amount may reduce these problems. SSA can be used as a cement replacement since its hydraulic properties and major elements are similar to cement i.e. silica, iron, calcium, aluminum, magnesium, phosphorus and oxygen. Numerous studies on the use of SSA as cement and fine aggregates (Sand) have been done [1]. Further study on SSA may lead to a breakthrough in the construction industry.

Concrete is the most preferred construction materials to build a structure since it is more economical, flexible, and durable compared to steel and timber. Concrete composed of water, cement, coarse and fine aggregates with a specific ratio of the components. Since one of the major component is cement which is increasing in price, substitute materials to replace it has been done. Examples of cement substitutes are hydraulic fly ash, activated slag, and calcium sulfoaluminate which have been researched to be a great alternative to cement [2]. The use of incineration for sewage sludge has been rising since the volume of the waste is reduced, and pathogenic organisms and volatile contents can be eliminated which leads to the disposing of SSA to be a new problem [3]. Since SSA can be reused for concrete, it is essential to further the research on the material.

SSA was suggested to be included since it exhibits pozzolanic properties similar to cement. However, the pozzolanic properties are lower compared to Portland cement which will reduce the strength of the concrete. To counter the low pozzolanic properties, replacing a percentage of cement in SSA may prove beneficial not only in terms of the concrete performance, but also helping the environment by providing a way to recycle the SSA [4]. Sewage Sludge was traditionally used as a fertilizer in agriculture however many countries reconsider its use because concern of accumulation of heavy metals in soil and pathogenic organisms. The increase in the production of sewage sludge has many countries resort to incineration, which led to the problems of disposing the SSA since landfills becomes less abundant [3]. Including SSA in concrete may provide an alternative to reuse this materials as well as reducing the construction price since most of this waste material is free. Researches need to be done before SSA can be included in concrete to observe the negative or positive effects.

This study is to determine the flexural strength of concrete where the percentage of cement is modified with the inclusion of SSA. The SSA has been thermally and mechanically treated and an optimum percentage of cement replaced has been determine based on past research which is 10% SSA replace as cement by weight. The SSA concrete is tested for its flexural strength and to see whether it is cost effective to be implemented.

Since countries like Spain and United Kingdom has problem disposing the sewage sludge, many studies have been done to utilize the SSA such as in lightweight concrete [5], bricks [6] and aerated concrete [7]. In the future, Malaysia will face a similar problem since Malaysia will continue to be modernized. Another problem is the rising price of cement due to the high demand of the construction industry [8].The optimum SSA replace in cement and the behavior of SSA such as water absorption, density and weight class has been studied. Question that needs to be answered are; what is the flexural strength of SSA replacing a percentage of cement? What is the flexural strength of SSA replacing a percentage of fine aggregates? Does SSA concrete will affect the concrete performance?

Thus, the objectives in this study are to determine the density and flexural strength of the concrete with 10% of the weight replace with cement and fine aggregate at 3, 7 and 21 days.

**Figure 1.**Process of preparation of SSA.

IOP Conf. Series: Materials Science and Engineering **371** (2018) 012010 doi:10.1088/1757-899X/371/1/012010

## 2. Methodology

*2.1. Preparation of SSA*

Figure 1 show the processes involved in SSA preparation. The Dried Sewage Sludge (DSS) was collected from Sewage Treatment Plant (STP). The collected DSS was incinerated using incinerator at 1000°C for 2 hours to produce the SSA and fast cooling in an open air. Then, some of the SSA was ground using the Ball Mill Machine for 12 hours at the speed of 30 rpm. Six (6) numbers of large size and fourteen (14) numbers of small sizes of ball mills was used. The grounded SSA will be used to replace a percentage of cement while the ungrounded SSA will be sieved and used to replace a percentage of sand during the preparation of the concrete beam. The SSA will be stored in separate containers until further testing.

*2.2. Preparation of concrete sample*

The concrete beam was designed according to British (DOE) method where the mix will be mixed using Concrete Mixer. The size of the beam used was 100mm x 100mm x 500mm. The design mix was calculated based on weight in accordance to DOE method and summarized in Table 1 and 2. The beam mould was cleaned properly and a layer of grease is smeared in the mould before concrete is to be poured. The concrete was poured in a mould and vibrated using a vibration table. The sample was stored in the laboratory for 24 hours until transfer to the curing tank.

**Table 1.**Concrete mix design per $m^3$ and per trial mix (water-cement ratio=0.47)

| Quantities | Cement(kg) | Water(kg) | Fine aggregate(kg) | Course Aggregate (kg) (Size =10mm) |
|---|---|---|---|---|
| Per $m^3$ (nearest 5 kg) | 436 | 205 | 587 | 1192 |
| Pertrial mix of 0.005 $m^3$ | 2.18 | 1.03 | 2.94 | 5.96 |

**Table 2.** Concrete sample mix design (w/c ratio = 0.47)

| Sample | No. of Sample | Cement (kg) | Water (kg) | Fine aggregate (kg) | Course Aggregate (kg) (Size =10mm) | SSA (kg) |
|---|---|---|---|---|---|---|
| C0 | 9 | 19.62 | 9.27 | 26.46 | 53.64 | - |
| C1 | 3 | 5.88 | 3.09 | 8.82 | 17.88 | 0.66 |
| C2 | 2 | 4.36 | 2.06 | 5.29 | 11.92 | 0.59 |
| **Total** | **14** | **29.86** | **14.42** | **40.57** | **83.44** | **1.25** |

(C0: Control concrete beam sample, C1: 10% of cement replaced by SSA concrete beam sample, C2: 10% of sand replaced by SSA concrete beam)

*2.3. Concrete testing*

*2.3.1. Density test.* The concrete beam size 100mm x 100mm x 500mm will be used in determining the density of the samples as in [9]. The mass of each concrete beam is determined by using weighing machine. The volume of the test specimen which is concrete beam should not less than 40d3, where d is the nominal size of the aggregate. The size aggregate used in this study is 10mm which means the concrete beam should not less than 40 000 mm3. The actual volume of the concrete beam is 5 000 000 mm3 which is more than 40d3. The density value will be determined by using Density equation (1) below.

$$\text{Density} = \frac{\text{Mass of Concrete}}{\text{Volume of Concrete}} \left(\frac{\text{kg}}{\text{m}^3}\right) (1)$$

*2.3.2. Flexural test.* The concrete beam size 100mm x 100mm x 500mm will be used in determining the flexural strength of the samples as in [10]. There are nine (9) samples for each type of concrete beam (3 types of concrete beam) and three (3) specimen of each beam will tested at 3, 7, and 21 days of curing time. Force (F) is applied to the concrete beam at an increment of 0.06 ± 0.04 N/(mm.s) as in BS 1881 Part 118 using the Universal Testing Machine until crack appears on the concrete or until the universal testing machine stopped. The flexural strength of each concrete can be calculated using the equation (2). Finally, the average flexural strength of each concrete beam tested is calculated and recorded.

$$\text{Flexural Strength} = \frac{\text{F x l}}{\text{w x d}^2} \left(\frac{\text{N}}{\text{mm}^2}\right) \quad (2)$$

## 3. Result and Discussion

*3.1. Density of concrete*

The average density of the control sample at 3, 7 and 21 days is 2412 kg/m$^3$, 2402 kg/m$^3$ and 2344 kg/m$^3$ respectively. The density for the concrete with 10% cement replaced with SSA at 3,7 and 21 days is 2436 kg/m$^3$, 2376 kg/m$^3$ and2 398 kg/m$^3$ respectively. The density for the concrete with 10% fine aggregate replaced with SSA at 3 and 7 days is 2456 kg/m$^3$ and 2464 kg/m$^3$ respectively. Figure 2 shows the densities of the different types of concrete for the specified curing time.

From this result, it is analysed that there is a small variation in the concrete density of the different types of concrete. It is to be expected that the density of the SSA concrete will be similar or higher compared to the control sample since the optimum incineration time of 1000°C was selected. The increase in density of SSA is directly proportional to the incineration temperature but drops off after the temperature exceeds 1000°C [7]. It is concluded that the concrete replaced with SSA incinerated at 1000°C has an insignificant impact in density.

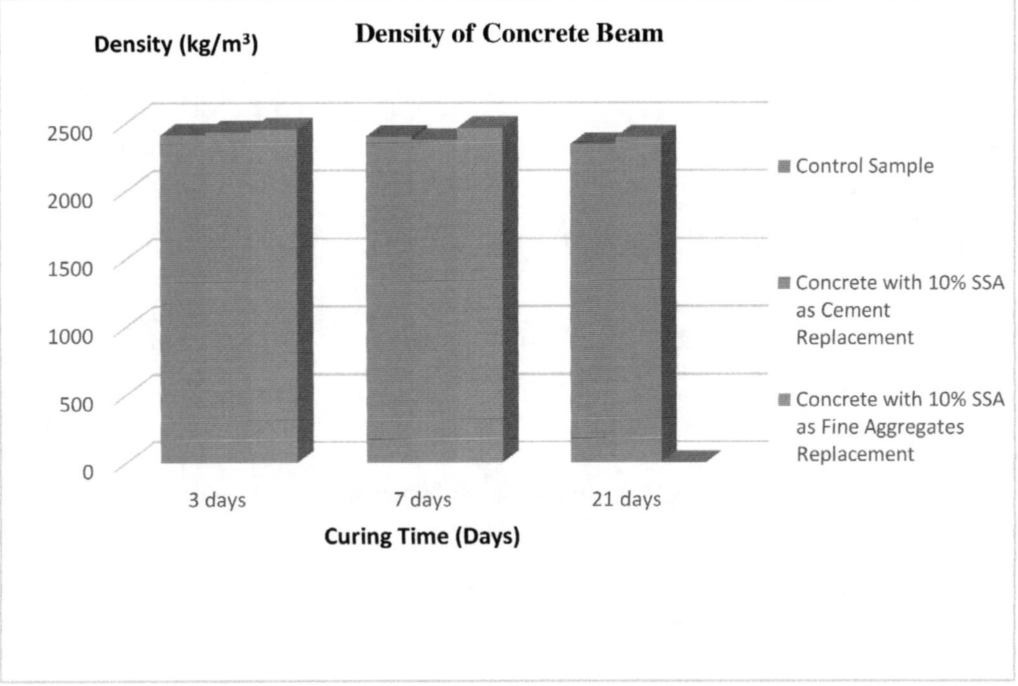


**Figure 2.**Relationship between density and curing time.

### *3.2. Flexuralstrength*

The average flexural strength of the controlled sample for 3, 7 and 21 days is 3.62MPa, 4.78 MPa and 4.84 MPa respectively as shown in Figure 3. The flexural strength of concrete with SSA replace 10% of cement for 3, 7 and 21 days is 4.63 MPa, 4.67 MPa and 4.73MPa respectively. The flexural strength of concrete with SSA replace 10% of fine aggregate for 3 and 7 days is 5.20 MPa and MPa respectively.

From the results, it is analysed that concrete replaced 10% of cement with SSA has a high early flexural strength with 4.63 MPa at 3 days compared to the control sample with a flexural strength of 3.62 MPa. The longer the curing time, the flexural strength of the SSA concrete increases rate has slowed down as shown in figure 4.2. The concrete control sample flexural strength surpasses the SSA concrete at 7 day onwards. This means that concrete replaced with 10% of cement with SSA has a high early flexural strength but lower flexural strength at increase curing time compared to a normal concrete.

Moreover, concrete replaced with 10% of fine aggregate with SSA has a higher early flexural strength compared to the control sample. The flexural strength of the SSA concrete at 3 day is 5.20 MPa surpasses the flexural strength of the concrete control sample for 21 day which is 4.84 MPa. This may have been a result of increase in density during the incineration process which results the SSA being more compact thus increase in strength. Besides, the different chemical composition of the SSA may result in the increase in strength of the concrete [11].

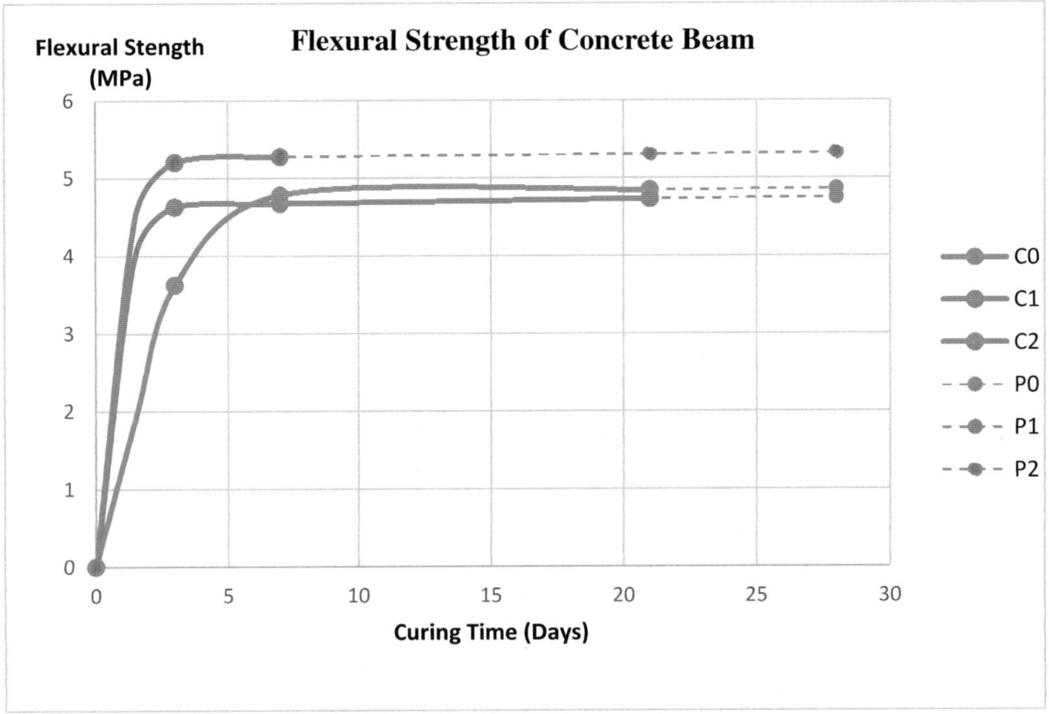


**Figure 3.**Relationship between flexural strength and curing time.
(C0: Control concrete beam sample, C1: 10% of cement replaced by SSA concrete beam sample, C2: 10% of sand replaced by SSA concrete beam)
(P0: Prediction of control concrete beam sample, P1: Prediction of 10% of cement replaced by SSA concrete beam sample, P2: Prediction of 10% of sand replaced by SSA concrete beam)

The results of SSA as cement may be lower since the SSA has a low pozzolanic activity compared to Ordinary Portland Cement (OPC). SSA as fine aggregate may produce different result due to the particle differs in size during sieving compared to the powdery SSA after grounding. The finer SSA reacts with water to become a cementitious material compared to the large particle size of the ungrounded SSA which is used as fine aggregates which limits its interaction with water. It can be concluded that SSA used to replace 10% of cement in concrete has a higher early flexural strength but lower end flexural strength. SSA used to replace 10% of fine aggregate in concrete has high flexural strength compared to the control sample.

For this study, we have limited time and resources therefore the flexural strength at 28 days for concrete with 10% cement replaced by SSA and flexural strength at 21 and 28 days for concrete with 10% fine aggregate replaced by SSA was predicted based on normal concrete behaviour which increases linear-elastically by a small amount nearing the end strength.

## 4. Conclusion and Recommendations

### *4.1. Conclusion*

From the investigation carried out, it can be concluded that:

1. The results obtained showed that the density of the concrete increase slightly or almost similar to the control sample. It can be concluded that SSA incinerated at 1000°C for 2 hours have an insignificant effect on the concrete.
2. The flexural strength of 10% of cement replaced with SSA in concrete shows a high early strength at 3 days with 1.01 MPa compared to the control sample but low strength with increase in the curing duration. The difference in the strength value of both of this concrete is very small that it can be concluded that the decrease in strength for the SSA as cement in concrete should be considered during design calculations but it does not affect the performance of the concrete.
3. The flexural strength of 10% of fine aggregates replaced with SSA in concrete shows high strength at any curing duration. This shows that the potential of SSA incinerated at 1000°C for 2 hours and sieved to the right particle size could be used to make concrete in the industry. The SSA as fine aggregate in concrete surpasses the flexural strength of the control sample with 1.58 MPa at 3 days and 0.49 MPa at 7 days. Therefore it can be concluded that there is an increase in the concrete performance with SSA replacing 10% of fine aggregates.

As a conclusion, all the objective of this study has been obtained to explain SSA as a partial replacement in concrete.

### *4.2. Recommendation*

Based on the results obtain from this study, the followings are recommended for future study to enhance the knowledge on SSA as a partial replacement in concrete:

1. It is recommended that future studies should be carried out to verify the prediction of the SSA concrete for 28 days of curing time.
2. Future studies can try using incineration temperatures ranging from 900°C - 1000°C to obtain a less dense concrete with the benefit of increase concrete strength.
3. Future studies to determine the relation and fine the optimum of incineration temperature, density and increase in concrete strength.
4. It is recommended that finding the relation of the chemical composition of sludge with the increase in concrete strength when SSA is used as fine aggregates should be carried out.

## 5. References

[1] Smol, M., Kulczycka, J., Henclik, A., Gorazda, K., & Wzorek, Z. (2015). The Possible Use of Sewage Sludge Ash (SSA) in the Construction Industry as a Way Towards a Circular Economy. *Journal of Cleaner Production* , 44-45.

[2] Hicks, J. K., & Caldarone, M. A. (2015). Opportunities from Alternative Cementitious Materials. *Concrete International* , 47-51.

IOP Conf. Series: Materials Science and Engineering **371** (2018) 012010 doi:10.1088/1757-899X/371/1/012010

[3] Naamane, S., Rais, Z., & Taleb, M. (2016). The Effectiveness of the Incineration of Sewage Sludge on the Evolution of Physicochemical and Mechanical Properties of Portland Cement. *Construction and Building Materials* , 783-789.
[4] Chang, F. C., Lin, J. D., Tsai, C. C., & Wang, K. S. (2010). Study on Cement Mortar and Concrete made with Sewage Sludge Ash. *Water Science and Technology* .
[5] Cheeseman, C. R., & Virdi, G. S. (2005). Properties and Microstructure of Lightweight Aggregate Produced from Sintered Sewage Sludge Ash. *Resources, Conservation and Recycling 45* , 18-30.
[6] Tay, J. H., & Show, K. Y. (1992). Utilization of Municipal Wastewater Sludge as Building and Construction Materials. *Resources, Conservation and Recycling* , 191-204.
[7] Lynn, C. J., Dhir, R. K., Ghataora, G. S., & West, R. P. (2015). Sewage Sludge Ash Characteristics and Potential for Use in Concrete. *Construction and Building Materials 98* , 767-779.
[8] Adnan, H. (2014, August 9). *Business News: Cement Prices Go Upwards*. Retrieved April 12, 2016, from The Star Online: http://www.thestar.commy/business/business-news/2014/08/09/cement-prices-go-upwards-current-prices-pegged-at-rm380rm400-per-tonne/
[9] British Standard – BS 1881: Testing Concrete (1983). *Part 114: Method for Determination of Density of Hardened Concrete.* London: British Standard Institution.
[10] British Standard – BS 1881: Testing Concrete (1983). *Part 118: Method for Determination of Flexural Strength.* London: British Standard Institution.
[11] Kazberuk, M. K. (2011). Application of SSA as Partial Replacement of Aggregate in Concrete. *Polish J. of Environment Stud. Volume 20* , 365-370.

## 6. Acknowledgement
The completion of this study would have been not possible if not dependent on the support of my undergraduate students Alister Saba anak Johnny and Sumba anak Ensanum from the Faculty of Civil Engineering, UniversitiTeknologi MARA, Shah Alam. They hence paid equal contribution to the study for which I always feel profound gratitude in my heart.

IOP Conf. Series: Materials Science and Engineering **371** (2018) 012011 doi:10.1088/1757-899X/371/1/012011

# Modified Pervious Concrete Containing Recycled Asphalt Pavement and Waste Tire Rubber

**Thevaneyan Krishta David[1]; RS Chandrasegaran[2] and Sivasan Karan Nair[3]**
1 Faculty of Civil Engineering, Universiti Teknologi MARA (UiTM), Shah Alam, Malaysia
2 Jabatan Kerja Raya, Daerah Batang Padang, Tapah, Malaysia
3 Department of Mechanical Engineering, Politeknik Premier Ungku Omar, Ipoh, Malaysia
Email: thevaneyan@salam.uitm.edu.my

**Abstract.** Portland Cement Pervious Concrete (PCPC) can reduce the risk of flash flood by letting the storm runoff to sip through the voids available in the PCPC hence infiltrate into the soil. This unique ability of pervious concrete offers advantages to the environment, public agencies, and building owners by controlling rainwater on site and addressing storm water runoff issues. Thus, this study was conducted to substantiate the usage of inorganic, cheap and reusable material namely Recycled or reclaimed asphalt pavement (RAP) and Fine Crumb Rubber (FCR) in pervious concrete. The primary objective of this study is to compare the compressive strength, density and the surface infiltration rate of pervious concrete containing waste tire rubber with pervious concrete that is made up by using natural aggregate without any enhancement. Standard pervious concrete will consist of natural coarse aggregates and cement. In this study, RAP will be replacing the natural coarse aggregates, the same size with the previous research. The modified pervious concrete will also be mixed with a certain percentage of FCR, replacing certain percentage of the coarse aggregate. The results of this study indicate that using recycled material could achieve similar performance as using natural aggregate, hence may reduce the construction cost.

## 1. Introduction

Pervious concrete is known to be a concrete mixture without any fine aggregate that has the ability to let fluid such as water, to flow through the concrete and infiltrate into the soil beneath. Portland Cement Pervious Concrete (PCPC) can reduce the risk of flash flood by letting the storm runoff to sip through the voids available in the PCPC hence infiltrate into the soil. This unique ability of pervious concrete offers advantages to the environment, public agencies, and building owners by controlling rainwater on site and addressing storm water runoff issues [3]. Based on [3], conventional pervious concrete has a flow rate of 3.4mm/s. With a rough calculation of $1m^2$ area, it can infiltrate 204mm of water in 1 minute. Other than infiltrating water from the surface into the soil, PCPC also can be used for the noise barrier because of the voids available that can reduce the sound wave.
Research works done in enhancing the performances of pervious concrete using certain percentage of waste tire rubber by [2] and properties of porous concrete from waste crushed concrete ( recycled

aggregate ) by [1] has set the scope and objectives of this study. Using recycled material may contribute to the economical and environmental factors positively. With the increasing number of vehicles in Malaysia, the waste tire rubber has also increased and rubber is difficult to be degradable. It is better to recycle these waste tire rubber by shredding it into small size and turn into tire chips (TC) or making smaller size into the Fine Crumb Rubber (FCR), as small as 5mm to enhance the concrete and at the same time, reduce the cost of producing pervious concrete.

Recycled or reclaimed asphalt pavement (RAP) is another recycled material which is available due to the extensive pavement maintenance could be used as an additional mixture or replacement to natural aggregate other than the FCR. This is because RAP consists of aggregate with different size that has been mixed with asphalt emulsion, producing a layer of asphalt emulsion binder. It has been heated at high temperature to make a tight bond between aggregates to make a flexible pavement which may contribute to unchanging strength of aggregate.

In this study, inorganic, cheap and reusable material which is RAP and FCR is used to make a modified pervious concrete to compare the compressive strength, density and the surface infiltration rate with pervious concrete using natural aggregate. Considering the above factors, the primary objective of this study is to determine the compressive strength, density and surface infiltration rate for the modified pervious concrete that contain recycled asphalt pavement (RAP) and waste tire rubber. Besides, the percentage amount of rubber to replace the coarse aggregate that give the best performance of the sample is another concern of this study. The performance of modified pervious concrete will then compared to natural aggregate pervious concrete.

Therefore, the percentage of the FCR and also the material used for coarse aggregate between natural aggregate and RAP will be the dependent variables, which is the limitation of the study. The tests that will be conducted for this research are as follows (Table 1):

- Compressive strength – to determine the compressive strength of the pervious concrete.
- Density – to determine the mass of the concrete with a specific volume of the specimen.
- Surface infiltration test – to determine the surface infiltration rate / rate of water to flow through the sample

**Table 1 .** Summary of the works that will be carried out in this study.

| Parameter | Testing | No of Sample | | | | Total no of Sample |
|---|---|---|---|---|---|---|
| | | Control sample | 0% FCR | 5% FCR | 15% FCR | |
| Compressive Strength | Compression Test (ASTM C109) | Fresh aggregates = 9 [ 3 sample for 7, 14 and 28 days] | RAP aggregates = 9 [ 3 sample for 7, 14 and 28 days] | RAP aggregates = 9 [ 3 sample for 7, 14 and 28 days] | RAP aggregates = 9 [ 3 sample for 7, 14 and 28 days] | 36 |
| Concrete Density | Hardened Concrete Density | | | | | |
| Surface Infiltration Rate | Surface Infiltration Test | | | | | |

## 2. Methodology

Methodology adopted for this study includes the following steps that taken for the study are explained from sample preparation, mixing design of the pervious concrete, preparing the mould, molding and curing, and sample testing. This are the steps required to produce the pervious concrete sample and also to determine the objective.

### *2.1. Sample preparation*

The preparation of samples for this study was done in reference to the work of [3] as shown in Table 2.

**Table 2**. Mix proportion according to [3].

| | Proportions, lb/yd$^3$ | Proportions, kg/m$^3$ |
|---|---|---|
| Cementitious materials | 450 to 700 | 270 to 415 |
| Aggregate | 2000 to 2500 | 1190 to 1480 |
| Water:cement ratio*** (by mass) | 0.27 to 0.34 | |
| Aggregate:cement ratio*** (by mass) | 4 to 4.5:1 | |
| Fine:coarse aggregate ratio**** (by mass) | 0 to 1:1 | |

*2.1.1. Natural aggregates.* Natural aggregate required to be washed and dried to make sure the aggregate does not have any other materials other than natural aggregate and to make sure the aggregates is dried from any moisture. The natural aggregate is dried by leaving it in an open space where the water can evaporate naturally. If it was dried using furnace, it can give an effect to the strength of the aggregate itself. After it is dried for 1 day, the natural aggregate is sieved using sieve shaker, to obtain a single size aggregates that is between 16-10mm. Since it is used for pervious concrete, it is better to use natural aggregate that is smaller size than 20mm.Therefore; 16mm was picked as the maximum size and 10mm for the minimum size. The reason of choosing a smaller size than 20mm is because it will give a slight effect to the strength of the concrete by the increase of the contact area between the cement pastes with the aggregate, more economical mixture since will be using less cement to achieve a high strength and producing smoother surface of the concrete. Thus it will increase the binding between cement paste and the aggregates and produce higher compressive strength.

*2.1.2. Recycled asphalt pavement.* The recycled asphalt pavement (RAP) was bought from the storage place where the entire RAP was compiled that is located at Kampung Jalan Kebun, Shah Alam, Selangor. The place is prepared for any RAP to collect and it is sold if anyone wants to buy. The RAP was left in an open area where it is already dried. According to Mr. Apai, the owner of the RAP collecting center, the latest RAP brought to the place is from a road around Kota Kemuning, Shah Alam, Selangor. The pavement was scrapped after 4 years since it is being placed. Since the RAP was brought in a bulk, therefore it required to sieve because of the RAP was made up of coarse and fine aggregates. Therefore the RAP was sieved using a sieve shaker and kept in a safe and dry place.

*2.1.3. Fine crumb rubber.* There are a few factories that could process the waste tire rubber. The waste tire rubber was collected and it is shredded into small pieces. In this study, 20kg of waste tire rubber that has shredded into a small size of 5-8mm size called as Fine Crumb Rubber (FCR) was bought from Yong Fong Rubber Industries Sdn. Bhd that is located at Port Klang, Selangor. The FCR should not be less than 5mm to make sure there are sufficient voids that still available in the pervious concrete. If it is smaller, it will be act similar to fine aggregate where it will reduce the void thus affecting the main purpose of pervious concrete that is to make water flows through it. The FCR was placed in a closed container to prevent for any water to enter into it. If there is water, the FCR will be moist and soon reduce its ductility.

*2.2. Mixing design of the pervious concrete*

Based on the literature review, there is no specific method for designing the concrete mix of the pervious concrete. Therefore, according to the research by [2], stated the specific material proportion used. The

IOP Conf. Series: Materials Science and Engineering **371** (2018) 012011 doi:10.1088/1757-899X/371/1/012011

research use a ratio of 0.27:1:3.57 (Water/cement: cement: Aggregate/cement) as shown in Table 2. In [3], given material proportion range that may found to be suitable for designing pervious concrete.

### 2.3. *Molding* and curing

There are several ways to molding the sample. However, in this study works done by [2] were followed. Pour 3 layers of concrete paste into the mould. For each layer, tamp the concrete paste using the 1.8mm diameter tapping rod, for 25 times. After that, the vibrators is turned on and make the sample vibrate for at least 5 seconds to make sure the aggregate was well connected. Continue the same steps for the other 2 layers.

After it was done, the sample is placed at a place where it is covered from rain or direct sun to prevent for any rapid hydrates or dehydrates process. If it is rapid hydrated, the samples will have more water and causes the samples to take more time to dry, However if it is rapid dehydrated, it is possible to see a visible cracks or hairline cracks on the sample after the sample is hardened,. Therefore, place the samples in a covered place and also place a piece of wet, clean cloth on top of the samples to make sure there are no rapid dehydration since Malaysia's average temperature is quite high.

The samples are left in the mould for 24 hours. Demould carefully to prevent any cracks on the samples. The samples were inspected to make sure the concrete paste was well dry, enough for curing. If it was dry, place the samples into a curing tank filled with water. Leave the samples in the curing tank according to the aging time required that is 7, 14 and 28 days.

### 2.4. *Sample testing*

In this study, 3 tests conducted namely compressive, density and surface infiltration test. The compressive test is to determine the strength of the concrete to resist the load applied. Meanwhile, for density test, it is to determine the category of the concrete's weight that is either lightweight or heavyweight concrete. Lastly the surface infiltration rate is to determine the surface infiltration rate of the sample.

*2.4.1 Compression test.* Compressive strengths of the concrete samples were determined according to BS: Part 116: 1983 (Method for determination of compressive strength of concrete cubes). Compression Machine Test was used to obtain the maximum loads and compressive strength. The loading to the concrete samples, were applied continuously at nominal rate within the range 0.2 $N/mm^2.s$ to 0.4 $N/mm^2.s$ until no greater load can be sustained. By using the compressive Machine Test, the maximum load is obtained and is recorded. The average of the concrete cube samples was calculated and recorded. By determining the compression strength of the specimen, the maximum strength of the concrete mix to resist the load is obtained. Thus it requires in designing either structural or non-structural component to make sure it is suitable to be used.

*2.4.2. Density test.* To determine the concrete's density, simply by placing the hardened concrete on a scale balance. By calculation using equation.1, measure the density by calculating the weight of the concrete divide with the volume of the concrete. The weight will be obtained in Kilogram (Kg)

$$Density\ (\frac{kg}{m^3}) = \frac{Mass\ (kg)}{Volume\ (m^3)} \tag{1}$$

Density also plays a big role in determining the load applied cause of its self-weight. It is a better mixture if it has higher compressive strength and at the same time have a low density. Therefore the pervious concrete requires determining it's self-weight to make sure it is adequate with the compression strength.

*2.4.3. Surface infiltration test.* This test is to determine the surface infiltration rate or rate of water to go through the voids of the sample. This is the main function of the pervious concrete to determine how fast water can go through it. The result may be considered good if the infiltration gets faster.

IOP Conf. Series: Materials Science and Engineering **371** (2018) 012011 doi:10.1088/1757-899X/371/1/012011

The test should be done according to ASTM C1871 – Surface Infiltration Rate. However since the material is insufficient to make a flat slab sample, approaching to do the test by modifying the equipment, using a similar concept as ASTM C1871. Using plastic container, the sample that already wrap with a plastic wrap is insert into the container (Figure 1). Time is taken for the water is drain out through the sample. The test only taken the vertical infiltration rate.

**Figure 1.** The modified equipment to test the surface infiltration rate.

### 2.5. *Composition of material*

The sample size used is 100mm x 100mm x 100mm and produced 9 samples for each mixture to check the average maximum load and stress, density and surface infiltration rate for each mixture. The previous research, with similar proportion of material (Table 3) used for the control sample, the researcher obtain a strength of 23.44 MPa of compressive strength after 28 days of curing in the water

The amount of material used for the control sample (M1) is using the same proportion by [2]. For M2, the coarse aggregate will be using the RAP instead of natural aggregate. For M3 and M4, 5% and 15% of the coarse aggregate will be replaced by the FCR.

**Table 3.** Material composition for this study.

| | Mix design | | | |
|---|---|---|---|---|
| | M1 - Control Sample | M2 – PC containing RAP with 0% rubber | M3 – PC containing RAP with 5% rubber | M4 – PC containing RAP with 15% rubber |
| Ratio (Water/Cement: Cement: Aggregate/Cement) | 0.27 : 1 : 4 | | | |
| Cement (kg/m$^3$) | 400 | | | |
| Water (kg/m$^3$) | 121.5 | | | |
| Natural Coarse Aggregate (kg/m$^3$) | 1600 | - | - | - |
| Recycled Asphalt Pavement (RAP) (kg/m$^3$) | - | 1600 | 1520 | 1360 |

IOP Conf. Series: Materials Science and Engineering **371** (2018) 012011 doi:10.1088/1757-899X/371/1/012011

| | | | | |
|---|---|---|---|---|
| Water Tire Rubber ($kg/m^3$) | - | 0 | 80 | 240 |
| Type of cement used | Ordinary Portland Cement (OPC) | | | |
| Aggregate size | 16mm – 10mm | | | |
| Rubber size | 5mm – 8mm | | | |
| Curing | In water | | | |

## 3. Result

Table 4 shows the results obtained in this study for compression and density. Based on the analysis that has been done, there are several discussion that can be done that is the physical appearance, amount of waste tire rubber effect the density of the sample, compression strength effected by the material and surface infiltration rate

### *3.1. Physical appearance*

There are many differences in aspect of physical appearance between each sample's mix, M1, M2, M3 and M4. For M1 shown in Figure 3, it is clearly seen that the surface is very rough even using a small size aggregate. This kind of roughness is not suitable to be used as side walk material or even the parking space. This is because of the uncomfortable surface of the pervious concrete. It is quite similar to M2 since both only using coarse aggregate, without any material that is smaller than the coarse aggregate. However when using RAP in mixture M2, the RAP that obtained for the study might slightly weak. Thus the RAP's outer surface is easily cracked. Therefore during the mixing, the surface might be taken off and the debris fills the gaps that available. Therefore, it is seen to be a little smooth surface of sample M2 compare to M1 as shown in Figure 3.

**Table 4.** Test results for density and compression

| Test | Permeability Test ( cm/s ) | Density Test ( kg/m3 ) | | | Compression Test ( MPa ) | | |
|---|---|---|---|---|---|---|---|
| Samples | 28 days | 7 days | 14 days | 28 days | 7 days | 14 days | 28 days |
| | Average | Average | | | Average | | |
| M1 - Control Sample | 0.615 | 1915 | 1893.333 | 1870 | 2.651 | 5.637 | 7.871 |
| M2 - RAP & 0% FCR | 0.493 | 1706.667 | 1691.667 | 1786.667 | 1.310 | 2.651 | 3.195 |
| M3 - RAP w/ 5% FCR | 0.522 | 1680 | 1703.333 | 1695 | 2.466 | 4.284 | 7.181 |
| M4 - RAP w/ 15% FCR | 0.735 | 1503.333 | 1508.333 | 1566.667 | 1.007 | 1.956 | 2.778 |

IOP Conf. Series: Materials Science and Engineering **371** (2018) 012011 doi:10.1088/1757-899X/371/1/012011

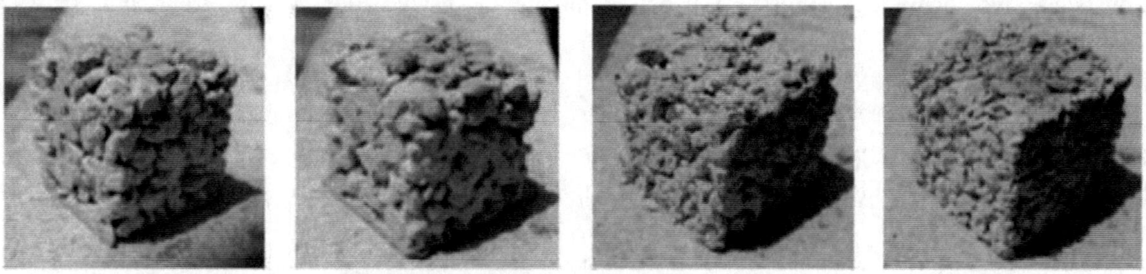

**Figure 2.** Sample M1, M2, M3 and M4.

Meanwhile for M3 in Figure 2, the surface of the concrete is slightly smoother after replacing 5% of the coarse aggregate with the FCR. This surface is more comfortable for parking space and sidewalk. Furthermore, for M4, it has the smoothers surface among the 4 mixture sample as seen in ture 3, since the coarse aggregate used that is the RAP is reduced to 15% and being replaced with FCR. In term of physical condition, M4 mixture sample is much more suitable to be used since it has less roughness.

After the samples were compressed to do the compression strength test, it also found that the FCR used for the sample, hold the sample and prevent from crumble. For a normal concrete sample, after the compression test, the sample will crumble into pieces. However with the presence of FCR, even it is well cracked, but the FCR still hold the pieces.

### *3.2. Amount of waste tire rubber effect the density of the sample*

When using the FCR to replace the coarse aggregate, theoretically it will be lighter since rubber is lighter compare with natural aggregate. Thus it is proven in Table 4 and Figure 3. This study indicates that not necessarily that a heavy material will have better strength compare with the lighter material. All mixture sample is in the suitable range of density as stated in [3] except for sample for mix M4. The sample is too light since it's containing 15% of FCR that replacing the weight of the course aggregate.

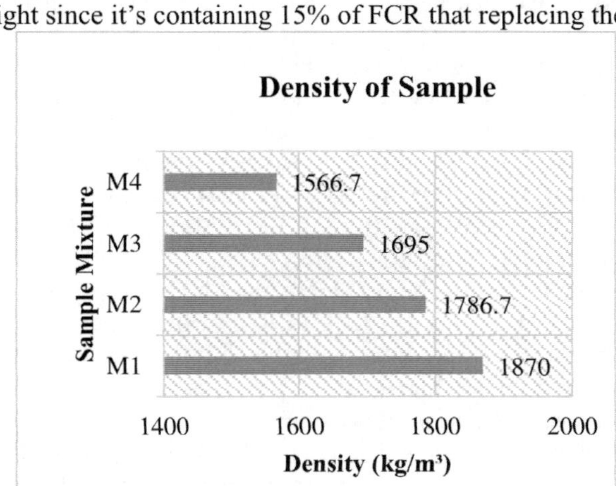


**Figure 3.** Density of samples.

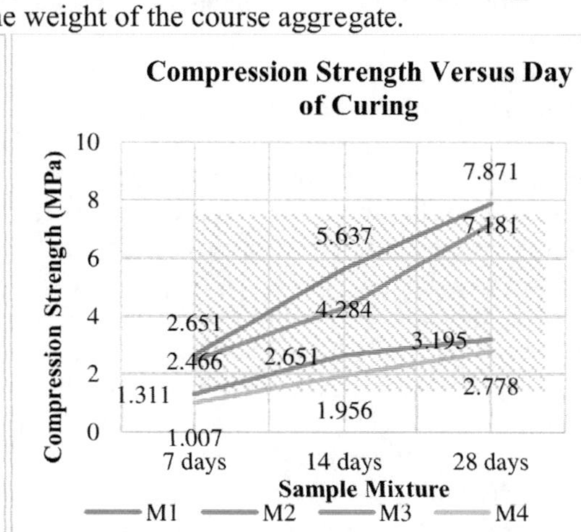


**Figure 4.** Compression strength vs curing.

### *3.3. Compression strength effected by the material*

According to the previous study by [2], the control sample that is only using natural aggregate as the coarse aggregate, cement and water, the compression strength obtain after 28 days is 23.44 MPa. Based on this study, the control sample that is M1, only obtaining 7.871 MPa after 28 days of curing as shown in Table 4 and Figure 4.

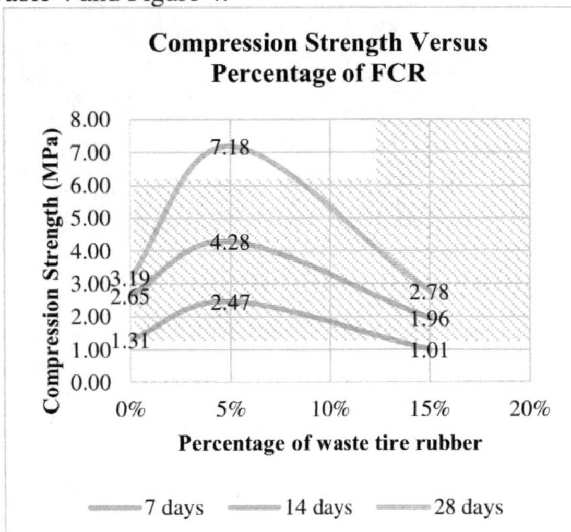


**Figure 5.** Compression strength vs. fcr percentages.

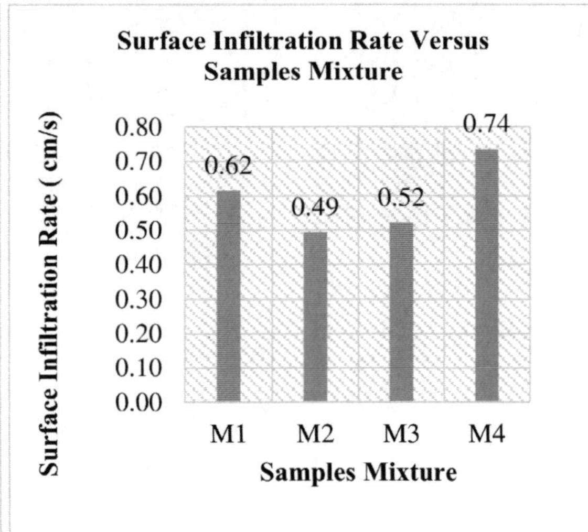


**Figure 6.** Surface infiltration rate.

Nevertheless, the sample's compression strength for 28 days was in the range 3 to 28 MPa as stated in [3]. Even M3 obtain only get 7.481 MPa, but with that strength, it still can resist the load from a vehicle that park on pervious concrete surface parking lot. Furthermore, the difference between mixture M1, that using natural aggregate and M2, that using RAP for the coarse aggregate. It shows that M2 is weaker compare to M1. The reason of this to be happen is probably the RAP that used for this study was weak than it is expected.

Therefore, to choose the right RAP to be used in the mixture, the condition of the RAP such as the period of time of exposure to the moving load on it, any water retained on the surface, type of moving vehicle that moves on it, may need to be considered. Even though it was coated with bitumen, it is unnecessarily to remaining its strength even after years of moving load applied on it. The RAP is originated from the stretch of road that sometimes heavy vehicles moves on it but there are no water usually retained on the surface. However it might be a little crack that makes the coarse aggregate of the RAP to be exposed to water and weaken the inner part of it, before it is bought to the collection center.

Moreover, the sample mixture for M3 was almost similar to the control sample and higher compression strength compare with M4. The possible reason behind of the result is 5% FCR is sufficient enough to make the concrete matrix to well bond between FCR and RAP and obtain a high strength. Meanwhile for M4, even it is far lighter, but the amount of FCR (Figure 5) is too many, at the same time make the harden concrete to be easily cracked during the compression test. Plus, as a rubber, when it is compressed, it will stretch. Therefore it cracked the bond of concrete between the FCR itself with the coarse aggregate. It might be stronger if the amount of cement used is more than 400 kg/m$^3$ or using superplastisizer. Results indicate that pervious concrete can be used as a non-structural material; it is more suitable for parking space or sidewalks. Further study requires producing more high compressive strength pervious concrete.

*3.4. Surface infiltration rate*

The surface infiltration rate varies for each sample (Figure 6). The surface infiltration rate is depending on the voids available in the pervious concrete mix. According to the result from Table 4, it shows that M4 has the highest surface infiltration rate among the entire mixture sample. This happens is probably because of the FCR that mixed in the previous concrete, producing much more voids with a larger surface area. Therefore, it is easier for water to flow through the pervious concrete sample. M2 sample has the lowest surface infiltration rate because of the broken outer surface of the RAP that turn into debris that fills between the coarse aggregate.

## 4. Conclusion

As a conclusion, sample mixture M3, that using recycled material that is RAP and FCR, has almost similar characteristics as sample mixture M1 that using new material that is natural aggregate. This shows that using recycled material could achieve the similar performance as using new material. At the same time, this could reduce the cost of material to be used for the pervious concrete, yet having the almost similar performance with the control sample.

Other than that, using RAP to replace natural aggregate as the coarse aggregate could be implemented in making pervious concrete. However the RAP needs to be in a good performance and condition. This also shows that there are significant differences when using RAP and natural aggregate as the coarse aggregate in the pervious concrete.

From this study, also shows that the amount of FCR required to get the best performance is 5%. Sample mixture that using 5% of FCR that is M3, obtaining almost similar compression strength and surface infiltration rate with control sample, M1 and lesser density compare to M1. Even the compression strength obtain from this study is very low, however it is still in the range that provided by [3].

Therefore, the objective of this study is achieved. However there are more studies on the usage of waste tire rubber and waste aggregate such as FCR and RAP, which is a recycled material, in making pervious concrete. Instead of this materials being decomposed, it can be used for the benefits of the society.

## 5. Recommendation

Based on the result acquired through this study, the following are recommended for future studies as to substantiate the modified pervious concrete containing RAP and FCR

i. Increase the amount of cement slightly higher from 450 kg/m3. According to [3], stated that aggregate cement ratio is in the range of 4-4.5. However the study's aggregate cement ratio is 3.55.
ii. Determine the RAP's strength that used for the experiment.
iii. Use percentage of FCR to be used between 5% and 15%. Since this study has the limitation of amount of FCR to be used, the future study should use various percentage of the rubber to determine the amount of FCR that give the best performance to the pervious concrete.
iv. Use the right equipment to determine the surface infiltration rate that is ASTM C1871. By producing the flat slab size, the result of the test would be more precise.

## 6. Reference

[1]. Bhutta, M. A. R., Hasanah, N., Farhayu, N., Hussin, M. W., Md Tahir, M., & Mirza, J. (2013). Properties of Porous Concrete from Waste Crushed Concrete ( Recycled Aggregate). Construction and Building Materials, 1243–1248.
[2]. Gesoğlu, M., Güneyisi, E., Khoshnaw, G., & İpek, S. (2014). Investigating properties of pervious concretes containing waste tire rubbers. Construction and Building Materials. http://doi.org/10.1016/j.conbuildmat.2014.04.046
[3]. Tennis, P. D., Leming, M. L., & Akers, D. J. (2004). Pervious Concrete Pavements EB202.02. Portland Cement Association, Skokie, Illinois, and National Ready Mixed Concrete Association, Silver Spring, Maryland, USA.

**Acknowledgement**

The completion of this study would have been not possible if not dependent on the support of my undergraduate student Adnin Razzaqi Awaludinfrom the Faculty of Civil Engineering, Universiti Teknologi MARA, Shah Alam. They hence paid equal contribution to the study for which I always feel profound gratitude in my heart.

IOP Conf. Series: Materials Science and Engineering **371** (2018) 012012 doi:10.1088/1757-899X/371/1/012012

# A Review of Agro-waste Materials as Partial Replacement of Fine Aggregate in Concrete.

**L N Somarriba Sokolova[1], E V Ermakova[1] and M Rynkovskaya[2*]**

1 Student, Department of Architecture and Civil Engineering, Peoples' Friendship University of Russia (RUDN University), Moscow, Russian Federation
2 Ph.D Associate professor, Department of Architecture and Civil Engineering, Peoples' Friendship University of Russia (RUDN University), Moscow, Russian Federation
E-mail: rynkovskaya_mi@pfur.ru

**Abstract.** An immense amount of agricultural waste is produced while growing, harvesting and processing goods; which should be treated as a resource for its prevalence and renewability. While developed countries are concerned with utilization and environmental issues, developing countries are focusing on the economic factors of social housing, especially in rural areas. Fortunately, environmental awareness has been raised in the construction industry by using agricultural waste as partial replacement for fine aggregate, coarse aggregate, reinforcing materials, cement and binders. This review is an attempt to collect world-wide data with references for future estimation possibilities of agro waste application focused on fine aggregate replacement in concrete.

## 1. Introduction

The use of vegetable fibers and other agricultural by–products materials such as sugarcane bagasse ash (SCBA), groundnut shell (GNS), oyster shell (OS), saw dust (SDA), oil palm shell (OPS), rice husk ash (RHA), cork waste ash (CPA) and coconut shell (CNS) in concrete leads to significant changes in the construction industry for house building and enriches the country's infrastructure with rapid development. This tendency in upper middle and high income countries is mainly viewed as an ecological disposal alternative; however, in the developing countries it is viewed as a cost reducing alternative. Additionally, in developing countries the safety issue is also a problem; properties are built without proper calculations, design and materials in earthquake-prone areas causing enormous human losses. For instance, as it happened in Haiti (2010) and Ecuador (2016). Consequently, the combination of local building materials with traditional building methods is rapidly gaining momentum worldwide which leads to an increase of research and experiments in countries as Brazil [1], Malaysia [2], India [3], as well as in Russia [4] and Belarus [5]. Currently, agricultural wastes are primarily used in concrete as a partial replacement for coarse and fine aggregates [6], admixtures [3], cement replacement [7] and even in ultra–high–strength concrete [8] and lightweight concrete [9]. Other uses encompass roof tile production [10,1], asphalt [11] and fibers for concrete reinforcement [12]. The most informative review was found by authors in [13], but as it is shown in the paper, this information is not enough. In this review paper, the data from various sources is gained for further possibilities examination of agro-waste materials as a replacement for fine aggregate; additionally, negative consequences are also taken into account.

## 2. Agricultural wastes production worldwide

Undoubtedly, biomass is an important and abundant resource, which can be utilized in different approaches. According to the information about the biomass and waste production of some countries annually, waste after biomass processing was approximately calculated in percentages.

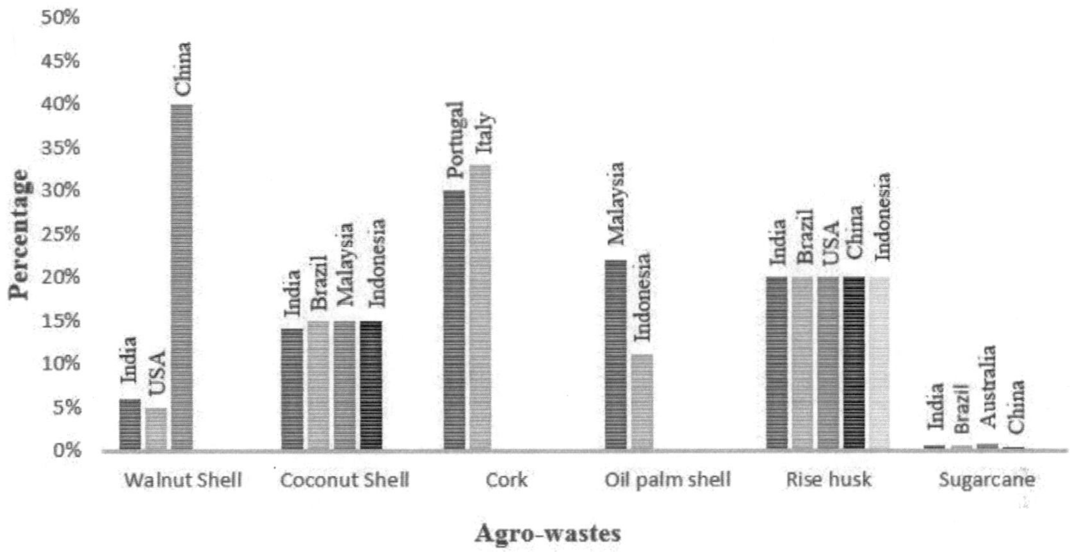


**Figure 1.** Annual waste balance per country

The information given in **figure 1** on the basis of Faostat (2016), Unimore (2017), Ricegrowers' Association, Indian production and trade statistics (2011) shows the need of recycling and biomass utilization; countries as India display an effective use of biomass with only 6% walnut shell waste, contrasting alongside China with an inefficient use of 40% walnut waste.

## 3. Physical and chemical properties of some agricultural wastes

In practice sand is usually used as a natural fine aggregate, but the necessity for alternative materials and need of sustainability has led to the use of agricultural waste as its replacement.

### *3.1 Physical properties*

The physical properties of biomasses are highly diverse as a consequence of moisture and size variations, ash yield and different genetic types of inorganic matter [8]. Due to this statement it is substantive to assemble agricultural wastes in order to comprehend their feasibility.

In **Table 1** the physical properties of agro-waste aggregates compared to sand are given. Bold figures were added by authors from various sources.

As it is seen in **Table 1** OPS[5] and CNS[8] have the highest fitness modulus making it challenging to use them as fine aggregate. These agro-wastes are used as rather coarse aggregate in experiments [21, 22]. OPS and CNS can also be used as aggregates when creating sidewalks in urban parks.

**Table 1.** Physical properties of sand and agricultural wastes

| Properties | Sand [7,14] | SCBA [13] | GNS [15,16] | OS [17] | SDA [18] | OPS [19,2] | RHA [20] | CPA [8] | CNS [21] |
|---|---|---|---|---|---|---|---|---|---|
| Specific gravity | 2.38– | 1.25– | **1.54** | 2.1– | **2.15-** | **1.17–** | 2.03 | **1.5** | **1.05–1.20** |

| | | | | | | | | |
|---|---|---|---|---|---|---|---|---|
| | 2.64 | 2.54 | | 2.48 | 2.5 | **1.37** | | |
| Fitness modulus | 2.21–3.44 | 1.42–2.12 | | 2–2.8 | 1.78 | **6.08** | | **6.26** |
| Bulk density (kg/m³) | 1428–1744 | 837–2040 | 254.55 | | 1250 | **590–600** | 11 2.4 | **650** |
| Water absorption (%) | 0.74–2.9 | 0.88 | 1.61 | 2.9–7.66 | | **14–33** | 0.1 | **24** |
| Moisture content (%) | 0.94–0.97 | | **0.42** | | **19.80** | **8-15** | | **4.2** |

[1]SCBA-Sugarcane bagasse ash
[2] GNS- Groundnut shell
[3] OS- Oyster shell
[4] SDA-Saw dust
[5] OPS- Oil palm shell
[6]RHA- Rice husk ash
[7]CPA Cork waste ash
[8]CNS- Coconut shell

### 3.2 *Chemical properties*

In the construction industry ash yield from agricultural wastes is used as an acidic additive in concrete mixtures but is not widely applied primarily due to a high degree of heterogeneity in the chemical composition. The chemical compound also varies according to specific species of each agro-waste [23]. Additionally, chemical properties are able to play a key factor in workability and overall performance of concrete. Bold figures were found in various sources pointed in **Table 2**.

**Table 2.** Some chemical properties of agro-wastes.

| Chemical compounds(%) | SCBA [13,24] | GNS [25,26,4] | OS [6,27] | SDA [28] | OPS [19,29] | RHA [30] | CPA [8] | WNS [31] |
|---|---|---|---|---|---|---|---|---|
| $SiO_2$ | 62.43–90 | **16.21** | 2–**20.06** | 65.3-**87** | | **62.5-97.6** | **38.15** | |
| $Al_2O_3$ | 2.85–4.28 | **5.93** | **0.5-5.85** | **2.5**-4 | | 0.14 | **3.65** | |
| $Fe_2O_3$ | 4.76–6.98 | **1.80** | 0.2-**3.05** | **2.0**-2.23 | | **0.54**-0.94 | **1.95** | |
| CaO | 1–11.8 | **8.69** | 51.06–**98.2** | **3.50**-9.6 | | **0.1 - 1.31** | **35.88** | **0.89** |
| MgO | 0.07–3.61 | **6.74** | 0.51-**0.93** | **0.24**-5.8 | | **0.01 - 1.96** | **1.41** | |
| $SO_3$ | 1.48 | **6.21** | 0.06–**2.71** | | | | | |
| $K_2O$ | 3.19–3.53 | **15.73** | 0.06–**0.97** | 0.11 | | **0.1-2.54** | **2.13** | |

| | | | | | | | |
|---|---|---|---|---|---|---|---|
| $P_2O_5$ | 0.23–2 | | | 0.43 | | **0.01 - 2.69** | **0.87** |
| $Na_2O$ | | **9.02** | **0.14, 0.58** | 0.07 | | **0.01 - 1.58** | **0.32** |
| $TiO_2$ | 0.02 | | **0.28** | | | | **0.28** |
| $Mn_2O_3$ | 0.02 | | **0.20** | | | | |
| $P_2O_5$ | 0.18 | | **0.17** | | | | |
| LOI | 1.86–4.73 | | 44.16 | **4.76** | **98-100** | 4.81 | **14.86** |

SCBA, SDA, RHA have the highest Silica content, while OS has mostly CaO. CPA has both $SiO_2$ and CaO as predominants.

## 4. Fresh properties of concrete

Fresh properties are used to determine the workability which by definition is a full compaction achievement in concrete mixtures that depends on several factors like chemical composition, slump test values, etc. But the main factor that affects workability is water. Unfortunately, there is no acceptable test which will measure directly this characteristic [23].

**Table 3.** Researched fresh properties of agro-waste as partial fine aggregate replacement.

| Agro-waste | Slump test | Compaction factor | Air content | Flow table test |
|---|---|---|---|---|
| SCBA [17,7] | + | + | - | - |
| GNS [16] | + | + | - | - |
| OS [6] | + | - | - | - |
| SDA [32] | + | - | + | - |
| OPS [33,14] | + | - | + | + |
| RHA [34] | + | - | - | - |

The most conducted test is the slump test due to the fact that it measures consistency of mixed proportions (see **Table 3**). Another test that helps us understand workability is the compacting factor which as we may see is not so widely used. CPA, WNS and CNS are not included since the information about the tests of these agro wastes as fine aggregate replacement has not been found.

## 5. Hardened properties of concrete

Studying hardened properties is the most important and necessary process in all experiments related to concrete. Researchers need to know how different aggregates in concrete significantly affect the compressive, flexural, tensile strength, etc. **Table 4** demonstrates the amount of researched hardened properties which were found for agro-wastes.

**Table 4.** Researched hard properties of agro-waste as partial fine aggregate replacement.

| Agro-waste | Compressive strength | Tensile strength | Flexural strength | Elastic modulus | Ultrasonic pulse velocity |
|---|---|---|---|---|---|
| SCBA [30] | + | + | + | - | - |
| GNS [26] | + | - | - | - | - |
| OS [28] | + | + | + | - | - |
| SDA [25] | + | + | + | + | - |
| OPS [33] | + | + | - | - | - |
| RHA [20] | + | + | - | - | - |
| CPA [11] | + | - | - | - | - |
| CNS [35] | + | + | - | - | + |
| WNS [31] | + | - | - | - | - |

According to **Table 4** the compressive strength test is the most common test, partly because of its simplicity and importance in structural engineering. However, the results for strength are not only affected by the aggregate, it has to be remarked that other factors as type of mould, curing, specimen size, rigidity of the testing machine and the likewise play an important role in the experiment; a thorough description should always be given. Tensile strength is of importance when estimating the load under which cracking will develop. This information is of vital importance since it helps maintain the continuity of concrete structures, and ever prevention of corrosion reinforcement. For unreinforced concrete beams flexural strength tests should be carried out and especially if the material is planned to be used in developing countries (see **Table 5**).

**Table 5.** Types of replacement researched in concrete.

| Type of replacement in concrete | SCBA [24;17;7; 36;37] | GNS [15;38; 39;26] | OS [28;30; 6] | SDA [32;41 42;43] | OPS [22;14; 44;45; 33;46] | RHA [47;5;3 4;48; 30;20] | CPA [49;8] | CNS [11;21; 50] | WNS [3;51;3 1] |
|---|---|---|---|---|---|---|---|---|---|
| Fine aggregate | + | + | + | + | + | + | + | + | + |
| Coarse aggregate | - | - | - | - | + | - | + | + | - |
| Cement | + | + | + | + | - | + | + | - | - |
| Admixture | + | - | - | - | + | + | - | - | - |

RHA had the lowest successful percentage replacement of sand with only 5% [47] followed by CNS and SCBA with 10% [35,17]. WNS was reported to have a 20% [51] replacement value along SDA with 25 %. It was reported that the partial replacement of sand with SDA did not affect the overall strength results of the specimen [42]. The most noticeable result was 80% with OS where the overall strength properties increased compared with the 0% OS specimen [28]. However, some tests showed how strength properties decreased as the amount of aggregate increased in the concrete mixture [11, 47, 15, 51]. Thus, as stated in [37] the overall improvement depends on the durability of aggregates in the cementious medium. Most of the aggregates are described as being suitable for the addition in conventional concrete to reduce the density of concrete which is useful to reduce weight of concrete and achieve light weight concrete [35,42,5,37,51] except for OS which according to [28] can be used as a structural element material in a single floor building.

## 6. Technological treatment processes

The described above aggregates are organic wastes, and therefore, are subject to varying degrees of biological decomposition in contact with an aggressive environment. A possible solution to this problem is the application of technological treatment processes. In [37] SCBA was dried at a temperature of 105-110C for 24 hours since not all the ash particles were equally burnt and mixed with water; this procedure was also applied to CPA [8]. On the other hand [52] GNS was only sun dried. In [44] OPS was washed with detergent in order to remove excess oil and dirt from the surface and then dried. But in [21] it is not indicated whether such procedures were carried out with CNS. Afterward, all the subjects were submitted to grinding with different types of machinery as rice milling machine [52] with GNS and planetary mill RETCSH PM 100 machine [8] with CPA. It is also required a more accurate description of the technological processes, since doubts arise in if it is possible to compare the obtained results or not. Moreover, the processing necessity has to be verified, since it is not stated whether it is absolutely necessary. No accounts on how labor-consuming or complex the technological processes were held.

## 7. Future recommendations

Concerning future recommendations, it would be useful to have some structural recommendations on agro-waste application possibilities for different conditions and climatic zones. For instance, which types of agro-waste should be used in earthquake-prone areas or areas with different moisture contents.

It considered that concrete based on natural aggregates or admixtures can only be used in economically underdeveloped areas as means to reduce costs, but it can also be used in upper middle or high income areas, if necessary, for the erection of temporary structures as pavilions, shelters, etc. The conclusion about expediency of application should be made not only based on economic issues, but also taking into account social factors as demands and environmental factors.

## 8. Conclusions

1. Not enough information was found about the tensile and bending tests, and an exact definition of workability was not given in research. Tensile and flexural tests are of importance due to the fact that some structures are designed to be used as structural elements in rural areas.
2. No information about the chemical composition of OPS, CNS and WNL was found. In addition, it is necessary to take into account the chemical heterogeneous properties mentioned before and their impact in the mixture. Furthermore, experiments are needed in order to bring the chemical composition of agro-wastes into homogeneity.
3. There were not found any specifications on how the technological processing of agro-wastes affects the overall mixture outcome or detailed information of costs, machinery, time and overall feasibility.
4. It is not widely observed how a final construction element has to be processed in order to improve its fire resistance, wear resistance, etc. For instance, the resistance to the elution of agro-waste particles, upon contact of the structure with water is of importance in tropical conditions with long rainy seasons.
5. There is a need on durability studies since most tests of hard properties were conducted on the $3^{rd}$, $7^{th}$,$14^{th}$ and $28^{th}$ day of concrete curing. Thus, there is an assumption that the strength properties may

IOP Conf. Series: Materials Science and Engineering **371** (2018) 012012 doi:10.1088/1757-899X/371/1/012012

decrease [11, 31, 27, 1], which is reassured by [37]; it is shown that on the samples tested on the 360 and 540 days how the strength was reduced with time.

## 9. References

[1] Jose A R, Sergio F S, Gustavo H D and Holmer S Jr Agricultural Wastes as Building Materials: Properties, Performance and Applications *Build. Mat.: Prop., Perf. and App.* **9** 1–44

[2] Alengaram U J et all Malaysia 2013 Utilization of oil palm kernel shell as lightweight aggregate in concrete–A review *Construction and Building Materials* **38** 161–172

[3] Akulova M V, Isakulov B R, Dzhumabaev M D, Imanbekova A M 2016 Getting binder of cement, fly ash and sludge, activated by the method of complex electromechanical activation for use in light arbolit concrete *Naykovedenie* **3** 1–9

[4] Asasutjarit C, Hirunlabh J, Khedari J, Charoenval S and Zeghmati B 2007 Development of coconut coir–based lightweight cement board *Construction Buildings Materials* **21** 277–288

[5] Obilade I O 2014 Experimental Study On Rice Husk As Fine Aggregate In Concrete *Int. J. of Eng. and Sci.* **3** 9–14

[6] Li G, Xu X, Chen E, Fan J and Xiong G 2015 Properties of cement–based bricks with oyster–shells ash *Journal of Cleaner Production* **91** 279–287

[7] Ramesh K V, Goutham R and Siva Kishore I 2017 An experimental study on partial replacement of bagasse ash in basalt concrete *Int. J. of Civ. Eng. and Techn.* **8** 335–341

[8] Ramos T, Matos A M and Sousa–Coutinho J 2014 Strength and Durability of Mortar using Cork waste Ash as Cement Replacement *Materials Research* **17** 893–907

[9] Anbazhagan A and Gopinath L 2017 Light Weight Concrete Using Coconut Shell SSRG *International Journal of Civil Engineering* **6** 73–75

[10] Saravanan J, ImthiyasAhamed S, Muniyasamy X, Muthu Ganesh P and Rawther Ibrahim Ali Sait Low Cost Roofing Tiles using Agricultural Wastes *International Journal of Civil Engineering* **4** 71–75

[11] Ramadhansyah P J, Nurfatin A M, Siti N A J, Norhafizah M, Norhidayah A H and Dewi S J 2016 Use of coconut shell from agriculture waste as fine aggregate in asphaltic concrete *Journal of Engineering and Applied Sciences* **11** 7457–462

[12] Sethunarayanan R, Chockalingam S and Ramanathan R 1989 Natural Fiber Reinforced Concrete *Transportation Research Record* **1226** 57–60

[13] Jnyanehdra K P, Sanjaya K P and Basarkar S S 2016 Concrete using agro–waste as fine aggregate for sustainable built environment *Int. J. of Sustainable Built Environment* **5** 312–333

[14] Shafigh P, Mahmud H B, Jumaat M Z and Zargar M 2014 Agricultural wastes as aggregate in concrete mixtures- A review *Construction and Building Materials* **53** 110-117

[15] Sada B H, Amartey Y D and Bako S Nigeria 2013 An investigation into the use of groundnut shell as fine aggregate replacement *Nigerian Journal of Technology* **32** 54–60

[16] Arshad H H and Kumar R D South India 2017 Performance of concrete properties by groundnut shell ash as a partial replacement of cement with sisal fiber *Int. J. of Civ. Eng.* **17** 23–28

[17] Prashant O M and Vyawahare M R 2013 Utilization of Bagasse Ash as a Partial Replacement of Fine Aggregate in Concrete *Procedia Engineering* **51** 25–29

[18] Kumar D, Singh S, Kumar N and Gupta A 2014 Low Cost Construction Material for Concrete as Sawdust *Int. J. of Current Engineering and Technology* **4** 3428–3430

[19] Teo D C L, Mannan Md A and Kurian J V 2006 Flexural Behaviour of reinforced lightweight concrete beams made with oil palm shell (OPS) *J. of Advanced Concrete Technology* **4** 1–10

[20] Padma Rao P, Pradhan Kumar A and Bhaskar Singh B 2014 A Study on Use of Rice Husk Ash in Concrete *International journal of education and applied research* **4** 75–81

[21] Gunasekaran K, Kumar P S and Lakshmipathy M 2011 Mechanical and bond properties of coconut shell concrete *Construcrion and Building Materials* **25** 92–98

[22] Khankhaje E, Salim M R, Mirza J, Hussin M W and Rafieizonooz M 2016 Properties of sustainable lightweight previous concrete containing oil palm kernel shell as coarse aggregate *Construction and Building Materials* **126** 1054-1065

[23] Neville A M 2011 Properties of concrete (London: Pearson) pp 1–2866

[24] Shafana T and Venkatasubramani R 2014 A study on the Mechanical Properties of Concrete with partial replacement of Fine aggregate with Sugarcane bagasse ash *International Journal of Advanced Structures and Geotechnical Engineering* **3** 34–39
[25] Olutoge F A 2010 Investigations on sawdust and palm kernel shells as aggregate replacement *Journal of Engineering and Applied Sciences* **5** 7–13
[26] Nwofor T C and Sule S 2012 Stability of groundnut shell ash (GSA)/ordinary portland cement (OPC) concrete in Nigeria *Adv. Appl. Sci. Res* **3** 2283–2287
[27] Hamester M R, Balzer P S and Becker D 2012 Characterization of Calcium Carbonate Obtained from Oyster and Mussel Shells and Incorporation in Polypropylene *Mater. Research* **15** 204–208
[28] Mohanalakshmi V, Indhu S, Hema P and Prabha V C 2017 Developing Concrete using Sea Shell as a Fine Aggregate International *J. for Innovative Research in Science &Techn.* **3** 282–286
[29] Abbas A and Ansumali S 2010 Global Potential of Rice Husk as a Renewable Feedstock for Ethanol Biofuel Production *BioEnergy Research* **3** 328–334
[30] Varshney H 2016 Utilisation of Rice Husk Ash in concrete as cement replacement *Journal of Mechanical and Civil Engineering* **16** 28–31
[31] Kamal I, Sherwani A, Ali A, Khalid A, Saadi I, Harbi A 2017 Walnut Shell for Partial Replacement of Fine Aggregate in Concrete: Modeling and Optimization *J. of Civ. Eng. Res.* **4** 109–119
[32] Oyedepo O J, Oluwajana S D and Akande S P 2014 Investigation of Properties of Concrete Using Sawdust as Partial Replacement for Sand *Civil and Environmental Research* **6** 35–42
[33] Muthusamy K, Zamri N, Zubir M A, Kusbiantoro A and Ahmad S W 2015 Effect of mixing ingredient on compressive strength of oil palm shell lightweight aggregate concrete containing palm oil fuel ash *Procedia Engineering* **125** 804-810
[34] Seyed A Z, Ameri F, Dorostkar F and Ahmadi M 2017 Rice husk ash as a partial replacement of cement in high strength concrete containing micro silica: Evaluating durability and mechanical properties *Case Studies in Construction Materials* **7** 78–81
[35] Anbazhagan A and Gopinath L 2017 Light Weight Concrete Using Coconut Shell SSRG *International Journal of Civil Engineering* **6** 73–75
[36] Dhegare S W, Dr Rauf S P, Bandwal N V and Khangan A 2015 Investigation into Utilization of Sugarcane Bagasse Ash as Supplementary Cementitious Material in Concrete *Int. J. of Emerg. Eng. Research and Techn.* **3** 109–116
[37] Bahurudeen A, Kanraj D, Gokul Dev V and Santhanam M 2014 Perfomance evaluation of sugarcane bagasse ssh blended cement in concrete *Cement and Concrete Research* **56** 32–45
[38] Arshad H H and Kumar R D South India 2017 Performance of concrete properties by groundnut shell ash as a partial replacement of cement with sisal fiber *Int. J. of Civ. Eng.* **17** 23–28
[39] Olutoge F A, Buari T A and Adeleke J S 2013 Characteristics Strength and Durability of Groundnut Shell Ash (GSA) Blended Cement Concrete in Sulphate Environment *International Journal of Scientific & Engineering Research* **4** 2122–134
[40] Erni S, Gagoek H and Purwanto K 2016 Green Concrete made of Oyster Shell Waste to Support Green Building Material *Jurnal Teknologi (Sciences & Engineering)* **78** 203–207
[41] Kumar D, Singh S, Kumar N and Gupta A 2014 Low Cost Construction Material for Concrete as Sawdust *Int. J. of Current Engineering and Technology* **4** 3428–3430
[42] Bdeir L M H Iraq 2012 Study Some Mechanical Properties of Mortar with Sawdust as a Partially Replacement of Sand *Anbar Journal for Engineering Sciences* **5** 22–30
[43] Auta S M et al 2016 Flexural strength of reinforced revibrated concrete beam with sawdust ash as a partial replacement for cement *Constr. of Unique Build. and Struct.* **5** 31–45
[44] Aslam M, Shafigh P and Jumaat M Z 2016 Oil–palm by–products as lightweight aggregate in concrete mixture: a review *Journal of cleaner production* **126** 56–73
[45] Mo K H 2015 Experimental investigation on the properties of lightweight concrete containing waste oil palm shell aggregate *Procedia Engineering* **125** 587–593

[46] Muthusamy K, Zulkepli N A, Mat Yahaya F 2013 Exploratory Study of Oil Palm Shell as Partial Sand Replacement in Concrete Research *J. of Applied Sc., Eng. and Techn.* **5** 2372–75

[47] Nithyambigai G 2015 Effect of Rice Husk Ash in Concrete as Cement and Fine Aggregate *International Journal of Engineering Research&Technology* **4** 934–936

[48] Alireza N G, Suraya A R, Farah N A A and Salleh M A M 2010 Contribution of Rice Husk Ash to the Properties of Mortar and Concrete: A Review *Journal of American Science* **6** 157–165

[49] Branco F G, Tadeu A and Reis M de L B G Can cork be used as a concrete aggregate? *ResearchGate* 1–8

[50] Yerramala A and Ramachandrudu C India 2012 Properties of concrete with Coconut Shells as Aggregate Replacement *International Journal of Engineering Inventions* **1** 21–31

[51] Mirza S H, Anwar A, Samarul H and Mohd A 2017 Cost Optimization of concrete by replacing fine aggregate with walnut shell powder *Int. J. of Civil Engineering and Technology* **8** 82–89

[52] Akeke G A, Ephraim M E, Akobo, I Z C and Ukpata J O 2012 Structural properties of rice husk ash concrete *International Journal of Engineering and Applied Sciences* **3** 57–62

**Acknowledgments**

This paper was financially supported by the Ministry of Education and Science of the Russian Federation on the program to improve the competitiveness of Peoples' Friendship University of Russia (RUDN University) among the world's leading research and education centers in the 2016-2020. This publication was prepared with the support of the "RUDN University Program 5-100".

IOP Conf. Series: Materials Science and Engineering **371** (2018) 012013 doi:10.1088/1757-899X/371/1/012013

# Numerical Analysis on Bond Strength of FRP Re-bars under Elevated Temperature

**M H Jamalan[1] and Dr F Fu[2]**

1 PhD Student, City University of London, London, UK
2 Lecturer in Structural Engineering, City University of London, London, UK
E-mail: abnz631@city.ac.uk

**Abstract.** The bond behaviour between fibre reinforced polymer (FRP) and concrete at a different temperature, ranging from room temperature (20°C) to high temperature of up to 300°C, are modelled using the FEA in this paper. Four different commercially produced FRPs with different bond treatments and the bond properties were study in this temperature range. It is observe that when the temperature increases from 20°C to 300°C the FRP's bond strength reduced between 80 and 90%, however, this reduction for steel is only 38% at this temperature range. Also as the temperature increased a reduction in the bond stiffness has occurred.

## 1. Introduction

In the past few years, the use of Fibre reinforced polymers (FRP) for reinforcing the concrete structures, where steel reinforcement is not suitable because of a highly corrosive environment or where electromagnetic transparency of the structure is required, has been increased. The properties of FRP re-bars are affected in high temperatures, such as where these re-bars face the fire or even in an extremely hot climate. As the temperature increased the mechanical properties of the polymers, especially their strength and stiffness decrease significantly and the polymer approaches its glass transmission temperature $T_g$ [1]. Different type of polymers are used as the matrix materials in FRP re-bars, the most common polymers are thermosetting polymers like; vinyl ester and polyester. The glass transition temperatures for these polymers are in a range of 60-130°C. Therefore it can be expected that if the temperature approaches or passes the glass transition temperature of the polymer resin the mechanical properties of the FRP re-bars will be affected.

High-performance inorganic and organic fibres, such as glass, carbon, and aramid fibres, are being used in FRP re-bars [2]. These fibres exhibit good mechanical property retention in the elevated temperature range considered in this work (up to 250°C) [3], and it is expected that failure of the re-bars at high temperature will occur first in the polymer matrix [4], studied the properties of the FRPs at high temperatures, as the result of these test on a variety of the FRP re-bars they found that the tensile strength of glass and carbon FRP re-bars approximately reduced 20% at a temperature of 250°C. However, this reduction was more significant for aramid FRP re-bars and it was about 60% at 250°C. Also, the tests showed that the effect of matrix composition is relatively small. Fujisaki et al. (1993) [5] reported a reduction of 40% in the tensile strength of glass-carbon FRP grids at a temperature of 100°C, probably due to changes in the properties of the polymer that forms the grid; however, no further reduction was seen for higher temperatures of up to 250°C. The impact resistance of concrete members (curtain walls) reinforced with the FRP grids at high temperature was good, and concluded that these concrete members would perform satisfactorily in the event of a fire. Wolff and Miesseler (1993) [6] proposed giving pre-stressed concrete members reinforced with glass FRP tendons the same

IOP Conf. Series: Materials Science and Engineering **371** (2018) 012013 doi:10.1088/1757-899X/371/1/012013

consideration for fire protection as that given for steel tendons, due to the small reduction in the tensile strength of glass FRP tendons, which was similar to that seen for steel tendons.

On the other hand, Wang and Evans (1995) [7] reported a reduction of 75% in the flexural strength of FRP beams as the temperature was increased from room temperature to 300°C. Okamoto et al. (1993) [8] investigated the performance of the pre-stressed concrete beams in a fire. In their test, the pre-stressing tendons were made of aramid FRP, and the reinforcement was made of either aramid or carbon FRP re-bars. As the temperature of the reinforcements reached approximately 200°C and 300°C for the aramid and carbon rods, respectively, Failure of the beams has occurred.

The variation in these results can be explained by the differences in the properties of FRP composites in different orientations, these discrepancies in orientation are because of the anisotropy of the composite material. In the direction of the fibres (i.e., the longitudinal or axial direction of the rebar), the properties are governed by the properties of the fibres, which exhibit good performance at high temperature. The polymer largely governs the properties in the transverse direction, such as shear in a plane parallel to the fibre direction. Therefore a reduction in the shear strength of an FRP rebar is expected as the temperature increases. This reduction explains the aforementioned results of Wang and Evans (1995) [7], as the flexural strength of a composite beam relies on the transfer of shear stress through the polymer.

## 2. Numerical programme

### *2.1. Rods*

In this paper four types of FRP re-bars were modelled and compared with steel re-bars. The used FRP re-bars are shown in Figure 1. Each FRP re-bars have a dimeter of 12.7 mm, and the diameter of the steel bar was 12.0 mm. The actual diameter of the FRP re-bars was slightly larger, depending on the surface treatment of each one. Properties of the FRP re-bars are given in Table 1.

Each FRP re-bars have different surface treatments in order to study the bonding between the FRP re-bars and concrete as can be seen in Figure 1. Surface of the CB re-bar is covered by a moulded polymer giving this bar lugs with similar geometry as those found on steel reinforcing bars. For other FRP re-bars a helical braid of glass fibres was wound to cover the surfaces of the bars. The braid was wider for CPI compared to CPH and NG re-bars as shown in Figure 1. In addition, for producing a large convex protrusions on the surface of the re-bars the braid was wrapped tightly around the NG bar. To increase the bond between the FRP re-bars and the concrete embedded sand particles on the surface of the CPH and NG re-bars was used. The sand was dispersed evenly on the surface of the CPH re-bar in a very thinner layer of polymer compared to NG re-bar.

### *2.2. Concrete*

Normal strength concrete was used in this study. Table 2 shows the general properties of concrete such as, mass density, young's modulus and Poisson's ratio. Also, concrete damage plasticity properties has been tabulated in Table 3 (a and b).

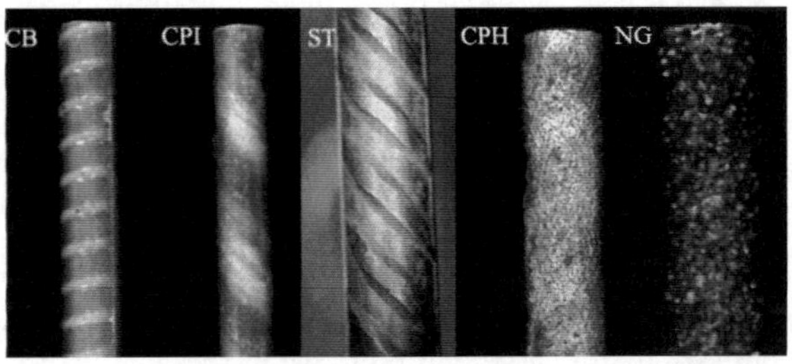


**Figure 1.** Tested re-bars.

**Table 1.** Properties of tested FRP re-bars.

| Bar Type | Diameter (mm) | Type of Resin | Physical bond improvement to surface of bars | Space between deformations (mm) | Glass transition temperature (°C) |
|---|---|---|---|---|---|
| **CB** | 12.7 | Urethane modified vinyl ester | Large deformations | 7.5 | 124 |
| **CPH** | 12.7 | Epoxy vinyl ester | Helix + sand coating | 27 | 122 |
| **CPI** | 12.7 | Epoxy vinyl ester | Helix + resin roughening | 25 | 95 |
| **NG** | 12.7 | Polyester | Helix + sand coating + deformations | 12 | 60 |
| **ST** | 12 | Steel | Large deformations | 7.7 | _ |

**Table 2.** General properties of concrete.

| **Mass Density [kN/mm$^3$]** | **Elasticity** | |
|---|---|---|
| | Young's Modulus [N/mm$^2$] | Poisson's Ratio |
| **2.36×10$^{-5}$** | 24855.578 | 0.2 |

**Table 3 (a).** Concrete damage plasticity properties.

| **Plasticity** | | | | |
|---|---|---|---|---|
| **Dilation Angle** | Eccentricity | $f_{b0}/f_{c0}$ | K | Viscosity Parameter |
| **36** | 0.1 | 1.16 | 0.666 | 0 |

**Table 3 (b).** Concrete damage plasticity properties.

| **Compressive Behavior** | | **Tensile Behavior** | |
|---|---|---|---|
| **Yield Stress** | Inelastic Strain | Yield Stress | Cracking Strain |
| 5.3 | 0 | 0.984219488 | 0 |
| 5.49 | 0.000222 | 1.524994267 | 9.6E-005 |
| 22.93 | 0.001083 | 1.296676701 | 0.000144 |
| 28.99 | 0.00174 | 1.155729537 | 0.000192 |
| 30 | 0.002219 | 1.057041912 | 0.00024 |
| 27.11 | 0.003355 | 0.982697179 | 0.000288 |
| 22.83 | 0.004438 | 0.923934094 | 0.000336 |
| 21.71 | 0.004719 | 0.875879203 | 0.000384 |
| | | 0.835570706 | 0.000432 |
| | | 0.801087969 | 0.00048 |
| | | 0.771122074 | 0.000528 |

| | |
|---|---|
| 0.744745197 | 0.000576 |
| 0.721278303 | 0.000624 |
| 0.700211106 | 0.000672 |
| 0.681151482 | 0.00072 |
| 0.663792309 | 0.000768 |
| 0.647889051 | 0.000816 |
| 0.63324418 | 0.000864 |
| 0.619696109 | 0.000912 |
| 0.607111153 | 0.00096 |
| 0.595377601 | 0.001008 |
| 0.584401251 | 0.001056 |
| 0.57410202 | 0.001104 |
| 0.564411316 | 0.001152 |
| 0.555269995 | 0.0012 |

## 3. Analysis

The pull-out simulation were performed in accordance to ACI440.3R-04 and CSA S806-02 standards in ABAQUS as shown in Figure 2. This is an implicit analysis procedure suitable for modelling both material and geometric nonlinearities was used, contact between different parts was also modelled. The FE analysis comprised an elastic-plastic simulation of the pull-out process and the determination of the load-displacement curve in the steel/GFRPs bars. Appropriate boundary conditions were applied to the model.

The rods were embedded vertically in a concrete block with a dimension of $200 \times 200\ mm$ and the length of $200mm$. Embedded length $(l)$ was $63mm$ (5 diameters) following a bond breaker zone of $63mm$. The layout of the concrete block has been shown in Figure 3.

The bottom surface of the model will be fully constrained along the concrete portion, but leave the steel/FRPs bars free. Far field displacements are also constrained due to limited influence of the pull-out loading.

The FE simulations were conducted for all the models and the FE results were compared with previous experimental works that have been done by (Baena .M. et al., 2009) [9] on bond behaviour between concrete and FRP bars, (Nanni .A. et al., 2005) [10] on thermal effects on bond between FRP re-bars and concrete and (Tastani .S. P and Pantazopoulou .S. J, 2006) [11] on bond of GFRP bars in concrete.

The FE mesh was uniform and was generated using a solid element which is an 8-node linear, the block after meshing has been shown in Figure 4. The concrete damaged plasticity model was assumed for the normal strength concrete. The concrete damaged plasticity model is based on the assumption of scalar (isotropic) damage, and is designed for applications in which the concrete is subjected to arbitrary loading conditions.

The average bond strength was determined by dividing the maximum pull-out load $(P_{max})$ by the embedded surface, assuming a uniform bond stress distribution along the rod, this is shown in equation 1.

$$\tau = \frac{P_{max}}{\pi d l} \tag{1}$$

Where $P_{max}$ is the peak load, $d$ the rod's diameter, and $l$ is the embedded length.

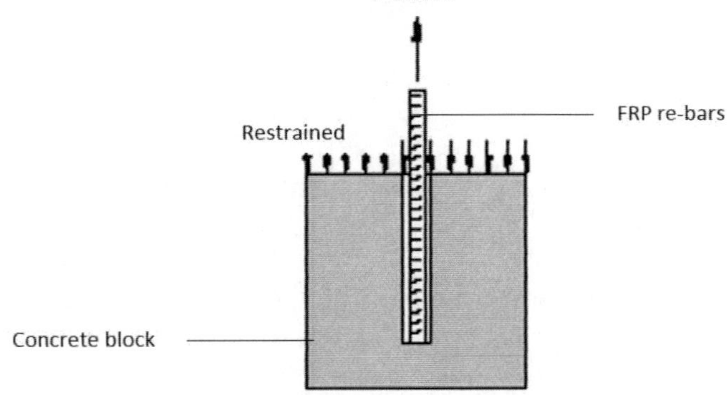


**Figure 2.** Pull-out test setup.

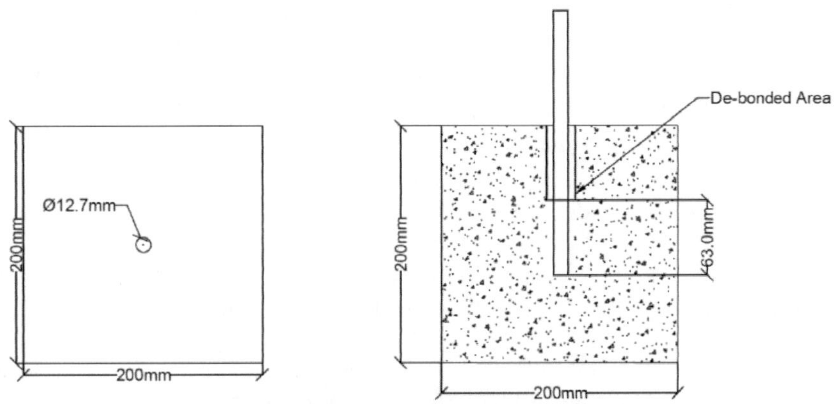


**Figure 3.** Concrete block layout.

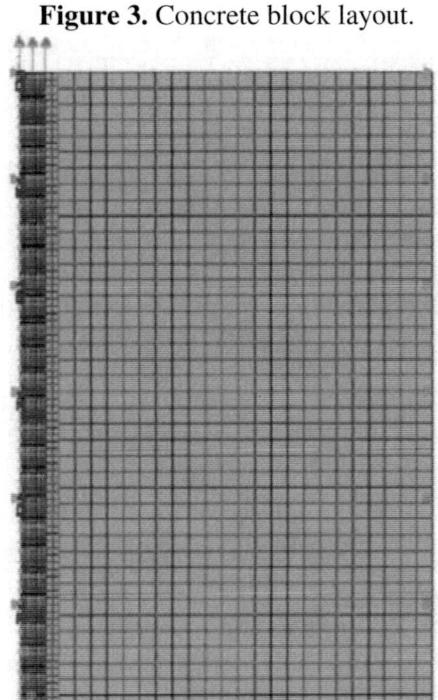

**Figure 4.** Mesh of the ABAQUS model.

## 4. Results

### *4.1. Bond strength*

As it shown in Figure 5, for all bars the effect of temperature on the bond strength calculated by using equation (1)

As it can be seen in Figure 5, all the FRP re-bars behave in a similar manner; except NG re-bar the other FRP re-bars shown a relatively high bond strength at room temperature. The calculated bond strengths for of CB, CPH, and CPI re-bars at room temperature were $13.1, 12.1, and\ 10.8\ MPa$, respectively. This value for steel re-bar was $11.2\ MPa$. A reduction in the bond strength was seen for all of the FRP re-bars at temperatures in the range of $80 - 160°C$. At nearly $200°C$ the bond strength has not changed and remained almost constant as the temperature was increased. Analysis were stopped at around $300°C$. According to (Gentry et al. 1999); at higher temperatures (400°C and above) the polymer matrix will begin to decompose no mechanical properties can be expected at these temperatures. Table 4 shows the values for the bond strength at room temperature, the residual bond at high temperature, and the degree of bond strength.

Comparing to other FRP re-bars NG bar has the weakest performance as the temperature increased and lost its bond strength quickly at low temperature. At a temperature of $92°C$, Loss of $55\%$ of the bond strength was seen, however, the entire loss of bond strength for NG re-bar was only $80\%$ compared with the other FRP re-bars $(85\text{–}92\%)$. For CB re-bars rapid reduction in bond strength was seen and most of the reduction was seen in a temperature ranging between $80\text{–}150°C$. This reduction for the CPH and CPI re-bars occurred in a wider range and extended up to $200°C$.

As the temperature increased the reduction in bond strength was also occurred in steel. Loss of bond strength was only $40\%$ for the steel re-bar as the temperature reaches $200°C$, this value for FRP re-bars was $80\%$ and more. As the temperature increased loss of bond strength for the steel re-bar increased gently. These findings are in good agreement with the findings of Diederichs and Schneidwe (1977) [12].

Effect of the temperature on the strength of the concrete was shown in Figure 6. Reduction in the strength of the concrete can affect the reduction in the bond strength of steel/FRP re-bars. As the temperature reaches $200°C$ the concrete strength reduced $15\%$, with some increase thereafter. These findings are typical of the behaviour of concrete at high temperature as described by Malhotra (1982) [13]. Therefore, some of the reduction in bond strength at high temperature for steel/FRP re-bars can be explained by a reduction in the concrete strength where the bond mechanism is mainly via the concrete.

Covering the surface of the FRP re-bars by inorganic components (i.e., sand particles and helical fibre) add extra resistance to the re-bars when the temperature rises. In CB and NG re-bars, where the bond relies on the properties of the polymer the effect of the elevated temperature was more critical, than for the CPH and CPI re-bars where the helical wrap of glass fibres was used. The performance of the CPH re-bar that had sand particles at the surface was similar to the CPI re-bars where no sand was used at the surface, therefore The effect of the sand particles is not clear.

### *4.2. Complete pull-out load-slip curve*

Figure 7 shows the load vs slip of the tested re-bars at different temperature. Load vs slip curves divided in two part for analysis: pre-peak and post-peak as the temperature rises.

### *4.3. Pre-Peak behaviour*

The slope of the load-slip curve in pre-peak stage can be considered to be the bond stiffness because it gives a relationship between load and deformation. This value has a great impact on the width of primary flexural cracks in reinforced concrete and on the deflection of beams and slabs. As the temperature rises for all the tested re-bars a reduction in the slope of the ascending curve was seen. For all of the FRP re-bars this reduction appears to be linear.

In most design codes and guidebooks, assumed that re-bars are a homogenous materials while describing the bond between the reinforcements and concrete, which means reinforcements are much

IOP Conf. Series: Materials Science and Engineering **371** (2018) 012013 doi:10.1088/1757-899X/371/1/012013

stiffer than the surrounding concrete. This assumption is appropriate for structures that have been reinforced by steel bars. However, in the case of using FRP re-bars as the reinforcing bars, this assumption is not true, because FRP re-bars are weaker and more flexible than the concrete, especially when the temperature increases.

### *4.4. Post-peak behaviour*

At room temperature the post-peak behaviour can be divided into two stages: (1) for a short distance after the peak Rapid reduction in the load bearing; and (2) more moderate reduction after that. Using different surface treatments affect the rate of reduction in the load in Stage 1 and the final values of the load and slip in Stage 2. Immediately after the peak load, a steep reduction of $30-40\%$ is seen for CB re-bars. A steep post-peak reduction can be still seen at the temperature of $130^{\circ}C$; however, after that a reduction occurs by a constant low slip. At a higher temperature of $200^{\circ}C$, a small peak is seen with a gradual decrease thereafter. In this type of re-bars (with lugs) the bond mechanism is rely on wedging of the lugs in the surrounding concrete. As the temperature raises, the surface layer with the lugs softens and becomes more flexible and weak, thus preventing the mechanism of wedging from being effective.

At room temperature CPH re-bar slips $5mm$ at its peak load. This is followed by a moderate reduction thereafter. A moderate reduction was seen as the temperature increased instantly after the peak to a complete reduction of the load. A similar behaviour was seen for CPI and NG re-bars; however, the peak load was lower than what was seen for the CPH rods at room temperature. The slope of the post-peak curve becomes more linear as the temperature increases, and the sudden drop right after the peak disappears. The helical wrapping on the rods and sand particles embedded on the surface of these re-bars caused the differences between the post-peak behaviour of them compared to CB re-bar. In these re-bars the bond mechanism is rely only on the additional features (i.e., the helical wrap and the sand particles), which are less sensitive to the temperature. Therefore the post-peak reduction is more moderate and exhibits constant bond strength to complete pull-out.

Steel re-bar does not show a significant change in behaviour in load-slip curve as the temperature increased. Generally, a rapid decrease of about 50% of the load was seen, after the peak load was reached, followed by a gradual decrease to complete pull-out. As the bond mechanism of steel re-bar are rely on mechanical anchoring to the concrete at all temperatures, so no significant change is expected as the temperature increased, however, for a moderate general reduction resulting from the reduction in the concrete strength as discussed previously.

**Table 4.** Bond strength value at room temperature, residual bond strength, and amount of bond strength loss

| Rod Type | Bond Strength (MPa) | | Loss of Bond (%) |
|---|---|---|---|
| | Room Temperature | Residual Bond | |
| **CB** | 13.1 | 1.2 | 91.6 |
| **CPH** | 12.1 | 1.8 | 86.2 |
| **CPI** | 10.8 | 1.7 | 85.2 |
| **NG** | 4.2 | 0.9 | 80.4 |
| **Steel** | 11.1 | 6.7 | 38.5 |

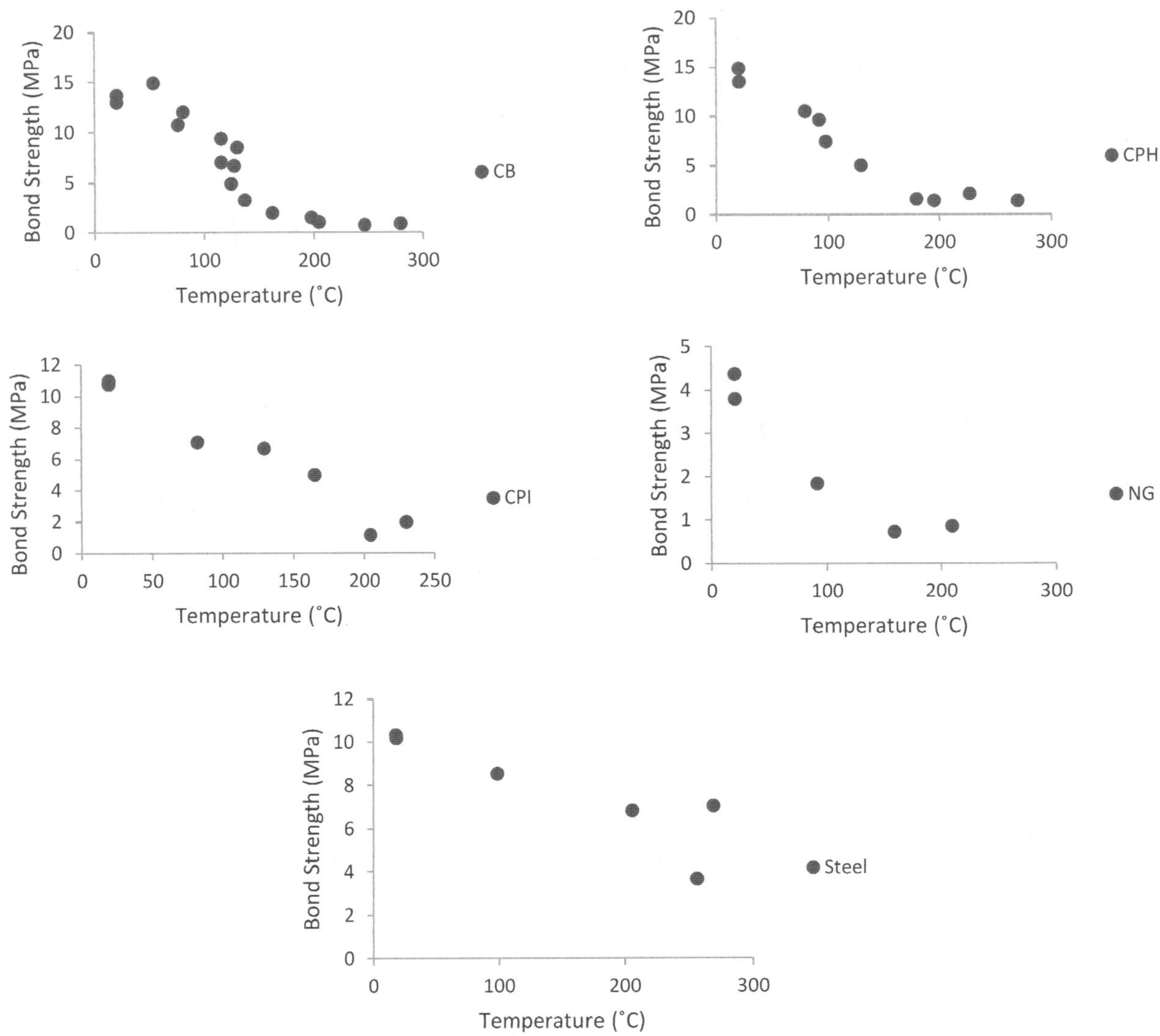


**Figure 5.** Effect of temperature on tested FRP re-bars.

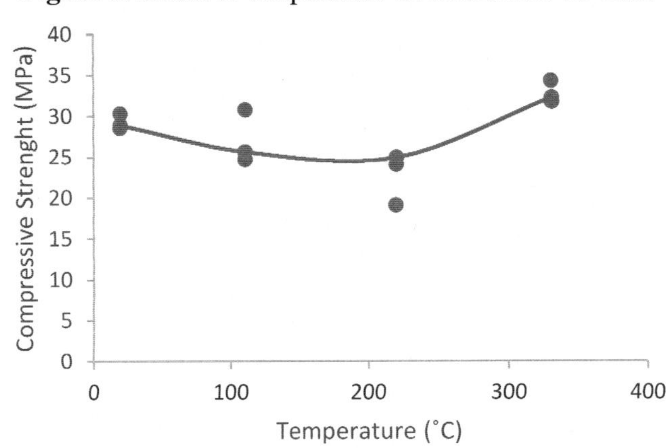


**Figure 6.** Effect of temperature on concrete compressive strength.

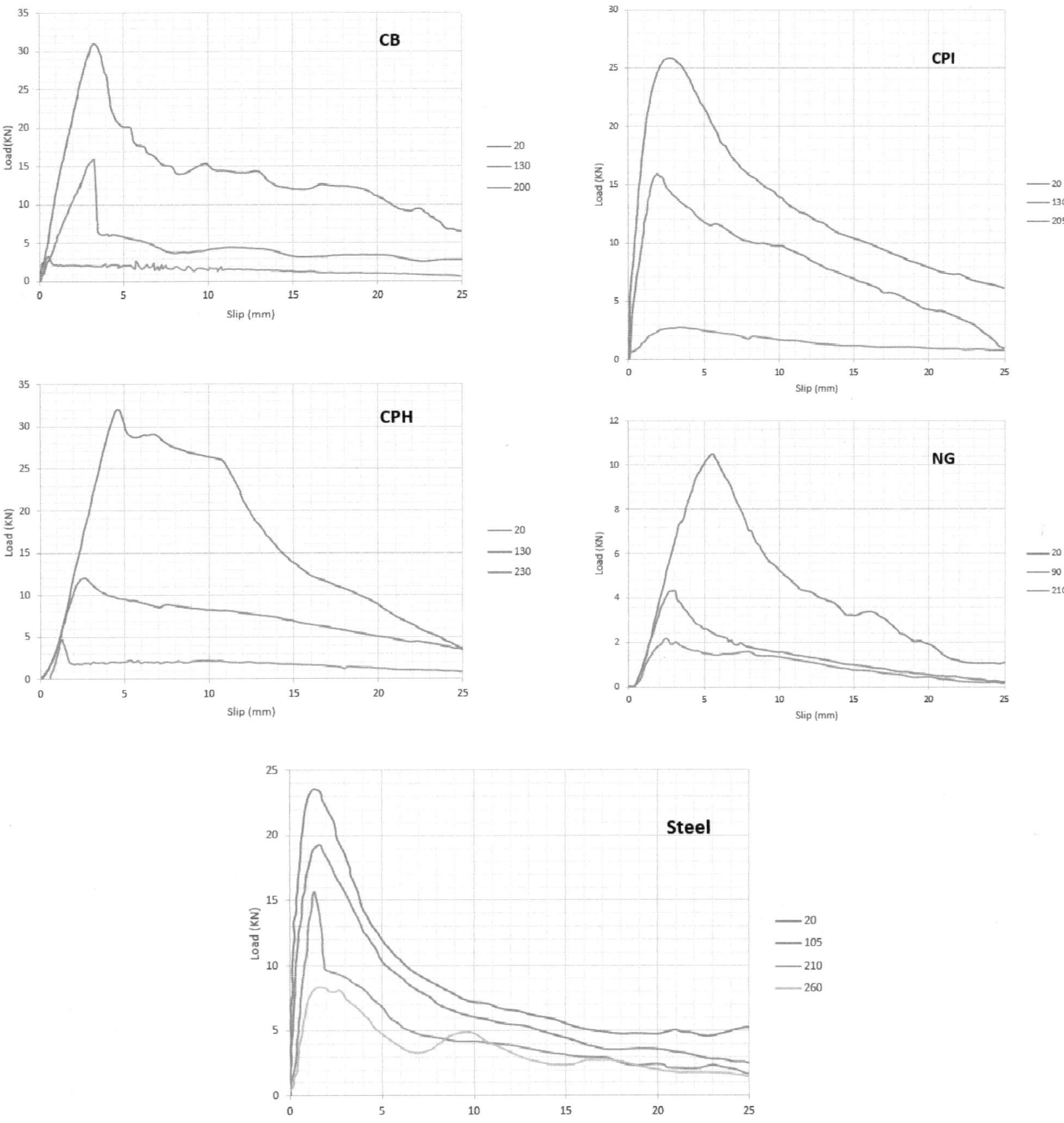


**Figure 7.** Load-Slip curves for tested FRP re-bars at different temperature.

## 5. Discussion

It appears that the bond strength of the FRP re-bars are depend on the bond mechanisms of the polymer at the surface of these re-bars at high temperatures. This study showed that in CB and NG re-bars the bond properties are only rely on the polymer. In CB rod a large stiff lugs at the surface provide the bond strength at room temperature, and in NG rod, a thick weak polymeric layer at the surface hides the deformations of the core and does not sufficiently support the sand particles embedded on the surface, which were presumably provided to enhance the bond. As the temperature increased a rapid reduction in bond strength was seen in both cases. However, in the post-peak behaviour the differences between these two re-bars can be seen.

CPH and CPI re-bars, where wrapped by helical glass fibre with a thin layer of polymer shown a better behaviour in bonding system. As the temperature increased the reduction in bond strength was more moderate, and in the post-peak stage the behaviour was more gradual reduction in the load bearing rather than a sudden drop, compared by the CB re-bar.

At elevated temperature concrete losses some of its strength due to internal vapour pressure, this reduction in strength can affect the reduction in the bond strength between the concrete and reinforcing bars as it was seen in steel.

## 6. Summery and conclusion

Different types of FRP re-bars were tested for pull-out at temperatures of up to $300^{\circ}C$. In all FRP re-bars high values of bond strength were obtained at room temperature, except for the one which had a weak external layer. The bond strength of all other FRP rods were as high as or higher than the steel rods.

The results showed that as the temperature increased, bond strength reduced massively. In CB rod this reduction was 92% and the bond strength dropped from 13.2 to 1.1 $MPa$. The polymeric lugs creates a bond between this type of rod and concrete, which appear to be better than other re-bars at high temperature. Having helical wrapping of glass fibre around the core of CPH and CPI re-bars, reduced the reduction in bond strength (from 12.2 to 1.7 $MPa$ and from 10.9 to 1.6 $MPa$, respectively), however, at high temperature the value of the bond strength was very low for all of the FRP re-bars. In the same temperature range the reduction in the bond strength of the steel re-bar, in comparison, was from 11.2 to 6.9 $MPa$.

As it seen in NG re-bar a thick polymeric layer at the surface appears to negatively affect the bond both at room temperature and at $200^{\circ}C$.

Temperature also affects the slope of the ascending load-slip curve (the bond modulus), which decreases as the temperature increases. As the temperature increases, the descending curve of the FRP re-bars becomes more linear, indicating a degradation in the polymeric surface treatments that support the bond, leaving the rebar with only a friction mechanism to create a bond.

## 7. References

[1]. Fried, J. R. (1995). *Polymer science and technology*. Prentice-Hall, Englewood Cliffs, N.J., 167–169.

[2]. Yamaski, Y., Masuda, Y., Tanano, H., and Shimizu, A. (1993). ''Fundamental properties of continuous fiber bars.'' *Proc., Fiber Reinforced Plastic Reinforcement for Concrete Struct.*, A. Nanni and C. W. Dolan, eds., American Concrete Institute, Detroit, 715–730.

[3]. Dostal, C. A., ed. (1987). *Engineering materials handbook, Vol. 1: Composites*. ASM International, Materials Park, Ohio.

[4]. Kumahara, S., Masuda, Y., and Tanano, Y. (1993). ''Tensile strength of continuous fiber bar under high temperature.'' *Proc., Fiber-Reinforced- Plastic Reinforcement for Concrete Struct.*, A. Nanni and C. W. Dolan, eds., American Concrete Institute, Detroit, 731–742.

[5]. Fujisaki, T., Nakatsuji, T., and Sugita, M. (1993). ''Research and development of grid shaped FRP reinforcement.'' *Proc., Fiber-Reinforced- Plastic Reinforcement for Concrete Struct.*, A. Nanni and C. W. Dolan, eds., American Concrete Institute, Detroit, 177–192.

[6]. Wolff, R., and Miesseler, H. J. (1993). ''Glass fiber prestressing system.'' *Alternative materials for the reinforcement and prestressing of concrete*, J. L. Clarke, ed., Chapman & Hall, London, 34–54, 131 132.

[7]. Wang, N., and Evans, J. T. (1995). ''Collapse of continuous fiber composite beam at elevated temperatures.'' *Composites*, 26(1), 56–61.

[8]. Okamoto, T., Matsubara, S., Tanigaki, M., and Hasuo, K. (1993). ''Practical application and performance of PPC beams reinforced with braided FRP bars.'' *Proc., Fiber-Reinforced-Plastic Reinforcement for Concrete Struct.*, A. Nanni and C. W. Dolan, eds., American Concrete Institute, Detroit, 875–894.

[9]. Baena, M. et al. 2009. Experimental Study of Bond Behaviour Between Concrete and FRP Bars Using a Pull-Out Test. Journal of Composites for Construction,pp. 784-797.
[10]. Nanni, A. et al. 2005. Thermal Effect on Bond Between FRP Rebars and Concrete. Journal of Composites for Construction.pp. 1223-1230.
[11]. Tastani, S.P., Pantazopoulou, S.J. 2006. Bond of GFRP Bars in Concrete: Experimental Study and Analytical Interpretation. Journal of Compusits for Construction. pp. 381-389.
[12]. Diederichs U, Schneidwe U. Untersuchung des Verbundverhaltens und der Verbund-Festigkeit von Rippenstraben und Glatten Rundstaber bei Hohen Temperaturen. Institute Fur Baustoffkundeund
[13]. Malhotra, L. (1982). *Design of fire-resistance structures*. Surrey University Press, London, 48–77.

IOP Conf. Series: Materials Science and Engineering **371** (2018) 012014 doi:10.1088/1757-899X/371/1/012014

# Properties of RPC with Calcium Carbonate Concrete Waste Powder

**Y W Sui[1,2,3], P F Qiao[2], Q B Tian[2], X T Yue[2] and J Li[2]**

1 Collaborative Innovation Center of green building of Shandong Province, Jinan 250101, China
2 School of materials sciences and Engineering of Shandong Jianzhu University, Jinan, 250101, China
3 Corresponding author, E-mail address: herrsui@sdjzu.edu.cn

**Abstract.** Calcium carbonate concrete is a normally concrete in the area with rich limestone. With the heavy use, a large of Calcium carbonate concrete waste is generated, how to reuse the waste is an important problem. The generation of reactive powder concrete (RPC) with silica fume is relative more expensive than normal concrete. Using the calcium carbonate concrete waste instead of silica fume in RPC can realize the recycling of construction and demolition waste, moreover, the high price of RPC manufacture can be decreased. The study mainly focused on the mechanical properties and durability as well as the microstructure of RPC with calcium carbonate concrete waste powder (CCWP) instead of silica fume, the results confirmed that the mechanical strengths of 3, 7, 28 days of RPC with CCWP did not obviously decrease, freezing and thawing test could attain to 500 recycles without obvious destruction and carbonation did not happen after 28 days. RPC with CCWP had no more amorphous material and good crystallinity, its microstructure was dense as that of RPC without CCWP.

## 1. Introduction

Reactive powder concrete (RPC) characterized by super-high strength, extreme durability and superior toughness [1], is a type of concrete with high doses of fine-grained cement and a low water–cement ratio, its compressive strengths can attain in the range of 200 ~ 800 MPa, compared with a maximum compressive strength in the range of 50 ~ 100 MPa for high strength concrete (HSC) [2,3], the Chinese standard GB/T 31387 also defines that RPC 100 has a compressive strength with more than 100 MPa [4].

The basic principles for the development of RPC, explained by many researchers, are characterized with no coarse aggregate, high compactness by powder, low water to cement ratio, use of suitable pozzolanic material, superplasticizer and steel fibers, treatment by heat or pressing during curing [5].

RPC can be used in the defense industry, in nuclear power plants, and in weapons factories. In Europe, RPC is also applied to the construction of bridges and viaducts to reduce their cross-sectional areas. In Turkey, RPC is used in the production of manhole covers and storm grates [2]. In China, RPC is used to produced cover plate for high speed railway, 200000 $m^3$ RPC were used in the construction of high speed railway in 2010, it was the most use in the world [6].

However, because of their ultra- high strengths and high manufacturing costs, the use of RPC has been questioned with concerns raised about possible ultra-brittle failure and unfavorable cost to performance efficiency [1].Silica fume is expensive as 6 ~ 10 times as the price of normal cement in China, therefore, RPC is only employed in restricted areas [2].

Nowadays, more and more concrete waste was accumulated in landfill, 500 Mio. Tons concrete wastes were produced in 2014 in China [7]. Most concrete wastes are difficult to reuse as aggregate to produce new concrete because of lower qualities of recycled concrete aggregate due to presence of residual mortar[8, 9], for this reason, when dealing with concrete recycling, a differentiation between coarse (nominal size > 5 mm) and fine aggregates (maximum size < 5 mm) is generally done. Coarse recycled concrete aggregates are commonly used in partial replacement of natural aggregates in concrete, however, the concrete mix design has to be adjusted in order to correct the worsening of final properties such as workability and durability, especially in respect to alkali – silica reaction, corrosion (due to chloride content) and freeze thaw resistance. On the contrary, fine recycled concrete aggregates are less useful as aggregates in concrete as they can be highly detrimental for what concern strength, workability and durability [10]. Therefore, at present, recycling industry has a very limited interest in fine concrete waste, even if they account at least for about the 30% of the entire building material waste [11].

Silica fume, a kind of pozzolanic material is used in RPC as a fine aggregate to play a micro filling role in no heat curing, however, because of the high price, silica fume is normally replaced by other pozzolanic material [12]. Limestone powder with the most content of calcium carbonate has low reactivity and is primarily used as a filler material [13-15]. The calcium carbonate concrete waste powder (CCWP) contains a lot of calcium carbonate and has reactivity material because of the rest cement without hydration [16], probably, CCWP has the same function in RPC like silica fume. Some studies certified that concrete waste powder had activity and could be used in Mortar replacing of 20% cement [17]. This study investigated the properties of RPC with CCWP instead of Silica fume in order to decrease the use quantity of silica fume.

## 2. Experimental Materials and Method

### *2.1 Experimental materials*

CCWP was made from a mass concrete with the aggregate of calcium carbonate, derived from a demolition site in China. CCWP was generated by the following methods:

- Direct generated grade (< 0.08 mm) by crushing,
- Grade (< 0.08 mm) after grinding the concrete waste grades more than 0.08 mm in ball mill.

The chemical elements of CCWP mainly contain CaO, $CO_2$ and $SiO_2$ due to the test results in table 1. According to the content CaO, the calculated content of calcium carbonate is about 76.4%.

The mineral composition of CCWP was mainly owned by calcium carbonate, DSC determination results showed that CCWP reacted with absorption heat in the temperature of 800 °C due to the decomposition of calcium carbonate (see figure 1).

**Table 1.** Chemical compositions of CCWP compared to silica fume and cement (%).

| Oxide | $SiO_2$ | $Al_2O_3$ | $Fe_2O_3$ | CaO | MgO | $K_2O$ | $Na_2O$ | $SO_3$ | $TiO_2$ | $CO_2$ |
|---|---|---|---|---|---|---|---|---|---|---|
| CCWP | 12.3 | 4.1 | 1.4 | 42.8 | 2.5 | 0.7 | 0.3 | 0.9 | 0.3 | 34.4 |
| Silica Fume | 77.9 | 0.7 | 1.6 | 1.9 | 2.1 | 6.4 | 1.1 | 1.5 | 0.1 | 4.6 |
| Cement [18] | 14.0 | 3.5 | 1.8 | 42.5 | 2.5 | 0.7 | 0.1 | 0.5 | 0.2 | 34.5 |

IOP Conf. Series: Materials Science and Engineering **371** (2018) 012014 doi:10.1088/1757-899X/371/1/012014

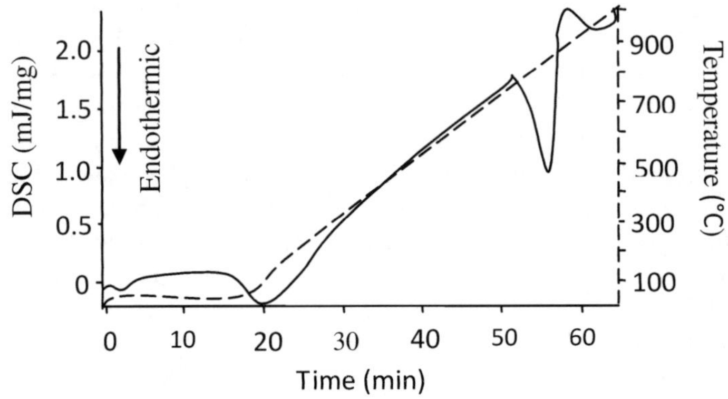


**Figure 1.** DSC determination for CCWP.

*2.2 Experimental method*

The test samples for mechanical properties of RPC with CCWP instead of silica fume contained cement, fine sand, silica fume, CCWP, Poly carboxylic acid and water, without steel fiber, the constitutes were detailed determined as follows:

- Cement: Portland cement (P.O 42.5) was manufactured in Shanshui Cement Company of Shandong.
- Silica fume (SF): was bought from Pengcheng silica fume company of Shandong with specific surface area of 20,700 $m^2/kg$.
- Polycarboxylic acid (PC): was produced by the Huadi of Shandong Company.
- Fine sand (FS): Quartz sand was obtained from the standard sand with maximum nominal sizes of 0.125 mm.

The concept of test samples preparation was listed in table 2, Silica fume in RPC was replaced by CCWP with 15%, 30% and 45%, and samples were cured in 65 °C warm water in a thermostat.

**Table 2.** Concept of RPC for experiments.

| | Cement | SF | FS | PC | Water | CCWP | W/C |
|---|---|---|---|---|---|---|---|
| Reference Sample | 1 | 0.250 | 1.1 | 0.018 | 0.28 | 0 | 0.28 |
| Replacement 15% | 1 | 0.213 | 1.1 | 0.018 | 0.28 | 0.037 | 0.28 |
| Replacement 30% | 1 | 0.175 | 1.1 | 0.018 | 0.28 | 0.075 | 0.28 |
| Replacement 45% | 1 | 0.138 | 1.1 | 0.018 | 0.28 | 0.112 | 0.28 |

Samples of Rectangular columns 40 × 40 ×160 mm for mechanical strengths were made according to the Chinese standard GB/T 17671 [19].The freezing and thawing experiment and carbonation resistance test were carried out with 100 × 100 × 400 mm prisms according to Chinese standard GB/T50082 [20].

The freezing and thawing experiment was determined the cycles of the samples in a fast freezing and thawing machine according to Chinese standard GB/T50082. The determination process of freezing and thawing depended on two indexes: mass loss (△W, as in equation (1)) and durability index (Pn, as in equation (2)). The freezing thawing test can be finished, when

- The test reaches the required cycles,
- △W > 5%, or
- Pn < 60%.

$$\Delta W = \frac{Wo - Wn}{Wo} \times 1\,0\,( \qquad (1)$$

where,

ΔW = weight change of specimen at n cycles of freezing and thawing in percent in %.

Wo = weight change of specimen at the beginning of the test in Kg.

Wn = weight of specimen after n cycles in Kg.

$$Pn = \frac{f_n^2}{f_o^2} \times 100 \tag{2}$$

where,
P= percentage of relative dynamic modulus of elasticity after n cycles of freezing and thawing in %.
n = number of cycles at the time of testing.
fn = fundamental transverse frequency after n cycles of freezing and thawing in Hz.
fo = fundamental transverse frequency at zero cycles in Hz.

## 3. Results and Discussion

### *3.1 Mechanical strengths of RPC with CCWP*

The 3, 7 and 28 days flexural and compressive strengths of RPC with CCWP were tested according to the GB/T17671 [19], results were indicated in tables 3 and 4.

**Table 3.** Flexural strengths of RPC with CCWP (MPa).

| Days | Reference (MPa) | Replacement of silica fume with CCWP | | |
|---|---|---|---|---|
| | | 15% | 30% | 45% |
| 3 | 10.90 | 10.35 | 10.40 | 10.15 |
| 7 | 15.05 | 15.75 | 14.80 | 15.95 |
| 28 | 12.75 | 11.40 | 11.10 | 11.75 |

According to test results, strengths of RPC with CCWP had little difference compared to those of reference, furthermore, the RPC with 45% CCWP instead of silicate fume indicated relatively higher compressive strengths than those of reference. Two obvious results could be come out: because of curing in 65 °C warm water, the RPC hydrated very quickly, 3 days compressive strength was 95.73 MPa, and had no big difference compared to 7 days and 28 days values. So far as to the flexural strength, 28 days value was smaller than that of 7 days. The placement of CCWP instead of silicate fume had little influence on the mechanical strength of RPC, especially, RPC with 45% CCWP had relative higher mechanical strengths (see figure 2).

**Table 4.** Compressive strengths of RPC with CCWP (MPa).

| Days | Reference (MPa) | Replacement of silicate fume with CCWP | | |
|---|---|---|---|---|
| | | 15% | 30% | 45% |
| 3 | 95.73 | 90.98 | 83.80 | 91.88 |
| 7 | 95.03 | 92.68 | 91.15 | 95.83 |
| 28 | 111.04 | 101.01 | 104.03 | 113.20 |

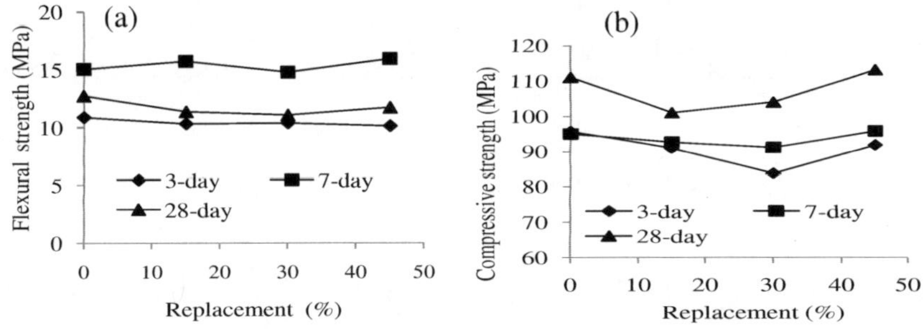


**Figure 2.** Influence of CCWP content on the flexural strengths (a) and compressive strengths (b) of RPC.

IOP Conf. Series: Materials Science and Engineering **371** (2018) 012014 doi:10.1088/1757-899X/371/1/012014

*3.2 Results of carbonation resistance and freezing and thawing*

The color of the RPC samples with different CCWP contents after 28 days carbonation test was lilac with the treatment of 1% Phenolphthalein, specially, the edge color did not obviously changed. Results showed that the samples did not be carbonized in the carbonation box after 28 days. RPC had dense structures, which was not changed with the addition of CCWP.

Freezing and thawing tests of RPC were determined until to 500 circles, results showed that all △W values were not more than 5%, all Pn values were not less than 60 % (see table 5). According to the tested data, the results could be confirmed that the samples of RPC with CCWP were not damaged after 500 cycles, which attained to the requirement of RPC Chinese standard GB/T 31387-2015.

*3.3 Microstructures of RPC with CCWP*

The crystal ingredient and microstructures of reference and RPC with CCWP instead of 45% SF were determined by the tests XRD and SEM. XRD determination Results showed that reference sample had amorphous materials and relative bad crystallinity, the amorphous material was active ingredients in silica fume , that meant silica fume in RPC did not completely take place the pozzolanic reaction. By contrast, the XRD photograph showed RPC with CCWP instead of 45 % SF had tiny amorphous materials and good crystallinity (Figures 3 and 4). The SEM test pictures showed that the microstructures had no obvious difference between reference and RPC with CCWP (Figure 5), they were both dense.

**Table 5.** Results of freezing and thawing of RPC with CCWP.

| Cycles | Index | Replacement of CCCWP instead of silicate fume | | | |
|---|---|---|---|---|---|
| | | 0 | 15% | 30% | 45% |
| 0 | Wo (kg) | 8.624 | 8.598 | 8.712 | 8.688 |
| | $f_o^2$ (GPa) | 4895807 | 4879682 | 5047069 | 4975143 |
| 100 | △W (%) | 0.46 | 0 | 0 | 0.88 |
| | Pn (%) | 96.54 | 100 | 94.34 | 100 |
| 200 | △W (%) | 1.27 | 0 | 1.29 | 1.30 |
| | Pn (%) | 96.11 | 100 | 100 | 100 |
| 300 | △W (%) | 0.20 | 1.30 | 0.61 | 0.12 |
| | Pn (%) | 96.90 | 100 | 100 | 100 |
| 400 | △W (%) | 0.98 | 1.68 | 2.34 | 2.66 |
| | Pn (%) | 94.56 | 95.33 | 97.69 | 94.38 |
| 500 | △W (%) | 2.19 | 3.78 | 3.56 | 4.03 |
| | Pn (%) | 89.79 | 87.29 | 90.24 | 88.55 |

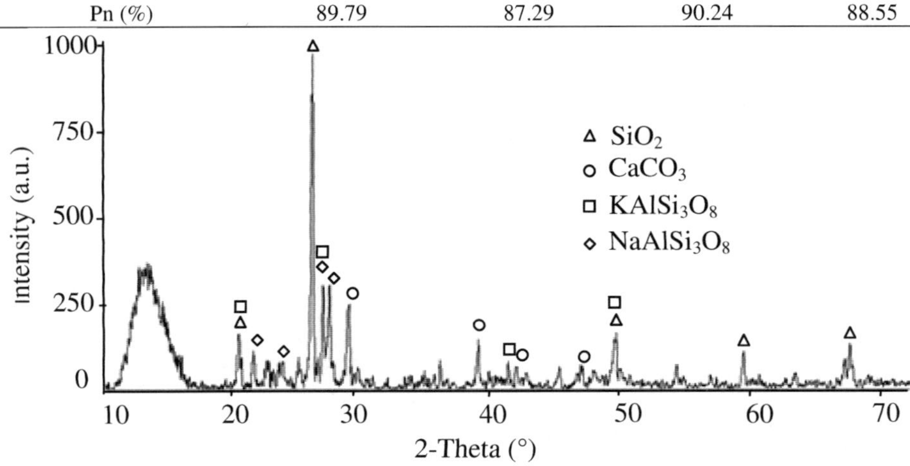


**Figure 3.** XRD-result for reference sample.

IOP Conf. Series: Materials Science and Engineering **371** (2018) 012014 doi:10.1088/1757-899X/371/1/012014

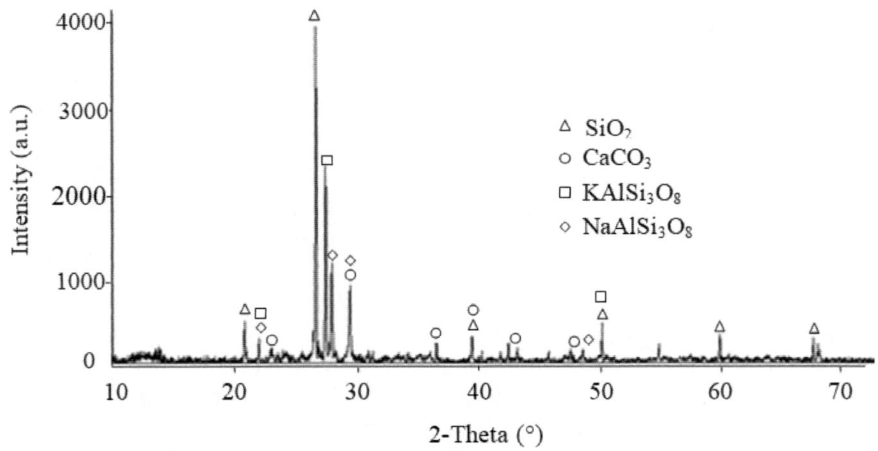


**Figure 4.** XRD-result for RPC with CCWP instead of 45% SF.

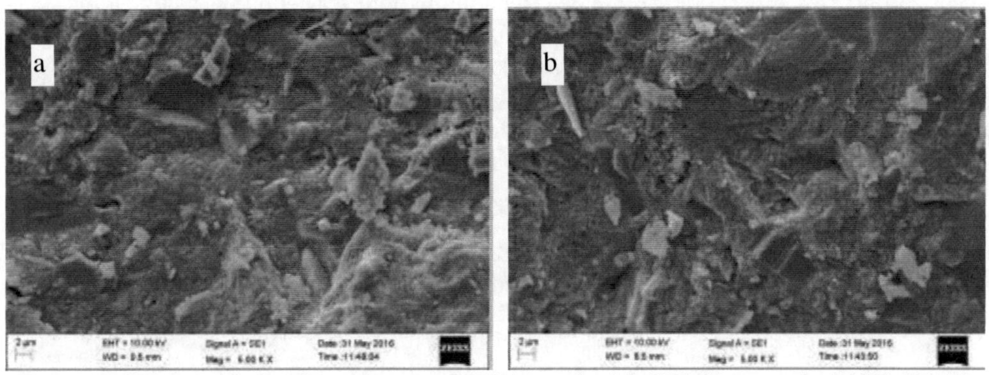


**Figure 5.** SEM-result for the reference RPC (a) and with CCWP instead of 45% SF (b).

## 4. Conclusions

This study examined the mechanical strengths, durability of RPC with CCWP instead of silica fume, according to the experiment data, the main conclusions were confirmed as follows:

1) The 3, 7 and 28 days flexural and compressive strengths of RPC with CCWP instead of 15%, 30% and 45% silica fume had no obvious difference with corresponding values of reference sample, the 28 days compressive strength of RPC with 45 % CCWP exceeded 100 MPa.

2) The freezing and thawing test indicated that the RPC with 15%, 30% and 45% CCWP instead of silica fume attained to 500 cycles. The carbonation reaction did not happen in the RPC with CCWP.

3) Silica fume in RPC had no entire pozzolanic reaction, and reference RPC has bad crystallinity, on the contrary, RPC with CCWP had no more amorphous materials and good crystallinity. The microstructure between reference sample and RPC with CCWP had no obvious difference.

## 5. References

[1] Yia N H, Kim J H J 2012 *Construction and Building Materials* **28** 694–707

[2] Canbaz M 2014 *Construction and Building Materials* **70** 508–13

[3] Zheng W Z, Luo B F, Wang Y, 2013 *Construction and Building Materials* **41** 844–51

[4] GB/T31387 2015 Reactive powder concrete (Beijing: Chinese quality supervision and administration bureau )

[5] Al-Hassani H M, Khalil W I, Danha L S 2014 *ACTA Ehnica Corvinensis-Bulletin of Engineering Tome VII* 47-58

[6] Compilers working group 2013 Explanation for test method of impact resistance super powder concrete (Solicit Opinions Manuscript)

[7] Guo S H 2015 Resource industry development report for 2014 on construction and demolition waste in China (Beijing: Products technology innovation strategic alliances of resourcezation of

construction and demolition in China)
[8] Sanchez de J M, Gutierrez P A 2009 *Constr. Build. Mater.* **23** 872-79
[9] Sui Y W, Tian Q B, Yue X T et al 2015 Study on influence factors of high quality recycled Concrete aggregates produced by thermal-grinding method. *Journal of Shandong Jianzhu University* **30** 544-49
[10] Zhao Z, Remond S, Damidot D, Xu W 2015 *Construct Build Mater* **81** 179-86
[11] Gastaldi D, Canonico F, Capelli L, Buzzi L, et al. 2015 *Cement & Concrete Composites* **61** 29-35
[12] He F, Yang J P, Ma S F 2007 Effect of Silica Fume Quantity on Performance in Reactive Powder Concrete *Journal of Guilin University of technology* **27** 77-80
[13] Suaiam G, Makul N 2013 *Constr. Build. Mater.* **38** 455-64
[14] Suaiam G, Makul N 2013 J. Clean. Prod. **57** 308-19
[15] Zhang Z Q, Wang Q, Chen H H 2016 *Powder Technology* **301** 16-25
[16] Zhang X, Wu Y S, Zhang J B, He J 2011 The Activity stimulated of construction wasted powder and its reuse feasibility analysis *Jiangxi Science* **29** 70-73
[17] Xu R L 2011 Activation and application of construction waste powders (Wuhan: Dissertation for master degree of Technology University of Hubei)
[18] Locher F W 2000 Zement-Grundlagen der Herstellung und Verwendung (Duesseldorf: Verlag Bau und Technik)
[19] GB/T17671 1999 Method of testing cements-Determination of strength (Beijing: Chinese quality Supervision and Administration Bureau)
[20] GB/T50082 2010 Standard for test method of long-term performance and durability of ordinary concrete (Beijing: Chinese quality supervision and administration bureau)

**Acknowledgments**
The project is sponsored by the Scientific Research Foundation for the Returned Overseas Chinese Scholars, State Education Ministry; the doctoral fund of Shandong Jianzhu University (XNBS1257), and the fund project of collaborative Innovation Center of green building of Shandong Province.

IOP Conf. Series: Materials Science and Engineering **371** (2018) 012015 doi:10.1088/1757-899X/371/1/012015

# Development of an Automatic Crack Inspection System for Concrete Tunnel Lining Based on Computer Vision Technologies

**C K Nguyen[1], K Kawamura[2], M Shiozaki[3], and A Tarighat[4]**

1 Graduate School of Science and Engineering, Yamaguchi University, Japan
2 Graduate School of Science and Technology for Innovation, Yamaguchi University, Japan
3 Sumitomo Mitsui Construction Co., Ltd., Tokyo, Japan
4 Dept. of Civil Engineering, Shahid Rajaee Teacher Training University, Iran

**Abstract.** This paper presents an automatic crack inspection system for concrete tunnel lining based on image processing technique and computer vision. The system includes a video image acquisition device, an image stitching software, and a crack detection software. Firstly, the video image acquisition device uses six video cameras and three illuminators mounted on a steel framework which is capable of sliding from side to top of the inspecting vehicle to shoot the full surface of the tunnel lining. Secondly, the image stitching software based on image matching technique was developed to create layout panorama from the tunnel lining surface images, making it easier to visualize a large and detailed section of the tunnel lining. Finally, we propose a semi-automatic crack detection software relied on a combination of image processing technique and interactive genetic algorithm (iGA) to crack detection for concrete surface images of the tunnel lining. Experimental results demonstrate the effectiveness of the proposed method.

## 1. Introduction

The number of deteriorated concrete components of existing civil infrastructure system has increased dramatically in many countries due to rapid aging. Therefore, the maintenance and management of existing concrete structures of the civil infrastructure system have become a major social concern worldwide. Especially, the existing tunnels were built over 50 years. To keep concrete components in good condition under all circumstances, the inspection plays an important role in the maintenance of infrastructure. Crack detection on concrete tunnel lining surface is one of the basic tasks of the inspection. Therein, the visual inspection is one of the most frequently adopted, which is carried out by inspector's observation. It has a number of limitations, such as the high labour costs associated with carrying out tasks, and inaccuracy due to subjectivity, safety, slow inspection speed and disruption to traffic. In recent years, automatic crack inspection system for large scale concrete structures is a very active area of research due to popular digital video camera and computer vision developments. Especially, for tunnel inspection, the layout panorama generation of the tunnel lining is one of the important tasks for preliminary tunnel inspection [1].

Altogether, many crack detection systems have been developed. Kawamura et al. [2] proposed interactive genetic algorithm is applied to adjust the image processing parameters. It is developed to extract crack pattern effectively. Fujita et al. involved preprocessing steps (median filter and background subtraction) [3] and post processing (Hessian matrix and threshold) to extract crack

features. Miyamoto et al. calculated crack widths based on the difference of pixel brightness of each row [4]. Nishinkawa et al. [5] designed System for Automatic Construction of Image Filter (SACIF) based on a genetic program to detect cracks on concrete structures. Stent et al. [6], [7] used clustering and ranking methods for detecting changes probability on tunnel lining surface. Yeum and Dyke [8] applied threshold of Frangi filter and Canny edge detector to detect bolt cracks.
Additionally, digital video and line-scan camera-based inspection devices have undergone substantial developments. However, the existing equipment requires a high level of expertise for effective inspection work to be performed. Therefore, the inspection cost is likely to be expensive, which the maintenance budgets of local governments often cannot handle.
This study contributes a system including automatic panorama generation and semi-automatic crack detection software with high accuracy and inexpensive inspecting cost because image processing expertise and exclusive vehicle are unnecessary. Moreover the system is based on computer vision technologies, to aid the visual inspection of tunnel lining.

## 2. Video image acquisition device

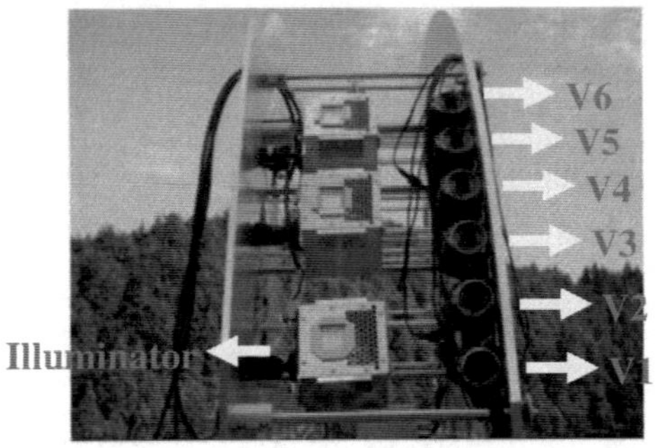


(a) Side view of the image acquisition device

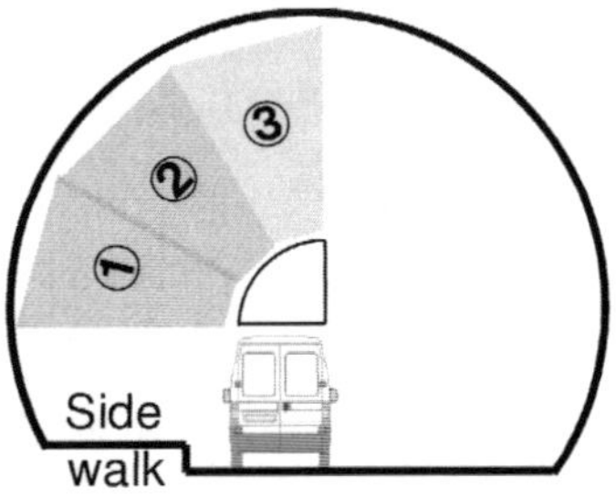


(b) Method for imaging a half of cross section of a tunnel

**Figure 1**. Video acquisition device.

The video images of the entire tunnel lining surface are continuously scanned by the video acquisition device assembled on a car shown in Figure 1. Figure 1 (a) exposes the system consisted of six digital video cameras (from V1 to V6) , three illuminators attached to the steel framework that its structure has shape appropriated with a half of cross-section of the tunnel, and a car. Further, this device is possible of sliding from the side to the top of the car in order to shoot the full tunnel lining surface under several passes through the tunnel. Figure 1 (b) shows how to obtain image data in moving forward three times of the inspection car in the longitudinal direction of the tunnel. To capture full cross-section of the tunnel, the inspection car has to run forward three times and backward three times. Quality of the image data depends on many factors such as distance between the tunnel lining and inspection car, the illumination, and specification of video camera. This device is simple and does not require an exclusive vehicle. The video data are converted into the consecutive images with the specific overlapped region depending on the scanning speed of the inspection car. Due to the characteristic feature of the imaging system, the consequence image motion is only the translation, so it is unnecessary to concern the rotation. Therefore, the direct method, which makes use of available information in the image-matching location, is applied to stitch images.

## 3. Image stitching method

Figure 2 shows an image stitching procedure for the entire tunnel lining including three steps. Captured image data are retrieved from six video cameras which each camera contains a number of images depended on the length of the tunnel. In Step 1, the dataset is connected to form a panoramic image in the longitudinal direction of tunnel for each camera. In Step 2, these panoramic images obtained in the step 1 are stitched in the circumferential direction of tunnel for each region corresponding to each pass.

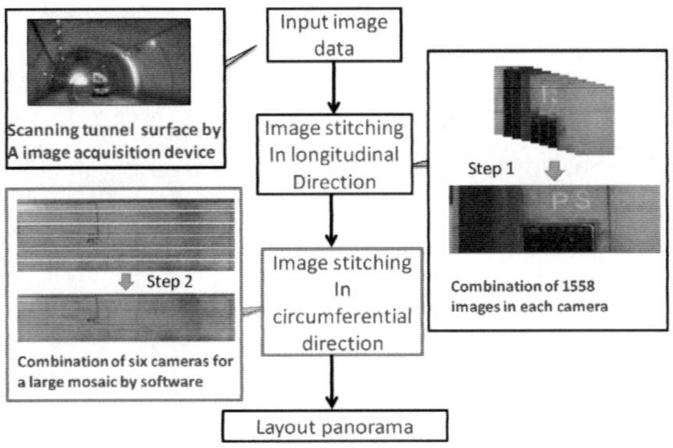


**Figure2.** The image stitching procedure for the entire tunnel.

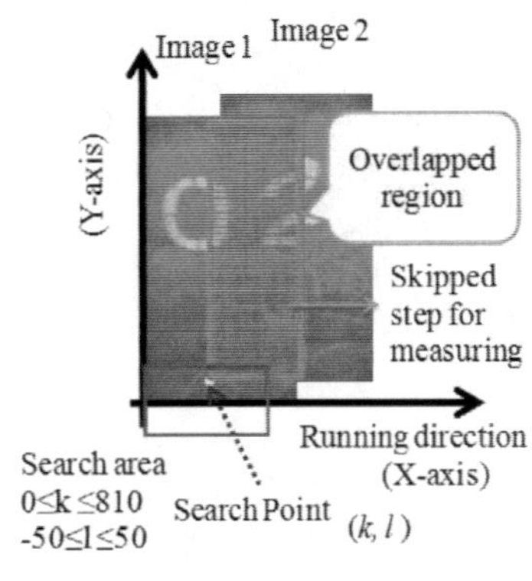


**Figure3**. Image matching based on similarity.

Subsequently, in Step 3, the connected images of the six regions (R1-R6) in step 2 are stitched together to make a layout panorama which consists of a full view of the entire tunnel lining surface. The layout panorama of the tunnel lining is produced from the combination of the automatic image stitching method of Step 1, and semi-automatic methods of Steps 2 and 3. In this paper, Step 1 is presented in detail.

### *3.1 Similarity metric*

Figure 3 shows a representation of automatic matching process of two consecutive images in the tunnel longitudinal direction (X-axis). Here, image 1 is the referenced image, and image 2 is the registered image of the image matching process. A search area is set in advance. Each the movement step of the search point is with respect to the location which enables image 2 to be shifted on image 1 to find the appreciate image-matching location by measuring the similarity and sharpness in the brightness of all pixel pairs in the overlapped region of the two images.

Furthermore, to accelerate the search and measurement process, the search point and similarity measurement are skipped in the search area and overlapped region with predefined values, respectively. Firstly, the matching cost functions are used to measure similarity of pixel-wise intensity in the overlapped region presented in detail:

$$SAD = \frac{\sum_{i=1}^{K}\sum_{j=1}^{L}\sum_{\{R,G,B\}} |(I_1 - I_2)|_{(i,j)}}{(K-k+1)\times(L-l+1)} \quad (1)$$

$$(k,l) = \{\underset{\in search\ area}{\arg\min}(SAD) \mid k \in [T_1\ T_2]\}$$

$$T_{1,2} = \frac{v\,\mathrm{m}\alpha}{u \times t} \quad (2)$$

Where, $I_1$ and $I_2$ are the intensity of pixels at coordinates *(i, j)* of images 1 and 2 respective in the overlapped region, *SAD* is determined from the sum of a difference between the pixel values of images 1 and 2 for each color channel (R, G, and B). Because standard *SAD* function has poor performance, image data are attributable to radiometric distortion and noise [9]. Authors propose a modified *SAD* function as follow: The numerator of the Equation (1) is divided by the area of the overlapped region (the total number of pixels) to normalize the overlapped region in each measurement. Furthermore, *(k, l)* and *(K, L)* are as the lower left and upper right coordinates (pixel) of the overlapped region of the image pairs, respectively. *k* coordinate is defined in image motion quantity (IMQ). Accordingly, the higher similarity of the overlapped region of two images obtains, the smaller score of *SAD* achieve. A

IOP Conf. Series: Materials Science and Engineering **371** (2018) 012015 doi:10.1088/1757-899X/371/1/012015

matching point at *(k, l)* coordinates is corresponding to the minimum cost value of *SAD* function with winner-take-all strategy.
$T_1$ and $T_2$ are the lower and upper threshold values determined from the (2). In that, $v$ is the initial setting speed of inspection car. It is equal to 30 Km/h. $\alpha$ is a parameter of the speed variance that is equal to 5 Km/h. In addition, $u$ is image resolution in unit mm/pixel, and $t$ defined frame rate is a number of frames per second in the scanning tunnel progress. To avoid too strict for keeping constant inspection speed, authors allow the speed change of inspection vehicle to 30 ± 5km/h.

### *3.2 Refinement*

We use median filter to smooth some local bad matches which appear corresponding to the inappropriate IMQ of the neighbor images as the following equations:

$$Q_i = \begin{cases} M_i \text{ if } |S_i - M_i| \le T_3 \\ S_i \text{ otherwise} \end{cases} \quad (3)$$

Where,

$$\begin{aligned} &S_i = med(M_{(i-j)}, \ldots M_i, \ldots, M_{(i+j)}) \\ &\text{if } i \le j \text{ then } M_{(i-j)} = 0 \end{aligned} \quad (4)$$

Here, $Q_i$ is IMQ in unit pixels after refinement, and $M_i$ is IMQ before refinement. $S_i$ is a median value obtained from (4), and $T_3$ is a threshold value depended on the speed change of the inspection car. The size of the median filter is $2j+1$ (where $j$ is an integer).

## 4. Experimental work

### *4.1 Case study*

The tunnel used in this experiment is an actual single-core circular tunnel with a length of 230 m in Yamaguchi Prefecture, Japan. Additionally, the image data sets captured for each camera from the image acquisition system consist of 1,558 images with a resolution of 1920 × 1080 pixels for each image. The initial setting inspection speed is $v$= 30 Km/h, and image resolution is $u$=0.231 mm/pixel. Further, the frame rate is $t$=60 fps, so the overlapped region of each image pair is over 50%. To ensure the resolution, a photographic laser distance meter is used to maintain the distance between the tunnel lining and the inspection car.
Further, a prototype software based on the C++ language was developed to implement the full image stitching algorithm. The search range is set in large enough space to ensure the search result of matching point objectively, as shown in Figure 3. The skipped step for measuring similarity and searching matching point is set to 11x4 pixels. Additionally, the parameter values of refinement step are chosen as follows: size of the median filter is 100, and the threshold value is 50.

### *4.2 Experimental results*

Figure 4 shows IMQ results of camera V1via the running direction of inspection vehicle before and after refinement. The IMQ result via the vertical direction (Y-axis) isn't presented because its variance is not much different. In Figure 4(a), some local points show inappropriate matching results before refinement. After refinement is applied, IMQ graph presents smoother shown in Figure 4(b). The IMQ results of the other cameras are similar to camera V1.

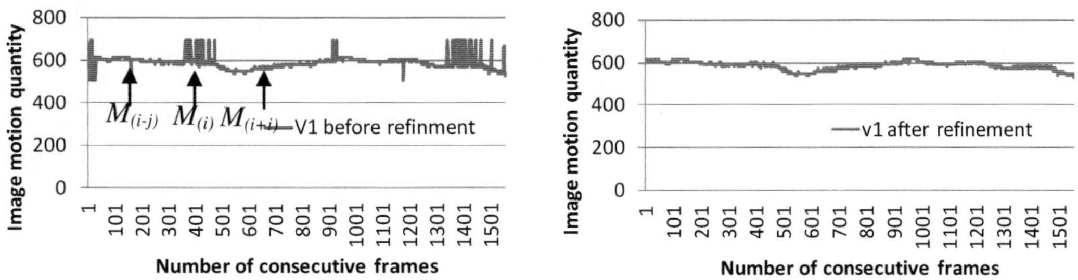


(a) The initial IMQ result before refinement (b) The IMQ result after refinement

**Figure** 4. The IMQ results before and after refinement.

Figure 5 shows a layout panorama segment resulted from the proposed image stitching method in both longitudinal and circumferential directions of a half of tunnel cross-section obtained from the image acquisition device. The panoramic pictures of the entire length of the tunnel lining include 22 segments. The length of each segment is 10.5 m excluded overlapped parts at the two ends of next segment. In addition, each mosaic region results from stitching panoramic image of each camera via circumferential direction. In that, each camera mosaic compressed with scale 1 per 32 has 45600x1920 pixels based on stitching 65 RGB (1080x1920) resolution images. The panorama shows high-accurate alignment in Figure 5. As a result, inspectors can detect tunnel lining abnormalities off-line easily.

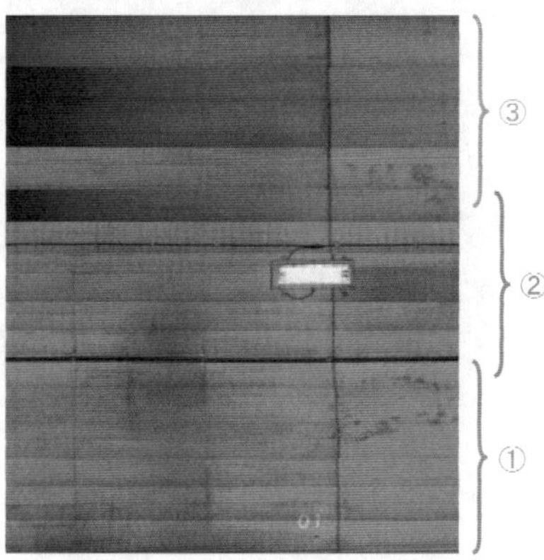


**Figure5**. A represented panorama segment of three imaging regions shown in Figure 1(b), and Step 3 in Figure 2.

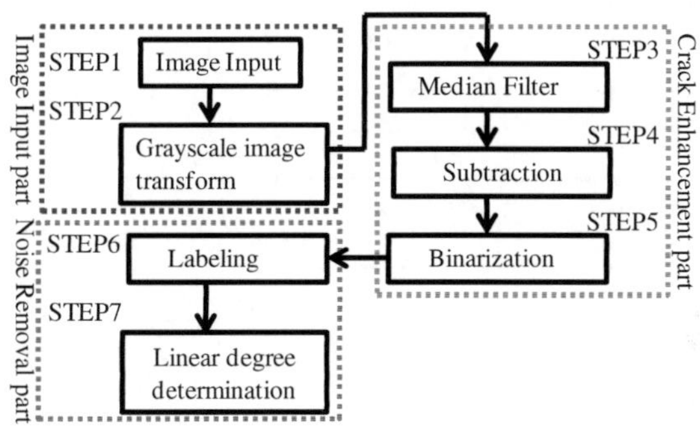


**Figure 6**. Image processing procedure.

## 5. Crack detection method

### *5.1 Image processing technique*

The algorithm for accurate crack detection has composed of three major process parts shown in Figure 6. The first part was image input part which included image input and gray-scale image transformation. The second part was crack enhancement part which included the following pre-processing steps: median filter, subtraction, binarization. The final part was noise removal part, there were two steps: labelling and linear degree determination. These parts are presented in [10].

### *5.2 Application of iGA to the optimization of image processing parameters*

In this study, the iGA is applied to searching optimal values of the image processing parameters (IPPs) because it has had many advantages such as the evaluation for each individual's fitness easier than the individual's fitness function of GA, small population size, less number of generation [11]. Figure 7 shows an operated mechanism of genetic algorithm including into the crucial three stages, namely the initial population generation, fitness evaluation of each individual in the current population, and evolution operation to create the next generation. Therein, the evolution operation consisted of selection, crossover, and mutation is repeated until finding best solution. Namely, the detailed steps are presented as follow:

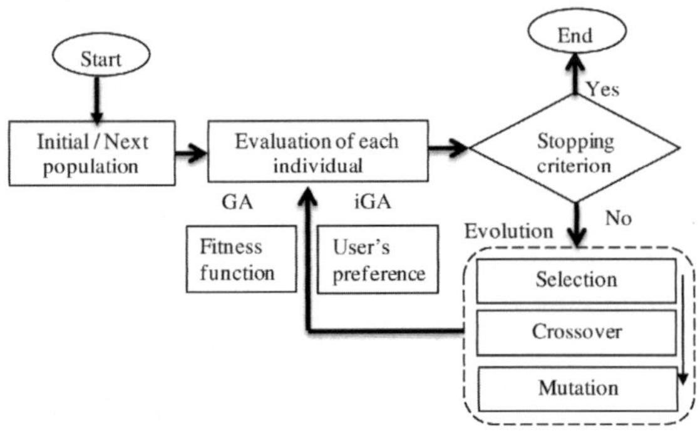


**Figure 7.** Operated mechanism of genetic algorithm.

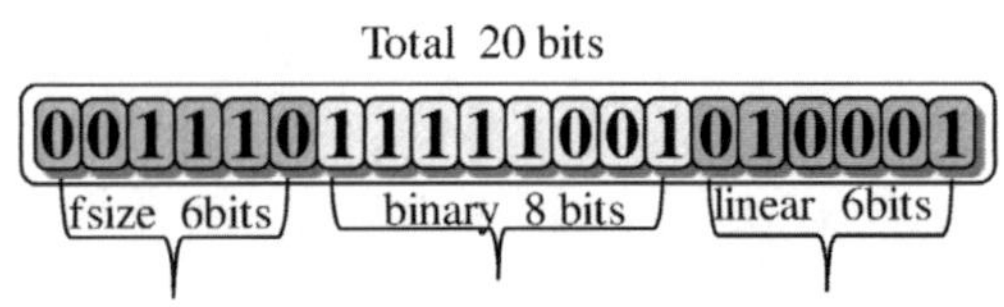


**Figure 8**. Coding presentation of a solution candidate.

**Table 1.** The solution ranges of IPP parameter.

| Variable | Range | Steps | bits |
|---|---|---|---|
| fsize | [1 127] | 2 | 6 |
| binary | [0 255] | 1 | 8 |
| linear | [0 32.5] | 0.5 | 6 |

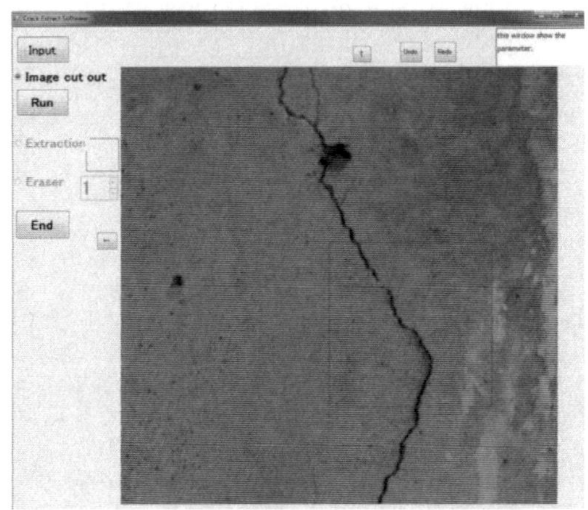


**Figure 9**. Input image window.

5.2.1 Initialize the first generation randomly.

Firstly, the IPPs are combined to create an individual in a population. Secondly, each individual is represented by a chromosome encoded to a binary string, as shown in Figure 8. Namely, the size of structuring element (fs) is assigned by 6 bits, the binary is expressed by 8 bits and the linear is expressed by 6 bits. Finally, an initial population was generated randomly with a predetermined size. Table 1 shows some parameter values based on the preliminary experiments.

5.2.2 Fitness evaluation of each individual.

The fitness of each individual (solution candidate) in a population is evaluated based on user's preference. Therefore, the evaluation result is often a relative fitness. This point was different from traditional GA.

5.2.3 Stopping criterion.

Evaluation of each individual is determined by the user until the user finds the best individual corresponding to the optimal parameters.

IOP Conf. Series: Materials Science and Engineering **371** (2018) 012015 doi:10.1088/1757-899X/371/1/012015

5.2.4 Evolution operation.

When stopping criterion is not satisfied, evolution procedure is implemented for operating selection, crossover, and mutation until the user chooses the best parameters.

## 6. Crack detection prototype software development

In this study, the image processing algorithm shown in Figure 6 is developed in the C program language. In addition, an interactive interface of the software for extracting crack and searching optimal parameter values are developed in C # program language (see Figures from 9 to 11). The following describes the prototype software that includes five STEPs of semi-automatic crack extraction.

### *6.1 STEP1*

Firstly, the user inputs an original digital image by clicking "Input" button in Figure 9. The software is able to divide a large size image into many subimages. In this paper, the original image size is 800x800 pixels. Secondly, the user checks "Image cut out" button to adjust the image processing parameters. Here, the size of the cut-out image is 300x300 pixels. Next, the user chooses an interested region of the cracks. Accordingly, a red frame appears automatically as Figure 9 The purpose of the crack region selection is to search optimal image processing parameter values for the cut-out image.

### *6.2 STEP2 (Adjustment of parameters)*

The interactive interface is presented as Figure 10 so that the user adjusts the parameters. The software generates 15 sample images (individuals) as an initial population. The user can keep elites (better individuals) to next new generation. If the user cannot keep any individuals, the user has to operate upon "Initialization" to generate a new population. In this window, the user evaluates the fitness of each individual with five buttons from 1 to 5 corresponding to the fitness from the worst to the best. These buttons are located below sample images. Subsequently, the user chooses "Evolution" button to perform gene operators (crossover, mutation).

### *6.3 STEP 3 (Crack extraction)*

To extract cracks from the original image smoothly, the user traces roughly the shape of the cracks by a finger on the touch panel as Figure 11 (a). This operation is also capable of processing with a computer mouse. The place in which the user traces on the touch panel is displayed as a red line. After limiting the crack region by a user's fingertip, the cracks are automatically displayed in the selected region with a blue line shown in Figure 11 (b). The user has to trace region of the crack to extract it from the background image. Hence, the crack of the final result image will be extracted with reduced noises.

As a matter of course, depending on the chosen parameter value, the cracks can be appeared with noise pixels or loss pixels. If some crack parts are not displayed, then the user returns STEP 2 to choose other parameter values. If many noises are also displayed as a portion of the crack, then the user moves to next STEP 4 for noise removal.

### *6.4 STEP4 (Noise removal)*

When some noises appear with the cracks are extracted from STEP3, the user selects the "Eraser" button in Figure 11 to carry out a process of noise removal. If the noises are not presented, the user moves to STEP5 without selecting the "Eraser" button.

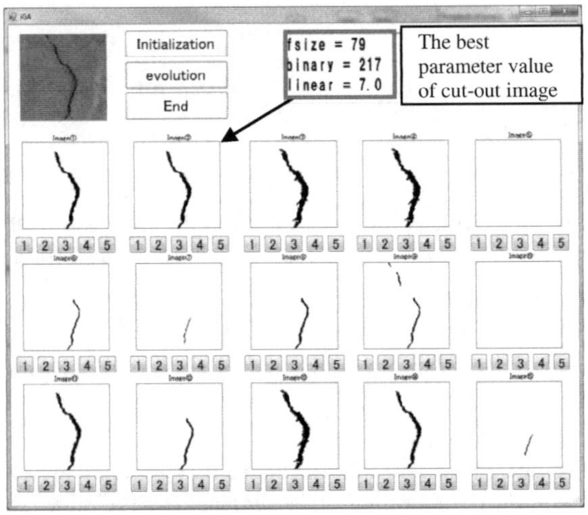


**Figure 10**. Parameter adjustment window.

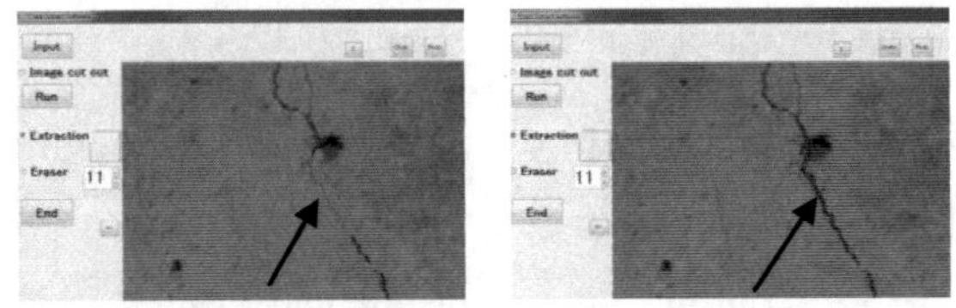

(a) The Tracing crack (b) The crack extraction

**Figure 11**. Traced crack on background

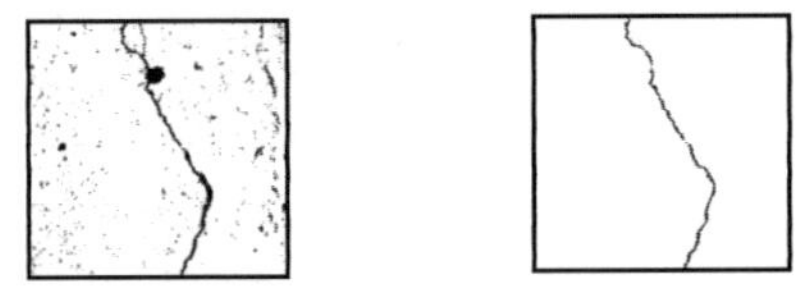

(a) Background image (b) Final crack image

**Figure 12**. The output of crack extraction.

### *6.5 STEP5 (End determination)*

When the user completes the crack extraction process, he/she presses "End" button at the bottom left of the window shown in Figure 11. On the other hand, when the user continues working crack extraction, he/she returns to STEP3. If the user wants to choose the parameters again, the user returns to STEP2. Consequently, the outputs of the software have two images automatically saved in the specified folder. Those are a crack extraction image as Figure 12 (a) and a crack map (final result image) as Figure 12 (b).

## 7. Evaluation and discussion

### *7.1 Evaluation method*

The proposed method is compared with the Haar wavelet transform (HWT) method with GA to optimize the parameters of HWT [12]. The HWT method is a useful tool for crack detection with many advantages as multi-resolution properties. They are compared with the benchmark image traced by human operator based on loss and noise rate values expressed as the (10) and 11 equations:

$$f_1 = \frac{m}{M} \times 100\%; f_2 = \frac{n}{N} \times 100\% \tag{10}$$

$$f = 1 - \sqrt{\frac{1}{2}(f_1)^2 + \frac{1}{2}(f_2)^2}\ f \in [0,1] \tag{11}$$

Where $m$ is the number of loss pixels, $n$ is the number of noise pixels. $M$, $N$ is the number of crack pixels and back pixels respectively. $f_1$, $f_2$ is percentage rate of loss and noise respectively. Additionally, $f$ is evaluation value. The higher value is obtained, the higher accuracy of the software is obtained.

The process of wavelet transform method is firstly the wavelet decomposition level which is considered as a parameter which needs to be found optimum value.

Secondly, hard threshold based all sub-band coefficients of HWT are adjusted. If the coefficient values are less than the hard threshold value, they are replaced by 0s (white pixels). Otherwise, they are kept unchanged values. The hard threshold also is taken account into the second parameter of solution candidates to search optimum value. Consequently, a filtering image is reconstructed by inverting wavelet transform. Next, the labelling is performed to connect 4-neighboring components. Finally the linear step is implemented as the part 3 of the proposed method, and it also is considered as the third parameter of solution candidate. As a result, the candidates of solution are encoded into binary string

16 bits which include the level the composition (2bits), the hard threshold (8 bits), the linear (6 bits). In this paper, the experiments are implemented for the two methods with 10 original images included various texture and uneven illuminating conditions as sample tests.

### *7.2 Discussion*

As a result, Table 2 shows HWT method and the proposed method. In this table, the accuracy of crack detection results of the proposed method outperforms the ones of the HWT method for all sample tests. The crack detection of the HWT method is limited with respect to low contrast images. For illustration, Figure 13 shows the crack detection results of orig. Image 10. The loss and noise rate in Figure 13.(c) of The HWT method are too much compared with the one of the proposed method. The crack detection accuracy using proposed method is $f$ equal to 0.9355. Meanwhile, the crack detection accuracy of the HWT method is $f$ equal to 0.8140.

## 8. **Conclusion**

In this study, the crack inspection system included the useful imaging device, image stitching and image processing software. Therein, the imaging device was capable of collecting full enough tunnel lining surface images without constraint traffic to create an enlarged long panorama of tunnel lining aided to visual inspection easily. The proposed image matching method in this study implements the similarity metrics and the dual threshold values based on the imaging speed. Moreover, median filter is applied to refine local inappropriate matching points of the images in the experiment.

Additionally, combination of image processing techniques and interactive genetic algorithm adjusted the optimized parameters of image processing procedure. The crack detection accuracy of the proposed method compared with the HWT method. The experimental results demonstrated that the effectiveness of the proposed method. The limitation of this system is that properties of the crack shapes and sizes aren't classified as well as measured automatically. These problems would be implemented in the future research.

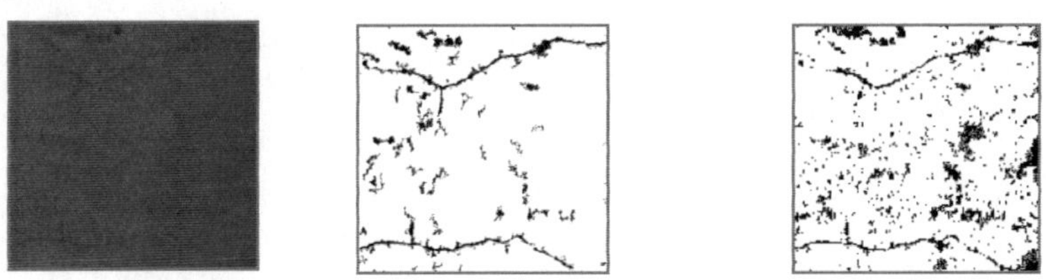

(a) (b) (c)

**Figure 13**. (a) Original mage No.10; (b): crack image result of proposed method; (c) crack image result of HWT method.

**Table 2.** Result of the accuracy analysis of hwt transform, the proposed method.

| | Proposed method | | | HWT method | | | | Proposed method | | | HWT method | | |
|---|---|---|---|---|---|---|---|---|---|---|---|---|---|
| No | $f_1$ | $f_2$ | $f$ | $f_1$ | $f_2$ | $f$ | No | $f_1$ | $f_2$ | $f$ | $f_1$ | $f_2$ | $f$ |
| 1 | 1.48 | 1.06 | **0.9887** | 1.06 | 1.54 | **0.9868** | 6 | 6.15 | 3.43 | **0.9502** | 11.71 | 5.48 | **0.9086** |
| 2 | 3.29 | 2.73 | **0.9698** | 2.36 | 3.04 | **0.9728** | 7 | 0.99 | 1.22 | **0.9889** | 6.5 | 10.11 | **0.915** |
| 3 | 1.67 | 1.86 | **0.9828** | 1.11 | 2.82 | **0.9786** | 8 | 2.06 | 2.45 | **0.9774** | 4.35 | 1.91 | **0.9664** |
| 4 | 2.72 | 2.55 | **0.9736** | 11.05 | 7.79 | **0.9044** | 9 | 10.07 | 12.87 | **0.8844** | 20.34 | 22.08 | **0.7877** |
| 5 | 6.98 | 10.16 | **0.9128** | 6.06 | 24.07 | **0.8245** | 10 | 7.79 | 5.26 | **0.9355** | 24.26 | 10.18 | **0.8140** |

## 9. References


[1] Zhu Z, Fu J, Zhang X, 2016 *Journal of Computer-Aided Civil and Infrastructure Engineering*, **00**, 1-19

[2] Kawamura K, Miyamoto A, Nakamura H, Sato R, 2003 *Proc. Japan Soc. Civ. Eng.* **742**, pp. 115-131

[3] Fujita Y, Mitani Y, and Y. Hamamoto 2006 *ICPR2006: IEEE 18th Int.Conf, on Pattern Recognition*, **3**, New York, 901-904

[4] Miyamoto A, Konno M, and Bruhwiler E 2007 *Asian J.Inf.Technol* **6(5)**, 553-561

[5] Nishikawa T, Yoshida J, Sugiyama T, and Fujino Y 2012 *Computer-Aided Civil and Infrastructure Engineering* **27(1)**, 29-47.

[6] Stent S, Gherardi R, Steneger B, Soga K, and Cipolla R 2013 *International association for pattern recognition (IAPR) Int. Conf. on Machine Vision Applications* Kyoto, Japan

[7] Stent S, Gherardi R, Steneger B, Soga K, and Cipolla R 2016 *Machine Vision and Applications* **27**:319–330

[8] Yeum C. M., and S.J. Dyke 2015 *Computer-Aided Civil and Infrastructure Engineering* 30,759-70

[9] Kumar S, and Chatterji N 2004 *ISSN:0377-2063 IETE Journal of Research*, **50**, No. I, pp 49-61

[10] Cuong Nguyen Kim, Kei Kawamura, Amir Tarighat, Junji Matsumoto, and Masando Shiozaki. Development of Semi-Automatic Crack Detection Software For Concrete Structures. *Proceedings of The 11th fib International PhD Symposium in Civil Engineering* , Tokyo, Japan. pp. 459-466, 2016-August

[11] Kawamura K, Miyamoto A, Nakamura H, Sato R 2003 Proc. Japan Soc. Civ. Eng.742, pp. 115-131

[12] Mukhopadhyay S, Mandal J 2013 Procedia Technology doi 10.1016/j.protcy.10.680-689



## Acknowledgments

The research in this article is supported by the financial support of JSPS KAKENHI Grant Number 15K0618.

IOP Conf. Series: Materials Science and Engineering **371** (2018) 012016 doi:10.1088/1757-899X/371/1/012016

# Lock-brick System for Sustainable and Environment Infrastructure Building Materials

**Susilawati[1] and Indah Wahyuning Tyas[2]**

1 Associated Professor, Lecturer of Flores University, Ende-Flores, Indonesia
2 Lecturer of Flores University, Ende-Flores, Indonesia
E-mail: susipi@uniflor.ac.id

**Abstract.** The community participation is expected to realize the appropriate housing. But the financial condition does not support, so that appropriate housing to be half constructed, which eventually became jammed unsustainable. Lock-brick system building construction can be implemented for appropriate housing and water infrastructure, such as small ponds and retaining wall construction. Lock-bricks are made from soil as the main ingredients added a little cement, thoroughly mixed in dry conditions, has dampened with water to a level sufficiently moist. Then, pressed with a pressing machine to 60%, forming interlocked bricks when arranged to form the walls of the liveable house. Columns, beams or retaining walls can be constructed with this system. Similarly, other water basin infrastructure such as rainwater catchers-canals can be constructed with this system. Lock-brick systems are used for a wide range of building infrastructure while considering the strength of the structure, so as to meet the standards. Various examples of usability have been done for the simple building liveable housing, student dormitories building, rain water reservoir, runoff catchment canals, soil retaining walls, and pavement floor. The usability, simple, easy to make and cheap, making this lock-brick system can support the development of sustainable infrastructure and built environment.

## 1. Introduction

Infrastructure development for raw water, such as ponds for conserving rainwater, which has been catch, or canal-traps to collect runoff water, are very important in managing rain water resource to meet the water requirement for live. It is also important to provide the liveable housing, which will support the health of the people. Everything requires a building structure that is quite resistant, strong and sustainable, not easily damaged quickly. This requirement requires an expensive building material, such as reinforced concrete iron, cement, sand and splits. The community encountered many difficulties to fulfil their need independently, even by little funding from the government. Programs have been launched, such as stimulant funding of Rp. 10.000.000.- for unliveable housing. The community participation is expected to realize the liveable housing, but their financial condition does not support, so that appropriate housing only could be half-constructed, which eventually jammed. It is hoped that this initial assistance can support the community to build their homes to be habitable and meet health requirements. But it is also still difficult for people to finish a half-finished house. These problems encourage being developed of an alternative building material that is lock-brick system technology, which can be used for various forms of buildings needed.

## 2. Lock-brick System Technology

Established in 1988, Habitech Centre has conducted research and developed various building components that are environmentally friendly, energy-efficient and cost-effective. This cost-effective building system is named Habitech Building System. This system has been tested, demonstrated and disseminated through various housing projects in Thailand and in other countries throughout Asia and Pacific [1].

Literature review of the lock-brick system shows that most of these systems are used for house building construction [2, 3, 4, 5]. The recapitulation of this literature review results is shown as in Table 1

**Table 1.** Recapitulation of articles literature review

| No. | Year | Author | Title | Note |
|---|---|---|---|---|
| 1 | 2014 | Anis Rahmawati | Lock-brick Modular as Innovation Wall Materials Residential Buildings | Research, funding by Ristekdikti |
| 2 | 2013 | M. Khasani, et.al. | Designing Pre-processed Components of Growing House | Vol 1, No 01 (2013) Jurnal Pendidikan Teknik Bangunan |
| 3 | 2011 | Sri Sumarni, dkk | Moulding Machine Development of Sloof and Beam component for Simple Growing House Earthquake Resistant Building | Draft scientific articles |
| 4 | 2009 | Simion Hosea Kintingu | Design of Interlocking Bricks For Enhanced Wall Construction Flexibility, Alignment Accuracy And Load Bearing | Dissertation |

Habsya, et al, in his article on the lock-brick system, "Lock-brick Concrete Modular for Alternative Wall Materials that Meets the Quality Standard of SNI at Low Cost", concludes that the cost per $m^2$ of modular lock-brick wall is 24% cheaper than the cost of the conventional brick wall [6]. From the reviews for various patents associated with the lock-brick system are presented as in Table 2 Patent development is widely available in the United States. That is also registered, the "Modular Lock-brick System for Constructing Buildings". Patent Registration Number, S00201401680, Date: March 21, 2014, Director General of Indonesian Law Ministry [7].

Habitech centre develops lock-brick base on soil material. The lock-bricks are made from the main ingredients soil, which be added a little cement, with the comparison of 1-part cement and 8-part soils. These materials thoroughly mixed in dry conditions, and be dampened with water to a level sufficiently moist. Then, poured to the pressing machine until 60% pressed, forming interlocked bricks. The lock-bricks arranged to form the walls of the liveable house. Columns and beams, could be constructed by this system also, with the special moulding, which form the place to lay the irons.

Lock-brick construction system could be an alternative solution for many problems, especially concerning of finance, environment and people participation. It also could be implemented for water infrastructure such as small ponds, channels and other infrastructure water buildings. Lock-brick systems are used for a wide range of building infrastructure but still considering the strength of the structure, according to standards [8].

**Table 2** Recapitulation of lock-brick patents review

| No. | Year | Inventor | Title | Note |
|---|---|---|---|---|
| 1 | 2013 | M.A. Kashinath | Interlocking Brick | No.: 3434/MUM/2013 |
| 2 | 2012 | Janssens et al | Building Blocks with Mating Coupling Means For Constructing Wall And Associated Method | US 2012/0102868 |
| 3 | 2012 | Nagy, John R and Krell, Clinton C. | Modular Concrete Building | Paten US8,132,388B2. |

| | | | | |
|---|---|---|---|---|
| 4 | 2009 | Steve Eugene Everett | Structural Building Block System and Method Comprising Same | Patent No.: US 7472520 |
| 5 | 2000 | Simmons, Scott. | Modular Building Materials | US Paten Number US006,088,987 |
| 6 | 1999 | John H | Uniform Building Construction Using Interlocking Connectors | Patent Number: 5901521 |
| 7 | 1997 | Williams et al | Flue Walls Using Interlocking | Patent Number: 5676540 |
| 8 | 1981 | Khoo Tian | Bricks | No. 4,299,071 |
| 9 | 1970 | Zagray | Interlocking building block construction | Patent US 3534518 |
| 10 | 1939 | A. J. Cilento | Interlocking Brick | Serial No. 251260 |
| 11 | 1934 | P. Brown | Interlocking Building Brick | US 1984393 |
| 12 | 1908 | J. Soss | Interlocking Brick | US 903906 |
| 13 | 1895 | George J. Herth | Lock-Brick | US Patent No. 535497 |
| 14 | 1892 | Louis A. Steiger | Interlocking Brick | Patent No. 468840 |

*2.1. Lock-brick System for livable, healthy and low-cost housing*
This lock-brick system could support the development of sustainable infrastructure and built environment because of simple, easy to make, cheap and multifunction of the usability. A design of appropriate technology by using the soil material for lock-brick becomes the building material of the residence. The Lock-brick system support people to build their own low cost healthy and livable house [9]. Some parts of the house construction are illustrated in Figure 1

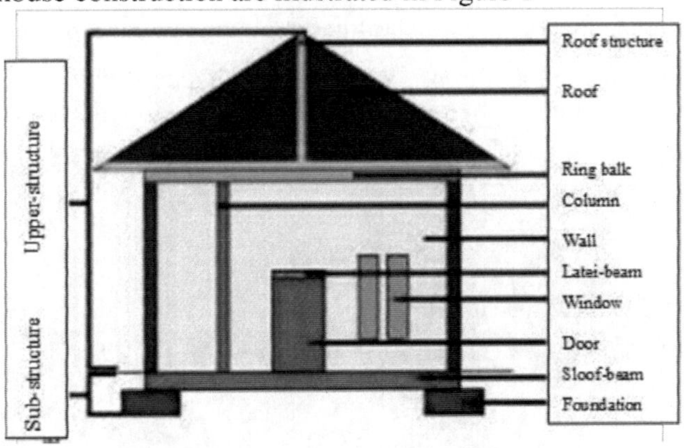


**Figure 1** Parts of the simple house.

Detailed stages description of the procedure:

1. The foundation is made of stone, or could be used of this lock-brick, which is arranged, and the hole's brick is filled by mortar.
2. Sloof beams are made of lock-bricks type of U-brick, where on top of brick form U to place in an iron bar, then casted as a concrete sloof (Figure 2).

**Figure 2** Foundation and installing iron on the sloof beam.

3. Practical columns structures are made by inserting an iron bar into the vertical hole of the lock-brick, then casted as concrete columns (Figure 3).

**Figure 3** Arrange the walls of the building and prepare the iron for the column

4. Walls are constructed using lock-bricks, which are arranged in accordance to the shape of the building, without adding mortar, but only arranged interconnected each other (Figure 4).

**Figure 4** Installation of lock-brick wall, door and window holes

5. Concrete beam wall hole, is a concrete latei-beam to hold the wall brick above the wall hole. The construct concrete reinforcement for latei-beam is shown on Figure 5.

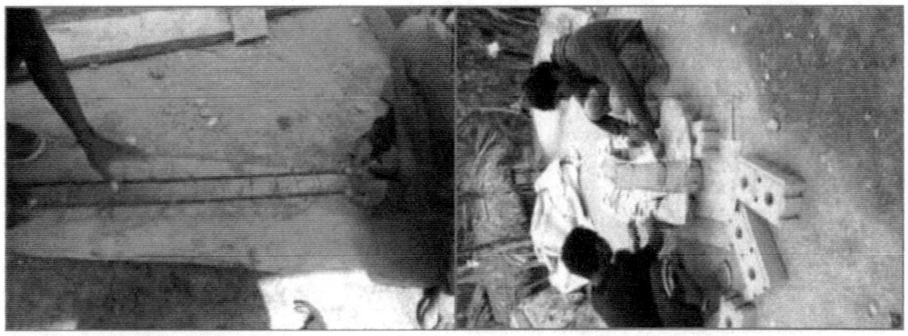

**Figure 5** Construct concrete reinforcement for latei-beam

6. Ring balk made as well as making sloof beam, as wall coverings and pedestal of the roof construction.
7. Roof structure is constructed from wood, and for the finishing by lock-brick also. The implementation of roof installation work is shown in Figure 6.

**Figure 6** Implementation of roof installation work

8. Building a simple house that has been completed. Figure 7 show the completed simple house lock-brick system in Tli'u and Nursing dormitory.

**Figure 7** The completed simple house lock-brick

*2.2. Lock-brick System for Various Building Infrastructure*

Lock-brick system is also suitable for various building structures, such as water structure, such as pond conservation, rain catcher channel, etc.

*2.2.1. Lock-brick System for Water Ponds*

To create a rain-water pond is required stages of work as follows:

1. Soil excavation
2. Lock-brick arrangement for the base of water pond
3. Lock-brick arrangement for the wall
4. Concrete slab for the top of water pond

The work stages are described as follow:

1. Soil excavation for the water pond. Excavation is done for the construction of existing tubs in the soil. For the construction of the container tub that does not require excavation, then directly arranged the retaining wall structure as a pond wall.
2. Lock-brick arrangement for the base of water pond. This arrangement is intended to stabilize the base of the pond, which can be directly casted, thereby reducing the volume of casting concrete that requires high cost. Figure 8 show the installation of sloof beam reinforcement floors and wall of the pond.

**Figure 8** Installation of sloof beam reinforcement floors and wall of pond.

3. Lock-brick arrangement for the wall. This lock-brick system intended to reduce the required concrete wall volume. The arrangement of this wall each brick hole given a rod iron as reinforcement of concrete. Figure 9 show the installation of up-ground pond wall.

**Figure 9** Installation of up-ground pond wall

4. Concrete slab for the top of water pond. The concrete slab structure is constructed for cover the water pond, and conventionally such as concrete slab construction (Figure 10).

**Figure 10** Concrete slab for the top of the pond

*2.2.2. Lock-brick System for Other Building Infrastructure*

Lock-brick system could be used for runoff water-trap channels (Figure 11).

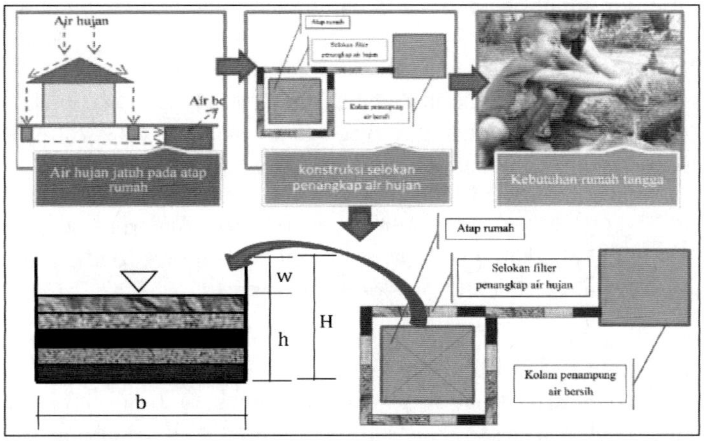


**Figure 11** Runoff water-trap channels concept

The rainwater runoff catcher channel is created to capture runoff flowing rainwater, and then flowed into a rainwater catchment pond, through this channel. Water runoff flowing through this channel is expected to experience of mineralization process, so that the quality of rainwater harvested is quite feasible for raw water. Figure 12 show the structure of rainwater runoff catcher and the connection to the pond.

**Figure 12** Rainwater runoff catcher channel and the connection to the pond

**3. The Lock-Brick System Can Be Environmentally, Friendly and Sustainable**

This lock-brick system is environmentally sound and suitable as a material that supports various building structures is a concern in this research. Several reasons that support this statement include: cement is about 1: 8-10. This indicates that most of the main materials used to form lock-brick are natural materials with no chemical processing or engineering. The lock-brick system used to build various structures is based on community participation system. Meaning that to make the lock-brick can be done by the community itself because it is simple and easy, as well as in arranging it into a building structure. More significant impact is the need for the required cost becomes very low. This provides an opportunity to build more people's needs for development to become more prosperous. Through various experiments performed and mention before, that can be said that this lock-brick system is environmentally sound, friendly and suitable as a material that supports various building structures

**4. Lesson Learned**

Some examples of usability have been done for the simple building liveable housing (Figure 7), student dormitories building (Figure 11), rain water ponds (Figure 9), runoff water-trap canals (Figure

IOP Conf. Series: Materials Science and Engineering **371** (2018) 012016 doi:10.1088/1757-899X/371/1/012016

12) and other construction. This lock-brick system could support the development of sustainable infrastructure and built environment because of simple, easy to make, cheap and multifunction of the usability.

## 5. Conclusion and Recommendation

From various studies conducted so far, and also the application of research results that have been done for various types of structural buildings using a lock-brick system, it can be concluded that lock-brick system is environmentally friendly and suitable as a material that supports various building structures. This lock-brick system is recommended to be applied to other building structures such as retaining wall structures, village road structures by adding plastic waste materials to increase its strength as filler fibres in lock-bricks, and other innovations, as it will be done for road construction on the way to Experimental garden in Agriculture Faculty of Flores University.

## 6. References

[1] Klong Neung, Klong Luang, Habitech Centre, Asian Institute of Technology, Pathumthani 12120 Thailand, 2016. Download at: http://www.habitech.ait.asia/

[2] Anis Rahmawati, Lock-brick Modular as Innovation Wall Materials Residential Buildings (Lockbrick Moduler Sebagai Inovasi Material Dinding Bangunan Rumah Tinggal), Penelitian Unggulan Perguruan Tinggi RISTEKDIKTI, 2014. Download at: http://jarlitbangnov.bappeda.jatengprov.go.id/index.php?ref=publication&mod=doc&id=19606

[3] M. Khasani, dkk., Designing Preprocessed Components of Growing House (Perancangan Komponen Prapabrikasi Rumah Tinggal Tumbuh), Jurnal Pendidikan Teknik Bangunan, Vol 1, No 01, 2013. Download at: https://eprints.uns.ac.id/1895/1/1375-3069-1-SM.pdf

[4] Sri Sumarni, dkk. Moulding Machine Development of Sloof and Beam component for Simple Growing House Earthquake Resistant Building (Pengembangan Mesin Pencetak Segmen Sloof dan Balok Moduler Rumah Sederhana Tumbuh Tahan Gempa (RT2G). Download at: http://lppm.uns.ac.id/kinerja/files/pemakalah/lppm-pemakalah-2011-0308201380248.pdf

[5] Simion Hosea Kintingu, Design of Interlocking Bricks For Enhanced All Construction Flexibility, Alignment Accuracy And Load Bearing, Dissertation, The University of Warwick, School of Engineering, May 2009.

[6] Habsya, et.al., Lock-brick Concrete Modular for Alternative Wall Materials that Meets the Quality Standard of SNI at Low Cost (Lockbrick Modular Beton Untuk Alternatif Bahan Dinding Yang Memenuhi Mutu SNI Dengan Biaya Murah), Sinektika Vol.14 No.2, 2014. Download at: http://journals.ums.ac.id/index.php/sinektika/issue/download/258/9

[7] Habsya C. 2014. Modular Lock-brick System for Constructing Buildings, Paten Register Nomor, S00201401680, Dirjen HKI Kemenkumham RI, Date 21 Mart 2014.

[8] Susilawati, System-Trap-Rainwater Engineering and Lockbrick Technology In Household Scope To Improve Quality Of Life, ARPN Journal of Engineering and Applied Sciences, Vol. **11**, No. 11, June 2016.

[9] Susilawati, Model Desain Prototipe Kombinasi Jebakan Air Berantai dengan Rumput Vetiver untuk Mengatasi Permasalahan Sumber Daya Air Pulau Kecil Daerah Kering Indonesia. Laporan Akhir Penelitian Hibah Bersaing, 2016.

IOP Conf. Series: Materials Science and Engineering **371** (2018) 012017 doi:10.1088/1757-899X/371/1/012017

# Development of Ultra High Performance Concrete and Reactive Powder Concrete with Nanosilica

**R Hela[1], L Bodnarova[2] and L Rundt[3]**

1 Professor, Brno University of Technology, Faculty of Civil Engineering, Veveri 95, 602 00 Brno, Czech Republic
2 Associated Professor, Brno University of Technology, Faculty of Civil Engineering, Veveri 95, 602 00 Brno, Czech Republic
3 Bc, Researcher, Brno University of Technology, Faculty of Civil Engineering, Veveri 95, 602 00 Brno, Czech Republic
E-mail: bodnarova.l@fce.vutbr.cz

**Abstract.** The article is dedicated to the design and production of Ultra High Performance Concrete (UHPC) and Reactive Powder Concrete (RPC) using silica fume and nanosilica. Nanosilica and fine steel fibres were used for the production of RPC. Compressive strengths of UHPC and RPC above 150 MPa have been achieved. It has been demonstrated that UHPC and RPC can be produced using standard concrete mixing system without the use of activating mixing and without a special treatment regime during maturing of the concrete. Aging of the concrete took place in a normal environment, without elevated pressure or temperature. The aging process at 20 °C allows the use of UHPC and RPC for the ready-mixed concrete when working on high volume construction projects. Even without thermal treatment, without the application of solidification pressure and without autoclaving, RPC reached a compressive strength of more than 180 MPa and a flexural tensile strength after 60 days greater than 22 MPa. The high tensile bending strength may be considered as the main advantage of RPC, as the RPC parameters allow, for instance, the use for pre-stressed structural elements where a high initial strength is also required.

## 1. Introduction

Ultra High Performance Concrete (UHPC) or Ultra High Strength Concrete (UHSC) refers to composite material with Portland cement based binder with a compressive strength of more than 150 MPa, high ductility and excellent durability. These properties are achieved by high content of cement, low water-cement ratio, using superplasticizers, additives (silica, fly-ash, blast furnace slag, metakaolin) and dispersed reinforcement. Reactive Powder Composite (RPC) is a cement binder material with compressive strengths between 200-800 MPa. It differs from UHPC by higher cement content and maximum grain sizes, which typically do not exceed 600 μm. The main principles of the RPC design include the elimination of coarse aggregate, thereby improving the homogenity of the mixture. Porosity is minimized by granulometry optimization. High strength is achieved by a very dense structure with optimal granulometry of all raw materials, thermal treatment at elevated pressure before and during setting and hardening, high proportion of siliceous extracts and dispersed reinforcement. The positive side of UHPC and RPC is the ability to produce supporting elements of very thin cross-sections, utilizing high strengths of these concrete, thereby relieving the whole construction and further positive is the high durability and associated sustainability and reduced need for repairs.

IOP Conf. Series: Materials Science and Engineering **371** (2018) 012017 doi:10.1088/1757-899X/371/1/012017

## 2. Specification of UHPC and RPC proposal in comparison with normal concrete

### *2.1. Input raw materials*

Ordinary dense concrete has an air content of about 2 %, with UHPC and RPC we are trying to achieve the maximum density, thereby increasing the strength and durability. In addition to cement, we select high doses of very fine admixtures; the maximum aggregate grain size for UHPC can be up to 16 mm, but usually up to 8 mm.

The amount of cement in the UHPC is between 700 and 1100 kg·m$^{-3}$, which is approximately three times higher than that of ordinary concrete. High dose of cement is one of the reasons for the high price of UHPC, greater development of hydration heat and volume changes during shrinking of concrete. Portland cement CEM I or mixed CEM II strength classes 42.5 or 52.5 are used [1]. Cement should contain a smaller proportion of $C_3A$ due to less consumption of mixing water, which also increases the fineness of grinding [2].

Due to the very low water-cement ratio (typically 14-20 %, compared to 40-50 % for ordinary concrete), UHPC contains less water than is needed for hydration of all cement grains (23-25 % by weight of cement). This way the cement grains also gain a filling function.

Superplasticizers (strongly water reducing) are necessary for the production of UHPC and RPC as the low water-cement ratio and fine impurities deteriorate, or even make the process impossible. A dose of superplasticizer in UHPC can be up to fifteen times higher than in conventional concrete [3]. Superplasticizers based on polycarboxylates proved to be the most effective.

An important component of UHPC and RPC are silica fume. According to various authors [1], [4] and [5], the optimal rate of silica fume in UHPC is between 20-35 %, but it is very dependent on the water-cement ratio, where the lower silica fume content decreases with lower water content. Round grains result in better workability, but mainly fill the gap between cement particles and contribute to a more dense structure. Silica fume reacts with free lime, in small amounts contained in dry cement, but mainly with calcium hydroxide resulting from hydration of cement during CSH gel production. The 18 % amount of silica from the cement weight is sufficient to react with all $Ca(OH)_2$. Not all grains of the cement get hydrated, so less silica fume is enough, however to fill the gap between the cement grains the optimal dose of silica is up to 30 %. In conventional concrete, the silica fume also has a stabilizing function – they reduce water separation. With higher proportions of silica, however, there's an increase in water consumption, resulting in the need to increase the dose of superplasticizers in the UHPC. Silica fume also plays an important role in the transit zone. The transition zone (the area between aggregate and cement paste) in conventional concrete is one of the weakest places because it contains pores and crystals of $Ca(OH)_2$ and ettringite. The layer thickness is 10-50 μm. Thanks to the low water-cement ratio in the production of UHPC and the pozzolan reaction between calcium hydroxide and mineral impurities, the CSH gel is formed and the transit zone then becomes almost as dense as the matrix itself.

The use of nanoparticles in UHPC seems also perspective. These are, for example, nano-silicon dioxide (nano-$SiO_2$), calcium nano-carbonate (nano-$CaCO_3$), nano-aluminum oxide (nano-$Al_2O_3$), nano-titanium dioxide (nano-$TiO_2$) and nano-iron oxide (nano-$Fe_2O_3$). Their features include a large surface area. Nanoparticles contribute to cement hydration due to their high reactivity, they can act as nano-reinforcement and as filler when they compact the microstructure and the transit zone, thereby decreasing porosity.

Nanosilica has a higher purity, a higher amount of non-crystalline silica, higher pozzolanic activity than silica. Nanosilica increases the amount of hydration products, thus decreasing the amount of portlandite. By adding nanosilica, consistency and water-consumption deteriorate, improving strength, especially at an early stage. The optimal amount to maintain acceptable spill values and at the same time the highest strengths was 3 % of the cement.

Aggregates and fillers. In ordinary concrete, the transition zone is the weakest point of the matrix, with UHPC it is the aggregate. It is therefore necessary to choose hard aggregates such as gabbro, granite, diabase or basalt. The largest grain size is 16 mm, but for strengths above 150 MPa, it is recommended to reduce the size. Larger fractions aggravate the homogenization of the mixture, on the other hand, it is one of the cheapest components of UHPC, therefore from and economic and ecological point of view there's an attempt to use it as much as possible. For example, quartz flour or other residues from aggregate grinding are used as fine fractions in UHPC.

UHPC containing a diffused reinforcement – fibres – is referred to as Ultra High Performance Fibre Reinforced Concrete UHPFRC. UHPC is a brittle material, so there short fibres are added to the mix to improve tensile strength, toughness, hardness and impact resistance. The reinforcement prevents the initiation and propagation of cracks, because it transmits tension through the fibre-to-concrete interface [6]. The most commonly used fibres are steel or carbon due to their high strength. Their content in the UHPC is between 0.5 and 3 % (by volume). High fibre content leads to worse UHPC workability and high cost because 1 % of the volume can be more expensive than the concrete itself. For RPC, a steel fibre dose of 1.5-3 % (volume) is recommended.

### *2.2. Homogenization*

During mixing of the concrete components, the mixture must be homogenized and the air pores eliminated as much as possible. Special procedures have been tried to achieve this, such as vacuum mixing or application of pressure before and during solidification. However, it is desirable to limit these processes because, in addition to the need for special technological equipment, they increase labour and cost compared to conventional strength concrete.

The dosing of components does not differ from conventional concrete; first there are binders, fillers, followed by water with superplasticizing additives (possibly with other additives) and finally dispersed reinforcement. Due to the high content of fines, the required mixing time may be increased.

### *2.3. Strength*

The minimum compressive strength varies according to different authors, but on average the UHPC should reach at least 150 MPa. Tensile strength for UHPC and RPC are between 8 and 15 MPa, and a tensile bending strength of between 30 and 60 MPa. Strengths are affected by composition, storage and care. Long-term strengths according to experiments [7] have shown that 3.5 years of strength were 40 % higher than in 28 days.

## 3. Experimental part

### *3.1. Goal of experimental work*

Given the current concrete technology, High Strength Concrete HSC with strengths of about 110 MPa is commonly used. However, there is also an area of so-called Ultra High Strength Concrete UHPC, which is characterized by strengths of about 160 MPa, and so-called reactive powder composites RPC with compressive strengths up to about 300 MPa. Until now, UHPC and RPC have not been commonly used in the industry for the construction of monolithic structures. UHPC and RPC are most commonly prepared under laboratory conditions, using a special treatment and homogenization. The realization of the preparation of large volumes of UHPC and RPC for the application on a construction site for the ready-mixed concrete is not very widespread precisely due to the challenging homogenization of the components of this special concrete, e.g. the use of activating mixers. In addition, special care is also frequently required in the course of aging of UHPC and RPC. The conducted experimental work has set out the goal of preparing UHPC and RPC using conventional mixers as they are used for the production of ready-mixed concrete. Also, the maturing conditions were set identically as in the case of treating the concrete after it was stored on the site, i.e. without the use of heat-curing modes.

Another objective of the experiment was to monitor the economic aspect of input materials for the production of UHPC and RPC. The price of these concrete types compared to ordinary concrete is increased by the high dose of cement and microfillers, and in the case of RPC, also the addition of dispersed reinforcement. For the production of UHPC and RPC, high-quality raw materials with well-defined properties and minimal variability of these properties need to be used, which represents a further increase in costs compared to ordinary concrete. The aim of the UHPC and RPC recipe design was, in addition to achieving the high strength required, also using common raw materials so that the UHPC and RPC prices were not too high in the comparison with ordinary concrete.

### *3.2. Raw materials*

*3.2.1. Cement.* CEM I 52.5 R cement was used. Medium grain is 10 μm, the remaining 20 μm on the mesh screen is 22.2 %. Specific weight is 3140 $kg{\cdot}m^{-3}$, specific surface is 501 $m^2{\cdot}kg^{-1}$, $Na_2O$ equivalent is 0.65 %.

*3.2.2. Silica fume.* Silica RW-Füller, RW silicium GmbH. It contains (96.0±1,5) % $SiO_2$ and not more than 0.9 % SiC, 0.9 % $K_2O$ and 0.12 % $Na_2O$. The loss on annealing is 1.2 %, pH 7.5. The specific surface area when using the BET method is 18-22 $m^2{\cdot}g^{-1}$. 95 % of the particles have a size of less than 10 μm and 70 % less than 1 μm.

*3.2.3. Nanosilica.* Nanosilica is from SkySpring Nanomaterials, Inc., Houston, Texas. Specific weight is 2160 $kg{\cdot}m^{-3}$, apparent density is only 100 $kg{\cdot}m^{-3}$, surface area 160 $m^2{\cdot}g^{-1}$ and content $SiO_2$ 98.7 %. The size of spherical particles is between 10 and 20 nm, but the grains are clustered into agglomerates having a medium particle size of 8.3 μm (measured in aqueous suspension).

*3.2.4. Ground limestone.* Ground Devonian limestone with a higher proportion of dolomite, specific surface area is 432 $m^2{\cdot}kg^{-1}$, apparent density 2160 $kg{\cdot}m^{-3}$ and specific weight 2540 $kg{\cdot}m^{-3}$.

*3.2.5. Aggregate.* UHPC samples contained 2 or 3 aggregate fractions of: 0-4, 4-8 and 8-16 mm. Mined siliceous sand was used for the 0-4 mm fractions. Basalt coarse crushed aggregates of fractions 4-8 mm and 8-16 mm were washed and dried to remove undesirable dust particles. In the RPC mixtures the aggregate was silica sand, the maximum grain was 0.135 mm.

*3.2.6. Superplasticizing additive.* Sika® ViscoCrete®-2700 superplasticizer based on ether-polycarboxylate. The bulk density at 20 °C is 1080 $kg{\cdot}m^{-3}$ and alkali ratio ($Na_2O$ equivalent) is less than 1.0 %.

*3.2.7. Water.* The mixing water was used from the water supply line.

*3.2.8. Fiber reinforcement.* The straight thin steel fibers from KrampeHarex were used in the RPC, diameter of 0.2 mm and a length of 6 mm. The tensile strength of steel fibers is 2100 MPa ± 15 %.

### *3.3. Design and production of UHPC*

The water-cement ratio (ratio of water to binder) was calculated in accordance with EN 13263 [8], when calculating the amount of silica fume to a maximum of 11 % of the cement quantity (70 kg for UHPC 1 and 82.5 kg for UHPC 2). The XC or XF environment according to standard EN 206 [9] was not specified, the k-value of silica fume was 2.0. The water-cement ratio for UHPC 1 was 0.18, with UHPC 2 at 0.16. The proportion of silica to cement is in both cases 0.15 and the superplasticizer dose is 2.5 % of the cement. The composition of UHPC is given in Table 1.

**Table 1.** UHPC composition.

| Material | UHPC 1 Quantity [$kg{\cdot}m^{-3}$] | UHPC 2 Quantity[$kg{\cdot}m^{-3}$] |
|---|---|---|
| CEM I 42.5 R | 700 | 750 |
| Silica fume | 105 | 112 |
| Siliceous sand 0-4 mm | 655 | 725 |
| Aggregate 4-8 mm basalt | 265 | 675 |
| Aggregate 8-16 mm basalt | 660 | - |
| Water | 155 | 150 |
| Superplasticizer | 17.5 | 18.75 |

Mixing: Dispersing into a forced mixer was carried out as follows: firstly, dry ingredients (aggregates, cement, silica fume) were mixed, and then gradually during the mixing the water with superplasticizer additive was added.
After mixing, a Slump test was performed and UHPC was then placed in 150x150x150 mm metal cubes. Compaction was carried out on a vibrating table. The samples were taken out of the forms after 1 day and stored in an aqueous environment until the compression tests were performed after 7 and 90 days [10].

### *3.4. Design and production of RPC*
Two RPC formulas were proposed, the first containing nanosilica (RPC 1), in the second mixture this nanosilica was replaced with ordinary silica fume (RPC 2). The composition is in Table 2 [10].

**Table 2.** RPC composition.

| Material | RPC 1 Quantity [kg·m$^{-3}$] | RPC 2 Quantity [kg·m$^{-3}$] |
|---|---|---|
| CEM I 42.5 R | 1000 | 1000 |
| Silica fume | 150 | 180 |
| Nanosilica | 30 | - |
| Ground limestone | 150 | 150 |
| Siliceous sand 0-4 mm | 1100 | 1100 |
| Water | 290 | 315 |
| Superplasticizer | 17.5 | 18.75 |
| Steel fibres | 110 | 110 |

The water-cement ratio was w = 0.23, for the RPC 2 mixture it was increased to 0.25 in order to achieve a higher consistency.
The ratio of sand to cement was 1.1, the amount of steel fibres being 4 % of the volume of the mixture.
The proportion of superplasticizing additive is 2.5 % by weight of cement.
Mixing: The nanosilica was first mixed with a total dose of water, superplasticizer and a smaller amount of silica sand for 10 minutes, and this suspension was then used as a dose of mixing water. The steel fibres were poured into the mixer at the end.

## 4. Tests carried out and results
The consistency of fresh UHPC was determined by the Slump test according to EN 12350-2 [11]. The consistency of fresh RPC was determined by a spill of mortar on the flow table test according to EN 1015-3 [12]. The volume weight was tested according to EN 12350-6 [13] and 12390-7 [14]. The results of those tests are shown in Table 3 [10].

**Table 3.** Consistency and Volume weight of UHPC and RPC.

| Mixture | Slump [mm] | Spill [mm] | Volume weight Fresh concrete [kg.m$^{-3}$] | Volume weight Hardened concrete [kg.m$^{-3}$] |
|---|---|---|---|---|
| UHPC 1 | 240 | - | 2550 | 2530 |
| UHPC 2 | 240 | - | 2500 | 2480 |
| RPC 1 | - | 190 | 2620 | 2610 |
| RPC 2 | - | 180 | 2590 | 2570 |

The compressive strength was tested according to EN 12390-3 [15]. The values of compressive strength on UHPC and RPC samples are shown in Table 4 [10]. Flexural strength was determined according to EN 1015-11 [16]. Flexural strength values are given in Table 5 [10].

IOP Conf. Series: Materials Science and Engineering **371** (2018) 012017 doi:10.1088/1757-899X/371/1/012017

**Table 4.** Compressive strength of UHPC and RPC.

| Mixture | Compressive strength [MPa] | | | |
|---|---|---|---|---|
| | 7 days | 28 days | 60 days | 90 days |
| UHPC 1 | 117.3 | 148.2 | 155.6 | 163.1 |
| UHPC 2 | 114.2 | 144.3 | 154.1 | 160.1 |
| RPC 1 | 113.3 | 156.9 | 191.9 | 202.1 |
| RPC 2 | 104.4 | 152.2 | 185.6 | 190.8 |

**Table 5.** Flexural strength of RPC.

| Mixture | Flexural strength [MPa] | | |
|---|---|---|---|
| | 7 days | 28 days | 60 days |
| RPC 1 | 17.2 | 22.3 | 24.4 |
| RPC 2 | 16.9 | 19.2 | 22.8 |

## 5. Discussion of results

Within the experimental work, UHPC and RPC recipes were designed and produced. Bulk density in fresh and hardened condition, fresh concrete consistency, compressive strength of concrete and tensile strength of concrete after bending have been determined.

The Slump was 240 mm for both UHPC recipes. Due to the high content of fines and the relatively high dose of superplasticizing additive, however, the concrete was very sticky.

The spill of RPC 1 was 190 mm, RPC 2 was then 10 mm smaller, although larger amount of mixing water was used. The content of the wires contributed to a stiffer consistency, as compared to the UHPC the mixture was less sticky. RPC often achieve fluid consistency, so it would also be appropriate to modify the design of the mixture, or to try another type of superplasticizing additive [10].

Volume weight in the fresh state of the UHPC1 mixture was 2550 $kg \cdot m^{-3}$, for UHPC2 2500 $kg \cdot m^{-3}$, in solid state it was lower, according to expectations - 2530 and 2480 $kg \cdot m^{-3}$

Compressive strengths of UHPC were tested after 7 and 90 days. Although UHPC1 contained also a larger aggregate fraction (8-16 mm), the compressive strength was slightly higher.

The compressive strength of 150 MPa was exceeded, which is usually referred to as the threshold for the classification of Ultra High Strength Concrete.

Due to the fact that the UHPC 1 recipe contained 50 $kg \cdot m^{-3}$ more cement than the UHPC 2 recipe and achieved sufficient strength, it would be possible to reduce the cement content, which would contribute to better workability and cost savings while maintaining the required high strength.

Two RPC recipes were designed and produced. The difference between them consisted in the use of nanosilica (RPC 1) and a slightly increased water dose (RPC 2).

A better RPC compaction would be contributed by the softer consistency, which was determined as a spill of mortar at the shaking table; the spill value was 190 mm for RPC 1 and 180 mm for RPC 2. Consistency, however, was very sticky („honey like") and would complicate use in both prefabrication and in situ casting. The solution could be to adjust the amount of superplasticizing additive, change its type, add water (water-cement ratio was very low) or replace part of the cement with another active ingredient. Increasing the water dose would, however, result in a decrease in strength, so the solution could consist in adjusting the dose of the superplasticizing additive or the change of the type of superplasticizing additive. In order to achieve the optimum consistency of RPC, different combinations of cement, silica dust, superplasticizing additives and limestone-free water, silica sand and steel wires will be tested in subsequent experimental work.

RPC compressive strengths were tested after 7, 28 and 60 days. Even without heat treatment, pressure application or autoclaving, RPC 1 and RPC 2 recipes achieved a compressive strength greater than 180 MPa. The bending tensile strength was 24.4 MPa for RPC 1 after 60 days, with the value of 22.8 MPa for RPC 2. The high tensile bending strength can be considered as the main advantage of RPC, which allows RPC parameters to be used, for example, for pre-stressed structural elements where a high initial strength is also required.

## 6. Conclusion

UHPC and RPC require careful selection of quality raw materials and their stable properties. When designing, grain aggregate curves have to be followed to achieve dense structure and to choose a superplasticizing additive compatible with the type of cement used, to achieve the desired workability. For optimal homogenization, it is usually necessary to prolong the mixing time.

The biggest disadvantage of UHPC and RPC is economic difficulty. The higher price is mainly due to the high content of cement and silica fume. The use of UHPC and RPC in building structures must be seen in the long term due to its excellent durability. The use of UHPC significantly reduces the need for maintenance of structures and often expensive repairs, especially for transportation infrastructure. In terms of mechanical properties, UHPC and RPC have the potential, in addition to special constructions (such as prestressed structures) especially when building high-rise buildings and bridges. It allows to realize thin cross-sections of the structural elements, which reduces the weight of the whole structure and therefore the loading into the foundations, thus increasing the usable surface. When producing subtle constructions, fuel is saved on aggregate transport, because there's a smaller need for it. The total consumption of cement depends on the particular building, theoretically it may be less when using UHPC, because smaller UHPC volume is needed than conventional concrete, which has a positive effect on the $CO_2$ emissions from the cement production. UHPC and RPC exhibit high initial strengths, allowing earlier stripping of formwork and acceleration of construction work.

The major contribution of the study is to demonstrate that UHPC and RPC can be produced using conventional raw materials and using conventional homogenization equipment and without special treatment regimes during the maturing of concrete. No special precautions have been taken with regard to concrete treatment; UHPC and RPC matured under normal laboratory conditions without heat curing. The maturing process at 20 °C allows the use of UHPC and RPC for ready-mixed concrete intended for high volume construction projects. The achieved results are beneficial for expanding UHPC and RPC for the ready-mixed concrete, allowing further application of these concrete types to monolithic structures. The positive environmental aspect of UHPC and RPC is also beneficial, as the application of these High Strength Concrete types with high utility properties may be used to build long-lasting structures with a smaller cross-section and thus save the volume of input raw materials.

## 7. References

[1] Shi C, Wu Z, Xiao J, Wang D, Huang Z and Fang Z 2015 A review on ultra high performance concrete: Part I. Raw materials and mixture design. *Construction and Building Materials.* **101** 742-749

[2] Strunge J and Deuse T 2008 Special cements for ultra high performance concrete. Ultra high performance concrete (UHPC). *Proc. of the Second International Symposium on Ultra High Performance Concrete.* 61-68

[3] Hirschi T, Wombacher E, Fehling M, Schmidt S and Stürwald 2008 Ultra high performance concrete (UHPC). Ultra high performance concrete (UHPC). *Proc. of the Second International Symposium on Ultra High Performance Concrete*

[4] Pierre R and Cheyrezy M 1995 Composition of reactive powder concretes. *Cement and Concrete Research.* **25- 7** 1501-1511

[5] Yang Y 2000 *Manufacturing reactive powder concrete using common New Zealand materials* Diploma thesis (Auckland: Dept. of Civil and Resource Engineering, University of Auckland, N.Z.) Supervisor: Jason Ingham

[6] Kotatkova J and Reiterman P 2014 Effects of different types of steel fibers on the mechanical properties of high strength concrete. *Advanced Materials Research.* **1054** 80-84

[7] Schachinger, Hilbig and Stengel 2008 Effect of Curing Temperature at an Early Age on the Long-Term Strength Development of UHPC Ultra high performance concrete (UHPC). *Proc. of the Second International Symposium on Ultra High Performance Concrete*

[8] EN 13263 Silica fume for concrete - Part 1: Definition, requirements and conformity criteria
[9] EN 206 Concrete - Specification, performance, production and conformity
[10] Rundt L 2016 *The Development of Ultra High Strength (UHPC) and Reactive Powder Composites (RPC)* Diploma thesis (Brno: Brno University of technology, Faculty of Civil Engineering, Czech Republic) Supervisor: R. Hela
[11] EN 12350 – 2 Testing fresh concrete - Part 2: Slump-test
[12] EN 1015 – 3 Methods of test for mortar for masonry - Part 3: Determination of consistence of fresh mortar (by flow table)
[13] EN 12350 – 6 Testing fresh concrete - Part 6: Density
[14] EN 12390 – 7 Testing hardened concrete - Part 7: Density of hardened concrete
[15] EN 12390 – 3 Testing hardened concrete – Part 3: Compressive strength of test specimens
[16] EN 1015 – 11 Methods of test for mortar for masonry - Part 11: Determination of flexural and compressive strength of hardened mortar

**Acknowledgment**
This outcome has been achieved with the financial support of project FV 10680, supported by Ministry of Industry and Trade, project No. LO1408 "AdMaS UP - Advanced Materials, Structures and Technologies", supported by Ministry of Education, Youth and Sport, and under the project GACR P104/15-23219S “Study of methods of nanoparticles dispersion, determination of conditions for preventing their re-agglomeration for application in cement composites”.

# Application of Simplified Deterioration Diagnosis Method to Existing Fishing Port Facilities

**T Fujita[1], K Nakamura[2], S Date[3] and T Kasai[3]**

1 Fisheries Civil Engineering, Fisheries Infrastructure Development Centre, Tokyo, Japan
2 Fisheries Civil Engineering, National Research Institute of Fisheries Engineering (Present affiliation: Fisheries Infrastructure Department, Fisheries Agency, Tokyo, Japan)
3 Development of Civil Engineering, Tokai University, Hiratsuka City, Kanagawa, Japan E-mail: t-fujita@fidec.or.jp

**Abstract.** For the efficient maintenance of a large number of fishing port facilities distributed over various locations, establishment of simple deterioration diagnosis method with acceptable precision is needed. In this study, the applicability of the rebound method and the mechanical impedance method to the existing fishing port facilities was examined with laboratory tests focusing on the effect of testing conditions, measurement positions, aggregate contents on compressive strength, and influence of polishing of the test surfaces followed by in-situ surveys. As a result, the mechanical impedance method showed an acceptable precision as compared with the rebound method and was able to easily estimate compressive strength without polishing the test surfaces. It was very likely that the mechanical impedance method could be highly applicable to deterioration diagnosis method for the existing fishing port facilities.

## 1. Introduction

A total 2866 fishing ports with huge fishing port facilities distribute over Japan [1] and their deterioration with time is a major concern [2].

Reflecting this condition, the Fishery Agency of Japan has started the Fishing Infrastructure Stock Management Project on April 2008 for realizing a longer life and lesser maintenance cost of existing fishing port facilities. Those who are responsible for the management of fishing port facilities are now executing repair constructions and performance maintenance planning on the basis of the project [1].

Fishing port facility is characterized by its volume, variety of structure types and local government control [3]. This leads to a shortage of maintenance/repair budget and difficulty in employing expertized staffs in the local governments [4]. Another problems with deterioration diagnosis method of fishing port facilities including development of simple method, avoidance of data scatter by human factors, improvement of precision in degradation prediction and LCC estimation have been left for the future study. The improvement of the precision in degradation diagnosis is important particularly in human factors at visual observation and in conditions of site such as immersion in the sea or masking by wave-dissipating constructions leading to a need of simple deterioration diagnosis method with an acceptable precision [5].

In this paper, simple deterioration diagnosis methods applicable to existing fishing port facilities were examined. In addition to the laboratory tests performed on factors affecting the compressive strength and testing conditions including testing position and surface pre-treatment, on-site testing was also carried out using rebound method that can estimate compressive strength by the rebound hardness as a

IOP Conf. Series: Materials Science and Engineering **371** (2018) 012018 doi:10.1088/1757-899X/371/1/012018

result of impacting concrete surface and mechanical impedance method that can also estimate compressive strength by the response waveform originated from an impact on the concrete surfaces by a hammer.

## 2. Fishing port facilities, rebound method and mechanical impedance method

### *2.1. State of the fishing port facilities*

Fishing port facilities in our country have been steadily developed since 1960's and accumulated total length became greater than 5000 km in 2005. Most of these facilities are expected to reach their service lives shortly (Figure 1.) and the deterioration problems may become apparent (Figure 2.).

Prefectural and city governments, whose estimated yearly repair cost for fishing port facilities in the twenty years ahead exceeds that of 2009 F.Y., are 24 out of 40 among which 12 are more than twice as much [2].

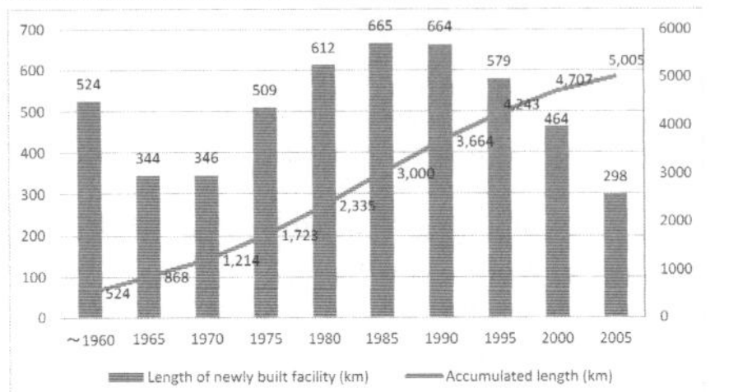


**Figure 1.** Changes in number of stocks in fishing port facilities.

**Figure 2.** Deterioration of fishing port facilities.

### *2.2. Characteristics of fishing port facilities*

Fishing port facilities are characterized by the following conditions for which any deterioration diagnosis method should pay particular attentions [4].

(1) Most of the structural elements are made of concrete particularly that without reinforcements.
(2) Lengthy structures such as breakwater form a major part.
(3) Structures are likely to be deteriorated due to ocean waves and seawater [6].

## 3. Simple degradation inspection method applicable to fishing port facilities

### *3.1. NDT methods applicable to fishing port facilities.*

Because major fishing port facilities are made of concrete without steel reinforcement, deterioration of concrete structure has been estimated in terms of cracking, flaking, internal flaws such as voids and compressive strength [7]. Among these measures, near-surface compressive strength of concrete may be a reliable deterioration index because non-reinforced concretes suffer from less effect of chloride attack and carbonation [8, 9].

Compressive strength of concrete in structure becomes most reliable when sampled from the targeted structures and tested according to JIS A 1108, while its destructive nature and time-cost from sampling to testing are known as major disadvantages [7]. Another compressive strength estimation methods include rebound method against concrete surface hardness [10, 11] and minute-destructive strength testing [7] and, for flaking and internal flaws, non-destructive methods using ultrasonic [12] and impact-echo methods or small core sampling method have been used [10, 13, 14]

Because repair of the targeted structure is not necessary and large number of data can be acquired, the non-destructive testing methods are favourably applied to the cases with the following conditions,

(1) Preliminary test for the subsequent detailed inspection,
(2) Cores necessary for compressive strength testing cannot be sampled,

IOP Conf. Series: Materials Science and Engineering **371** (2018) 012018 doi:10.1088/1757-899X/371/1/012018

(3) Number of tests for the strength estimation are huge, where statistical treatment or calibration curves shall be accompanied.

On the other hands, several problems with the non-destructive testing methods including low precision and limitation in applied ranges have been pointed out [7].

In this study, rebound method and mechanical impedance method which is a kind of the impact echo method, were focused as possible deterioration diagnosis methods applicable to existing fishing port facilities [4, 15, 16, 17]. Reasons of the selection are as follows.

(1) Easy operation

(2) Fast, continuous and multiple measurements.

*3.2. Characteristics of rebound method and mechanical impedance method.*

The rebound method estimates compressive strength of concrete with rebound number resulting from an impact on the concrete surfaces by a rebound hammer, as shown in Figure 3. It is much easier in operation and less destructive to targeted structures compared with core sampling method, and favourably applied as a supplement of visual inspection and preliminary tests [4]. However, several disadvantages, including large variation of data by instruments, impossibility of repeated test at the same point and large variation by the type of material, have been reported [18, 19].

Mechanical impedance method estimates compressive strength of concrete with contact impedance resulting from response waveform of an impact on the concrete surfaces by an impulse hammer [18, 19] as shown in Figure 3. More precisely, the measured response waveform, as shown in Figure 4, is divided into two parts: time domain when a hammer presses concrete and that concrete elastic response thrusts back the hammer. These measurements provide compressive strength estimation, deterioration degree of concrete surfaces and flaking/separation of near-surface concrete [18, 19].

**Figure 3.** Measurement of rebound method (left) and mechanical impedance method (right).

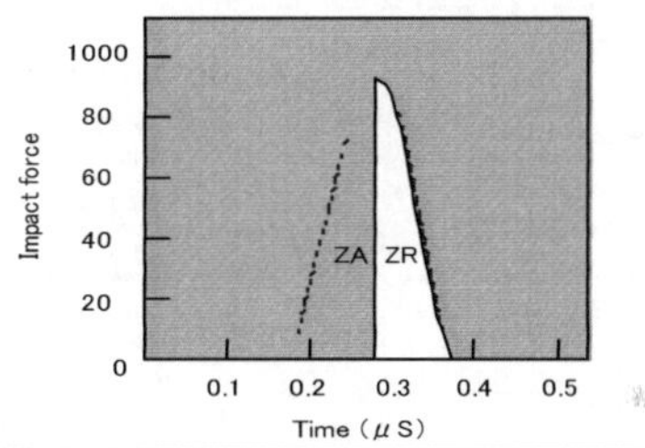


**Figure 4.** An example of impact response waveform of concrete surface.

## 4. Laboratory and in-situ testing methods

*4.1. Laboratory tests: Comparison of compressive strengths obtained with rebound method, mechanical impedance method and standard core testing method.*

In order to compare the results obtained with the rebound method, mechanical impedance method and core compressive strength, and to examine influences of the testing position on the compressive strength, prism specimens of 300 × 300 × 600 mm with a water cement ratio (W/C) of 35, 50 and 65% were prepared and subjected to measurements by the rebound method and the mechanical impedance method to estimate compressive strength. Appearance of the prism specimen is shown in Figure 5. Materials used for the prism specimens and mix proportions are listed in Tables 1 and 2.

IOP Conf. Series: Materials Science and Engineering **371** (2018) 012018 doi:10.1088/1757-899X/371/1/012018

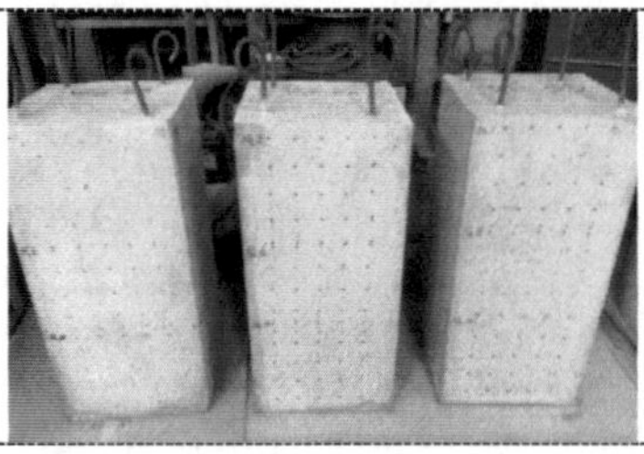

**Figure 5.** Concrete prism specimens.

**Table 1.** Materials used for the prism specimen

| | Type | Notation | Characteristics |
|---|---|---|---|
| Cement | Ordinary portland cement | C | Density: 3.16 g/cm3, Blaine specific surface area: 3300 cm2/g |
| Fine aggregate | Pit sand from Kikugawa influent | S | Density: 2.59 g/cm3, Water absorption: 2.18%, Fineness modulus: 2.42 |
| Coarse aggregate | Hard sandstone from Oume | G | Density: 2.70 g/cm3, Water absorption: 0.62%, Maximum aggregate size: 20 mm |
| Admixture | AE water reducing agent | Ad1 | Lignin sulfonic acid compound and polyol complex |
| | AE auxiliary agent | Ad2 | Modified rosin acid compound anionic surfactant |
| | Superplasticizer | SP | Polycarboxylic acid ether compound |

**Table 2.** Concrete mix proportion of the prism specimen

| W/C (%) | s/a (%) | Air content (%) | Unit content (kg/m3) | | | | Admixture (kg/m3) | | | Measured value | |
|---|---|---|---|---|---|---|---|---|---|---|---|
| | | | W | C | S | G | Ad1 | Ad2 | SP | Air content (%) | Slump (cm) |
| 35 | 42.8 | 4.5 | 162 | 463 | 717 | 998 | − | 0.009 | 3.70 | 4.5 | 15.5 |
| 50 | 45.8 | | 167 | 334 | 809 | 998 | 1.67 | 0.003 | − | 4.0 | 15.8 |
| 65 | 47.2 | | 172 | 265 | 854 | 998 | 1.06 | 0.004 | − | 4.3 | 11.0 |

*4.2. Laboratory tests: Effects of surface deterioration on the compressive strengths estimated with rebound method and mechanical impedance method.*

In order to confirm the influence of difference in test surface conditions on the measured value, one side of the prism specimen, which was not used for tests as described in the previous section 4.1, was dipped in a hydrochloric acid (diluted concentration: 3.3%) for 5 to 10 mm in depth. After deteriorating the surface for 24 hours, it was washed with a water jet and used as a test surface. The appearance of the prism specimen with a deteriorated surface layer is shown in Figure 6.

After completion of these measurements, a core with a diameter of 75 mm was taken at each measurement point and subjected to the compressive strength test (JIS A 1108).

The test surfaces were prepared in three conditions (without deterioration, deterioration without polishing, polishing after deterioration) and then measured with the rebound method and mechanical impedance method. These measurement positions were at the centre portion of 30 cm from the upper surface, and the number of measurements was 50 times. For all the tests, measured values are all used without invalidating ± 20% or more unlike the conventional procedure of the rebound method.

Polishing of the specimen surfaces was performed using a motorized disc polisher with a disc diameter of 100 mm and a grain size of # 60. (surface layer about 1 mm).

**Figure 6.** Concrete prism specimens with a deteriorated surface.

*4.3. In-situ concrete strength estimation with rebound method and mechanical impedance method at the existing fishing port facilities.*

In order to estimate concrete strength with the rebound method and mechanical impedance method for the application to existing fishing port facilities, the measurement with both methods and concrete compressive strength test using sampled cores (JIS A 1108) were carried out.

Structure type, year of completion, measurement position, measurement items, number of measurement points at existing fishing port facilities are listed in Table 3.

In these fishing port facilities, measurement for the test surface with or without polishing was executed with the rebound method and the mechanical impedance method, and at the same position, cores were taken and the compressive strength test (JIS A 1108) was carried out. The number of measurements was 25, and polishing was performed using an electric disk grinder (100 mm disk grinder, grain size # 60) to a degree (approx. 1 mm) that it can be judged that the concrete surface had no irregularities by visual observation as in the previous report.

Although each facility had a wide range of the year of completion as listed in the fishing port ledger, the history of repair was unknown or judged as having no achievement. Neither facility has left design documents, while it was judged from the structure type to be a non-reinforced concrete structure although the mix proportions of the material was unknown.

**Table 3.** Measurement positions at existing fishing port facilities

| Prefecture | Fishing port | Facility | Structure type | Year of completion | Location | Measurement items and number of tests |
|---|---|---|---|---|---|---|
| Aomori | Port U | Breakwater | Concrete single block | 1983 | 2 locations (No.1, 40) | Core compressive strength test: 1<br>Rebound test: each 25 points (3cm mesh)<br>Mechanical impedance test: each 25 points (3cm mesh) before and after polishing |
| | | wharf | | 1982 | 1 location (No.2) | |
| Niigata | Port N | Sea wall | Retaining wall | 1980-2006 | 1 location (NN.1) | Core compressive strength test: 3<br>Rebound test: each 25 points (3cm mesh)<br>Mechanical impedance test: each 25 points (3cm mesh) before and after polishing |
| | Port T | Breakwater | Retaining wall | 1980-2005 | 2 locations (TB1, TB3) | |
| Kanagawa | Port M | Breakwater | Concrete debris with on-site concrete top | 1953 | 2 locations (NO9, 16) | |
| | | Mooring pier | Concrete block masonry | 1961 | 1 location (NO.11) | |

IOP Conf. Series: Materials Science and Engineering **371** (2018) 012018 doi:10.1088/1757-899X/371/1/012018

**Table 4.** Estimated compressive strength of rebound and mechanical impedance methods as compared with average core compressive strength (Unit: N/mm$^2$)

| W/C=35% | | | W/C=50% | | | W/C=65% | | |
|---|---|---|---|---|---|---|---|---|
| Rebound method | Mechanical impedance method | Average core compressive strength | Rebound method | Mechanical impedance method | Average core compressive strength | Rebound method | Mechanical impedance method | Average core compressive strength |
| 27.4 | 31.4 | 50.8 | 25.1 | 27.5 | 45.7 | 22.0 | 27.1 | 29.5 |

**Figure 7.** Overview of breakwaters at fishing port U (Upper left), fishing port T (Upper right), fishing port M (Lower left) and mooring pier of fishing port M (Lower right).

## 5. Results and discussion

### *5.1. Laboratory test: Comparison of compressive strength estimated with rebound method, mechanical impedance method and core strength test*

Results of compressive strengths estimated with rebound and mechanical impedance method, and measured with sampled core are listed in Table 4. Rebound test was performed 25 times at each point and the averaged rebound number (R-value) was used for the estimation of compressive strength.

Estimated strength = $-18.04+1.27R_o$ [20] (1)

Measurement and estimation of compressive strength with mechanical impedance method were performed according to the Proposed Operation Manual for Assessing Concrete Strength using Mechanical Impedance Method at Fishing Port Facilities and strength was estimated using 25 impacts per point.

Estimated strength = 2.98 $ZR$ (mechanical impedance value) -13.35 [21] (2)

By comparing the compressive strengths obtained with the three methods, the magnitude of compressive strength tends to be in the order of, the rebound method <mechanical impedance method <average core compressive strength, at any W / C (see Table 4), and the mechanical impedance method showed closer results to the average core compressive strength than that of the rebound method.

### *5.2. Laboratory tests: Effects of surface deterioration on the compressive strength estimated with rebound method and mechanical impedance method*

Estimated strengths and coefficient of variation by the rebound method and mechanical impedance method are shown in Figure 8. for specimens of three different surface conditions: without deterioration, deterioration without polishing, and polishing after deterioration. The estimated strength of the mechanical impedance method tended to be larger than the rebound method in any W/C or surface condition, showing similar tendency to the laboratory test result as shown in section 5.1.

The estimated strength of the rebound method tended to decrease for specimens with a surface condition "deterioration without polishing" compared to that of the mechanical impedance method leading a remarkably large coefficient of variation in this case. This may imply that the polishing of test surface is necessary for the rebound method when applying to existing fishing port facilities.

IOP Conf. Series: Materials Science and Engineering **371** (2018) 012018 doi:10.1088/1757-899X/371/1/012018

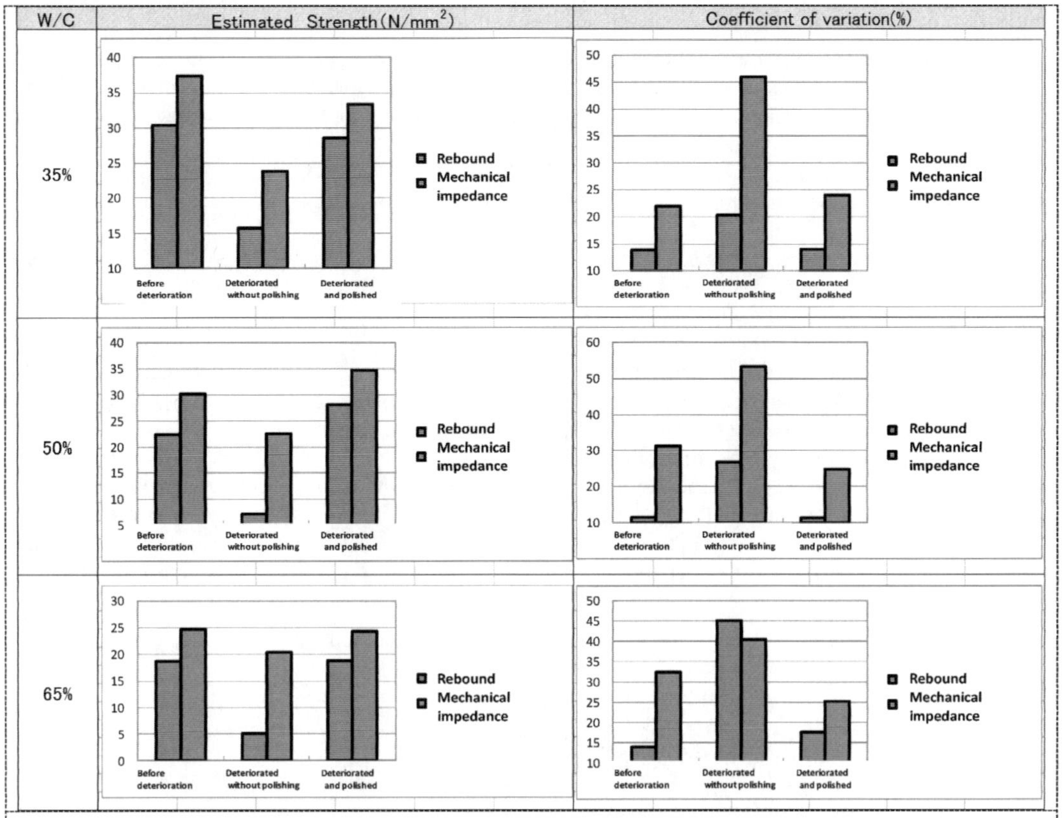


**Figure 8.** Effects of surface conditions on the estimated concrete strength and coefficient of variation measured with rebound method and mechanical impedance method.

*5.3. In-situ concrete strength estimation with rebound method and mechanical impedance method*

Results of the in-situ measurements with the rebound method, mechanical impedance method and core compressive strength test at each fishing port are shown in Table 5. The strength estimation by the rebound method was performed in the same way as described in sections 5.1 and 5.2, while deviation larger than ± 20% was invalidated [20], while in the mechanical impedance method, those with the asymmetrical ZA and ZR and disturbed waveform in the impact response waveform were invalidated [21].

The estimated strength of the rebound method in any of the fishing port approached the average core compressive strength when specimens were polished. Comparing the estimated strength after polishing (a) and the average core compressive strength (c) at each fishing port, the values ranged from 80.3 to 102.9% which were nearly the same as the average core compressive strength, while those of the breakwater NO. 1 and 40 at the fishing port U, breakwater No. 9 and No. 11 mooring pier of fishing port M showed considerably lower ranging from 37.3 to 70.1% (see Table 5). These support the fact that the rebound number shows considerably scatter depending on the type of the targeted material, the surface condition, and the aggregate beneath the test surfaces [10]. It should be noted that the application of the rebound method to the existing fishing port facility needs extensive number of measurements, polishing of test surfaces as well as core compressive strength tests as conventionally believed [10, 20].

Also, the mechanical impedance method tended to show the estimated strength close to the average core compressive strength (c) after polishing like the rebound method, and much closer to the average

IOP Conf. Series: Materials Science and Engineering **371** (2018) 012018 doi:10.1088/1757-899X/371/1/012018

core compressive strength than that of the rebound method as b/c of 61.0 to 94.9% before polishing and 67.8 to 117.4% after polishing.

Mikami et al. compared the average core compressive strength and the measured value by the mechanical impedance method at the existing levees (20 points) of the coastal structures [5], and proposed a correction formula (3) for applying this method to the levee facility in order to estimate the core average compressive strength from the mechanical impedance measurement.

$$Y = 1.07\,X + 8.46\ (R^2 = 0.73) \qquad (3)$$

where $Y$: average core compressive strength N/mm2 and X: estimated strength by mechanical impedance method N/mm2.

Correlation between the estimated compressive strength obtained with the mechanical impedance method before polishing and the average core compressive strength at 9 fishing ports in this study was given by the equation (4).

$$Y = 1.2\,X + 0.71\ (R^2 = 0.80) \qquad (4)$$

The correlation coefficient is higher than that of the correction formula (3) of Mikami et al. [5], and it is considered that the proposed method could estimate the compressive strength with a high precision using conversion equation (4), even applied before polishing. In the future, continuous data accumulation is needed by types of facility and structure.

Because the proposed method tends to show closer values to the average core compressive strength than that of the rebound method, it is confirmed that simple estimation of compressive strength close to that the core compressive strength test without polishing the test surfaces could be possible when data accumulation is repeated associated with the waveform analysis of the impact response waveform

**Table 5.** Comparison of the measured data with tree methods at each fishing port facility

| | | | | | | | |
|---|---|---|---|---|---|---|---|
| Prefecture | | Aomori | | | | | |
| Fishing port | | Fishing port U | | | | | |
| Facility | | Breakwater | | | | Wharf | |
| | | NO. 1 | | NO. 40 | | NO. 2 | |
| Polishing of the test surface | | Before | After | Before | After | Before | After |
| Core compressive strength | Averaged compressive strength: c ($N/mm^2$) | 42.3 | | 23.4 | | 25.2 | |
| Rebound | Estimated Strength: a ($N/mm^2$) | 18.8 | 22.6 | 7.8 | 16.4 | 11.1 | 22.6 |
| | a/c (%) | 44.4% | 53.4% | 33.3% | 70.1% | 44.0% | 89.7% |
| Mechanical impedance | Estimated Strength: b ($N/mm^2$) | 33.6 | 36.3 | 14.3 | 15.9 | 20.51 | 25.06 |
| | b/c (%) | 79.3% | 85.8% | 61.0% | 67.8% | 81.4% | 99.4% |
| Prefecture | | Niigata | | | | | |
| Fishing port | | Fishing port N | | Fishing port T | | | |
| Facility | | Sea wall | | Breakwater | | | |
| | | NN1 | | TB1 | | TB3 | |
| Polishing of the test surface | | Before | After | Before | After | Before | After |
| Core compressive strength | Averaged compressive strength: c ($N/mm^2$) | 27.3 | | 23.3 | | 22.0 | |
| Rebound | Estimated Strength: a ($N/mm^2$) | 11.3 | 28.1 | 11.6 | 18.7 | 14.4 | 21.9 |
| | a/c (%) | 41.4% | 102.9% | 49.8% | 80.3% | 65.5% | 99.5% |
| Mechanical impedance | Estimated Strength: b ($N/mm^2$) | 21.6 | 26.2 | 22.1 | 25.3 | 19.7 | 23.6 |
| | b/c (%) | 78.9% | 96.0% | 94.9% | 108.8% | 89.6% | 107.4% |
| Prefecture | | Kanagawa | | | | | |
| Fishing port | | Fishing port M | | | | | |
| Facility | | Breakwater | | Mooring pier | | | |
| | | NO. 9 | | NO. 16 | | NO. 11 | |
| Polishing of the test surface | | Before | After | Before | After | Before | After |
| Core compressive strength | Averaged compressive strength: c ($N/mm^2$) | 30.3 | | 20.8 | | 38.8 | |
| Rebound | Estimated Strength: a ($N/mm^2$) | 7.8 | 11.3 | 14.6 | 16.7 | 15.4 | 22.4 |
| | a/c (%) | 25.7% | 37.3% | 70.2% | 80.3% | 39.7% | 57.7% |
| Mechanical impedance | Estimated Strength: b ($N/mm^2$) | 26.7 | 28.3 | 19.1 | 24.4 | 27.5 | 35.0 |
| | b/c (%) | 88.2% | 93.2% | 91.7% | 117.4% | 70.8% | 90.2% |

in the future.

## 6. Conclusions

As a simple deterioration diagnosis method, applicability of rebound and mechanical impedance methods to fishing port facilities was examined and following conclusions were obtained.

(1) Compressive strength estimated with the rebound method was lower than that obtained with the averaged value of core compressive strength and showed large scatter with or without polishing test surface. It was confirmed that polishing of the test surfaces as pointed out in JIS, larger number of measurements and combined core compressive strength test are needed when applied to fishing port facilities.

(2) Even though test surfaces were not polished, concrete strength estimated with mechanical impedance method tended to be closer to mean core compressive strength unlike that estimated with rebound method. The mechanical impedance method could be more favourably applied to simple on-site strength estimation when relationship between impact response waveform and compressive strength is further accumulated.

As shown above, application of mechanical impedance method to fishing port facilities, forming large volume and huge stock, is a reasonable option as a simple estimation method of compressive strength. Based on these methods, further research shall be made by accumulating data from various locations, structure types and date of completion, and by taking into account effects of concrete age and surface deterioration conditions. The inspection of existing fishery port facilities was a result of authors' compilation of the inspection data based on the Infrastructure Inspection Project sponsored by the Fisheries Agency of Japan.

## 7. References

[1] National Association of Fisheries Infrastructure 2016 Pocket Handbook of Fishing Port Fishery and Fishing Village Tokyo pp 21-69

[2] Umezu T, Hayashi H and Mikami N 2009 Development of Fishing Port Facilities and Necessity of Subsequent Maintenance and Upgrading Measures Fisheries Engineering 46(2) pp 181-186

[3] Fishery Agency of Japan and Fishing Port, Fishery and Fishing Village Research Institute 2006 Investigation Report on the Efficient Maintenance and Upgrading Measures pp 31-51

[4] Fujita T, Kaneda T, Tomikawa Y and Date S 2014 Study on the Application of Simplified Deterioration Diagnosis Method to Fishing Port Facilities Proceedings of the Japan Concrete Institute 36 (2) pp 1303-08

[5] Mikami N, Fujita T, Gokudan K and Kasai T 2009 Application of Impact-Echo Method to Degradation Inspection of Surface Layer of Fishing Port Concrete Structures Proceedings of the Japan Concrete Institute 31(2) pp 1555-60

[6] Mehta P K and Gerwick B C Jr 1982 Cracking-Corrosion Interaction in Concrete Exposed to Marine Environment Concrete International 4 pp 45-51

[7] Fujita T, Yoshizuka Y, Kaneda T, Fudo M and Date S 2014 Features and Problems of Function Conservation Project in Fishing Port Facilities IACSIT International Journal of Engineering and Technology 6(5) pp 409-13

[8] Mikami N, Fujita T, Hosaka M, Gokudan K, Shigematsu H and Kasai T 2010 Application of Simple Degradation Diagnosis Method to Maintenance and Control of Coastal Structures Journal of Ocean Development 26 pp 165-170

[9] Kim J K, Moon Y H and Eo S H 1998 Compressive Strength Development of Concrete with Different Curing Time and Temperature Cement and Concrete Research 28(12) pp 1761-73

[10] Japan Concrete Institute 2014 Concrete Diagnosis Technology 14' Basics pp 97-109

[11] British Standard Institution 1986 Testing concrete Recommendations for Surface Hardness Testing by Rebound Hammer BS pp 1881-202

[12] Trtnik G, Kavcic F and Turk G 2009 Prediction of Concrete Strength Using Ultrasonic Pulse Velocity and Artificial Neural Networks Ultrasonics 49 pp 53-60

[13] Basu A and Aydin A 2004 A Method for Normalization of Schmidt Hammer Rebound Values International Journal of Rock Mechanics and Mining Sciences 41 pp 1211-14

[14] Hedayati M R, Sadr S A, Shirazi S A G 2010 Acoustic Surface Wave Condition Monitoring of Subsea Concrete Structure International Journal of Simulation Systems Science & Technology 11(3) pp29-33
[15] Giurgiutiu V and Zagrai 4-8 March 2001 Newport Beach CA A Electro-Mechanical Impedance Method for Crack Detection in Metallic Plates SPIE's 8th Annual International Symposium on Smart Structures and Materials and 6th Annual International Symposium on NDE for Health Monitoring and Diagnostics paper SS02 pp 4335-22
[16] Soh C K and Bhalla S 2005 Calibration of Piezo-Impedance Transducers for Strength Prediction and Damage Assessment of Concrete Smart Materials and Structures 14(4)
[17] Abdulkader E M and Nehme S G 2017 Repeatability of the Rebound Surface Hardness of Concrete with Alteration of Concrete Parameters Construction and Building Materials 131 pp 317-26
[18] Gokudan K, Kubo M, Sakai T and Kubo H 2004 Estimation of Compressive Strength of Cylinder Specimen by Mechanical Impedance Measurement Proceedings of Japan Concrete Institute 26(1) pp 1995-2000
[19] Kubo M, Gokudan K, Morihama K and Sakai T 2005 Assessment of Soundness of Full-Scale Retaining Wall by Mechanical Impedance and Impact-Echo Measurement Proceedings of Japan Concrete Institute 27(1) pp 1711-16
[20] Japan Materials Society 1958 Guideline of Compressive Strength Estimation of Hardened Concrete by Schmidt Hammer (Proposal) Materials Testing 7(59) pp 426-30
[21] http://nrife.fra.affrc.go.jp/seika/inpidansuhoumanyuaru/inpitannsu_index.html

**Acknowledgement**
Authors thank those who involved in this research at Fisheries Agency of Japan, Aomori, Niigata and Kanagawa Prefectures, and Mikuniya Corporation for their generous cooperation and support.

IOP Conf. Series: Materials Science and Engineering **371** (2018) 012019 doi:10.1088/1757-899X/371/1/012019

# Study on Compressive Strength Prediction for Steam Curing

**N Nakajima [1], H Koyama [2], S Hashimoto [3] and S Date [4]**

1 Tokai University, Development of Civil Engineering, Hiratsuka city, Japan
2 BASF Japan Ltd, Development of Civil Engineering, Chigasaki city, Japan
3 Fukuoka University, Development of Civil Engineering, Fukuoka city, Japan
4 Tokai University, Development of Civil Engineering, Hiratsuka city, Japan
E-mail: hayabusa_ripobitann_0613@yahoo.co.jp

**Abstract.** Steam curing process is generally adopted for precast concrete products to improve productivity. For setting reasonable production cycles and selecting economical formulation, it is important to predict the strength of mortar / concrete at early stage. Compressive strength estimation by the Maturity method is being widely used in conventional research. However, it is pointed out that accuracy is relatively low for estimating the strength of concrete under steam curing condition. Therefore, in this study, a relation between effective material ages based on the Maturity and Arrhenius's law and both initial and long-term strength were evaluated. As a result, it was confirmed that the Arrhenius's law shows higher estimation accuracy than the Maturity method. Furthermore, it was confirmed that the prediction method using effective material ages based on the Arrhenius's law which takes the activation energy into consideration has higher estimation accuracy.

## 1. Introduction

In recent years, the population of Japan is on a downward trend and the declining birthrate and the aging of the population are advancing. Also, at the construction site, the declining birthrate and aging population are progressing, hence lack of skilled workers is being concerned. It is worried that current skilled workers account for 33% of the total over the age of 45 and 1.1 million people may leave the job after 10 years. The labor on the construction site is required for certain skills to set up forms, to assemble reinforcing bars and to perform complex tasks. Also, the Tokyo Olympic Games will be held in Japan in 2020. Therefore, improvement of the productivity of the structure is in social demand in the construction industry.

Precast concrete products attract attention to shorten the construction term of the structure. Since precast concrete is fabricated and managed in the factory, it doesn't influence of the weather, it is manufactured in a high accuracy. Therefore, precast concrete products are said to be of high quality. On the other hand, as a countermeasure to cost savings, blast furnace slag and fly ash are used as supplemental cementitious materials [1]. Also, to improve the productivity of precast concrete products, two-cycle production system using same formwork is being applied, in case of need [2]. In the two-cycle production system, it is necessary to apply the steam curing process to remove the product from the mold within adequate period. Prediction of the strength of the product makes it possible to clarify the removal timing of the mold. Therefore, strength prediction is key point from the viewpoint of productivity improvement demolding strength from the viewpoint of installation and dismantling of formwork, cost aspect, quality and quality assurance. Also, it is important to grasp the control of concrete products [3]. concrete with the same material and mix proportions, generally, it is said that "the compression strength obtained is equal if the Maturity are the same even when the temperature changes "[5]. However, the difference in

IOP Conf. Series: Materials Science and Engineering **371** (2018) 012019 doi:10.1088/1757-899X/371/1/012019

cement types is not considered, and if the type of cement and the curing temperature change, the hydration reaction rate also differs. Therefore, the strength development time varies depending on the curing temperature. Therefore, there is strength prediction formula based on the effective material age using the Arrhenius' law as another compressive strength predicting expression expressing the curing time for obtaining the predetermined strength. Since this formula is based on the hydration reaction of cement, its reaction rate is to express the temperature dependence of the chemical reaction. The Maturity method and the Arrhenius' law have equations for predicting the time, as for effective material age, the prescribed strength development, respectively.

Generally, the strength prediction formula used for predicting the compressive strength development time is mostly a Maturity method where calculation with effective material age is easy [6-10]. Many studies on the prediction of the initial strength development of concrete under various curing conditions are reported using the Maturity method. However, there are few reports on research on Maturity method under steam curing. A few reports on the concrete strength development time using the cumulative temperature method and the Arrhenius law are issued so far. In general, it is said that both the Maturity method and the Arrhenius' law do not change the strength development time. However, there have been few reports on studies investigating the relationship between strength development time and compressive strength under the different temperature conditions using both equations. Therefore, in this study, this experiment used three types of curing conditions, two kinds of cement and blast furnace slag fine powder and mixing of mortar. Then, then, this experiment measured the relationship between compressive strength and curing time with respect to the strength development time of the Maturity method and the Arrhenius' law, the Young's modulus and the propagation velocity with respect to the change in the physical properties on the curing temperature, and applicability of the range of use of the strength prediction formula, were investigated

In general, the strength of concrete increases with the age, but it is known that there is a correlation with the curing temperature, and the time to reach a predetermined strength differs depending on the curing temperature [4]. As a factor influencing the strength development of concrete, the curing condition including steam curing program of concrete in the early age can be mentioned. The strength development of concrete depends on the curing temperature and curing time, and in the case of.

## 2. Outline of experiment

### *2.1. Material use and mix proportions*

Materials used and mix proportions are shown in table 1 and table 2 respectively. As a material, Ordinary portland cement, high early strength cement, fine blast furnace slag powder were used. As the admixture, high range water reducing agent for concrete products was used, and an admixture was
added for each mortal specimen so as to have adequate workability. In recent years, cases of use of blast furnace cement B type are increasing in Japan from the viewpoint of durability improvement. Blast

**Table 1.** Matertials used.

| Materials | | Properties | Density ($g/cm^3$) |
|---|---|---|---|
| Cement | N | Ordinary Portland cement | 3.16 |
| | H | High-early-strength cement | 3.14 |
| Mineral admixture | Sg | Blast furnace slag fine powder ($4000cm^2/g$) | 2.90 |
| Fine aggregate | S | River sand from Yamakitamati Kanagawa | 2.69 |
| Chemical admixture | Ad | High range water reducing agent | - |

furnace cement B type (hereinafter referred to "BB", c.f. table 2 is one in which 45% of ordinary portland cement is replaced with blast furnace slag fine powder. In this research, this cement was also used to confirm an influence of type of cement on strength development. was prepared. The ratio of sand and cement was set to 1: 1 by setting the water cement ratio to 3 levels of 30, 40, and 50%, and the sand cement ratio (S / C) to 2.0. Also, in this experiment, the amount of cement was made constant, weighed value was obtained, afterwards, the mixing was carried out.

### *2.2. Curing conditions*

All the material used were controlled and managed temperature so that the mixing temperature became the same as the curing temperature, then mixing was carried out. However, if the mixing temperature is set to 45℃ or higher, there is concern of false set, so the maximum mixing temperature must be 40°C or less. After preparing the specimen, it was cured in a steam chamber. The curing parameters are shown in Figure 1. Curing was performed at designated temperature until a predetermined time, and then the specimen was moved to a constant temperature chamber kept at 20°C.

**Table 2.** Mix proportions.

| Type of cement | W/C (%) | S/C | C (g) | Ad (C×%) |
|---|---|---|---|---|
| N | 30 | 2.0 | 7000 | 0.7 |
| | 40 | | | 0.5 |
| | 50 | | | 0 |
| H | 30 | | | 0.7 |
| | 40 | | | 0.5 |
| | 50 | | | 0 |
| BB* | 30 | | | 0.7 |
| | 40 | | | 0 |
| | 50 | | | |

*Replaced 45wt% of N with Sg

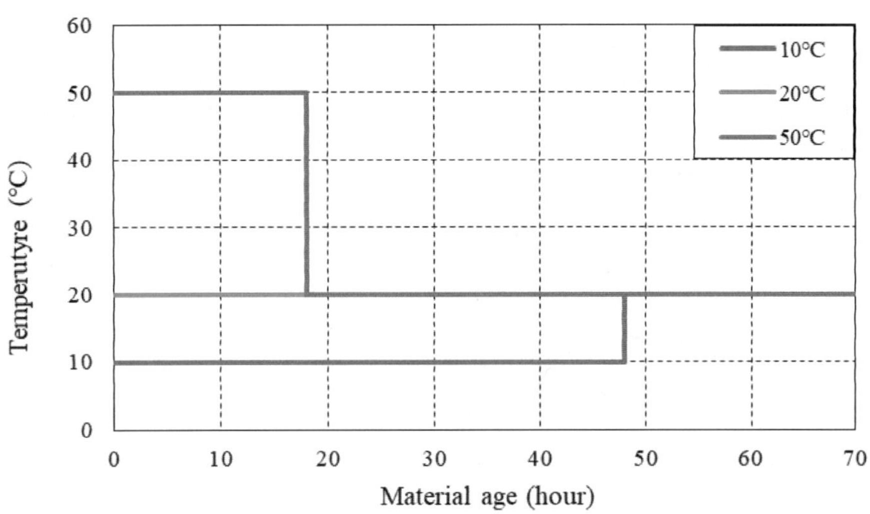


**Figure 1.** Curing temperature.

### *2.3. Experiment combinations*
Three kinds of cements used in this experiment, the mixing conditions of mortar are as shown in table 3. Since the curing conditions are shown in Figure 1, it was decided to consider a combination of blending conditions of 26 patterns in total.

### *2.4. Test items*
For the test items, both 0 hit and 15 hits mortal flow value and the mixing temperature were measured in the fresh mortar. After curing, compressive strength, Young's modulus, and ultrasonic wave propagation velocity were measured respectively. Cylindrical specimens with φ50 × 100 mm were used for compressive strength and Young's modulus, and specimens for ultrasonic wave propagation velocity of 50 × 55 × 500 mm were prepared and measured. Since the age of the test material varies depending on the curing temperature, it is shown in table 3.

## 3. Method for predicting strength development
Evaluation of compressive strength prediction at initial material age of mortar under steam curing was conducted by using Maturity method generally known as conventional prediction expression of strength as followings.

$$M = \int_0^t (T_k + 10)dt \tag{1}$$

here,
M: Maturity [°C · hr]
Tk: environmental temperature [°C]
Δt: curing time [hr]

### *3.1. Effective material age*
Effective material age was considered to effective for evaluating the mechanical properties of concrete subjected to unsteady temperature history. Generally, effective age refers to the age obtained by converting the influence of the curing temperature on the hydration reaction so as to be equivalent to the degree of hydration when the curing temperature was 20 °C.
In this study, applicability was evaluated by using already proposed effective material age ($t_e$). The existing effective material age calculation formula is shown below.

#### *3.1.1. Effective material age①*
Effective material age expression (1) is calculated based on the Maturity method as an empirical rule, and is obtained by dividing the constant by the value of the Maturity method. Therefore, the Maturity method and the effective material age formula ① have the same tendency. Therefore, in this study, the examination by the accumulated temperature method was omitted and decided to consider the effective material age formula ① shown in (2) as a representative te.

$$t_e = \int_0^t \frac{(T_k + 10)}{30} dt \tag{2}$$

**Table 3.** Curing temperature and material age.

| Curing temperature (°C) | Material age (hour) |
|---|---|
| 10 | 48, 168, 336, 672 |
| 20 | 10, 18, 168, 336, 672 |
| 50 | 5, 10, 18, 168, 336, 672 |

### *3.1.2. Effective material age②*

The effective material age formula ② is expressed as the temperature dependence of hydration reaction of concrete on the basis of the Arrhenius' s law as shown in equation (3). This formula is derived by assuming the activation energy to 33 kJ / mol.

$$t_e = \int_0^t \exp\left[13.65 - \frac{4000}{273 + T_k}\right] dt \tag{3}$$

### *3.1.3. Effective material age③*

The effective material age formula ③ is also based on the Arrhenius' s law of the formula (4). However, regarding the calculation of activation energy (Uh), numerous studies including experimental studies have been reported so far. On the other hand, in fib Model Code 2010, it is pointed out that relational expressions should be determined separately by experiments when cement with a high mixing ratio of admixtures showing pozzolanic activity is used or when influence of temperature is important at designing are doing. Therefore, in this experiment, since it gives a high temperature history due to steam curing, the influence due to being different from the assumed activation energy is concerned, therefore in calculating the activation. energy, in order to modify the effective material age formula②, it was obtained by the rule of thumb advocated by Saetta's calculated [13] by the following equation (5).

$$t_e = \int_0^t \exp\left[\frac{U_h}{R}\left(\frac{1}{293} - \frac{1}{T_k}\right)\right] dt \tag{4}$$

$$\frac{U_h}{R} = 4600\left[\frac{30}{T_k - 263}\right]^{0.39} \tag{5}$$

## 4. Result and discussions

### *4.1. Compressive strength*

The results of the compressive strength test are shown in Figure 2. The vertical line shows the compressive strength and the horizontal line shows the test material age logarithmically. Figure 2 shows 3 results at deferent curing temperatures.

Comparing the compressive strength with W/C, compressive strength tended to be higher as W/C was smaller. In addition, when comparing with the same mix proportions, the initial compressive strength was higher as the curing temperature was higher in all case. However, in the case of high-temperature steam curing, the initial strength shows relatively high, but the long-term strength is compared, the compressive strength is reduced by about 20% on average as compared with the 10 °C curing, hence the internal structure can be considered to sparse. This is because in the case of high temperature steam curing, the pore structure inside the mortar was rapidly built at the initial stage, so the internal structure is considered to be sparse. On the other hand, for the case of curing at 10 °C and 20 °C, it is considered that the internal structure is dense because slow pore structure is constructed.

Therefore, when considering the compressive strength for a long time, it can be said that the compressive strength is higher as the curing temperature is lower. However, since the state of the pore structure cannot be obtained exactly by only the compressive strength, the relationship between compressive strength and Young's modulus is as follows.

### *4.2. Ultrasonic wave propagation velocity*

The test results of the ultrasonic wave propagation velocity are shown in Figure 3. The vertical line shows compressive strength and the horizontal line shows logarithmic representation of the test material age. Figure 3 shows curing at 10℃. from the left, curing at 20℃, and curing at 50℃. For all mix proportions, the ultrasonic wave propagation velocity increases as the compressive strength increases. As the compressive strength increases, the internal structure becomes dense, so the ultrasonic wave propagation velocity is considered high

### *4.3. Yong's modulus*

The Young's modulus results are shown in Figure 4. indicating 3 results at deferent curing temperatures. As the ultrasonic wave propagation velocity increases in all formulations, the value of Young's modulus also increases. It is known that when the value of Young's modulus is large, the object is hardly deformed, so that the compressive strength is large. Therefore, since the fact that the Young's modulus is rising means that the compressive strength is also rising, it can be considered that the internal structure is dense. Since it is known from the whole formulation that the ultrasonic wave propagation velocity increases in proportion to the Young's modulus, the internal structure of the specimen can be considered to be dense.

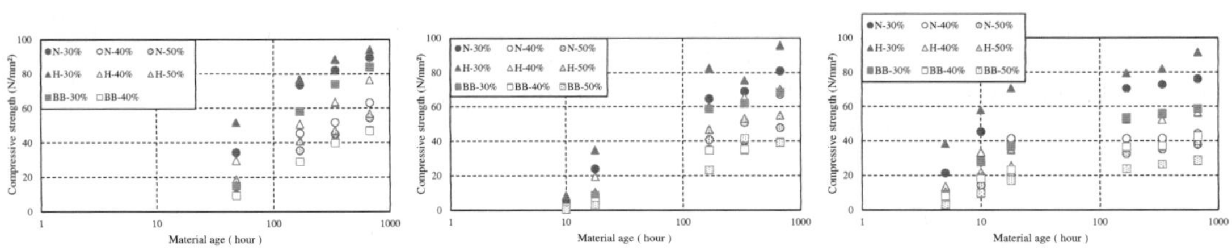


Curing temperature (10°C) Curing temperature (20°C) Curing temperature (50°C)

**Figure 2.** Relationship between material age and compressive strength.

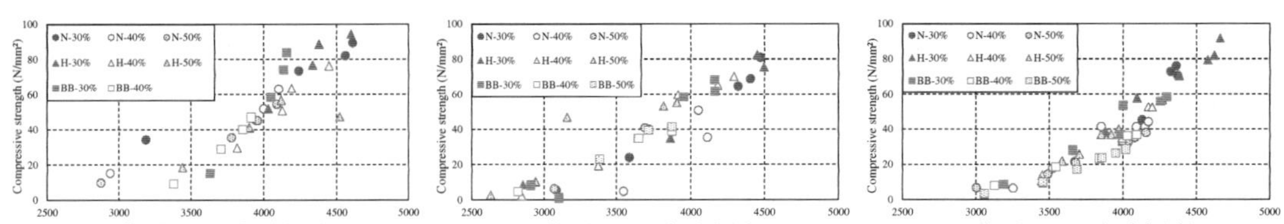


Curing temperature (10°C) Curing temperature (20°C) Curing temperature (50°C)

**Figure 3.** Relationship between compressive strength and ultrasonic wave propagation velocity.

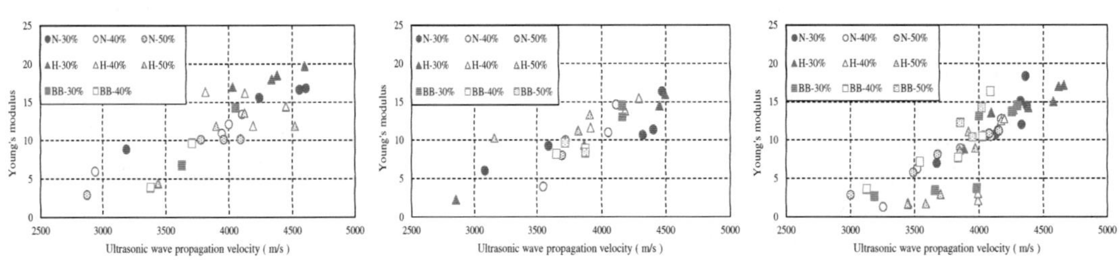


Curing temperature (10°C) Curing temperature (20°C) Curing temperature (50°C)

**Figure 4.** Relationship between Young's modulus and ultrasonic wave propagation velocity.

### *4.4. Comparison result of strength development prediction using three kind of "Effective material age"*

The relationship between effective material age by 3 formulas and compressive strength are shown in Figure 5 and Figure 6. As in consequence, the estimated strength at 20 °C, which is the reference temperature of the effective material age formula, and the compressive strength at 10 °C and 50 °C are compared at the time of effective age of each calculated by the Maturity method and the Arrhenius' s law. Since both the Maturity method and the Arrhenius's law used at this time set the reference temperature to 20°C, hence there is no difference in effective material age at 20°C under curing. Each regression line ($\sigma = a \cdot \log(t_e) + b$) are calculated by using the compressive strength at 20°C and at the age of measurement and reflect reference temperature of 20°C.

Overlaps on the regression line when the temperature is the same 20°C, even if the age of age changes. Also, when the temperature is different, the effective material age is corrected based on the effective material age formula. At that time, substitute the effective material age ($t_e$) corrected for the case where the temperature is different to 20°C, and calculate the estimated strength. When the estimated strength is close to the actual measurement value, the applicable range of the effective material age formula is wide. Also, when the estimated strength is far from the measured value, the applicable range of the effective material age formula is narrow. In addition, by plotting the effective material age and the measured value on the regression line, when the deviation from the regression line is large, the application range is narrow, and the estimation accuracy of the intensity compression strength is low.

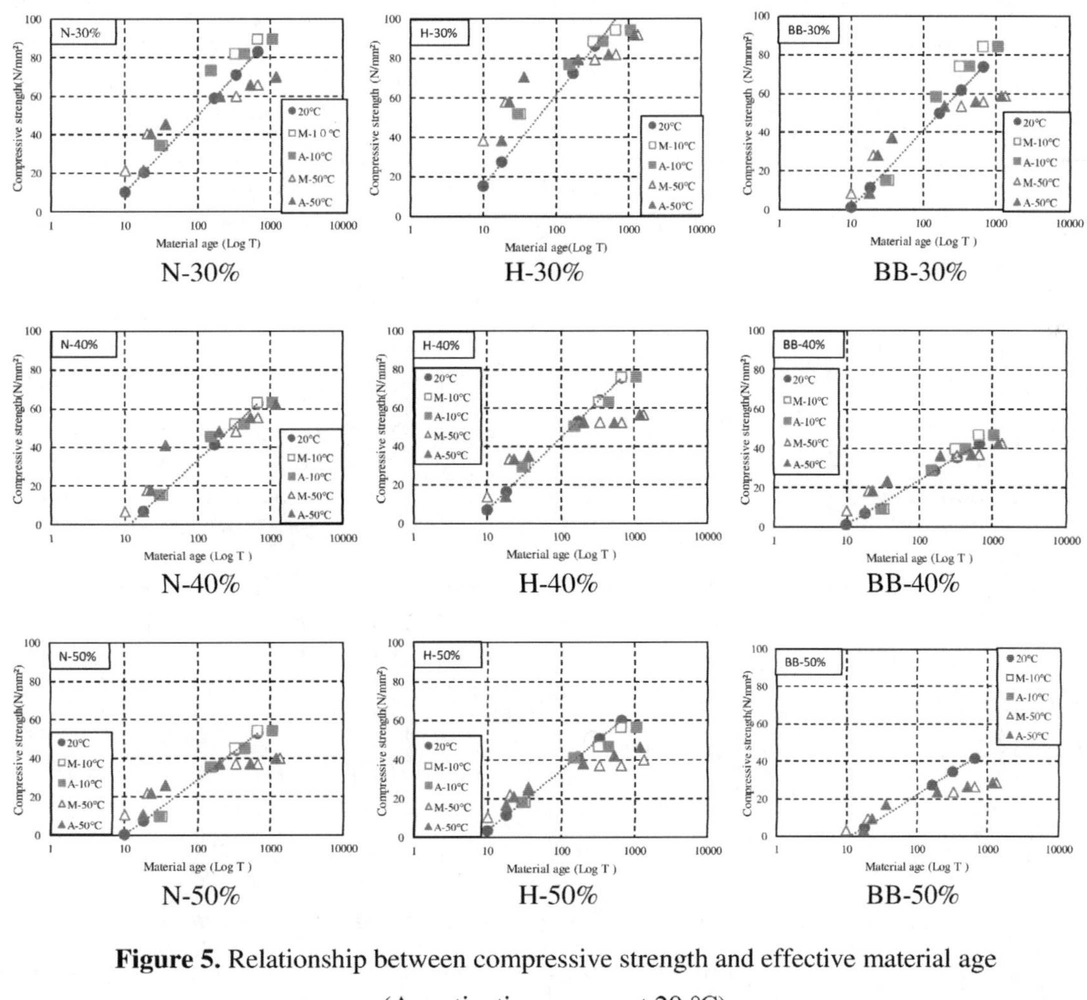


**Figure 5.** Relationship between compressive strength and effective material age (As activation energy at 20 °C).

IOP Conf. Series: Materials Science and Engineering **371** (2018) 012019 doi:10.1088/1757-899X/371/1/012019

Therefore, the nearer to the regression line of 20, it can be considered that it is higher estimation accuracy of the compressive strength than other curing temperatures, and it can be said that versatility is high.Figure 5 shows those using the effective material age formula with activation energy at 20℃ of the Arrhenius's law, Figure 6 shows the effective material age formula of the Arrhenius' s law, of which is considering adequate activation energy based on each curing program. table 4 shows accuracy of strength prediction based on the results of both Figure 5 and Figure 6. Those that are not filled in the plot show the maturity method, and those filled with paint indicates the Arrhenius law. The standard deviation of prediction errors by 3 methods are shown in table 4, and from Figure 5, it is confirmed that the Arrhenius's law is the one with small deviation of the regression line in all formulations. Therefore, when comparing the maturity method and the Arrhenius's law as in the past research, it can be said that the Arrhenius's law is better for the strength estimation accuracy [14].
These results also show that the Arrhenius's law shows better accuracy of estimating the compressive strength than the maturity method. Also, when the curing temperature is compared between 10 °C and 50 °C, the deviation from the regression line is larger at 50 °C, indicating that the compressive strength estimation accuracy is poor. Furthermore, when the prediction accuracy in Figure 5 and Figure .6 is compared, it was confirmed that estimation accuracy improves with consideration of activation energy. In consequence, it was found that the Arrhenius's law has better accuracy of the estimation of the

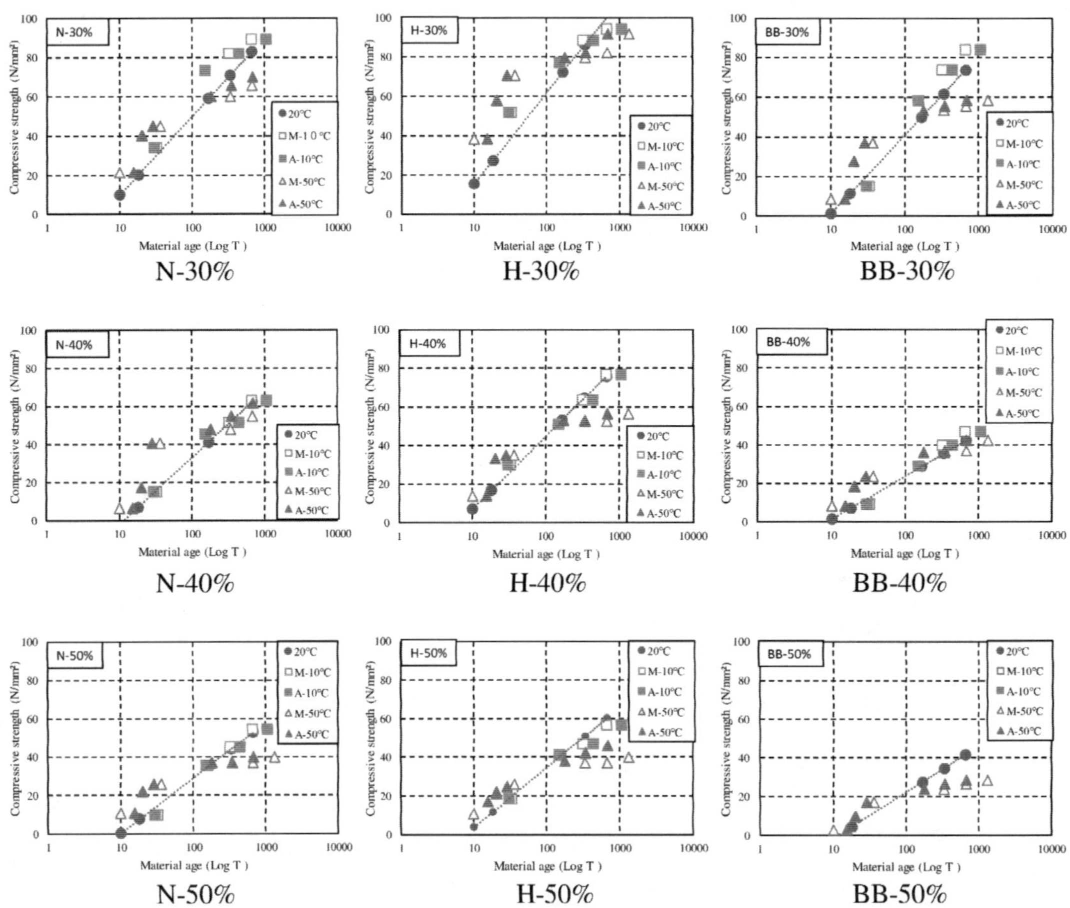


**Figure 6.** Relationship between compressive strength and effective material age (Considering adequate activation energy based on each curing program).

**Table 4.** Accuracy of strength prediction based on three kinds of effective material age.

| Type of cement | Temperature (°C) | W/C (%) | Standard deviation of prediction error (N/mm²) | | |
|---|---|---|---|---|---|
| | | | Maturity method | Arrhenius's law | |
| | | | | Activation energy at 20°C | Consider adequate activation energy |
| N | 10 | 30 | 6.3 | 8.3 | 5.5 |
| | | 40 | 3.3 | 5.1 | 2.8 |
| | | 50 | 1.6 | 3.2 | 0.9 |
| H | | 30 | 2.7 | 3.8 | 2.2 |
| | | 40 | 1.8 | 0.8 | 2.4 |
| | | 50 | 1.9 | 3.2 | 1.5 |
| BB | | 30 | 4.8 | 7.1 | 3.9 |
| | | 40 | 1.1 | 1.9 | 1.0 |
| N | 50 | 30 | 8.1 | 7.3 | 7.0 |
| | | 40 | 10.8 | 9.8 | 9.8 |
| | | 50 | 4.9 | 4.1 | 3.7 |
| H | | 30 | 9.9 | 9.6 | 9.4 |
| | | 40 | 7.9 | 8.0 | 7.4 |
| | | 50 | 5.2 | 5.1 | 4.7 |
| BB | | 30 | 9.2 | 9.2 | 8.7 |
| | | 40 | 5.3 | 4.7 | 4.2 |
| | | 50 | 4.4 | 4.0 | 3.9 |

compressive strength than the Maturity method, and the activation energy is deeply related to the compressive strength estimation.

## 5. Conclusion

In this study, the influence of the curing temperature by the Maturity method and Arrhenius's law was investigated and the relationship between the cumulative temperature and the effective material age was examined. As a result, the following was found out.

(1) It was found that the ultrasonic propagation speed increases in proportion to the Young's modulus. And internal structure of the specimen can be considered to be dense.

(2) About the accuracy of estimating the compressive strength, the Arrhenius's law has better accuracy of estimating the compressive strength than the Maturity method

(3) Accuracy of compressive strength estimation based on effective material age the greater the difference between the curing temperature and the reference temperature (20°C), the poorer the accuracy of estimating the compressive strength.

(4) Among the active ingredient age formula of the Arrhenius's law, it was found that the accuracy estimation accuracy becomes even better when the effective material age formula considering the activation energy is used.

## 6. References

[1] Ilwoo Won, Youngju Na, Jeong Tai Kim, Sunkuk Kim, "Energy-efficient algorithms of the steam curing for the insitu production of precast concrete members", Volume 64, pp.275-284, September 2013

[2] Elmo C. Higginson, "Effect of Steam Curing on the Important Properties of Concrete", Volume 58, Issue 9, pp.281-291, September 1961.

[3] A.A.Ramezanianpour, M.H.Khazali, P.Vosoughi, "Effect of steam curing cycles on strength and durability of SCC: A case study in precast concrete", Construction and Building Materials, Volume 49, pp. 807-813, December 2013

IOP Conf. Series: Materials Science and Engineering **371** (2018) 012019 doi:10.1088/1757-899X/371/1/012019

[4] G. A. Khoury, "No Access Compressive strength of concrete at high temperatures: a reassessment", Magazine of Concrete Research, Volume 44 Issue 161, pp. 291-309, December 1992
[5] Franz-Josef Ulm, "Modeling of Thermochemomechanical Couplings of Concrete at Early Ages", Journal of Engineering Mechanics, Volume 121, Issue 7, July 1995
[6] V. Walle, L.d'Aloïa, F.Cussigh, S.Lecruxa, "Using the maturity method in concrete cracking control at early ages", Cement and Concrete Composites, Volume 26, Issue 5, pp. 589-599, July 2004
[7] Guo Chengju, "Maturity of Concrete: Method for Predicting EarlyStage Strength", ACI Materials Journal, Volume 86, No.4, pp.341-343, July-August 1989
[8] Carino, N., "The Maturity Method: Theory and Application", Cement, Concrete and Aggregates, Vol. 6, No. 2, pp. 61-73, 1984
[9] Carino, N. J. and Tank, R. C., "Maturity functions for concretes made with various cements and admixtures." ACI Materials Journal, 89(2), pp.188-196, 1992
[10] J.-KKim, Y.-HMoon, S.-HEo, "Compressive strength development of concrete with different curing time and temperature", Cement and Concrete Research, Volume 28, Issue 12, pp.1761-1773, December 1998
[11] V.Waller, L.d'Aloïa, F.Cussigh, S.Lecruxa, "Using the maturity method in concrete cracking control at early ages", Cement and Concrete, Volume 26, Issue 5, pp.589-599, July 2004
[12] fib, "Model Code 2010, Final Draft", Vol.1, fib bulletin Vol.65, 2013
[13] Anna V. Saetta, Bernhard A, Schrefler and Renato V. Vitaliani: The Carbonation of Concrete and the Mechanism of Moisture, Heat and Carbon Dioxide Flow Through Porous Materials, Cement and Concrete Research, Vol.23, pp.761-772, 1993
[14] Nozomi Nakajima, Takayoshi Maruyama, Hiromitsu Koyama, Shigeyuki Date, "Effect of Steam Curing on Initial Strength Development of Mortar", International Journal of Engineering and Technology, Vol.9, No.6, pp.468-473, December 2017.

IOP Conf. Series: Materials Science and Engineering **371** (2018) 012020 doi:10.1088/1757-899X/371/1/012020

# Convergence Analysis of a Markov Chain Monte Carlo Based Mix Design Optimization for High Compressive Strength Pervious Concrete

**Jiaqi Huang**
Faculty of Engineering and Mathematical Sciences,
The University of Western Australia, WA, Australia
E-mail: 21971574@student.uwa.edu.au

**Lu Jin**
Faculty of Engineering and Mathematical Sciences,
The University of Western Australia, WA, Australia
E-mail: lu.jin@uwa.com

**Abstract.** Compared with conventional concrete products, pervious concrete usually features with high water permeability rate and low compressive strength due to the lack of fine aggregates. Thus the determination of optimal mix design of ingredients has been recognized as an effective mechanism to achieve the trade-off between compressive strength and permeability rate. In this paper, we proposed a Markov Chain Monte Carlo based approach to approximate the optimal mix design of pervious concrete to achieve a relatively high compressive strength while maintaining desired permeability rate. It is proved that the proposed approach effectively converges to the optimal solutions and the convergence rate and accuracy rely on a control parameter used in the proposed algorithm. A number of simulations are carried out and the results show that the proposed system converges to the optimal solutions quickly and the derived optimal mix design.

## 1. Introduction

Pervious concrete is formed by mixing cement, water and coarse aggregates [1]. Due to the excellent permeability of pervious concrete, it has been widely used in urban construction and pavement. The component of cement paste usually forms a thick layer of coating around the coarse aggregate, the small particles of coarse aggregate are strongly combined with the cement paste to enhance stability and provide the desired mechanical properties [1,2]. Due to the absence of fine aggregates, a large amount of vacancies between coarse aggregates can be observed inFigure 1. These vacancies result ina much higher permeability rate of pervious concrete compared with the conventional counterparts that is shown inFigure 2. Contrarily, the lack of fine aggregate weakens the pervious concrete on compressive strength. Therefore, an effective and efficient pervious concrete mix design optimization that achieves the maximum compressive strength while maintaining a suitable permeability rate to meet the construction requirement is desirable [1].

In this paper, a design of pervious concrete as a Markov Chain Monte Carlo (MCMC) process is modeled [2,3,4,5,6,7] and an optimization approach to approximate the optimal mix design based on Gibbs Sampling is proposed [2]. In order to achieve the aim of maximize the compressive strength while maintaining a relatively high permeability rate, the key problem of this paper is to find out an optimal mix design to obtain those features (i.e., optimal relative proportions of ingredients). As a result, a global optimal solution that effectively and efficiently converged by the proposed algorithm is discussed. The rate and accuracy of convergence is dominated by a control parameter employed in the approach. In addition, a number of simulations are carried out to implement the Makov Chain Monte Carlo based approach. From the simulation results, we could see that the system converges to the optimal mix design globally effectively and efficiently by using the Makov Chain Mnte Carlo based approach. Also the convergence of iteration is accurate and fast when the optimal system coefficients are used.

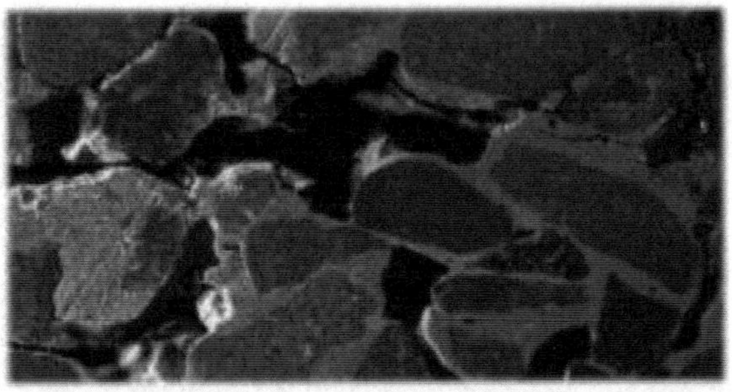

**Figure 1.**Material structure of pervious concrete.

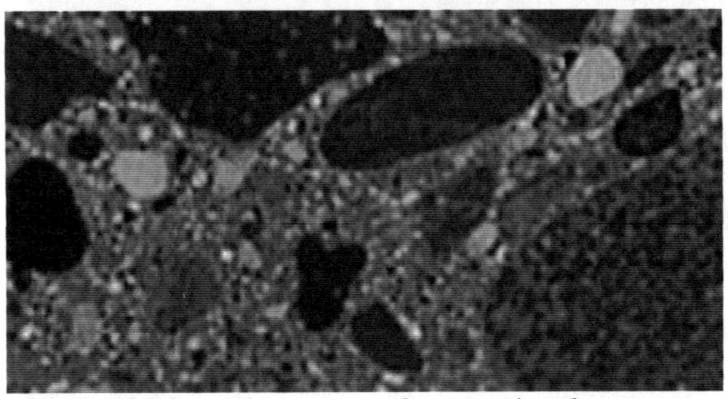

**Figure 2.**Material structure of conventional concrete.

The rest of this paper is organized as follows. In Section 2, the markov chain monte carlo technique and the Gibbs Sampling method are briefly reviewed. Section 3 describes a MCMC based pervious concrete model and a Gibbs Sampling method based optimization approach. The convergence analysis of the proposed algorithm is also depicted in Section 3. In Section 4, a number of simulations are carried out to show the effectiveness and efficiency of the proposed approach. Section 5 concludes this paper.

## 2. Methodology

The Markov Chain Monte Carlo (MCMC) technique is a general computing technique that has been widely used in physics and computer science, which include a family of algorithms for sampling from probability distributions based on constructing a Markov chain whose stationary distribution is the desired distribution. MCMC solve the complicated or high dimension problems by constructing a Markov chain having $\pi(\cdot)$ as its stationary distribution, and carry on the Monte Carlo method to the samples collected according to the $\pi(\cdot)$. Initially, the prior distribution may be arbitrary, the system transits its states randomly with small steps to generate and filter samples. Once the qualified samples are generated for the Markov chain, this chain will be executed for a relatively long time to reach its stable state, which is referred as burn-in process. Then, the stable samples will be used in the Monte Carlo integration to approximate the expectation of function.

The Gibbs Sampling method is one of the well-known MCMC sampling methods which carry out the Monte Carlo method to the samples collected from a stationary distribution $\pi$. In a given system, system variable vector $x$ is a $N$-dimension row vector with element $x_n$, $n = 1, \ldots, N$ and the distribution of interest $F(x)$ is a function of vector $x$ and can be of any form. The key idea of the Gibbs Sampling method is that the value of each random variable in the vector $x$ is updated iteratively and asynchronously according to a probability distribution.

In each iteration, a sample is drawn from the conditional distribution $P(x_n \mid x_{-n}, F(x))$ where $x_{-n} = (x_1, \ldots, x_{n-1}, x_{n+1}, \ldots, x_N)$, $n \in [1, N]$. It can be seen that the Gibbs Sampling method defines a Markov Chain on variable $x$. Assuming $\chi = (\chi_1, \chi_2, \ldots, \chi_N)$ is the feasible real domain corresponding to random variable vector $x$, the detailed Gibbs Sampling method is given in Algorithm shown below.

**Algorithm 1**: Gibbs Sampling Algorithm

- Initialization: Randomly select an initial point $x \in \chi$;
- while $n \in N$ do
  - the element $x_n$ is updated by a sample from the probability distribution $P_n(x_{-n}) = (P_n(x_n \mid x_{-n}), \forall x_n \in \chi_n)$ with

$$P_n(x_n \mid x_{-n}) = \frac{\exp(\frac{-b}{F(x_n, x_{-n})})}{\sum_{x_n' \in \chi_n} \exp(\frac{-b}{F(x_n', x_{-n})})} \quad (1)$$

where $b$ is a positive constant.

  - increment $n$ by one;
  - if $n > N$, then $n = 1$;

According to (2), $x_n$ which produces a higher value of objective function deserves a higher probability to be selected. In light of the Gibbs Sampling method, it becomes possible to estimate highly complicated models.

The weighted product method (WPM) is employed in our system to generate a utility function. WPM is one of the widely used multi-criteria decision making (MCDM) methods which allow one to carry out comprehensive analysis by taking account of multiple aspects of systems[1,2]. In a WPM system consisting of $k$ system variables (e.g. $n_1, n_2, \ldots, n_k$) and $l$ system measures (e.g. $m_1, m_2, \ldots, m_l$), a weighting coefficient $\alpha_i$, $i \in [1, l]$ is assigned to each of system performance measures. Given a variable vector $a$, the weighted product $P$ is calculated as follows

$$P(a) = [m_1(r)]^{\alpha_1} [m_2(r)]^{\alpha_2} \ldots [m_l(r)]^{\alpha_l} = \prod_{j=1}^{l} [m_j(r)]^{\alpha_j} \quad (2)$$

For any two variable vectors $a_1$ and $a_2$, if $P(a_1) > P(a_2)$ can be obtained, it can be concluded that $a_1 > a_2$ i.e. the variable vector $a_1$ is better than $a_2$.

## 3. Pervious Concrete

### 3.1. System descriptions

A mix design of concrete is usually given as the relative ratios of the weights of all ingredients, i.e. $w = 1 : r_{wc} : r_{a_1c} : r_{a_2c} : \ldots : r_{a_nc}$ where $r_{wc}$ is water-to-cement ratio and $r_{a_ic}$, ($i \in [1, n]$), is the weight ratio between the $i$-th aggregate and cement. The ratio $1 : r_{wc} : r_{a_1c} : r_{a_2c} : \ldots : r_{a_nc}$ implies one part (by weight) of cement, to $r_{wc}$ parts of water, to $r_{a_1c}$ parts of aggregate $a_1$, until $r_{a_nc}$ parts of aggregate $a_n$. For simplicity, in the proposed porous concrete system, the mix design w is expressed as a ratio vector $r = (r_0, r_1, r_2, \ldots, r_n)$ where $r_0 = r_{wc}$ and $r_i = r_{a_ic}$. Note that, because the ratio of cement to itself is constantly one, its value is not included in the ratio vector $r$. Then the ratio vector $r$ is used as the system variable vector and the system utility function is denoted by $U(r)$ [3,4].

Now the problem seeking the optimal mix design $w^*$ has been converted into the problem finding the optimal ratio vector $r^*$ which maximizes the overall system utility, $U(r^*)$. Moreover, it is assumed that each ratio $r_i$, ($i \in [0, n]$), can only take finite discrete values within the given space $R_i$, i.e., $r_i \in R_i$ where $R_i = (r_{i1}, r_{i2}, \ldots r_{ik})$, $r_{i1}, r_{i2}, \ldots r_{ik} \in [r_i^{\min}, r_i^{\max}]$ and $r_i^{\min}$ and $r_i^{\max}$ are the minimum and maximum boundaries of space $R_i$. Thus the

IOP Conf. Series: Materials Science and Engineering **371** (2018) 012020 doi:10.1088/1757-899X/371/1/012020

space of ratio vectors is $R = R_1 \times R_2 \times L \times R_n$. Once the optimal ratio vector $r^*$ is determined, the optimal value of utility function is derived and the weights of cement and other components can be easily derived.

### *3.2. Implementation of the Gibbs sampling method*

Let $T = \{1, 2, L\}$ denote the index of a sequence of iterations. At the first step $t = 1$, an arbitrary ratio vector denoted by $r(t)$, ($r(t) \in R$), is select. When the iteration moves from $t$ to $t+1$, ($t \in T$, $t+1 \in T$), the ratio vector is updated from $r(t)$ to $r(t+1)$, i.e., the system state transits from $r(t)$ to $r(t+1)$. In this process, there is one and only one element $r_i$ in $r(t)$, is updated according to the probability distribution $\Lambda_i(r_{-i}(t))$ as given below.

$$\Lambda_i(r_{-i}(t)) = \Lambda_i(r_i \mid r_{-i}(t)) = \frac{\exp(\frac{-b}{U(r_i, r_{-i}(t))})}{\sum_{r_i' \in R_i} \exp(\frac{-b}{U(r_i', r_{-i}(t))})}, \forall r_i \in R_i \qquad (3)$$

where $\beta$ is a positive constant and

$$r_{-i}(t) = \begin{cases} r_0(t) & \text{if } i = 1, \\ r_1(t) & \text{if } i = 0. \end{cases} \qquad (4)$$

according to the observation of $r_{-i}(t)$. After updating the ratio from $r_i(t)$ to $r_i(t+1)$, the value of utility function $U(r(t+1))$ is evaluated for the new generated ratio vector $r(t+1)$. From (4), the value of $r_i$ inducing a high $U(r_i, r_{-i}(t))$ deserves a high probability to be employed. Then the process moves forward to next iteration. Please note that, at any step $t$ of the iteration process, system keeps tracking the average value of utility function $\overline{U}(t)$ and compare it with $U(r(t))$. If the difference between these values is equal or smaller than a pre-defined value $\varepsilon$ and remains stable, the system converges. The average value of utility function $\overline{U}(t)$ is calculated as follows.

$$\overline{U}(t) = \frac{1}{t} \sum_{t'=1}^{t} U(r(t')) \qquad (5)$$

In the design of utility function $U(r)$ of pervious concrete, the pervious concrete system can be treated as a multi-criteria decision making (MCDM) problem which consists of $l$ system performance measures. These $l$ desired performance measures must be taken into account so that the system can reach the tradeoff among them. As mentioned in previous sections, the porosity in pervious concrete is produced by the reduction or the elimination of fine aggregate from the mix design of general concrete. As a result, the pervious concrete is featured with a high permeability rate. On the other hand, a high compressive strength is one desirable property of pervious concrete. Therefore, by using the weighted product model (WPM), the utility function $U$ of the proposed system is defined as a function of compressive strength ($f_c(r)$ in $MPa$) and permeability rate ($K(r)$ in $mm/\sec$), i.e.,

$$U\left(f_c(r), K(r)\right) = \left[f_c(r)\right]^{\alpha_1}\left[K(r)\right]^{\alpha_2} \qquad (6)$$

where $\alpha_1 \in (0,1)$ and $\alpha_2 = 1 - \alpha_1$.

### *3.3. Convergence Analysis of the proposed system*

The proposed mix design optimization approach of pervious concrete can be modelled as a Markov chain. Given $R$ as the space of the ratio vectors of the system and let $r_{init}$ and $r_{end}$ denote the initial and the end ratio vectors, respectively, the transition matrix is $P=\sum_{i=1}^{n} P_i$ where $P_i = p_i(r_{init}, r_{end})$ , $\forall r_{init}, r_{end} \in R$, with

$$p_i(r_{init}, r_{end}) = \begin{cases} \Lambda_i(r_{end,i} \mid r_{init,-i}) & \text{if } r_{init,-i} = r_{end,-i}, \\ 0 & otherwise. \end{cases} \quad (7)$$

where $r_{init,-i} = (r_{init,1}, \cdots, r_{init,i-1}, r_{init,i+1}, \cdots, r_{init,n})$ is defined as the same manner. In (8), $p_i(r_{init}, r_{end})$ indicates the transition probability between $r_{init}$ and $r_{end}$ of an iteration step with respect to the *i*-th element in the ratio vector. Based on this definition, it could be induced that there must exist such an integer $k$ , $k>0$ , that, after $k$ iterations, all ingredients (i.e. all elements in the ratio vector) have updated their ratios at least once. In other words, all the entries in $P^{k'}$ , $k'>k$ , are strictly positive, which corresponds to a regular Markov chain. Because a regular Markov chain is actually also an ergodic Markov chain, the Markov chain of the proposed system is an irreducible, aperiodic and positive recurrent Markov chain and has the convergence properties of an ergodic Markov chain. Moreover, the Markov chain of the proposed system must converge to the unique stationary distribution $\pi = (\pi_1, \pi_2, \cdots, \pi_{|R|})$ in which $|R|$ is the cardinality of $R$ and $\pi_r$ , $\pi_r \in \pi$ is defined as follows.

$$\pi_r = \frac{\exp(\frac{-b}{U(r)})}{\sum_{r' \in R} \exp(\frac{-b}{U(r')})} \quad (8)$$

Let $r^*$ denote optimal ratio vector maximizing the utility function $U$ . For simplicity of notation, we denote $U(r^*)$ by $U^*$ which indicates the value of the utility function when the variable is the optimal ratio vector $r^*$ . Note that, because $r^*$ is the optimal ratio vector, the corresponding $U^*$ is the maximum value compared with the values of $U(r)$ when $r \neq r^*$ . Moreover, if there are more than one optimal ratio vector, we define $R^*$ as the set of all optimal ratio vectors, i.e. $r_i^* \in R^*$ , $i \in [1, |R^*|]$ .As a result, $U^* = (U_1^*, U_2^*, \cdots)$ is the collection of maximum values of utility function corresponding to $r_1^*, r_2^*, \cdots$ , $(r_1^*, r_2^*, \cdots \in R^*)$.

Dividing the numerator and the denominator of (9) by $\exp(\frac{-b}{U(r^*)})$ gives

$$\pi_r = \frac{\exp[-b\delta(r)]}{\exp[-b\delta(r_1)] + \exp[-b\delta(r_2)] + \cdots + \exp[-b\delta(r_M)]} \quad (9)$$

where $\delta(r) = (\frac{1}{U(r)} - \frac{1}{U^*})$ .

Note that, when $r = r^*$ , we could induce that $\delta(r^*) = (\frac{1}{U(r^*)} - \frac{1}{U^*}) = (\frac{1}{U^*} - \frac{1}{U^*}) = 0$ . As a result, $\exp[-b\delta(r^*)] = \exp(0) = 1$ . Thus,

$$\pi_r = \begin{cases} \dfrac{\exp[-b\delta(r)]}{1+\sum_{r'\in R, r'\notin R^*}[-\beta\delta(r')]} & \text{if } r^* \text{ is unique,} \\ \dfrac{\exp[-b\delta(r)]}{|R^*|+\sum_{r'\in R, r'\notin R^*}\exp[-b\delta(r')]} & \text{otherwise.} \end{cases} \quad (10)$$

where $|\cdot|$ indicates the cardinality of $(\cdot)$. In (11), it can be observed that the value of $b$ has great impact on the stationary distribution. Considering the general cases where $|R^*|>1$, we have

$$\lim_{b\to\infty}\pi(r) = \lim_{b\to\infty}\frac{\exp[-b\delta(r)]}{|R^*|+\sum_{r'\in R, r'\notin R^*}\exp[-b\delta(r')]} \quad (11)$$

Note that $\lim_{\beta\to\infty}\exp[-b\delta(r')] = 0$ , $r'\in R$ and $r'\notin R^*$. If $r'\notin R^*$, then $\delta(r)=0$, and

$$\lim_{b\to\infty}\pi(r) = \lim_{b\to\infty}\frac{\exp[0]}{|R^*|+\sum_{r'\in R, r'\notin R^*}\exp[-b\delta(r')]} = \frac{1}{|R^*|} \quad (12)$$

otherwise,

$$\lim_{b\to\infty}\pi(r) = \lim_{b\to\infty}\frac{0}{|R^*|+\sum_{r'\in R, r'\notin R^*}\exp[-b\delta(r')]} = 0 \quad (13)$$

From (13) and (14), it can be found that the value of each element in the stationary distribution is primarily dominated by the coefficient $b$. It can be observed that the Markov chain of the proposed system must converge to a stationary distribution when the coefficient $b$ approach infinite, as shown in (13) and (14).

## 4. Simulation Configurations and Results

### *4.1. Configurations*

As a primary interest of pervious concrete research, the permeability is measured as the water penetrating through pervious concrete samples which is expressed in millimeters per second ( $mm/\sec$ ). In our proposed system, we will use the experimental data provided in [] to build up the MCMC system model. According to the experimental data, the maximum compressive strength and the maximum permeability rate are $f_{c,\max} = 19.8573MPa$ and $K_{\max} = 26.6630mm/\sec$, respectively. In all simulations, it is assumed that, except cement, only water and one type of coarse aggregate sizes between $4.75mm$ and $9.5mm$ are used to make the pervious concrete samples. The water to cement ratio and the aggregate to cement ratio are denoted as $r_0$ and $r_1$, respectively. More specifically, the minimum and maximum values of these parameters are summarized in Table 1. Please refer to [3] for the detailed experimental data and more information on the configurations and preparations of the specimen.

By following (7), the maximum value of utility function can be obtained as follows

$$U_{\max} = U_{\max}(f_c(r), K(r)) = \max_{r\in R}\{[f_c(r)]^{\alpha_1}[K(r)]^{\alpha_2}\} \quad (14)$$

Based on the configurations given above, a number of simulations are carried out to verify the effectiveness and efficiency of the proposed Gibbs Sampling approach.

**Table 1**Initial relative ratios of components

| Components | Ratio | Min value | Max value | Steps |
|---|---|---|---|---|
| Cement | 1 | NA | NA | NA |
| Water | $r_0$ | $r_0^{min} = 0.3$ | $r_0^{max} = 0.4$ | 1000 |
| Aggregate | $r_1$ | $r_1^{min} = 3.1$ | $r_1^{max} = 5.7$ | 1000 |

### *4.2. Simulation results*

We first observe the effect of the controlling parameter $b$ on the performance of the Gibbs Sampling method. Figure 3compares the compressive strength of the Markov Chain Monte Carlo based approach with the other two approaches, each of which focuses on maximizing either compressive strength or permeability rate only. It is shown that the Gibbs Sampling optimal mix design achieves the trade-off between a high compressive strength while maintaining a desired permeability rate. The figure suggests that a large value of $b$ is preferable when it comes to the optimality of the solution.

Next, we evaluate the effect of the weighting coefficient $\alpha_1$ on the performance of the system. The value of $\alpha_1$ varies from 0 to 1 with 0.1 increment. Figure 4 shows the comparison of the compressive strength at different values of $\alpha_1$. It can be observed that, when $\alpha_1 \in [0, 0.45]$, the compressive strength remains stable regardless the value of $b$. Thereafter, the value of compressive strength dramatically and almost linearly increases with the increase of $\alpha_1$. Figure 5 shows that the permeability rate remains stable when $\alpha_1 \le 0.45$. When $\alpha_1 > 0.45$, the permeability rate significantly decreases with the increase of $\alpha_1$.

## 5. Conclusions

This paper proposed a Markov Chain Monte Carlo algorithm-based optimization algorithm to approximate the optimal mix design of pervious concrete. The derived optimal mix design achieves the trade-off between the maximum compressive strength and the desirable permeability rate of pervious concrete. We also proved that the proposed method effectively and efficiently converges to the optimal solutions. A control parameter $b$ employed in the proposed algorithm dominates the speed and accuracy of convergence, specifically, a large value of $b$ leads to a fast convergence speed and a high convergence accuracy. Simulations show the consistent results.

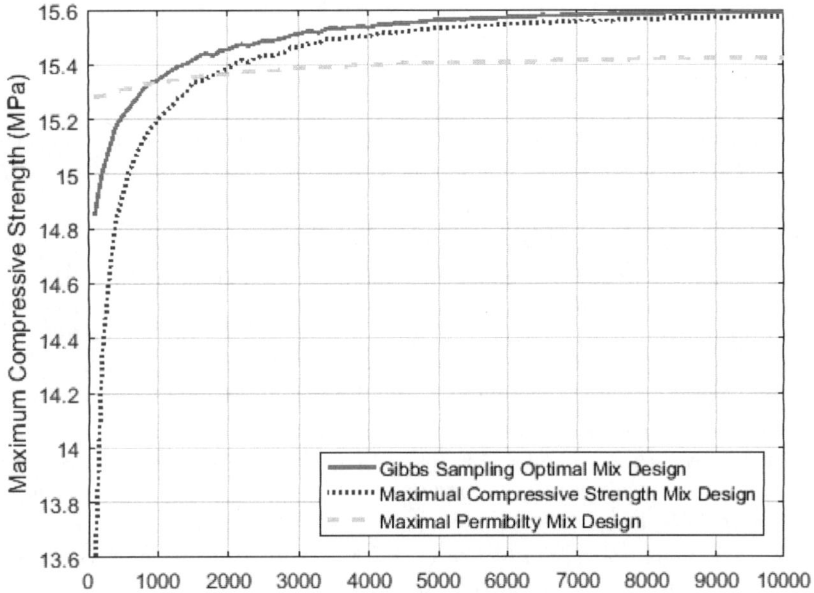


**Figure 3.** Compressive strength for different ***b***.

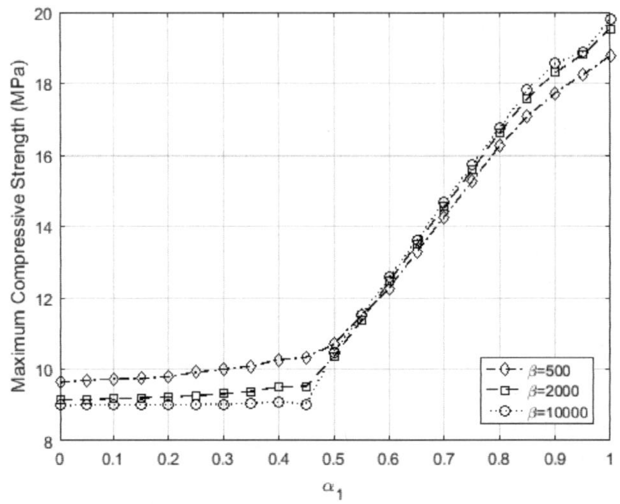


**Figure 4.**Compressive Strength for different $\boldsymbol{\alpha_1}$.

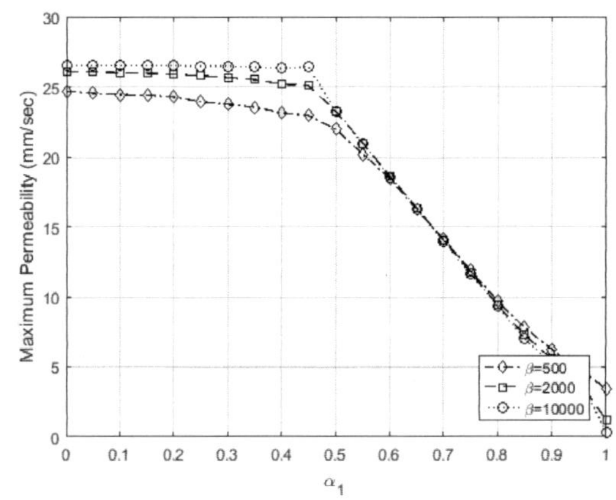


**Figure 5.** Permeability for different $\boldsymbol{\alpha_1}$.

## 6. References

[1] Alireza J., Ali Akbar R. and Mohammad J. 2014, Mechanical Characteristic of Pervious Concrete Considering the Gradation and Size of Coarse Aggregates, *Research Journal of Environmental and Earth Sciences*vol 6, no. 9, pp. 437-442.

[2] Tattersall G 1991 Workability and Quality Control of Concrete. E &Fn Spon.

[3] Ghafoori N and Dutta S, 1995 Laboratory investigation of compacted no fines concrete for

[4] paving materials, *Journal of Materials in Civil Engineering*, vol. 7, no. 3, pp. 183–191.

[5] Rajasekhar K and Spandana K 2016, Strength Properties of Pervious Concrete Compared with Conventional Concrete, *IOSR Journal of Mechanical and Civil Engineering*, vol. 13, pp 97-103.

[6] Hastings W 1970 Monte Carlo Sampling Methods Using Markov Chains and Their Applications, *Biometrika*, vol. 57, no. 1, pp. 97–109.

[7] Kantas N, Doucet A, Singh S, and Maciejowski J 2009 An overview of sequential montecarlo methods for parameter estimation in general state-space models in *Proc. IFAC System Identification (SysId) Meeting*.

[8] Cao X, Zhang W, Song J, Zhu X, and Wang S 2009 Estimating parameters of airsea oscillator with MCMC method," *Acta Phys. Sin.*, vol. 58, no. 9, pp. 6057–6057.

[9] Metropolis N, Rosenbluth A, Rosenbluth M. Teller A, and Teller E 1953 Equation of state calculations by fast computing machines," *The Journal of Chemical Physics*, vol. 21, no. 6, pp. 1087–1092.

[10] Casella G and George E 1992 Explaining the Gibbs Sampler," *The American Statistician*, vol. 46, no. 3, pp. 167–174, 1992.
[11] Gilks W and Spiegelhalter D 1996 *Markov Chain Monte Carlo in Practice*.Chapman & Hall/CRC.
[12] Vignesh Kumar R., Vivek Kumar G., Vineet V. and Freida F., 2016 An Experimental Study on Properties of Pervious Concrete, *International Journal of Advances in Engineering Research*, vol. 11, pp 50-63.
[13] Triantaphyllou E and Mann S 1989 An examination of the effectiveness of multi-dimensional decision-making methods: A decision-making paradox," *An Examination of the Effectiveness of Multi-Dimensional Decision-Making Methods: A Decision-Making Paradox*, vol. 5, pp. 303–312.
[14] Triantaphyllou E 2000 Multi-criteria decision making methods: a comparative study. Kluwer Academic Publishers.
[15] Triantaphyllou E and Mann S 1989 An examination of the effectiveness of multi-dimensional decision-making methods: A decision-making paradox," *An Examination of the Effectiveness of Multi-Dimensional Decision-Making Methods: A Decision-Making Paradox*, vol. 5, pp. 303–312.
[16] Triantaphyllou E 2000 Multi-criteria decision making methods: a comparative study. Kluwer Academic Publishers.

IOP Conf. Series: Materials Science and Engineering **371** (2018) 012021 doi:10.1088/1757-899X/371/1/012021

# Influence of Stone Powder on Properties of Concrete with Low Cementitious Materials

**ZHU Yun, WEN Chaokai and CHEN Jing**

China West Construction Southwest Group Co., Ltd., Chengdu 610052, China
E-mail: zhuywh712@163.com

**Abstract.** With the growing shortage of natural sand, mechanism sand become an important source of sand. And the large amount of low quality is a common phenomenon of the mechanism sand, which mainly reflecting in its bad gradation and grain shape and the results will lead to the cementitious materials and water consumed greatly. The influence of stone powder of high quality mechanism sand on C30 concrete properties is studied. The results show that the proper amount of stone powder could improve the gradation of mechanism sand, increase the paste volume and fill in the void among particles. The workability, compressive strength, permeability and frost resistance of the mechanism sand concrete will be improved on the basis of the content of stone powder, and the optimum range of stone powder content of mechanism sand is 9% to 15% for the C30 concrete.

## 1. Introduction

It is important to develop and research the mechanism sand, called artificial sand, since the natural riverine sand has become increasingly strained. Even though the mechanism sand is adequate, the quality of that is so poor that it owns harmful grain gradation and grain shape, others are worth exploring. A number of studies have shown that the quality of mechanism sand affects the amount of cementitious materials and water consumption of concrete, especially influence of the low strength concrete[1]. Cementitious materials of the C30 fresh concrete generally above 330kg/m$^3$, part of the region and even reach 380kg/m$^3$, and improve the amount of cementitious materials is an effective measure to reduce the risk that the strength of concrete is less stable. Choosing the high-quality aggregate is so significant that it can reduce water consumption and cementitious materials, but improve the quality of concrete[2].

Less than 0.075mm partial is known as the stone powder in mechanism sand. There are different opinions for the influence of stone powder content on concrete performance of different scholars[3,4], someone hold that it will reduce the performance of the concrete and the others consider that it has the best content. Whether the results are accurate or not, it is necessary to verify[5]. Because, after all, the studies are researched based on the ordinary concrete with the high-quality aggregate, low cementitious materials, low water consumption and the concrete low strength is no systematic studied. Therefore, it is of important significance to carry out effects of influence of stone powder content of high-quality mechanism sand on C30 concrete with cementitious materials in the 300kg/m$^3$ or even lower.

## 2. Experiment

### *2.1. Raw materials*

IOP Conf. Series: Materials Science and Engineering **371** (2018) 012021 doi:10.1088/1757-899X/371/1/012021

(1) Cement: P.O42.5 cement produced by Sichuan Lafarge are used in the test. Their basic performance is shown in Table 1.

**Table 1.** Basic performance of cement.

| Cement | Specific area($m^2$/Kg) | Flexural Strength(MPa) | | Compressive strength(MPa) | | Density (g/$cm^3$) | Stability |
|---|---|---|---|---|---|---|---|
| | | 3d | 28d | 3d | 28d | | |
| P.O42.5 | 349 | 6.1 | 9.3 | 28.1 | 49.6 | 3.13 | Conforming |

(2) Fly ash: Grade I fly ash are used in the test. Their basic properties are as follows: fineness of 7.6%, $SO_3$ addition of 0.54%, loss on ignition of 2.01%, water requirement 95%.
(3) Gravel: 5~25mm gravel are used in the test, mud addition of 0.1%, apparent density 2740kg/$m^3$, crushing value of 4.9%.
(4) Mechanism sand: Fineness modulus of 2.6, mud addition of 0.2%, apparent density 2740kg/$m^3$, grade of Machine-Made Sand is shown in Table 2.
(5) Superplasticizer: Polycarboxylate superplasticizer are used in the test. Solid addition of 12%.

**Table 2.** Grade of mechanism sand.

| Screen size(mm) | 4.75 | 2.36 | 1.18 | 0.60 | 0.30 | 0.15 | 0.075 |
|---|---|---|---|---|---|---|---|
| Cumulative sieve residue(%) | 1.2 | 9.4 | 32.7 | 56.1 | 75.3 | 86.8 | 89.4 |

*2.2. Mix design*
The influence of stone powder of mechanism sand on the workability, mechanical properties and permeability of the concrete is studied for the purpose of analyzing stone powder of mechanism sand of its influence on concrete.
The mechanism sand used in the experiment is to remove the particles which is less than 0.075mm, by designing the amount of stone powder is 0%, 3%, 6%, 9%, 12%, 15%, 18% and 21% respectively to replace mechanism sand. The influence of stone powder content of mechanism sand on properties of C30 concrete with low cementitious materials is researched. Test mix proportion is shown in Table 3.

**Table 3.** Mix proportion of concrete.

| Sample No. | Stone powder | Cement | Fly ash | Mechanism sand | Gravel | Water | Superplasticizer |
|---|---|---|---|---|---|---|---|
| JP-1 | 0 | 240 | 60 | 940 | 1030 | 135 | 3.4 |
| JP-2 | 28 | 240 | 60 | 912 | 1030 | 135 | 3.9 |
| JP-3 | 56 | 240 | 60 | 884 | 1030 | 135 | 4.3 |
| JP-4 | 85 | 240 | 60 | 855 | 1030 | 135 | 4.6 |
| JP-5 | 113 | 240 | 60 | 827 | 1030 | 135 | 4.7 |
| JP-6 | 126 | 240 | 60 | 814 | 1030 | 135 | 5.2 |
| JP-7 | 169 | 240 | 60 | 771 | 1030 | 135 | 5.5 |
| JP-8 | 197 | 240 | 60 | 743 | 1030 | 135 | 5.9 |

(kg/$m^3$)

## 3. Results and analysis

### *3.1. Influence of stone powder on workability of concrete*

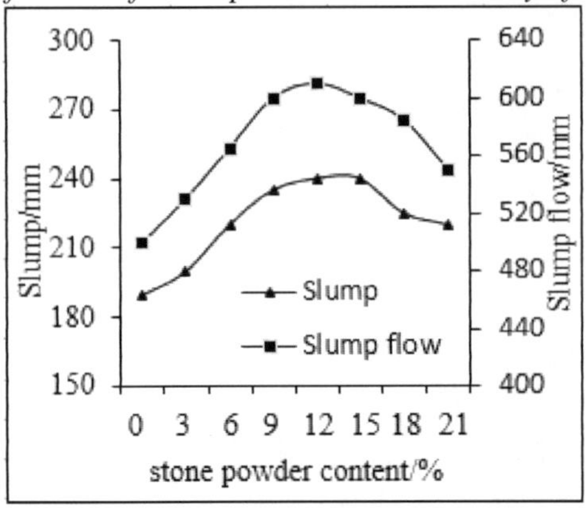


a. Slump flow

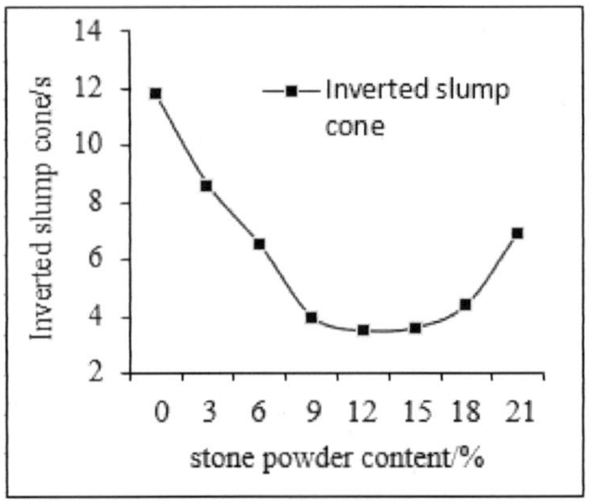


b. Inverted slump cone

**Figure** 1. Workability of C30 concrete.

The workability test results is shown in Figure 1. When the mechanism sand has no limestone powder, there has a poor workability and easy bleeding isolation of concrete. With the stone powder content of mechanism sand increasing, the working performance of concrete mixture is gradually improved, as it is between 9%~15%, the concrete mixture has good workability and excellent various aspects performance. Accounting for 12% of mechanism sand, the slump and slump spread of fresh concrete are 240 mm and 610 mm respectively, delivery time of inverted slump cone of 3.5s and the working performance of C30 concrete with low cementitious material content of 300kg/m$^3$ has reached or exceeded the performance of ordinary C30 concrete. The working performance of concrete increasingly become so bad that too much stone powder of mechanism sand is not conductive to concrete with the increase of stone powder content of mechanism sand. So, the stone powder content of mechanism sand should be range of 9%~15% on C30 concrete with low cementitious materials.

Complete removal of stone powder particles in mechanism sand will also affect the whole graded of concrete system and make graded appeared interrupted, and the cement slurry will be reduced, likewise, the working performance of concrete mixture is detrimental with low cementitious materials. To change the gradation of concrete system and the slurry of concrete mixture, the stone powder content of mechanism sand should be improved appropriate. As it is too much stone powder in mechanism sand, it will adsorb more water and superplasticizer, resulting in high viscosity and content superplasticizer, which is adverse to the workability of concrete mixture.

### *3.2. Influence of stone powder on compressive strength of concrete*

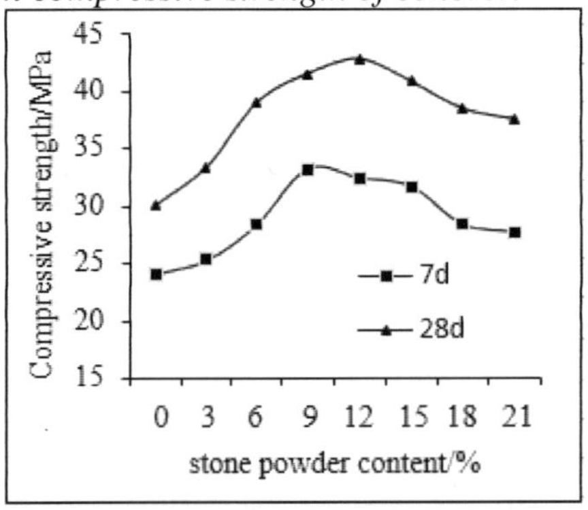


**Figure 2.** Compressive strength of C30 concrete.

The test results shown in Figure2, which indicates that it is detrimental to the compressive strength of concrete with the mechanism sand has no limestone powder, and it is also especially obvious for concrete with low amount of cementitious materials that the 28d compressive strength of concrete is only 30.1MPa. Increasing the dosage of stone powder of mechanism sand, the compressive strength of concrete is improved. When the stone powder content is 12%, the 28d compressive strength can reach up to 42.8MPa. When the content reaches 21%, the compressive strength of concrete decreased, and lower than the group JP-1. Therefore, in order to ensure the compressive strength of concrete with low amount of cementitious materials, the stone powder content of mechanism sand should be 9%~15%.
Effects of stone powder content of mechanism sand on the mechanical properties of concrete is mainly reflected in the following aspects:
The stone powder of mechanism sand can improve the gradation system of concrete;
The stone powder of mechanism sand can accelerate the hydration of cement and mineral admixture;
The stone powder of mechanism sand can react to some substance in cement and promote the degree of concrete density, as a certain active substance.
The test results also show that too much stone powder of mechanism sand is not conductive to the compressive strength of concrete, for it can undermine whole graded system and absorb more water, thereby limiting the hydration of cement and mineral admixture.

### *3.3. Influence of stone powder on hydraulic permeability of concrete*

Hydraulic permeability is one of the most important indexes to evaluate the compactness and permeability of concrete, and it is also an important index to evaluation low strength concrete. This section researches the influence of stone powder content on permeability of C30 concrete with low amount of cementitious materials. Test results are shown in Table 4.

**Table 4.** Test results of hydraulic permeability.

| Stone powder (%) | 0 | 3 | 6 | 9 | 12 | 15 | 18 | 21 |
|---|---|---|---|---|---|---|---|---|
| Relative permeability coefficient/ ($10^{-6}$mm/s) | 10.34 | 8.99 | 3.13 | 1.22 | 0.98 | 1.33 | 1.94 | 4.12 |

The test results show that the permeability of concrete falls after increases first along with the stone powder content of mechanism sand increased. C30 concrete with low amount of cementitious materials and progressively increased stone powder content has a poor permeability and the gradation system and compactness will be improved. If the stone powder content of mechanism sand, such as more than 15%, up to 21%, this need for more cement wrapped in powder particle surface that can destroy the concrete system and affect the compactness and permeability of concrete. In order to ensure hydraulic anti-permeability of concrete with low amount of cementitious materials, the stone powder content of mechanism sand should be controlled in the range of 9%~15%.

### *3.4. Influence of stone powder on frost resistance of concrete*

The influence of stone powder content with 0%, 6%, 12% and 18% on frost resistance of C30 concrete with low amount of cementitious materials is researched. Test results are shown in Table 5.

**Table 5.** Influence of stone powder on frost resistance of concrete.

| Sample No. | Stone powder (%) | Relative dynamic elastic modulus (%) | | | Frost resistance grade |
|---|---|---|---|---|---|
| | | 50 | 100 | 150 | |
| JP-1 | 0 | 99.2 | 75.9 | 50.9 | F100 |
| JP-3 | 6 | 99.1 | 85.9 | 68.9 | >F150 |
| JP-5 | 12 | 99.0 | 94.2 | 75.6 | >F150 |
| JP-7 | 18 | 99.0 | 86.1 | 61.3 | >F150 |

IOP Conf. Series: Materials Science and Engineering **371** (2018) 012021 doi:10.1088/1757-899X/371/1/012021

In our-in this particular study we have some things good and some things bad, but overall, this-these would relatively good with the addition of stone powder in the mechanism sand. As the stone powder content of mechanism sand increased, the frost resistance of concrete is aggrandized first and then decreased. When the stone powder content is 12%, the frost resistance of concrete is best. Continue to improve the stone powder content of mechanism sand, the frost resistance of concrete becomes worse. More holes and poor frost resistance will be generated of C30 concrete with few or no stone powder. Gradually improve the stone powder content can make gradation more reasonable and more compact of concrete, and react with some substance in cement and promote the structure of concrete more compact stone powder, as a certain activity. But the filling effect will be weakened with too much stone powder in the mechanism sand, and more water absorbed that will undermine whole graded system and limit the hydration of cement and mineral admixture to reduce the frost resistance of concrete.

## 4. Conclusion

Footnotes should be avoided whenever possible. If required they should be used only for brief notes that do not fit conveniently into the text.

- The influence of stone powder of high quality mechanism sand on properties of C30 concrete is studied. The results show that the proper amount of stone powder could improve the gradation of concrete system, increase liquidity, improve the overall performance of concrete and make the structure more compact. When the stone powder content of mechanism sand is 9%~15%, the concrete has better working performance and mechanical performance;
- The stone powder of mechanism sand is good for compact and hydraulic anti-permeability of concrete. The optimal hydraulic anti-permeability properties of concrete will be reached as the stone powder content of mechanism sand is 12%;
- The frost resistance and strength of concrete can be increased with the proper amount of stone powder, and the best properties of concrete react as the amount of stone powder in mechanism sand is 12%.

## 5. References

[1] Zhao, S., & Yu, J. (2016). Effect of mechanism sand quality on concrete quality. Construction Technology.
Zhao Shemin and Yu Jun 2016 Effect of Mechanism Sand Quality on Concrete Quality (China: Construction Technology) pp 577-579

[2] Cai, J. W., Bei-Xing, L. I., Zhou, M. K., & Xiao-Man, H. U. (2006). Effects of crusher dust on properties of low/medium strength concrete with manufactured sand. Journal of Wuhan University of Technology, 28(4), 27-30.

[3] Song, S., Bao, W., & Jin, D. (2013). Research on performance of concrete with low cementitious material content based on the high-quality aggregate. Concrete, 30(10), 157-159.

[4] Yan, Y., Wang, Y. L., & Zhou, M. K. (2007). Effects of the amf contents of mfa on performances of the c30 pump concrete. Journal of Wuhan University of Technology.

[5] Wang, J., Zhou, M., Tusheng, H. E., Guoju, K. E., & Gong, C. (2008). Effects of stone dust on resistance to chloride ion permeation and resistance to freezing of manufactured sand concrete. Journal of the Chinese Ceramic Society, 36(4), 482-486.

IOP Conf. Series: Materials Science and Engineering **371** (2018) 012022 doi:10.1088/1757-899X/371/1/012022

# Effect of Shrinkage Reducing Admixture on the Strength and Shrinkage of Alkali Activated Cementitious Mortar

**C V Nguyen[1,2,3], P S Mangat[2] and G Jones[3]**

1Faculty of Civil Engineering, DaNang University of Science and Technology, 54 Nguyen Luong Bang, Danang, Vietnam
2Centre for Infrastructure Management, Materials and Engineering Research Institute Sheffield Hallam University, Howard Street, Sheffield, S1 1WB, UK
3C-Probe Systems Ltd, UK Unit 2, Sherdley Road Industrial Estate, Wharton Street, St Helens, WA9 5AA, UK
E-mail: chinhx1a@gmail.com

**Abstract.** The effect of a shrinkage reducing admixture (SRA) on the mechanical properties and drying shrinkage of a proprietary alkali activated cementitous material (AACM) was investigated. Five AACM mortar mixes were prepared. SRA replaced the liquid activator at a dosage of 0%, 1%, 2%, 4% and 7%. The liquid/binder ratio was 0.38. The samples were cured in water. The flexural and compressive strengths of all mixes were determined at 1day, 2 days, 7 days and 28 days and the drying shrinkage was determined up to 80 days. The results show that SRA reduced the shrinkage of AACM mortar by up to 69% after 80 days at 7% dosage. However, SRA also reduced the compressive and flexural strengths of AACM mortars. Increasing dosages of SRA reduced the compressive and flexural strength while recording less shrinkage. Regardless of the dosage of SRA, a unique relationship exists between flexural strength and compressive strength, which correlates with the data of previous research. The AACM mortar maintained strength of over 43MPa (greater than 75% of the control mix at 0% SRA dosage) at 7% SRA dosage, which is classified as high strength in accordance with British Standard PAS 8820:2016.

## 1. Introduction

Alkali activated cementitious material is developing significantly as a global sustainable material for the construction industry. Alkali activation is the chemical reaction between a solid aluminosilicate precursor and an alkaline activator which produces hardened products [1, 2]. The main precursors used for producing alkali activated materials are industrial by products which are high aluminates such as granulated blast furnace slag, fly ash and natural clays (metakaolin) [3].

The drying shrinkage is a key factor causing the early age cracking of mortar. There are many causes of shrinkage such as materials constituents, curing methods, etc. Previous research has revealed that the characteristics of the hydrated calcium silicate gel and pore size distribution have a direct effect on drying shrinkage [4, 5]. Kutti [6] stated that the alkali activation of slag produced two main hydration products, a C–S–H gel and a Si-rich gel. The latter consists of higher uncombined water content that is eliminated during the drying process, causing substantial shrinkage and, therefore, microcracking.

The shrinkage can be reduced by using low liquid activator/slag ratio, low activator concentration or high aggregate/slag ratios [7]. There are also some possible solutions to reduce the shrinkage such as using of polypropylene fiber [8] or admixtures [9]. The effect of shrinkage reducing admixtures (SRA) has been studied in Portland cement systems [10-14], where they decrease capillary stress in pore water, which in turn reduces cement paste shrinkage. However, the effect of SRA on the mechanical properties and dry shrinkage of alkali activated slag has not been fully investigated. This paper reports the investigation on the effect of a shrinkage reducing admixture on the strength and shrinkage of a proprietary AACM developed by Sheffield Hallam University [15] and produced under licence by C-Probe Systems Ltd, UK.

## 2. Experimental Programme

The test programme was planned according to the British Standard for AACMs, PAS 8820:2016 [16].

### *2.1 Material and mix composition*

Five AACM mortar mixes were prepared. The binder /sand ratios were kept constantly for all mixes. The mix proportions were 1:1:0.38 for AACM binder: sand: liquid respectively. The liquid is defined as the total of shrinkage reducing admixture (SRA) and liquid activator. Five different dosages of SRA (0%, 1%, 2%, 4% and 7% by weight of binder) replaced the liquid activator, so that the liquid/ binder ratio remained constantly for all mortar mixes. The mix proportions are presented in Table 1.

**Table 1.** Mixture proportion of AACM mortar

| Sample ID | AACM binder (kg) | Sand (kg) | SRA (kg) |
|---|---|---|---|
| 0%SRA | 5.5 | 5.5 | 0 |
| 1%SRA | 5.5 | 5.5 | 0.055 |
| 2%SRA | 5.5 | 5.5 | 0.110 |
| 4%SRA | 5.5 | 5.5 | 0.220 |
| 7%SRA | 5.5 | 5.5 | 0.385 |

### *2.2 Sample preparation and curing*

The samples of dimensions 40x40x160 mm were cast in steel moulds. After casting, all samples in the moulds were covered by plastic sheets and left in the laboratory ($20^oC$, RH 65%) for 24 hours. Then, all the samples were demoulded and cured in water at $20^oC$. The shrinkage samples were cured in water for 7 days before recording the first reading (datum) as detailed in section 2.3.

### *2.3 Flexural and compressive strength*

The flexural and compressive strengths were determined in accordance with BS EN 196-1: 2005 [17]. The three point bending test method was used to determine the flexural strengths. The two halves of the broken prisms from the flexural strength test were used to determine the compressive strengths of mortar. The strength measurements of AACM mortar were made at 1, 2, 7 and 28 days age. Each result of flexural strength and compressive strength was the average value of three specimens and six specimens respectively.

### *2.4 Drying shrinkage*

The drying shrinkages of different compositions of AACM mortar mixes (Table 1) were determined according to BS ISO 1920-8:2009: Testing of concrete - Part 8: Determination of drying shrinkage of concrete for samples prepared in the field or in the laboratory [18]. Prism specimens of 160mm length and cross section 40x40 mm were cast and demoulded after 24 hours. Two demec points were attached on each face of the prism at a gauge length of 100 mm to measure the distance (strain) between the demec points with an extensometer. The samples were then cured in water at $20^oC$. The samples were removed from water at 7 days after casting, dried with a cloth and the first (datum) strain reading was taken with a demec extensometer. The shrinkage specimens were then cured in air at $20^oC$, 65% RH. Subsequent shrinkage readings were taken with the extensometer up to 80 days. The shrinkage strain and time relationships were plotted.

## 3. Results and discussion

### *3.1. Flexural and compressive strength*

The flexural strength and flexural strength activity index of all samples are shown in Table 2 and plotted in Figure 1. The flexural strength activity index is defined as the ratio (in percent) of flexural strength of the SRA replacement sample to the corresponding control sample (0%SRA). Similarly, the compressive strength and compressive strength activity index of all mixes are shown in Table 2 and plotted in Figure 2. The compressive strength activity index is defined as the ratio (in percent) of compressive strength of the SRA replacement sample to the corresponding control sample (0%SRA).

**Table 2.** Flexural strength and compressive strength of AACM mortars

| Sample ID | Flexural strength (MPa) (Flexural strength activity index) | | | | Compressive strength (MPa) (Compressive strength activity index) | | | |
|---|---|---|---|---|---|---|---|---|
| | 1 day | 2 days | 7days | 28 days | 1 day | 2 days | 7 days | 28 days |
| 0%SRA | 3.72 (100) | 5.36 (100) | 7.73 (100) | 8.25 (100) | 20.79 (100) | 28.64 (100) | 44.68 (100) | 57.90 (100) |
| 1%SRA | 3.00 (80) | 4.79 (89) | 6.92 (90) | 7.63 (92) | 17.76 (85) | 25.49 (89) | 40.36 (90) | 50.62 (87) |
| 2%SRA | 3.28 (88) | 4.70 (88) | 7.35 (95) | 7.98 (97) | 16.04 (77) | 26.13 (91) | 36.96 (83) | 52.52 (91) |
| 4%SRA | 3.76 (101) | 5.74 (107) | 7.27 (94) | 7.31 (89) | 19.13 (92) | 27.21 (95) | 37.46 (84) | 47.03 (81) |
| 7%SRA | 3.80 (102) | 5.07 (95) | 5.86 (76) | 6.96 (84) | 18.11 (87) | 21.09 (74) | 32.59 (73) | 43.20 (75) |

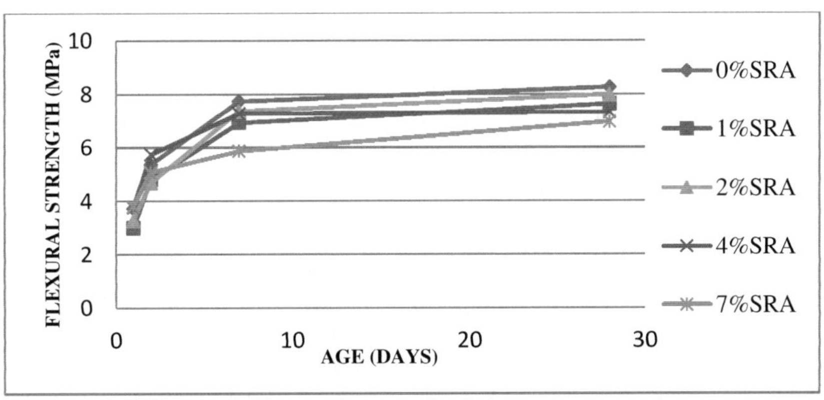


**Figure 1.** Flexural strength of the mortar samples.

Figure 1 shows that at early age the flexural strength of AACM mortar is more than 3MPa at 1 day and more than 4.7MPa at 2 days regardless of the dosage of SRA replacement of liquid activator. At 28 days age the flexural strength of AACM mortar is about 7MPa regardless of the dosage of SRA replacement. Table 2 shows that at 1 day the flexural strength of 1%SRA and 2%SRA samples reduced compared with the control sample (0%SRA). The flexural strength activity index values of 1%SRA and 2%SRA are 80% and 88% respectively. However the flexural strengths of 4%SRA and 7%SRA increased slightly compared with the control samples (0%SRA). The flexural strength activity index values of 4%SRA and 7%SRA are 101% and 102% respectively. Similar trends occur with flexural strengths at 2 days age. The flexural strength activity indexes of 1%SRA, 2%SRA and 4%SRA are 89%, 88% and 107% respectively, however the flexural strength activity index of 7%SRA reduced to 95%. Although the 1%SRA and 2%SRA reduced early age flexural strengths, these were still maintained at over 80% of the control samples (0%SRA). At 28 days age, reduction in flexural strength was recorded for all mortar samples with SRA replacement. The flexural strength activity index values of 1%, 2%, 4% and 7%SRA are 92%, 97%, 89% and 84% respectively. It shows that at later age (28 days), an increasing dosage of SRA replacement causes more reduction in flexural strength, although these flexural strengths are still maintained above 84% of the control sample even with 7% of SRA addition.

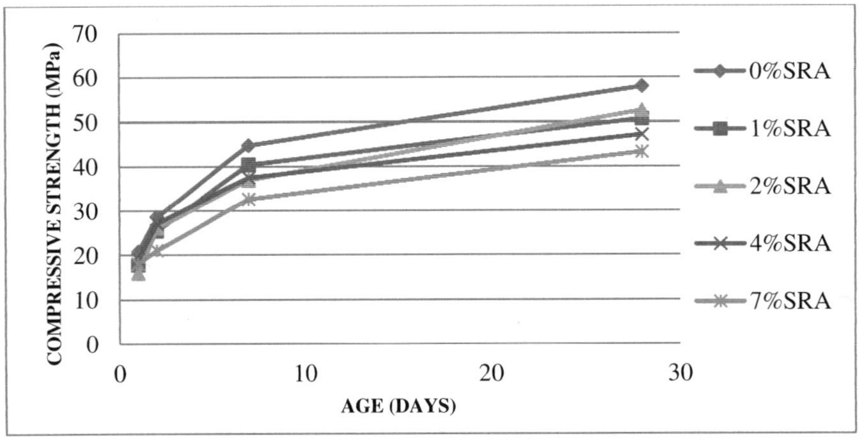


**Figure 2.** Compressive strength of the mortar samples.

Figure 2 and Table 2 show that at early age (1 day) the compressive strengths of all AACM mortar mixes reduced with increasing SRA content, similar to the case of flexural strength. The compressive strength activity index values of 1%, 2%, 4% and 7%SRA mixes are 85%, 77%, 92% and 87% respectively. Similar to the flexural strength, the compressive strengths of 4% and 7%SRA mixes were higher than that of 1% and 2%SRA mixes at 1 day. At 2 days the compressive strengths of SRA replacement samples reduced compared with the control mix (0%SRA), the 7%SRA mix showing the most reduction with the compressive strength activity index being 74% compared with 89%, 91% and 95% for 1%, 2% and 4%SRA mixes respectively. At later age of 28 days, the compressive strength again reduced compared with the control sample (0%SRA), giving the compressive strength activity index values of 87%, 91%, 81% and 75% for 1%, 2%, 4% and 7%SRA respectively. However, the compressive strength of all SRA mixes still remained greater than 75% of the control samples at 28 days. Regardless of the dosage of SRA the compressive strengths of AACM mortar are more than 16MPa, 21MPa, 32MPa and 43MPa at 1, 2, 7 and 28 days respectively, classifying as 42.5R in accordance with PAS 8820:2016 [17]. The compressive strengths of AACM mortar without SRA gained a high value of nearly 58MPa at 28 days age.

### *3.2. Relationship between flexural and compressive strengths*

The relationship between flexural and compressive strength is plotted in Figure 3 for the experimental data of this investigation including summary data of all AACM mortar samples with and without SRA and the results of other researcher [19-21] whose mixes did not contain a shrinkage reducing admixture (SRA). The ratio of flexural strength and compressive strengths of all AACM mortar samples is in the range of 0.14 to 0.24 with the higher ratios occurring at higher dosage of SRA addition. It means that when SRA was added to the mix, the reduction in compressive strength appears to be higher than the reduction in flexural strength. A power relation of summary data with and without SRA of the form $f_f = 0.398 f_c^{0.77}$ fits all data with a coefficient of correlation of $R^2$=0.99. Where $f_f$ is the flexural strength in MPa; $f_c$ is the compressive strength in MPa; and R is the correlation coefficient of the equation.

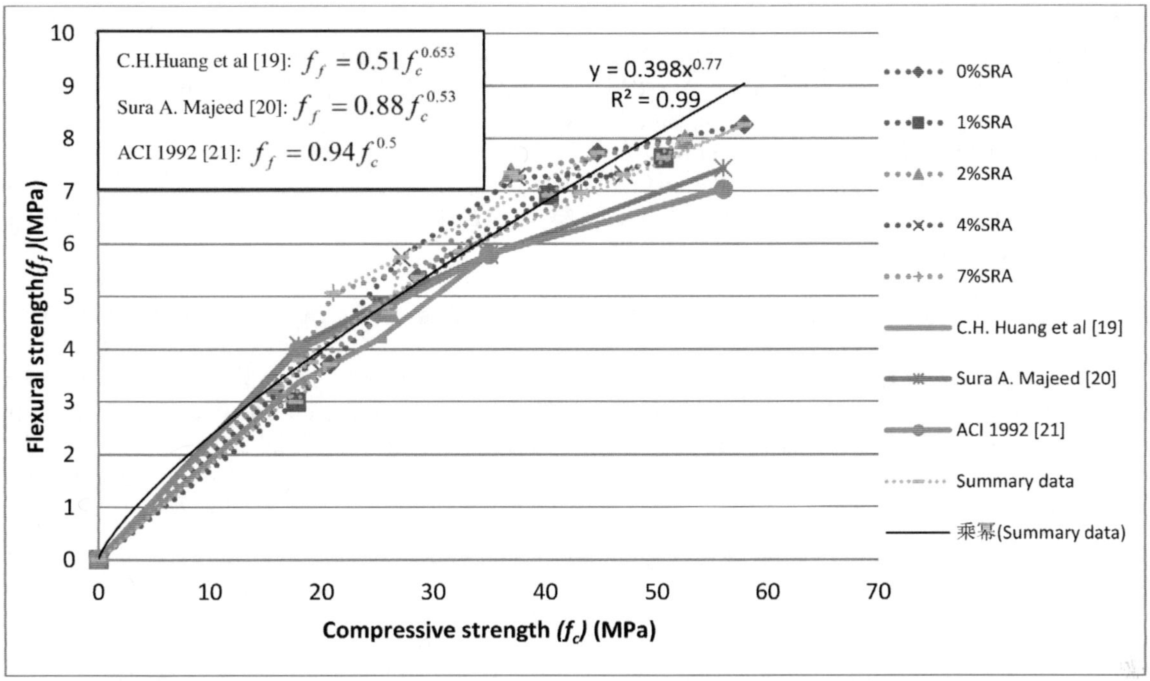


**Figure 3.** Relationship between flexural and compressive strength of AACM mortar.

### *3.3. Drying shrinkage*

The drying shrinkage of all AACM mortar mixes with and without SRA is presented in Figure 4. Each result presented is the average value of three specimens with 6 readings. It is clear that SRA reduced drying shrinkage with increasing dosage of SRA. At 80 days the drying shrinkage of all mortar mixes were 2194, 2079, 1835, 1274 and 679 microstrain for 0%, 1%, 2%, 4%, 7%SRA respectively. The shrinkage of AACM mortar reduced by 5.2%, 16.4%, 41.9%, and 69.1% when 1%, 2%, 4% and 7%SRA (by weight of binder) were used to replace the liquid activator.

At 28 days, the drying shrinkages of all mortar samples with and without SRA are 1390, 1302, 956, 680 and 330 micro strain representing reductions of 6.3%; 31%, 51%, and 76% with 1%, 2%, 4% and 7% replacement by SRA. In comparison, the flexural and compressive strengths of SRA samples at 28 days were reduced by less than 15% and 25% respectively. It can be said that depending on the purpose of AACM application, we can select the reasonable dosage of SRA so that meeting the required strengths and reducing of shrinkage preventing surface cracking. It was also observed that after 80 days, the surfaces of all AACM samples were still in very good condition without any sign of cracking on the surface.

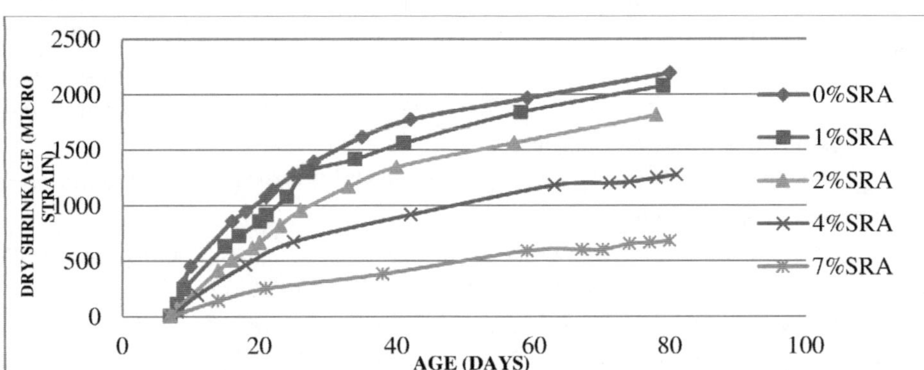


**Figure 4.** Shrinkage of AACM mortar.

IOP Conf. Series: Materials Science and Engineering **371** (2018) 012022 doi:10.1088/1757-899X/371/1/012022

## 4. Conclusions

Based on the results reported in this paper, the following conclusions can be made:

- The shrinkage reducing admixture (SRA) significantly decreases the drying shrinkage of AACM mortar while still maintaining its strength.
- The drying shrinkage is reduced by 76% at 28 days and 69% at 80 days with 7% of SRA (by weight of binder) addition to the AACM mortar mixes.
- The surfaces of AACM samples (with and without SRA) were still in very good condition without any signs of cracking after 80 days.
- SRA reduces both the flexural and compressive strength of AACM mortar. These reductions at 28 days are less than 15% and 25% for flexural and compressive strength respectively with 7% of SRA content.
- Regardless of the dosage of SRA, the flexural and compressive strengths of AACM mortar exceed 7MPa and 44MPa respectively at 28 days. These values are considered as high strength and classified as 42.5R according to British standard PAS 8820:2016.
- The flexural strength ($f_f$) and compressive strength ($f_c$) of AACM mortar are related by the following equation $f_f = 0.398 f_c^{0.77}$ with a coefficient of correlation of $R^2$=0.99.
- The microstructure of AACM mortar samples (with and without SRA) should be investigated in future researches.

## 5. References

[1] Shi C, Krivenko P, Roy D 2006 *Alkali- Activated Cements and Concretes, Abingdon* UK: Taylor & Francis.

[2] Mangat PS, Lambert P 2016 Sustainability of alkali-activated cementitious materials and geopolymers *Sustainability of construction materials*, ed Jamal M. Khatib, Woodhead Publishing Series in Civil and Structural Engineering 70

[3] Duxson P, Provis J 2008 Designing precursors for geopolymer cements *Journal of the American Ceramic Society* 91 pp 3864-3869

[4] Wittmann F H 1982 Creep and shrinkage mechanisms *Creep and Shrinkage in Concrete Structures*, ed Bazant ZP, Witmann FH, Wiley Chichester pp129–161.

[5] Young J F 1988 Physical mechanisms and their mathematical descriptions *Mathematical Modelling of Creep and Shrinkage of Concrete*, ed Bazant ZP, Wiley Chichester pp 63–98

[6] Kutti T 1992 Hydration products of alkali-activated slag, *Proc. Int. Conf. on the Chemistry of Cement* (New Delhi) vol 4 pp 468–474

[7] Collins F, Sanjayan JG 2000 Effect of pore size distribution on drying shrinking of alkali-activated slag concrete *Cem. Concr. Res* 30

[8] Puertas F, Amat T, Vázquez T 2000 Behaviour of alkaline cement mortars reinforced with acrylic

[9] Bakharev T, Sanjayan JG, Chen Y B 2000 Effect of admixtures on properties of alkali-activated

[10] Berke NS, Dallaire MP, Hicks MC, Kerkar A 1997 New Developments in Shrinkage ReducingAdmixtures, *Proc. Int. Conf. on Superplasticizers and Other Chemical Admixtures in Concrete*. Supplementary Papers. Ed. Malhotra pp 971–998

[11] Mora J, Aguado A, Gettu R 2003 The influence of shrinkage-reducing admixtures on plastic shrinkage, *Mater. Constr.* 53 pp71–80

[12] Shah SP, Karaguler ME, Sarigaphuti M 1992 Effects of shrinkage-reducing admixtures on restrained shrinkage cracking of concrete *ACI Materials Journal* 89

[13] Bentz D, Geiker M, Hansen K 2001 Shrinkage-reducing admixtures and early age desiccation in cement pastes and mortars. Cem and Con Research 31 pp1075–1085

[14] Gettu R, Roncero J, Martin M, Agull´o A 2002 Experimental study of concretes incorporating shrinkage reducing admixtures. *Report C-4098/1* (Universitat Polit´ecnica de Catalunya, Barcelona, March)
[15] Mangat PS 2009 Liquid Granite: Building Material of the Future Unveiled *Science Daily,* Sheffield Hallam University, UK pp2016-2017
[16] British Standard 2016 'PAS 8820:2016: Construction materials- Alkali activated cementitious material and concrete- Specification
[17] British Standard 2005 BS EN 196-1:2005: Method of testing cement- Part 1 Determination of strength- Compressive strength'
[18] British Standard 2009 BS ISO 1920-8:2009: Testing of concrete -- Part 8: Determination of drying shrinkage of concrete for samples prepared in the field or in the laboratory
[19] Huang CH, Lin SK, Chang CS, Chen HJ 2013 Mix proportions and mechanical properties of concrete containing very high-volume of class F fly ash *Const and Build Mater* 46 pp71-78
[20] Sura A M 2009 Predicting The Relationship Between The Modulus Of Rupture and compressive strength of mortar *Al-Rafidain Engineering* vol 17 (5)
[21] ACI 363 1992 State-of-the art report on high strength concrete.

**Acknowledgements**
The authors gratefully acknowledge the support provided by Innovate UK, C-Probe Systems Ltd, UK, Sheffield Hallam University, UK and Danang University of Science and Technology, Vietnam.

IOP Conf. Series: Materials Science and Engineering **371** (2018) 012023 doi:10.1088/1757-899X/371/1/012023

# Development of the Structural Uses of Date Palm Rachis for Cheap Wide Span Construction for Rural Communities in the Middle East

**E A Darwish[1,4], Y Mansour[2], H I Elmously[3]**
1 Teaching Associate, Department of Architecture, Faculty of Engineering, Ain Shams University, Cairo, Egypt.
2 Professor, Department of Architecture, Faculty of Engineering, Ain Shams University, Cairo, Egypt.
3 Professor, Department of Design and Production, Faculty of Engineering, Ain Shams University, Cairo, Egypt.
4 E-mail: eman.atef@eng.asu.edu.eg

**Abstract.** Date palm rachis is one of the most abundant agricultural wastes in Egypt that is being used traditionally in making furniture and sheathing by rural communities in Egypt. This familiarity attracted previous researchers to develop the use of date palm rachis for cheap construction in order to introduce palm rachis as a local building material which contributes to empowering rural communities in Egypt. Most of the wide-span date palm rachis based previous studies lacked simplicity and flexibility that is required to encourage communities to exploit the structural abilities of palm rachis. This paper aims to exploit the philosophy and potentials of palm rachis by combining traditional concepts with modern techniques in order to increase the available covered spans, which would promote the potentials of Date Palm Rachis. This aim is served by introducing a new concept of using palm rachis for simple wide span construction utilizing arched truss Form-Active structural systems. This primary design is built in an experimental scaled model to investigate the durability and details of the suggested system. The findings encourage researchers and local builders to investigate the potentials of date palm rachis as a cheap and local building material for rural communities in Egypt.

## 1. Introduction

The demand for shaded areas is increasing rapidly under the circumstances of Global Warming that make activities in exposed areas almost impossible. However, the choice of conventional steel trusses is not always available for poor rural communities under the circumstances of the high inflation rate, which increases the cost of the expensive imported materials.

Therefore, natural local materials are now the new trend to be used as a substitute for the expensive conventional building materials [1]. Natural local materials enjoy the privilege that most rural communities around the world have a strong technical heritage in using them in a wide range of inherited techniques, from making furniture and ornaments, to building their own houses [1]. This heritage qualifies natural local materials such as timber and bamboo to replace conventional shades and steel trusses that are used to cover wide spans to host flexible activities such as markets and community centres for poor rural communities [1].

Date Palm Rachis ranks the first among the amounts of the annual pruning residues of Date Palm Trees in Egypt. This gave Date Palm Rachis a huge part of the craftwork such as ornaments, chairs, fences and simple sheathing in rural communities in the western oases and Nile Delta as illustrated in Figures 1 and 2 [3].

IOP Conf. Series: Materials Science and Engineering **371** (2018) 012023 doi:10.1088/1757-899X/371/1/012023

**Figure 1.** Ornamented partition from Date Palm Rachis in Egypt.

**Figure 2.** Roofing of Date Palm rachis for roofing in Egypt.

This undeniable familiarity between Date Palm Rachis and the rural communities in the Middle East encouraged several previous researches to investigate the potentials of Date Palm Rachis as durable and versatile independent structural elements [2].

## 2. Techniques of utilizing date palm rachis in construction

### *2.1 Bundling technique*

Collecting thin elements into bundles has been a common way to store and move agricultural residues. Moreover, shaping those bundles with ropes within a fixed distance along the bundles adds stability and durability to the bundles, that they can be used for construction instead of depending entirely on wooden columns. This technique created the vault in Mudhif in Iraq as illustrated in Figure 3 [4]. Reeds are shaped into 70cm diameter bundles that are planted in the soil on each side of the proposed vault, and then the bundles are bent in order to tie the ends of every bundle with its counterpart on the other side [5]. The bundle purlins are assembled with robes every four inches, and then they are forced through the compressed arches using friction to add stability to the structure and unite the arches into one structure, as illustrated in Figure 4 [5]. The span of the result vault is about 6-13 m, and the vault itself is fast and easy to erect and move, as it depends only on the reeds bundles [4].

**Figure 3.** Mudhif Structure [4].

**Figure 4.** Inserting the purlins to unite the arches [5].

This technique inspired [6] to try the same system using Date Palm Rachis. Unlike reeds, Date Palm Rachis is not hollow and has higher density, which makes bundles with 70cm diameter made of Date Palm Rachis very heavy to erect. Therefore, Date Palm Rachis elements were shaped to make long bundles of 20 cm diameter. One end of each bundle was planted along one side of the proposed vault, and then each bundle is bent where the other end is to be planted along the other side to form an arch, as illustrated in Figure 5. However, palm trunks were added at distance 3.25 m along every arch to support the arches and decrease the deformation as illustrated in Figure 6. Hence, the actual span decreased to only 3.25m, which is less than what is needed to cover flexible and various activities [6].

**Figure 5.** The intermediate columns and sheathing mats [6].

**Figure 6.** The intermediate palm trunks [6].

*2.2 Truss and space truss*

Lately, [7] began the first attempt to use Date Palm Rachis for wide span construction using Space truss structural system without depending on palm trunks or wooden columns. A geodesic dome was designed and analysed to cover a diameter of 13m. Due to the shortage of resources, Ref. [7] built an experimental space truss to cover an area of 3x3 m to test the joinery that was proposed in the geodesic dome.

The whole date palm rachis elements were shaped to 10x10mm strips [7]. The strips are gathered to make 1m bundled squared members of cross section 10cm2 [7]. The space truss was a 4-faced triangular truss system with its diagonal length equals to one meter [7]. The joints of the structure were designed as a steel plate with 2 U-shaped high strength bolts holding down the members to the steel plate, as illustrated in Figure 7 [7].

**Figure 7.** The U-shaped bolt connection in the experimental Space Truss [7].

However, the proposed structure suffers of the high cost of the joinery due to the high complexity of the non-standard joints. In addition, the difference in the coefficient of expansion between metal and Date Palm Rachis decreases the durability and stability of the joints.

## 3. Objectives and methodology

Based on the previous literature review, it can be concluded that most of the previous researches utilized Date Palm Rachis to act as elements among a specific structural system, regardless of the nature of the material or the traditional techniques of building using the material.

This paper aims to introduce a new understanding of the nature of Date Palm Rachis, in order to employ this understanding to propose a structural concept that is based on a fusion between the traditional techniques and the works of the previous researchers. This structural concept should be characterized with the following:

- Cheap components: consists mostly of whole Date Palm Rachis with minimum dependence on manufacturing.
- Simple joints: utilizes minimalist connections that depend on friction between the elements with the minimum use of ropes or intermediate elements.

The objectives are served by the following methodology:

- Studying the nature of the material through personal experience building a simple shade.
- Proposing the structural technique via analysis and a scaled physical model in order to study the practicality of the details corresponding with the nature of the material. This defines the limitations of the proposed concept for further researches.

## 4. Discussion

*4.1 Nature of date palm rachis*

Date Palm Rachis, illustrated in Figures 8 and 9, ranks the first in the amount of the pruning residues of Date Palm Trees Phoenix Dactylifera in Egypt, with the annual rate of 105 ton/year, which equals approximately 30% of the total annual amounts [8.] Each mature Palm tree produces 10 to 26 rachises annually with total number that ranges from 100 to 125, with only half of them photosynthetic-active while the rest are required to prune [9]. Each date bunch is carried by 8 rachises, where each rachis is naturally curved to support the weight of 1-1.5 kg. This weight gives Date Palm Rachis its natural high strength at bending, where the value of the Modulus of Elasticity of bending is approximately 102 N/mm2 [10]. Depending on the age, the surrounding environment and species, the length of the members, which vary between 3-6m [9].

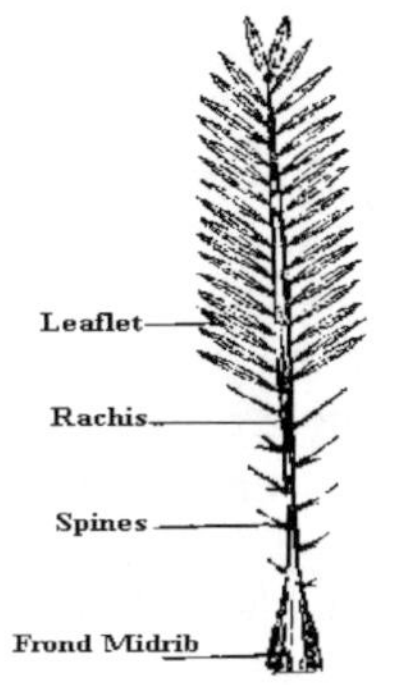


**Figure 8.** Anatomy of Date Palm Leaf [9].

**Figure 9.** Dried and leafless Date Palm Rachis.

The density of the fiber bundles in the rachis peaks right at the periphery layer which conceals the fibers inside and decreases at the core zone [11]. Previous measurements showed that the density in the periphery layer equals 1.14gm/cm3, with Young Modulus of 15.4 kN/mm2, while the density decreases in the core zone to be 0.8gm/cm3, with Young's Modulus of 4.9 kN/mm2 [8]. Most of the traditional techniques use whole rachises to make use of its high strength and high resistance to humidity, dryness and direct sunlight [2].

*4.2 Structural concept for date palm rachis*

*4.2.1 Structural deformation of date palm rachis.* The Bundling technique was used by the authors and a group of local crate makers to build a 2x3 shade in less than 4 hours in a Qayat village in Menya Governorate in Upper Egypt. The technique was found to be very simple, cheap and fast. However, the shade suffered major deflection in the 5cm diameter beams and buckling in the 10cm diameter columns over the small spans of the shade, in addition to the entire dependence on ropes for the joints, which decreased the overall stability and increased the threat of failure under the force of wind.

It can also be realized that the natural bent shape of Date Palm Rachis contradicts post-beam structural concept, which causes the instantaneous deformation and buckling of the structure, as illustrated in Figure 10. Moreover, the perpendicular connection of the bundles depended entirely on ropes to hold the bundles which do not intersect at any point, but rather overlap within the space where the ropes are tied repeatedly to secure the connections.

**Figure 10.** Deformations in the shade in Qayat.

In addition, the connections are clearly unstable and ineffective for homogenous load distribution. This obvious deformation did not occur in the bundles shaped as arches that were discussed previously by [6]. The arches fitted the natural curved shape of the bundles, decreasing the straight members that face the threat of buckling in in the structure. In addition, the natural mild curvature of the bundles made the formed arches consistent and spontaneous with homogeneous overlaps that are fastened using ropes over adequate distances, which required less complicated connections along the length of the arches. Hence, the structural behaviour of Date Palm Rachis is better distributed within systems that respect their natural curved shape.

*4.2.2 Compressive form active elements: concept of arches.* It is clear that the simplicity and workability of the arched bundles concept are higher than those of the space truss systems. However, the Date Palm Rachis space truss in the geodesic dome theoretically provided wider spans without depending on intermediate columns [6]. Therefore, the arched bundles concept requires improvement to increase the covered span while sustaining maximum practicality and simple connections.
The main concept of arches is to turn all the internal forces and bending moment to compression, as much as possible, unlike a planar slab which contains internal forces of tension and compression from moments [12]. Single span arches without moment-resistant special bases with coplanar supports and uniform load distribution, conform all the internal forces to compression, following a parabolic shape. Parabolic arches are more significant when using building materials that are subject naturally to bending and shear, than when using stiffer materials such as wood, steel and reinforced concrete [12].

*4.2.3 Semi form-active structures.* In the Mudhif structure, the 70 cm diameter parabolic arches were used to cover the span of 13m [4]. As previously stated, Ref. [6] added intermediate columns to replace the large cross section area. Therefore, the idea of semi Form-Active Structures emerges. A Semi Form-Active structure is used when it is needed to cover a long span using more efficient members than post-beam structures while using light members with less weight without using fabrics or tent like fully form-active structures [13]. This can be realized in Bridge engineering which utilizes more efficient shapes in cross-sections and longitudinal profiles instead of simple Post-Beam structures for long spans [13].

*4.2.4 Proposed structure: arched space truss.* The use of parabolic arches in the previous examples confirms the significance of parabolic arches in shaping the chords of the bridge space trusses. Hence, using parabolic arches that contain only compression will exert compression on any purlins or additional members that are inserted perpendicularly into the arches. Thus, the threat of disassembly due to tension is minimized, where friction is theoretically large enough to hold the members together without the dependence on complicated steel connection like in the space truss concept, or using robes to wrap the joint repeatedly like in the local shade experiment discussed previously.
Accordingly, arched space trusses solve the challenges of covering long spans while remaining light for shades. The proposed structure is triangular arched space truss, as the most simple space truss configuration [14], where the three parabolic chords are assumed to be made of bundles of vertically dried Date Palm Rachis with a diameter of 20cm as in [6]. The bracings are assumed to be regular

gable to simplify the structure, with bundles of horizontally dried Date Palm Rachis with a diameter of 10cm diameter similar to the columns in the Qayat shade that is discussed previously (Figure 10). The basic span chosen is 8m, and the height of the chords depends on the parabolic equation. The bundles are stacked with ropes on moderate intervals along the bundles. The connections depend on the friction of the compression of inserting the endings of the bracing members inside the continuous bundles of the parabolic truss chords. The proposed design is illustrated in the Figures 11 and 12.

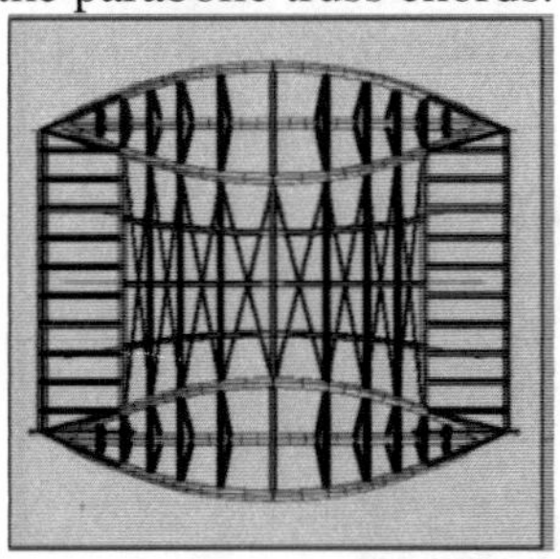

**Figure 11.** Plan of the proposed system.

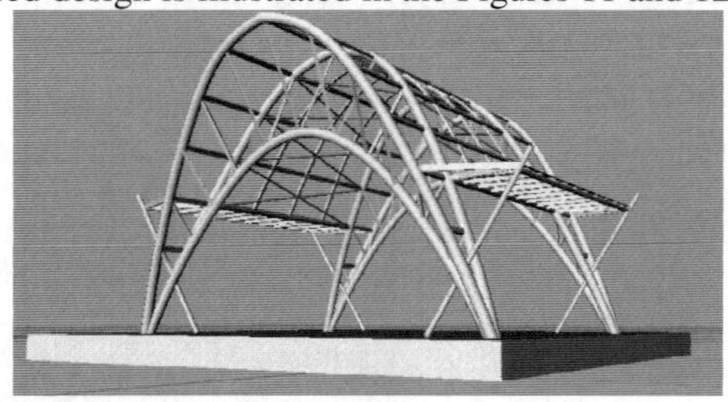

**Figure 12.** Perspective of the proposed structure.

4.3 Procedures of the physical model

Physical models help to address the consecutive challenges of the construction process. A scaled physical model was built to study the details of the structure system in terms of the bundles assembly, bracings and joints. Dried whole rachises were sliced to smaller strips, where every strip imitates 4 actual rachises. The physical model was constructed with a scale of 1:20 due to the limitation of available space and time. A short brief about this model was published in a previous paper by the authors [15]. Following are the details and analysis of the model:

*4.3.1 Dimensions of the structure.* The main chords were assumed to be 20cm diameter according to [6], the intermediate bracings were assumed to be 10cm diameter according to the Qayat shade that was discussed previously in Figure 10. The span was designed to be 8m with parabolic chords.

*4.3.2 Shaping the bundles.* Firstly, the bundles were assembled to make the arches. The assumed cross section was made for the entire length of each bundle which is 15m, where each bundle consisted of three overlapping groups to imitate the real length of the rachises which is limited to 6m. However, the bundles upon manual bending were fractured in the middle where the curvature is the most acute along the arches, as illustrated in Figure 13. Therefore, the bundles were modified to be more flexible at bending. The cross section of the middle group was reduced as it can be assumed that the compression loads of the own weight in the peak of the parabolic arches are the least. Those new bundles were flexible at bending and no fracture occurred, as illustrated in Figure 14.

A plywood base was used and drilled at the bases of the arches to create the target span, and by adjusting the length of the arch to be planted; the parabolic shape could be achieved. The bundles were bent and inserted into the drilled holes. In addition, an inclined post was used to assess the versatility of the system. That inclined post was held only by the friction between the three arches that share the same bases, as illustrated in Figure 15.

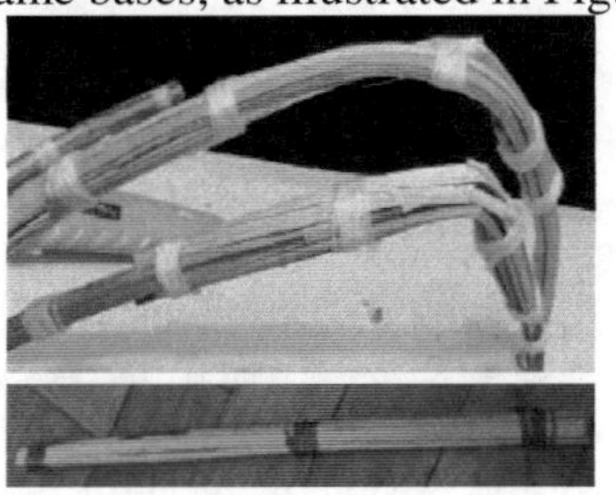

**Figure 13.** Fractured full cross section arch.

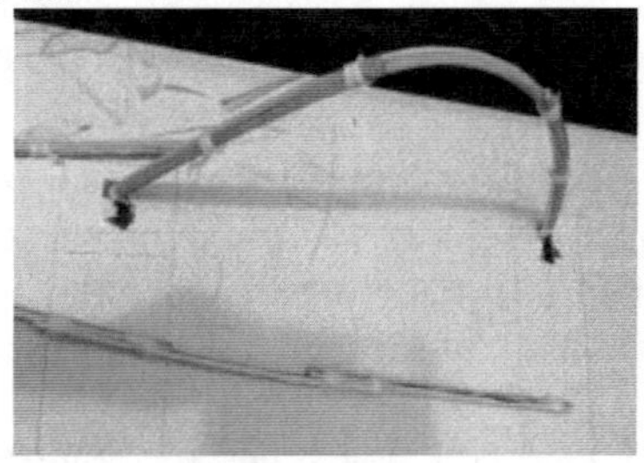

**Figure 14.** The decreased middle cross section arch.

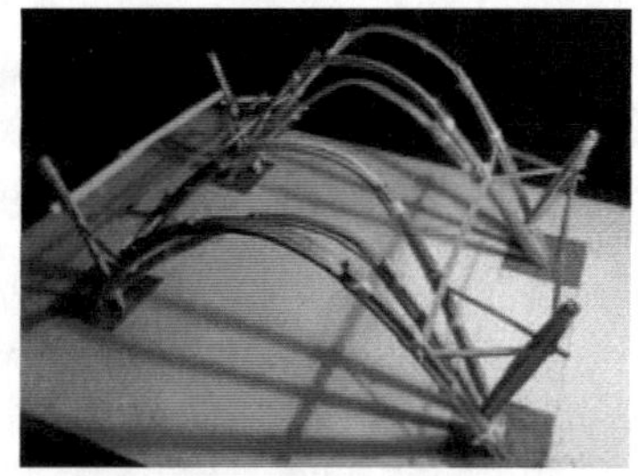

**Figure 15.** The structure before bracings.

IOP Conf. Series: Materials Science and Engineering **371** (2018) 012023 doi:10.1088/1757-899X/371/1/012023

*4.3.3 Inserting the bracings and purlins.* The bracings at the bottom of the arches were forced into the bundles, and the bracings at the peaks of the arches were rested on the arches, and then the rest of bundles that were subtracted from the middle group were put and tied tightly to enclose the inserted bracing using the pressure of friction in the parabolic bundles. In addition, the purlins between the two arched space trusses were also inserted through the bundles, as illustrated in Figure 16. Those purlins were assumed to only carry the loads of the wind and their own weight, as the cover is out of the scope of the research. After completion, the physical model was found to be stable and the bracings and the chords were well attached.

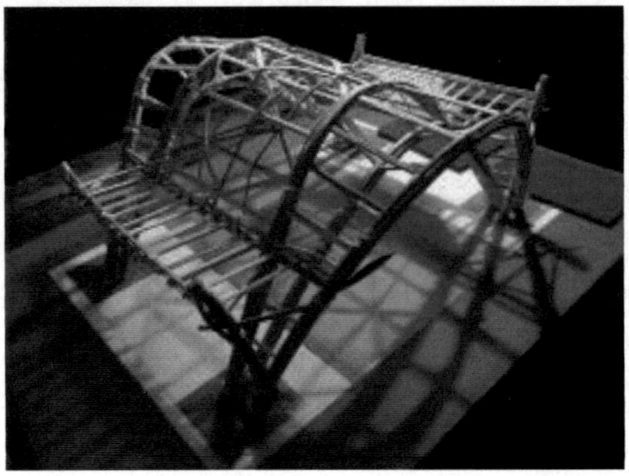

**Figure 16.** The physical model after completion.

*4.4 Analysis of the physical model*

Under the available conditions, the scale of physical model was insufficient to undergo physical load testing. However, it was sufficient to understand the behaviour of the elements of the structure. The inspection of the stability of the physical model during the building process proved that the core strength of this system depends on the bracings that unite the 3 arched bundles to act as a triangular arched space truss. The single arches only stood up due to the fixed foundation which resisted the thrust occurring at the ends of the arches. However, the least pressure would break the reduced middle section of the bundles.

On the other hand, the unified action of the arches increased the stability of the structure and reinforced the resistance to deformation under pressure applied by hand, especially when compared to the stability of the structure before inserting the bracings. Moreover, tying the subtracted elements of the middle section of the arches over the ends of the bracings exposed the bracing joints to the compression inside the parabolic arches, which ensured the decrease of the threat of failure under applied pressures by hand on all the joints, especially in the lower parts of the arches, where the compression loads are the maximum. However, the purlins between the two arched space trusses suffered deformations because of their low weight and the lack of connections to fix the purlins along the spacing between the 2 space trusses. Moreover, the long beams on the side suffered minor deflection over the spacing of 6m.

## 5. Conclusion

Date Palm Rachis is a natural agricultural pruning residue that is abundant in Egypt and the Middle East. Date Palm Rachis enjoys familiarity in the rural communities as a material for hand crafts and simple sheathing for shades and simple structures. This paper analysed traditional techniques of utilizing Date Palm Rachis and recent researches that exploit the potentials of Date Palm Rachis for modern structure systems. In order to introduce the fusion concept that merges the simplicity of the traditional techniques and the practicality and structural integrity of the modern systems, the paper proposed arched space truss, where the form active parabolic arches offer compressive friction that sustain stable connections between the arched chords and the bracings without foreign metal joints that were used in the previous Date Palm Rachis trusses.

A physical scaled model was constructed to study the full behaviour of the structure. The scaled physical model proved to be durable and structurally stable and the threat of disassembly of the bracings were minimized. However, the intermediate purlins between the two arched space trusses were found to need more improvement in order to reduce their deformation to carry the weight of the sheathing. In addition, the side post and beam shades suffered deformations, which confirm that form active systems are more suitable than post and beam structures for Date Palm Rachis construction.
The results of this paper call for additional studies to build the model in adequate scales for physical testing of the structural performance. Moreover, additional studies are needed to define the most suitable light covering technique and the fixation of the covering to the intermediate purlins. Also, natural builders and social workers are invited to study the uses of the structure in order to be introduced to the rural communities as fast, simple and cheap shades that can be built for rural communities in Egypt and the Middle East.

## 6. References

[1] Kennedy J 2004 *Building Without Borders: Sustainable Construction for the Global Village*, (Ontario: New Society Publishers) pp30-45
[2] Elmously H 2001 The Industrial Use of The Date Palm Residues An Eloquent Example Of Sustainable Development *Proc. 2nd International Conference on Date Palms (Cairo),* (Dubai: United Arab Emirates University) pp 866-886
[3] Ahmed R M 2014 Lessons Learnt from the Vernacular Architecture of Bedouins in Siwa Oasis, Egypt *Proc. The 31st Int. Sym. on Automation and Robotics in Const. and Mining* (London: ISARC).
[4] Almssad A and Almusaed, A 2015 Building materials in eco-energy houses from Iraq and Iran, *Case Studies. in Const. Mat.* Ed 2 Grantham M, ( Amsterdam: Elsevier) pp42-54.
[5] Ochsenschlager E L 2004 *Iraq's Marsh Arabs in the Garden of Eden*. (Philadelphia: University of Pennsylvania Museum of Archaeology and Anthropology)
[6] Piesik S 2012 *Arish: Palm-Leaf Architecture* Ed 2 (London: Thames & Hudson.) pp60-127
[7] Hassan N 2001 *Use of Local Raw Materials in Low Cost Roofing* Ph.D Institute of Environmental Studies & Research Ain Shams University Cairo Egypt
[8] ElMously H 2005 The Palm Fibers for the Reinforcement of Polymer Composites: Prospects and Challenges *Proc. 1st Ain Shams Int. Conf. on Enviro. Eng. (Cairo)* vol 1 (Cairo: Ain Shams University) p 14.
[9] Zaid A and De Wet P 1999 *FAO plant production and protection papers*, Chapter I pp.1-28 http://www.fao.org/docrep/006/y4360e/y4360e05.htm, last accessed Aug.2017
[10] Darwish E A, Mansour Y, Elmously H 2017 Date Palm Rachis as a local and renewable structural material for rural communities in Egypt, *Al Azhar's 14th International Conference on: Engineering, Architecture and Technology (Cairo),* In press.
[11] Megahed M, and Elmously H 1995 Anatomical Structure of Date Palm Leave's Midrib And Its Variation Across And Along The Midrib *IUFRO XX 191World Congress (Tampere)*
[12] Ambrose J and Tripney P 2012 *Building Structures* Ed 3 (London: John Wiley & Sons) pp60-65.
[13] Macdonald A 2001 *Structure and Architecture* Ed 2 (Oxford: Reed Educational and Professional Publishing) pp110-125.
[14] Lan T 2005 *Space Frame Structures, in Handbook for Structural Engineering*, Ed 2 W. Chen, (Tokyo: CRC Press) pp 24-50
[15] Mansour Y, Darwish E A and Elmously H 2017 Utilizing Date Palm Rachis for Flexible and Eco-friendly Construction in Egypt, *Proc. World Sustainable Built Environment (Hong Kong)* vol 1 (Hong Kong: HKGBC) pp 2824-33.

IOP Conf. Series: Materials Science and Engineering **371** (2018) 012024 doi:10.1088/1757-899X/371/1/012024

# Full-Scale Standard Fire Experiment and Numerical Modelling Behaviour of Non-Load Bearing Calcium Silicate Partition Drywall

**Ching-Yuan Lin[1], Wang Yinuo[2] and Ying-Ji Chuang[1]**

1 Department of Architecture, National Taiwan University of Science and Technology, Taiwan.
2 China Academy of Building Research, Beijing 100013, China.
E-mail: wyn_up@foxmail.com

**Abstract.** The study uses the non-load bearing calcium silicate partition drywall as the test specimen, with full-scale standard fire test and the CFD finite element numerical modelling simulation on 300 cm × 300 cm area specimen. The whole process of experiment fire test and the numerical modelling simulation result is highly correlated. This numerical modelling simulation can be applied to other researches on the other wall systems. This is an innovative and valuable research. Since they are highly correlated, the CFD finite element numerical modelling can be processed to work out the predicted results before the test, which can greatly save the cost of test. When the predicted results is satisfying, the standard fire test can be processed to verify the results.

## 1. Introduction

The firewalls installed in fire compartment should be possessed of flame retardant effectiveness to limit fire growth and forestalling structural failure. But whether the building component reach the fire resistance, is requires to be testing with the standard furnace fire. This approach, leading to a ranking of products as a function of the building materials and structures behaviour under a standard fire exposure is use in a lot of countries. (e.g. NFPA 251[1], ISO 834-1[2] and CNS 12514[3]). Nowadays, various countries have adjusted the specimen sizes of firewall to at least more than 3m high and 3m wide. For the trend of building engineering is towards increasing in dimension and high-rise, the conventional heavy building materials and high labour intensive methods are descending. For example, different non-load bearing drywall systems which are expected to replace the traditional thick and heavy building materials appear constantly. These systems have features of optimization constructional methods, shortening the constructional duration, various constructional methods, light materials and the quality is more stable than concrete. The surface of the non-load bearing partition drywall is boards. Generally, there are calcium silicate boards, gypsum boards, cement board and magnesium oxide boards, with light steel frames of 0.8mm~1.2mm serving as the supporting framework. Between two plates, the common rock wool of $60kg/m^3$~$100kg/m^3$ are stuffed.

There are many researchers study on the performance issues of non-load bearing partition drywall. Chuang et al. [4] had proposed influences exerted by the environmental temperature during the relevant firewall tests on the specimen temperature of the non-fire exposure side. Collier and Buchanan [5] conducted researches related to the fire rating of constitutions of partition walls. Ho and Tsai [6] proposed that the quality of boards had a great effect on the fire resistance of wall. Wang et al.

[7] once proposed that installing devices, such as embedded junction box in the wall could influence the fire resistance of the wall.

The above researches are focus on the fire resistance of non-load bearing drywall based on a standard fire test experiments. With CFD (Computational Fluid Dynamics) is increasingly applied in various engineering fields, numerical simultaneous aided researches in building field by making use of CFD is becoming widespread. For example, Do et al. [8] came up with that the thermal conductivity of porous material is mainly related to the thermal conductivity of its components and spatial arrangement of its complex structure by formula, microstructure observation and experiments.

At present, there is no study focus on comparison of the performance of calcium silicate drywall by full-scale standard fire experiment and numerical modelling, which this research will study on it. The research will use quantitative analysis of a fire resistance through a standard fire test experiment and CFD numerical modelling. Aim to explore the correlation between the full-scale standard fire test and CFD numerical modelling simulation. The test uses the standard in ISO 834-1[2] to perform on a test specimen which size is 300cm (width) ×300cm (height). Proposed the numerical model to simulate the process of transient heat transmission, compare the results of computer simulation and the test results.

## 2. Test methodology

### *2.1. Standard test furnaces*

In this research, the test equipment is applied and can conduct material testing horizontally or vertically. The standard test furnace of equipment is 300cm (width) ×300cm (height) ×240cm (depth). The furnace adopt the electronic ignition and the control system is the computer PID temperature controller. There are 8 burners in the furnace which only 4 are switched on for wall test. Ceramic wool with a thickness of 300mm and a density of 160 kg/m$^3$ was paved within the furnace. The furnace floor was composed of firebricks (size: 230mm×114mm×65mm) and gaps among firebricks were coated with refractory mortars. Such bricks were evenly allocated in nine thermocouples furnaces (Type K) (the distance from the fire exposure side of specimen was 100mm) (see figure 1), and 2 furnace pressure meters and T-type pipes (the heights of T-type pipes were respectively equal to the 150 cm and 250cm from the bottom of furnace).At the back of the furnace, there is an air outlet for exhaust air and it is connected to the outdoor chimney. All thermocouples are 10cm away from the fire testing area of specimen. The temperature in the furnace is measured by K-type thermocouple of which specification conforms to the regulation of CNS 5534 [9] that the thermocouple shall possess property above 0.75 Grade. The heat-resistance stainless steel pipe is placed in the insulated stainless steel pipe of which inner diameter is 14mm and front end is open and the hot junction of front end extrudes 25mm. All thermocouples in the furnace have been placed in the environment of which temperature is 1000℃ for 1 hour [2] before the first use to increase the sensitivity of measuring the temperature and the accuracy requirement is ±3%.The Type K thermocouples in furnaces and Type K thermocouples on the un exposure side of the specimen utilize the compensating lead wires of thermocouples to connect to the data miner.

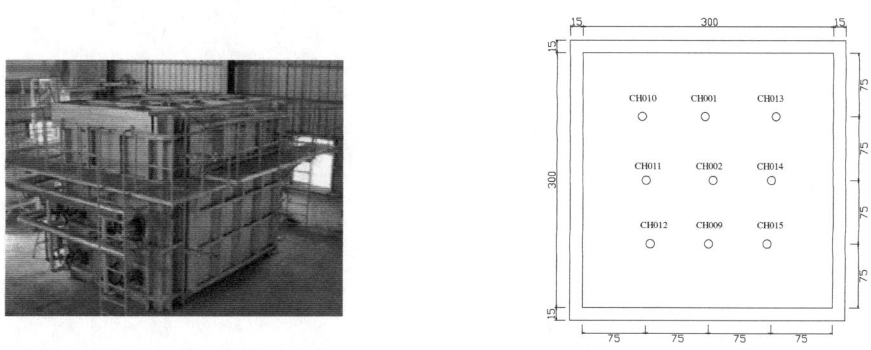


**Figure 1.** Full-scale standard fire test furnace and location of furnace thermocouples.

The data miner is made by Japanese YOKOGAWA Company. All equipment signals will firstly connect to DS 600 data recorder. And then, DS 600 will deal with the signals before transferring them to DC 100. In the end, the signals will be converted to output to the computer. In the test, the researcher heated the specimen respectively based on the heating curve of ISO 834-1 [2] standard. The relationship formula between temperature and time is shown as below:

$$T=20+345\log_{10}(8t+1) \tag{1}$$

where
T: is the average furnace temperature, in degree Celsius;
*t*: is the time, in minutes.

*2.2. Test specimen materials*

Thickness of specimen framework is 15cm and the framework welded with steel plates of 5mm thick. Its inner part is made of concretes with 210kg/cm$^2$ compressive strength. In the concretes, juxtapositions of spot welding reinforcing mesh with a diameter of 5mm and an interval of 150mm are set. The specifications conform to CNS 6919 [10]. In short, specimen frameworks and the contact surfaces of specimens shall be similar to the practical installation situations to the greatest extent. Test materials including light steel frames, rock wool, calcium silicate boards. As to the C-type internal framework, the size is 65×35×0.8mm@406mm. The size of C-type peripheral framework is 67×30×0.8mm. The size of flying shore is 38×12×0.95mm@1200mm. The thickness and density of rock wool is 50mm and 60kg/cm$^3$ respectively. The calcium silicate boards conforms to provisions of the first level, with white colour, a thickness of 9mm, a density of 1.31 g/cm$^3$. The thickness of wall being composed of all materials is 83mm. Test materials are shown as figure 2, figure 3. As the test expects to study the similarity of results of numerical modelling and test.

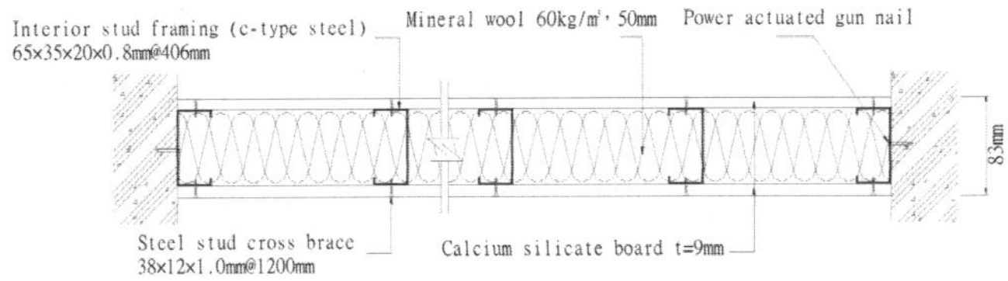


**Figure 2.** Horizontal cross section of wall.

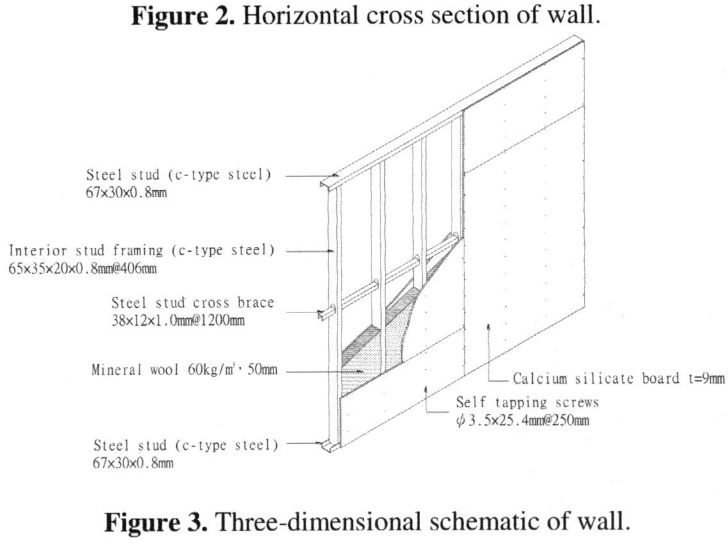


**Figure 3.** Three-dimensional schematic of wall.

*2.3. Record the temperature of the test specimen*

Test sets 8 thermocouples on the unexposed surface of the specimen, as shown in figure 4. According to the requirements of the ISO 834-1[2], it observes the temperature distribution. Because it is required to compare the similarity of results of numerical modelling and tests as well as optimize the computer model, three measuring points are set in the middle layer of wall and measuring positions are respectively in the position of 9mm, 41mm and 74mm, as shown in figure 5. The temperature measurement is recorded once every 6 seconds by computer and recorded by photos during the test.

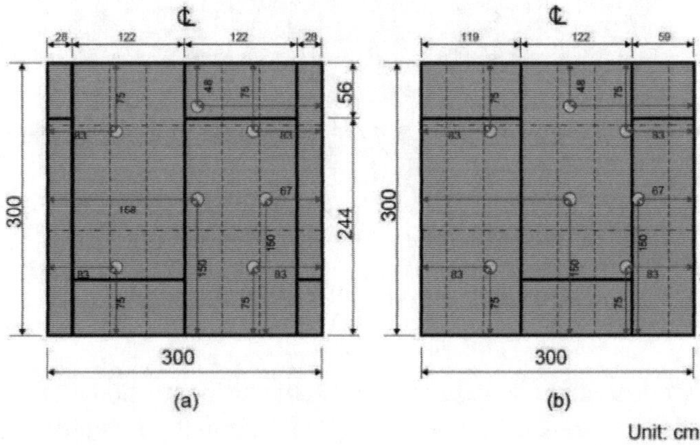


**Figure 4.** The position of test thermocouples.

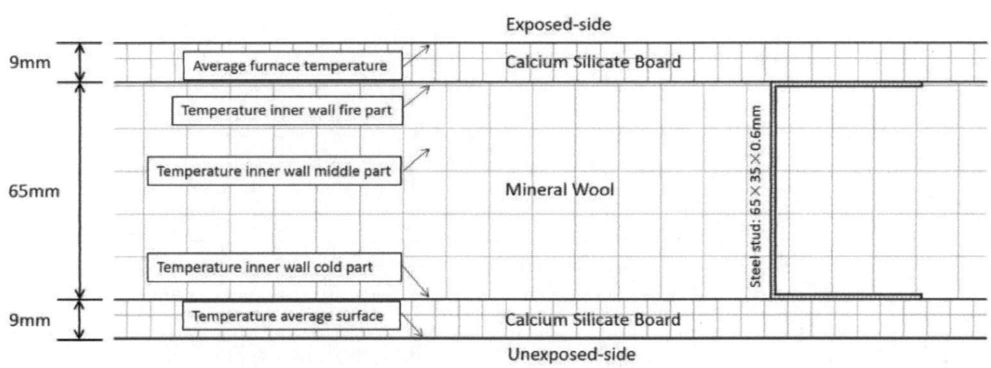


**Figure 5.** Cross-section of the wall with indication of the Thermocouple position in the Test.

## 3. Numerical modelling methodology

### *3.1. General modelling*

In this research, the numerical modelling takes the Finite Element Numerical Analysis of which principle is to divide the solution domains to several interrelated sub-domain units, assume a proper approximate solution to every unit, deduce the general satisfied conditions of the domain and work out the solution of the question. Numerical modelling is based on CFD technology for a series of computer simulation analysis, the use of software fluent to solve. It can be roughly divided into three parts: Pre-Processing, Solve and Post-Processing. Pre-Processing mainly focuses on how to build geometric model and mesh. This research needs to get the computer model which matches the standard fire test result, so the geometric model is corresponding with the standard fire test of specimen. Therefore, it's necessary to build the geometric model, and mesh it into several units. Based on the shape and size of building components, the test meshes it by hexahedron. The meshing size of calcium silicate board, steel stud and fireproof cotton are respectively 9×9×9mm, 0.6×0.6×5mm and 5×5×5mm, as shown in figure 6. Solve mainly includes how to set up the related material parameters,

IOP Conf. Series: Materials Science and Engineering **371** (2018) 012024 doi:10.1088/1757-899X/371/1/012024

set boundary conditions and select the mathematics model and calculation methods. Post-Processing aims to analyze the data solved by modelling.

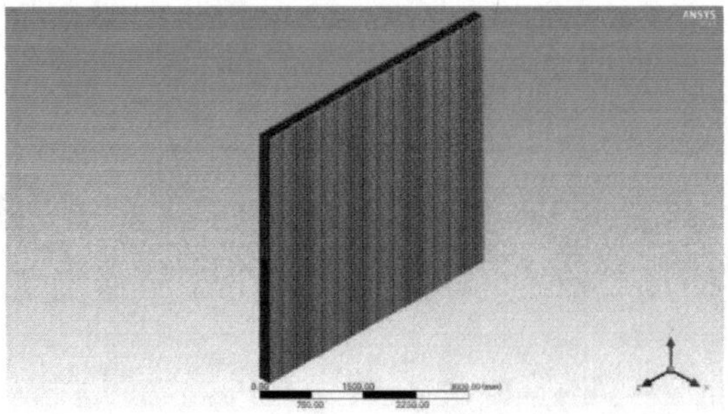

**Figure 6.** Numerical model of meshing.

*3.2. Parameter setting*

The materials used in the test include steel, calcium silicate board and rock wool. The parameters involved in modelling material are specific heat (J/kg℃), Thermal conductivity (W/m℃) and Density (kg/m$^3$). Parameters used in steel refer to the regulations of BS EN 1993 [11]. In addition, the specific heat of steel linearly increases as the temperature rising, as shown in figure 7.

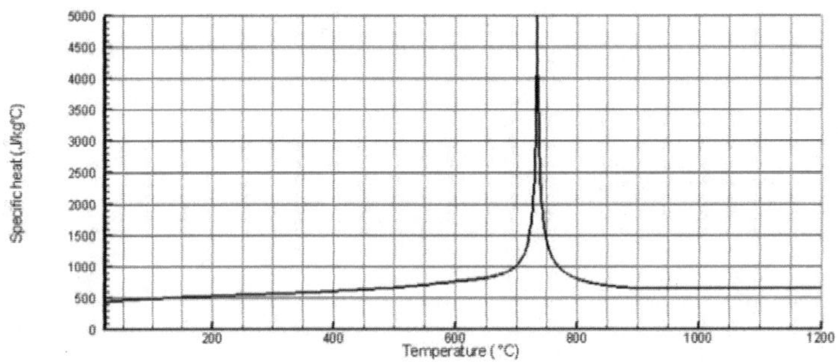


**Figure 7.** Specific heat for steel at elevated temperatures. BS EN 1993.

The thermal conductivity of steel is 53.3 W/m℃ when the temperature is 20℃ and it rises to 27.4 W/m℃ when the temperature is 800℃. However, the thermal conductivity is stable when the temperature is over 800℃. The density of steel is 7850 kg/m3. Rosanne et al. [12] performed research on the thermal performance of a series of insulation materials. According to their related data, the specific heat of calcium silicate board is 819.4J/kg℃. The regulation of GB/T 10699—1998[13] stipulates the thermal conductivities of calcium silicate board at various temperatures and other parameters. Its thermal conductivity as follow.

$$\lambda=\begin{cases}0.0564+7.786\times10^{-5}\times t+7.8571\times10^{-8}\times t^2 & (t\leq500℃)\\ 0.0937+1.67397\times10^{-10}\times t^3 & (500<t\leq800℃)\\ 0.179 & (t>800℃)\end{cases} \quad (2)$$

where

$\lambda$: Thermal conductivity (W/m℃)

$t$: temperature (℃)

In the research made by Nassif et al. [14], it presents some material property parameters related to rock wool. Combining the specifications of rock wool used in this research, the final specific heat, thermal conductivity and density of rock wool are respectively 840 J/kg℃, 0.035 W/m℃ and 60 kg/m$^3$. The specific setting is shown as Table 1.

IOP Conf. Series: Materials Science and Engineering **371** (2018) 012024 doi:10.1088/1757-899X/371/1/012024

**Table 1.** Thermal properties.

| Material | Specific heat (J/kg℃) | Thermal conductivity (W/m℃) | Density (kg/m$^3$) |
|---|---|---|---|
| Steel | According to Fig.7 | 20℃,53.3 T≥800℃,27.4 | 7850 |
| Calcium Silicate | 819.4 | According to (2) | 1350 |
| Mineral Wool | 840 | 0.035 | 60 |

## 4. Results and discussion

### *4.1. Experiment results*

The test time lasts 60 minutes. After 9 minutes of the test, trace smokes with abnormal taste burst out above the unexposed surface of specimen and the seams of framework. At this time, all temperatures of measuring points obviously tend to rise. Until 14$^{th}$ minute, the temperature of measuring point on unexposed surface tends to decline until the 35$^{th}$ minute. From the 35$^{th}$ minute to the end, the temperatures rise all the time. The temperature inner wall fire part rises rapidly after 9 minutes and it slowly rises to the end after 22 minutes. At the end of test time, the temperature of measuring point is 738.1℃. The temperature inner wall middle part rises rapidly after the test begins 9 minutes and it begins to slowly rise to the end after 38 minutes. At the end of test time, the temperature of measuring point is 487.8℃. The situation of temperature inner wall cold part is generally the same to the temperature side wall within the first 18 minutes. After that, the temperature gradually goes up to the end and the final temperature is 316.5℃ (see figure 8). At 21$^{th}$ minute, a transverse crack appears on the upward side of left board of unexposed surface and the crack extends to the center at 38$^{th}$ minute. When the test time is over, the temperature upper left center is the highest one among temperatures on unexposed surfaces, is 104.7℃ and the highest average temperature is 97.5℃ (see figure 9) which is not in excess of stipulated fire resistance in regulations of ISO 834-1[2]. After the test, the integrity of unexposed surface of specimen is still good (see figure 10). Therefore, the specimen meets the demand of 60 minutes fire resistance.

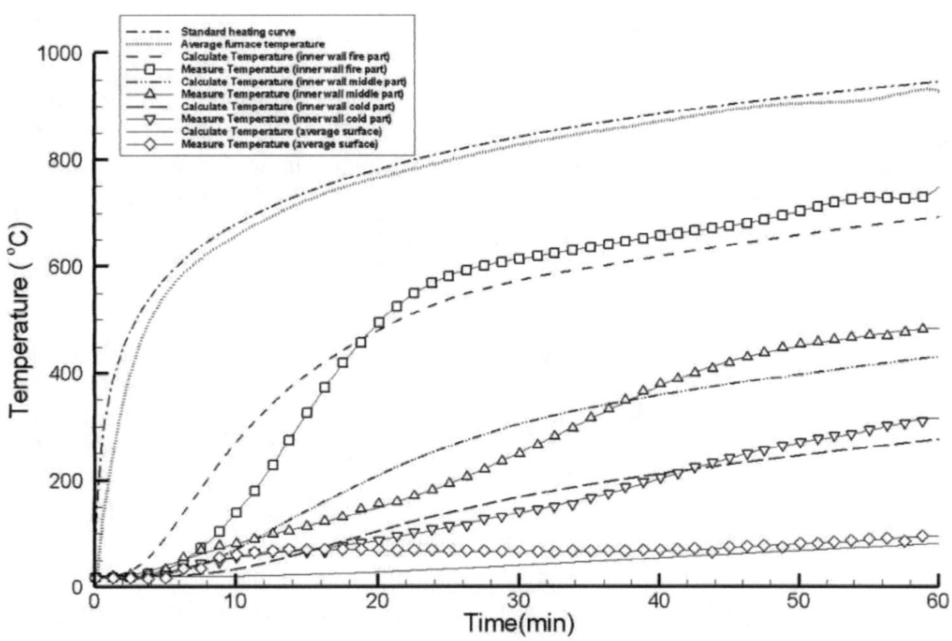


**Figure 8.** Measure temperatures compared to simulation values.

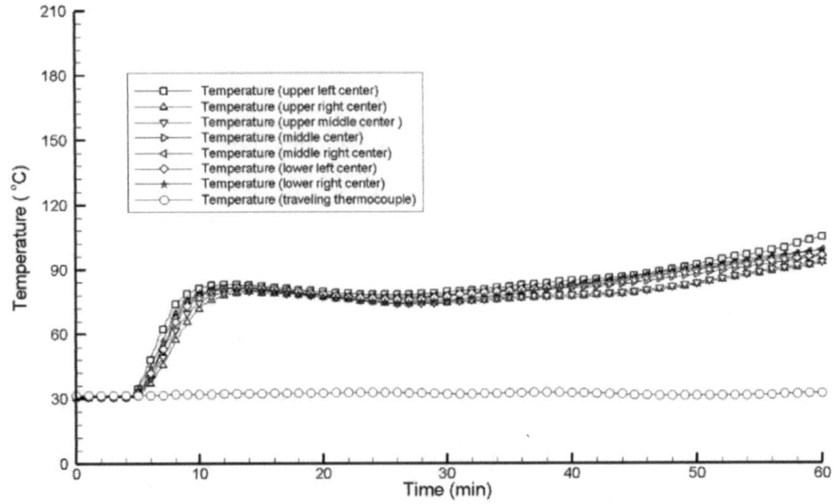


**Figure 9.** Time-temperature chart for the Test.

**Figure 10.** The result of the specimen after 60 min standard fire test.

*4.2. Comprehensive discussion*

Although the test boards have some cracks on the surface facing the fire during the experiment, the surface does not explode, and the integrity is good by visual inspection. After testing it for 60 minutes, the fire resistance meets the requirements of ISO 834-1 [2] and the 60 min of fire ratings. From the 14th to the 35th minute, the temperature shows a steady decrease, indicating that there is some moisture inside the board and the rock wool to absorb the heat. The temperature at the surface on the backside then starts going up only after the material itself is totally dried. This often occurs in firewall tests when the material is more consistent. For example, the metal sandwich wall shows such phenomenon. The metal surface does not get burned, and the insulation layer (rock wool) in between can consistently absorb heat for a while. Only when the heat reaches saturation will the temperature on the surface not facing the fire continue to go up.

The full-scale standard test specimen and the CFD finite element numerical modelling have the same location of measuring points. The results by comparing numerical modelling and standard fire test is shown as figure (see Figure 8). The heating curve of test furnace is responding to the modelling standard heating curve, shows that the standard fire test temperature conforms to the regulation of ISO 834-1[2]. In the standard fire test, the temperature inner wall fire part rises faster in the first 22 minutes of test and rises slower from 22nd minutes to 60th minutes. Similarly, although the heating curve of numerical modelling on this measuring point is relatively ideal, it can be found that the escalating trend of temperature in first 20 minutes is obviously higher than that in last 40 minutes, basically coincident with the escalating trend of this measuring point in standard fire test. In first 20 minutes, the temperature inner wall middle part rises slower, and after that it rises faster. At 21th minutes, a transverse crack appears on the upward side of left board of unexposed surface, so that causes that the temperature rises faster. Because the numerical modelling is in an ideal condition, the numerical modelling temperature on this point rises slowly all the time, and the situation that its temperature slope and temperature slope of standard fire test alternate with each other may occur. However, the overall escalating trends are highly coincident. The temperature inner wall cold part is away from the fire source, so its temperature heating curve is gentler than previous two temperatures and it is highly correlated to the numerical modelling temperature. In standard fire test, trace smokes with abnormal taste burst out above the unexposed surface of specimen and the seams of framework at 9th minute, so the average temperature on unexposed surface greatly rises but tends to rise slower, corresponding with the numerical modelling temperature. Through the comparison, it can be found that the temperature ascending curves of different measuring points of numerically predicted values are highly correlated to the data of all temperature measuring points in standard fire test in 60 minutes.

## 5. Conclusion

This is an innovative research. It successfully builds the CFD numerical models which is corresponding with the specimens, successfully predicts the effect of reinforcement by modified model parameters and verifies it in the following standard fire test. It's proved that the numerical modelling corresponding with the standard fire test is effective. Similar patterns can be applied to other researches on the wall systems and that can greatly save the cost of test. Before the test, the numerical modelling can be processed to work out the predicted results.

## 6. References

[1] NFPA 251, Standard Methods of Tests of Fire Resistance of Building Construction and Material; 2006.
[2] ISO 834-1, Fire-resistance tests-Elements of building construction: Part1: General requirements, 1999.
[3] CNS 12514, Method of fire resistance test for structural parts of building, 2010.
[4] Ying-Ji Chuang, Wu-Tai Wu, Hung-Yuan Chen, Chieh-Hsin Tang and Ching-Yuan Lin, Experimental investigation of fire wall insulation during a standard furnace fire test with different initial ambient air temperatures, Journal of Applied Fire Science, 15(1), pp. 41-55, 2006.
[5] P.C.R. Collier and A.H. Buchanan. (2002) Fire Resistance of Lightweight Timber Framed Walls, Fire Technology, Vol. 38(2), pp.125-145.
[6] M.-C. Ho and M.-J Tsai (2010), Relative importance of fire resistance performance of partition walls, Journal of Marine Science and Technology, Vol. 18(3), pp. 430-434.
[7] Wang Yinuo., Chuang, Y. J., & Lin, C. Y. (2015). The performance of calcium silicate board partition fireproof drywall assembly with junction box under fire. Advances in Materials Science and Engineering, Volume 2015, Article ID 642061, 12 pages.
[8] C.T. Do, D.P. Bentz, and P.E. Stutzman, Microstructure and Thermal Conductivity of Hydrated Calcium Silicate Board Materials, Journal of Building Physics, 31(1), pp. 55-67, 2007.
[9] CNS 5534, "Thermocuples," 1982.
[10] CNS 6919, Welded steel wire fabrics, 1999.
[11] BS EN1993-1-2. Eurocode 3: Design of steel structures—Part 1–2: general rules structural fire design.
[12] Rosanne Walker, Sara Pavía, Thermal performance of a selection of insulation materials suitable for historic buildings, Building and Environment, 94 (2015) 155-165.
[13] GB/T 10699—1998, Calcium silicate insulation, 1998.
[14] A.Y. Nassif, I. Yoshitake and A. Allam, Full-scale fire testing and numerical modelling of the transient thermo-mechanical behaviour of steel-stud gypsum board partitionwalls, Construction and Building Materials, 59(30), pp. 51-61, 2014.

**Acknowledgement**
The authors would like to thank TFPT Laboratory for technically supporting this research.

IOP Conf. Series: Materials Science and Engineering **371** (2018) 012025 doi:10.1088/1757-899X/371/1/012025

# The Effects of Carbon Fiber Reinforced Polymer Strengthening on Cylindrical Steel Storage Tanks under Bending Shear Load

**P V Nhut[1] and Y Matsumoto[2]**

1 Graduate student, Department of Architecture and Civil Engineering, Toyohashi University of Technology, Toyohashi, Japan
2 Associate Professor, Department of Architecture and Civil Engineering, Toyohashi University of Technology, Toyohashi, Japan
E-mail: p175510@edu.tut.ac.jp

**Abstract.** Large thin-walled cylindrical steel storage tanks play important roles in the development of economy and infrastructures. Storage tanks subjected to internal pressure can cause buckling modes, especially elephant foot bulge at the bottom of the tanks. The design of tanks against buckling is very necessary. In this paper, the load-carrying capacity and the failure modes of the storage tanks subjected to internal pressure under bending shear load will be calculated by the finite element analysis. The results will be compared with the design method from Architectural Institute of Japan. Moreover, this paper will also consider the effects of the use of carbon fiber reinforced polymer layers to strengthen storage tanks against buckling and the increase of the load-carrying capacity of the tanks under bending shear load.

## 1. Introduction

Large thin-walled cylindrical steel storage tanks (CSSTs) have an extremely important role in the development of economy and infrastructures. They have been widely used in fluid tanks, chemical, electric power, food engineering, etc. CSSTs are also often used as key facilities in the whole production system. The destruction of CSSTs will cause the interruption of the production process, even lead to disaster. Therefore, the evaluation the load-carrying capacity of CSSTs is very necessary. In addition, the most common failure mode of CSSTs subjected to internal pressure is buckling and capacity to buckling resist is an important evaluation item in the design of cylindrical storage tanks. Buckling is a serious failure mode for tanks, which drastically lowers their structural strength and collapses their geometries [1,3]. One of the design standards, which related to cylindrical steel tanks against buckling was introduced by the design recommendation of Architectural Institute of Japan [2]. However, the small-scale tanks were used in the experiments and the accuracy of these small-scale tanks' results with the large real tanks has not been verified. Actually, the performance of the large real tank's experiments against buckling is very difficult and costly. In this paper, the finite element analysis (FEA) will be used to analyze the large real CSSTs subjected to internal pressure and the results will be compared with small-scale ones in the design recommendation [2].
Another problem is that a large number of CSSTs have appeared increasing signs of deterioration and reducing load-carrying capacity because of corrosion or aging degradations. For achieving sustainable development of infrastructure and economy, these structures need to be repaired to improve the performance. The conventional method of repairing or strengthening these aging structures often involves bulky and heavy plates that are difficult to fix and prone to corrosion. Moreover, this method

could cause business interruption and the damage of materials. In recent years, the use of carbon fiber reinforced polymers (CFRPs) with many outstanding characteristics such as lightweight, high strength, and high corrosion resistance to increase the load capacity and ductility of steel structures proved to be an economical and reliable strengthening solution [5,6]. In this paper, the authors will consider the effectiveness of CFRP strengthening on buckling restraining of CSSTs subjected to internal pressure under bending shear load using finite element analysis. CFRP layers are unidirectional materials; therefore, they will be applied to the circumferential direction of the tanks. CSSTs strengthened by CFRP layers in this way can restrain the deformation of the tanks in the horizontal direction and limit the elephant foot bulge in the bottom of the tanks.
Three models with different geometrical parameters with two values of internal pressure will be surveyed by finite element analysis to investigate the load-carrying capacity and buckling modes of CSSTs. After that, these models with CFRP strengthening will also be simulated by FEA to consider the effects of CFRP layers on buckling protest of CSSTs under bending shear load.

## 2. Material properties of CFRP layers and steel storage tanks

### 2.1. CFRP layers
CFRP layers contain CFRP fiber layer and matrix. The volume fraction of fiber $v_f$ and of matrix $v_m$ are 50% and 50% respectively. One layer of CFRP and matrix has the thickness of 0.444 mm. The longitudinal modulus of the CFRP fiber and matrix are $E_f$ = 245000 (MPa) and $E_m$ = 4100 (MPa) respectively. These properties are calculated by typical carbon fiber properties [4]. From these parameters, the material properties of CFRP layers are determined by the composite rules. Table 1 shows the evaluated properties of CFRP layers. In the table, $E_x$ and $E_y$ are the longitudinal and transverse elastic modulus of CFRP layers, $G_{xy}$ is the shear modulus of CFRP layers. $E_x$ and $E_y$ are also elastic modulus of CFRP layers in the circumferential ($x$) and vertical ($z$) direction in FEA.

**Table 1.** The properties of CFRP layers.

| $E_x$ (MPa) | $E_y$ (MPa) | $G_{xy}$ (MPa) |
|---|---|---|
| 124550 | 8065 | 2946 |

### 2.2. Steel storage tanks
The tanks use the SS400 steel with the material properties as shown in Figure 1. The yield stress of SS400 steel is 235 (MPa). This is the lowest value of SS400 steel. The stress-strain relationship of steel used for the tanks has the tri-linear shape.

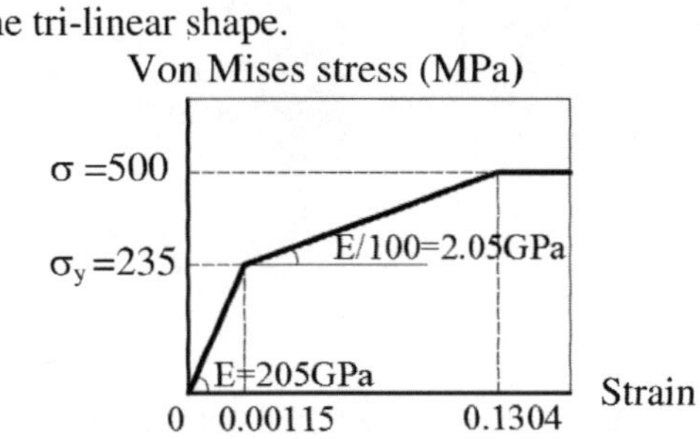


**Figure 1.** Material properties of steel.

## 3. Theory of buckling of CSSTs under bending shear load
According to design recommendation, the buckling of CSSTs subjected to internal pressure depends on the ratio of tensile hoop stress $\sigma_h$ caused by internal pressure to the yield point stress of the tank's material $\sigma_y$ ($\sigma_h/\sigma_y$). In the region below $\sigma_h/\sigma_y$ = 0.3, the buckling occurs in inextensional modes, whereas beyond $\sigma_h/\sigma_y$ = 0.3, the buckling occurs in elephant foot bulge mode. In this paper, the design of tanks against the elephant foot bulge will be focused because this kind of failure is very dangerous and remarkably affects the load-carrying capacity of CSSTs.

In the case of the ratio of $\sigma_h/\sigma_y$ over 0.3, the equivalent bending stress $\sigma_{cr}$ of cylindrical steel storage tanks is determined by equations (1) to (4). In these equations, $r$ is the inside radius of the tank, $t$ is the thickness of the tank, and $E$ is the elastic modulus of the tank's material.

$$\sigma_{cr} = \sigma_{cr0}\left(1 - \frac{\sigma_h}{\sigma_y}\right) \tag{1}$$

$$\frac{r}{t} < 0.069\frac{E}{\sigma_y} \rightarrow \sigma_{cr0} = \sigma_y \tag{2}$$

$$\frac{r}{t} > 0.807\frac{E}{\sigma_y} \rightarrow \sigma_{cr0} = \frac{0.8E}{\sqrt{3(1-\nu^2)}}\left(\frac{t}{r}\right) \tag{3}$$

$$0.069\frac{E}{\sigma_y} < \frac{r}{t} < 0.807\frac{E}{\sigma_y} \rightarrow \text{interpolating } \sigma_{cr0} \text{ from (2) and (3)} \tag{4}$$

## 4. Finite element analysis

### *4.1. Description and generation of 3D finite element models*

3D nonlinear analysis (LUSAS package) was used to analyze the failure buckling modes and the ultimate strength of cylindrical steel storage tanks subjected to internal pressure under bending shear load with two cases: without CFRP strengthening and having CFRP strengthening. The simulated models in FEA were followed to the experimental tests of Akiyama in recommendation book [2] so that the results can be compared and evaluated. The configuration of the models is described in Figure 2.

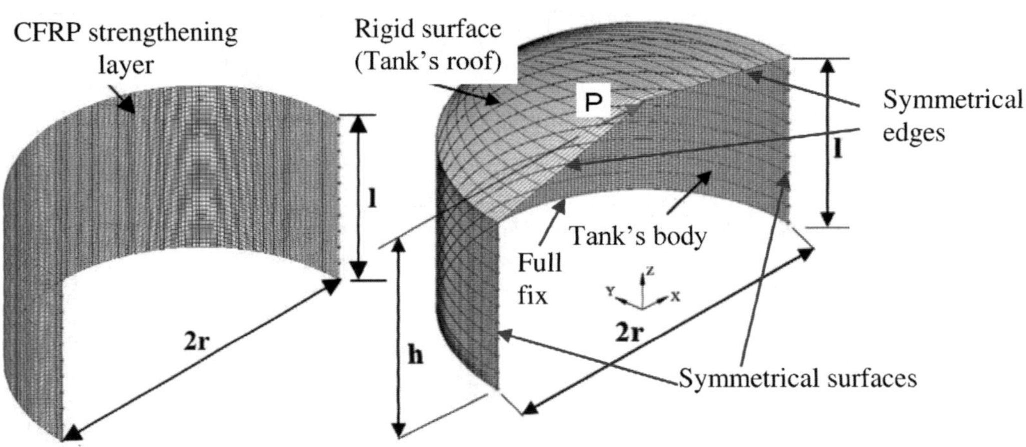


**Figure 2.** Half model in finite element analysis and boundary conditions.

The simulated tanks had two parts: the tank's wall and the tank's roof. In the cases of tanks with CFRP-strengthened layers, CFRP layers are modelled as surface elements at the outside of the tank's walls as shown in Figure 2. In this paper, three kinds of tanks were modelled with different ratios of inside radius to the thickness of the tank ($r/t$), the height of the tank to the inside radius ($l/r$), and the distance between the lower edge of tank where horizontal loading is applied to the inside radius ($h/r$) (Table 2).

**Table 2.** Three kinds of models in FEA.

| Parameter | r/t | l/r | h/r | t(mm) | l(mm) |
|---|---|---|---|---|---|
| Tank 1 | 999 | 0.5 | 0.621 | 10 | 5000 |
| Tank 2 | 749 | 0.668 | 0.828 | 10 | 5000 |
| Tank 3 | 499 | 1 | 1.24 | 10 | 5000 |

Every kind of tank subjected to internal pressure under bending shear loads was considered in three cases: without CFRP strengthening, 5-ply CFRP strengthening, and 12-ply CFRP strengthening. The internal pressure with the ratios $\sigma_h/\sigma_y$ = 0.5 and 0.7 was focused on the analysis because the main buckling failure is elephant foot bulge in these cases. In total, there were 18 models in finite element analysis.

In these models, the tank's walls were simulated with 3D quadratic solid elements and the tank's roofs were used the surface elements with rigid materials. These rigid elements can help the load transfer from the top of the tanks to the tanks easily. Mesh sizing and the number of analysis iterations play important roles in the analysis. The mesh sizing and the analysis iteration increment must small enough so that the convergence can occur. Also, small mesh sizing assures the accuracy of the analysis results and collapse behavior of the tanks. In this analysis, half models were applied for all the tanks because of symmetrical conditions. In the tank's walls, there were 170 elements in the circumferential direction, 80 elements in the height direction, and 3 elements in the thickness direction.

### *4.2. Boundary conditions*

The tanks were fixed at the bottom surface, two surfaces on the tank's walls and two lines on the tank's roofs were applied symmetrical conditions as shown in Figure 2.

### *4.3. Loading conditions*

The loads were applied to two steps. In the first step, only the internal pressure was applied to the analysis. After finishing this step, the results were fixed and then the horizontal loading was assigned. The values of internal pressure were defined according to the ratios of the tensile hoop stress $\sigma_h$ to the yield point stress of the tank's material $\sigma_y$ ($\sigma_h/\sigma_y$). Two levels of $\sigma_h/\sigma_y$ were surveyed 0.5 and 0.7. From the tensile hoop stress $\sigma_h$, the values of the internal pressure were defined by equation (5). The values of internal pressure were presented in Table 3.

$$p = \frac{\sigma_h t}{r} \tag{5}$$

**Table 3.** The values of internal pressure in three kinds of tanks.

| Internal pressure (MPa) | Tank 1 | Tank 2 | Tank 3 |
|---|---|---|---|
| $\sigma_h/\sigma_y$=0.5 | 0.11762 | 0.1569 | 0.2355 |
| $\sigma_h/\sigma_y$=0.7 | 0.16466 | 0.21963 | 0.32966 |

## 5. Buckling modes and ultimate strength of cylindrical steel storage tanks in FEA

### *5.1. Ultimate strength*

Table 4 shows the maximum horizontal loading $P_{max}$ obtained from FEA in the cases of tanks without CFRP strengthening and with 5-ply and 12-ply CFRP-strengthened layers. In the case of tanks without CFRP strengthening, the maximum horizontal loads decreased when the values of the internal pressure increased. When strengthening with CFRP layers, the maximum horizontal loads increased comparing with the cases without strengthening that having the same values of internal pressure. Moreover, the increase of maximum loading is corresponding to the increase of the number of CFRP layers.

Figure 3 and 4 present the relationships between the horizontal loading and the displacement at the top of the tank's wall for all the models. In the cases of tanks without CFRP strengthening, the loading decreased rapidly after the tanks reached the maximum load capacity. This is very dangerous for the working of the structures. However, when strengthening with CFRP layers, the loading decreased slowly and more stable. This is because the CFRP layers helped to keep the loading decrease slowly. This is very important thing for the buckling restraining of the tanks under the loading, especially seismic loading.

In this paper, the analysis will be stopped when getting one of two conditions. Firstly, the maximum displacement at the top of the tank's wall was constrained to the value of one percent of the tank's height (50mm). This is because the big displacement may affect the equipment connected to the tanks. Secondly, in the cases of strengthening with CFRP layers, the analysis results will consider the damage of CFRP layers. The analysis will be stopped when the strain in the CFRP layers reaches the breaking strain of CFRP layers in fiber direction (1.4%). This breaking strain value was calculated from the properties of CFRP layer (TORAYCA) [4].

**Table 4.** The maximum horizontal loading obtained from FEA.

| $P_{max}$ (kN) | $\sigma_h/\sigma_y=0.5$ | | | $\sigma_h/\sigma_y=0.7$ | | |
|---|---|---|---|---|---|---|
| | Tank 1 | Tank 2 | Tank 3 | Tank 1 | Tank 2 | Tank 3 |
| Without strengthening | 18226 | 13833 | 8799 | 16186 | 11417 | 6356 |
| 5-ply CFRP strengthening | 19183 | 14720 | 9871 | 18658 | 13862 | 8098 |
| 12-ply CFRP strengthening | 20091 | 15926 | 10718 | 20743 | 15903 | 10356 |

### *5.2. Buckling modes*

Figure 5 and 6 show the buckling modes and Von Mises stress of tanks without CFRP strengthening and the tanks with 5-ply CFRP-strengthened layers and 12-ply CFRP-strengthened layers in two levels of internal pressure. The main buckling failure of tanks without CFRP strengthening was elephant foot bulge. When strengthening with CFRP layers, the elephant foot bulge at the bottom of the tanks was limited and the shear buckling occurred at the tank's walls.

## 6. The correspondence of analysis results compared with theory and strengthening effects of CFRP layers on storage tanks

### *6.1. The correspondences of analysis results compared with theory in the cases of tanks without CFRP-strengthened layers*

In the cases of tanks without CFRP strengthening, the main buckling modes were elephant foot bulge. These results are good correspondence with the design recommendation.

Equations (6) and (7) show the calculation process of the equivalent bending stress $\sigma_{cr}$ of cylindrical steel storage tanks, which was determined from FEA. In these equations, $P_{max}$ is the maximum horizontal loading obtained from FEA and $A$ is the sectional area of the half tank.

$$\tau = \frac{P_{\max}}{A} \tag{6}$$

$$\sigma_{cr,FEA} = \tau \frac{h}{r} \tag{7}$$

Table 5 shows the equivalent bending stress calculated from FEA compared with these values calculated by theoretical equations from the design recommendation. There were large differences in the results between FEA and theory in the case of the ratio $\sigma_h/\sigma_y$ = 0.5. With the ratio $\sigma_h/\sigma_y$ = 0.7, the results were more accurate with the theoretical equations. This is because the tanks, which were used in the experimental tests in design recommendation were made of very thin shells, whereas, the thickness of the tanks in EFA is much larger. This causes the differences in the behavior under buckling modes. In the case of the ratio $\sigma_h/\sigma_y$ =0.7, the analysis results were approximately same with the results of experimental tests.

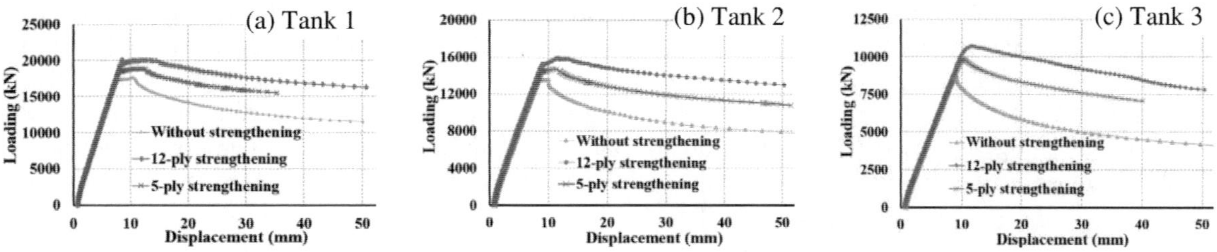


**Figure 3.** The displacement - loading relationship in the case of $\sigma_h/\sigma_y$ =0.5.

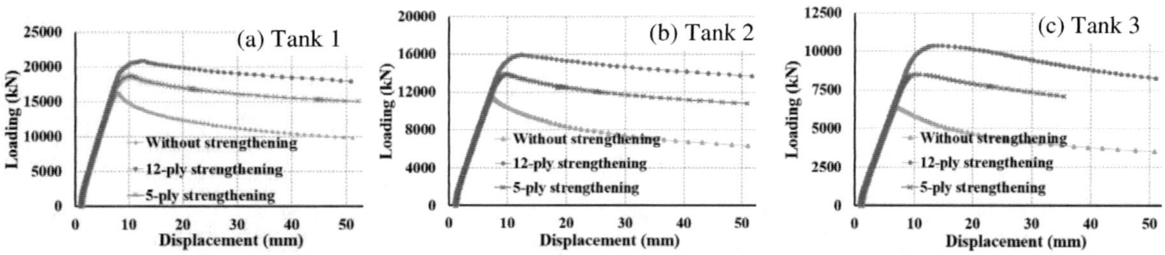


**Figure 4.** The displacement - loading relationship in the case of $\sigma_h/\sigma_y$ =0.7.

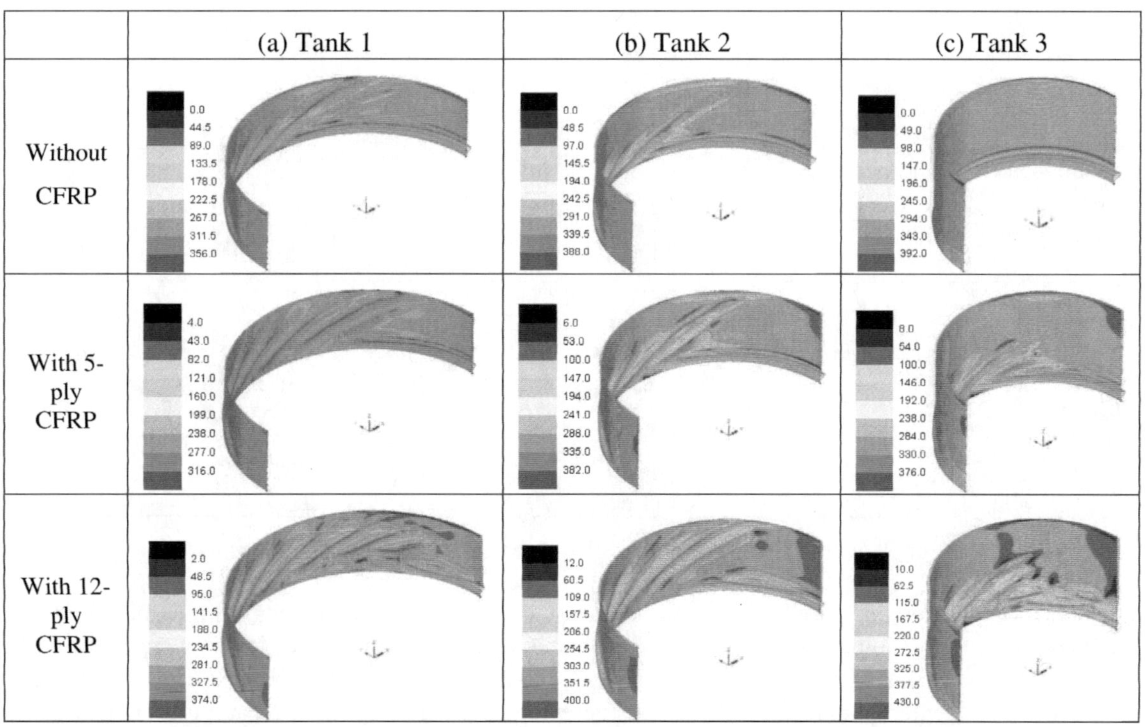


**Figure 5.** The Von Mises stress of tanks in the case of $\sigma_h/\sigma_y$ =0.5. Unit: MPa.

### *6.2. The strengthening effects of CFRP layers on storage tanks*

From Figure 3 and 4, it is clear that when strengthening the tanks with CFRP layers, the ultimate strength of the tanks was increased and the increased level of ultimate strength was corresponding with the increase of the numbers of CFRP - strengthened layers. On the other hand, the strengthening would be more effective if the ratio *r/t* was smaller and the ratio *l/r* was larger as shown in Figure 7.

IOP Conf. Series: Materials Science and Engineering **371** (2018) 012025 doi:10.1088/1757-899X/371/1/012025

*6.3. The impact of CFRP strengthening on the decrease of internal pressure inside the storage tanks*

When strengthening storage tanks by CFRP layers, the value of internal pressure inside the tanks will be decreased. This is the reason for the increase of the ultimate strength in the tanks with CFRP-strengthened layers. The decrease value of internal pressure can be calculated from the decrease of tensile hoop stress $\sigma_h$ as shown from equations (8) and (9). In these equations, $p$ is internal pressure, $t$ is the thickness of the tank's wall, $E_{FRP}$ is the elastic modulus of CFRP layers, $t_F$ is the thickness of CFRP layers, and $\sigma_{h,eq}$ is the equivalent tensile hoop stress.

$$\sigma_h = \frac{p.r}{t} \tag{8}$$

$$\sigma_{h,eq} = \frac{p.r}{t + \frac{E_{FRP}}{E} t_F} \tag{9}$$

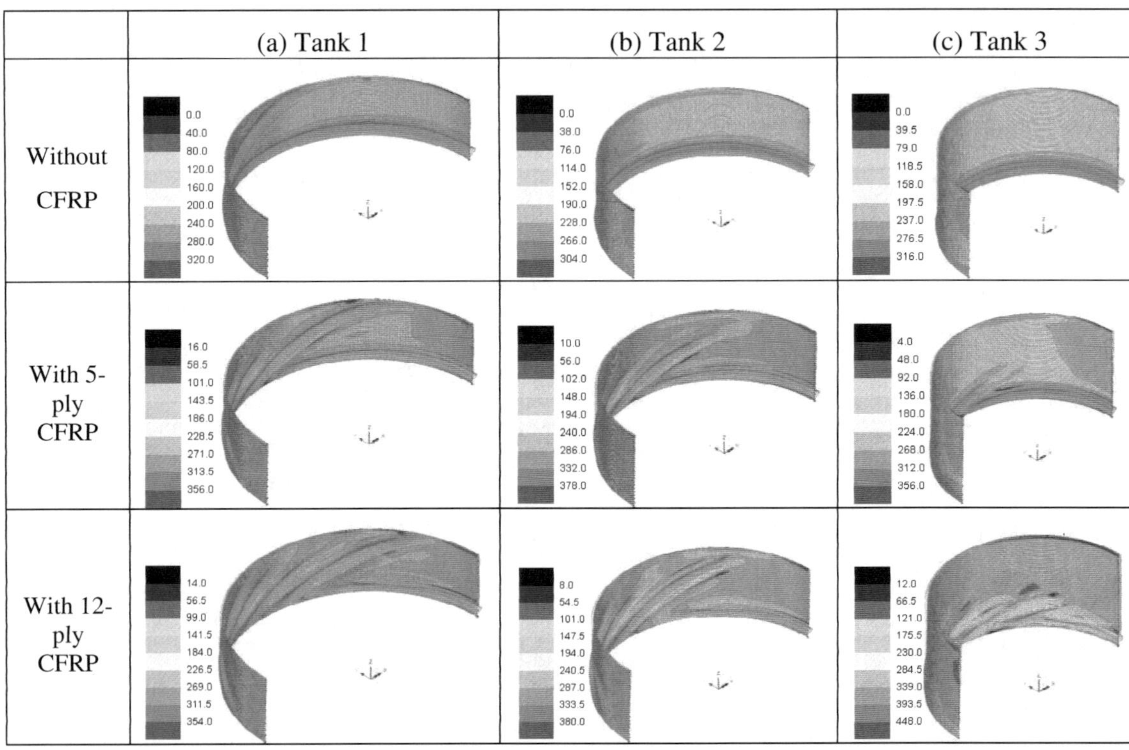


**Figure 6.** The Von Mises stress of tanks in the case of $\sigma_h/\sigma_y$ =0.7. Unit: MPa.

Figure 8 shows the decrease of the value of internal pressure inside the tanks when strengthening with CFRP layers. In finite element analysis, the same tanks with and without CFRP layers were be applied the same value of internal pressure according to the ratio $\sigma_h/\sigma_y$. After analysis completion, the decreased value of internal pressure will be considered and presented by equivalent tensile hoop stress $\sigma_{h,eq}$. From Figure 8, it can be seen that when strengthening with CFRP layers, the ratios $\sigma_h/\sigma_y$ of 0.7 and 0.5 were decreased to ratios $\sigma_{h,eq}/\sigma_y$ of 0.53 and 0.38 respectively. Along with the decrease, the loading was increased with various levels depending on the kinds of models. The increase of the loading was higher with the models which have the ratio $r/t$ smaller and the ratio $l/r$ larger. Besides, the strengthening effects of CFRP layers in the case of $\sigma_h/\sigma_y$=0.7 was remarkably higher compared with the case of $\sigma_h/\sigma_y$ =0.5. If using the theoretical equations from design recommendation, the increase of loading when the values of the ratio $\sigma_h/\sigma_y$ decrease from 0.7 to 0.53 and 0.5 to 0.38 would be higher compared with the results in FEA

**Table 5.** The values of equivalent bending stress from FEA compared with theoretical equations.

| | $\sigma_h/\sigma_y=0.5$ | | | $\sigma_h/\sigma_y=0.7$ | | |
|---|---|---|---|---|---|---|
| Equivalent bending stress (KN/cm$^2$) | Tank 1 | Tank 2 | Tank 3 | Tank 1 | Tank 2 | Tank 3 |
| $P_{max}$ (KN) | 18226 | 13833 | 8799 | 16186 | 11417 | 6356 |
| Theory | 4.968 | 6.625 | 10.52 | 2.981 | 3.975 | 6.132 |
| FEA | 3.604 | 4.865 | 6.97 | 3.201 | 4.016 | 5.035 |
| Variation (%) | -27.46% | -26.57% | -33.75% | 7.38% | 1.03% | -17.89% |

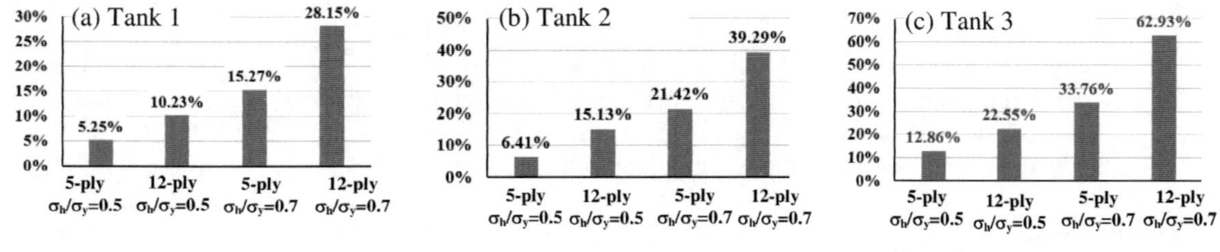


**Figure 7.** The strengthening effects of CFRP layer on the increase the ultimate strength of tanks.

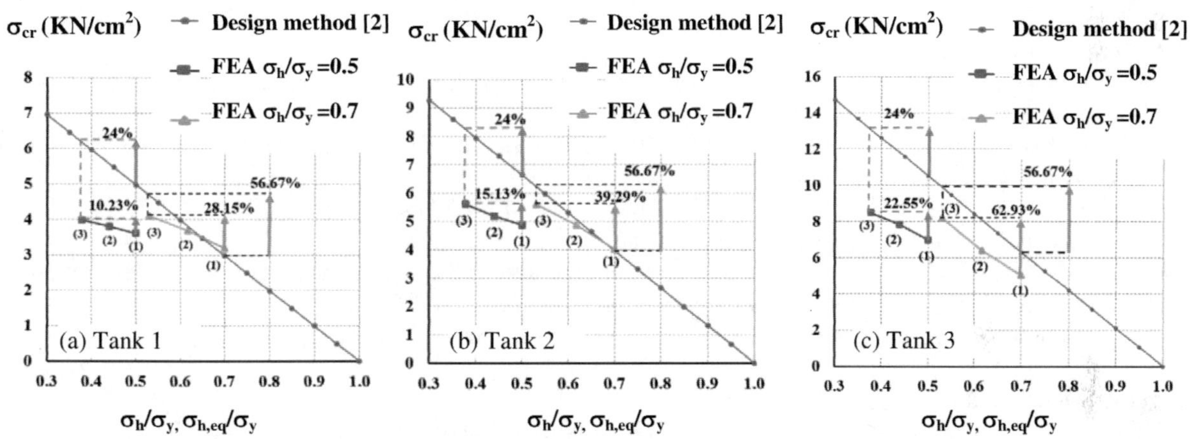


**Figure 8.** The impact of CFRP strengthening on the decrease of internal pressure inside the storage tanks. (1) Without CFRP strengthening, (2) 5-ply CFRP strengthening, and (3) 12-ply CFRP strengthening.

## 7. Conclusion

In this paper, finite element analysis was used to survey the load-carrying capacity and buckling modes of cylindrical steel storage tanks under bending shear load with and without CFRP strengthening. Three models with different geometry parameters were simulated with two levels of internal pressure. Moreover, the strengthening effects of CFRP layers on the ultimate strength of the tanks was also considered. Based on the analysis and calculated results, some following main points may be highlighted:

- In the case of without CFRP strengthening, the main buckling modes of storage tanks obtained from FEA were elephant foot bulge. These failure modes were very corresponding with the design recommendation of Architectural Institute of Japan. However, there were some differences in ultimate strength of the tanks because the differences in steel materials used. When the ratio of the tensile hoop stress to the yield point stress of the tank's material was 0.5, the results had about 26% to 33% variation, whereas if this ratio increased to 0.7, the analysis results were more accurate with the design recommendation.

IOP Conf. Series: Materials Science and Engineering **371** (2018) 012025 doi:10.1088/1757-899X/371/1/012025

- The strengthening effects of CFRP layers on the increase of ultimate strength of storage tanks were clear in this study. The increase of ultimate strength of the tanks corresponds to the increase of the numbers of CFRP layers. On the other hand, the effectiveness of strengthening depends on the ratio of the tank's radius to the tank's thickness and the ratio of the height of the tanks to the tank's radius. If the first ratios were lower and the second ratios were higher, the strengthening by CFRP layer would be remarkably effective. Another issue when strengthening the tanks by the CFRP layers is about the buckling modes of tanks after strengthening. The elephant foot bulge was limited and the shear buckling occurred in the tank's wall. This is because the CFRP-strengthened layers are unidirectional materials; therefore, the strengthening effects of CFRP layers on the circumferential directions were clear. This restrained the elephant foot bulge in the tanks. However, the effectiveness of strengthening against shear buckling was limited.
- When strengthening the tanks by CFRP layers, the impact of internal pressure on the tanks was decreased. This is also the reason for the increase of ultimate strength of the tanks and makes them more safety in the working and production.

## 8. Future works
The authors will consider the effective and reasonable height of CFRP layers for the strengthening of CSSTs. The results will have important contributions to enlarge this strengthening method in the future.

## 9. References
[1] A Maekawa 2012 *Earthquake-Resistant Structures - Design, Assessment and Rehabilitation*, ed A Moustafa (In Tech) chapter 12 pp 307-336
[2] Architectural Institute of Japan (AIJ) 2010 *Design Recommendation for Storage Tanks and Their Supports* (Tokyo, Japan)
[3] NASA SP-8007 1968 *Buckling of Thin-walled Circular Cylinders*
[4] High-performance carbon fiber Torayca, Torayca cloth (Online)
[5] T G Teng, T Yu, and D Fernando 2012 Strengthening of Steel Structures with Fiber-Reinforced Polymer Composites *Journal of Construction Steel Research* **78**
[6] X Zhao *FRP-Strengthened Metallic Structures* (CRC Press) chapter 1 pp 1-11

## Acknowledgments
This work was supported by JSPS KAKENHI Grant Number 17K06640.

IOP Conf. Series: Materials Science and Engineering **371** (2018) 012026 doi:10.1088/1757-899X/371/1/012026

# Cyclic Loading Tests of Composite Moment Frames Using Octagonal CFTs

**J J Lim[1] and T S Eom[2]**

1 Ph.D. candidate, Dept. of Architectural Engineering, Dankook Univ., 152 Jukjeon-ro, Gyeonggi-do, Korea
2 Associate Prof., Dept. of Architectural Engineering, Dankook Univ., 152 Jukjeon-ro, Gyeonggi-do, Korea
E-mail: doublej17@naver.com

**Abstract.** This research investigates the seismic performance of two interior beam-to-column connections with floor slabs. Concrete-filled octagonal steel tube column, H-section steel beam, and U-section steel beam were used for composite moment frames, respectively. To release the stress concentration at the column face, two design methods were proposed. First, H-section steel beam passed through the column, directly. Stopper plates were also used at the column faces to decrease the slip deformation. Second, top and bottom flanges of U-section beam were not welded to the column face. Instead, to transfer the tensile force of the beam flange, reinforcement bars passing through the column were welded to the beam flanges. The test results showed that the ductility of moment connections was sufficient in the case of through details; However, the load-carrying capacity was significantly affected by the details of beam-to-column connections.

## 1. Introduction

Concrete filled tube (CFT) has a good flexural capacity, because the steel plates which resist the tension or compression are placed along the perimeter of the column section. Also, the steel tube confines the concrete core; at the same time the concrete inhibits the buckling of steel tube, which increases the stiffness and ductility. Therefore, CFT columns are used for many high rise buildings in highly seismic region. Recently, to maximize the composite effect, the octagonal steel tube was fabricated by bending a thin steel plate of thickness of 6 mm. The octagonal steel tube is fabricated by welding by two C-shaped plates and two planer plates (see Fig. 1). The rib plates at the corner increase the buckling resistance of steel tube. Also, octagonal section similar to circular section has a better concrete confinement effect than that of rectangular section. These guidelines, written in the style of a submission to *IOP Conference Series: Materials Science and Engineering*, show the best layout for your paper using Microsoft Word. If you don't wish to use the Word template provided, please use the following page setup measurements.

In the connection where the beam frames into the column, particular attention should be paid to the connection details to secure the seismic performance required by AISC 341 [1]. This is due to the stress concentration followed by weld joint fracture at the beam-column connections. Such weld joint fracture can be easily more occur in the case of octagonal CFT column with thin steel plates.

Previous studies on the seismic performance of steel beam-CFT column moment connection have focused on reducing stress concentration at the weld joint and transferring beam flange force to column. Special details at the steel beam-CFT column connection can be divided into three categories. First, additional stiffener, cover plate, and diaphragm are welded at the connection [2, 3, 4, 5, 6]. Such details increased the stiffness and ductility of connections, effectively. However, in some cases where the connection details of the additional plates and diaphragms were inappropriate, the brittle rupture at the weld joint occurred. Second, end-plate and T-stub welded to the beam flanges were bolted to the flange of CFT column [7, 8]. In the beam-column moment connection with such details, the tensile force of beam flange transferred to the column plate without weld joint fracture. However, in the case where the stiffness of column flange was not sufficient, the load-carrying capacity and the deformation capacity decreased due to the out-of-plane deformation of column plate. Third, additional steel plate, reinforcement bar, and diaphragm were passed completely through the column [9, 10, 11, 12, 13]. In this case, the beam-column connections showed superior ductility over the other comparable connections without a through detail.

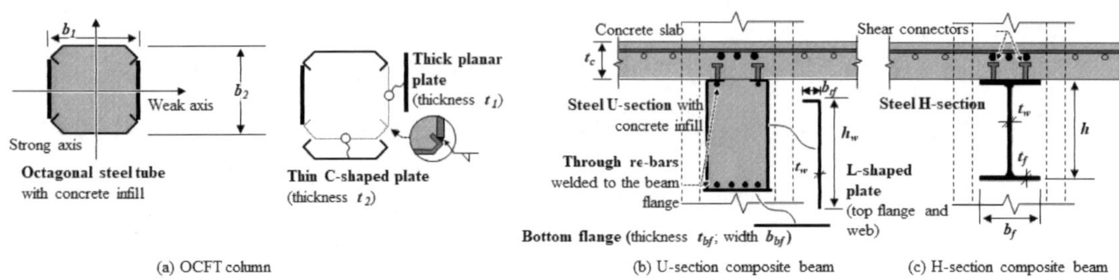


**Figure 1.** Composite member sections

In the present study, seismic performance of composite moment connections using octagonal concrete filled tubes (OCFT) was investigated as follows.

1) Two full-scale beam-column connection specimens were tested, including with OCFT columns, with H-section beams, and with U-section composite beams.
2) H-section beams were directly passed through the OCFT column as the beam plates welded to the column flange. In the U-section beam-OCFT column connection, beam was not passed through the column and welded to the column plate. The beam flanges were not welded to the column to reduce a stress concentration. Instead, reinforcement bars were passed through the column and welded to the flanges of through beam.
3) Concrete slab was installed at the top of beam. Sufficient studs were also installed for full composite action between the concrete slab and the steel beam.
4) To evaluate the compressive resistance of slab, effective width of slab was designed to have sufficient transverse reinforcement bars. A transverse beam which passes through the column was also installed.
5) For each type of specimen, deformation capacity, load carrying capacity, energy dissipation capacity and failure mode were evaluated.

## 2. Moment connection details

Fig. 2 (a) shows the moment connection detail using conventional H-shaped steel beam and OCFT column. To transfer the beam force to the concrete infill of OCFT without a severe out-of-plane deformation of column plate, the entire section of steel beam is passed through the column with welding. Additional steel plates are also welded to cover the holes made by through beam. To increase the negative moment strength at the connection, holes are drilled in the column flange, and slab reinforcement bars pass through the column with sufficient anchorage length. Transverse beam passes through the column and longitudinal beam, respectively, with the area loss of the web of longitudinal beam.

Fig. 2 (b) shows the hybrid moment connection of welding and through reinforcement bars for an interior beam-column connection. The U-section composite beam was not passed through the column. Instead, to transfer the beam flange force to the column and the opposite beam, reinforcement bars are passed through the column and welded to the top and bottom flanges of U-section beams. To secure the yield strength of through reinforcement bars, the weld size and length should be sufficient. Note that, to secure a plastic region without welding, the welding between the beam flanges and through reinforcement bars is performed far away from the column face. Holes are also drilled at the column flange for through reinforcement bars in slabs and U-section beams.

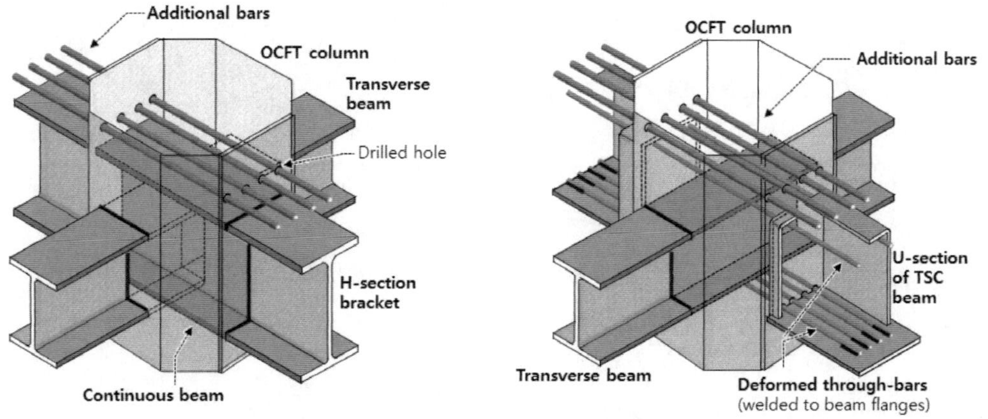


(a) Moment connection with through beam (b) Hybrid moment connection of welding and bar reinforcement

**Figure 2.** Details of interior composite beam-OCFT column connections

## 3. Test program

*3.1 Tests parameters and section details*

In the present study, as indicated in Fig. 3, two full-scale interior beam-column connections, J1 and J2, were fabricated. The dimension symbols in cross section of the OCFT column, H-section beam, and U-section beam were indicated in Fig. 1. The sectional dimension of OCFT column (b1 b2) was 650 mm 650 mm. The thickness of the planer plate and C-shaped plate of OCFT column was 6 mm. The specimens were designed for strong column-weak beam condition as the bending moment strength of OCFT column was $M_{nc}$ = 2898 kN. The clear height of OCFT column was 3030 mm and the shear span between the column face and the loading point was 3055 mm. The entire length, width, and thickness of concrete slab were 7400 mm, 3000 mm, and 180 mm, respectively. The dimensions of the column, beam, and slab were same in both specimens. However, details of the beam-column connections differed in each specimen as follows.

In J1 (see Fig. 3(a)), H-section steel beam ($h$= 482 mm, $b$= 300 mm, $t_w$= 11 mm, $t_f$= 15 mm) passed through the OCFT column with welding at the column face. For a transverse beam, H-section steel beam ($h$= 400 mm, $b$= 200 mm, $t_w$= 9 mm, $t_f$= 11 mm) was used. Considering a required beam moment strength, beam splices with inclined end-plates were installed at a distance of 1200 mm. The cover plates were welded for continuity of column flanges. The continuity plate was designed to have the same thickness as the column flange. The concrete slab was installed at top of the steel beams. For full composite action between the steel beam and concrete slab, 2-$\phi$22 shear studs were placed at a spacing of 140 mm. Twenty SD400 D13 bars were used for longitudinal slab reinforcement bars, while four SD400 D22 bars passed through the column were used. The transverse slab reinforcement bars (D10 SD400) were installed closely to prevent the buckling of through slab-reinforcement bars in the vicinity of the connection. To decrease the slip deformation between the beam flange and the concrete infill of OCFT column, stopper plates were used at the column face. The stopper plates were welded only to the beam bottom flanges. The welding details were designed to resist the yield strength of beam bottom flanges (= $b_f t_f F_y$).

In J2 (see Fig. 3(b)), the slab details were the same as those of J1. U-section composite beam was used instead of H-section steel beam. The beam flanges were not welded to the column plates to avoid stress concentration. Instead, through reinforcement bars welded to the beam flanges were used. The welding between the through reinforcement bars and beam flanges was provided at a distance of 250 mm for a plastic hinge region without welding. The yield strength of bottom through reinforcement bars ($f_{yj}A_{sj1}$ = 699 kN) was about 24% less than that of bottom flange ($F_y b_{bf} d_{bf}$ = 916 kN).

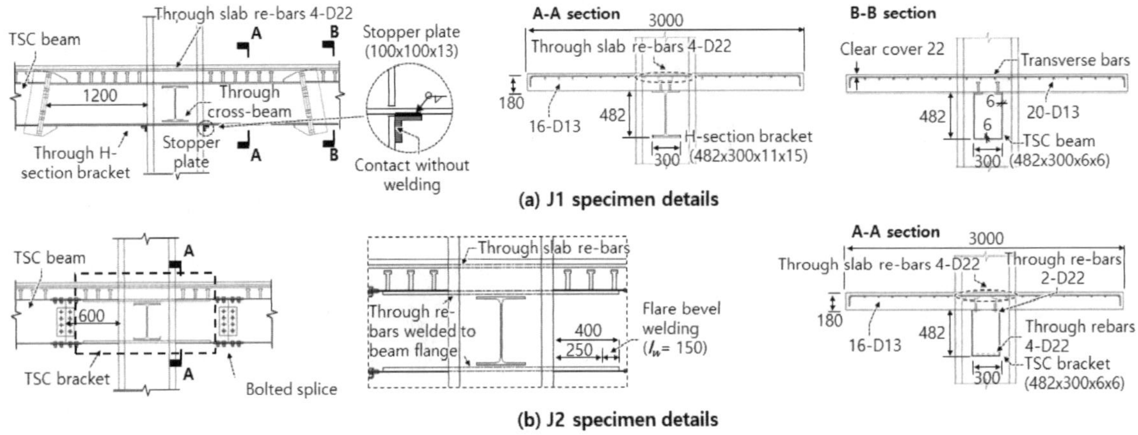


**Figure 3.** Connection details of specimens

The material strengths of the steel plates, reinforcement bars, and concrete were represented in Table 1. The actual yield strengths of 6 mm- and 15 mm-thick plates used for OCFT column, U-section beam and H-section beam were $F_y$= 509 and 424 MPa, respectively. The yield strengths of slab reinforcement bars were $f_y$= 535 MPa for D10 and $f_y$= 476 MPa for D13. For the through reinforcement bars located in the concrete slab and the U-section beam, the yield strength of D22 bar was $f_y$= 460 MPa. The compressive strength of the concrete was $f_{ck}$= 38 MPa. Compression tests of three concrete cylinders were performed on the first day of cyclic tests.

### *3.2 Loading and measurement plan*

Fig. 4 shows the test set-up for cyclic loading test of interior beam-column connection. The OCFT column was erected vertically and pinned at the bottom, while the beam with concrete slab was horizontally placed with pin supports at both ends. The transverse beam was not supported as the column and beam was supported laterally to prevent the transverse displacement.

Displacement history for cyclic loading was planned in accordance with the SAC protocol in AISC 341-02 [1]. Note that an axial load was not applied to the OCFT column because it was not feasible in the laboratory to apply a meaningful axial force to the full-scale column. Axial load is beneficial in decreasing diagonal shear cracking in the beam-column connection, but it can accelerate the buckling of the steel tube.

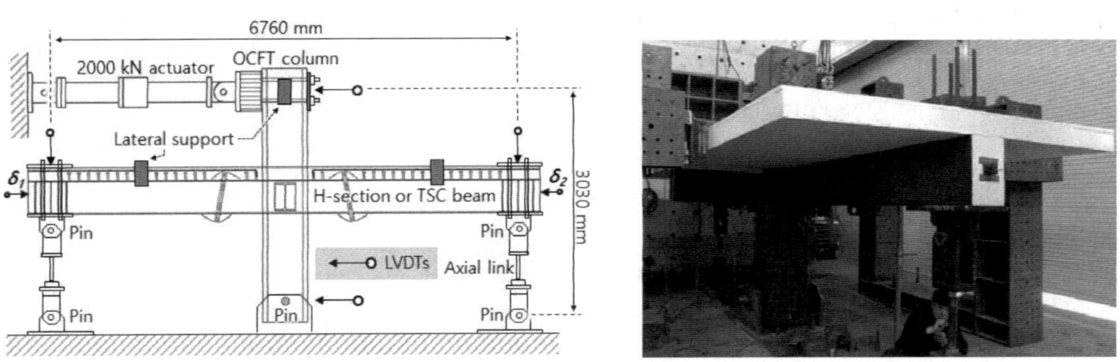


**Figure 4.** Test set-up and measurement

## 4. Test results

*4.1 Load-displacement relationships and failure modes*
Fig. 5 shows the lateral load-drift ratio ($P$ –$\delta$) relationships of the connections under cyclic loading. The lateral drift ratio $\delta$ was calculated by dividing the lateral displacement at the loading point by the clear height of the OCFT column (= 3060 mm). Initial slip deformation at the beam end with pin connection occurred in both specimens. Cyclic loading plan was modified by considering the initial slip deformation of 6.5 mm (see Fig. 5).
In J1 with through H-section beam and stopper plates (see Fig. 5(a)), flexural cracks transverse to the beam span occurred at $\delta = \pm 0.75\%$ and the concrete crushing at the column face occurred at $\delta = \pm 3.0\%$ ~ 4.0%. The stiffness of connection decreased severely at $\delta = \pm 2.0\%$ as the beam flanges yielded in tension. Also, the weld joint fracture at the column face occurred at $\delta = \pm 3.0\%$ (see Fig. 6(a)). However, the load-carrying capacity maintained even after the weld joint fracture. This indicates that the stopper plates can effectively prevent the slip deformation between the beam flange and the concrete infill after weld joint fracture. The buckling of beam bottom flanges and weld joint fracture between the cover plates and column plates occurred at $\delta = \pm 7.0\%$ (see Fig. 6(a)), which resulted in significant strength degradation.

In J2 with U-section beam (see Fig. 5(b)), the peak strengths were $P_u^+$ = 741 and $P_u^-$ = 704 kN at $\delta = \pm 4.0\%$, respectively. After the peak strengths, the load-carrying capacity decreased gradually during repeated load cycles. The maximum lateral load at $\delta = +6.0\%$ was 598 kN greater than 80 % of peak strength $P_u^+$(= 741 kN). Weld joint fracture between the beam webs and column plates also occurred about $\delta = \pm 4.0\%$. This result indicates that the through reinforcement bars welded to the beam flanges successfully transferred the flexural moment of the beam to column.

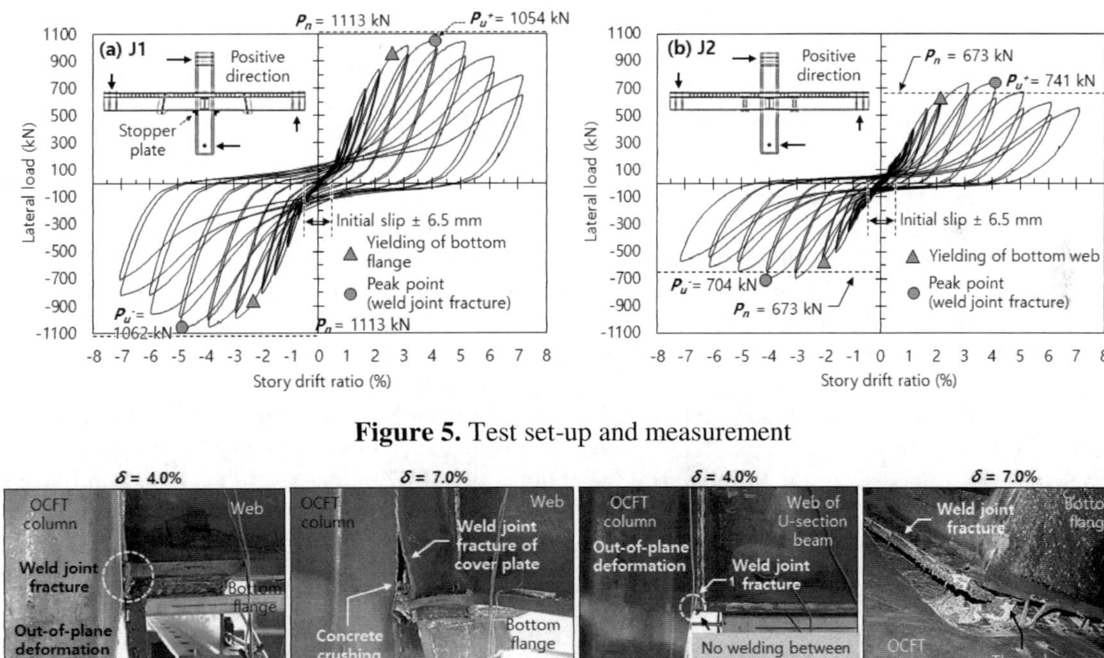


**Figure 5.** Test set-up and measurement

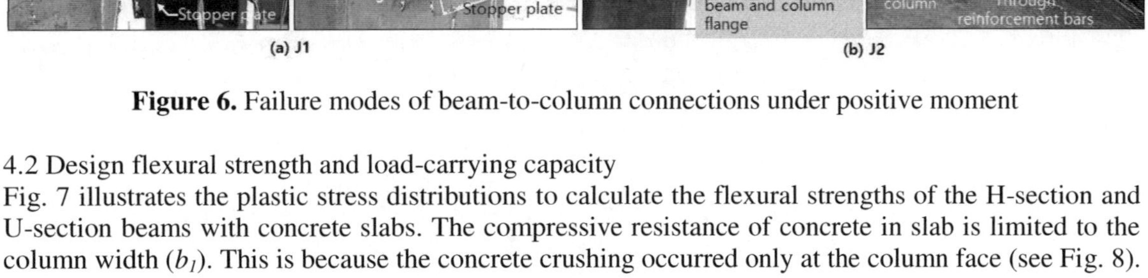


**Figure 6.** Failure modes of beam-to-column connections under positive moment

4.2 Design flexural strength and load-carrying capacity
Fig. 7 illustrates the plastic stress distributions to calculate the flexural strengths of the H-section and U-section beams with concrete slabs. The compressive resistance of concrete in slab is limited to the column width ($b_1$). This is because the concrete crushing occurred only at the column face (see Fig. 8).

IOP Conf. Series: Materials Science and Engineering **371** (2018) 012026 doi:10.1088/1757-899X/371/1/012026

The slab reinforcement bars passing through the column develop their yield strength $f_y$ in compression and tension, respectively.

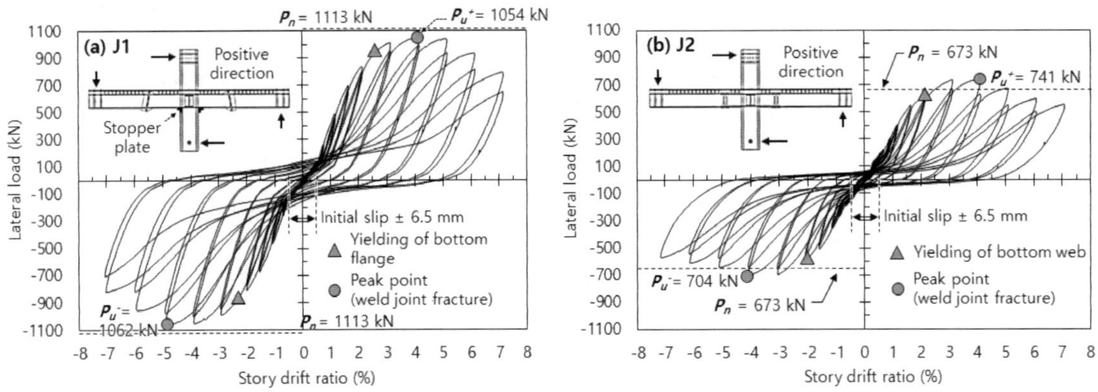


**Figure 7.** Test set-up and measurement

In the case that H-section beam passes through the column with welding and stopper plates, the beam web plates welded to the tube wall do not develop their full yield stress, since the thin steel plates were used for OCFT column. In addition, since the web plate should carry shear force of the beam to the column, its contribution to the flexural strength can decrease. Therefore, the design plastic stress of the is reduced to 0.5 $F_y$. In the case of a hybrid moment connection using U-section beam and through reinforcement bars, the top and bottom flanges developed their yield strength $F_y$ only in compression.

As indicated in Fig. 5 and Fig. 7, the lateral strengths calculated by plastic stress distribution. In J1, the peak strengths were similar to the design lateral strengths ($P_u/P_n$= 1.03). In J2, the peak strengths $P_u$ were also similar to the design lateral strengths $P_n$ ($P_u/P_n$= 1.04).

4.3 Strain measurement results

Fig. 9 illustrates the peak strain distributions at each load cycles. In J1 with the through H-section beam reinforced with stopper plates, the strain of bottom flange was significantly greater than the yield strain (=0.002) and large strain reversals occurred during load cycles. However, the strain of bottom web was nearly constant after yielding. This indicates that the slip deformation of bottom flange was limited due to the stopper plates. The neutral axis calculated from the stress distributions was located above due to composite action between the steel beam and the concrete slab. The strains of through reinforcement bars in slab were significantly greater than the yield strain (=0.002). This indicates that the through slab reinforcement bars yielded experiencing large yield strains, thereby contributing to the negative moment strength of TSC beam at the column face. In J2, since the top and bottom flanges were not welded to the column plate, the tensile strains were nearly zero. The strain of bottom web welded to the column plate was constant after yielding due to the out-of-plane deformation of column plate and weld joint fracture..

IOP Conf. Series: Materials Science and Engineering **371** (2018) 012026 doi:10.1088/1757-899X/371/1/012026

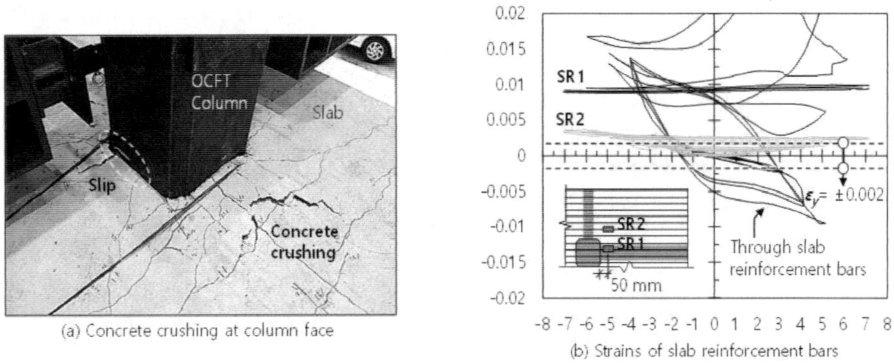


**Figure 8.** Test set-up and measurement

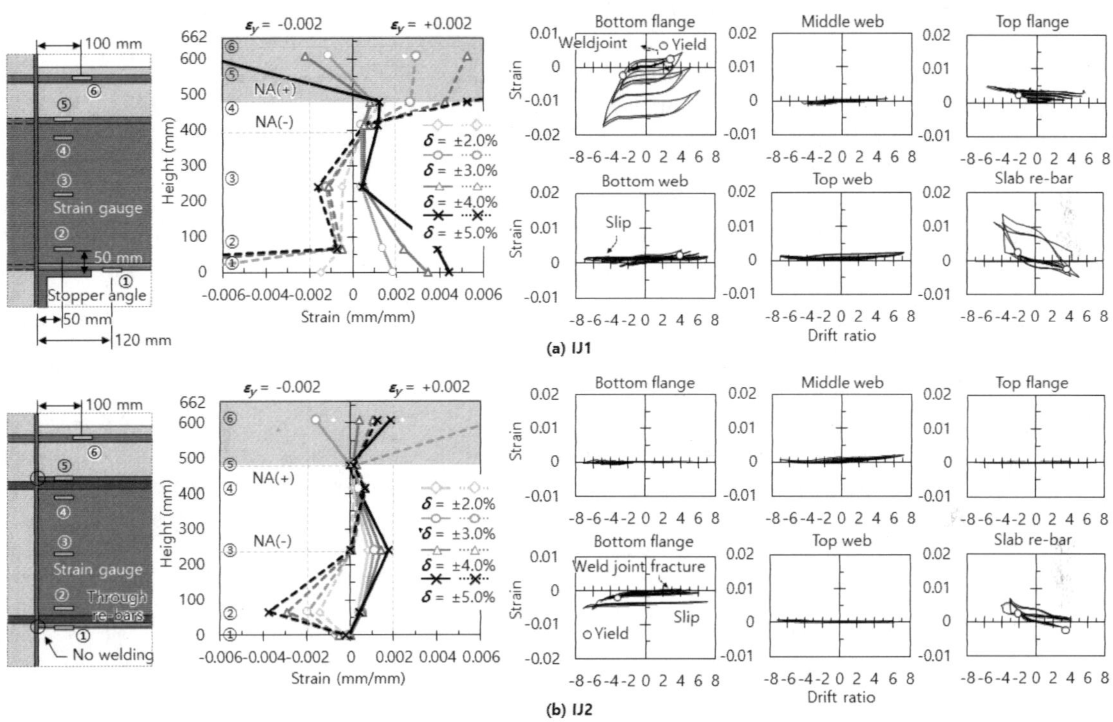


**Figure 9.** Test set-up and measurement

## 5. Summary and concrete

In the present study, full-scale interior beam-column connections using OCFT column, H-section beam and U-section were fabricated. Cyclic loading tests were performed to investigate the load-carrying capacity, deformation capacity and failure modes. The major findings of the present study can be summarized as follows:

1) In J1, although the through H-section beam was used with welding, the out-of-plane deformation of steel tube occurred at the weld joint. However, the stopper plates successfully decreased the slip deformation between the beam bottom flanges and the OCFT column. Therefore, the load carrying capacity was maintained even after the weld joint fracture at the column face and the deformation capacity reached $\delta = \pm 7.0\%$
2) In J2 with hybrid connection details, the through reinforcement bars successfully the flexural moment of the beam to column. The tensile force of the beam flange was directly transferred to

the column concrete and the opposite beam flange through the reinforcement bars. Therefore, ductile behavior occurred without welding between the beam flange and column plate.

3) When the concrete slab was subjected to compression under positive bending, the compressive resistance of concrete was limited to the column width. Through reinforcement bars in concrete slab underwent significant plastic strains during cyclic loading and thus contributed to the positive and negative moment strength of the beam. However, other slab bars bypassing the column underwent relatively small elastic strains. To evaluate the compressive resistance of slab, effective width of slab was designed to have sufficient transverse reinforcement bars. A transverse beam which passes through the column was also installed.

## 6. References

[1] ANSI/AISC 341-02 2010 *Seismic Provisions for Structural Steel Buildings* (Chicago)
[2] Morino S, Kawaguchi J, Yasuzaki C and Kanazawa S 1993 *In Composite Construction in Steel and Concrete II* 726
[3] Schneider S P and Alostaz Y M 1998 *J. of Constructional Steel Research* **45** 726
[4] Ricles J M, Peng S M and Lu L W 2004 *J. of Structural Engineering* **130** 223
[5] Qin Y, Chen Z and Wang X 2014 *J. of Constructional Steel Research* **98** 35
[6] Shin K J, Kim Y J, Oh Y S and Moon T S 2004 *J. of Engineering Structure* **26** 1877
[7] Wu L Y, Chung L L, Tsai S F, Shen T J and Huang G L 2005 *J. of constructional steel research* **61** 1387
[8] Azizinamini A and Schneider S P 2004 *J. of Structural Engineering* **130** 213
[9] Mirghaderi S R, Torabian S and Keshavarzi F 2010 *J. of Engineering Structure* **32** 2034
[10] Jeddi M Z, Ramli Sulong N H and Arabnejad Khanouki M M 2017 *J. Engineering Structure* **131** 477
[11] Beutel J, Thambiratnam D and Perera N 2002 *J. of Engineering Structure* **24** 29
[12] Korol R M, Ghobarah A and Mourad S 1993 *J. of Structural Engineering* **119** 3463.

**Acknowledgments**
The present research was conducted by the research fund of Transportation and Maritime Affairs of Korea (Code 17CTAP-C129746-01, Infrastructure and transportation technology promotion research Program). The authors are grateful to the authorities for their support.

IOP Conf. Series: Materials Science and Engineering **371** (2018) 012027 doi:10.1088/1757-899X/371/1/012027

# Study on Mechanical Behavior of Self-tapping Screws Connection Using Washers in Single-Lapped Glass Fiber Reinforced Plastic Plates by Experiment and Finite Element Analysis

**N N Duong[1], P V Nhut[1], C Satake[1] and Y Matsumoto[2]**

1 Graduate student, Department of Architecture and Civil Engineering, Toyohashi University of Technology, Japan
2 Associate Professor, Department of Architecture and Civil Engineering, Toyohashi University of Technology, Japan
E-mail: n165504@edu.tut.ac.jp

**Abstract.** In recent years, mechanical characteristics of Glass Fiber Reinforced Plastic (GFRP) connections using bolts/rivet have been researched in detail, and they are used in many GFRP structures. However, the connection lack bearing strength compared with material strength and they need the prepared holes. In this paper, we surveyed the self-tapping screws connection strength and effects of washers on connection strength of single-lapped GFRP plates using self-tapping screws under tensile-shear loading by experiment and finite element analysis (FEA). The strength of self-tapping screws connection in GFRP plates depends on the washer's diameters. When increasing washer's diameter, the bearing strength of connections increased. Hence, increasing washer's diameter at reasonable levels is an effective method to increase the load carrying capability for self-tapping screws connection in GFRP plates. Moreover, FEA was proposed to investigate the bearing strength of connections and there were good agreements between FEA and experimental results.

## 1. Introduction

Fiber Reinforced Plastic (FRP) has currently been widely used in the fields of vehicles, aircraft, because of their have many good characteristics when comparison with other materials such as light weight, high strength, high corrosion resistance, good freedom of molding, etc[1]. These characteristics lead to high seismic resistance, long life, and ease in construction; therefore, FRPs have also been used in architecture and civil engineering field[2]. At present, because the connection behavior of FRP connection is relatively complex, it is impossible to construct freely like concrete and steel. If that problem is solved, FRP will be a material which can change significantly the construction structure. Unlike steel, the connection of FRP is difficult to welding, so mechanical connection, adhesive connection, or both of them have been used. However, the mechanical properties of connection, which are indispensable for constructing structures by FRP material, are not clarified sufficiently. Therefore, accumulation of research results to improve design method are necessary. There have been many researches about mechanical connection for FRP by steel bolts/rivet[3]. This connection requires drilling to create the holes in FRP material first and bearing strength is lower than the material strength due to the influence of connection clearance. Hence, solving this problem to improve initial stiffness of connection and bearing strength by tapping screw should be considered carefully. Furthermore, the screw has a drill at the tip, so there don't need hole processing, increasing construction speed. However, because of self-tapping screw's rotation in single-lapped connection, it cannot mobilize all the material strengths inside connection together, lead to the partial damage occurred. Therefore, the bearing strength of single-lapped GFRP plates connection using self-tapping screw was decreased when comparing with double-lapped GFRP plates connection [4]. Reducing the rotation of self-tapping screws while bearing load by increasing the diameter of the washer, thereby increasing the connection strength of single-lapped GFRP plates using self-tapping screws should be considered carefully. In previous study, we investigated the self-tapping screws connection strength and effects of washers on connection strength of single-lapped GFRP plates using self-tapping screws with the end distance was 25mm. The load early reached the maximum load and rapidly decreased in all specimens due to the occurrence of shear-out failure. This problem leads to the sudden collapse of the construction structure, danger for the users because we cannot recognize the symptoms of failures. Increasing end distance in order to increase connection strength, structure stability, and investigate other self-tapping screws connection's failure modes should be researched in detail. In this study, we surveyed the self-tapping screws connection strength and effects of washers on connection strength of single-lapped GFRP plates using self-tapping screws with end distance was 50mm, under tensile-shear loading by experiment and finite element analysis (FEA).

## 2. Experiment

In this study, GFRP which was used in the experiment is pultruded GFRP (AGC Matex manufactured "Plalloy" C100B), fiber is mainly roving, resin uses unsaturated polyester. The mechanical properties of the GFRP materials are shown in Table 1.

**Table 1.** Mechanical properties of GFRP materials

| GFRP materials | Value |
|---|---|
| Tensile strength (MPa) | 446 |
| Longitudinal elastic modulus (GPa) | 30 |
| Glass content rate (weight %) | 55 |

GFRP plates used in the experimental test were cut from original C-shape GFRP materials with the 175mm length, 50mm with and 6.5mm thickness. Self-tapping screws (Japan Power Fastening manufactured "MB TEKS" HEX#5), which were used in the experimental tests, are shown in Figure 1 and Figure 2. Self-tapping screws were made of stainless steel. Self-tapping screws were tightened by an electric screwdriver, tightening torque of the self-tapping screws was controlled as 15 Nm in order to prevent the damage to the specimen material.

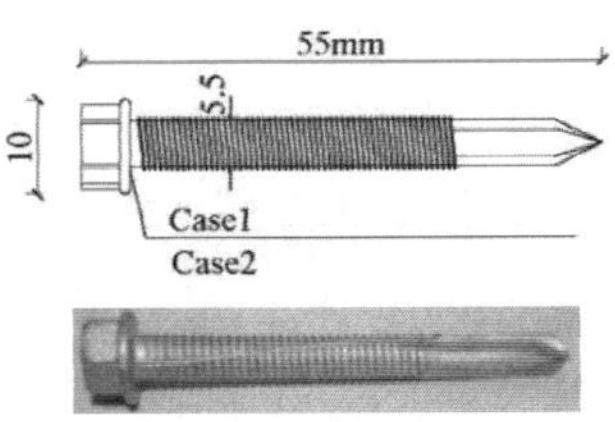


**Figure 1.** Self-tapping screws without alternative washer (Unit: mm).

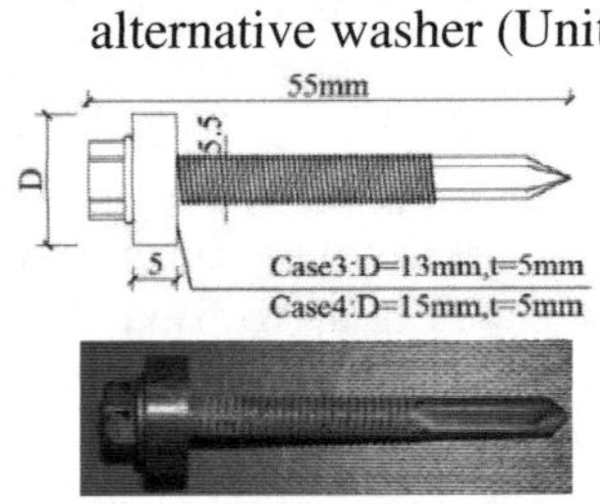


**Figure 2.** Self-tapping screws with alternative washer (Unit: mm).

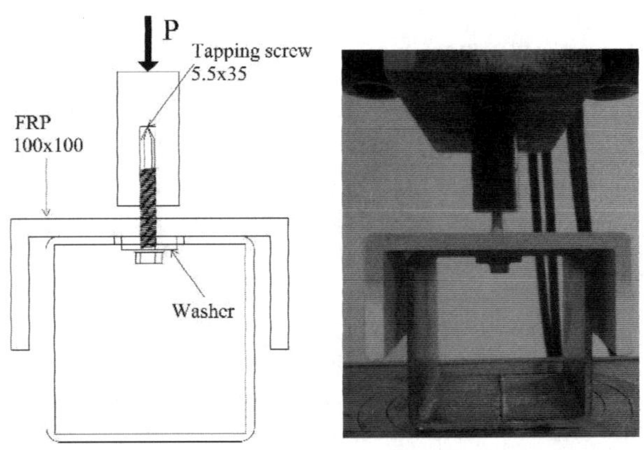


**Figure 3.** Experimental method and setup of Pull-out test (Unit: mm).

### 2.1. Experimental methods

*2.1.1. Pull-out tests.* Figure *3* shows experimental method and setup. Because the shaft neared tapping screw-head does not have threads, we put 5mm thickness-washer between self-tapping screws-head and FRP plates in order to accurately survey bearing strength in the pull-out test. Specimens were cut from the long C-shaped material into 100mm segments with an average thickness of 6.53mm.

*2.1.2. Single-lapped tensile-shear tests.* The experiment was conducted with 4 different cases to survey the influence of the washer on the connection strength. Case 1 is self-tapping screws without the washer's effect. Case 2 is original self-tapping screws with washer's diameter 10mm. Cases 3, 4 include self-tapping screws with alternative washer with the diameters 13mm and 15mm, respectively. Figure 4 shows the experimental methods with specimen's dimension, we have five specimens for each case, each specimen includes two FRP plates and two added plates for fixing in the testing machine. Figure 5 shows the experimental setup. All the tests were conducted by a universal testing machine, the top and bottom of the specimens were clamped onto the machine, data recording includes load and crosshead displacement. In case 1, in order to release the effect of the washers on GFRP plates, the distance between the washers and GFRPs plate is 5mm. The connection strength of single-lapped GFRP plates was investigated through single-lapped tensile-shear loading test.

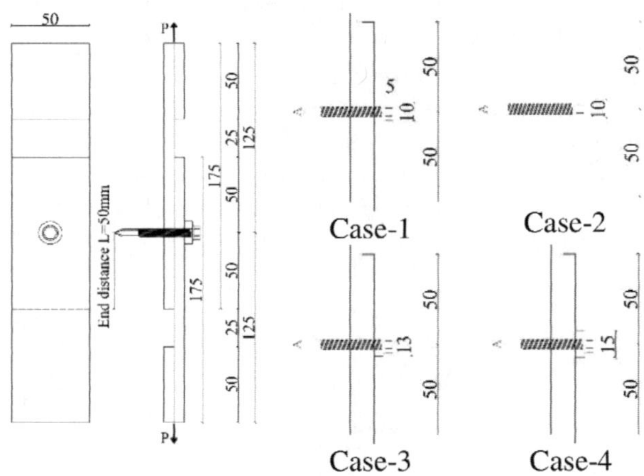


**Figure 4.** Experimental method (Unit: mm).

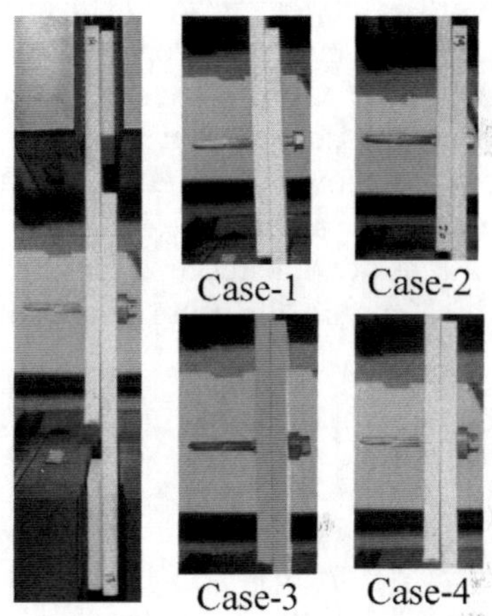


**Figure 5.** Experimental setup.

### 2.2. Experimental results and discussion

*2.2.1. Pull-out tests*

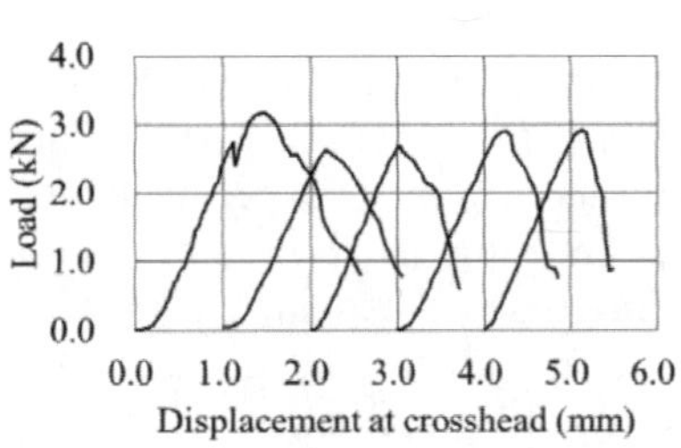


**Figure 6.** Load-displacement relations.

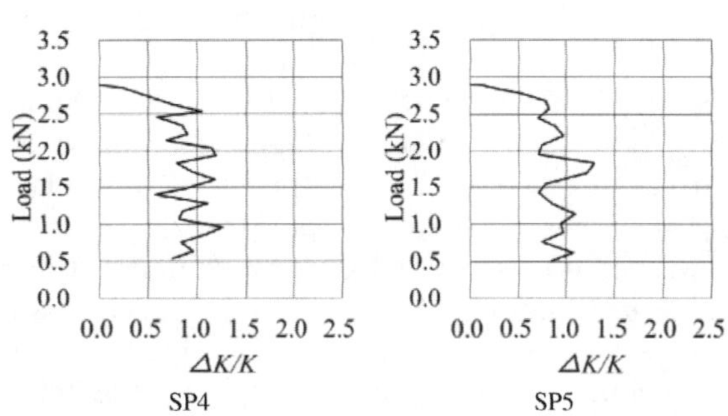


**Figure 7.** Load-ΔK/K relation (specimen 4, 5).

Figure 6 shows the load-displacement relations, the loading was gradually decreased after reaching the maximum load in all specimens. Figure 7 shows Load-*ΔK/K* curves (specimen 4, 5). From this Figure, we can know the levels of the load which load-displacement curves is still linear. $K$ is the stiffness (inclination until linear completed), $K$ was calculated from $0.2P_{max}$ to $0.5P_{max}$ of the load-displacement curves. $\Delta K$ is the inclination of previous plot in each experimental plot. The elastic limit is defined as the loading when $\Delta K/K$ reaches 0.5[5] then the loading level of elastic limit is calculated to be $0.96P_{max}$. Table 2 summarizes results obtained by the pull-out tests.

**Table 2.** Summarization of pull-out test results.

| $N_o$ | SP1 | SP2 | SP3 | SP4 | SP5 |
|---|---|---|---|---|---|
| Thickness (mm) | 6.54 | 6.53 | 6.52 | 6.55 | 6.52 |
| $P_{max}$ (kN) | 3.18 | 2.64 | 2.70 | 2.91 | 2.92 |
| Stiffness K (kN/mm) | 3.04 | 2.75 | 3.33 | 2.97 | 2.94 |
| Average thicness (mm) | | | 6.53 | | |
| CV of thickness (%) | | | 0.18 | | |
| Average $P_{max}$ (kN) | | | 2.87 | | |
| CV of $P_{max}$ (%) | | | 6.65 | | |
| Average Stiffness K (kN/mm) | | | 3.00 | | |
| CV of K (%) | | | 6.28 | | |

*2.2.2. Single-lapped tensile-shear test.* Figure 8 shows load-displacement relations. Figure 9 shows the comparison of maximum loads between case-end distance 25mm[6] and case-end distance 50mm. The deformation of GFRP plates at the contact position between GFRP plates and washers are shown in Figure 11, the deformation modes of specimens and the rotation restriction of self-tapping screws by washers are shown in Figure 12. Self-tapping screws in single-lapped connection is rotated by shear loading as shown in Figure 12, lead to the occurrence of partial bearing-failure. Shear-out failure didn't occur as case-end distance 25mm, so maximum load of case-end distance 50mm higher than case-end distance 25mm in all specimens as shown in Figure 9.

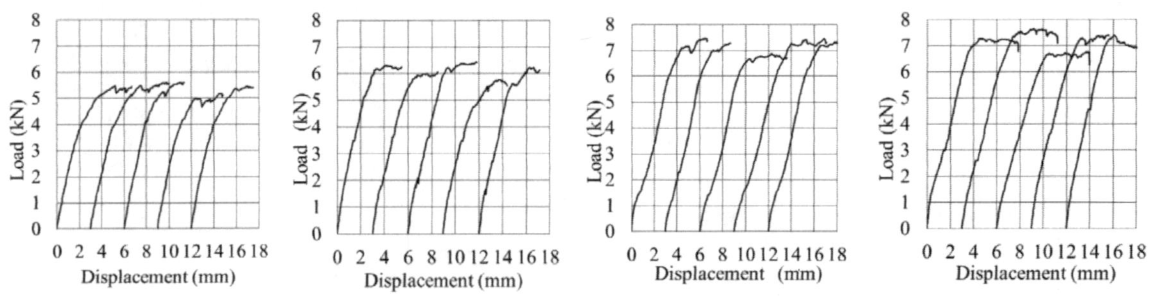


**Figure 8.** Load-displacement relations.

The max load increased when increasing the diameter of the washer because the rotation was restrained by washer as shown in Figure 12. This rotation restriction helps to reduce partial bearing-failure in the connections. The coefficient of variations (CV) are around 5% in all experimental cases as shown in Figure 9. It can be said that the experimental results are quite stable. With specimens having washer in case of end distance 25mm, the load linearly increased when increasing washer diameter from 10mm to 15mm. However, specimens having washer in case of end distance 50mm, when increasing washer's diameter from 10mm to 13mm, load increased significantly, and increasing washer's diameter from 13mm to 15mm, load slightly increased as shown in Figure 9. In case end distance 50mm, comparing with bearing strength in double-lapped connection (240MPa) [7], bearing strength in single-lapped connection is still lower; however, the bearing strength has increased significantly in case washers diameter 15mm (201MPa) compared with case no washers influence (150MPa), approximately 35% as shown in Figure 10. Increasing end distance lead to bearing-failure occurred, shear-out failure didn't occur. After reaching maximum load, the load remained approximate

IOP Conf. Series: Materials Science and Engineering **371** (2018) 012027 doi:10.1088/1757-899X/371/1/012027

maximum load level for long periods while displacement still increased as shown in Figure 8. This helps the user can see the signs of damage before the collapse of the structure, increasing the structure safety. Because the stiffness of steel washers is higher than GFRP plates, rotation restriction of self-tapping screws by washers lead to the deformation of GFRP plates at the contact position between GFRP plates and washer as shown in Figure 11.

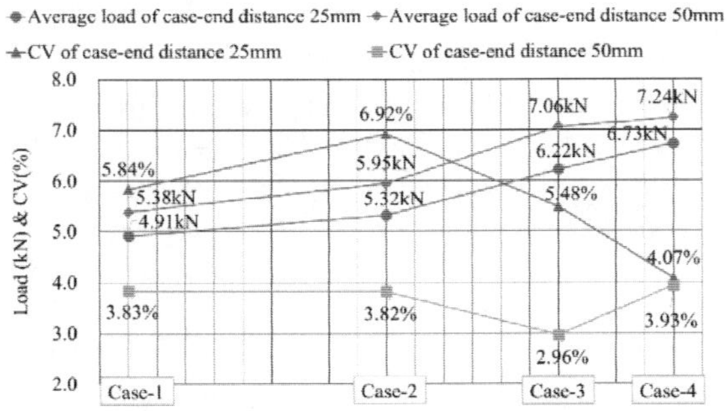


**Figure 9.** Comparison max load between case end distance 25mm and case end distance 50mm.

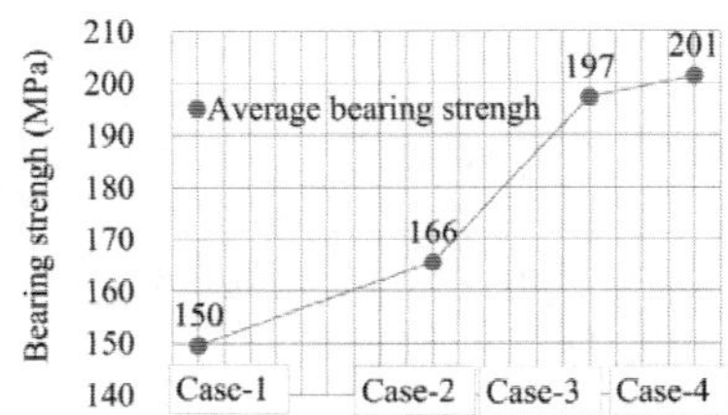


**Figure 10.** Average bearing strength.

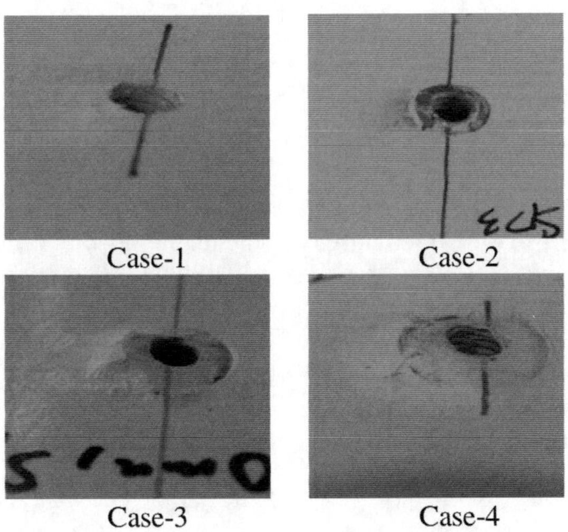


**Figure 11.** The deformation of GFRP plates at the contact position between GFRP plates and washers.

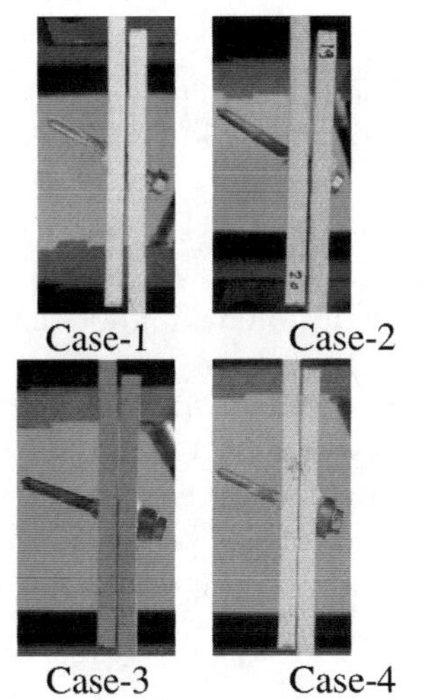


**Figure 12.** Deformation modes of specimens

## 3. Finite element analysis

### *3.1. Finite element analysis*

In FEA, important FEA features were the initial contacts between component parts. The contacts of self-tapping screws connection in GFRP plates include initial contact between washer with GFRP plates and contact between self-tapping screws with GFRP plates. The contact between the washer and GFRP plates was taken easily by available frictional contact properties. Because the contact was only the frictional contact. However, the simulation of contact between self-tapping screws and GFRP plates were not only frictional contact but also pull-out contact. The simulation of pull-out contact through self-tapping screw's threads is necessary to analysis by FEA, but it is very complicated. Hence, we used truss elements instead of this pull-out contact properties in simulation.

3.1.1. *Pull-out test.* In experiments, the self-tapping screws and GFRP plates were held by the screw thread. In this research, we modified the contact between self-tapping screws with GFRP plates by

IOP Conf. Series: Materials Science and Engineering **371** (2018) 012027 doi:10.1088/1757-899X/371/1/012027

four truss elements for every GFRP plates with the approximate stiffness with self-tapping screw threads. We determined when these truss elements were yielding, it means the connections were failed. The materials of these truss elements are assumption with the elastic modulus E=205 (GPa). In order to define the stiffness as well as other properties of the truss elements, we conducted the experimental pull-out tests. From the experimental pull-out tests, the sectional area for the truss elements were calculated following to (1).

$$A = \frac{K_{ave} \times t_{FRP}}{4E} \tag{1}$$

Here, $K_{ave}$=3 (kN/mm) is the average stiffness obtained from pull-out tests. $t_{FRP}$ is the average thickness of FRP plates, this is also the length of the truss elements. A is the sectional area of the truss elements. From (1), the sectional area that was calculated for every truss element was 23.9x10-3 ($mm^2$). Also, the initial uniaxial yield stress for every truss element was 30 (GPa), which was calculated from (2)

$$\sigma = \frac{P_{\max}}{4A} \tag{2}$$

These values will be assigned into FEA to check the accuracy of the results of pull-out tests. The model of pull-out tests in FEA is shown in Figure 13. The model of truss elements in the pull-out test is shown in Figure 14.

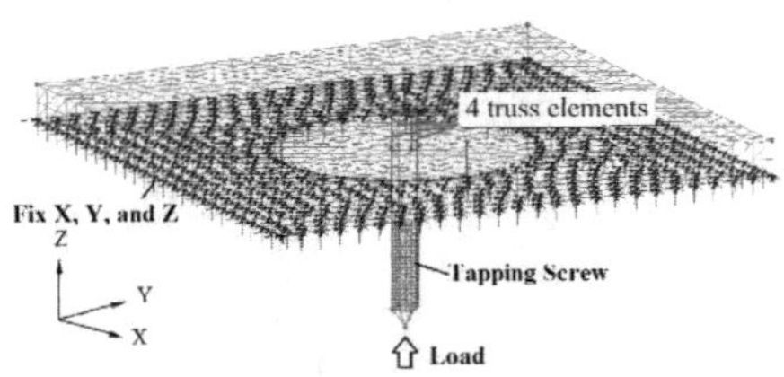


**Figure 13.** Model pull-out tests in FEA.

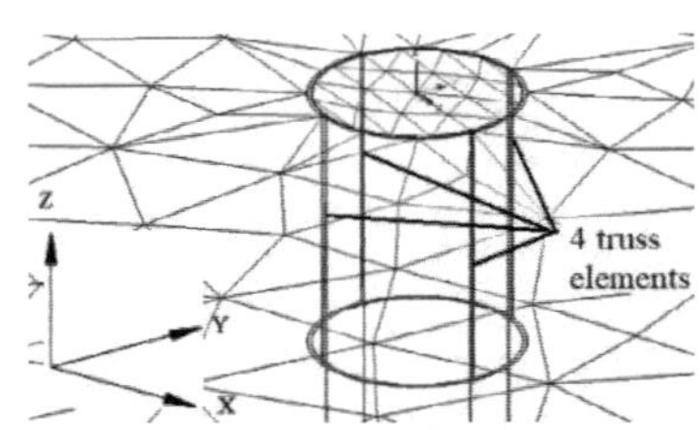


**Figure 14.** Model truss elements in pull-out tests.

*3.1.2. Single-lapped tensile-shear test.* Single-lapped tensile-shear test was carried out using nonlinear 3-D analysis in FEA to analysis the strength of the connections with 4 different cases. Case 1 is tapping screws without the washer. Cases 2, 3, 4 include washer diameter with 10mm, 13mm, and 15mm respectively. The overview and truss element's positions of single-lapped tensile-shear test model are shown in Figure 15, Figure 16. The model of single-lapped tensile-shear test in all case are shown in Figure 17.

**Figure 15.** Overview of Single-lapped tensile-shear test model

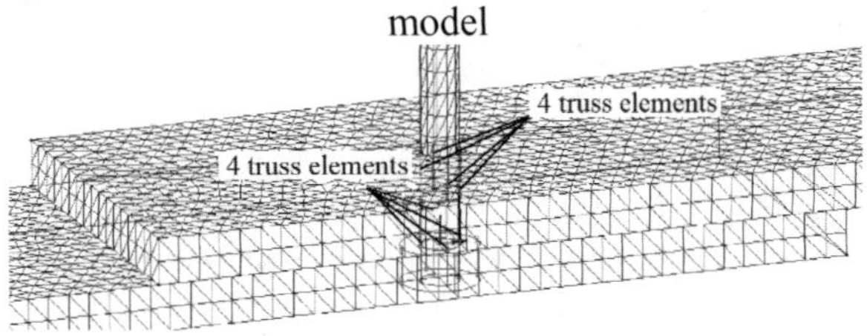


**Figure 16.** Truss elements of single-lapped tensile-shear test model.

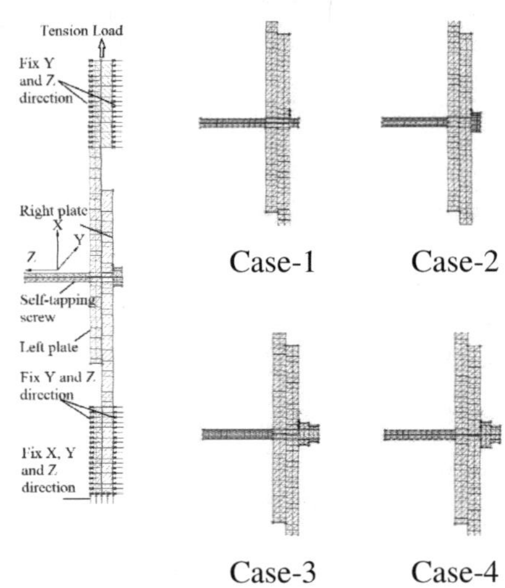


**Figure 17.** Single-lapped tensile-shear test model in case 1, 2, 3, and 4.

### *3.2. Finite element analysis results*

*3.2.1. Pull-*out *test.* Figure 18 shows the relationships between the loading and crosshead displacements that obtained from both experiments and FEA. The red line shows the result in FEA. There was good agreement between experimental data and FEA results. Therefore, the present analysis model using truss elements could be used for the simulation of contacts between GFRP plates and tapping screws replace for tapping screw's threads.

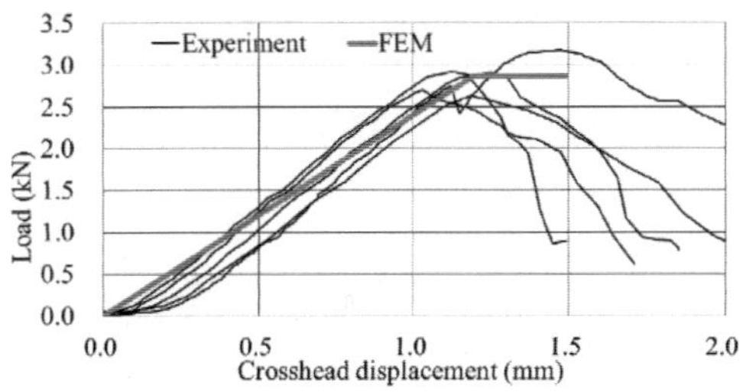


**Figure 18.** Experimental data and FEA results of Pull-out test.

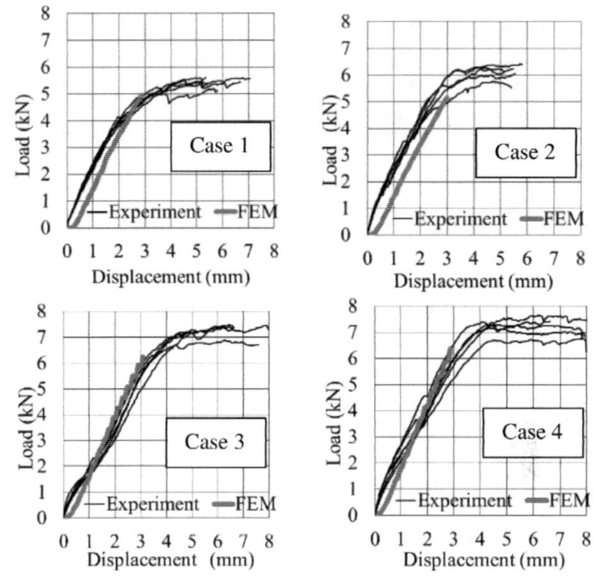


**Figure 19.** Load-displacement relations in experiments and FEA.

*3.2.2. Single-lapped tensile-shear test.* Figure 19 shows the relationships between the load and cross-head displacement obtained from both experiments and FEA. The red line shows the result in FEA. The maximum load in FEA is the load at the time in which the truss elements reaches the initial uniaxial yield stress.

The slope of load and cross-head displacement curve in FEA results corresponding with experimental results as shown in Figure 19. Figure 20 shows the comparison between average maximum load of experiments and FEA. There were good agreements between experimental data and FEA results. We simulated the distance between self-tapping screws and holes in GFRP plates is 0.1mm in order to simulate four truss elements. Because of the effect of this clearance and the limitation of the number of truss elements, the bearing strength of connection in FEA analysis was decreased, so the results of FEA analysis in all case are lower than experimental data as shown in Figure 20. According to FEA results shown in Figure 20, in case of specimens having a washer, increasing washer diameter from 10mm to 13mm, load increased significantly, but increasing washer diameter from 13mm to 15mm, load slightly increased, so it can be said that FEA results closely relate with experimental results. Figure 21 shows the stress distribution in self-tapping screws with the load level 4.9 (kN). In finite element analysis, the compressive stress appeared in the tangential positions between GFRP plates and tapping screw. When the diameter of the washer increased, the compressive stress decreased in the tapping screw and the tensile stress increased in the intersection positions between tapping screw and washer with the same load level. This is because when the washer's diameter increased, the constrained rotation of tapping screw was increased and this leads to the decrease of compressive stress concentration. Figure 22 shows the longitudinal stress distribution in FRP plates with the load level 4.9 (kN). In the areas between two FRP plates, the concentration of compressive stress occurred higher than other areas; therefore, the damage also occurred first in these areas. This damage was good

IOP Conf. Series: Materials Science and Engineering **371** (2018) 012027 doi:10.1088/1757-899X/371/1/012027

correspondence with the bearing-failures that occurred in the same areas in the experiments. Moreover, the concentration of compressive stress decreased corresponding to the decrease of compressive stress in the tapping screw when the washer's diameters increased in the same load level as shown in Figure 22. Figure 23 shows the load and self-tapping screw's rotation relation, θ is the rotation angle of self-tapping crews under tensile-shear load testing. Figure 24 shows the deformation of self-tapping screws and GFRP plates at the load level 4.9 (kN). The rotation of the self-tapping screws were constrained by the washers, so the self-tapping screws using larger washer's diameter have smaller rotational angle at the same level of load as shown in Figure 23 and Figure 24. The difference of slopes of load and self-tapping screw's rotation relation curve between all cases corresponding to the difference of maximum load in experiment and FEA results.

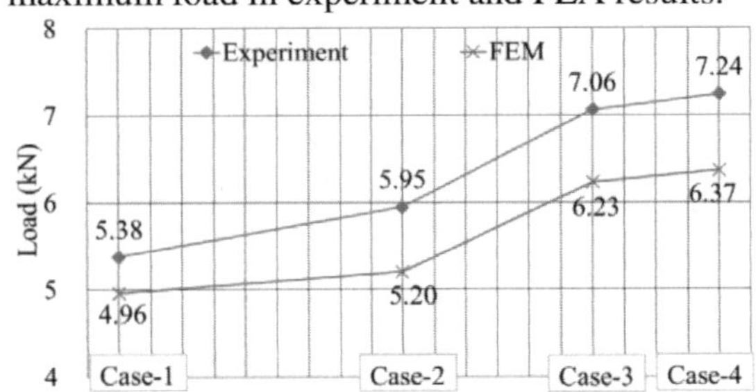


**Figure 20.** Comparison between average maximum load of experiments and FEA.

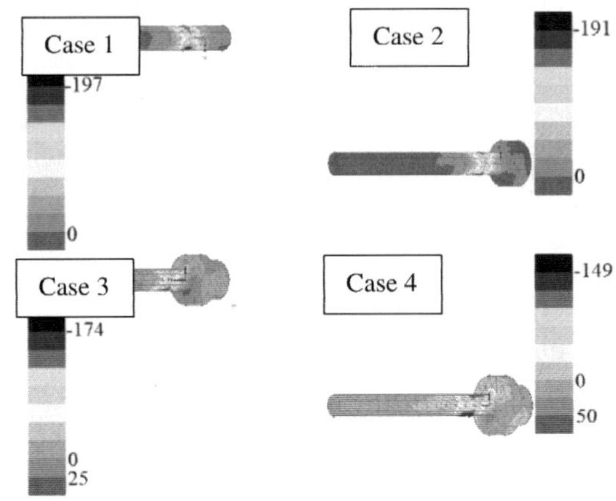


**Figure 21.** The stress distribution in tapping screw with the load level 4.9 (kN) (Unit: MPa).

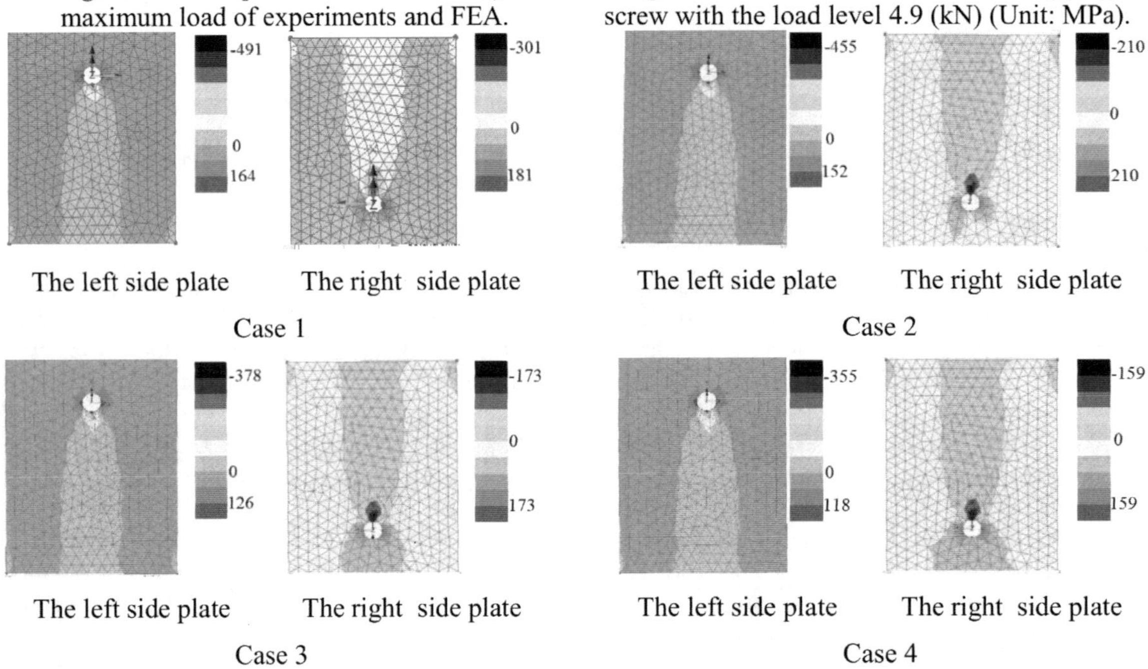


**Figure 22.** The longitudinal stress distribution in FRP plates with the load level 4.9 (kN) (unit: MPa)

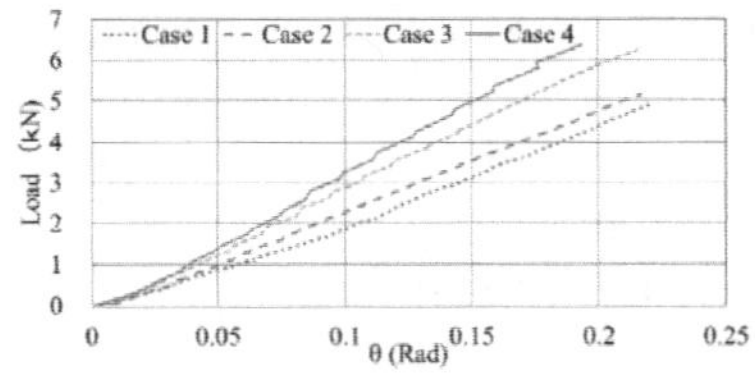


**Figure 23.** Load and Self-tapping screw's rotation relation.

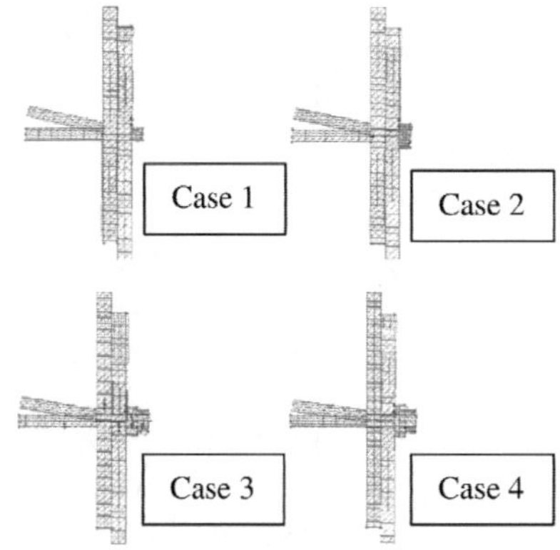


**Figure 24.** Deformation modes in FEA with the load level 4.9 (kN).

## 4. Conclusion

In this study, we surveyed the self-tapping screws connection strength and effects of washers on connection strength of single-lapped GFRP plates using self-tapping screws with end distance was 50mm, under tensile-shear loading by experiment and finite element analysis (FEA).

- Then, the following conclusive remarks are obtained:
- The connection strength in GFRP plates using self-tapping screws depends on washer's diameter; when increasing washer's diameter, the connection strength increased. Hence increasing washer's diameter to a reasonable level is an effective method to increase the load carrying capability for self-tapping screws connection in GFRP plates.
- The occurrence of shear-out failure is the reason load early reached the maximum load and rapidly decreased then. Hence, increasing the length of end distance to a reasonable level to avoid the shear-out failure is a good method to increase connection strength and connection stability, avoid the sudden damage.
- We proposed the truss elements model to simulate the pull-out behavior of connection in FEA. We conducted experimental pull-out test in order to determine truss element properties in finite element analysis model. There were good agreements between FEA results and experimental results in pull-out test and single-lapped tensile-shear test. Hence, the pull-out behavior of connection can be simulated by truss elements and failure of connection is determined by yielding of truss elements. This simulation method can be used to investigate mechanical behavior of single-lapped and double-lapped GFRP plates connection using single or multiple self-tapping screws in the future research.

## 5. References

[1] JEC Group：THE CHALLENGING WORLD OF COMPOSITES，2013.6
[2] Alfred Kofi GAND, Tak-Ming CHAN, James Toby MOTTRAM, "Civil and structural engineering applications, recent trends, research and developments on pultruded fiber reinforced polymer closed sections: a review," Front. Struct. Civ. Eng. 2013, 7(3): 227–244. DOI 10.1007/s11709-013-0216-8.
[3] Yukihiro Matsumoto, Seishi Yamada and Iwao Komiya: Nonlinear Failure Behavior and Bearing Strength of Bolted Joints in Fiber Reinforced Polymer Plates, Proceeding of The SAMPE Tech 2012 (SAMPE Tech 2012) (North Charleston, USA), 2012.10
[4] Nguyen Ngoc Duong, Phan Viet Nhut, Chito SATAKE, Yukihiro MATSUMOTO: Effect of Washers on Mechanical Behavior of Single-lapped Connections in GFRP Plates using Tapping Screws, The 12th Symposium on Research and Application of Hybrid and Composite Structures- Architectural Institute of Japan (7), 2017.11.

[5] Yukihiro MATSUMOTO, Yuya INOUE and Kazunari MATSUNO: Fundamental Study on Joint Strength using Tapping Screw for GFRP, Proceeding of the 12th International Symposium on Fiber Reinforced Polymers for Reinforced Concrete Structures (FRPRCS-12) & the 5th Asia-Pacific Conference on Fiber Reinforced Polymers in Structures.

[6] Nguyen Ngoc Duong, Phan Viet Nhut, Chito SATAKE, Yuya INOUE, Yukihiro MATSUMOTO: Influence of Washers on Mechanical Behavior of Single Lap Connection in GFRP Plates using Tapping Screws, Summaries of Technical Papers of The 72nd Annual Meeting-Japan Society of Civil Engineers (CS5-020), 2017.9.

[7] Yuya Inoue, Nguyen Ngoc Duong, Chito Satake and Yukihiro Matsumoto: Mechanical Characteristics of Connection for GRFP Plates using Tapping Screws, International Conference of Global Network for Innovative Technology & AWAN International Conference in Civil Engineering (AICCE) 2017.

**Acknowledgment**

In this study, GFRP materials were supplied by AGC MATEX CO., LTD and Self-tapping screws were supplied by Japan Power Fastening Co Ltd.

IOP Conf. Series: Materials Science and Engineering **371** (2018) 012028 doi:10.1088/1757-899X/371/1/012028

# Seismic Strengthening Design and Performance Analysis of an Existing RC Frame

**Sun Yong [1], Chen Xin [2], Zhu Yan-qing[2], Yu Wei-gen[1], Li Jia-qing[1], and Su Guan-xing[1]**

1 Jiangsu Research Institute of Building Science co., Ltd, Nanjing 210008, China
2 Jiangsu Province Key Laboratory of Structure Engineering, Suzhou University of Science and Technology, Suzhou 215011, China
E-mail: tmxsuny@gmail.com

**Abstract.** The reinforcement design and seismic performance analysis of the Teaching Building of Jiangyan Experimental Elementary School are investigated in this paper. Firstly, according to the seismic assessment report provided by an inspection company, the seismic strengthening design is conducted by using the methods of enlarging sections and sticking CFRP. Secondly, a three-dimensional model is established by using the structural analysis software OpenSees, and both the dynamic characteristics and seismic responses of the structure under frequent earthquakes are analysed. Lastly, the dynamic elastic-plastic time history analysis of the structure is conducted under rare earthquakes, then seismic performance of the structure with and without strengthening are compared under far-field earthquakes. The results show that the seismic performance of the structure is effectively improved, as the maximum reduction of the storey drifts can reach 35%, when the structure is subjected to rare earthquake.

## 1. Introduction

China is a country with severe earthquakes which have the characteristics of great frequency, high intensity, shallow seismic focus and wide distribution. Only in 2016, there were 17 earthquakes, which induced 119 people killed, 662 people wounded and a direct economic loss of 6.67 billion RMB. Therefore, earthquake retrofitting plays an important role in the disaster prevention of the urban and rural area in China.

Most buildings in middle and primary schools in China are composed of multi-storey masonry structure and reinforced concrete frame structure, many of which are running the risks of damaging due to the low earthquake resistance level or exceeded design life. During Wenchuan earthquake in 2008, the collapsing of school buildings had led to serious death and injury of teachers and students, which accounts 7% of total deaths. Therefore, many school buildings in China have been assessed and strengthened since 2008. During these projects, several common methods are used in reinforcement concrete structures retrofitting, such as enlarging concrete section, bonded with steel plate or CFRP, and strengthening with hysteretic dampers [1,2] etc. A lot of theoretical and experimental research has been conducted relating to the retrofitting methods for existing RC frame structures by now [3-5]. Xu Cheng-xiang et al. [6] summarized the quasi-static loading simulation method of earthquake damage of concrete structures and the simulation method of that by inputting seismic wave to vibration table and then compared the major reinforcement methods to damaged concrete structures after earthquakes. Wang Su-yan et al. [7,8] studied the seismic performance of damaged high-strength square columns after strengthened with CFRP. The effects of different reinforcement methods on the improvement of displacement ductility coefficient, energy dissipation capacity and stiffness degradation of high-strength concrete columns were discussed. The study indicated that the seismic performance of repaired earthquake damaged members has been significantly improved. The ductility was far better than that of the comparative members but still slightly lower than that of members directly reinforced without seismic damage. Wang Xin-ling [9] applied a horizontal low-cycle repeated load on a 1/3 scale reinforced concrete frame model to study the effect of the degree of damage on the seismic performance of the strengthened structure. The experimental results showed that with the increase of the damage, the greater the maximum crack width of the frame strengthened by CFRP, the larger the strain hysteresis of CFRP. When the degree of damage became severe, the ductility of the strengthened structure was reduced.

This paper mainly focuses on the seismic strengthening design and performance analysis of a teaching building in the Experimental Primary School of Jiangyan District, Taizhou City, China. The structural characteristics of the RC frame are firstly introduced and then the seismic strengthening design is carried out according to the identification results. A three-dimensional finite element model is established to conduct the dynamic characteristics analysis and the dynamic time history analysis of the structure under the frequent earthquake. Finally, the elastic-plastic seismic behaviour of the structure under rare earthquakes is analysed, and the seismic performance of structure with and without reinforcement under far-field earthquakes is compared.

## 2. Project Overview and Strengthening Design

### *2.1 Project overview*

The teaching building in Jiangyan Experimental Primary School of Taizhou City (Figure 1) was built in 2007 which is a five-story reinforced concrete frame structure with a total area of 2500$m^2$, a height of 14.400m and an independent foundation. The structure is located in Jiangyan District, Taizhou City. The building is in class B of seismic fortification and the structural safety level is 2. The fortified earthquake group is the second group and the seismic fortification intensity is 7th degree. The design basic seismic acceleration is 0.1g, and the site category class is III, with a site characteristic period of 0.55s. It is also assumed that the structural damping ratio is 5%.

(a) side view (b) Backside

**Figure 1.** The teaching building.

### *2.2 Comprehensive reinforcement design*

According to the assessment report for the structure, some of the concrete components can't meet the anti-seismic requirements of the Chinese Codes. Therefore, in order to improve both the rigidity and bearing capacity of the structure, the section enlargement method was used to strengthen the columns, and the enlarged section and reinforcement were designed according to Chinese Specification JGJ 116-2009. As shown in Figure 2, The cross section of reinforced column JKZ1 is increased from 400mm x 400mm to 550mm x 550mm, that means the four sides are increased by 75mm, 75mm, 0mm, and 100mm respectively. The reinforcements of enlarging-section column section are 4C22+(2C18+2C22) +(2C18+2C22) with stirrups of C8@100/200. The cross section of reinforced column JKZ2 is increased from 350mm x 350mm to 500mm x 450mm, that means the four sides are increased by 75mm, 75mm, 0mm, and 100mm respectively. The reinforcements of enlarging-section

IOP Conf. Series: Materials Science and Engineering **371** (2018) 012028 doi:10.1088/1757-899X/371/1/012028

column section are 4C22+(1C18+C22) +2C18) with stirrups of C8@100/200. The C40 grouting material is adopted to increase column section.

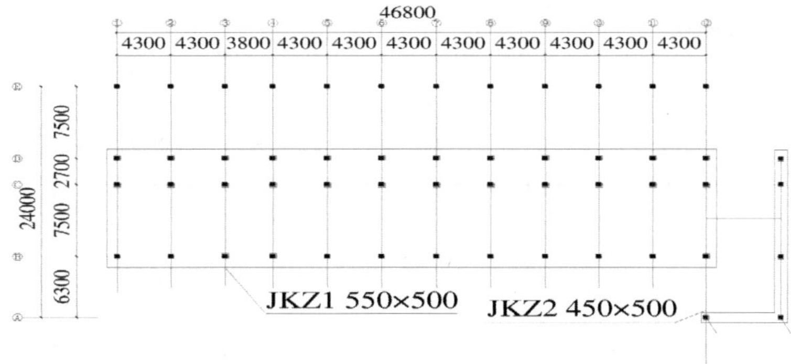


(a) Column floor plan

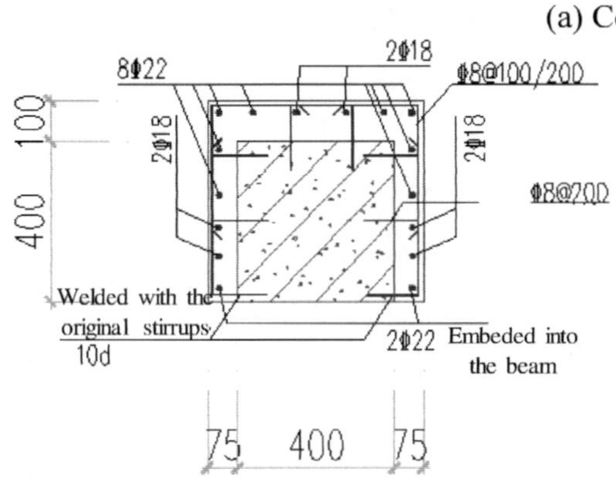


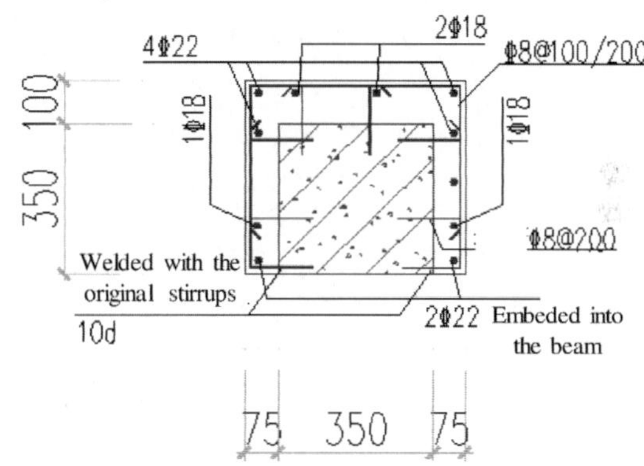


(b) detail drawing of column 1 (c) detail drawing column 2

**Figure 2.** Reinforcement method of the frame columns.

Concrete frame beams are bonded with CFRP to improve their ductility, as shown in Figure 3. The detailed method is to stick one-layer CFRP with the width of 200mm or 300mm along the bottom of the beam, meanwhile, the side of the beam affixes one-layer layering and U-shaped hoops of CFRP, in which, the width of the strip and the U-type hoop are 100mm and 200mm respectively. The spacing of U-shaped hoops of CFRP is 400mm and the specification of all the CFRP is $200g/m^2$.

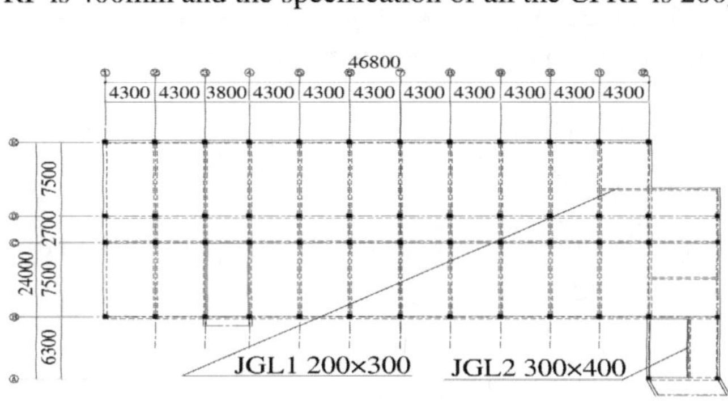


(a) layout of beams

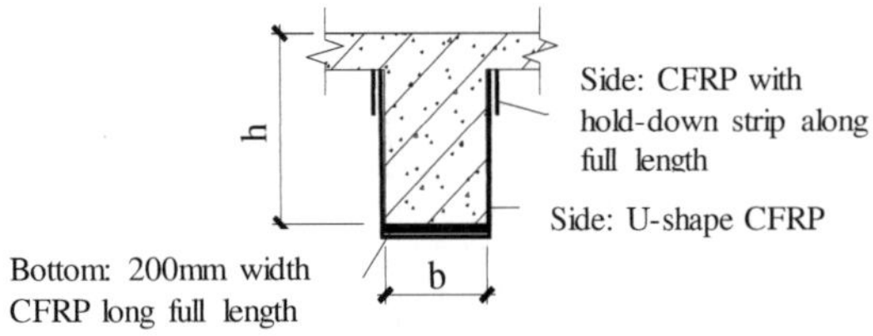


(b) detail drawing of beam 1

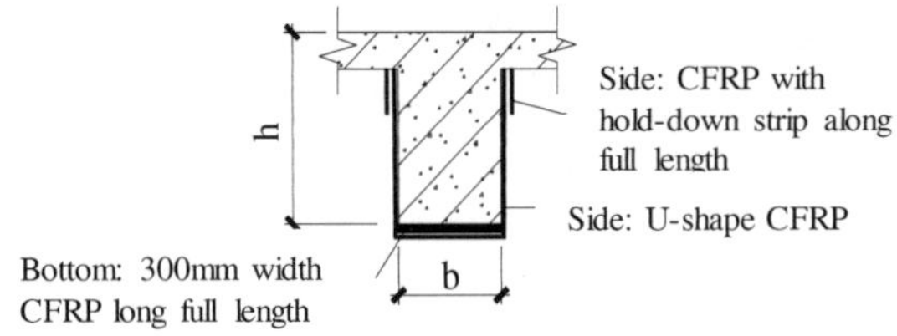


(c) detail drawing of beam 2

**Figure 3.** Reinforcement method of the frame beams.

## 3. Finite element model and dynastic characteristics

### *3.1. Finite element model*

The finite software OpenSees is used to establish a three-dimensional finite element model of the teaching building [10]. As shown in Figure 4, the Elastic-Beam-Column element is adopted to simulate beams and columns under the frequent earthquakes. Connections between Beams and columns are rigid and the bottom of columns are fixed.

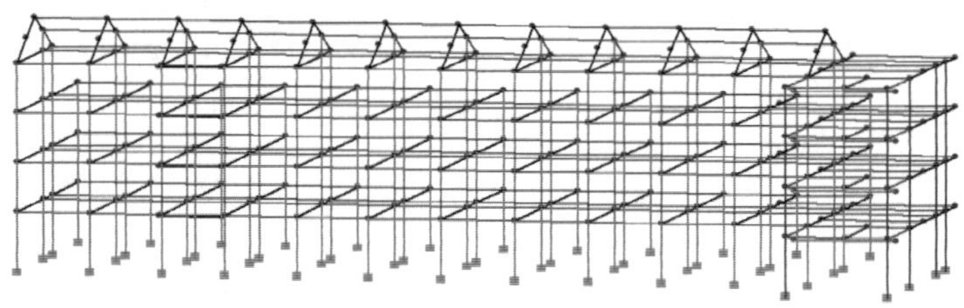

**Figure 4.** Finite element model.

The original part of the columns adopts C20 concrete and HRB400 steel bars and the enlarged part uses C40 grouting material and HRB400 steel bars. Concrete and reinforcement are modelled using Concrete02 and steel 01 model respectively, and CFRP is simulated by elastic material with elastic modulus of 2.4x105 MPa. Where, Concrete02 is based on the Kent-Scott-Park model without considering the restraint effect of steel bars. The main parameters of this constitutive model are shown in Figure 5(a). Where, $E_0$ is the initial elastic modulus of concrete, $f_{c,m}$ and $\varepsilon_0$ are the peak strain and stress of concrete respectively, $\varepsilon_u$ and $f_u$ are the ultimate strain and ultimate stress of concrete, $f_t$ is the peak stress of concrete under tension, $E_t$ is the degradation modulus of concrete in tension zone, and $\lambda$ is the reduction coefficient of initial elastic modulus at the ultimate compressive stress of concrete. Steel01 uses the ideal elastic-plastic model as shown in Figure 5(b). Where, $f_y$ is the yield strength, $E_0$ is the modulus of elasticity and $b$ is the coefficient of hardening.

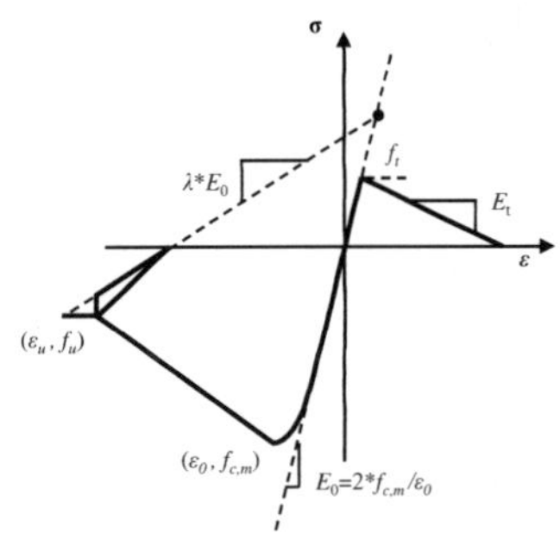


(a) concrete material

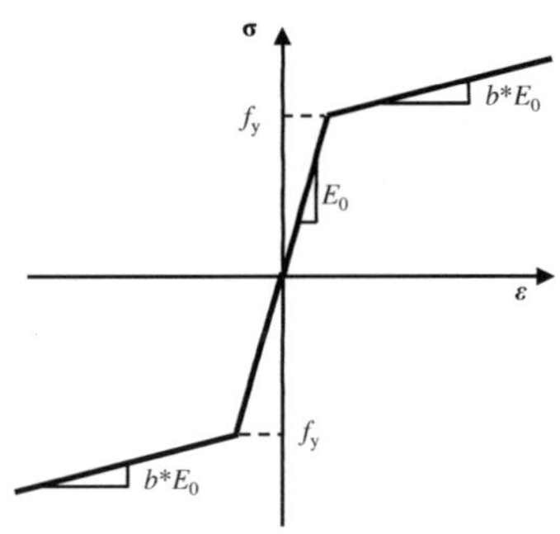


(b) steel material

**Figure 5.** Stress-strain relationship of the materials' constitutive model.

IOP Conf. Series: Materials Science and Engineering **371** (2018) 012028 doi:10.1088/1757-899X/371/1/012028

*3.2. Dynamic characteristics*

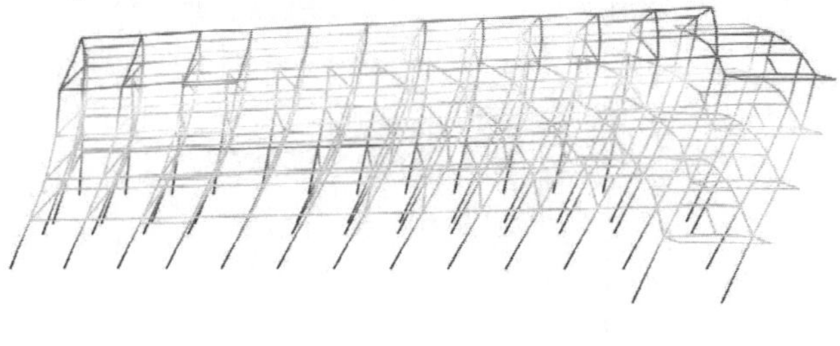

(a) 1st mode

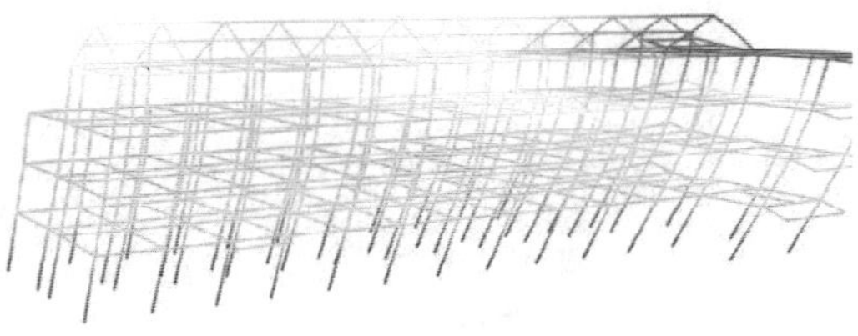

(a) 2nd mode

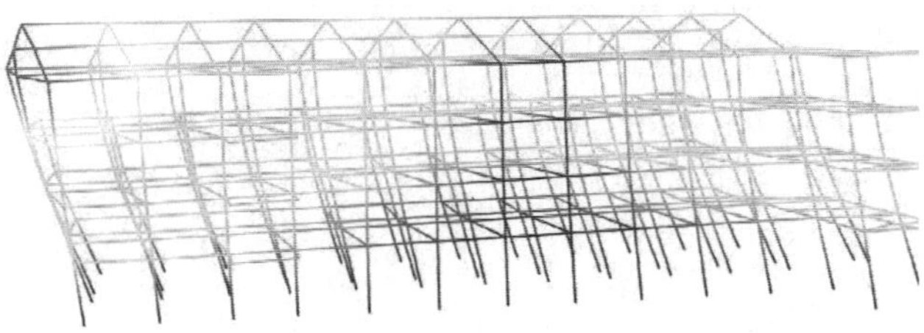

(c) 3rd mode

**Figure 6.** Vibration mode shape of the first three modes.

In order to verify the model and analyse the dynamic characteristic of the structure, two structural models are established by using software PKPM and ETABS respectively. The first three natural period of the original structure are shown in Table 1. It can be seen that they are very close to each other. The first natural period of the vibration differs by only 3.3% and the maximum difference between the third natural period is 7.8%, so it can be concluded that the finite element model using OpenSees is reliable. The first three vibration modes of the structure are shown in Figure 6. It can be seen that the first and second modes are the translational mode and the third mode is the torsional vibration mode. Although the structure has some corner planes, it still reveals the vibratory features of regular frameworks.

**Table 1.** First three modes of the structure.

| Number of stages | Natural period of vibration/ s | | |
|---|---|---|---|
| | PKPM model | ETABS model | OpenSees model |
| 1 | 1.094 | 1.0577 | 1.0590 |
| 2 | 1.016 | 1.0385 | 1.0326 |
| 3 | 0.942 | 0.997 | 1.0154 |

## 4. Elastic seismic behaviour of structures subjected to earthquakes

*4.1. Seismic waves*

Three seismic ground motions are selected for dynamic time history analysis. The response spectrum curves of these three seismic waves are shown in Figure 7 (a). By comparing with the normalized Chinese Code response spectrum curves, the selected seismic waves meet the code's requirements of statistical significance in the frequency spectrum energy. The acceleration time history curve of Kern County ground motion is shown in Figure 7 (b). By using these earthquake waves, the structural responses under both frequent earthquake and rare earthquake are analysed respectively. Here, frequent earthquake is defined as the earthquake with a 50 years exceedance probability of 63%, while rare earthquake is the earthquake with 50 years exceedance probability of 2-3%.

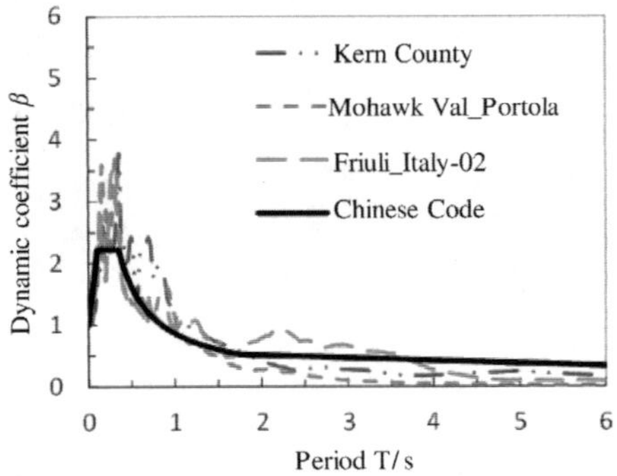


(a) Response spectrum curves of seismic waves

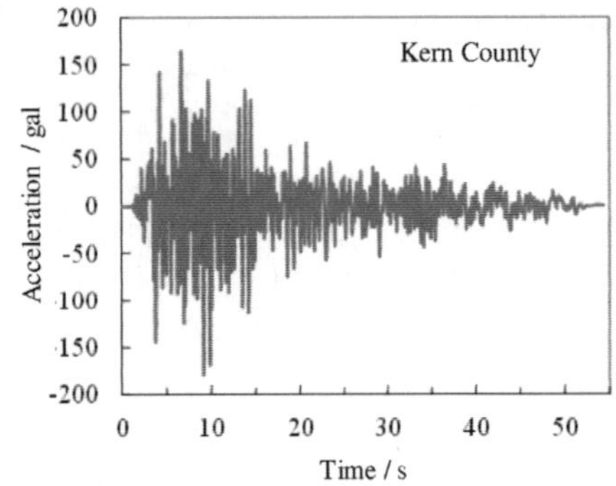


(b) Kern County wave

**Figure 7.** Seismic wave for time history analysis.

*4.2. Structural response under frequent earthquakes*

*4.2.1. Base shear force*

**Table 2.** Base shear force under frequent earthquakes.

| Direction | Horizontal | Longitudinal |
|---|---|---|
| Response spectrum of PKPM | 1588.10kN | 1622.20kN |
| Response spectrum of ETABS | 1573.09kN | 1488.53kN |
| Kern County | 1191.28kN | 1242.91kN |
| Mohawk Val Portola | 1135.19kN | 1258.38kN |
| Friuli Italy-02 | 1294.70kN | 1320.97kN |

Response spectrum analysis is proceeded by using PKPM and ETABS, in which the accidental eccentricity and 15 modes of vibration are taking into account. The three seismic waves are amplitude-modulated to 35gal and the elastic time-history analysis is conducted by using OpenSees. As shown in Table 2, under the same conditions, the values of the base shear force in the longitudinal and transverse directions of the structure are close to each other. Combined with the results in Table 1, it shows that due to the square cross-section of the frame columns, the stiffness in both directions and the mass source of each storey of the structure are equal, which leads to the relatively close dynamic characteristics of the two directions in the elastic state.

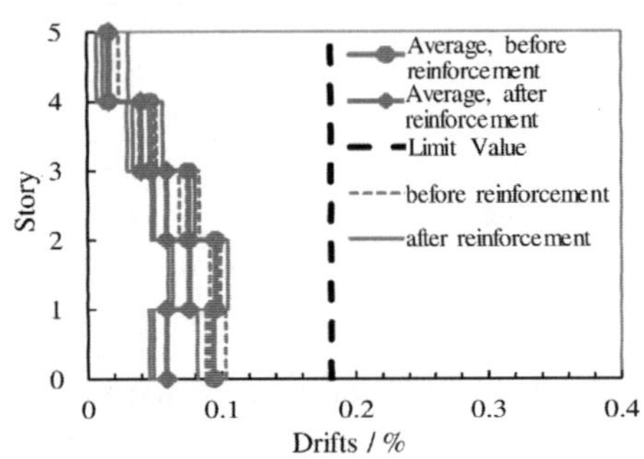


(a) Storey drifts

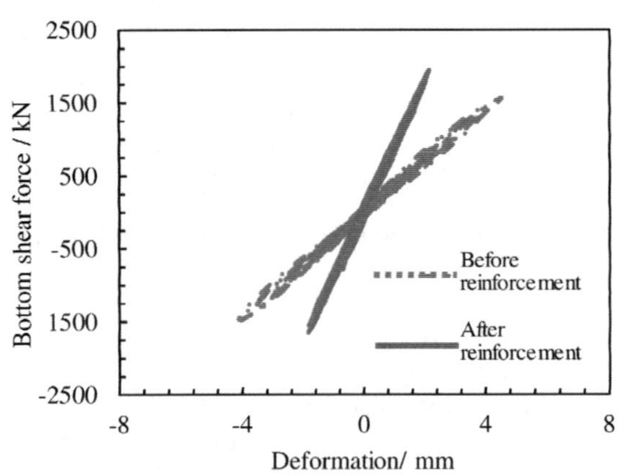


(b) Base shear force-displacement curve under Friuli Italy-02 earthquake

**Figure 8.** Lateral displacement responses of the structure under frequent earthquakes.

### *4.2.2. Structural displacement response*

Figure 8 shows the displacement responses of the structure under frequent horizontal earthquakes. It can be seen that: (1) In Figure 8 (a), the drift of the first storey is the largest before retrofitting, and decreases obviously after retrofitting. However, the weak storey is transferred from the first one to the second one. (2) As shown in Figure 8 (b), the degradation of interlayer stiffness of first storey is not obvious under frequent earthquakes. So the bottom storey with the largest structural deformation is still in elastic state, and its residual deformation is very small.

## 5. Structural elastic-plastic seismic behaviour under earthquakes

### *5.1. Structural displacement response*

The seismic waves are exemplified to a peak acceleration of 220gal for rare earthquakes, and then structural elastic-plastic time-history analysis under rare earthquakes is carried out. Figure 9 shows the displacement responses of the structure under rare earthquake. By comparing Figure 9 (a) and (b), it can be seen that: (1) the average of maximum storey drifts of rare earthquakes before strengthening is only 45.8% of the code's limit value; (2) After the strengthening, the storey drift of the bottom floor of the structure has a large attenuation; (3) After strengthening, the storey drifts of the 1st, 2nd and 3rd storeys of the structure tend to be close, and the structural weak storey under earthquakes transforms to the 2$^{nd}$ storey which is similar to that of elastic time history analysis.

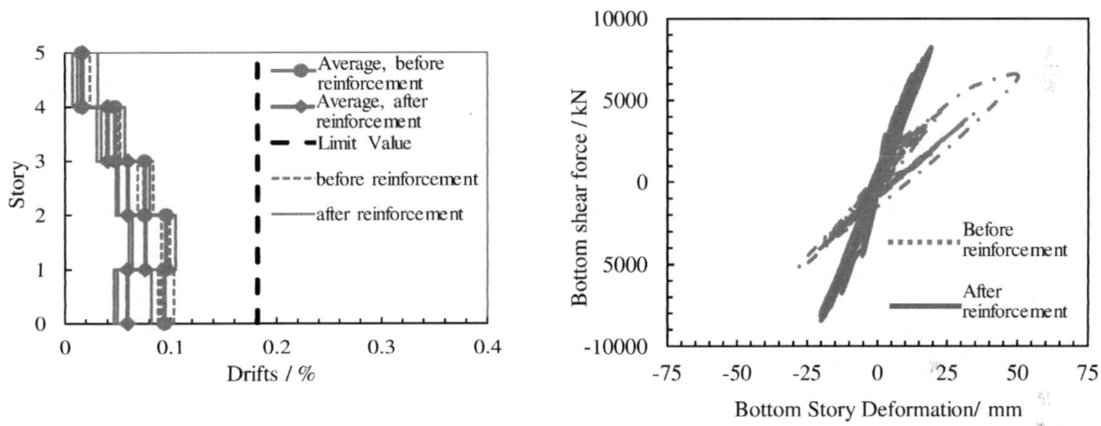


(a) Storey drifts

(b) Base shear force-displacement curve under Friuli Italy-02 earthquake

**Figure 9.** Lateral displacement responses of the structure under rare earthquakes.

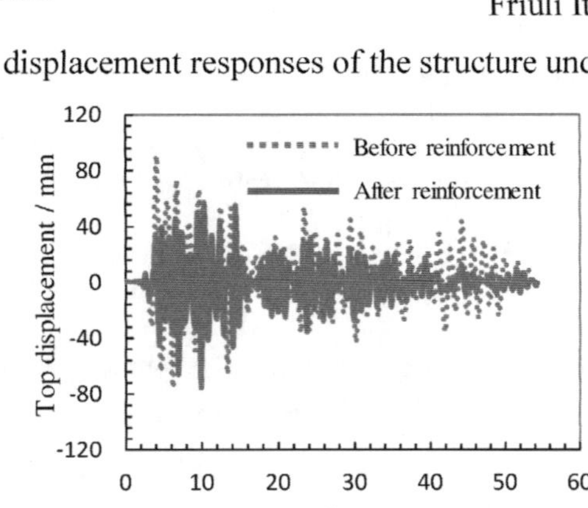


**Figure10.** Lateral top displacement time history curves of the structure under rare earthquakes.

Figure 10 shows the time history curves of the structural top displacement under rare earthquakes. It can be seen that under the same seismic ground motions, both the top displacements of the structure are small before and after strengthening, and the time history curve is flat under far-field earthquakes. After strengthening, the top displacement of the structure at the pulse time decreases by 23.6% compared with that before strengthening. So, the reinforcement scheme greatly reduces the damage probability of the structure at this moment and improves the aseismic performance of the structure.

### *5.2. Overall structural deformation*

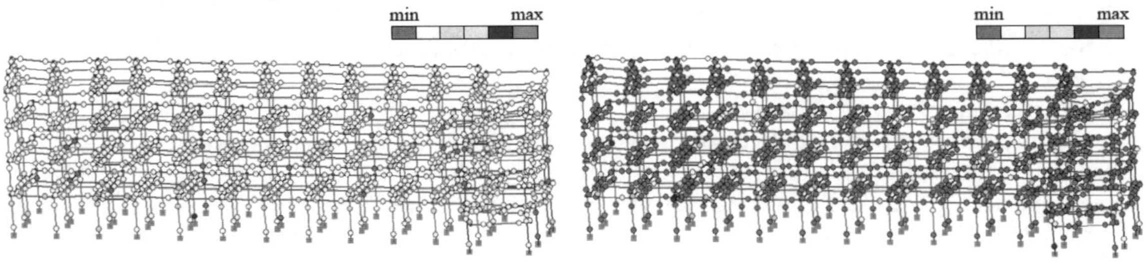


(a) Before strengthening (b) After strengthening

**Figure 11.** Structural damage under rare earthquakes.

Further analysis is carried out on the overall damage distribution of the structure under rare earthquakes. As shown in Figure 11, the structural damages of all structural members are small and the changes before and after strengthening are significant. In general, the comprehensive reinforcement scheme in this paper improves the seismic performance of the reinforcement concrete frame structure obviously.

## 6. Conclusions

In this paper, the seismic strengthening design and the seismic performance analysis of a teaching building in Jiangyan Experimental Primary School are presented. The main conclusions are as follows:

1) The method of enlarging section effectively increases the stiffness of the bottom columns, so that the displacement angle of the bottom storey of the structure decreases obviously comparing with that of the other storeys. The method of bonding CFRP can effectively improve the bearing capacity and ductility of the frame beams.

2) After strengthening, the structural performance parameters, such as the storey drifts, the top displacement and the base shear of the structure, have been attenuated in different degree, and the maximum attenuation of storey drifts can reach 35%.

## 7. References

[1] Benavent-Climent A., Mota-Páez S. Earthquake retrofitting of R/C frames with soft first story using hysteretic dampers: Energy-based design method and evaluation[J]. Engineering Structures, 2017, 137: 19-32.

[2] Khampanit A., Leelataviwat S., Kochanin J., et al. Energy-based seismic strengthening design of non-ductile reinforced concrete frames using buckling-restrained braces[J]. Engineering Structures, 2014, 81: 110-122.

[3] Merczel D. B., Aribert J. M., Somja H., et al. Plastic analysis-based seismic design method to control the weak storey behaviour of concentrically braced steel frames[J]. Journal of Constructional Steel Research, 2016, 125: 142-163.

[4] Qu Z., Kishiki S., Maida Y., et al. Seismic responses of reinforced concrete frames with buckling restrained braces in zigzag configuration[J]. Engineering Structures, 2015, 105: 12-21.

[5] Stefani L D, Scotta R, Lazzari M. Optimal design of seismic retrofitting of RC frames with eccentric steel bracing[J]. Bulletin of Earthquake Engineering, 2015, 13(2):613-633.

[6] Xu Cheng-xiang, Peng Wei, Xu Kai-long. Research status and prospect of the reinforcement of seismic damaged concrete frame structure [J]. Journal of Yangtze University(Natural Science Edition), 2014, 11(2): 68-71(in Chinese)

[7] Wang Su-yan, Cao Huai-chao, Liu Yi. Experimental investigation on seismic behavior of damaged high strength concrete columns repaired with CFRP sheets[J]. Journal of Railway Science & Engineering, 2012, 9(3): (in Chinese).

IOP Conf. Series: Materials Science and Engineering **371** (2018) 012028 doi:10.1088/1757-899X/371/1/012028

[8] Wang Su-yan, Yu Wen-hua. Experimental research on ductility behavior of souare high concrete columns retrofitted with FRP hybridized methods[J]. China Civil Engineering Journal, 2010, (S1): 429-435(in Chinese).
[9] Wang Xin-ling, Fan Jian-wei, Yao Zhang-tang, et al. Damaged degree effect on seismic behavior of RC frames strengthened by CFRP sheets[J]. Building Structure, 2011, 41(6): 94-97(in Chinese).
[10] Qi Hu, Sun Jing-jiang, Lin Lin. Research on fiber model of OPENSEES[J]. World Earthquake Engineering, 2007, (04): 48-54(in Chinese).

**Acknowledgments**
The authors are grateful for the joint supports of the National Natural Science Foundation of China (No. 51408389), Natural Science Foundation of Jiangsu Province (No. BK20140281, No. BK20161581) and Qing Lan Project.

IOP Conf. Series: Materials Science and Engineering **371** (2018) 012029 doi:10.1088/1757-899X/371/1/012029

# A Review of Geometry Investigations of Helicoids

**V Jean Paul**

MSc, PhD student, Department of Architecture and Civil Engineering, Peoples' Friendship University of Russia (RUDN University), Moscow, Russian Federation
E-mail: jeanpaulvladimir@yahoo.fr .

**Abstract.** This paper presents research on geometry of helicoids which occurred in different papers. The purpose of the paper is to show a variety in types of helical surfaces which are not well known, and discuss their possible application in architecture and civil engineering. This review is a first stage of a wide investigation on stress-strain analysis methods as applied to helical structures. Author's attention is paid to the differences in terms, geometry and materials used for these shells, and future possibilities of application of five types of helicoid are also discussed. The work would be interesting for architects, civil engineers, mathematicians and designers, and could lead to a progress in material usage efficiency.

## 1. Introduction

Helical shells have been known since the time of Archimedes, who proposed to use the so-called 'Archimedes screw' to raise water up to 0.5 m in height. Later, these shells were used as auger blades for drilling wells of various purposes: exploration, hydrogeological, seismic, drainage and other technical wells. According to [1] there are five types of helicoids: Right helicoid, Oblique helicoid, Developable helicoid, Convolute helicoid, Pseudo-developable helicoid. These types with different names were found by author in several sources. One of the goals of this paper is to investigate the terms and geometry of different types of helicoids.

After the destruction of the Twin Towers in New York, architects and engineers have set of much thinking which was written in the Y.S. news and world report. Despite the temporary doubts, the demand for skyscrapers has not decreased. One of the reasons of popularity of skyscrapers is the non-availability of the lands for construction and its high cost; especially it concerns the large cities as, the smaller areas are necessary for multistoried buildings, they are very relevant for modern megalopolises in which millions people live. Similar architectural concepts reduce the price of the cost of inhabited office and other rooms.

Besides high-rise buildings in the modern world, there is some kind of polish up for megalopolises which emphasizes their prestige and wealth. William Mitchell, the dean of faculty of architecture and planning of the Massachusetts Institute of Technology, says [2]"Giant buildings are tribute to fashion, prestige and similar."

Due to the continuous scientific and technical progress, it has become possible to embody almost any innovative ideas in architecture. Scientific and technical progress promotes search of new forms of the screw line surfaces applicable for construction combining such qualities as beauty, reliability and technological effectiveness.

Today one of the fashionable architectural concepts is used in construction of screw surfaces which not only have attractive design, but also reduce the area used for construction and efficient material usage.

IOP Conf. Series: Materials Science and Engineering **371** (2018) 012029 doi:10.1088/1757-899X/371/1/012029

As the world of surfaces is various and boundless, the innovative ideas in the sphere of architecture are inexhaustible, and geometry of different types of surfaces is worth being investigated from several points (less material usage, optimization of the structure according to stress-strain analysis, etc.).
The helicoid was first ever documented by J. Meusnier [3] around 1775, it was the first unbounded non-periodic minimal surface to be found. Besides the plane, it is the only known ruled minimal surface (zero mean curvature). Important contribution to the theory of helical shells was made by L. Euler (Switzerland) [4], G. Monge (France), E. Catalan [5], G. Darboux [6], D. Hilbert [7] and many others.
The Minimal Surface has a Helix as its boundary. For many years, the helicoid remained the only known example of a complete embedded Minimal Surface of finite topology with infinite Curvature. However, in 1992 a second example, known as Hoffman's Minimal Surface and consisting of a helicoid with a Hole, was discovered [2].
In the 1960s, Scientists began to be interested in calculating the helical shells, and in 1960 the first reinforced concrete helix was built in Poland. In the first half of the 20th century, only single objects were realized, but presently, helical shell structures are becoming popular again [8], as well as helicoids in mathematics [9].
The main idea of this paper is to investigate the differences in helicoids geometry and start discussing the possibilities of their future shape optimization.

## 2. Geometry investigations

### *2.1. General information*

Geometrically, the helicoid is defined by simultaneously rotating and translating a line at constant speed at an axis to which it is perpendicular [10]. According to [11], the helicoid is the surface generated by the lines parallel to the *xy* plane which go through each point of the helix and intersect the *z* axis.
Helicoid is a ruled surface swept out by a line which always intersects a fixed axis at right angles and which rotates uniformly as its point of intersection moves uniformly along the axis and it intersects any cylinder concentric with the axis in a helix. The helicoid has the shape of the thread of a screw. The helicoid is a minimal surface. The intersection of the helicoid with a cylinder produces a helix, which would be equivalent to the railing of a spiral staircase. The ruled helicoid that is obtained when the generating curve is a straight line intersecting the axis at right angles is called a right conoid.
The helicoid has a wide variety of shapes, depending on the pitch, the proximity of the lines and the points, and whether the pairs of points connected by the rulings are in fact not quite diametrically opposite.
The helicoid and the catenoid are parts of a family of helicoid-catenoid minimal surfaces [12].
Helicoids play a very important role as minimal surfaces. In the technical area generalized helicoids are used for staircases, screws and helical pipes [13, 14].
A generalized helicoid is a surface in Euclidean space generated by rotating and simultaneously displacing a curve, the profile curve, along a line, its axis. Any point of the given curve is the starting point of a circular helix. If the profile curve is contained in a plane through the axis, it is called the meridian of the generalized helicoid. Simple examples of generalized helicoids are the helicoids. The meridian of a helicoid is a line which intersects the axis orthogonally.
Very important types of generalized helicoids are:

- ✓ The ruled generalized helicoids. Their profile curves are lines and the surfaces are ruled surfaces.
- ✓ The circular generalized helicoids. Their profile curves are circles.

The helicoid is shaped like Archimedes screw, but extends infinitely in all directions. According to [10], it can be described by the following parametric equations in Cartesian coordinates:

$$x = \rho \cos(\alpha\theta) \tag{1}$$

$$y = \rho \sin(\alpha\theta) \tag{2}$$

$$z = \theta \tag{3}$$

where $\rho$ and $\theta$ are range from negative infinity to positive infinity, while $\alpha$ is a constant. If $\alpha$ is positive, then the helicoid is called right-handed; if negative then it is called left-handed.
The helicoid has principal curvatures

$$\pm\frac{\alpha}{(1+\alpha^2\rho^2)}. \tag{4}$$

The sum of these quantities gives the mean curvature (zero since the helicoid is a minimal surface) and the product gives the Gaussian curvature [15].
The helicoid is homeomorphic to the plane $R^2$. To see this, let alpha decrease continuously from its given value down to zero. Each intermediate value of $\alpha$ will describe a different helicoid, until $\alpha = 0$ is reached and the helicoid becomes a vertical plane.
Conversely, a plane can be turned into a helicoid by choosing a line, or axis, on the plane, then twisting the plane around that axis:

$$\pi\ [R\sqrt{(R^2+h^2)} + h^2 * In(\frac{R+\sqrt{R^2+h^2}}{h})] \tag{5}$$

### *2.2. Five types of helicoid*

The helicoid is a special case of an ordinary screw surface. The ordinary screw surface is formed by the ordinary screw movement and characterized by points trajectory which represent the cylindrical screw lines with a constant step lying on coaxial round cylinders. As stated earlier, this paper is focusing on five types of helical surfaces which are shown on figures 1-5.

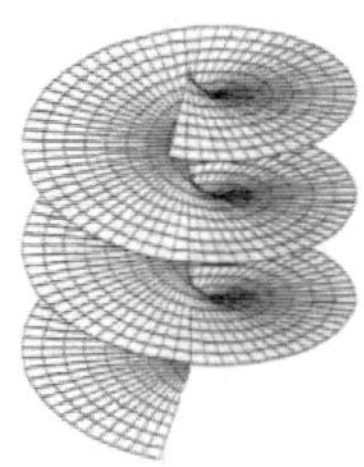

**Figure 1.** Right helicoid.

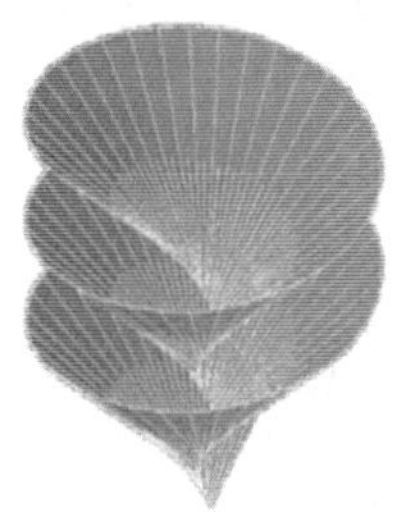

**Figure 2.** Inclined helicoid.

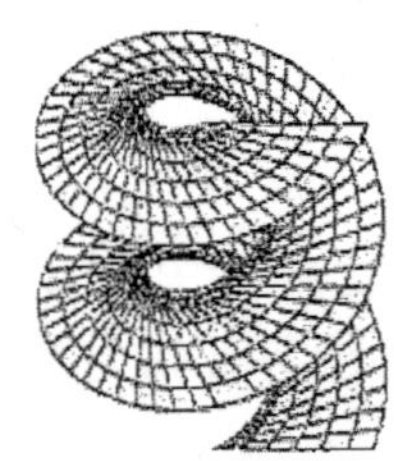

**Figure 3.** Deployable helicoid.

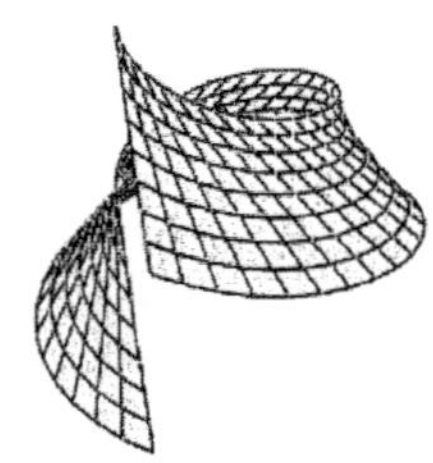

**Figure 4.** Convoluted helicoid.

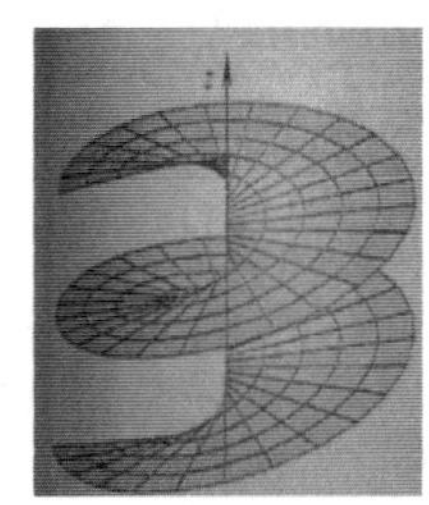

**Figure 5.** Elliptic helicoid.

*2.2.1. Right helicoid.* Right helicoid is called the screw line surface described by a straight line which crosses a helicoid axis at right angle rotates with a constant angular speed around this axis and at the same time moves progressively along the same axis [10]. Sometimes it is also called 'straight', as in [16].
Parametric form of right helicoid (**figure 1**) and its coefficients of quadratic forms [10]:

$$x = (r, v) = rcosv \tag{6}$$
$$y = (r, v) = rsinv \tag{7}$$
$$z = c \tag{8}$$
$$A = 1 \tag{9}$$
$$F = 0 \tag{10}$$
$$B^2 = r^2 + c^2 \tag{11}$$
$$L = N = 0 \tag{12}$$
$$M = {}^{-C}/_{B} \tag{13}$$

*2.2.2. Inclined helicoid.* Inclined helicoid is called 'oblique' helicoid in [1, 10]. Formation of inclined helicoid is similar to formation of a screw conoid: formation moves on the screw line and on his axis, remaining at the same time, all the time parallel consecutive forming, a direct circular cone (figure 2).

IOP Conf. Series: Materials Science and Engineering **371** (2018) 012029 doi:10.1088/1757-899X/371/1/012029

If the height of the directing cone equal to zero, then the inclined helicoid will turn into a screw conoid. Thus, the screw conoid is a special case of an inclined helicoid; forming a screw conoid are perpendicular to a surface axis, with reference to this surface, is called differently a direct or right helicoid.
Parametric form of inclined helicoid (figure 2):

$$x = x(r, v) = rcosv \tag{14}$$

$$y = y(r, v) = rsinv \tag{15}$$

$$z = cv + kr \tag{16}$$

Coefficients of the basic quadratic forms of the surface:

$$A^2 = 1 + k^2 \tag{17}$$

$$F = ck \tag{18}$$

$$B^2 = r^2 + c^2 \tag{19}$$

$$L = 0 \tag{20}$$

$$M = \frac{-C}{\sqrt{B^2 - F^2}} \tag{21}$$

$$N = \frac{kr^2}{\sqrt{B^2 - F^2}} \tag{22}$$

$$k_r = 0 \tag{23}$$

$$k_v = \frac{N}{B^2} \tag{24}$$

where $k < 0$

$$cos\chi = \frac{ck}{\sqrt{r^2 + c^2}} \tag{25}$$

where k = angular coefficient, $k = ctg\alpha$, $\chi$= angle between coordinate lines r and v. The end section of the helicoid at z = 0 yields ρ = $cvtg\alpha$, that is, the Archimedean spiral.

*2.2.3. Deployable helicoid.* The torso surface formed by tangents to the screw line of a constant step on the circular cylinder is called the developable helicoid [1, 10, 17] and torso helicoids in [10], but may be also called deployable helicoid (figure 3). There were found several forms of parametric equations of this helicoid.

a. Parametric form of deployable helicoid.

$$x = x(u, v) = \frac{acosv - au\, sinv}{m} \tag{26}$$

$$y = y(u, v) = \frac{asinv + au\, cosv}{m} \tag{27}$$

$$z = z(u, v) = \frac{bv + bu}{m} \tag{28}$$

$$\text{where } m = \sqrt{a^2 + b^2} \tag{29}$$

b = pitch of helix u = 0 (return edges), v = angle, counted from the axis $Ox$.
The coefficients of the basic quadratic forms of the surface and its principal curvatures:

$$A = 1 \tag{30}$$

$$F = m \tag{31}$$

$$B^2 = \frac{m^2 + u^2 a^2}{m^2} \tag{32}$$

IOP Conf. Series: Materials Science and Engineering **371** (2018) 012029 doi:10.1088/1757-899X/371/1/012029

$$N = \frac{uab}{m^2} \tag{33}$$

$$L = M = 0 \tag{34}$$

The coordinate lines coincide with the rectilinear generators of the helicoid, and the lines v represent coaxial helical lines. A conjugate but non-orthogonal system of curvilinear coordinates.

b. Parametric form of deployable helicoid (figure 3):

$$x = x(u, v) = acosv - ucos\varphi\ sinv \tag{35}$$

$$y = y(u, v) = asinv + ucos\varphi\ cosv \tag{36}$$

$$z = z(u, v) = avtg\varphi + usin\varphi \tag{37}$$

where φ = the angle of inclination of the rectilinear generators of the helicoid to the *xOy* plane, then

$$tg\varphi = \frac{b}{a} \tag{38}$$

The coefficients of the basic quadratic forms of the surface:

$$A = 1 \tag{39}$$

$$F = \frac{a}{cos\varphi} \tag{40}$$

$$B^2 = F^2 + u^2cos^2\varphi \tag{41}$$

$$B^2 - F^2 = u^2cos^2\varphi \tag{42}$$

$$L = M = 0 \tag{43}$$

$$N = usin\varphi cos\varphi \tag{44}$$

*2.2.4. Convoluted helicoid.* The convoluted helicoid is formed by means of the straight line of AB moving in space, and which all the time is crossed with the screw line (figure 4), concerning at the same time the side surface of the direct circular cylinder with radius a. Axes of the screw line and the cylinder coincide and form straight line and axis are not crossed at the right angle.

Parametric form of convoluted helicoid (fig d.).

$$x = x(t, v) = \mathrm{a} \cos v - t\ sin\ \gamma \sin v \tag{45}$$

$$y = y(t, v) = \mathrm{a} \sin v + t\ sin\gamma\ cosv \tag{46}$$

$$z = z(t, v) = pv + t\ cos\gamma \tag{47}$$

where a = the shortest distance between the line *AB* and the axis *Oz; γ*= the angle between the generator of the straight line *AB* and the screw axis *Oz*; The parameter t determines the position of the point M on the rectilinear generator. The equations give both cavities of the helicoid with

$$t > 0\ and\ t < 0 \tag{48}$$

The coefficients of the basic quadratic forms of the surface:

$$A = 1 \tag{49}$$

$$F = a\ sin\gamma + p\ cos\gamma \tag{50}$$

$$B^2 = a^2 + p^2 + t^2\ sin^2\gamma \tag{51}$$

$$L = 0 \tag{52}$$

$$M = \frac{sin\gamma(actg\gamma - p)}{\sqrt{(actg\gamma - p)^2 + t^2}} \tag{53}$$

$$N = \frac{[a(actg\gamma - p) + t^2\ sin\gamma\ cos\gamma]}{\sqrt{(actg\gamma - p)^2 + t^2}} \tag{54}$$

$$K = -\frac{(actg\gamma - p)^2}{[(actg\gamma - p)^2 + t^2]^2} < 0 \tag{55}$$

The angle χ between the coordinate lines t and v is calculated by equation

$$cos\chi = \frac{(a\ sin\gamma + p\ cos\gamma)}{\sqrt{a^2 + p^2 + t^2\ sin^2\gamma}} \tag{56}$$

Straight-line generators will be orthogonal to the helix lines $v$, if the constant angle $\gamma$ is taken by equation

$$tg\gamma = -\frac{p}{a} \tag{57}$$

A direct convolutional helicoid can be obtained by taking $\gamma = \frac{\pi}{2}$. (58)

*2.2.5. Elliptic helicoid.* The elliptic helicoid – is a linear surface of negatively Gaussian curvature. This type of a helicoid can be carried also to a class of spiral surfaces (figure 5), and to a class of linear surfaces to group of line surfaces of negative Gaussian curvature.
Parametric form of elliptic helicoid (figure 5) [12]:

$$x = x(u, v) = av\ cosu \tag{59}$$

$$y = y(u, v) = bv\ sinu \tag{60}$$

$$z = z(u) = cu \tag{61}$$

where a and b = constants.
Coefficients of the basic quadratic forms of the surface and its curvature:

$$A^2 = a^2v^2sin^2u + b^2v^2cos^2u + c^2 \tag{62}$$

$$F = v(b^2 - a^2)sinu\ cosu \tag{63}$$

$$B^2 = a^2cos^2u + b^2sin^2u \tag{64}$$

$$A^2B^2 - F^2 = a^2b^2v^2 + c^2(a^2cos^2u + b^2sin^2u) \tag{65}$$

$$L = 0 \tag{66}$$

$$M = \frac{abc}{\sqrt{a^2b^2v^2+c^2(a^2cos^2u+b^2sin^2u)}} \tag{67}$$

$$N = 0 \tag{68}$$

$$k_u = k_v = 0 \tag{69}$$

$$K = \frac{-a^2b^2c^2}{(A^2B^2-F^2)^2} < 0 \tag{70}$$

$$H = -\frac{abcF}{(A^2B^2-F^2)^{\frac{3}{2}}} = -\frac{abcv(b^2-a^2)sinu\ cosu}{(A^2B^2-F^2)^{\frac{3}{2}}} \tag{71}$$

*2.3. Helicoids transformations*
Helicoid and Catenoid are members of the same associate family of surfaces therefore, one can bend a catenoid into a portion of a helicoid without stretching [12,18]. In other words, one can make a (mostly) continuous and isometric deformation of a catenoid to a portion of the helicoid such that every member of the deformation family is minimal (having a mean curvature of zero). A parametrization of such a deformation is given by the system below:

$$x(u, v) = \cos\theta \sinh v \sin u + \sin\theta \cosh v \cos u$$

$$y(u, v) = -\cos\theta \sinh v \cos u + \sin\theta \cosh v \sin u$$

$$z(u, v) = u\cos\theta + v\sin\theta$$

For $(u, v) \in (-\pi, \pi] * (-\infty, \infty)$, with deformation parameter $-\pi < \theta \leq \pi$, where $\theta = \pi$ corresponds to a right-handed helicoid $\theta = \pm\frac{\pi}{2}$ corresponds to a catenoid, and $\theta = 0$ corresponds to a left-handed helicoid.

## 3. Discussion and future recommendations

Despite developed geometrical characteristics of five types of helicoids mentioned above, there are still some gaps that could lead to misunderstandings from other researchers. For example, during the process of review some differences in determinations of helicoids named by separate authors were found out. In this paper, the most mentioned and logical (in author's opinion) terms are used, however more research is needed to prove it.

Thus, the helicoid on figure 1 is called 'right' helicoid in [1,10,19,20], while the same type of helicoid is mentioned as 'straight' helicoid in [12]. The helicoid on figure 2 is generally called 'oblique' helicoid [1,10], but it is described as 'inclined' helicoid in [21]. As the helicoid on figure 3 is concerned, despite the term suggested in this paper ('deployable' helicoid) it is usually called 'developable' helicoid [1,10], because it logically coincides with the term 'developable surface' that is widely used in practice and research papers [22], and it is often mentioned as 'torso' helicoid [23]. Some researchers call helicoids as 'pre-twisted beams' [24].

It may be caused by difficulties in translation or differences in understanding, but it is obviously required to put all terms in accordance with some etalon on the base of mathematical or engineering logic. However, the terms of two other types of helicoids which are not widely used are even less clear for researchers, and there were found much more terms for different types of helicoids in [10,12]. That is why it is needed to create a clear classification of helicoid types according to their geometry before starting investigation on searching the optimal shape for specific parameters.

In recent times, the most complete overviews on the geometry, calculation and application of helical shells of a general type were presented by Krivoshapko [23] and Rynkovskaya [8], and a review for linear helical shells is given in [25], where, in particular, it is shown that the geometry of all five types of helicoids has been studied quite well. At the same time, not all types of ruled helical shells are used in practice, which is associated, in a number of cases, with insufficient information. Investigators are aware of the state of affairs in the area of determining the stress-strain state of these shells. That is why it would also be necessary to investigate these helicoids stress-strain state for future choice of the type of helicoid and widen their application in architecture and civil engineering.

In mechanical engineering, right, oblique and deployable helicoids are often used, however, their strength is determined in most cases on the basis of experiments, in particular, due to the relatively small size of the screws, augers and threads used, but some research is also done [26-30]. Obviously, to use these surfaces in architecture and civil engineering, it is required to conduct more research [31] on methods of stress-strain calculations of helicoids.

## 4. Conclusions

The geometry of five types of helicoids are presented in the article. Some of them are well developed and some of them were not investigated too much, and most of differences are unknown by architects, civil engineers and designers. However, these differences could lead to impact on the stress-strain state of a real structure, as well as help to use material more efficiently. Besides, knowledge of the differences in geometry could provide a proper information for parametric investigations according to maintenance of a real structure.

The investigation on stress-strain state of these five types of helicoids is also required to supply other researchers, architects and engineers with data for choosing an optimal type of helicoid for determined parameters in practice.

As it is shown in the article, some differences in terms were found by author which could lead to misunderstanding between other researchers, and it is suggested to put all terms into an order according to a determined logic.

## 5. References

[1] Krivoshapko S N and Rynkovskaya M 2017 Five Types of Ruled Helical Surfaces for Helical Conveyers, Support Anchors and Screws *MATEC Web of Conferences* **95**, 06002 (2017)
[2] Indiana University 2006 Getting A Handle On Minimal Surfaces *ScienceDaily*, 3 January 2006
[3] Meusnier J B Mémoire sur la courbure des surfaces 1776 *Mém. des savans étrangers* **10** 477-510
[4] Euler L 1771 Novi commentarii Academial Scientiarum Petropolitanae pp 3-34
[5] Catalan E 1842 Sur les surfaces réglées dont l'aire est un minimum *J. Math. Pure Appl.* Pp 203-11
[6] Darboux G 1914 Theorie des surfaces V.1
[7] Hilbert D and Cohn-Vossen S 1932 Anschauliche Geometrie p 310
[8] Rynkovskaya M I 2012 Application and calculation of helicoidal shells in architecture and civil engineering *RUDN Journal of Engineering Researches* **4** pp 84-90

[9] MeeksIII W H and Weber M 2007 Bending the helicoid *Mathematische Annalen* V.339 **4** pp 783–98
[10] Krivoshapko S N and Ivanov V N 2015 Encyclopedia of Analytical Surfaces *Springer* p 752
[11] http://www.geom.uiuc.edu/zoo/diffgeom/surfspace/helicoid/
[12] Weisstein E W 2003 CRC Concise Encyclopedia of Mathematics, Second Edition, Chapman & Hall/CRC p 3252
[13] Nield D A and Kuznetsov A V 2004 Forced connection in a helical pipe filled with a saturated porous medium *Int. J. of Heat and Mass Transfer*. V. 47 **24** pp 5175-80.
[14] Vasudevaiah M and Patturaj R 1994 Effect of torsion in a helical pipe flow *Int. J. of Mathematics and Mathematical Sciences* V. 17 **3** pp 553-60
[15] https://en.wikipedia.org/wiki/Helicoid
[16] Hazewinkel M 2013 Encyclopaedia of Mathematics: V. 6 *Springer Science & Business Media* p 732
[17] Lewis J R, Mccurdy A and Gaudet A D 1985 Laser printing system with developable helicoids reflector *US Patent 4494821*, 01.22.1985.
[18] Fomenko A T and Tuzhilin A A 2005 Elements of the Geometry and Topology of Minimal Surfaces in Three-dimensional Space *American Mathematical Soc.* 2005 p 142
[19] Hirashima M and Iura M 1979 A geometrically nonlinear theory of right helicoidal shells *Theor. and Appl. Mech.* **27** pp 155-67
[20] [20] Rynkovskaya M I 2008 Rekach's method of calculation as applied to right helicoid *Structural Mechanics of Engineering Structures and Constructions* **3** pp 23-29
[21] Condra G E and Elias M K 1944 Study and Revision of Archimedes (Hall) *Geological Society of America* p 243
[22] Rynkovskaya M 2017 Analysis of displacements in beam structures and shells with middle developable surfaces *MATEC Web of Conferences* **108** 16001
[23] Krivoshapko S N 1999 Geometry and strength of general helicoidal shells *Applied. Reviews* V. 52 **5** pp 161-175
[24] Filiz S, Bediz B, Romero L A and Ozdoganlar O B 2014 Three dimensional dynamics of pretwisted beams: A spectral-Tchebychev solution *Jour. of Sound and Vibration* V. 333 **10** pp 2823-39
[25] Krivoshapko S N 2006 Calculation and design of screw-like structures used in construction and construction machines: overview information, VNIINTPI **1-2,** Build. Struct. and Mater. p 68
[26] Hoyashita S 1995 Calculation method for profile of helical gear with tooth-trace modification finished by worm-shaped tool or form grinding wheel *Jap Soc. Mech. Eng.* **61** pp 1149-56
[27] Taylor D W 1933 The Speed and Power of Ships *Ransdell Incorporated* p 130
[28] Cohen J W 1955 On stress calculations in helicoidal shells and propeller blades *Walman* p 100
[29] Grigorenko Y M, Gulyaev V I, Gozulyak E A and Ashuri K 1983 Stress-strain state of tubular shells under action of uniform surface pressure *Prikl. Mechanika* **8**
[30] Dabrowski O and Sapian C 1987 Loading of a central screw chute in a coal storage container *Pr. nauk. Inst. beed. pwrocl.* **51** pp 125-30
[31] Knabel J and Lewinski T 1999 Selected equilibrium problem of thin elastic helicoidal shells *Arch. Civil Eng.* **42(2)** pp 245-57

**Acknowledgments**
This paper was financially supported by the Ministry of Education and Science of the Russian Federation on the program to improve the competitiveness of Peoples' Friendship University of Russia (RUDN University) among the world's leading research and education centers in the 2016-2020.

IOP Conf. Series: Materials Science and Engineering **371** (2018) 012030 doi:10.1088/1757-899X/371/1/012030

# Experimental Investigation on Inelastic Cyclic Behaviour of Bracing Member with Wide-flange Section

**Dong-Hyeon Shin[1] and Hyung-Joon Kim[2]**

1 Ph.D Student, University of Seoul, Seoul, Korea
2 Associate Professor, University of Seoul, Seoul, Korea
E-mail: hyungjoonkim@uos.ac.kr

**Abstract.** Braced frames have been frequently employed as the lateral force resisting system of low- to mid-rise buildings. However, available test results are relatively limited to establish analysis models of braces with wide-flange that are required to capture their inelastic cyclic response. To address this, a full scale quasi-static test of a brace with wide-flange section was carried out to investigate the important structural characteristics which significantly influence its inelastic cyclic behaviour. This paper focuses on the buckling strength, tensile yield strength and local buckling of the test specimen and compares them with those predicted using an existing physical theory model. The physical model well predicts the hysteretic curve within tensile regions. However, the analytical compressive resistance of the specimen after buckling is overestimated compared to that obtained from the test results.

## 1. Introduction

Braced frames have been frequently used for the lateral force resisting systems of low- to mid-rise buildings due to their relatively simple design, fabrication and erection (Tremblay et al., 2008) Wide-flange sections have been generally implemented as braces. The cyclic response of such bracing members should be reasonably reflected into their analysis models to properly estimate the seismic performance of braced frames. Many researchers (Astaneh-Asl and Goel, 1984; Black et al., 1980; Gugerli and Goel, 1982; Jain and Goel, 1978; Zayas et al., 1981) have conducted a large number of tests with the purpose of investigating the hysteretic behaviour of bracing members. Most of previous tests were performed with hollow sections that can effectively resist to applied axial forces. However, existing cyclic tests of bracing members with wide-flange section are relatively limited and most of test specimens, if any, were scaled according to available testing environments. In practices, longer and/or larger bracing members have been commonly used in braced frames (Oliveira et al., 2008; Azad and Topkaya, 2017) and full scale cyclic tests have been required to neglect size effects and, ultimately, to establish analysis models that can properly capture the hysteresis of bracing members.

To figure out this issue, a full scale quasi-static test of a wide-flange section brace was carried out to investigate the important structural characteristics which significantly influence its inelastic behaviour. An example braced frame was designed to size a test specimen with a wide-flange section brace. The hysteretic behaviour of the test specimen was predicted using a physical theory model (Remennikov and Walpole, 1997). Cyclic quasi-static test results and analysis results are compared.

## 2. Design of a test specimen and its analytic cyclic behaviour

### *2.1. Selection of a test specimen*

A test specimen was chosen from bracing members used for the main lateral load-carrying structural members in an example braced frame as shown in Figure 1. The 3-story example braced frame was designed with seismic design parameters such as a response amplification factor, $R = 3.25$, a deflection amplification factor, $C_d = 3.25$, an overstrength factor, $\Omega_0 = 2$ and an important factor, $I = 1.0$. It was assumed that the frame is located in Seoul, Korea and its site class is assigned to SD. 5 $kN/m^2$ of dead load and 4 $kN/m^2$ of live load are assigned.

Table 1 summarizes geometrical properties of the brace which was selected as a test specimen for this study. For the specimen, it is assumed that the steel grade of SS400 with the nominal yield strength, $F_y$ of 235 MPa and ultimate strength, $F_u$ of 400 MPa. The brace length between working points is 3974mm. The expected axial compression strength, $P_n = 1331$ kN and the expected axial tensile yield strength, $P_{yt} = 1438$ kN is calculated. More detailed information for the seismic design of the example braced frame can be found in Shin et al. (2017).

*2.2. Prediction of cyclic behavior of the test specimen*

The cyclic behaviour of a brace is dependent on quite complex mechanisms including the global buckling phenomena, the yielding of material, the local buckling effects, and the post-buckling deterioration of compressive load capacities due to the Bauschinger effects and the tangent modulus deterioration (Remennikov and Walpole, 1997). Phenomenological, finite element and physical theory models have been suggested by various researchers (Ikeda and Mahin, 1986; Zayas et al., 1981; Haddad, 2017) to properly reflect the complicate mechanism of braces subjected to cyclic loading. Of them, this study employs a physical theory model to predict the hysteretic behaviour of the selected brace.

The physical theory model is developed based on analytical expressions for axial force-axial deformation behaviour of a pin-ended bracing member with a plastic hinge located at mid-span. As shown in Figure 2, the axial force-moment interaction and tangent modulus history should be defined to predict the hysteretic behaviour of a brace subjected to cyclic loading. Table 2 summarizes parameters required for defining the physical theory model for the brace. The hysteretic behaviour of the brace under a qualified testing protocol prescribed in the AISC seismic provision (AISC, 2010) as presented in Figure 3 is found in Figure 4. This hysteretic behavior will be comparatively described with test results in following chapter.

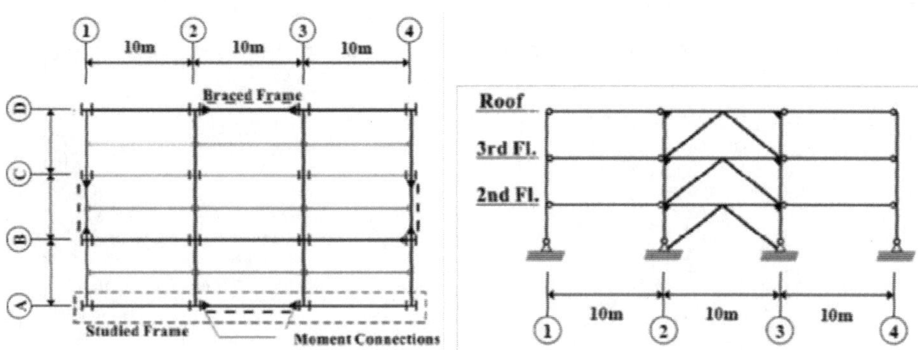


**Figure 1.** Plan and elevation of an example braced frame.

**Table 1.** Geometrical properties of a test specimen.

| **Specimen** | **Width, B, mm** | **Height, H, mm** | **Web thickness, $t_w$, mm** | **Flange thickness, $t_f$, mm** | **Slenderness ratio, KL/r** | **Area, A, $mm^2$** | **Width to thickness ratio** |
|---|---|---|---|---|---|---|---|
| WF 175 | 175 | 175 | 8 | 12 | 39 | 5542 | 14.58 |

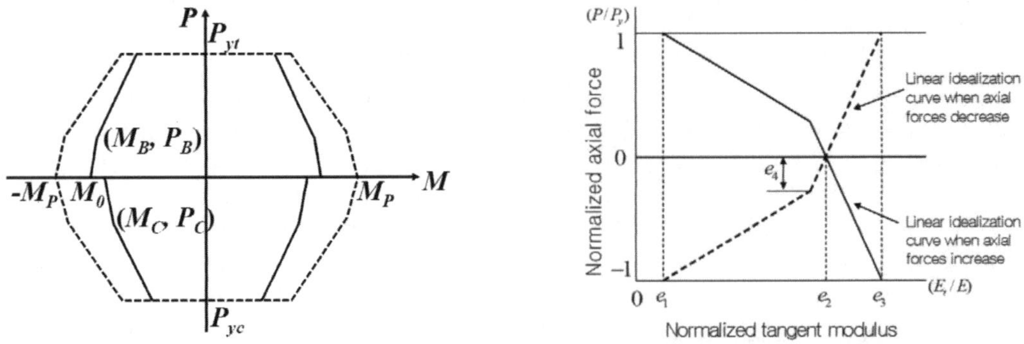


(a) Axial force-moment interaction (b) Tangent modulus history

**Figure 2.** Parameters to predict hysteretic behaviour using a physical theory model.

**Table 2.** Parameters required for defining a physical theory model.

| Specimen | Axial force-moment interaction curve | | | | | | | Normalized tangent modulus | | | |
|---|---|---|---|---|---|---|---|---|---|---|---|
| | **$P_{yc}$, kN** | **$P_{yt}$, kN** | **$M_0$, kNm** | **$P_B$, kN** | **$M_B$, kNm** | **$P_C$, kN** | **$M_C$, kNm** | **$e_1$** | **$e_2$** | **$e_3$** | **$e_4$** |
| WF 175 | 1331 | 1438 | 36.9 | 266 | 33.2 | 719 | 18.4 | 0.05 | 0.5 | 0.8 | -0.7 |

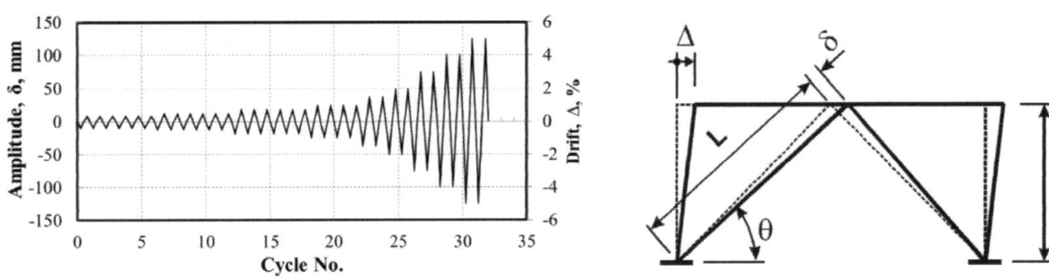


**Figure 3.** Loading protocol and expected deformed shape of the test specimen.

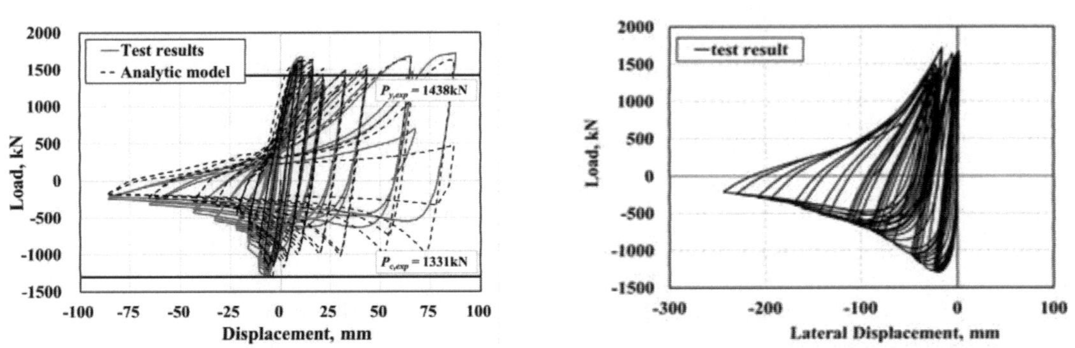


(a) Axial force – axial deformation relation (b) Axial force – lateral displacement relation

**Figure 4.** Hysteretic behaviour of the test specimen.

## 3. Full-scale test results of the test specimen

### *3.1. Test results*

The full-scale test was conducted with the brace selected from the example braced frame and gusset pl ates are neglected to concentrate on the cyclic response of pure axial loaded steel members. A 10MN universal testing machine was used to apply a displacement-controlled loading protocol described in the AISC seismic provision (AISC, 2010). The applied load and member deformations were monitored using the actuator's load cell and two cable position transducers, respectively. One of the transducers was installed for measuring the axial deformations along the loading direction, while the other for measuring the out-of-plane deformations at the mid-length of the test specimen. With considering the global buckling of the specimen, lateral displacements at its mid-span were observed up to maximum 250mm which is over 2-times larger than its imposed axial deformation.

The test results shown in Figure 4 are well-matched with the expected yield and buckling strengths. Figure 5 shows deformed shapes and failure modes of the specimen. The elastic global buckling of the specimen was observed at a 0.5% drift and minor inelasticity was developed when a 0.8% drift was imposed. Dominant local buckling at the mid-span and both ends of the specimen were measured at a 2% drift. As imposed drifts are increased, the fractures of the locally buckled flanges at the mid-span were expanded. Figure 6 shows the tearing of the flanges at the mid-span and the local buckling at the end of the specimen after the test.

(a) 0.5% drift (b) 0.8% drift (c) 1% drift (d) 2% drift (e) 3% drift (f) 4% drift

**Figure 5.** Deformed shapes of the specimen under compressive loading.

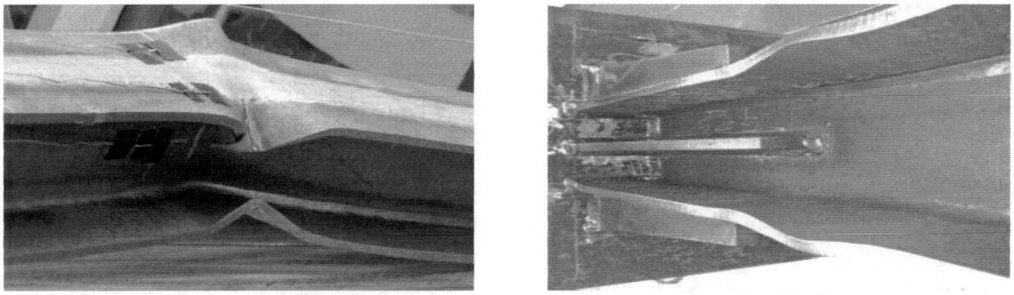

(a) Fracture at the mid-span (b) Local buckling at ends

**Figure 6.** Failure modes of the specimen.

### *3.2. Comparison of analytic and experimental results*

Figure 4 presents the direct comparison of the analytic axial force-axial displacement relation with the test results of the specimen. The analytic hysteresis behaviour is reasonably well matched with hysteresis behaviour obtained from the test. The analysis well predicts the hysteretic curve within tensile regions. However, the analytical compressive resistance of the specimen after buckling is overestimated compared to that obtained from the test results. More specific comparison is summarized in Figure 7 where tensile, compressive strength, dissipated energy and tangent modulus at various imposed axial displacements are presented. For the tensile strength of the specimen, the analytic model well captures with average 2.3% relative differences regardless of the imposed axial displacements. Although the analytic compressive strengths under the lower imposed axial displacements agree with the test results, those measured at the larger axial displacements overestimate up to 38%. Such big differences in the compressive strengths, the differences in energy dissipated during a cycle are increased with the increase in the imposed axial displacements. Also, the changes of tangent modulus of the brace tracked by the analysis are different with those obtained from the test results.

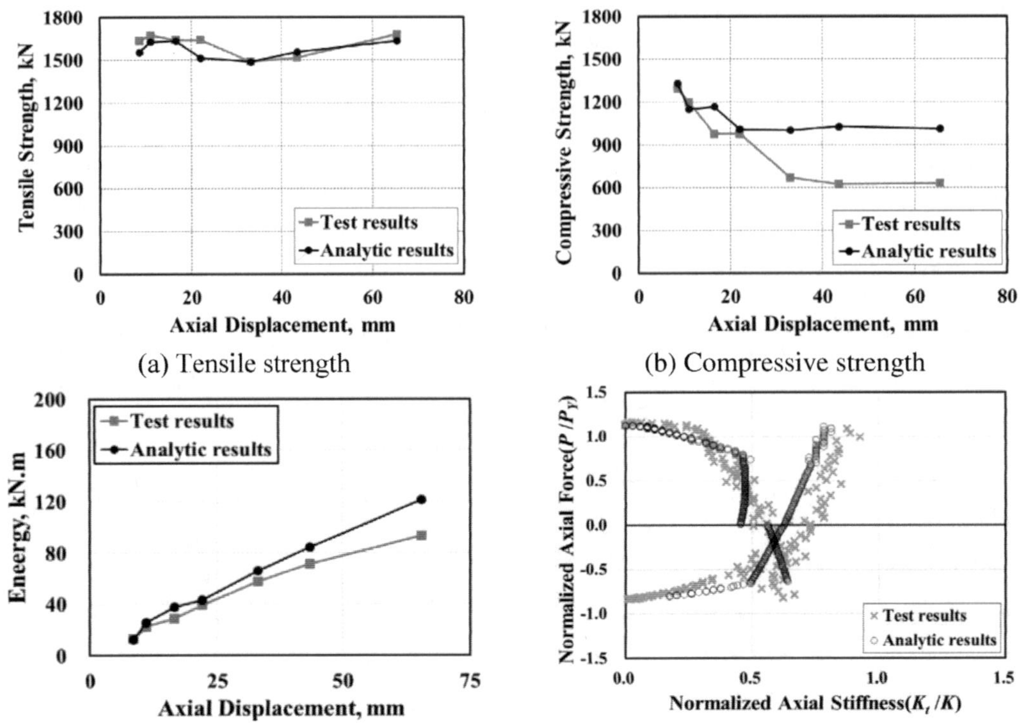

(a) Tensile strength (b) Compressive strength

(c) Dissipated Energy (d) Tangent Modulus

**Figure 7.** Comparison of analytic and test results.

## 4. Conclusion

This study experimentally investigates the inelastic cyclic behaviour of a bracing member with wide-flange section since available test results are relatively limited to establish analysis model of it. To investigate the important structural characteristics which significantly influence cyclic behaviour, the full scale cyclic quasi-static test was carried out with the brace selected from an example braced frame. The brace specimen fractured at the mid-span at an axial displacement corresponding a 4% story drift of the example braced frame.

Analytic results obtained from the physical theory model and the test results are compared in terms of compressive, tensile strength, dissipated energy and tangent modulus. Although the analytic hysteresis behaviour is reasonably well matched with the hysteresis behaviour in the tensile excursion lower than 3% difference, the analytic cyclic response after buckling in compressive regions overestimates up to 38%. Such big differences in compressive regions result in the difference in energy dissipated and changes of tangent modulus of the brace during loading cycles.

For this reason, additional researches on the physical model for braces with wide-flange section are needed to properly capture their hysteretic curves in compression regions. Furthermore, it is expected from the improvement of analytic model for brace that the seismic performance of braced frames, which are frequently employed as the lateral force resisting system of low- to mid-rise buildings, can be more clearly evaluated.

IOP Conf. Series: Materials Science and Engineering **371** (2018) 012030 doi:10.1088/1757-899X/371/1/012030

## 5. Reference


[1] R Tremblay, M Haddad, G Martinez, J Richard, K Moffatt 2008 Inelastic cyclic testing of large size steel bracing members *The 14th World Conference on Earthquake Engineering* October 12-17, Beijing, China.

[2] Astaneh-Asl A, Goel S C 1984 Cyclic in-plane buckling of double angle bracing *J. Struct. Eng.* 110(9)

[3] Black R G, Wenger W A B, Popov E P 1980 *Inelastic buckling of steel strut under cyclic load reversals Rep. No. UCB/EERC-80/40* (Berkeley: Earthquake Engineering Research Center, Univ. of California)

[4] Gugerli H, Goel S C 1982 *Inelastic cyclic behavior of steel bracing members Rep. No. UMEE 82R1* (Michigan: Dept. of Civil Engineering, Univ. of Michigan, Ann Arbor)

[5] Jain A K, Goel S C 1978 *Hysteresis models for steel members subjected to cyclic buckling or cyclic end moments and buckling UMEE Rep. No. 78R6* (Michigan: Univ. of Michigan, Ann Arbor)

[6] Zayas A Z, Shing P B, Mahin S A, Popov E P 1981 *Inelastic structural modeling of braced offshore platforms for seismic loading Rep. No. UCB/EERC-81/04* (Berkeley: Earthquake Engineering Research Center, Univ. of California)

[7] J C de Oliveira, C Christopoulos, J A Packer, R Tremblay, M G Gray 2008 Full-Scale Experiment Validation of Standardized Seismic-Resistant Cast Steel Brace Connectors *The 14th World Conference on Earthquake Engineering* October 12-17, Beijing, China.

[8] S K Azad, C Topkaya 2017 A review of research on steel eccentrically braced frames *Journal of Constructional Steel Research* 128

[9] Remennikov A M, Walpole W R 1997 Analytical prediction of seismic behavior for concentrically-braced steel systems *Earthquake Eng. Struct. Dyn.* 26(8)

[10] D H Shin, S J Hong, H J Kim 2017 Risk-Targeted Seismic Performance of Steel Ordinary Concentrically Braced Frames Considering Seismic Hazard *J. Comput. Struct. Eng. Inst. Korea* 30(5)

[11] K Ikeda, S A Mahin 1986 Cyclic response of steel braces *J. Struct. Eng. ASCE* 112.

[12] M Haddad 2007 Cyclic behaviour and finite element modelling of wide flange steel bracing members *Thin-Wall Structures* 111

[11] Zayas A Z, Shing P B, Mahin S A, Popov E P 1981 *Inelastic Structural Modeling of Braced Offshore Platforms for Seismic Loading, Report No. UCB/EERC-81/04* (Berkeley: Earthquake Engineering Research Center, Univ. of California)

[12] American Institute of Steel Construction (AISC) 2010 *ANSI/AISC 341-10, Seismic provisions for structural steel buildings, including Supplement No. 1* (Chicago: AISC)



**Acknowledgments**

This work was supported by the National Research Foundation of Korea (NRF) grant funded by the Korea government (MSIP) (No.NRF-2016R1A2B4011900).

IOP Conf. Series: Materials Science and Engineering **371** (2018) 012031 doi:10.1088/1757-899X/371/1/012031

# Can Existing Design Codes Be Used to Design Flexural Reinforced Concrete Elements Strengthened with Externally Bonded Novel Materials?

**C G Papakonstantinou[1] , C Kakae[2], N Gryllakis[2]**

1 Assistant Professor, Dept of Civil Engineering, Univ of Thessaly, Greece
2 Dept of Civil Engineering, Univ of Thessaly, Greece
E-mail: cpapak@uth.gr

**Abstract.** The use of composite overlays to strengthen or rehabilitate existing structural elements has been investigated over the past few decades. The most common composites that were used are based on Glass or Carbon fibers impregnated in an organic polymeric matrix, known as GFRP or CFRP, respectively. Thus, many design codes throughout the world provide methodologies to calculate the flexural strength of reinforced concrete members strengthened with externally bonded Carbon or Glass fiber composites. It is essential to keep in mind that the experimental databases that were used in these codes were based on experiments conducted on these materials. However over the last years new materials have emerged, such as steel fibers (SRP/SRG) and basalt fibers (BFRP), as well as textile reinforced mortars (TRM). In this study the target was to investigate if the analytical models described in the codes can effectively be used to describe these novel materials as well. In order to do so, a statistical analysis was conducted to compare experimental results to results obtained using the methodologies described in the codes. It was found that most codes perform surprisingly well and in only few cases they do not provide acceptable results.

## 1. Introduction

Over the last decades due to aging infrastructure and an increase in design loads several researchers studied the possibility to rehabilitate or strengthen existing reinforced concrete structures. The answer was given by the use of "exotic" materials such as fiber-reinforced polymers (FRP) that up to that time were very expensive and were used mainly in the aerospace industry [1]. Since then their prices have dropped significantly, and thus their affordability and availability increased. In the first studies Glass, Aramid or Carbon fibers were utilized. The resins used were organic polymers, such as epoxy, polyester, vinylester resins etc. Based on these materials a significant number of experiments were conducted. Due to increasing interest in the aforementioned strengthening systems several organizations decided to develop design codes in order to provide guidance through design tools for civil engineers. These design codes were based on an experimental database containing the aforementioned materials. Some of the codes actually provide specific coefficients for Glass and Carbon fibers. However, during the last decade, new materials have been investigated. These materials include fibers such as basalt [2] and steel [3] and inorganic matrices such as Geopolymer [3] and cementitious mortars [4]. None of the existing codes discuss the use of steel or basalt fibers and even more the use of inorganic matrices.

The codes that were examined in this study include the American Concrete Institute ACI 440 [5], the international federation for structural concrete (fib) [6], the Canadian standard (ISiS) [7], the Japanese code (JSCE) [8], the Egyptian code [9], and the design recommendation presenter by the British

Concrete Society (TR55) [10]. In all these codes several modes of failure are taken in consideration. In normal reinforced concrete design there are mainly two failure modes, yielding of the steel reinforcement and crushing of concrete. When a structural element is strengthened with externally bonded composite materials some additional modes of failure can be identified, such as fracture of the fibers, debonding of the composite, shear failure of the concrete cover. In most codes the failure mode is dictated by the maximum strain that can be developed in the FRP. It was noticed that brittle failures that were based on the delamination of the FRP or the shear failure of the composites occur at much lower strains compared to the characteristic maximum strain of the composite material. Therefore the design methodologies suggest the use of reduction factors in order to drive the design in more ductile modes of failure. However, the new materials mentioned before differ in terms of either the matrix used or the fibers compared to the GFRPs and the CFRPs. The different material properties can very well affect the performance of the strengthened elements. Thus the question that this study tries to answer is if these older codes can be used to predict the flexural capacity of reinforced concrete structural elements strengthened with externally bonded novel composites.

## 2. Design models

In the following paragraphs a short description of the examined design models is presented.

### *2.1. American concrete institute (ACI 440.2R-08)*

The American Concrete Institute Committee 440 design guidelines [5] are based on strain compatibility as well as force and moment equilibrium. ACI suggests that strain at the time of strengthening, at the concrete substrate, must be considered. FRP is not permitted to carry compressive stresses, according to ACI 440.2R-08. ACI introduces two safety factors $\varphi$ and $\Psi_f$. The first depends upon the strain in the steel reinforcement at the time of failure, thus it depends upon the mode of failure. More specifically $\varphi$ equals 0.9 for a tension-controlled failure (ductile) while a value of 0.65 is recommended for a compression-controlled failure (brittle without yielding). In the transition zone between the previous two extremes, $\varphi$ can be determined from the linear interpolation. The second factor $\Psi_f$ is a strength reduction factor for the FRP, which is multiplied by the flexural contribution of FRP reinforcement ($M_{nf}$), as shown in equation (1):

$$M_{nf} = \Psi_f A_f f_{fe}\left(d_f - \frac{\beta_1 c}{2}\right) \qquad (1)$$

where:

$A_f$ is the FRP cross sectional area

$f_{fe}$ is the FRP effective strength

$d_f$ is effective depth of FRP shear reinforcement

$\beta_1$ is factor taken as 0.85 for concrete strength $fc'$ up to and including 28 MPa

$c$ is the distance from extreme compression fiber to the neutral axis, in. (mm)

The, $\Psi f$ , factor is taken as 0.85 based on the reliability analysis of the experimentally calibrated statistical values to mainly account for the less predictable failure mode of delamination of FRP reinforcement [11].

For a FRP-strengthened reinforced concrete section, the following failure modes are likely to take place in flexure: (i) Ductile crushing of concrete: crushing of concrete in compression after yielding of tensile steel (desirable failure mode), (ii) Brittle crushing of concrete: crushing of concrete in compression before yielding of tensile steel (undesirable failure mode) , (iii) Rupture of FRP: rupture of FRP reinforcement after yielding of tensile steel (desirable failure mode) and (iv) Debonding of FRP: The intermediate crack-induced debonding of FRP reinforcement.

It is important to note that a failure mode which was mentioned by several researchers [12], is not covered by ACI 440. This failure mode is a shear delamination of the concrete cover initiating at the FRP curtailment, and is typically mentioned as cover delamination.

ACI introduces an FRP strength reduction coefficient, $C_E$, which is related to the type of fibers and the enviroment that the FRP will be used. Values of this factor range from 0.5 to 0.95.

Finally ACI provides equation (2) in order to calculate the design strain of the FRP overlay:

IOP Conf. Series: Materials Science and Engineering **371** (2018) 012031 doi:10.1088/1757-899X/371/1/012031

$$\varepsilon_{fd} = 0.41\sqrt{\frac{f_c'}{nE_f t_f}} \leq 0.9\varepsilon_{fu} \quad (2)$$

where:
$\varepsilon_{fd}$ is the design FRP strain,
$\varepsilon_{fu}$ is the ultimate FRP strain,
$E_f$ is the FRP modulus of elasticity,
$t_f$ is the FRP thickness,
$f_c'$ is the concrete compressive strength

### 2.2. ISIS Canada (design manual)

The ISIS Canada Research Network provided a design manual for the design of strengthened reinforced concrete structural elements with fiber reinforced polymers. According to the manual the most commonly used fibers are Glass and Carbon, while the matrices are epoxy, vinyl ester or polester resins. The internal force due to tension in the FPR reinforcement can be expressed using equation (1).

$$T_{frp} = \varphi_{frp} A_{frp} E_{frp}\, \varepsilon_{frp}\; \mu\varepsilon\; \varepsilon_{frp} \leq \; \varepsilon_{frpu} \quad (3)$$

where:
$T_{frp}$ is the internal force due to tension in FRP reinforcement,
$\varphi_{frp}$ is a safety factor related to the FRP material used,
$A_{frp}$ is the cross sectional area of the FRP
The safety factor $\varphi_{frp}$ depends upon the types of the fibers and the type of structure. ISIS makes a distinction between bridges and buildings. For buildings and Carbon fibers the suggested value is 0.75 while for Glass fibers 0.6.

### 2.3. TR 55

TR 55 [10] suggests that partial safety factors to be applied to the characteristic mechanical properties are a function of the type of fiber and the manufacturing/site application process. Thus:

$$\gamma_{mF} = \gamma_{mf} \times \gamma_{mm} \quad (4)$$

where $\gamma_{mf}$ depends on the type of fiber and $\gamma_{mm}$ depends on the manufacturing and/or site application process. Specific values are given in [10]. The accuracy with which the properties obtained from test samples reflect the overall properties of the material will depend on the method of manufacture, the level of quality control and application. It is essential that material properties used in design are replicated in the site application. It should be taken into consideration that the many products (fibers and matrices) offered in the market place possess different properties.

### 2.4. Fib bulletin 14

fib Bulletin 14 [6] proposes the use of FRP material safety factors in order to reduce the stresses in the FRP. The design guidelines recommend an even further reduction of the allowable FRP strain at ultimate when the design is governed by concrete crushing or bond failure.
In this situation the FRP design stress $f_{fd}$ becomes:

$$f_{fd} = \frac{f_{fk}}{\gamma_f}\frac{\varepsilon_{fue}}{\varepsilon_{fum}} \quad (5)$$

where:
$f_{fk}$ is the FRP characteristic strength,
$\gamma_f$ is a safety factor which varies from 1.35 to 1.50 depending upon the type of fibers (1.35 for Carbon, 1.40 for Aramid and 1.50 for Glass), values are presented in Table 3.1 of [5],
$\varepsilon_{fue}$ is the effective ultimate FRP strain expected to be reached in situ, and
$\varepsilon_{fum}$ is the mean strain obtained through uniaxial tensile testing
Typically for flexural strengthening cases the ratio:

$$\frac{\varepsilon_{fue}}{\varepsilon_{fum}} = 1 \quad (6)$$

It should be mentioned that fib Bulletin 14 [5] points out that the $\gamma_f$ factors are subject to further study, due to the limited numbers of experimental results.

### *2.5. Japanese society of civil engineers*
Japanese Society of Civil Engineers provides recommendations for the design of reinforced concrete structures with externally bonded FRPs [8]. The design guidelines also recommend an even further reduction of the allowable FRP strain at ultimate to account for the case of premature delamination. More specifically the allowable FRP strain is reduced by 30%. There is no distinction between the different FRP types.

### *2.6. Egyptian code of practice (ECP)*
The Egyptian Code of Practice (ECP) provides recommendations for the design of reinforced concrete structures with externally bonded FRPs [9]. The design guidelines also recommend an even further reduction of the allowable FRP strain at ultimate to account for the case of premature delamination. Initially the allowable FRP strain is divided by 1.50. Furthermore, the allowable FRP strain is multiplied by a safety factor that varies from 0.75 for Glass fibers to 0.95 for carbon fibers. The design code does not offer alternatives for other types of fibers or externally bonded overlay systems such as the TRM.

## 3. Experimental data
A series of experimental data were collected from different experiments [4, 13-32] conducted on reinforced concrete beams strengthened for flexure with different externally bonded FRP systems. The systems that were examined were based on Glass, Carbon, Aramid and Basalt fibers. Additionally TRM systems were examined. An experimental database was created that included all relevant information regarding the tested beams, such as beam dimensions, reinforcement, properties of materials used, thickness and width of FRP system, maximum load capacity, type of failure. A total of 290 beams are included out of which 159 are strengthened with CFRP, 20 with GFRP, 8 with BFRP, 37 with SRP and 66 with TRM. The experimental data were used to run an analysis using the aforementioned six design codes.

## 4. Analysis and discussion
A statistical analysis was conducted to compare analytical flexural beam strengths to experimental strengths. More specifically correlation analysis was performed (Pearson and Spearman's correlation coefficient) for each design code. Pearson correlation coefficient indicates a measure of strength of a linear association between the two variables (in our case experimental beam strength vs. analytical beam strength) a while Spearman's rank correlation coefficient is a nonparametric measure of rank correlation. Figure 1 shows experimental vs. theoretical values that were plotted using the design methodology described previously. In these graphs all available experimental data were used regardless of FRP type. In each graph one main diagonal line was plotted which indicates the exact linear relationship between the experimental and the theoretical flexural strength. A point above the diagonal line means that an analytical value is higher than the corresponding experimental, and therefore one can conclude that the model overestimates the beam's flexural strength. On the contrary points below the diagonal indicate that the model underestimates the beam's strength, and thus provides safe predictions. The percentages of beams that failed at bending moments lower than that estimated by the codes are: 90% for JSCE, 84% for ISIS, 81% for TR55, 80% for ECP, 79% for ACI, and 65% for fib. It is evident that if we exclude the fib design recommendations the remaining codes perform very well. It should be noted though, that the fib design recommendations were drafted using much smaller experimental databases than the one used in this article, since at the time the available experimental data were significantly fewer than today.
As can be seen in Table 1. the design code that exhibits the highest Spearman correlation coefficient is JSCE followed by ISIS and ACI, while JSCE exhibits also the highest Pearson correlation coefficient, followed again by ACI. Based on these results it can be concluded that the JSCE design recommendation provides both safety and great correlation between experimental and analytical values. However in cases where the flexural strength was high (more than 200kNm) it provided significantly lower values than the experimental.

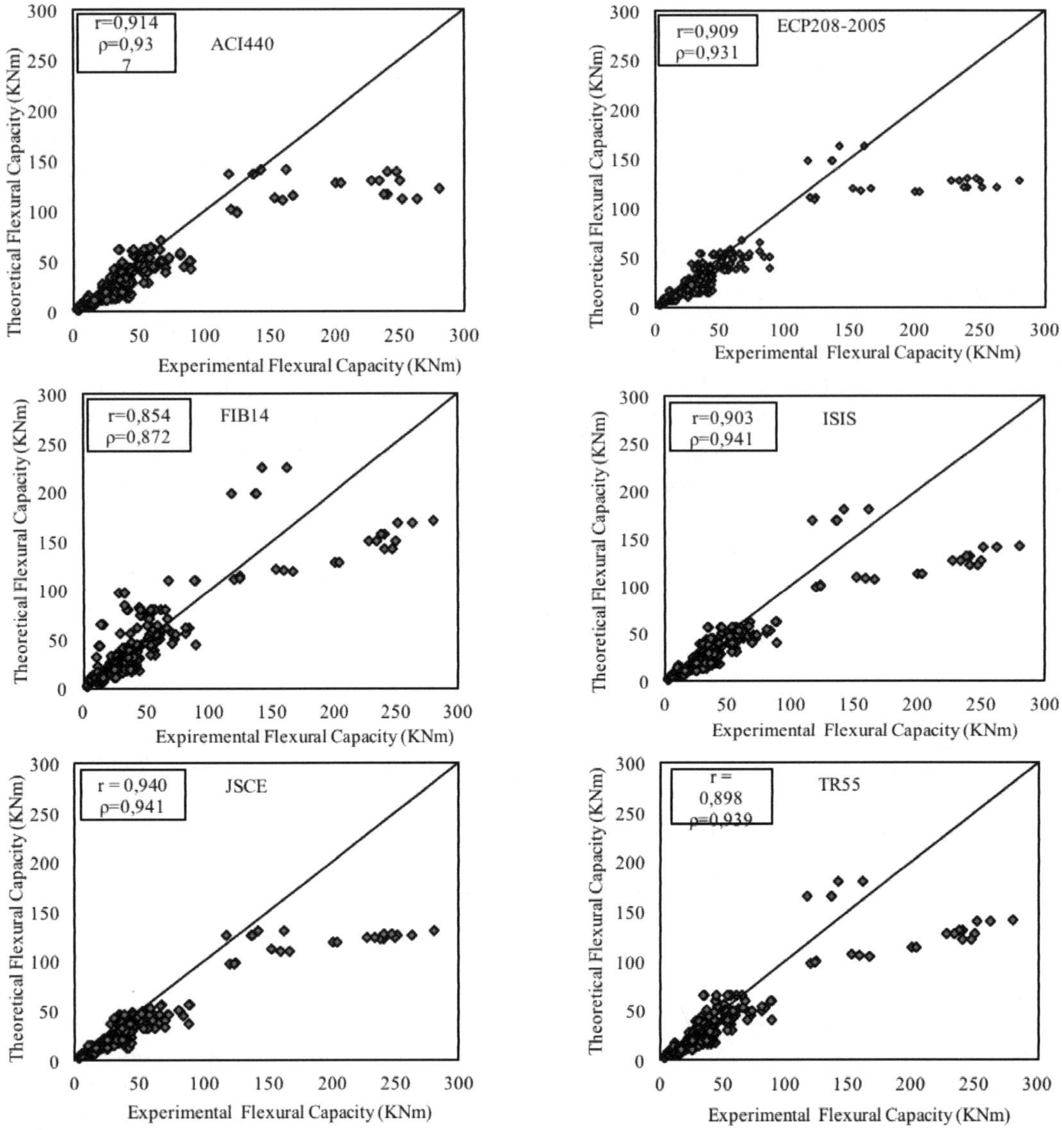


**Figure 1.** Experimental vs theoretical moment capacity values for the six examined codes.

Figure 2. shows similar graphs but for reinforced concrete beams strengthened with Textile Reinforced Mortars that are not covered by any design code. In order to obtain the analytical results, the same equations recommended for CFRP/GFRP were used. Even though some assumptions were made, it can be noticed that all codes perform very well in terms of safety. JSCE underestimated all beam strengths, while ACI and fib, which were the most unsafe, proved to be unsafe only in 15% of the cases. Strong Spearman coefficients (shown in the graphs) display that there is a very good correlation between experimental and analytical values. However, Pearson correlation coefficients in this case were relatively low, ranging from 0.68 to 0.70 (while in the previous case the coefficients were higher than 0.85). Similar graphs were obtained for other types of composite materials as well such as SRP/SRG and BFRP. The difference was that the Pearson correlation coefficient was higher than 0.95 when SRP/SRG was used for all codes. A fact that indicates, that existing codes can be very efficient to describe the SRP/SRG strengthening technique.

**Table 1.** Pearson correlation coefficients for examined design manuals.

| Code | Spearman Correlation Coefficient (ρ) | Pearson Correlation Coefficient (r) |
|---|---|---|
| ECP208-2005 [9] | 0,931 | 0,909 |
| FIB14 [6] | 0,872 | 0,854 |
| ISIS [7] | 0,941 | 0,903 |
| TR55 [10] | 0,939 | 0,898 |
| JSCE [8] | 0,941 | 0,940 |
| ACI440 [5] | 0,937 | 0,914 |

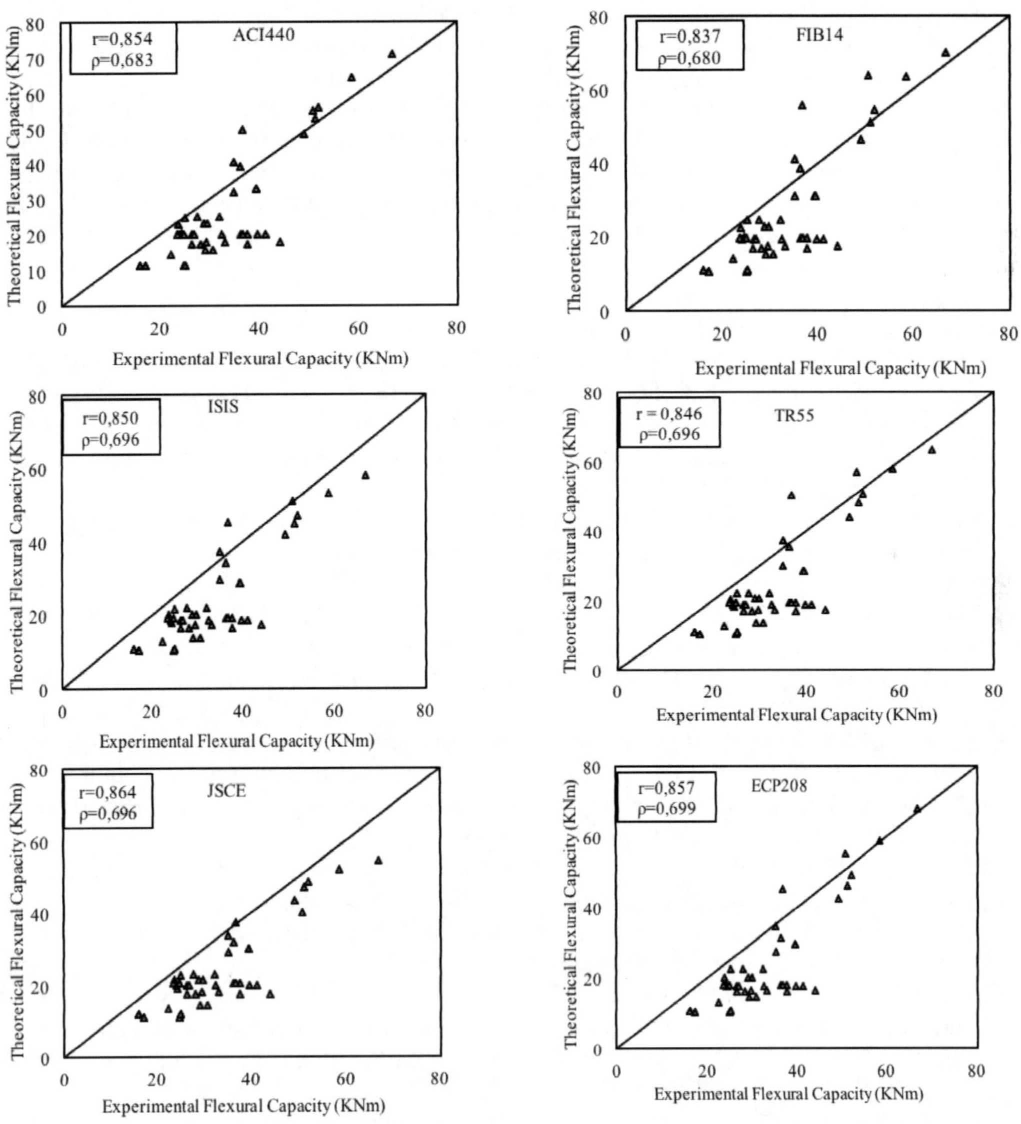


**Figure 2.** Experimental vs. theoretical moment capacity of RC beams strengthened with TRM.

**Table 2.** Performance of JSCE and ACI for five different types of composite overlays.

| | | CFRP | GFRP | SRP/SRG | BFRP | TRM |
|---|---|---|---|---|---|---|
| JSCE | Safety Ratio | 83% | 91% | 94% | 100% | 98% |
| | Pearson Coef. | 0.949 | 0.938 | 0.946 | 0.98 | 0.864 |
| ACI | Safety Ratio | 69% | 86% | 94% | 100% | 85% |
| | Pearson Coef. | 0.916 | 0.947 | 0.981 | 0.947 | 0.854 |

The performance of two best performing overall codes is provided in Table 2. More specifically, both safety ratio (% of beams failed at load over the design load), and Pearson's correlation coefficient are presented for JSCE and ACI for all five different types of composite overlays. It is clear that both codes can be used regardless of the composite material.

## 5. Conclusions

Based on the performed statistical analysis it can be concluded that almost all codes (other than fib) can be successfully used to describe the flexural strength of reinforced concrete beams strengthened with externally bonded materials regardless of their specific type. However in some cases the available experimental database is quite small and definitely more experiments are necessary in order to tweak and verify the equations proposed by the codes. All codes should take in consideration the different types of composite materials, available in the market place, that can be used as externally bonded strengthening systems. Finally the design methodology that performed the best in terms of both safety and correlation is the one described by JSCE. The authors believe that some changes in the JSCE code may provide even better results.

## 6. References

[1] Nanni, A 1997 *Cfrp strengthening* Concrete International, **19** 19-23

[2] Di Ludovico, M, A Prota, and G Manfredi 2010 *Structural upgrade using basalt fibers for concrete confinement* Journal of Composites for Construction, **14** 541-552

[3] Papakonstantinou, C G and K Katakalos 2009 *Flexural behavior of reinforced concrete beams strengthened with a hybrid inorganic matrix - steel fiber retrofit system* Structural Engineering and Mechanics, **31** 567-585

[4] Ebead, U, K C Shrestha, M S Afzal, A El Refai, and A Nanni 2017 *Effectiveness of fabric-reinforced cementitious matrix in strengthening reinforced concrete beams* Journal of Composites for Construction, **21**

[5] ACI440.2R-08 2008 *Guide for the design and construction of externally bonded frp systems for strengthening concrete structures*. American Concrete Institute.

[6] fib 2001 *Externally bonded frp reinforcement for rc structures*. International Federation of Structural Concrete: Lausanne.

[7] Canada, I 2002 *Strengthening reinforced concrete structures with externally—bonded fibre reinforced polymers (frps)* in *Design Manual no. 4*. Canadian Standards Association.

[8] JSCE 2001 *Recommendations for upgrading of concrete structures with use of continuous fiber sheets*. Japan Society of Civil Engineers: Tokyo.

[9] ECP-208", E C o P 2005 *Use of fiber reinforced polymer (frp) in the construction fields*. Egyptian Housing and Building National Research Center: Cairo.

[10] (UK), T C S 2000 *Technical report no. 55 – design guidance for strengthening concrete structures using fibre composite materials*.

[11] Okeil, A M, Y Bingol, and T AlKhrdaji 2007 *Analyzing model uncertainties for concrete beams flexurally strengthened with frp laminates* in *TRB 86th Annual Meeting*. Washington.

[12] GangaRao, H V S V, P. V. 1998 *Bending behavior of concrete beams wrapped with carbon fabric* Journal of Structural Engineering, **124** 3-10

[13] Akbarzadeh, H and A A Maghsoudi 2010 *Experimental and analytical investigation of reinforced high strength concrete continuous beams strengthened with fiber reinforced polymer* Materials & Design, **31** 1130-1147

[14] Pecce, M, G Manfredi, and E Cosenza 2000 *Experimental response and code models of gfrp rc beams in bending* Journal of Composites for Construction, **4** 182-190

IOP Conf. Series: Materials Science and Engineering **371** (2018) 012031 doi:10.1088/1757-899X/371/1/012031


[15] Aram, M R, C Czaderski, and M Motavalli 2008 *Debonding failure modes of flexural frp-strengthened rc beams* Composites Part B-Engineering, **39** 826-841

[16] Ashour, A F 2002 *Size of frp laminates to strengthen reinforced concrete sections in flexure* Proceedings of the Institution of Civil Engineers: Structures and Buildings, **152** 225-233

[17] Kara, I F and A F Ashour 2012 *Flexural performance of frp reinforced concrete beams* Composite Structures, **94** 1616-1625

[18] Babaeidarabad, S, G Loreto, and A Nanni 2014 *Flexural strengthening of rc beams with an externally bonded fabric-reinforced cementitious matrix* Journal of Composites for Construction, **18**

[19] Loreto, G, L Leardini, D Arboleda, and A Nanni 2014 *Performance of rc slab-type elements strengthened with fabric-reinforced cementitious-matrix composites* Journal of Composites for Construction, **18**

[20] Ombres, L 2011 *Flexural analysis of reinforced concrete beams strengthened with a cement based high strength composite material* Composite Structures, **94** 143-155

[21] Elsanadedy, H M, T H Almusallam, S H Alsayed, and Y A Al-Salloum 2013 *Flexural strengthening of rc beams using textile reinforced mortar - experimental and numerical study* Composite Structures, **97** 40-55

[22] Jung, K, K Hong, S Han, J Park, and J Kim 2015 *Prediction of flexural capacity of rc beams strengthened in flexure with frp fabric and cementitious matrix* International Journal of Polymer Science,

[23] Aravind, N, A K Samanta, J V Thanikal, and D K S Roy 2017 *An experimental study on the effectiveness of externally bonded corrugated gfrp laminates for flexural cracks of rc beams* Construction and Building Materials, **136** 348-360

[24] Napoli, A and R Realfonzo 2015 *Reinforced concrete beams strengthened with srp/srg systems: Experimental investigation* Construction and Building Materials, **93** 654-677

[25] Kim, Y J, A Fam, and M F Green 2010 *Flexural strengthening of reinforced concrete beams with steel-reinforced polymer composites: Analytical and computational investigations* Journal of Reinforced Plastics and Composites, **29** 2141-2155

[26] Prota, A, T Kah Yong, A Nanni, M Pecce, and G Manfredi 2006 *Performance of shallow reinforced concrete beams with externally bonded steel-reinforced polymer* ACI Structural Journal, **103** 163-70

[27] Barros, J A O and A S Fortes 2005 *Flexural strengthening of concrete beams with cfrp laminates bonded into slits* Cement & Concrete Composites, **27** 471-480

[28] Balsamo, A, F Nardone, I Iovinella, F Ceroni, and M Pecce 2013 *Flexural strengthening of concrete beams with eb-frp, srp and srcm: Experimental investigation* Composites Part B-Engineering, **46** 91-101

[29] Rahimi, H and A Hutchinson 2001 *Concrete beams strengthened with externally bonded frp plates* Journal of Composites for Construction, **5** 44-56

[30] Gopinath, S, A R Murthy, N R Iyer, and M Prabha 2015 *Behaviour of reinforced concrete beams strengthened with basalt textile reinforced concrete* Journal of Industrial Textiles, **44** 924-933

[31] Tomlinson, D and A Fam 2015 *Performance of concrete beams reinforced with basalt frp for flexure and shear* Journal of Composites for Construction, **19**

[32] Raoof, S M, L N Koutas, and D A Bournas 2017 *Textile-reinforced mortar (trm) versus fibre-reinforced polymers (frp) in flexural strengthening of rc beams* Construction and Building Materials, **151** 279-291

IOP Conf. Series: Materials Science and Engineering **371** (2018) 012032 doi:10.1088/1757-899X/371/1/012032

# Load Rating and Buckling of Circular Concrete-Filled Steel Tube (CFST): Simulation and Experiment

**L H Vu, N C Duc, L V Dong, D L Truong, N M T Anh, H Q Hung and P V Hue**

Mien Trung University of Civil Engineering, Ministry of Construction of the Socialist Republic of Vietnam
E-mail: lehoangvu@muce.edu.vn, nguyencongduc@muce.edu.vn

**Abstract.** The paper studies estimate ultimate strength and local buckling of circular concrete filled steel tube (CFST) specimens subjected to axial load using strain gage, force transducer and displacement sensor. The strain responses of CFST column in testing procedure and the results of finite element modelling will be used to predict failure modes when cracks occur in concrete core and deformation increases on steel tube's surface. Besides, the principles of Eurocode 4 and AISC 360-10 are used to calculate the critical buckling stress in comparison with experimental approach and numerical evaluation. Combining experimental and simulation data can be very useful for diagnostic load rating and limit, and effective to evaluate the damage on structural health monitoring of CFST members.

## 1. Introduction

It is undeniable that concrete structures with CFST members play a major role in modern construction structures and materials for coastal and island construction buildings in recent years. CFST column using high-strength steel is one of the most promising new developments to increase fatigue life of concrete and reduce corrosive effects of seawater for concrete structures in coastal areas. The behavior of CFST column under compressive axial load is an important factor in limit load analysis for engineers that will be calculated by using finite element method (FEM), can be estimated via testing methods and measurements. The familiar research of this problem is the response of circular and square thin-walled CFST specimens with length-to-diameter ratios subjected to axial loading through finite element analysis (FEA) model to predict bearing capacity and the failure modes, experimental analysis using displacement transducer and strain gauges to plot load versus strain curve [2].

In addition, Investigating failure under flexural buckling of CFST columns by an experimental study was compared with compression load limits in the South African code (SANS 10162-1) and Eurocode 4 (EC4) [3]. Experimental behavior of circular CFST columns with different L/D ratios and D/t ratios are based on EC4 and AISC 360-10 predictions [4]. Also, Experimental investigations with displacement transducers using the Design of Experiments (DOE) approach on circular CFT samples with D/t ratio of 22.3-50.8 and L/D ratio of 7.8-22.5, were compared with AISC-LRFD-2005 and EC4-1994 [1].

Besides, there are many studies for experimental approaches and some other papers only focus on theoretical methods: Computer-aided engineering (CAE) and Design Codes. To illustrate, new theoretical model developed to analyze the axial and lateral stress–strain response of CFT columns by integrating a lateral-to-axial strain model of confined concrete, an axial stress–strain model of confined concrete, and a biaxial stress–strain model of steel tube [5]. A parametric study was investigated on the effects of concrete strength, steel pile, FRP wraps and steel ring properties on the structural capabilities of external CFST columns [6]. The F.E modelling of confined concrete and steel

tube (concrete-steel interface) was analyzed and compared with CFST specimens tested by other published researches [7].
Similarly, the behavior of concrete-encased CFST columns, concrete-encased CFST, hollow RC and conventional CFST with square and circular section under combined compression and torsion were investigated according to the Chinese design guideline for concrete structures GB50010-2010, experimental method using strain gages to predict shear strain and failure modes [8]. Axial compressive strength and ductility of CFST stub columns using new approach for reducing the axial stress in the steel tube were experimentally studied [9]. Tensile properties of steel in CFST columns applied to a cyclic loading and corrosion test [10].
There has been no shortage of heated discussion on the issues from experiment to computation, and the purpose of the present works was to study the behavior of CFST specimens under axial load. The specimens were analyzed theoretically and experimentally to estimate the capacity of CFST and to calculate ultimate compressive strength, and determine positions of local buckling through the 3D-finite element modelling of steel tube and concrete core, together with testing procedures via measuring strain, displacement and force.

## 2. Experimental investigation and computational model evaluation

An overall view of the test facilities in the experimental study will be as shown in Figure 1 and 2. The system consists of stand-alone Mobile Base Station along and three four-channel Nodes, which can communicate through wireless with the user's PC. WinSTS software controls over the wireless structural testing system (STS-WiFi) to monitor Node status, signal strength, and record sensor signals. The wireless structural testing system was supplied by Bridge Diagnostics - USA [17].

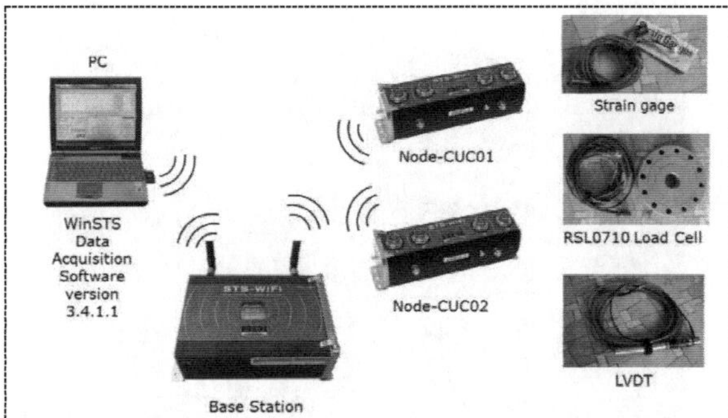


**Figure 1.** Wireless structural testing system (STS-WiFi).

**Figure 2.** Overall view experimental equipment and instruments for CFST.

The mix proportions of 01 cubic metre concrete are: PCB 40 Cement (Nghi Son, Vietnam): 395 kilograms; water: 210 litres; river sand: 530 kilograms; coarse aggregate: 1150 kilograms. The concrete strength B25 with slump of 12 centimeters follows to TCVN 5574:2012, [13], equal to C20/25 according to Eurocode 2 (EC2), [11] which is compressive strength for cylindrical concrete specimens $f_{c,cyl}$ of 20 MPa and modulus of elasticity of concrete $E_c$ of $29x10^3$ MPa at 28 days. Material properties of steel follow to BS 1387/1985, ASTM A53 with specified minimum yield stress $F_y$ of 240 MPa and modulus of elasticity of steel $E_s$ of $2x10^5$ MPa. Three CFST specimens are shown in Figure 3. It is assumed that there is sufficient interaction between the steel pipe and the concrete core, then the ultimate bearing strength of CFST, $P_{u,cal}$ is determined according to Eurocode 4 (EC4) as follows, [12]:

$$P_{u,cal} = P_{a,cal} + P_{c,cal} = f_y A_a + f_{c,cyl} A_c \tag{1}$$

where: $P_{a,cal}$ is the nominal yielding strength of steel material; $P_{c,cal}$ is the nominal yielding strength of concrete material; $f_y$ is specified minimum steel yield stress; $f_{c,cyl}$ is compressive strength for cylindrical concrete specimens; and $A_a$, $A_c$ is are cross-sectional areas of the steel pipe and concrete core, respectively.

IOP Conf. Series: Materials Science and Engineering **371** (2018) 012032 doi:10.1088/1757-899X/371/1/012032

AISC 360-10: The specimens are classified as compact sections for filled composite sections subjected to axial compression, Width (*D*) to Thickness (*t*) Ratio is checked as follows, (Table I1.1A – ANSI/AISC 360-10 Pages 16.1-84), [14]:

$$( D/t ) < \lambda_p = (0.15\times E_s) / F_y \tag{2}$$

The available compressive strength of axially loaded doubly symmetric filled composite members for compact sections shall be taken as, [14]:

$$P_p = F_y A_s + C_2 f_c' (A_c + A_{sr} \times E_s/E_c) \tag{3}$$

In which: coefficient of $C_2$ is 0.85 for rectangular sections and $C_2$ is 0.95 for round sections; $A_{sr}$ is equal to 0.0 with no steel reinforcement (rebar) in composite members; $F_y$ is specified minimum yield stress of steel section; $A_s$ is area of steel; $A_c$ is area of the concrete section; $f_c'$ is specified compressive strength of concrete.

The mounting of the test specimen and the arrangement of Spring Return Linear Variable Differential Transformer (LVDT), strain gages are shown in Figure 4 and more clearly in Figure 5. The loading apparatus consists of compression testing machine pressure machine set with 100 tons, Model: WE-1000B produced by Wuxi Xiyi - China; LoadCell with 120000 lbs of RDP Electronics Ltd, UK (Model: RSL0710/120000LB) is used to measure compression force signal and installed at bottom of CFST. Strain gauge PL-60-11(TML Tokyo Sokki Kenkyujo co., Ltd. - Japan) in concrete core is 60 mm length, 120Ω gauge resistance, 2.13 gauge factor. Strain gauge PLA-5-11 (TML Tokyo Sokki Kenkyujo co., Ltd. - Japan) on steel surface is 5 mm length, 120Ω gauge resistance, 2.13 gauge factor.

**Figure 3.** Manufacturing of CFST specimens.

**Figure 4.** Test setup for CFST column.

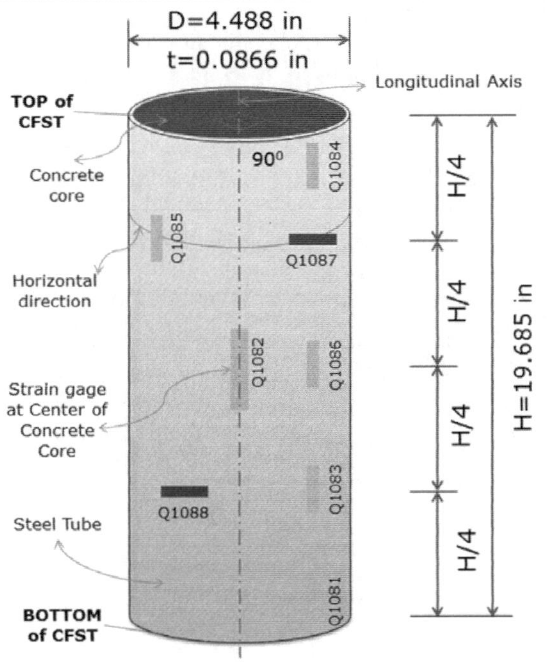


**Figure 5.** Schematic diagram of positions and identifications for strain gages.

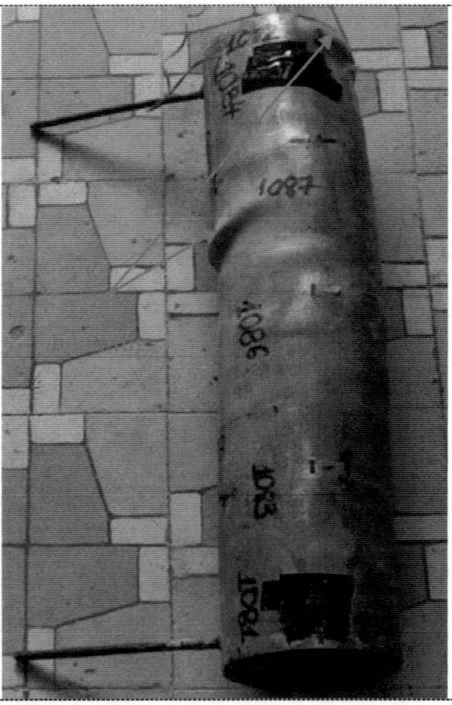

**Figure 6.** Buckling failure mode of the steel tube surface.

The numerical results in CAE are shown in Figure 7 for observing and predicting failure mode of modelling CFST, and large deformation under compressive stresses on the steel tube's surface near the concrete top surface also is described in this Figure. However, Three test specimens have different failure modes compared with numerical simulation of CFST model under axial load. Specimens, namely MUCE-01 is Euler buckling [15]; MUCE-02 is local buckling near the column top; The clearest result is MUCE-03 with outward local buckling in the region near the top and middle of CFST column, are shown in Figure 8.a, 8.b, 8.c respectively.

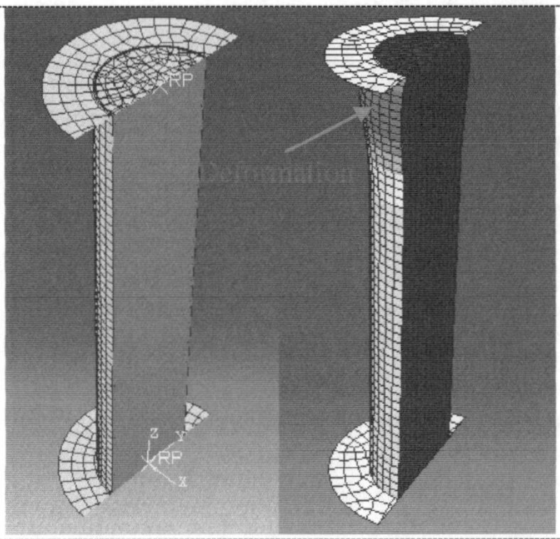


**Figure 7.** Finite element modelling of CFST column: steel tube and concrete core.

**Figure 8.** Tested results and failure modes of three CFST specimens.

Figure 9 illustrates the behavior of CFST specimen (MUCE-03) subjected to compression force. Figure 9.b shows force versus strain of strain gages at the top (Q1084), the bottom (Q1081), the middle (Q1086), H/4 (Q1085 and Q1083) of steel tube surface parallel to the longitudinal direction. It can be seen that loading is about 101100 lbf, compression strain at position H/4 (Q1085) decreases rapidly from 0.1104 to 0.05, part of the explanation lies in local outward buckling. Two strain gages including Q1087 and Q1088 are mounted at H/4 of steel surface in the horizontal direction for monitoring local buckling as shown in Figure 9.c. The axial load versus displacement (LV3552) with approximately maximum values: 116600 lbf and 1.119 inches in the Figure 9.d. Moreover, Figure 9.a demonstrates crack occurring in concrete core (installed strain gage - Q1082) due to local buckling, signal curve of Q1082 is interrupted, as well as some other signals can be errors at this time.

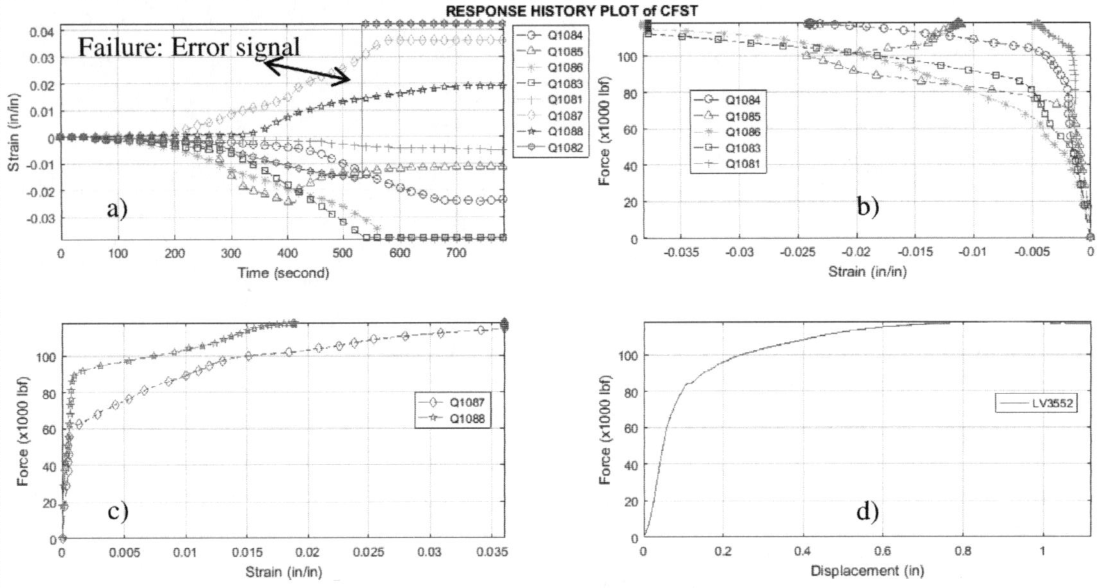


**Figure 9.** Force-displacement and force-strain curves of tested specimen (MUCE-03).

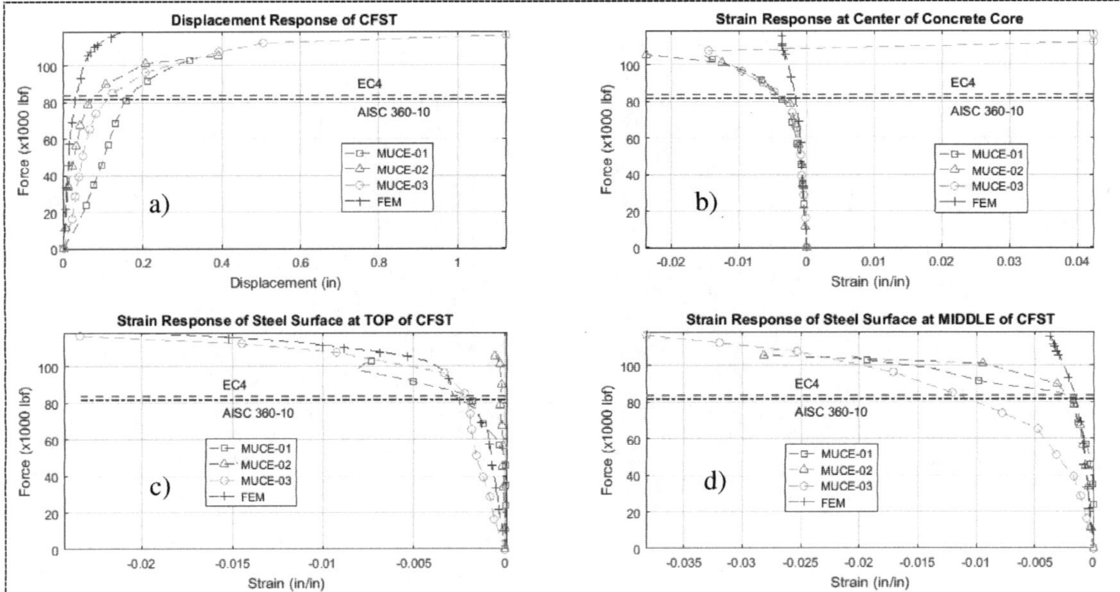


**Figure 10.** Response history curves of three tested specimens.

As shown in Figure 10 above and Figure 11 below, displacement and strain versus axial force graphs are measured in different positions of three specimens. As can be seen, MUCE-01, MUCE-02, MUCE-03 correspond to dashed line with square markers, dashed line with upward-pointing triangle markers, dashed line with circle markers, respectively. Load test records are compared with data in CAE for fracture simulation of CFST model applied to longitudinal compression and CAE result is dashed line with plus-sign markers. In this paper, ultimate strength of concrete-filled composite steel column is calculated by equation (1) according to EC4 with dashed line of this figure. Moreover, the ultimate capacity prediction of CFST is based on AISC 360-10 from the equation (3) with dash-dot line under line of EC4. According to EC4 and AISC 360-10 standard are about 84190.949 lbf (374.5 kN), 82504.882 lbf (365 kN), respectively, while the real axial load capacity of CFST specimen is approximately equal to 96430 lbf (nearly 429 kN) under failure modes higher than design standards. From Figure 10.a, it is observed that critical load of CFST specimens reaches a peak at just over 101200 lbf (450 kN) and then loading increases very slowly while strain signal is error in Figure 10.b, partly because of failure criterion for concrete core, partly because local buckling occurs in the region near the top of column in Figure 10.c and the region near the mid-height of specimen in Figure 10.d. Strain values in CAE modelling are less than results in experiment of figures because CFST modelling is not calibrated, steel and concrete's modulus of elasticity can be changed to compare with design standards. From CAE modelling of CFST can be used to simulate failure modes, and then support in planning and optimisation for installing sensors, and can be used to calculate ultimate strength calculations of CFST column under axial compression.

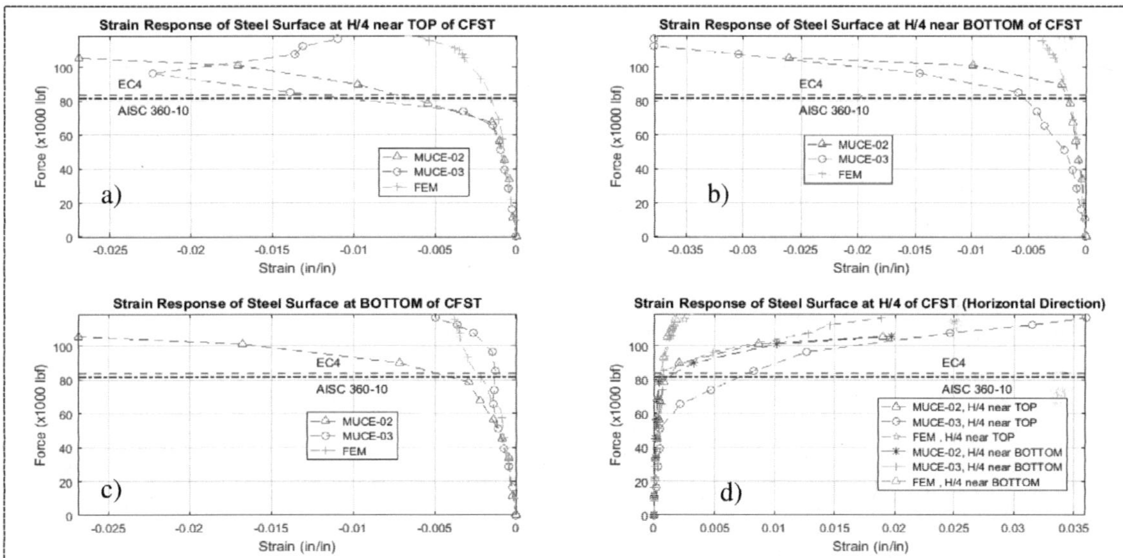


**Figure 11.** Behavior curves at different locations.

In Figure 11.a, strain data at region of H/4 rises significantly (MUCE-02), MUCE-03 jumps sharply then goes down moderately. This means that, large deformation on steel surface corresponding to region of local buckling is unstable (Figure 11.b, c, d). Buckling modes can be monitored by strain gages at weak locations near top and bottom of CFST column.

### 3. Conclusions

The following principal conclusions are drew from the results that were obtained in the study:

- The ultimate strength of CFST specimen under axial load was determined by EC4, AISC 360-10 and strain measurement.
- The location of local buckling failure of CFST column was observed clearly through testing procedure, and also showed in CAE modelling. Comparison between measured results and predicted by uncalibrated model was moderately error because material properties, friction coefficients of steel and concrete were obtained as described in design standards and studied papers. In order to solve

IOP Conf. Series: Materials Science and Engineering **371** (2018) 012032 doi:10.1088/1757-899X/371/1/012032

limitations of this paper, the modelling of CFST member can be updated by data of strain and force load overtime in the further work.
- The wireless structural testing system (STS-WiFi) in this study was correctly measured and recorded dynamic and static time historical responses of CFST specimens. This experimental equipment and instruments can be used in long-term structural health monitoring (SHM) for CFST members and other structures including: buildings, bridges, industrial structures based on Non-Destructive Testing-NDT in the field.

## 4. References


[1] Manojkumar V. Chitawadagi, Mattur C. Narasimhana and M. Kulkarni 2010, Axial strength of circular concrete-filled steel tube columns - DOE approach, *Journal of Constructional Steel Research*ed G. A. R. Parke (England: Elsevier) volume 66 issue 10 pp 1248-1260.

[2] Y F Yang and L H Han 2012, Concrete filled steel tube (CFST) columns subjected to concentrically partial compression, *Thin-Walled Structures* ed N. Silvestre (Portugal: Elsevier) volume 50 issue 1 pp 147-156.

[3] M. Dundu 2012, Compressive strength of circular concrete filled steel tube columns, *Thin-Walled Structures* ed N. Silvestre (Portugal: Elsevier) volume 56 pp 62-70.

[4] Talha Ekmekyapar and Baraa J.M.AL-Eliwi 2016, Experimental behavior of circular concrete filled steel tube columns and design specifications, *Thin-Walled Structures* ed N. Silvestre (Portugal: Elsevier) volume 105 pp 220-230.

[5] A. K. H. Kwan, C. X. Dong and J. C. M. Ho 2016, Axial and lateral stress–strain model for concrete-filled steel tubes, *Journal of Constructional Steel Research*ed G.A.R. Parke (England: Elsevier) volume 122 pp 421-433.

[6] C. X. Dong, A. K. H. Kwan and J. C. M. Ho 2017, Effects of external confinement on structural performance of concrete-filled steel tubes, *Journal of Constructional Steel Research*ed G. A. R. Parke (England: Elsevier) volume 132 pp 72-82.

[7] Y. Ouyang, A. K. H. Kwan, S. H. Lo and J. C. M. Ho 2017, Finite element analysis of concrete-filled steel tube (CFST) columns with circular sections under eccentric load, *Engineering Structures* ed P.L. Gould (USA: Elsevier) volume 148 pp 387-398.

[8] Qing-Xin Ren, Lin-Hai Han, ChaoHou, Zhong Tao and ShuaiLi 2017, Concrete-encased CFST columns under combined compression and torsion: Experimental investigation, *Journal of Constructional Steel Research* ed G. A. R. Parke (England: Elsevier) volume 138 pp 729–741.

[9] Liusheng He, Yangang Zhao and Siqi Lin 2017, Experimental study on axially compressed circular CFST columns with improved confinement effect, *Journal of Constructional Steel Research* ed G. A. R. Parke (England: Elsevier) volume 140 pp 74–81.

[10] Fang Yuan, Mengcheng Chen, Hong Huang, Li Xie and Chao Wang 2017, Circular concrete filled steel tubular (CFST) columns under cyclic load and acid rain attack: Test simulation, *Thin-Walled Structures* ed N. Silvestre (Portugal: Elsevier) volume 122 pp 90–101.

[11] Eurocode 2, 2004, Design of concrete structures, *European Committee for Standardization* EN1992-1-1.

[12] Eurocode 4, 2004, Design of composite steel and concrete structures, part 1.1: General rules and rules for buildings, *European Committee for Standardization* EN1994-1-1.

[13] TCVN 5574-2012, Concrete and reinforced concrete structures - Design standard, Vietnam.

[14] Specification for Structural Steel Building 2010, American Institute of Steel Construction.

[15] Timoshenko S. P. and Gere J. M. 1961. Theory of Elastic Stability, 2 ed., McGraw-Hill.

[16] Amar Khennane 2013, Introduction to Finite Element Analysis Using MATLAB and Abaqus, *Taylor & Francis Group,* USA.

[17] Bridge Diagnostics Inc 2015, Wireless Structural Testing System STS-WiFi Operations Manual (Colorado: USA).



**Acknowledgments**

The authors would like to thank the Mien Trung University of Civil Engineering (MUCE, LAS-XD 162) in Phu Yen province, Ministry of Construction of the Socialist Republic of Vietnam (MOC) for supporting this research project.

IOP Conf. Series: Materials Science and Engineering **371** (2018) 012033 doi:10.1088/1757-899X/371/1/012033

# Cyclic Behaviour of HPFRCC Coupling Beams with Diagonal Reinforcement

**SW Han[1] and SB Kim[2]**

1 Professor, Department of Architectural Engineering, Hanyang University, Seoul, Korea
2 Graduate Student, Department of Architectural Engineering, Hanyang University, Seoul, Korea
E-mail: swhan82@gmail.com

**Abstract.** Coupled shear walls are efficient in resisting lateral forces induced by winds and earthquakes. However, it is difficult to construct coupled shear walls particularly because current design codes require complex reinforcing details within coupling beams. The objective of this study was to investigate the cyclic behaviour of HPFRCC (high performance fiber reinforced cement composites) diagonally reinforced concrete (DRC) slender coupling beams. For this purpose, HPFRCC and conventional reinforce concrete DRC coupling beams were made and tested. This experimental study shows that HPFRRCC DRC coupling produced better seismic behaviour than corresponding conventional DRC beams.

## 1. Introduction

Coupled wall systems consisting of separate shear walls linked together by coupling beams at floor levels are effective in resisting wind- and earthquake-induced forces in high-rise buildings. When subjected to design-level earthquakes, coupling beams are expected to suffer substantial inelastic deformations due to the geometry of adjacent walls (MacGregor and Wright 2009); they play a key role in the performance of the entire system. Therefore, coupling beams should be equipped with adequate reinforcement details that perform well during seismic events, providing adequate ductility and energy dissipation, as well as strength and stiffness.

Historically, Paulay and Binney [1] first developed diagonal reinforcement for coupling beams. Diagonally reinforced coupling beams strongly resist sliding and have ductility, energy dissipation, and stiffness retention capacities superior to those conventionally reinforced coupling beams [1]-[ 3].

Based on the previous studies mentioned above, ACI 318-14 [4] specifies two confinement options for coupling beams with diagonal reinforcement. In the first option, each group of diagonal reinforcement comprises at least four longitudinal bars enclosed by densely spaced transverse reinforcement. This confinement method entails a very intricate arrangement of reinforcement, especially at the mid-span of the beam where opposite diagonal reinforcement groups meet each other. Also, a small amount of additional transverse reinforcement should be applied around the beam perimeter. According to Harries et al. (2005), the first confinement option is practically difficult to construct

Due to such shortcomings, a second confinement option is allowed by ACI 318-14, in which transverse reinforcement required for beams and columns of special moment frames is provided for the entire cross section of the beam, as a replacement for transverse reinforcement directly enclosing diagonal bars. However, reinforcement congestion remains a problem when he second option is used, especially near the midspan of the beam, where horizontal crossties conflict with diagonal bars.

During the last decade, research groups [5]-[8] have investigated the effectiveness of high performance fiber-reinforced cement composites (HPFRCCs) with respect to earthquake resistance of structures. HPFRCCs are characterized by strain-hardening behaviour in uniaxial tension by developing numerous micro-cracks with the assistance of a small portion of engineered fibers [6]. HPFRCCs generally show much higher ductility than normal concrete, under both tension and compression. Thus, the use of HPFRCC may relieve the confinement requirements of members with highly congested reinforcement. When subjected to seismic forces, HPFRCCs are deemed to improve energy dissipation by means of fiber bridging over micro-cracks and excellent bonding between the reinforcing steel and cement within the composite.
The objective of this study was to evaluate the cyclic behaviour of HPFRCC DRC beams. Experimental tests were conducted using slender concrete and HPFRCC DRC beams. Six test specimens were made, which included three concrete DRC specimens and three HPFRCC specimens with the same reinforcement details as the corresponding concrete DRC beams. The aspect ratio of the specimens was set to 3.5, and the reinforcement detail of the specimens follows the second detail option provided by ACI 318-14. The test variable was the application of HPFRCC

## 2. Experimnets

### *2.1. Test specimens*

Experimental tests were conducted using four diagonally reinforced coupling (DRC) beam specimens subjected to reversed cyclic loads. Test variable was material types (HPFRCC and regular concrete). The aspect ratio of all the specimens was 3.5.
In this study, precast DRC beam specimens were made and tested for high quality. It is noted that constructing HPFRCC DRC beams at the construction site and controlling their quality is a difficult task. Shear keys and U-shaped dowel bars were provided at the interface between the beam and stub. Concrete with a compressive strength of 60 MPa was used to construct the stubs and sufficient reinforcement was placed in the stubs to avoid failure before the DRC beam specimens failed.
Among the six specimens, one normal concrete DRC beam specimen (RC-110) was made as a standard specimen, which was designed and detailed according to ACI 318-14 (Fig. 1a). Figures 1 and 2 shows six specimens tested in this study. Please follow these instructions as carefully as possible so all articles within a conference have the same style to the title page. This paragraph follows a section title so it should not be indented.

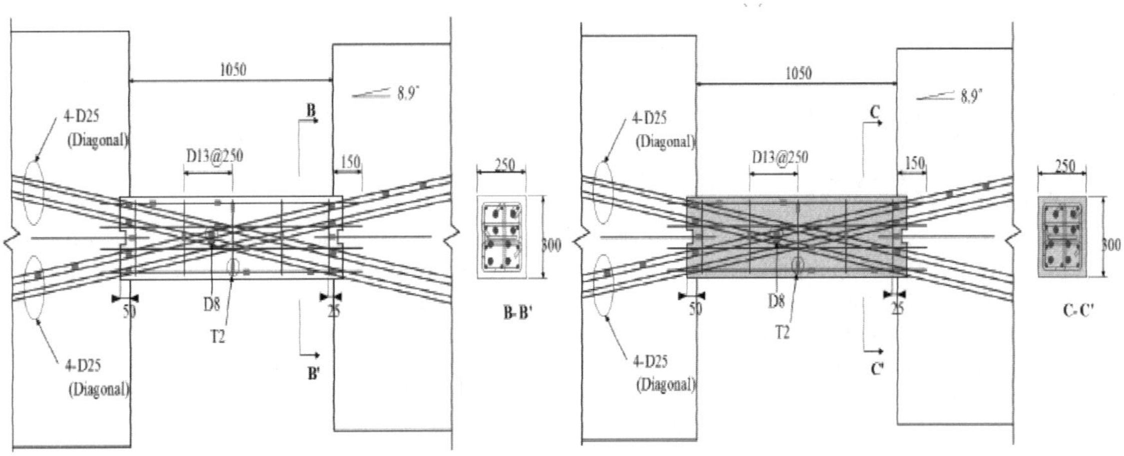


(a)RC-250 (b)HRC-250

**Figure 1.** Specimens

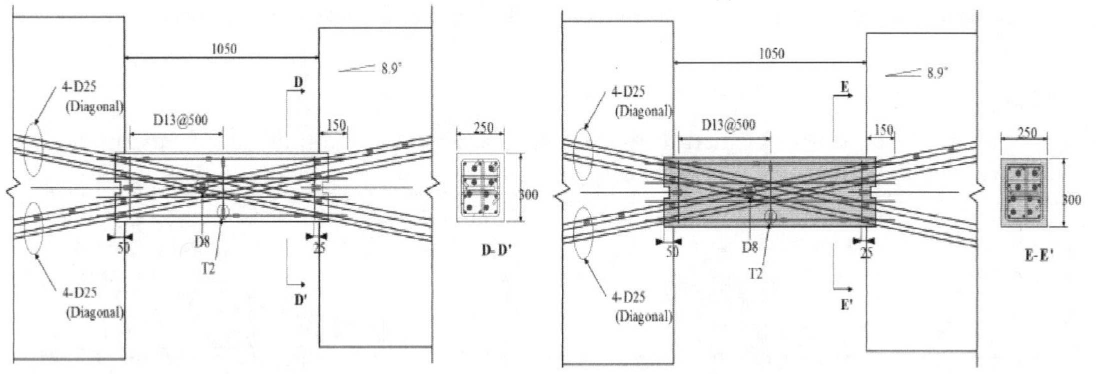

(c)RC-500 (b)HRC-500

**Figure 2.** Specimens.

*2.2 Test setup*

Figure 3 shows a test set-up with a specimen that is placed vertically. The bottom stub was fixed to a strong floor with anchors. Loading was imposed horizontally by a hydraulic actuator on the steel frame connected to the top stub to simulate the cyclic behaviour of coupling beams under later loads. Two roller supports (Fig. 3b) were installed at the ends of the steel frame to prevent relative rotation between the ends of the coupling beam (i.e., to prevent the top stub from rotating), but allowing for horizontal translation only. The roller supports restrained the axial elongation of the coupling beam specimens as well, which was similar to the restraint provided by adjoining shear walls due to their large in-plane stiffness [9]. Stoppers (Fig. 3d) were provided at the bottom stub to prevent the specimen from sliding.

Quasi-static displacement-controlled cyclic loading was used (Fig. 3). The drift loading amplitude was increased by 0.25% up to a 1.5% drift ratio, 0.5% up to a 4% drift ratio, 1% up to an 8% drift ratio, and 2% after an 8% drift ratio. The drift ratio ($\theta$) was defined as the lateral drift divided by the span length of the beam. For each drift loading amplitude, two same-drift consecutive cycles were applied to evaluate strength and stiffness degradations. The same loading sequence was used in all tests. The applied load was measured using a load cell installed at the actuator, and the shear in the coupling beam was taken to be equal to the applied load.

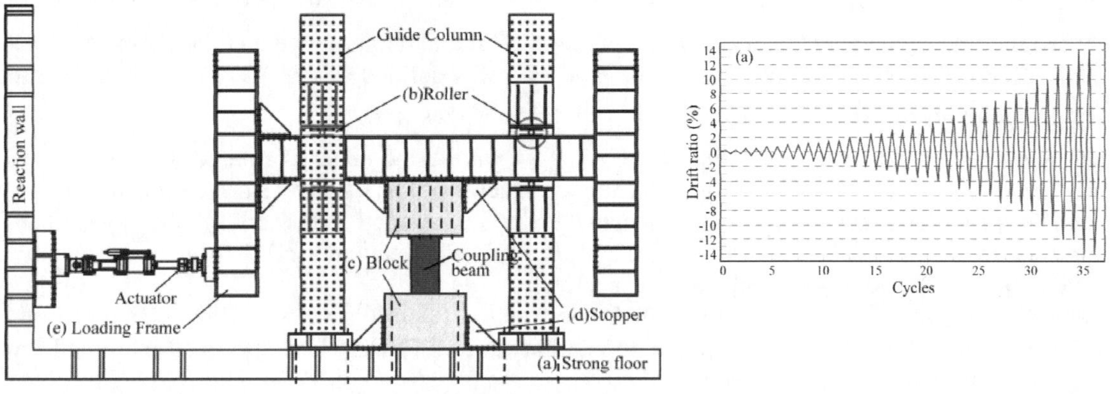


**Figure 3.** Test setup.

The drift in the specimen was measured using LVDTs installed horizontally at the top stub. Another horizontal LVDT was installed at the bottom of the top stub to check the possible rotation of the specimen. To measure the flexural deformation and shear distortion of the coupling beam, vertical and diagonal LVDTs were installed at one side of the beam, and the fixed-end rotation at the interface between the beam and stub was monitored by vertical LVDTs at the top and bottom faces of the beam (Fig. 3).

## 3. Cyclic behaviour

Figure 4 shows cyclic curves obtained from the test. The abscissa and ordinate of this figure are the drift ratio ($\theta$) and shear force ($V$), respectively.

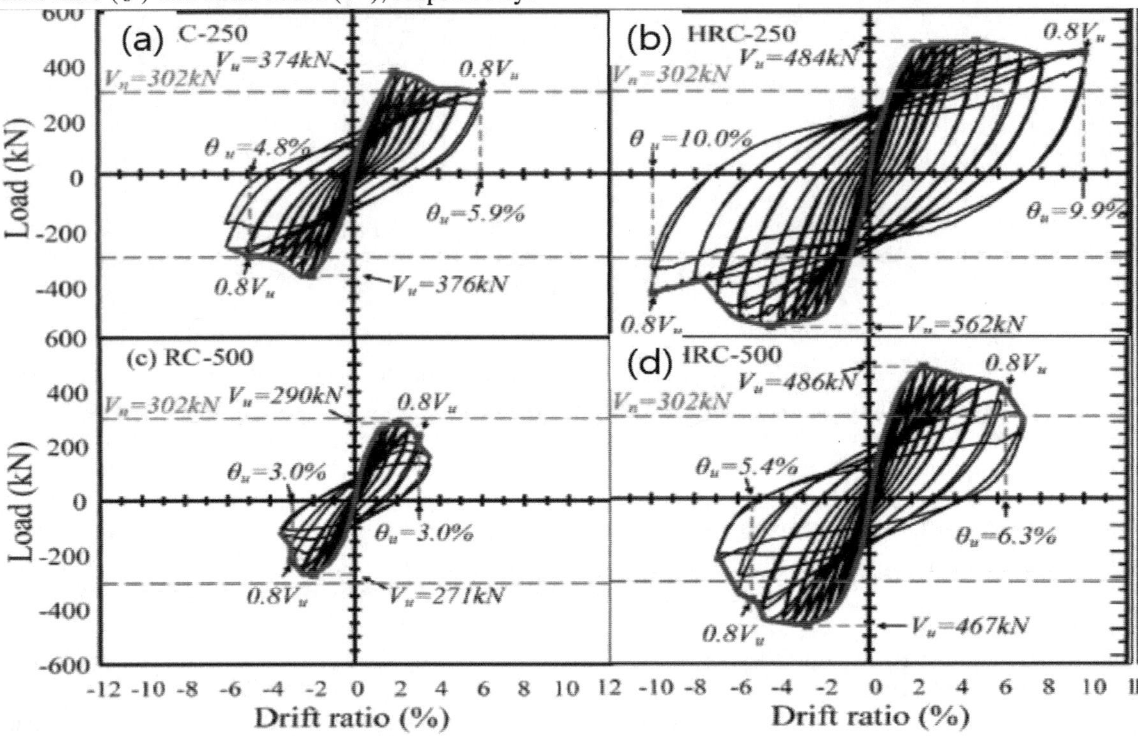


**Figure 4.** Cyclic curves of the specimens.

Table 1 summarizes the important properties obtained from cyclic curves, including shear force ($V_y$) and drift ratio ($\theta_y$) at yielding, maximum shear force ($V_u$), maximum drift ratio ($\theta_u$), drift ratio at failure ($\theta_f$), and ductility ratio ($\mu$). The yield point for $V_y$ and $\theta_y$ was taken as the point of intersection between the line connecting the origin to the point at 2/3 the maximum load and the horizontal line at the maximum load [10]. The maximum drift ratio ($\theta_u$) was defined as the drift ratio when the shear strength decreases by 20%. The ductility ratio was calculated as $\theta_u / \theta_y$.

The shear strength calculates according to ACI 318-14 with tested material strength for all specimens were 302 kN. The calculated shear strength for each specimen is marked by a dotted line in Fig. 4 All specimens produced shear strength larger than that calculated strength except for specimen RC-500.

Figure 4a shows the cyclic curve of specimens RC-250. The strength and deformation capacity of specimen RC-250 were 374 kN and 5.9%, respectively.

In Fig. 4b, the cyclic curve of HPFRCC DRC specimen HRC-250 is shown, which was an identical specimen to specimen RC-250 except for the application HPFRCC. This specimen exhibited more stable cyclic behaviour than specimen RC-250 due to the beneficiary effect of HPFRCC. The strength and deformation capacity of this specimen were 484kN and 9.9%, respectively. Thus, the seismic behaviour of slender DRC beams could be significantly improved by using HPFRCC.

The cyclic curves of specimens RC-500 and HRC-500 were plotted in Figs. 4c and 4d. These specimens had transverse reinforcement as small as 25% of that required by ACI 318-14. As shown in Fig. 4c, specimen RC-500 exhibited the poorest seismic behaviour among the specimens: $V_u$ and $\theta_u$ of

this specimen were 290 kN and 2.0%, respectively. However, the seismic behaviour of the specimen with limited transverse reinforcement was much improved by using HPFRCC (Fig. 4d).
The maximum shear force ($V_u$) and maximum drift ratio ($\theta_u$) of specimen HRC-500 were 486 kN and 6.3%, which were 167% and 315% of those of specimen RC-500, respectively.

**Table 1.** Summary of experimental test results.

| Specimen | | $V_y$ (kN) | $\theta_y$ (%) | $V_u$ (kN) | $\theta_u$ (%) | $V_f$ (kN) | $\theta_f$ (%) | $\mu$ ($\theta_u / \theta_y$) |
|---|---|---|---|---|---|---|---|---|
| RC-250 | (+) | 347 | 1.7 | 374 | 5.9 | 256 | 6.0 | 3.5 |
| | (-) | 341 | 1.7 | 376 | 4.8 | 181 | 5.9 | 2.8 |
| HRC-250 | (+) | 457 | 2.0 | 484 | 9.9 | 358 | 10.0 | 4.9 |
| | (-) | 531 | 2.5 | 562 | 10.0 | 340 | 10.1 | 4.0 |
| RC-500 | (+) | 266 | 1.2 | 290 | 2.0 | 204 | 3.0 | 1.7 |
| | (-) | 245 | 1.2 | 271 | 2.0 | 165 | 3.0 | 1.7 |
| HRC-500 | (+) | 465 | 2.0 | 486 | 6.3 | 247 | 7.0 | 3.2 |
| | (-) | 434 | 2.0 | 467 | 5.4 | 226 | 7.0 | 2.7 |

## 4. Conclusions

The cyclic behaviour of HPFRCC DRC beams was investigated. Four specimens were made and tested. The seismic behaviours of concrete and HPFRCC specimens with different transverse reinforcements were compared with that of the corresponding reinforced concrete DRC beams. The seismic behaviour of HPFRCC specimen HRC-250 was superior to that of concrete specimen RC-250, which had the same amount of transverse reinforcement as specimen HRC-250. Similar observations were made for specimens HRC-500 and RC-500. Thus, the use of HPFRCC for DRC beams could improve the seismic behaviour of slender DRC beams.

## 5. References

[1] Aderhold J, Davydov V Yu, Fedler F, Klausing H, Mistele D, Rotter T, Semchinova O, Stemmer J and Graul J 2001 *J. Cryst. Growth* **222** 701

[2] Paulay T and Binney JR 1974 Diagonally reinforced coupling beams of shear walls, ACI special publication SP-42, 579–598, American Concrete Institute, Detroit, MI, U.S.A.

[3] Barney GB, Shiu KN, Rabbit BG, Fiorato AE, Russell HG, and Corley WG 1980 Behavior of coupling beams under load reversals, RD068.01B, Portland Cement Association Skokie, IL, U.S.A.

[4] Naish D, Wallace JW, Fry JA, and Klemencic R 2009 Reinforced concrete link beams: alternative details for improved construction, UCLA-SGEL Report 2009-06, University of California at Los Angeles, CA, U.S.A.

[5] ACI Committee 318 2014.Building code requirements for structural concrete and commentary, ACI 318-14, American Concrete Institute, Farmington Hills, MI, U.S.A.

[6] Canbolat BA, Parra-Montesinos, GJ and Wight JK 2005 Experimental study on seismic behavior of high-performance fiber-reinforced cement composite coupling beams, ACI Structural Journal 102(1), 159–166.

[7] Naaman AE and Reinhardt HW 1996 Characterization of high performance fiber reinforced cement composites-HPFRCC, High performance fiber reinforced cement composite 2(HPFRCC 2), Proceedings of the 2nd international RILEM workshop, Cachan Cedex, France, 1-24.

[8] Lequesne RD, Setkit M, Parra-Montesinos GJ and Wight JK 2011 Seismic detailing and behavior of coupling beams with high-peformance fiber-reinforced concrete, ACI special publication SP-272, 205–222. Berg, G.V., and Stratta, J.L., 1964. Anchorage and the Alaska earthquake of March 27. American Iron and Steel Institute, Washington, D.C., U.S.A.

[9] Han SW, Lee CS, Kwon HW, Lee KH and Shin M.S 2015 Behavior of fibre-reinforced beams with diagonal reinforcement, Magazine of Concrete Research 67(24), 1287-1300.

[10] Barbachyn SM, Kurama YC, and Novak LC 2012 Analytical evaluation of diagonally reinforced concrete coupling beams under lateral loads, ACI Structural Journal 109(4), 497-508.
[11] Pan A and Moehle JP 1989 Lateral displacement ductility of reinforced concrete flat plates, ACI Structural Journal 86(3), 250-258.

**Acknowledgement**
A The research was supported by grants from the National Research Foundation of Korea (NRF-2017R1A2B3008937).

IOP Conf. Series: Materials Science and Engineering **371** (2018) 012034 doi:10.1088/1757-899X/371/1/012034

# Application of HPFRCC on Coupling Beams with Bundled Diagonal Reinforcement

**SW Han[1], CS Lee[2], KH Lee[3], MS Shin[4]**

1 Professor, Department of Architectural Engineering, Hanyang University, Seoul, Korea E-mail: swhan82@gmail.com
2 Graduate Student, Department of Architectural Engineering, Hanyang University, Seoul, Korea
3 Professor, Department of Architectural Engineering, Sejong University, Seoul, Korea
4 Professor, School of Urban and Environmental Engineering, Ulsan National Institute of Science and Technology (UNIST), Ulsan, Korea
E-mail: mtsonicc@naver.com

**Abstract.** The objective of this study was to develop simple reinforcement details for diagonally reinforced coupling beams; reducing transverse steel by use of high-performance fiber-reinforced cementitious composites (HPFRCCs) and bundling diagonal bars are explored. Four coupling beam specimens with length-to-depth aspect ratios of 2.0 or 3.5 were fabricated and tested under cyclic lateral displacements. The test results revealed that HPFRCC coupling beams with bundled diagonal bars and widely spaced transverse reinforcement (one-half the amount of reinforcement required by current seismic codes.) exhibited excellent seismic performance compared with ordinary concrete coupling beams to bundled diagonal bars and code-required transverse reinforcement.

## 1. Introduction

Coupling beams reinforced with longitudinal bars parallel to the span of the beam, may experience sliding shear failure near the beam ends at which flexural cracks caused by reversed cyclic loading come across one another [1]. Transverse reinforcement is not capable of preventing sliding shear failure when flexural cracks propagate across the entire depth of the beam between stirrups [2]. Many studies have been conducted with the aim of resolving this problem. Paulay and Binney [3] first developed diagonal reinforcement for coupling beams. Diagonally reinforced coupling beams strongly resist sliding and have ductility, energy dissipation, and stiffness retention capacities superior to those conventionally reinforced coupling beams [4, 5, 6].

According to Harries et al. [7], the confinement option provided in ACI 318-14 [8] is practically difficult to construct when the average shear stress in the beam is greater than $0.5\sqrt{f'_c}$ MPa, where $f'_c$ is the concrete compressive strength in MPa, especially near the mid-span of the beam, where horizontal crossties conflict with diagonal bars.

In efforts to resolve the difficulty of fabricating diagonal reinforcement in coupling beams, various reinforcing details have been investigated and proposed to date [4, 9] Han et al. [9] recently tested the efficiency of bundled diagonal reinforcement in reinforced concrete (RC) coupling beams. Bundled diagonal reinforcement allows more internal space, enhancing workability and allowing simple construction, compared with code-specified diagonal reinforcement in which spacers are needed to

IOP Conf. Series: Materials Science and Engineering **371** (2018) 012034 doi:10.1088/1757-899X/371/1/012034

maintain the gaps between separate diagonal bars. Bundling also increases the angle of diagonal reinforcement measured from the longitudinal axis of the beam, thereby increasing both flexural and shear strength. Han et al. [8] reported that coupling beams having bundled diagonal reinforcement achieved greater strength and energy dissipation than, and a similar displacement ductility to, those with code-specified diagonal reinforcement.

Several researchers [10-14] have reported that the use of high performance fiber-reinforced cement composites (HPFRCCs) in structural members improved their seismic resistance. HPFRCCs produce strain-hardening behaviour in uniaxial tension by developing numerous micro-cracks with the assistance of a small portion of engineered fibers [15]. HPFRCCs generally show much higher ductility than normal concrete, under both tension and compression. Thus, the use of HPFRCC may relieve the confinement requirements of members with highly congested reinforcement [10]. When subjected to seismic forces, HPFRCCs are deemed to improve energy dissipation by means of fiber bridging over micro-cracks and excellent bonding between the reinforcing steel and cement within the composite [15].

The aim of the present work was to investigate the effect of HPFRCC on the hysteretic behaviour of coupling beams with bundled diagonally reinforcement. For this purpose, four specimens were made and tested.

## 2. Experimental Program

### *2.1. Specimen details*

Figure 1 illustrates the dimensions and reinforcing details of the specimens. All specimens were reinforced with bundled diagonal reinforcement and also used the second confinement option illustrated in Fig. 1c.

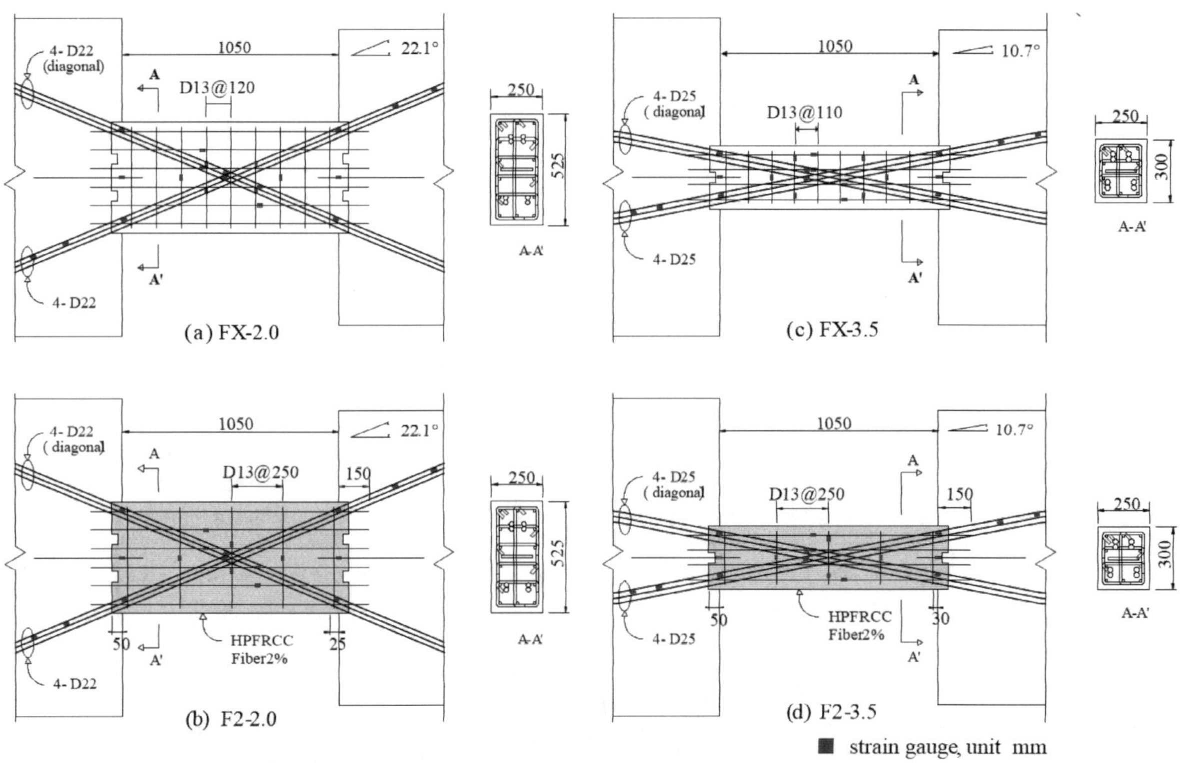


**Figure 1**. Dimensions and reinforcing details of coupling beam specimens.

The main test variables were the use of conventional RC or the use of HPFRCC in combination with reduced transverse reinforcement, and the use of two different length-to-depth aspect ratios ($l_n/h$). Table 1 summarizes the test variables and specimen dimensions. The HPFRCC used in this study

contained polyvinyl alcohol (PVA) fibers. The PVA fiber content in the HPFRCC was 2% by volume. The reduced amount of transverse reinforcement was about one half that required by ACI 318-141 [8]. The beam aspect ratio (defined as the span length divided by the total depth of the beam) was selected to be either 2.0 or 3.5 to respectively represent typical deep or slender coupling beams in high-rise residential buildings [14].
The beam length was 1050 mm in all specimens, and the beam depth was 525 mm or 300 mm. The total area of diagonal bars was determined so that the average shear stress in the coupling beam would be limited to approximately $0.5\sqrt{f'_c}$ MPa. The spacing of transverse reinforcement was 120 and 110 mm in the specimens with the aspect ratios of 2.0 and 3.5 respectively, not exceeding six times the diagonal bar diameter as required in ACI 318-14. The inclination angles of bundled diagonal reinforcement in the specimens with the aspect ratios of 2.0 and 3.5 were about 22.1° and 10.7°, respectively. ACI 318-14 specifies that horizontal reinforcement in diagonally reinforced coupling beams mainly used to provide anchorage for horizontal crossties, shall not develop the yield strength at walls. In this study, the embedment length of horizontal reinforcement into the top and bottom stubs was 50 mm in all specimens.
To prevent any damage to the stubs, they were built using concrete of a compressive strength of 60 MPa and included ample reinforcement details.

**Table 1.** Specimen details and test variables

| Specimen | Material | Width (mm) | Depth (mm) | Span length (mm) | Length-to-depth ratio ($l_n/h$) | Angle α (∘) | Transverse reinforcement spacing (mm) |
|---|---|---|---|---|---|---|---|
| RC-2.0 | RC | 250 | 525 | 1050 | 2.0 | 22.1 | 120 |
| HC-2.0 | HPFRCC | 250 | 525 | 1050 | 2.0 | 22.1 | 250 |
| RC-3.5 | RC | 250 | 300 | 1050 | 3.5 | 10.7 | 110 |
| HC-3.5 | HPFRCC | 250 | 300 | 1050 | 3.5 | 10.7 | 250 |

*2.2 Loading Frame, Protocol, and Measurements*
Figure 2 illustrates the test setup and loading protocol. The test setup was intended to represent the behavior of coupling beams subjected to lateral loading. The coupling beam was arranged vertically, and the steel frame rigidly connected to the top stub was loaded horizontally by a hydraulic actuator. The bottom stub was fixed to the strong floor with anchors. In order to enforce zero moment at the mid-span of the coupling beam, the axis of the actuator was arranged to pass through the mid-span. Also, two roller supports were installed near the ends of the steel frame to prevent the top stub from rotation and vertical translation but allow horizontal translation. Stoppers were installed at the ends of the bottom stub to prevent the specimen from sliding.
Quasi-static reversed cyclic loading was applied with controlled displacement as shown in Fig. 4b The drift ratio is defined as the lateral displacement between the ends of the coupling beam divided by the beam length. For each drift ratio, two same-drift consecutive cycles were applied to assess strength and stiffness degradations. The same loading protocol was used in all tests.

*2.3 Material tests*
Compression and uniaxial (direct) tension tests were conducted to examine the properties of the normal concrete and HPFRCC used to construct the coupling beam specimens. PVA fibers in the HPFRCC were 2.0% by volume. The water/PCM ratio was approximately 20%, where "PCM" stands for dry premixed cement mortar consisting of binder, fillers, and chemical admixtures. Table 2 summarizes the physical properties of the PVA fibers used in the HPFRCC. In the normal concrete, the maximum aggregate size was 25 mm, whereas the HPFRCC contained neither coarse nor fine aggregates.
For the compression tests, three cylindrical specimens having 100 mm in diameter and 200 mm in

IOP Conf. Series: Materials Science and Engineering **371** (2018) 012034 doi:10.1088/1757-899X/371/1/012034

height were fabricated according to ASTM C39, and cured under the same condition as the coupling beam specimens. Three LVDTs in parallel were installed around the perimeter of the specimen in the loading direction to estimate the average compressive strain, with the gage length of approximately 90 mm.

Fig. 3a shows the compressive stress-strain curves of the HPFRCC and normal concrete acquired after 28 days of curing. Both the HPFRCC and normal concrete showed slightly higher strengths than the design compressive strength of 40 MPa. The compressive stress-strain relationships indicated that the HPFRCC was much more ductile than the normal concrete; the strain measured at failure was approximately 67% greater for the HPRCC. On the other hand, the secant modulus of elasticity of the normal concrete was about 24% greater than that of the HPFRCC. The secant modulus was calculated according to ACI 318-14, which is the slope of a line passing through 45% of the maximum compressive strength in the stress-strain curve.

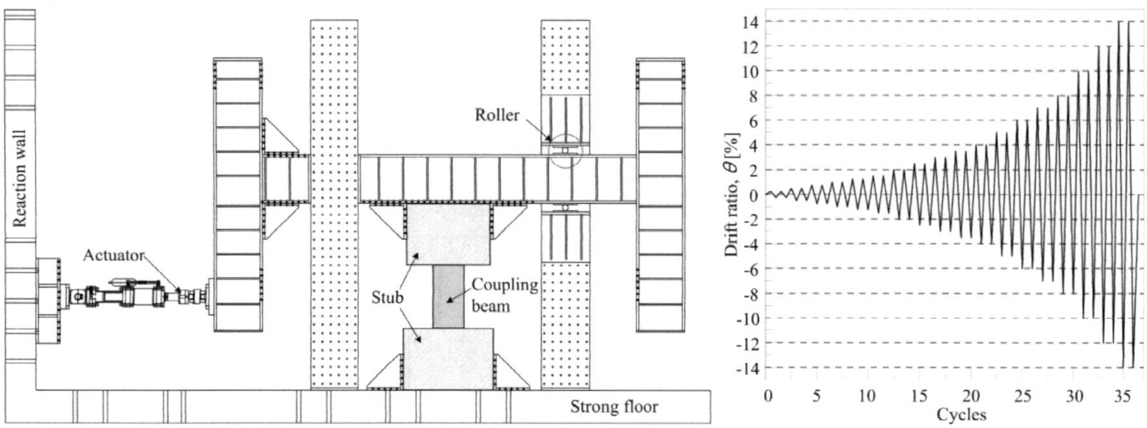


(a) (b)

**Figure 2**. Test setup and loading protocol.

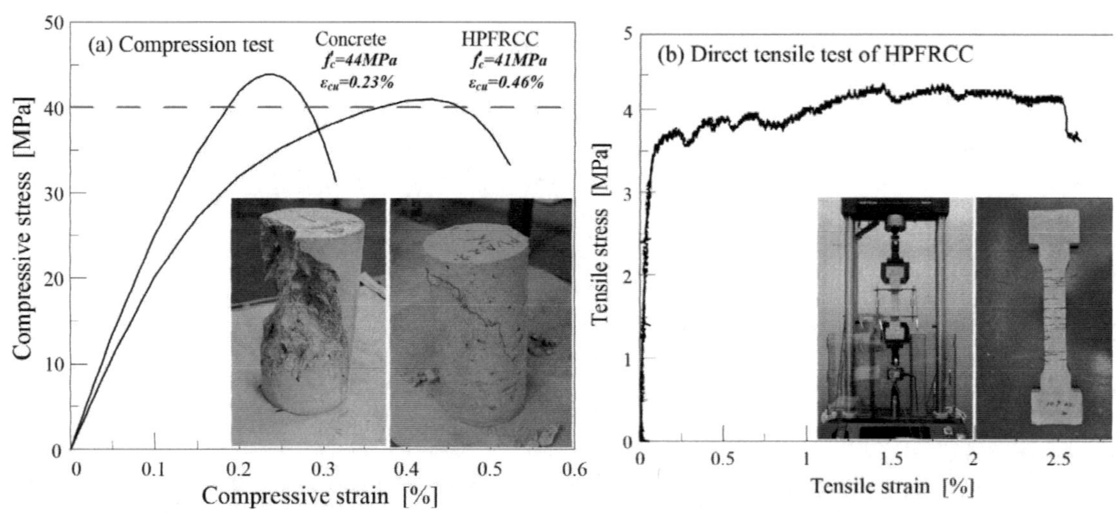


(a) Compression test (b) Direct tension test of HPFRCC

**Figure 3**. Stress-strain curves of concrete tests.

For direct tension tests, three dog-bone shaped specimens were fabricated, similar to those used by the

University of Michigan research group (Parra-Montesinos 2005). Two LVDTs in parallel were mounted along the sides of the specimen to estimate the average tensile strain, with the gage length of approximately 180 mm.

Fig. 3b shows the tensile stress-strain curve and cracking pattern of the HPFRCC used in this study. Under tension, the HPFRCC specimens showed ductile behavior by means of strain hardening, developing numerous well-distributed micro-cracks with the effect of fiber bridging [13, 15, 16]. The maximum tensile strain exceeded 2.5%, and the tensile strength was approximately 4.3 MPa. Table 2 summarizes the compressive and tensile strengths of the HPFRCC and normal concrete measured at the curing age of 28 days. Table 3 summarizes the mechanical properties of steel acquired from the tension tests.

**Table 2.** Properties of normal concrete and HPFRCC

| Type | Compressive strength $f_{cu}$ (MPa) | Maximum compressive strain $\varepsilon_{cu}$ (%) | Direct tensile Stress (MPa) | Maximum tensile Strain (%) |
|---|---|---|---|---|
| Normal concrete | 44 | 0.23 | - | - |
| HPFRCC | 41 | 0.46 | 4.3 | 2.5 |

**Table 3.** Properties of normal concrete and HPFRCC

| Type | Compressive strength $f_{cu}$ (MPa) | Maximum compressive strain $\varepsilon_{cu}$ (%) | Direct tensile Stress (MPa) | Maximum tensile Strain (%) |
|---|---|---|---|---|
| Normal concrete | 44 | 0.23 | - | - |
| HPFRCC | 41 | 0.46 | 4.3 | 2.5 |

## 3. Test Results and Observations

### *3.1 Load-displacement responses*

Figure 4 shows the cyclic shear-drift responses of the coupling beam specimens. The right side vertical axis of this figure indicates the normalized shear stress, namely, the ratio of shear force in the coupling beam, taken equal to the actuator load, to the product of the beam cross-sectional area ($A_{cw}$) and the square root of the concrete compressive strength ($f'_c$). Table 4 summarizes the yield load ($V_y$), yield drift ratio ($\theta_y$), maximum load ($V_u$), maximum drift ratio ($\theta_u$), and ductility ratio ($\mu$) of each specimen. The yield and maximum drift ratios were determined according to the work of Pan and Moehle (1989). The yield drift ratio corresponds to the point of intersection between the secant line connecting the origin to the point of 2/3 of the maximum load and the horizontal line at the point of the maximum load. The maximum drift ratio was measured when the strength reduced to 80% of the maximum load. The ductility ratio is defined as the maximum drift divided by the yield drift.

Specimens RC-2.0 and RC-3.5 of normal concrete, composed of standard RC, having code-specified transverse reinforcement exhibited stable load-drift behavior up to about 4% and 8% drift respectively, without considerable drop in strength. Between the two, the slender beam specimen showed much more ductility and had full load-drift loops. The RC-2.0 specimen suffered significant and successive strength drops from the second cycle to 5% drift, when some diagonal and transverse reinforcement ruptured. In contrast, the higher-aspect-ratio specimen RC-3.5 sustained more than 80% of its maximum load up to the first positive loading cycle of 10% drift ratio. Thus, the ductility ratios of Specimens RC-2.0 and RC-3.5 for loading in the positive direction were approximately 3.0 and 5.7 in, respectively.

Specimens HC-2.0 and HC-3.5, composed of HPFRCC, having one half the amount of code-specified transverse reinforcement, showed load-drift responses that were generally similar to those of the normal concrete specimens discussed above. HC-2.0 and HC-3.5 exhibited stable behavior up to 5% and 8% drift, respectively. HC-2.0 underwent large strength drops from the first cycle up to 6% drift, and completely collapsed during the 7% drift cycles, owing to rupture of the diagonal reinforcement. On the other hand, HC-3.5 started to become unstable only from the 10% drift cycles. The ductility ratios of Specimens HC-2.0 and HC-3.5 for loading in the positive direction were approximately 3.7 and 5.6, respectively.

It is noted that the normalized shear stress greatly exceeded the design limit of 0.83 for diagonally reinforced coupling beams as specified in ACI 318-14. Therefore, the test results suggest that bundled diagonal reinforcement was applicable in both deep and slender coupling beams, and that the HPFRCC was effective in allowing the amount of transverse reinforcement required by ACI 318-14 to be reduced by half.

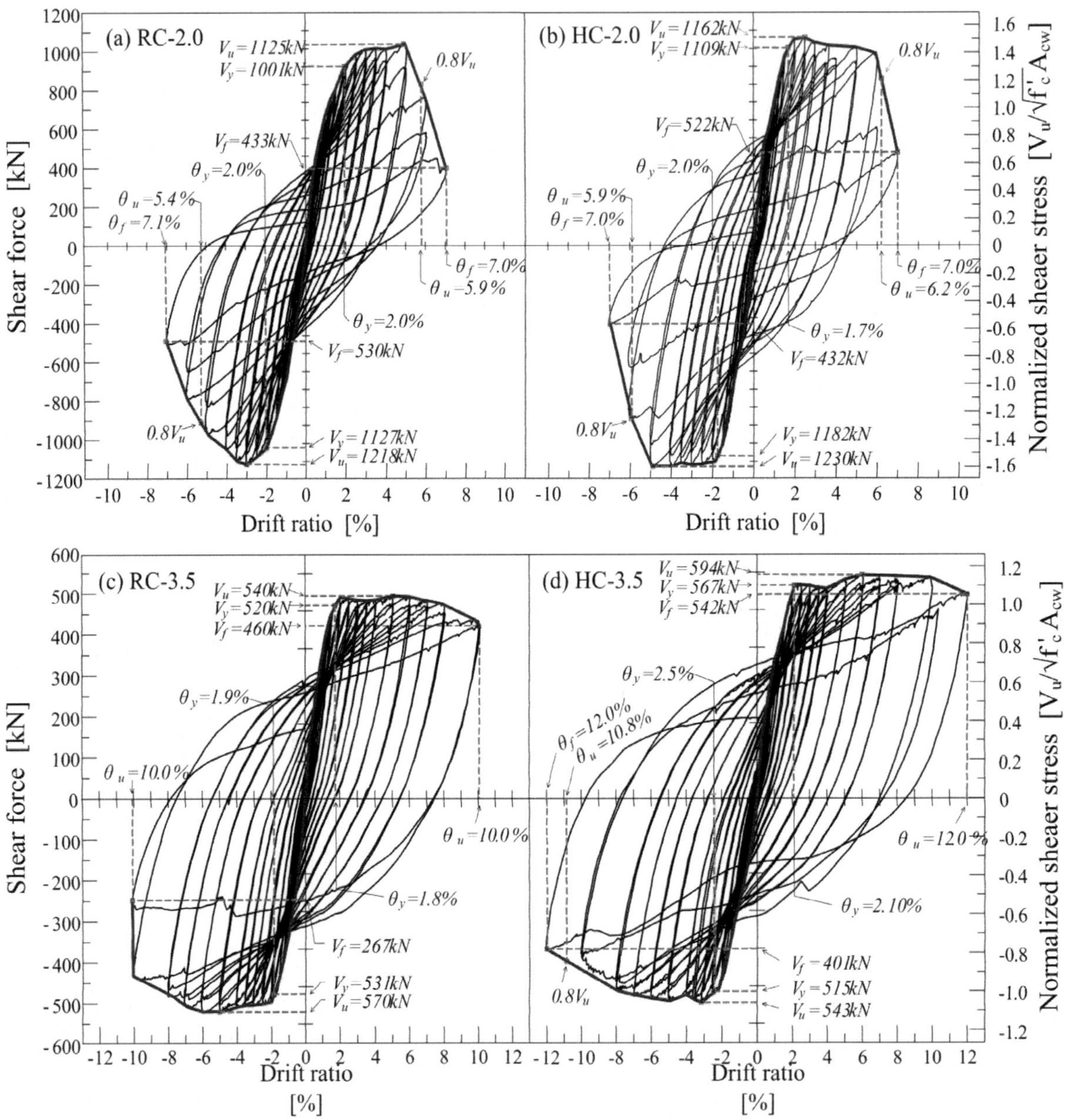


**Figure 4**. Hysteretic shear-drift responses.

**Table 4.** Test results for critical strengths and drift ratios

| Specimen | | $V_y$ (kN) | $\theta_y$ (%) | $V_u$ (kN) | $\theta_u$ (%) | $\mu$ $(\theta_u/\theta_y)$ | $V_{n-ACI}$ (kN) | $M_{n-ACI}$ (kN-m) |
|---|---|---|---|---|---|---|---|---|
| RC-2.0 | (+) | 1001 | 1.98 | 1125 | 5.88 | 2.96 | 490 | 259 |
| | (−) | 1127 | 2.00 | 1218 | 5.37 | 2.68 | 490 | 259 |
| HC-2.0 | (+) | 1109 | 1.70 | 1162 | 6.24 | 3.67 | 490 | 259 |
| | (−) | 1182 | 1.76 | 1230 | 5.88 | 3.34 | 490 | 259 |
| RC-3.5 | (+) | 532 | 1.76 | 540 | 10.04 | 5.70 | 316 | 167 |
| | (−) | 531 | 1.88 | 570 | 10.02 | 5.34 | 316 | 167 |
| HC-3.5 | (+) | 567 | 2.15 | 594 | 12.02 | 5.60 | 316 | 167 |
| | (−) | 515 | 2.53 | 543 | 10.79 | 4.27 | 316 | 167 |

Note: $V_y$ = yield load; $V_u$ = maximum load; $\theta_y$ = yield drift ratio; $\theta_u$ = maximum drift ratio; $\mu$ = ductility ratio equal to $\theta_u / \theta_y$ ; $V_{n-ACI}$ , $M_{n-ACI}$ = shear and moment strength calculated by ACI 318-11 (ACI 2014)

Figure 5 shows the envelopes of the cyclic shear-drift response curves for the four coupling beam specimens. Among both pairs of specimens with the same aspect ratio, the RC and HPFRCC specimens not only achieved similar maximum loads, but also underwent similar strength degradation histories. In addition, the primary cause of sudden strength drops in all four specimens was the rupture of diagonal reinforcement with no buckling. Therefore, combining the use of HPFRCC and half the amount of code-required transverse steel appeared to provide confinement equivalent to that of the code-required transverse reinforcement.

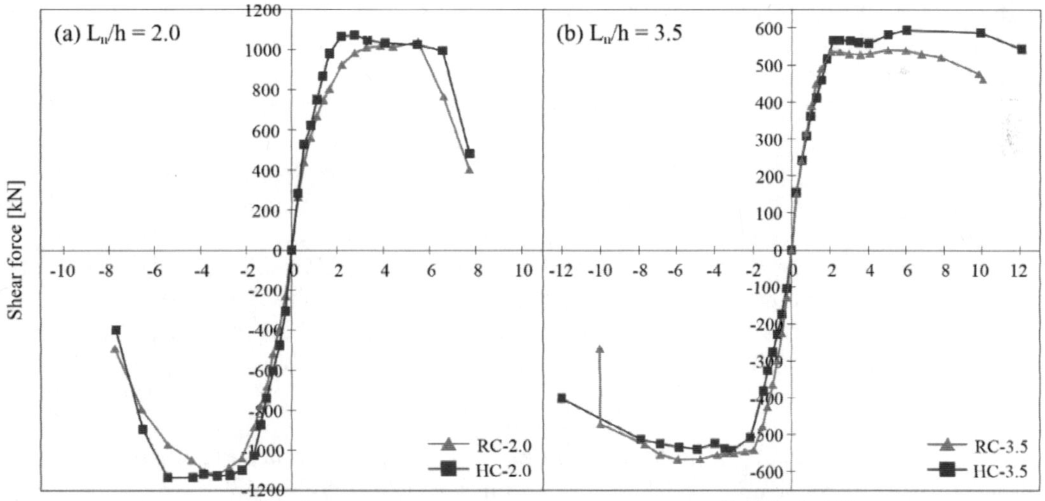


**Figure 5.** Envelope curves.

## 4. Conclusions

In this study, HPFRCC and RC coupling beam specimens with bundled diagonal reinforcement were tested under cyclic lateral loading in order to assess the effectiveness of HPFRCC. The RC specimens were reinforced with the code-required amount of transverse steel per ACI 318-14, whereas the HPFRCC specimens were reinforced with about half the code-required amount of transverse steel.

The HPFRCC coupling beams had similar lateral strengths to those of the RC coupling beams and

IOP Conf. Series: Materials Science and Engineering **371** (2018) 012034 doi:10.1088/1757-899X/371/1/012034

superior seismic capacities, even though the HPFECC specimens were reinforced using only half the code-required amount of transverse steel. This suggests that the amount of transverse reinforcement required by ACI 318-14 could be reduced by half with the use of HPFRCC, which supplied adequate confinement for the bundled diagonal bars.
The maximum drift ratio achieved was at least about 5.37 and 10.0 in the specimens of aspect ratios 2.0 and 3.5, respectively. Also, no substantial strength drop occurred until the bundled diagonal bars ruptured in all specimens. This indicates that the use of bundled diagonal reinforcement was effective in ensuring suitable seismic performances of both deep and slender coupling beams; this design also enables simplified construction.

**5. Reference**

[1] Berg VB and Stratta JL 1964 Anchorage and the Alaska Earthquake of March 27, 1964, AISC.
[2] Paulay T and Priestley MJ 1992 Seismic design of reinforced concrete and masonry buildings, John Wiley & Sons, N.Y.
[3] Paulay T and Binney JR 1974 Diagonally reinforced coupling beams of shear walls. Amer. Conc. I., 42, 579-598.
[4] Tassios TP, Moretti M and Bezas A 1996 On the behavior and ductility of reinforced concrete coupling beams of shear walls. ACI Struct. J., 93(6), 711-720.
[5] Galano L and Vignoli A 2005 Seismic behavior of short coupling beams with different reinforcement layouts. ACI Struct. J., 97(6), 876-885.
[6] Fortney PJ, Rassati GA and Sharooz BM 2008 Investigation on effect of transverse reinforcement on performance of diagonally reinforced coupling beams. ACI Struct. J., 105(6), 781-788.
[7] Harries K. A., Fortney P. J., Shahrooz B. M., and Brienen P. J. (2005). "Practical design of diagonally reinforced concrete coupling beams-critical review of ACI 318 requirements." ACI Struct. J., 102(6), 876-882.
[8] ACI (American Concrete Institute). (2014). Building code requirements for structural concrete. ACI 318-14, Farmington Hills, MI, USA.
[9] Han SW, Lee CS, Shin M and Lee K 2015 Cyclic performance of precast coupling beams with bundled diagonal reinforcement. Eng. Struct., 93, 142-151.
[10] Canbolat BA, Parra-Montesinos GJ and Wight JK 2005 Experimental study on seismic behavior of high-performance fiber-reinforced cement composite coupling beams. ACI Struct. J., 102(1),159-166.
[11] Parra-Montesinos GJ 2005 High-performance fiber-reinforced cement composites: an alternative for seismic design of structures. ACI Struct. J., 102(5), 668-675.
[12] Parra-Montesinos GJ, Peterfreund SW and Chao SH 2005 Highly damage-tolerant beam-column joints through use of high-performance fiber-reinforced cement composites. ACI Struct. J., 102(3), 487-495.
[13] Naaman AE, Likhitruangsilp V and Parra-Montesinos GJ 2007 Punching shear response of high-performance fiber-reinforced cementitious composite slabs. ACI Struct. J., 104(2), 170-179.
[14] Shin M, Gwon SW, Lee K, Han SW and Jo YW 2014 Effectiveness of high performance fiber-reinforced cement composites in slender coupling beams. Constr. Build. Mater., 68, 476-490.
[15] Li VC 2003 On engineered cementitious composites (ECC). J. Adv. Concr. Technol., 1(3), 215-230.
[16] Yang EH and Li VC 2010 Strain-hardening fiber cement optimization and component tailoring by means of a micromechanical model. Constr. Build. Mater., 24(2), 130-139.

**Acknowledgement**

A The research was supported by grants from the National Research Foundation of Korea (NRF-2017R1A2B3008937).

# Seismic Performance of Steel Tube-high Strength Concrete Squat Walls

**Z Z Zhao[1,2] , G Z Fan[1] , X G He[3] and X Liu[1]**

1 Key Laboratory of Civil Engineering Safety and Durability of the China Education Ministry, Tsinghua University, Beijing 100084, China
2 Beijing Engineering Research Center of Steel and Concrete Composite Structures, Tsinghua University, Beijing 100084, China
3 Architectural Design and Research Institute of Tsinghua University, Tsinghua University, Beijing 100084, China
E-mail: fgz14@mails.tsinghua.edu.cn

**Abstract.** In order to investigate the seismic behaviour of the high strength concrete squat wall with distributed steel tubes, Three squat walls with steel tubes and one RC squat wall were tested. The strength, shear deformation, and energy dissipation capacity of these specimens were analyzed. The test results showed that the load-carrying capacity of the high strength concrete squat wall can be decreased after inserting steel tubes. However, the deformation capacity and the energy dissipation capacity of the squat wall can be improved dramatically when steel tubes are inserted into the section of the wall, especially the squat wall with steel tubes evenly distributed in the section, and as a result, it can be concluded that they are very suitable for application in the conversion layer to improve the collapse resistant capability of structures.

## 1. Introduction

Squat shear wall refers to the wall which the aspect ratio is less than 1.5. This type of wall is considered to exhibit poor ductile behavior due to the premature shear failure. Squat walls have been widely used in practical engineering, especially in walls that directly connected with foundation, that located on the upper part of the conversion layer, and that used in the nuclear engineering buildings [1, 2]. Because of the requirement for practical applications, many research works were carried out on the following related squat walls. Su et al. [3, 4] summarized and discussed the experimental data of four shaking table tests and numerical findings, and found that the phenomenon of shear concentration appeared at exterior walls above the transfer floor due to the local flexural deformation of transfer structures. The effect of shear concentration made the squat shear walls more prone to the occurrence of brittle failures, leading to the interrupt of the vertical load path and then the local collapse.

Based on the investigation of earthquake disaster and related researches [5-7], a new type squat shear wall with steel tubes distributed in the section was proposed in this study. This type of squat wall is supposed to enhance the deformation capacity and collapse resistant capability of structures making use of the advantages of concrete-filled steel tubes (CFSTs). In this paper, four high strength concrete squat shear walls with distributed steel tubes are designed and tested to investigate the seismic performance of this type wall. The shaped steel ratio and the arrangement of steel tubes are taken as major parameters. Based on the test data, an analysis of the failure modes, strength, stiffness degradation, ductility, and energy dissipation capacity of those specimens was conducted. The work in this paper also provides a foundation for further development of a theoretical analysis, which will be discussed in another paper.

## 2. Experimental program

### *2.1. Test specimens*

Four squat walls (SW1-SW4) with the same dimension were tested under cyclic loading. All the specimens had a height of 1080mm and a thickness of 160mm. The aspect ratio of the specimens was 0.95. The overall geometries of the squat walls are shown in Figure 1. The axial compression ratio of the specimens was the same, which was 0.35. Figure 2 shows the detailed sectional configurations of the specimens. SW1 was a RC squat wall taken as a comparative specimen. SW2 and SW3 were the squat walls with steel tubes arranged at both sides and the middle section of walls, respectively. SW4 was a squat wall with steel tubes evenly distributed in the cross-section of the specimen.

The outer diameter of the steel tube was 89mm, which was selected to match the size of specimens. The thickness of steel tubes was 4mm to ensure that the steel tube had sufficient constraint on the concrete filled in it. The diameter-to-thickness ratio of the steel tubes was 22.3. Steel tubes and vertical rebars were inserted into the foundation beam and top beam to ensure sufficient anchorage. The anchoring lengths of the vertical rebars and steel tubes were 650mm and 480mm, respectively. Four annular bars were welded at the bottom of each steel tube to anchor the steel tube into the foundation for the purpose of transferring the loads, as shown in Figure 1.

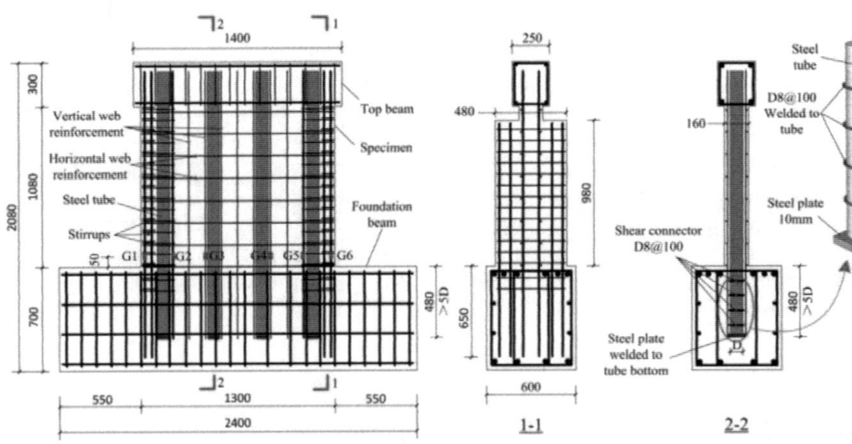


**Figure 1.** General layout dimensions and reinforcement arrangement of the specimens (Unit: mm).

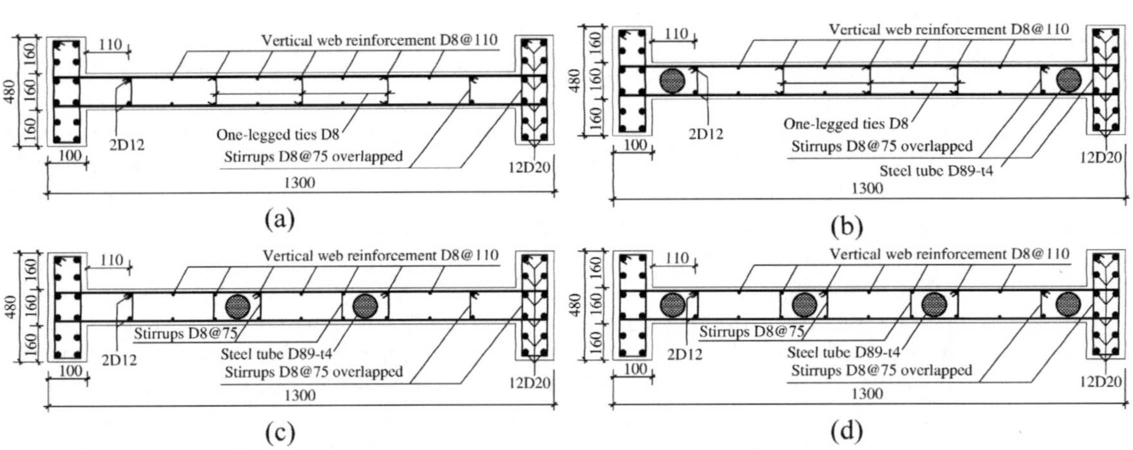


**Figure 2.** Sectional dimensions and reinforcement details of specimens (Unit: mm): (a) SW1; (b)SW2/SW3; (c) SW4/SW5; (d) SW6.

### *2.2. Material properties*
The measured yield and ultimate strengths of steel rebars are summarized in Table 1. The steel tubes were made up of Grade Q345 steel. The material properties of steel tubes, used in the tests, are also reported in Table 1. The concrete that filled in the steel tube had the same strength grade with that outside the tube. The average cube strength ($f_{cu}$) at the time of tests is 66.7 MPa.

**Table 1.** Material properties for steel. (Unit: MPa).

| | Rebar D8 | Rebar D12 | Rebar D20 | Steel tube |
|---|---|---|---|---|
| Yield strength $f_y$ | 410 | 483 | 443 | 348 |
| Ultimate strength $f_u$ | 587 | 659 | 600 | 416 |

## 3. Experimental observations and test results

### *3.1. Damage and failure mode*
Figure 3 shows eventual failure of the four specimens. According to the test of four specimens, the failure model of the high strength concrete squat shear walls with distributed steel tubes was analysed.
SW1 was a RC shear wall. The first crack occurred at the upper left corner of the specimen during the controlled rotation 1/1000, and then extended to the lower right corner. When reaching the peak load (the controlled rotation was 1/100), SW1 failed with the main diagonal crack coalesced which inclined at an angle around 45° to the horizontal direction. Simultaneously, the horizontal reinforcement in the central of the wall fractured, and the longitudinal reinforcement at the lower right corner buckling. The concrete around the main diagonal crack spalling severely and the concrete at the lower right corner crushed. The eventual failure was the brittle shear failure mode along the main diagonal crack, as shown in Figure 3(a).
SW2 was the squat wall with steel tubes arranged at both sides. The first crack occurred at the upper right corner of the specimen during the controlled rotation 1/2000. When the controlled rotation was 1/250, large numbers of 45° skew cracks occurred at the middle of the web of the wall, some of which developed as the main diagonal cracks at further stage. When the controlled rotation was 1/100, two main diagonal cracks appeared apparently with the concrete at the intersection crushed severely. When the controlled rotation was 1/75, the concrete at the intersection of two main diagonal cracks was almost completely peeled off, and the vertical web reinforcement buckling, as shown in Figure 3(b). Compared with SW1, the damage of SW2 was more homogeneous before the peak load of SW1, and the damage was mainly concentrated in the middle part of the specimen, as shown in Figure 4.
SW3 was the squat wall with steel tubes arranged at the middle section. The first crack occurred at the controlled rotation 1/2000. In the initial stage of 1/500 cycle, several skew cracks appeared on the left side of the wall. However, the propagation of the cracks was suppressed because of the existence of steel tubes, and there were some cross bond cracks at the position of steel tubes at the meantime. When the controlled rotation was 1/75, large area of concrete spalling between the two steel tubes occurred, and the spalling area of concrete made up around 1/3 of the total area of the wall. When the controlled rotation was 1/50, the root of the left flange of the wall damaged severely, along with significantly load-carrying capacity deterioration, as shown in Figure 3(c).
SW4 was a squat wall with steel tubes evenly distributed in the cross-section of the specimen. The first crack occurred at the upper right corner of the specimen during the controlled rotation 1/1000. When the controlled rotation was 1/500, the development of inclined cracks was obviously suppressed by the steel tubes. However, the bond crack zone along the height direction of the steel tube formed apparently. During the 1/50 cycles, concrete along the height direction of steel tubes spalling severely, and four steel tubes exposed. When the controlled rotation was 1/30, the concrete cover in the web of the wall was almost completely lost, along with slight damage of flanges, as shown in Figure 3(d). There was no obvious shear failure in the specimen. SW4 exhibited excellent deformation capacity and stability and worked in a way similar to the rocking wall [8] at the later stage.

IOP Conf. Series: Materials Science and Engineering **371** (2018) 012035 doi:10.1088/1757-899X/371/1/012035

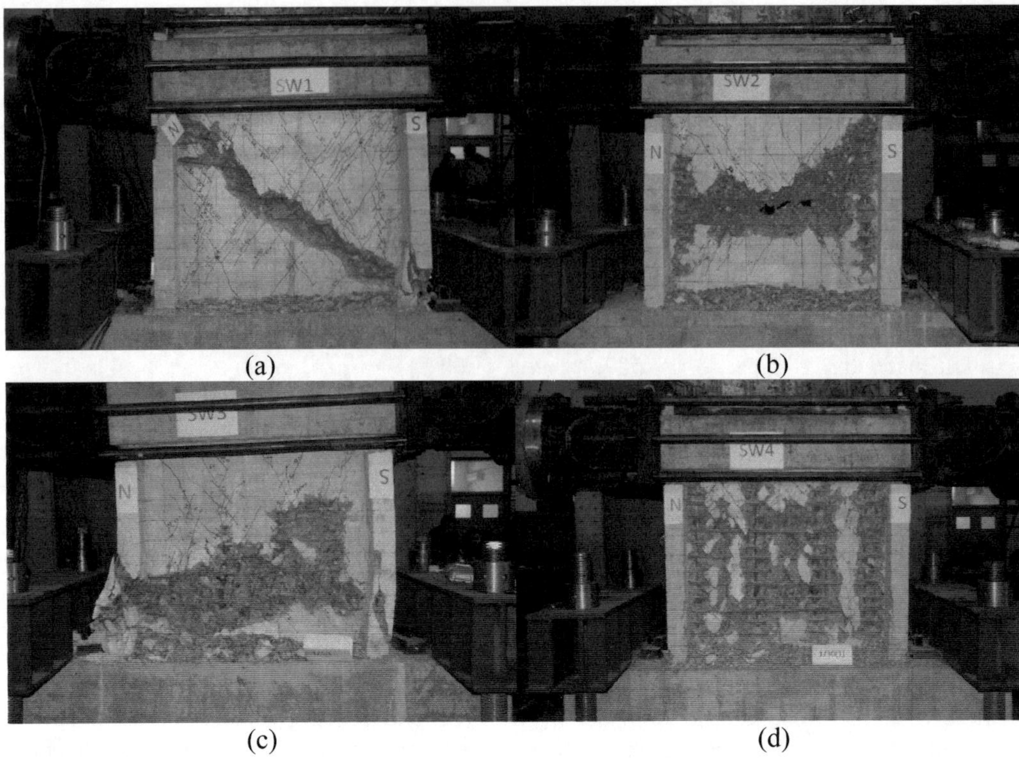


**Figure 3.** The failure modes of the specimens: (a) SW1; (b)SW2; (c) SW3; (d) SW4.

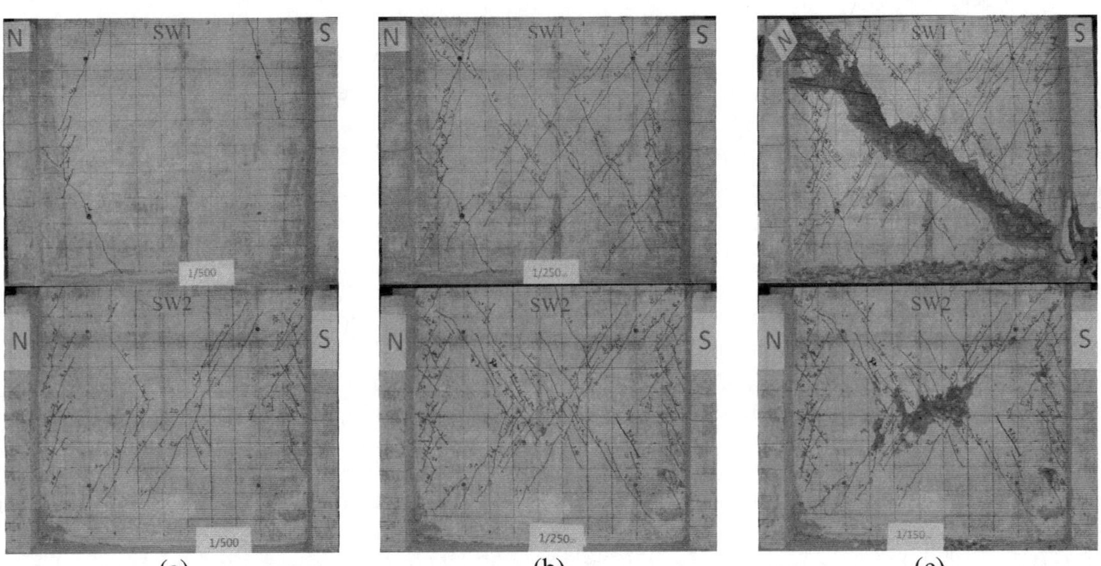


**Figure 4.** Comparison of damage evolution between SW1 and SW2: (a) 1/500; (b) 1/250; (c) 1/150.

As seen from the above description, steel tubes could not delay the time that the first crack occurred. After inserted steel tubes, the development of inclined cracks was suppressed obviously, thus avoiding the brittle failure. However, the vertical bond slip cracks would be presented in the connections between steel tubes and the concrete after inserted steel tubes. By comparing the failure mode of SW2 with that of SW3 could be found, steel tubes which inserted into the both sides of the wall could play a role in protecting the flanges, so that the specimen was more stable. The failure mode of the specimen was no longer a shear failure after steel tubes were evenly inserted into the wall section. SW4 combined the advantages of SW2 and SW3 effectively, and showed a good deformation capacity and stability.

### *3.2. Force-displacement relationship*

The recorded curves of lateral force (P) versus lateral displacement (Δ) for all specimens are shown in Figure 5. It was found that the lateral force versus lateral displacement response of SW1 cycled almost linearly. The specimen behaved approximately elastically with little residual displacement occurred. The energy dissipation and ductility of SW1 were very poor, and the specimen showed the characteristics of brittle failure.

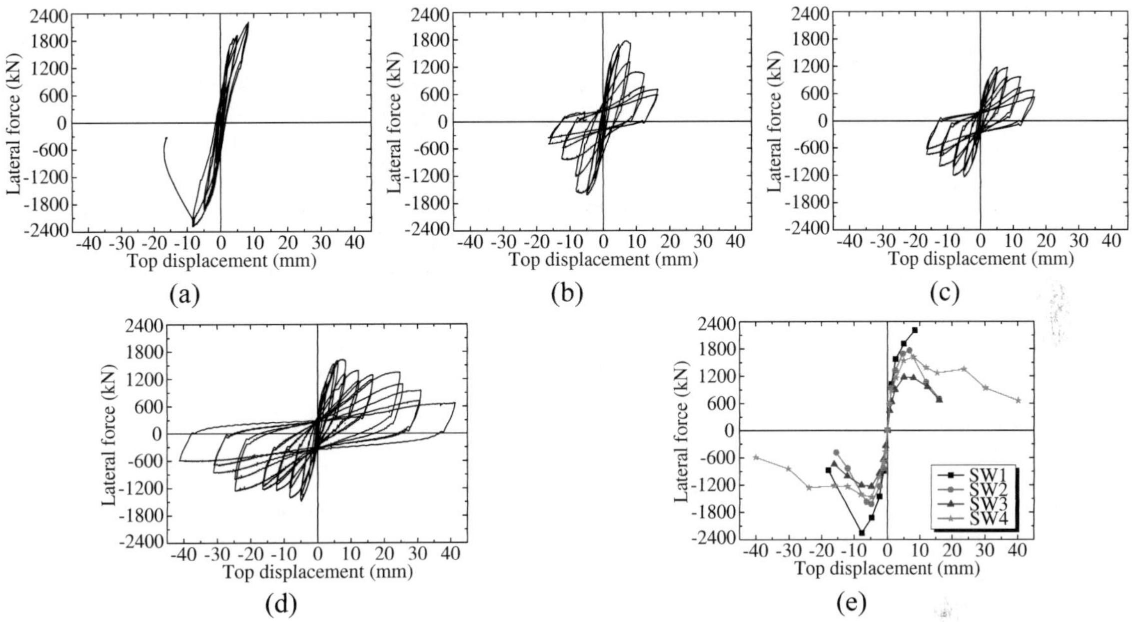


**Figure 5.** Hysteresis and Skeleton curves: (a) SW1; (b) SW2; (c) SW3; (c) SW4; (e) Skeleton curves.

Compared with SW1, SW2 had plumper hysteresis curve, better ductility and plastic deformation capacity. However, load-carrying capacity of SW2 was lower than that of SW3. The damage evolution of SW4 was more homogeneous than that of SW1 due to the insertion of steel tubes. And this caused the damage of SW2 was more severe than that of SW1 at the same controlled rotation before the peak load. This in turn led to the reduction of load-carrying capacity.

The deformation capacity of SW3 was similar to that of SW2. Nevertheless, the load-carrying capacity of SW3 was lower than that of SW2. Firstly, the damage of specimen with steel tubes inserted in the middle section was more severe than that of specimen with steel tubes inserted in the both sides during the same controlled rotation, especially in the flange, and then, the moment capacity of SW2 was higher than that of SW3 due to the steel tubes on both sides. Both contributed to the fact that SW2 had a higher load-carrying capacity.

As shown in Figure 5(d), SW4 exhibited excellent deformation capacity and stability with plump hysteresis curve. Compared with SW2, the load-carrying capacity of SW4 was decreased slightly. It could be noted that the damage at the peak load could be aggravated when the steel tubes inserted into the middle section of squat walls.

Figure 5(e) shows the hysteresis skeleton curves of four specimens. It could be seen that inserted steel tubes, especially in the middle section, would decrease the load-carrying capacity of the squat wall, which could significantly improve the deformation capacity at the same time. The deformation capacity and ductility of the squat wall with steel tubes evenly distributed in the cross-section (SW4) was dramatically improved compared with the other specimens.

## 4. Analysis of test results and discussions

### *4.1. Dissipated energy*

The dissipated energy in each cycle could be calculated from the lateral load (P) versus lateral displacement (Δ) curve as the area bounded by the hysteretic hoop of that cycle. The single-cycle energy dissipation Es of each specimen is shown in Table 2. The cumulative energy dissipation Ec of each specimen is shown in Figure 6.

The cumulative energy dissipation capacity of specimens, from high to low, was in the following order: SW4, SW2, SW3 and SW1. The single-cycle energy dissipation capacity of SW3 was similar to that of SW1 at the early stage. However, the ultimate cumulative energy dissipation capacity of SW3 was much better than that of SW1 because of the better ultimate deformation. And this phenomenon also occurred between SW4 and SW2. According to observations of the test data, the conclusions are as follows: (1) the effect of steel tubes on the energy dissipation capacity in the squat wall with steel tubes arranged at both sides was slightly better than that of the squat wall with steel tubes arranged at the middle of the section. And the ultimate cumulative energy dissipation capacities were similar because of the similar deformation capacity and single-cycle energy dissipation capacity; (2) the cumulative energy dissipation capacity of the squat wall with steel tubes evenly distributed in the cross-section (SW4) was about three times higher than that of the squat wall with steel tubes arranged at both sides (SW2) or the squat wall with steel tubes arranged at the middle of the section (SW3), in spite of the doubling of the amount of steel used. And the deformation capacity of SW4 was also improved significantly.

**Table 2.** The single-cycle energy dissipation $E_s$. (kN·m).

| Rotation angle/rad | SW1 | SW2 | SW3 | SW4 |
|---|---|---|---|---|
| 1/2000 | 0.296 | 0.419 | 0.258 | 0.326 |
| 1/1000 | 0.713 | 1.043 | 0.805 | 0.946 |
| 1/500 | 2.026 | 2.957 | 2.175 | 2.703 |
| 1/250 | 4.862 | 7.872 | 5.922 | 7.437 |
| 1/250 | 4.683 | 7.178 | 5.356 | 7.985 |
| 1/150 | 8.353 | 12.915 | 9.368 | 12.200 |
| 1/150 | 8.216 | 10.120 | 7.719 | 10.077 |
| 1/100 | | 11.396 | 11.839 | 15.725 |
| 1/100 | | 7.868 | 8.857 | 13.512 |
| 1/75 | | 9.475 | 11.309 | 19.950 |
| 1/75 | | 7.469 | 8.941 | 18.612 |
| 1/50 | | | | 31.248 |
| 1/50 | | | | 25.280 |
| 1/40 | | | | 26.783 |
| 1/40 | | | | 21.120 |

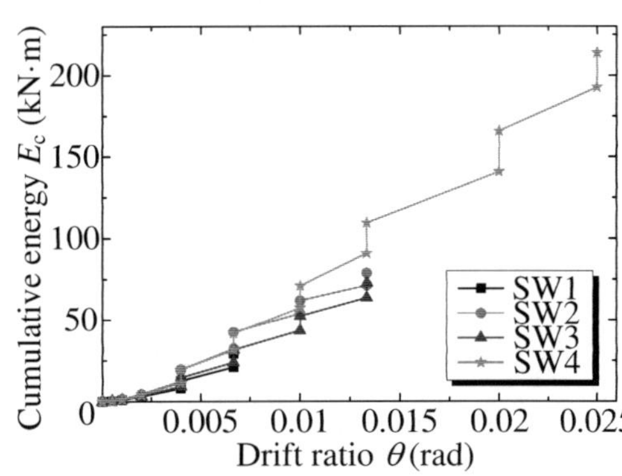


**Figure 6.** Cumulative energy dissipated of the specimens.

### *4.2. Rigidity degradation*

The stiffness of specimens decreased with the cyclic loading for the reason of cumulative damnification. The stiffness of specimens under cyclic loading can be evaluated by the index-cyclic stiffness, which can be determined as Equation (1):

$$K_l = \sum_{i=1}^{n} P_j^i / \sum_{i=1}^{n} \Delta_j^i \qquad (1)$$

IOP Conf. Series: Materials Science and Engineering **371** (2018) 012035 doi:10.1088/1757-899X/371/1/012035

In which, $K_l$ is the cyclic stiffness; $P_j^i$ is the peak load of the $i$ th cycle when the deformation is controlled as $\Delta_j$ ; $\Delta_j^i$ is the peak deformation of the $i$ th cycle when the deformation is controlled as $\Delta_j$ ; and n= the number of cycles when the deformation is controlled as $\Delta_j$ .

The $K_l$–Δ curves of specimens are shown in Figure 7. These cyclic stiffness degradation curves are compared, and the analysis can be made as follows:

(1) The initial rigidities of specimens in a lower to higher order were the first for SW3, the second SW1, the third SW4 and the fourth for SW2. The initial rigidity of the squat wall could be improved by inserting steel tubes into both sides of the wall, and could be reduced by inserting steel tubes into the middle section of the wall. The initial stiffness of SW4 was higher than that of SW1, which indicated that the enhancing effect of steel tubes arranged at both sides to the initial stiffness of squat walls was higher than that of steel tubes arranged at the middle section.

(2) Due to the interaction between steel tubes and the concrete, the positive initial stiffness of specimens was larger than the negative initial stiffness. After several hysteresis hoops, the positive stiffness was almost the same as the negative stiffness due to the concrete spalling. In SW4, this characteristic was especially obvious.

(3) In the case of large deformation, the stiffness degradation of SW4 was flat without sundden drop in the stiffness. This kind of wall exhibited excellent deformation capacity and collapse resistant capability.

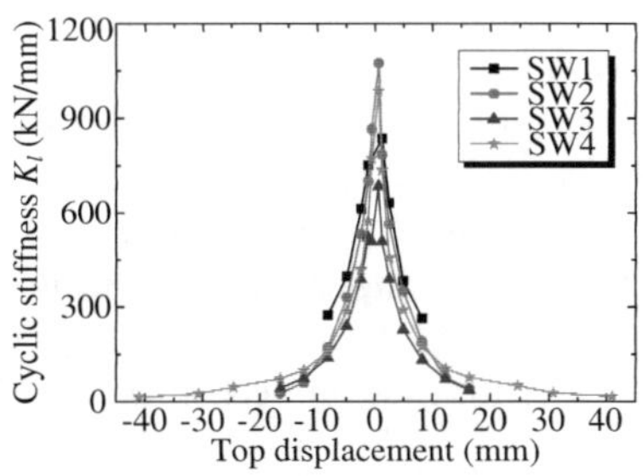


**Figure 7.** Comparison of $K_l$–Δ curves of specimens.

### *4.3. Strain distribution of the section*

In order to study the strain distribution of the section of specimens in the process of loading, 6 strain gauges (G1~G6) were arranged at the bottom section of each squat wall, and the distance from the strain gauges to the foundation beam was 50mm, as shown in Figure 1. The strain distributions of states of axial force exerted, cracking, yield, and peak load were shown in Figure 8. The ordinate axis represents the strain measured by strain gauges, and the abscissa axis represents the distance from the strain gauge to the left edge of the wall. Because of the problem of the strain gauge data acquisition, the strain distribution of SW4 was not presented in this paper. Based on the illustrations, the following conclusions could be made: (1) After the axial force was exerted, the strain distributions of SW1 and SW2 were basically in accordance with the plane section assumption. However, due to the existence of steel tubes in the middle section of SW3, the strain of the middle section was smaller than that of the two ends; (2) Before the cracking, the strain distributions of specimens basically obeyed the plane section assumption. However, after squat walls cracking, the strain distributions of specimens was no longer consistent with the plane section assumption.

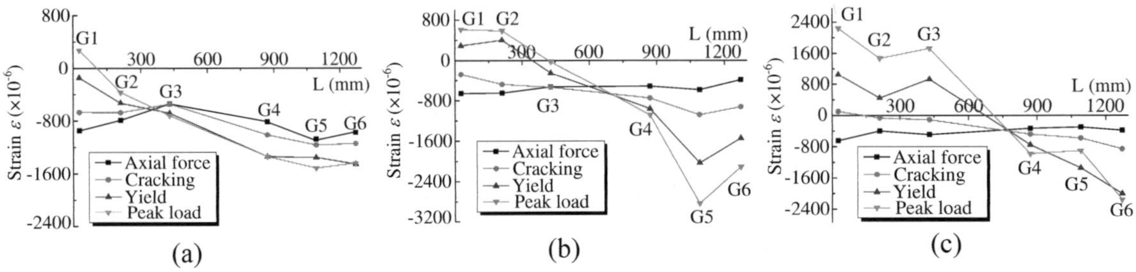


**Figure 8.** Strain distributions along the bottom sections of the specimens: (a) SW1; (b) SW2; (c) SW3.

## 5. Conclusions

Three squat walls with steel tubes and one RC squat wall were tested and analysed. The main conclusions can be summarized as follows:

(1) The ordinary high strength concrete squat wall showed a poor ductility and energy dissipation, with obvious brittle shear failure.

(2) The deformation capacity of the high strength concrete squat wall was improved dramatically after the insertion of steel tubes. After inserted steel tubes, the development of inclined cracks was suppressed obviously, and then, the damage evolution of squat walls was more homogeneous at each stage. The deformation capacity and energy dissipation capacity of the squat wall with steel tubes evenly distributed in the section (SW4) were much better than that of any of the first three squat walls. The cumulative energy dissipation capacity of SW4 was about three times higher than that of SW2 or SW3, although the amount of steel tubes doubled.

(3) The initial rigidity of the squat wall could be improved by inserting steel tubes into both sides of the squat wall, and could be reduced by inserting steel tubes into the middle section of the squat wall. And the enhancing effect of steel tubes arranged at both sides to the initial stiffness of squat walls was larger than the weakening effect of steel tubes arranged at the middle section to the initial stiffness of squat walls.

(4) The strain distribution of bottom section of the squat wall was no longer consistent with the plane section assumption after the squat wall cracking. Thus, the plane section assumption was no longer suitable for the calculation of the peak load of the squat wall.

## 6. References

[1] Parulekar YM, Reddy GR, Vaze KK, Pegon P, and Wenzel H 2014 Simulation of reinforced concrete short shear wall subjected to cyclic loading. *Nuclear Engineering and Design* **270(5)**: 344-350

[2] Abdlebasset YM, Sayed-Ahmed EY, Mourad SA 2016 Seismic analysis of high-rise buildings with transfer slabs: state-of-the-art-review. *Electronic Journal of Structural Engineering* **16(1)**: 38-51.

[3] Su RKL and Cheng MH 2009 Earthquake-induced shear concentration in shear walls above transfer structures. *The Structural Design of Tall and Special Buildings* **18(6)**: 657-671

[4] Elawady AK, Okail HO, Abdelrahman AA and Sayed-Ahmed EY 2014 Seismic Behaviour of High-Rise Buildings with Transfer Floors. *Electronic Journal of Structural Engineering* **14(2)**: 57-70.

[5] Chen L, Mahmoud H, Tong SM and Zhou Y 2015 Seismic behaviour of double steel plate-HSC composite walls. *Engineering Structures* **102(8)**: 1-12

[6] Hu HS, Nie JG, Fan JS, Tao MX, Wang YH and Li SY 2016 Seismic behaviour of CFST-enhanced steel plate-reinforced concrete shear walls. *Journal of Constructional Steel Research* **119(3)**: 176-189

[7] Qian JR, Jiang Z and Ji XD 2012 Behaviour of steel tube-reinforced concrete composite walls subjected to high axial force and cyclic loading. *Engineering Structures* **36(4)**: 173-184

[8] Nazari M, Sritharan S, Aaleti S 2017 Single precast concrete rocking walls as earthquake force-resisting elements. *Earthquake Engineering & Structural Dynamics* **46(5)**: 753-769.

**Acknowledgments**
This work presented herein was supported by National Natural Science Foundation of China (Grants No. 51478243). The writers wish to express their sincere gratitude to the sponsor. The writers would also like to thank the anonymous reviewers for their valuable comments and suggestions.

IOP Conf. Series: Materials Science and Engineering **371** (2018) 012036 doi:10.1088/1757-899X/371/1/012036

# Evaluation of Limit State for Pier According to Scour and Peak Ground Acceleration

**M. H. Kwon[1], W. Y. Jung[2], D. H. Shin[3] and K. Y. Kim[4]**

1 Dept. of Civil Engineering, ERI, Gyeongsang National University, Jinju, South Korea
2 Dept. of Civil Engineering, Gangneung-Wonju National University, Gangneung, South Korea
3 K-water Institute, Daejeon, South Korea
4 K-water Institute, Daejeon, South Korea
Email: kwonm@gnu.ac.kr, kwonm@gnu.ac.kr, cute_lion@hanmail.net, kky0606@kwater.or.kr

**Abstract.** Recently, materials and structure analysis techniques used in various structures have been developed and have made remarkable progress. As the design method for reinforced concrete structures is shifting to the limit state design method, it is necessary to conduct the limit state assessment. So far, several researchers have evaluated the safety of many structures covered by the civil engineering field by introducing the limit state assessment. However, there is a lack of research on bridge piers considering the level of scour. Therefore, in this study, T-shaped bridge piers with a height of 11.5 m and a width of 2.8 m installed in underwater ground were selected and the critical condition of bridge pier for 20 seismic waves was evaluated using LS-DYNA. The presence or absence of scouring was also set as a variable and the level of scour is changed from 0 to 5 meter. As a result of the analysis, the main fracture was tensile failure at the maximum ground acceleration of 0.2g to 0.4g, and the maximum ground acceleration after the earthquake was found to be weak as a whole except compression. Finally, as a result of comparing analysis results according to the presence or absence of scour, the limit state reaching ratio showed a similar pattern without significant change.

## 1. Introduction

In addition to material and structure analysis techniques, design methods have been developed for high-rise and large-scale structures. Recently, a design method for reinforced-concrete structures has been shifted by the limit state design method, and several researchers have evaluated the safety by introducing the limit state evaluation for many structures. However, the evaluations of structures and study on bridges, which play a large role as social infrastructures, remain insufficient. Because the bridge structure is designed for various conditions, the structure is safe under normal extreme conditions. Generally the structure, which is designed with seismic load, is safe to the targeted earthquake. However, since the frequency and magnitude of earthquakes are increasing every year, the risk of earthquakes is also increasing. Therefore, it is necessary to evaluate the marginal condition of bridge structures for earthquakes. In this study, the limit states of bridge structures for various seismic loads and PGA were evaluated. To evaluate the limit state of the bridge structure according to the scour, the pier is analysed with the difference level of scour.

## 2. Analysis and evaluation plan

The bridge pier in the evaluation is T-shaped bridge pier with pit bases, 11.5 m in column height, and 2.8 m in width. A structural analysis was performed using infinite ground conditions. LS-DYNA which is a finite-element analysis program to simulate the dynamic behaviour of three-dimensional inelastic structures, reinforced-concrete and soil material model and contact boundary condition model in the analysis code, is used in the analyses. When the seismic waves were applied, infinite ground was implemented using the PML model to absorb the energy and control the reflected wave. The modelling results and real structural analysis model results using PML are shown in Fig. 2. Table 1 summarizes the finite-element analysis model.

**Table 1.** Finite element analysis plan.

| Material model | | | Contact condition | Damping | Element | |
|---|---|---|---|---|---|---|
| concrete | soil | Unbounded Soil | 2D-AUTOMATIC _NODE_TO_SURFACE | FREQUENCY _RANGE | 3-NODE BEAM | 4-NODE SHELL |
| 159-CSCM | 147-FHWA_SOIL | 230-PML | | | | |
| FPC=24MPa | K=65MPa, G=186MPa | E=1.32GPa | Soil-concrete(FS=0.5) | CDAMP=5% | Number of = 752 | Number of = 5609 |

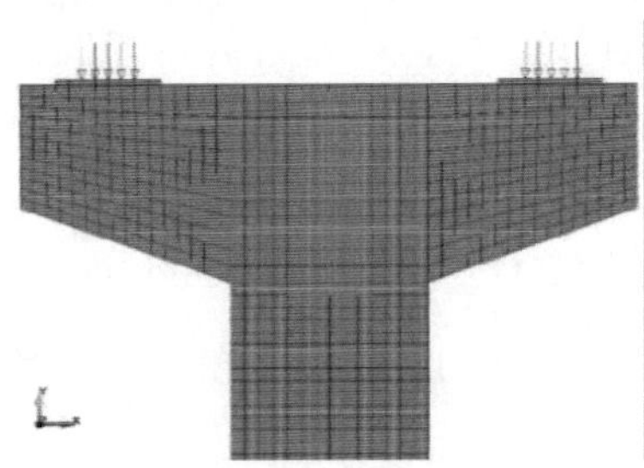

**Figure 1**. Load condition.

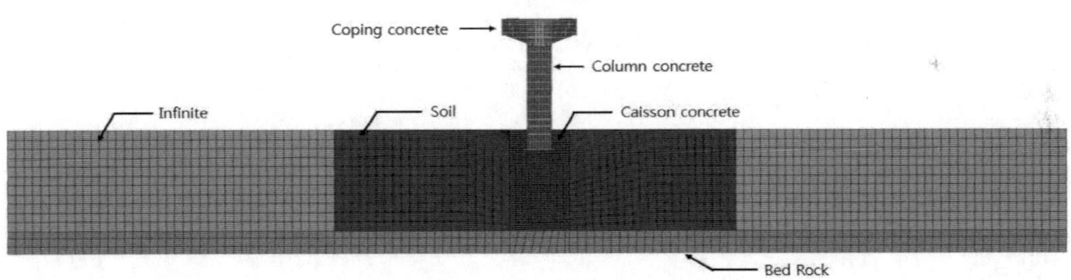


**Figure 2.** Whole model (scour 0m).

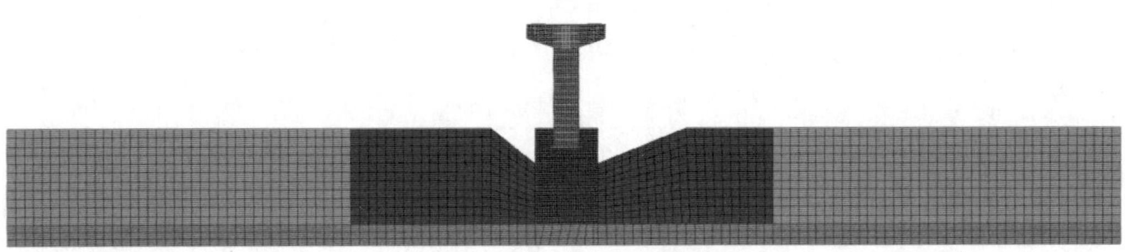

**Figure 3.** Scour 5m model.

Natural waves were selected to account for the uncertainty of seismic waves. Due to that, 20 natural seismic waves are selected from the seismic records in the recent 20 years worldwide. The natural wave enables one to derive the effects of different characteristics of seismic waves. Among various factors that can characterize seismic waves, the peak ground acceleration is used the main characteristic of the seismic since it is widely used in practice. As another variable, scour was created

IOP Conf. Series: Materials Science and Engineering **371** (2018) 012036 doi:10.1088/1757-899X/371/1/012036

at 1-m intervals from 0 m to 5 m around the pier in different conditions. The upper load is shown in Fig. As shown in Fig. 1, the dead load of 7.5 $kN/m^2$ was applied to both bearing supports. The limit state of the concrete structure was set to the tensile and compressive failure of the concrete. The maximum relative displacement between the right top of the pier and the bottom of the column is set to displacement limit state, and the von-Mises stress is used as the limit stat of reinforcing bar.

## 3. Results and discussion

The Tables 2-7 show the probability of the bridge failure to the maximum ground acceleration with difference level of scour. First, Table 2 shows the analysis results with the 0m of scour. It indicates that there is no concrete compression failure at 0.2g and the probability of failure in reinforcement and drift ratio is 5%, but the probability of failure about concrete tension failure is relatively high as 40%. At 0.4 g, the failure probability for the tensile, rebar and maximum displacements increased to 100%, 60% and 75%, respectively, which can be treated as tensile failure. However, after reaching the ground acceleration of 0.6 g to 1.5 g, the probability of failure for the tensile, reinforcing bar, and maximum displacements exceeded 90% except for concrete compression. Second, Table 3 shows the analysis results with the 1m of scour. It indicates that the probability of concrete compression failure, reinforcement failure and drift failure are very similar with the case of scour with 0 m and the probability of concrete tension failure is increased as 45% at PGA 0.2g. At 0.4g, all limit state reaches to the failure except concrete compression failure. The probability of concrete tensile, reinforcement, and drift failure are sharply increased to 95%, 65%, and 75% respectively.

The main failure can be considered a tensile failure. Next, from 0.6 g to 1.5 g, the probability of failure for the tensile, reinforcing bar, and maximum displacements exceeded 90% except for concrete compression. The scouring range of 2-5 m was similar. Therefore, there is no large change in tendency of failure probability depending on the presence and depth of the scour since the bridge pier possesses the strong caisson concrete foundation. Fig. 4 shows the results of the tensile stress and compressive stress of bridge pier subject to PGA 1.5g seismic load with change of the scour level.

**Table 2.** Probability of failure with maximum ground acceleration (scour 0m).

| PGA | Limit state | | | |
|---|---|---|---|---|
| | Tension of Concrete | Compression of Concrete | Steel | Drift |
| 0.2g | 40% | 0% | 5% | 5% |
| 0.4g | 100% | 0% | 60% | 75% |
| 0.6g | 100% | 0% | 90% | 90% |
| 0.8g | 100% | 0% | 95% | 100% |
| 1.0g | 100% | 0% | 100% | 100% |
| 1.5g | 100% | 0% | 100% | 100% |

**Table 3.** Probability of failure with maximum ground acceleration (scour 1m).

| PGA | Limit state | | | |
|---|---|---|---|---|
| | Tension of Concrete | Compression of Concrete | Steel | Drift |
| 0.2g | 45% | 0% | 5% | 5% |
| 0.4g | 95% | 0% | 65% | 75% |
| 0.6g | 100% | 0% | 90% | 90% |
| 0.8g | 100% | 0% | 95% | 100% |
| 1.0g | 100% | 0% | 100% | 100% |
| 1.5g | 100% | 0% | 100% | 100% |

IOP Conf. Series: Materials Science and Engineering **371** (2018) 012036 doi:10.1088/1757-899X/371/1/012036

**Table 4.** Probability of failure with maximum ground acceleration (scour 2m).

| PGA | Limit state | | | |
|---|---|---|---|---|
| | Tension of Concrete | Compression of Concrete | Steel | Drift |
| 0.2g | 40% | 0% | 5% | 10% |
| 0.4g | 90% | 0% | 75% | 75% |
| 0.6g | 100% | 0% | 90% | 85% |
| 0.8g | 100% | 0% | 95% | 100% |
| 1.0g | 100% | 0% | 95% | 100% |
| 1.5g | 100% | 0% | 100% | 100% |

**Table 5.** Probability of failure with maximum ground acceleration (scour 3m).

| PGA | Limit state | | | |
|---|---|---|---|---|
| | Tension of Concrete | Compression of Concrete | Steel | Drift |
| 0.2g | 35% | 0% | 0% | 0% |
| 0.4g | 95% | 0% | 65% | 70% |
| 0.6g | 100% | 0% | 90% | 90% |
| 0.8g | 100% | 0% | 100% | 95% |
| 1.0g | 100% | 0% | 100% | 100% |
| 1.5g | 100% | 0% | 100% | 100% |

**Table 6.** Probability of failure with maximum ground acceleration (scour 4m).

| PGA | Limit state | | | |
|---|---|---|---|---|
| | Tension of Concrete | Compression of Concrete | Steel | Drift |
| 0.2g | 25% | 0% | 0% | 0% |
| 0.4g | 95% | 0% | 60% | 95% |
| 0.6g | 100% | 0% | 90% | 95% |
| 0.8g | 100% | 0% | 95% | 100% |
| 1.0g | 100% | 0% | 100% | 100% |
| 1.5g | 100% | 0% | 100% | 100% |

**Table 7.** Limit state arrival rate with maximum ground acceleration (scour 5m).

| PGA | Limit state | | | |
|---|---|---|---|---|
| | Tension of Concrete | Compression of Concrete | Steel | Drift |
| 0.2g | 35% | 0% | 0% | 0% |
| 0.4g | 85% | 0% | 50% | 60% |
| 0.6g | 100% | 0% | 90% | 90% |
| 0.8g | 100% | 0% | 95% | 95% |
| 1.0g | 100% | 0% | 95% | 100% |
| 1.5g | 100% | 0% | 100% | 100% |

IOP Conf. Series: Materials Science and Engineering **371** (2018) 012036 doi:10.1088/1757-899X/371/1/012036

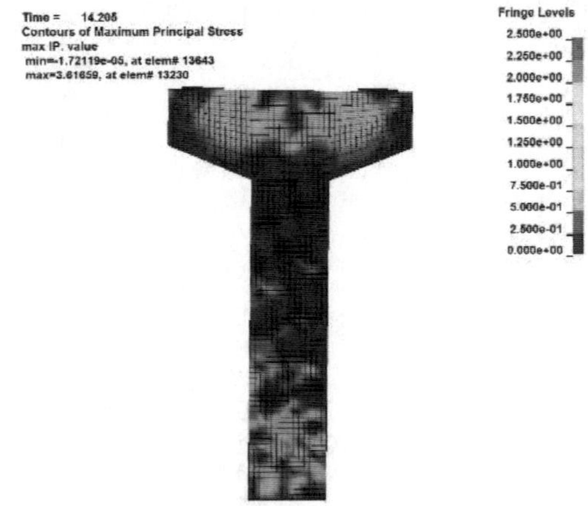


(a) 1.5g Tension of concrete (scour 0m)

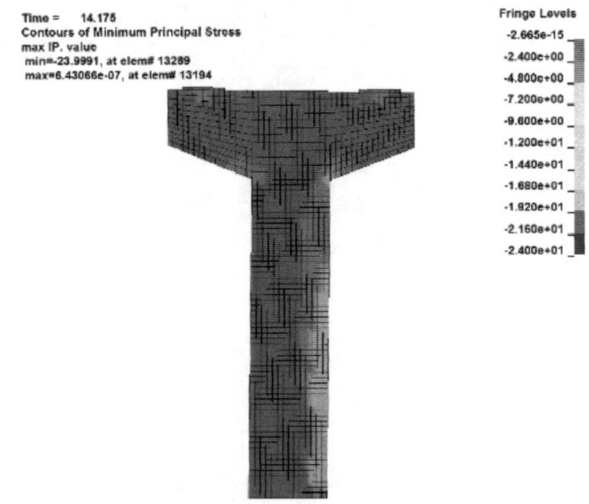


(b) 1.5g Compression of concrete (scour 0m)

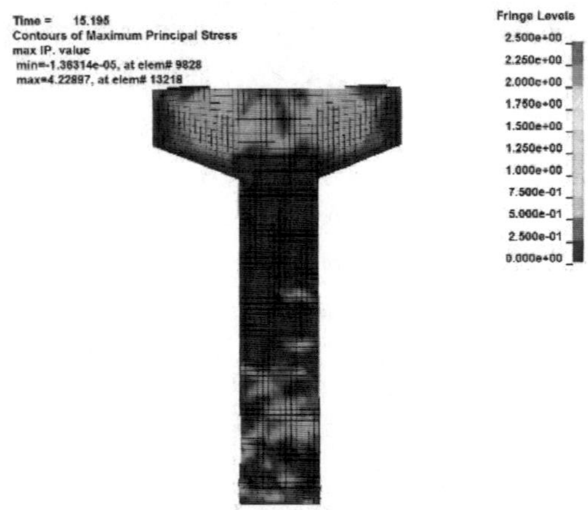


(c) 1.5g tension of concrete (scour 1m)

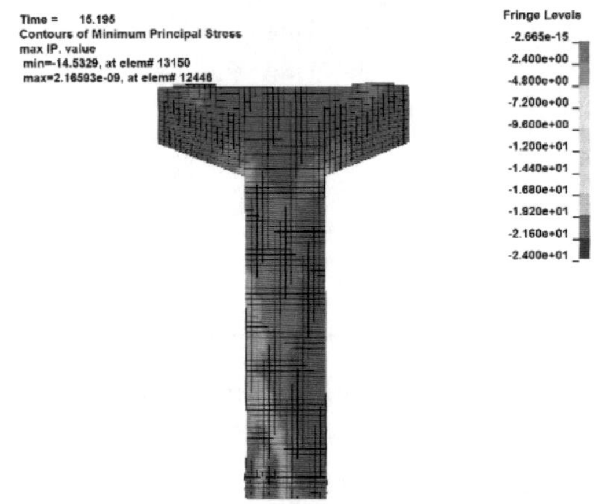


(d) 1.5g compression of concrete (scour 1m)

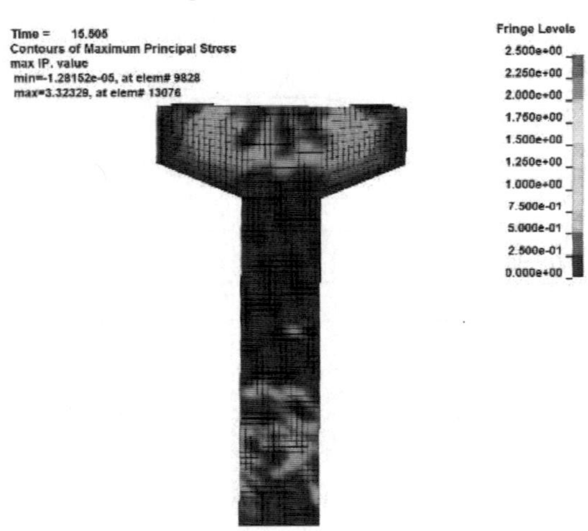


(e) 1.5g tension of concrete (scour 2m)

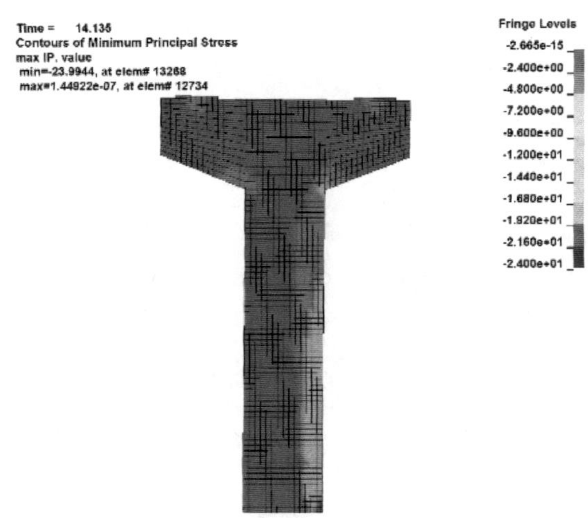


(f) 1.5g compression of concrete (scour 2m)

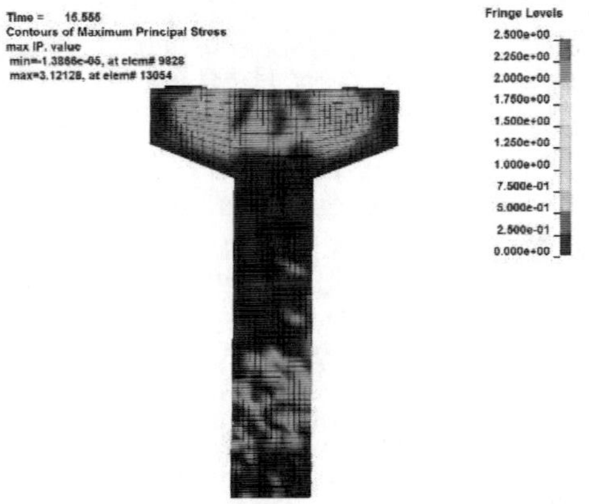


(g) 1.5g tension of concrete (scour 3m)

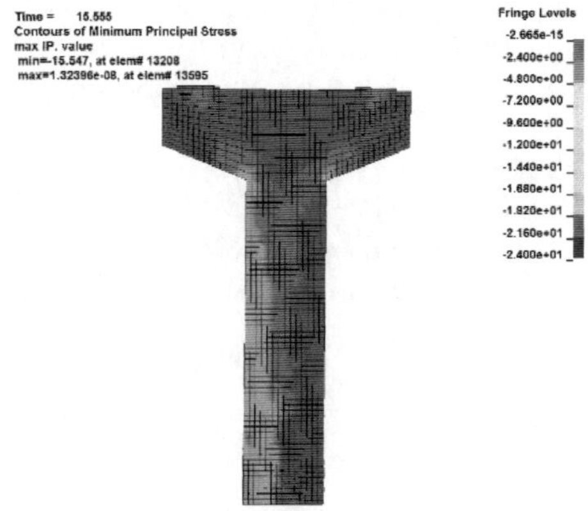


(h) **1.5g compression of concrete (scour 3m)**

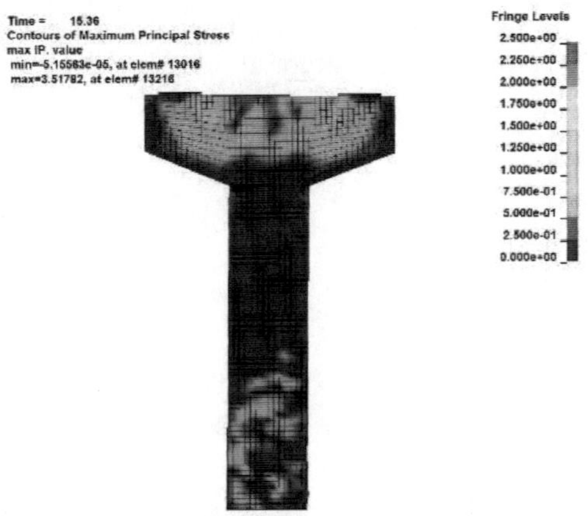


(i) 1.5g tension of concrete (scour 4m)

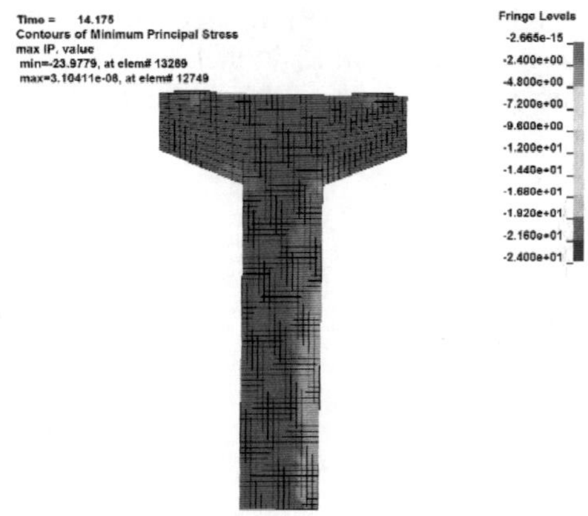


(j) 1.5g compression of concrete (scour 4m)

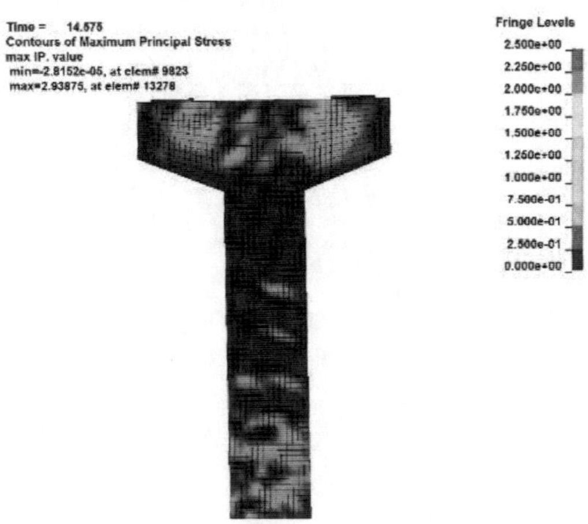


(k) 1.5g tension of concrete (scour 5m)

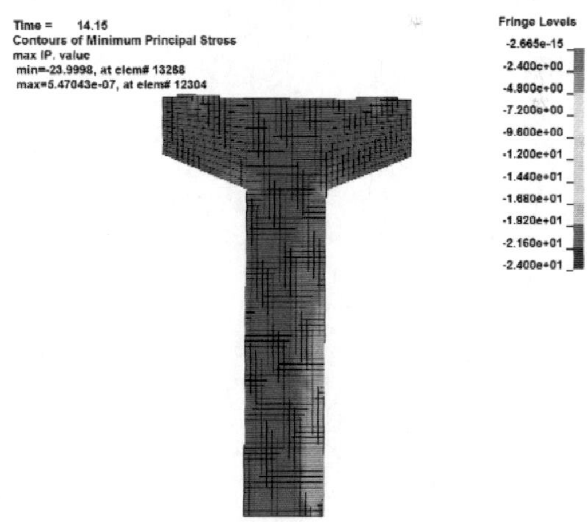


(l) 1.5g compression of concrete (scour 5m)

**Figure 4.** Results (tension, compression).

## 4. Conclusion

The evaluation results of the limit state of the bridge with variety natural seismic loads are as follows. When PGA is 0.2 g, the failure probability of reinforcement and drift condition is 0~5% and the probability of concrete tension failure is 25~45%. With PGA 0.4 g, the probability of failure for all limit states are increased by 50% more. When PGA is greater than 0.4g, probability of failure is more

IOP Conf. Series: Materials Science and Engineering **371** (2018) 012036 doi:10.1088/1757-899X/371/1/012036

than 90%, which indicates that the usability is weak as a whole such as tension and compression limit state. In addition, it is observed that the failure probability of the pier is not effected much as the increasing the level of scour due to the strong caisson concrete foundation.

## 5. References

[1] Kotsvos, Michael D and M Pavlovic 1995 Structural concrete: finite-element analysis for limit-state design (Thomas Telford)
[2] Anastasopoulos I, Anastasopoulos P.C, Agalianos A., and Sakellaridis L. 2015 Simple method for real-time seismic damage assessment of bridges, *Soil Dynamics and Earthquake Engineering,* 78, pp 201-212
[3] LS-DYNA Keyword User's Manual 2007 vol. 1.
[4] Design standards for highway bridges (Limit state design method) Jan 2015.

## Acknowledgment

This research was supported by a grant(17SCIP-B065985-05) from Smart Civil Infrastructure Research Program (SCIP) funded by Ministry of Land, Infrastructure and Transport of Korean government and by the National Research Foundation of Korea[NRF] grant funded by the Korea Government [MEST] [No. 2017RIA2B3008623].

IOP Conf. Series: Materials Science and Engineering **371** (2018) 012037 doi:10.1088/1757-899X/371/1/012037

# Ledras (Ledra Street): *Discovering it's Visible and Invisible Multiple Layers*

**G Ozturk**[1]
1Lecturer, Faculty of Architecture, Architecture Department, Near East University, Nicosia, Cyprus
E-mail: gizem.ozturk@neu.edu.tr

**Abstract.** Today, historic preservation and re-use of historic sites are common techniques in urban development process of cities with rich historical background, which creates many contradictions and architecturally aesthetical problems. Architects should try to find ways to understand the problems underneath them and suggest a technique with a continuous outcome. Researching the multiple layers of this study area will help to discover the *time-space relationship* of the existing environment.
*Prof. Dr. Semra Aydinli states at her presentation in Boston "Enmeshed experience motivates us to grasp the multi-sensory qualities of the existing environment. Every place that has specific spatiality in Istanbul establishes a different integrity for production of sensation because of its major changes in life and enmeshed experience."* [1]
The disconnection between heritage and contemporary leads this research to search for possible ways to maintain the environmental value of the Cyprus' architecture in terms of quality, continuity, and aesthetical experience. Every historic building, their surrounding, and society carry its own has its own unique spatiality in its context. Therefore it is inevitable to not to come across with enmeshed experiences through a detailed research at any of the historic quarters of Cyprus. Places contains *enmeshed experiences* and *cultural codes*, with the intention that in this research the chosen method is the *bodily experiencing* to point out the identity of the area to identify a new conception of sense and to produce new events. Bodily experience of Ledra is supported with photography, and maps aesthetical value and identity. The island's built environment is rich in variety and each nation.

## 1. Introduction

In Ledras case, it is worth to discover the layers for a continual successful re-use and future. In this research Ledra is narrated with photos and bodily experiences. Here the layers are not only tangible and but also intangible. Bringing out both tangible and intangible layers of Ledra and its enmeshed experiences is necessary for future becoming.
Throughout history Cyprus has gone under control of different ethnics such as Byzantine, Romans Lusignan, Venetian, Ottoman and British, and as a result many architectural types and cultures codes have formed the island. Even at present Cyprus remains very rich with its built environment and architectural context. Footprints of each nation are sensed in different quarter of the island and each has their own unique spatiality in the island's context.
In fact the importance of historical identity and the value of the island's architecture have been ignored for several decades as the public and the authorities were busier with the conflict between the two communities and other political reasons. The division had a huge impact on the island's urban transformation and architectural development. The demand for housing was high after 1974 war, as a result of immigration between both sides and rapid population growth and immigration from Turkey and Far East countries.

IOP Conf. Series: Materials Science and Engineering **371** (2018) 012037 doi:10.1088/1757-899X/371/1/012037

As a result, box like concrete blocks were emerged and the historical quarters left to deteriorate and therefore faded away. Since then the historic quarters have been occupied by low income families. As they could not afford to repair these buildings and the necessities of historic buildings the hurting fact is collapsed roof tops, broken windows, decaying the stone facades. Nowadays restoration projects are going off at these sites, however, is the re-use of these buildings taken into consideration? Or just facade repair is main aim and still remains as no man's zone. No matter how the European Communities' are interested and aware of how important and unique the Island's architecture and history is, the Cypriots in contrary is reckless to this afflictive situation. The first duty is to bring social awareness and inform the occupying inhabitants for sustainable results and continuing surrounding. Then the re-use of old buildings cannot be considered without social awareness, it should be avoided to allow occupancy of these spaces.

> "*………The richness of art is manifested in contrast: contrast with other works of art, other features of human experience, other times and places, and other forms of judgment and understanding. The possibilities of contrast are inexhaustible. Every being shares this inexhaustibility of openness to novel possibilities, although inexhaustibility is most fully realized in art. Art is the manifestation of inexhaustibility, achieved through intensity of contrast.*[2]
>
> *…………, history and tradition are essential components of complex contrast, providing a background against which minimalist works inhabit more complex contrast in proportion to their purity and simplicity.* [3]

The above citations indicate why the other times and places, historical background and buildings are important for Art and aesthetic value. To have an environment with full of aesthetical value, ethic built environment and responsible surrounding, an essential attention should be given to contrast. The newly build technologic environment must learn to work together with the existing historical environment. Ignorance should be avoided; instead integrity of those two should be encouraged, so that perfection might be achieved.

David Seamon stated in his article Goethe's Way of science as a Phenomenology of Nature, that;

> *…a deeper sense of responsibility and care for the natural world should be fostered. Animals, but also its plants, its terrestrial forms, its ecological regions, its formations of earth, sky and water, its sensual presence as expressed, for example, through light, darkness, and colour.*[4]

This theory is somehow proves that the existing history is also a part of our nature and should be involved in every step that will be taken in terms of environmental and urban development. Without history the contemporary architecture lacks of identity, contrast and aesthetic value. Perfection which is one of the elements of intense contrast can only be perceived here if only history is attached with the present-day. Both past and present day should be analyzed in advance to avoid failure and embrace the three dimensions of aesthetic value- Perfection, Invention and Celebration.

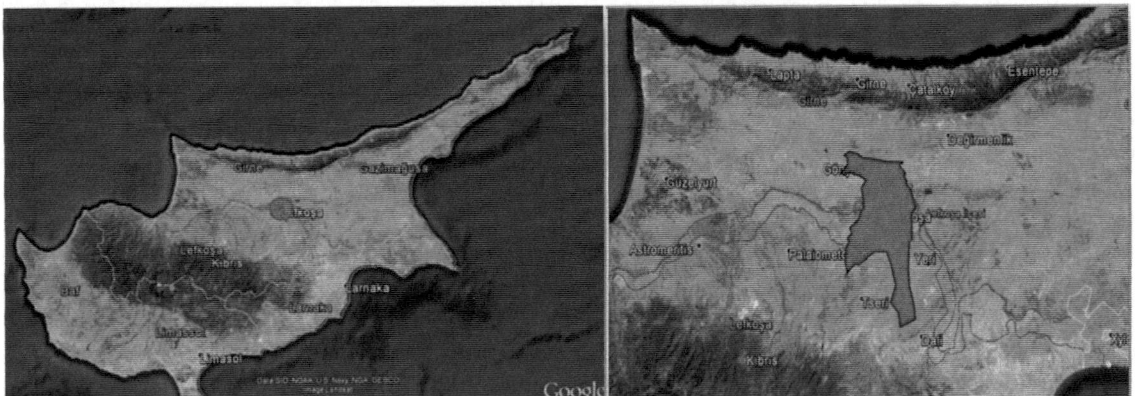

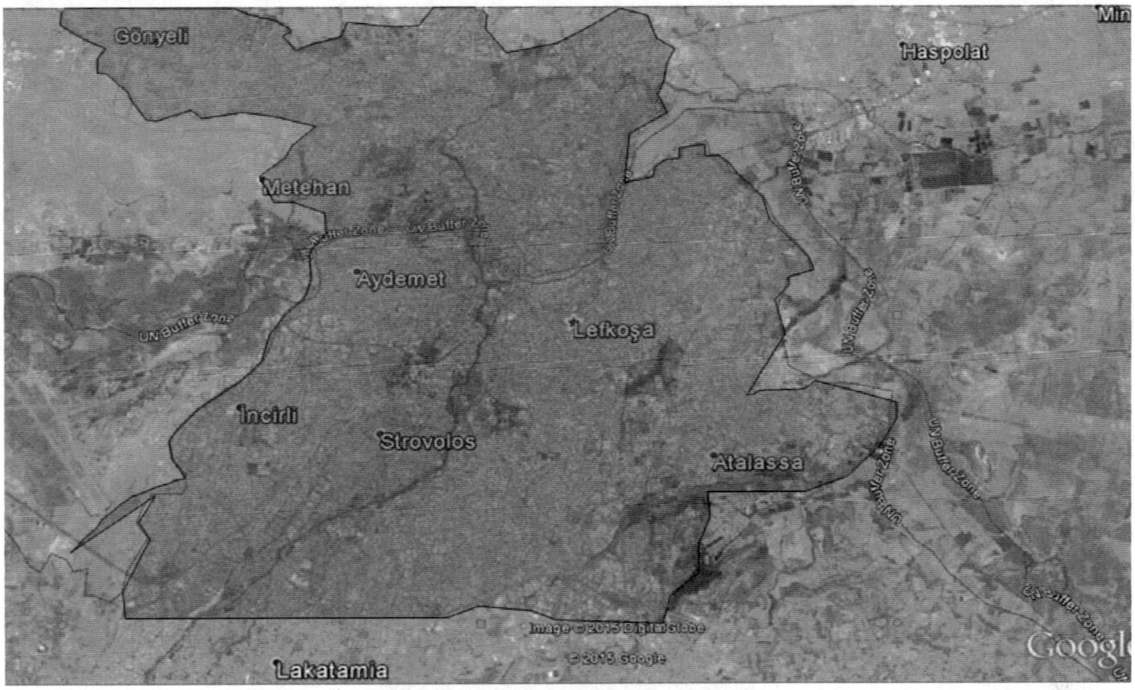


**Figure 1.**Geographical location of Nicosia showed in different.

## 2. Location- Ledra

The name of the ancient city-kingdom of Ledra, established in 1050 BC located in the centre of the island where the capital Nicosia is today. Today Ledras is the name of a street that that leads of Elefteria square runs almost in the centre of the walled part of Nicosia in a South to North direction and is about 1 km long. Most of it lies within the area of Greek part of Nicosia while the short stretch at the northern end lies between the UN Buffer Zone and the intersection with Arasta Street and Kyrenia Street falls within the Turkish part of Nicosia. Ledras plays a very important role in the urban context of Nicosia as it is one of the only two pedestrian crossings between the divided Nicosia and Cyprus.

**Figure 2.**Ledra street connection routes.

IOP Conf. Series: Materials Science and Engineering **371** (2018) 012037 doi:10.1088/1757-899X/371/1/012037

## 3. Character – Ledra

Traditionally, Ledra was the main commercial of the capital, although in recent years it has been suppressed by more accessible streets further out from the centre. Buildings lying on both sides of the street are largely for commercial use on street level; however the upper levels are serving as residential. On this street while shoppers are eying boutiques, couple of people sip there coffee or dinner and armed soldiers and policemen guards the barricade. Yet this feeling could not be encountered at any European city.

**Figure 3.**Soldiers walking up and down Ledra Street and communicating with tourists.

The decaying no-man's land zone and crumbling buildings slicing Ledra into Greek and Turkish interrupts the vibrant Ledra, disturbs the continuity of this unique medieval core. Ledra street barricade, acts as one of the symbols of the city enduring separation between two communities, therefore attracting tourists from different part of the world. Yet it is possible that this symbol of division may now turn out to become a symbol of reunification.

**Figure 4.**The diagram narrates the division and unity of Ledras.

IOP Conf. Series: Materials Science and Engineering **371** (2018) 012037 doi:10.1088/1757-899X/371/1/012037

Besides hope, the crossing offers tangible benefit injecting new life in the old town nestled with 15th century Venetian walls, as tourists and locals eager to satisfy their curiosity, commerce is increased, therefore socio-economic standard is boosting and Ledra is bustling more and more day by day.

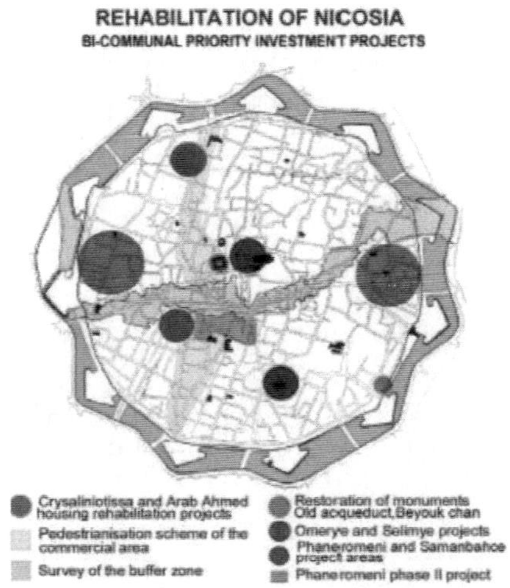


**Figure 5.**Bi-communal rehabilitation project map of Nicosia old town.

As it was and still is the main touristic quarter of the city with its coffee shops, restaurants, and small boutiques, in April 2008 the street reopened and encouraged to return to normal civilian and pedestrian activities. This area was subject to physical decay and socio-economic decline conditions for many years. The proximity of the Green Line was the primary reason for the lack of maintenance and the subsequent conditions. The urban upgrading of these areas was the priority of authorities under Nicosia Master Plan supported by UNDP.

## 4. Time – Space Relationship

Before understanding a problem its origin should be probed, in Ledra's case one approach was to define the layers, examine the facts then to bring them out to daylight for further debate and analysis. As mentioned previously the first but not the only reason of the deterioration, is the division. Secondly it is observed that some of the historic buildings have been wiped out and new buildings aroused instead. The reasons behind this method might be it might have been believed that this was a cheaper approach. However there is one thing that authorities failed to notice the upcoming disastrous result - the link between the past and the present was detached as a result the continuity of the space was lost as well as the originality. The importance of layers and the past have been described profoundly at Merleau Ponty's Phenomenology of Perception and can be summed up briefly as;

Space maybe defined as a form of external experience, rather than as a physical setting in which external objects are arranged. The relationships between objects in space are revealed by the experience of the perceiving subject. A perceptual field is a field in which perceptions are present in time and space. Space is modified and restructured by time.

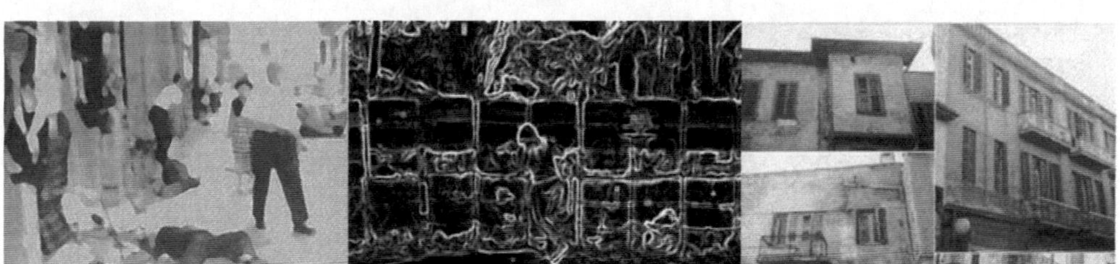

**Figure 6.** Picture narrates time-space relationship of Ledras.

## 5. Revitalization- Ledra

At stage 1 the revitalization project of Phaneromeni Area covering Ledra Street has been completed in 2004 and stage 2 was completed in 2005. The project focused on upgrading urban context as the strengthening of buildings overlooking the street on both sides foreseen. The works included the restoration of foundations and walls, the reconstruction of intermediate floors, the restoration of facades, and the replacement of rain water pipes, external doors and windows. It included transportation upgrading, partial pedesterianization, improving existing road, buildings with architectural value and facade restoration and street lighting. In order to create a safe pathway external cladding and bracing of the buildings in the northern and southern part of the street away from the buildings for the pedestrians an also new public space was created where all Cypriots can come together. And then the result was satisfying; proper conservation of the buildings overlooking the street on both sides was ensured, safety for pedestrians was increased. Pedesterianization of the commercial streets of Ledra upgraded its functionality and environment and allowing the area to compete with the new modern commercial centers outside the walled city. During planning and implementation of the project an effort was made to enrich the area functionally, aesthetically and architecturally, rendering it more attractive to residents, shoppers and visitors. Visible and invisible changes carry occurring at Ledra since the re-use.

**Figure 7.**This picture shows the present pathway with safe facades lying on the both sides of the street.

**Figure 8.**In this picture the upgraded quality of Ledras is presented.

IOP Conf. Series: Materials Science and Engineering **371** (2018) 012037 doi:10.1088/1757-899X/371/1/012037

## 6. Urban Transformation – Ledra

The urban and architectural development of Ledra did not occur instantaneously, or within just several years, it happened in a continuous change throughout decades and this change will continue for forever, it will never stops. As it continues o change new layers are introduced, each year put many things on top of the previous year, therefore creating visible and invisible layers and cultural codes. There has been a continuous difference at Ledra Street albeit it manages to keep up with the present conditions and retain honest with its past. The transparency can be grasped on even a routine visit to the site. The metamorphosis of urban transformation is described as the following in the study - Experiencing the place- memory of Kuzguncuk, Avsar 2014;

> *....Transformation does not refer to a phenomenon that occurs in an instant or within a couple of years; instead it denotes a kind of continuous change that would last forever. Transformation is not a continuity that produces sameness, but it generates differences. Here, transformation is used in the sense Deleuze uses the term metamorphosis. In his work Difference and Repetition (1994) Deleuze links metamorphosis to the eternal return of Nietchze, in the way that eternal return exists in every metamorphosis. Moreover he states that what causes to return is also subject to metamorphosis, which he refers as the metamorphosis of the agent. He understands metamorphosis as a difference in itself. Regarding Deleuze's metamorphosis, the urban transformation could be understood as a continuous production of difference via everyday life routines of the users in a certain locality. This production of difference that depends on the continuous social relations of the users with the place, transforms both the users and the locality simultaneously, like in metamorphosis.* [5]

**Figure 9.** The multiple street layers of Ledras and the urban transformation from the past to present.

## 7. Inhabitants – Visitors:

Ledras has got visitors from all over the world. It is alive for 24/7. As it is the main artery of the walled and the divided city and visitors use Ledra's road every day. The shop owners are mainly Greeks and Turks and the working class is from Far East, Georgia, Bulgaria, and Turkey etc. The population is very hybrid and the temporary population changing of the street is changing for every day. It is a multicultural street. The diversity and the unity between the communities despite the division of the street make Ledra very unique. There is a variety of cultures, religion, and tradition. In every corner of the street mind comes up with different feeling and imagination, one discovers new fact at each step through Ledra. At this street many states occur: the past-present relationship, time-space connection, visible- invisible cultural codes, love- hater between communities, crowded- but quite road, narrow streets- but wide content. Even though Ledra is narrow in structure, it shelters a global world here; therefore a wide variety of human beings and phenomena is inevitable.

IOP Conf. Series: Materials Science and Engineering **371** (2018) 012037 doi:10.1088/1757-899X/371/1/012037

**Figure 10, 11.** In these photographs diverse cultural codes of Ledras and both blending - clashing immigrants within the scene is narrated.

It is used in diverse ways by the locals and visitors, which is another generator of tension. Ledra is a preferred place for the visitors every day. People from different part of the world visit Ledra or use Ledra Street every day either for exploring the historic environment or use to cross over from north to south. Locals from the other parts of the island visit here for shopping, breakfast, lunch or dinner or taking photographs of the historic buildings, or students for research purposes. Tourist purpose may only be taking pictures of the border. Activists use this street to protest and camp on the UN line to take attention of the authorities to unification. Film producers also use this street because of its mystic and layered character. It consists of everything that one can ask on a photograph or on a movie scene. Limestone buildings of the Colony period, the buffer zone, the check point etc. The divided scenery might be the part of daily life of the inhabitants and locals but this is very unusual situation for the rest of the visitors.

**Figure 12.** In this photograph few green texture of the street is captured to indicate the dominancy of manmade environment.

## 8. Man Made vs. Nature– Ledra

Unfortunately there is very little green tissue and vegetation at Ledra Street; few trees are spread along the road and parks within couple of minutes reach. On the south side at the very far end of the east side of the walled city there is a park called Athens Park. And on the north side there is Kugulu Park that is right at the far end of the walled city and perpendicular to Ledras. The Public area right at the end of Ledra surrounded by trees where people may sit under shade at hot summer days for relaxing and chatting and also people can feed pigeons encourages the body and mind to question the link between nature- men even in contrast to each other. This physical environment causes the body to sense the tangible and intangible aspects of the scenery.

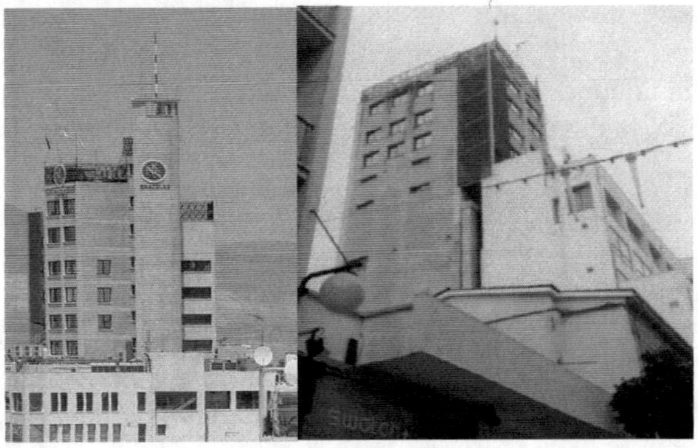

**Figure 13.** The Schaloas Tower- the manmade observation hill of Ledras - is shown in the above picture.

The street is naturally straight and plain; there are no slopes at all in contrast to the road reaching up to other bastions. And this fact makes it impossible to perceive the city and the street from different levels. However as a solution a tower called Scholas was built as an observation tower as the demand of the locals and the visitors to capture different vistas of the walled city and the last divided capital and lessen their curiosity about the forbidden zone. Observation request of buffer zone provides another source of tension, curiosity: gives birth to necessity to bodily experience this outrageous fact. The tower made it possible to have various vistas and to capture unforgettable scenery of life. But this man made block is not the only level difference at Ledras, the hundreds of years old bastion at the far end of the street provides a different vista of this interesting street. Going on top of this observation tower and bastion recalls history (memories) and evokes imagination.

**Figure 14, 15.**Ledras from different angles and street levels.

IOP Conf. Series: Materials Science and Engineering **371** (2018) 012037 doi:10.1088/1757-899X/371/1/012037

It has become obligatory to boundary the shop fronts with flower pots as the smell and visual beauty of green is missing at Ledras. It should be obligatory to the residents of the houses as well to have plantation at their balconies facing the street for providing at least a little amount of green tissue and bringing an aesthetical look to the facades of the buildings. The nature has a direct relationship with social and cultural life. If greenery is provided, the soul of a body will rest from the surrounding man made environment. The connection between the outdoors and indoors of buildings should be provided to avoid disconnection between the human and nature.

## 9. Scale & Architecture

Ledras might seem as an attractive centre for outsider and they may want to live in but it does not have a potential to develop horizontally any longer, luckily because of Nicosia walls surrounding it and putting limitations to its boundaries. Only vertical development is possible if few recently built apartment blocks aredemolished and new high rise technological buildings arose.

**Figure 16.**Ledras architectural layers.

However the demand is low to live in this area as it is a commercial busy street. And its geographical and historical constrains become an advantage for it, preventing it to expand and became a victim of regeneration projects and new development sites and thus get over populated and disturb the historic texture.

**Figure 17**.The roof scale of the street.

The buildings of Ledras are many two, three and even up to five storey houses or apartments. On the street level of Ledras there are only as commercials and used by authorities, such as restaurants, jewellery shops, souvenir shops, restaurants, cafes, bars, school and banks etc. The upper levels are mainly inhabited by low income immigrants from Far East and few locals or by local law firms and dentistry or medical clinics.

The dominating architecture of the area dates back to British Colonial Period. It is observed that there are late 20th century's architecture extensions on the historic buildings and also again late 20th century concrete tower blocks with marble cladding and it is notable that there are few replica facades of British period with modern structure and elements which is not an honest approach. It is seen that Ledra preserves its overall impression of a place that hosts good examples of British colonial architecture, evoking common memory and successfully integrating past with present architecture. Albeit these buildings were abandoned and decayed till the revitalization project, they survived and did not deprive Nicosia from their beauty. And also the low come inhabitants of the area could not cope well with the necessities of the old houses After the gate were open the shops were renovated, new cafes and restaurants were open therefore triggering its cultural, social and economic transformation, through these different ways of conservation old and abandoned buildings have been reoccupied and the population of the area increased.

**Figure 18.**Ledra time-space relationship-the honest development of the street in the whole time of becoming.

The inhabitants of the Ledra Street despite religious and ethnic differences and despite the division and hater between them still continue to have strong social and cultural relations with each other. Today not only the unique physical appearance but also the social relations that are departing from it make Ledra a desirable place to live, study or visit. In this architectural transformation process the old buildings of Ledra have not been replaced with new ones but instead the present day conditions are integrated with the past. The physical conditions of the buildings are improved or modified, and some of their functions have been changed, according to different demands. New meanings have been introduced to the street; inhabitants adapted their own needs to the existing. Both the inhabitants and the buildings integrated into a new look of Ledra, forming future of it. The past, the present and the future are integrated transparently in Ledra. The life styles of inhabitants have not been changed instead they are reflected to space adding something from themselves. The past, present and future cannot be separated. Without past there is no present, and without the present there will be no future.

## 10. Conclusion

This study focused on covering and discovering the layered visible and invisible character of the Ledras Street – linkage between the divided districts of Nicosia city. As this road links people from one side to another, it also links religion, ethnics and architectural environment of both sides. The link between the communities was detached but through this road visible and invisible attachment is

IOP Conf. Series: Materials Science and Engineering **371** (2018) 012037 doi:10.1088/1757-899X/371/1/012037

provided once again. The intangible and tangible shifts can be grasped via this zone of Phaneromeni quarter of old Nicosia. The architectural character of the street remained honest in comparison to the rest of the conservation projects going around the island. The integration of the past and present is honestly and ethically applied at Ledras. The rhythm between the street and the community has been kept. The street is not left as a ghost zone, today it lives and it will live in the future. Sustainable and continual architectural conditions have been developed on this street. The street has been looked by different ranges of perspectives and the multi layered character of Ledra brought up to daylight. This diversity should be preserved well for an infinite future. We are responsible for forming lively environments and read our cities, towns, village and streets as narratives. By doing so, it is possible to obtain holistic, transparent, and honest places to live in.

**Figure 19.** Diagram narrates the desired form of Nicosia and Ledras.

## 11. References


[1] Aydinli S, 2014 *Lecture notes Cyprus International University Boston Presentation*

[2] Ross S D, 1982 *A Theory of Art: Inexhaustibility by ContrastUSA***94** 246

[3] Ross S D, 1982 *A Theory of Art: Inexhaustibility by Contrast New York* **120** 246

[4] Seamon D, Zajonc A, 1998 *Goethe's Way of Science: A Phenomenology of Nature USA*

[5] Aydinli S, 2014 *Lecture notes Cyprus International University Experiencing the Place-Memory in Kuzguncuk: A. Phenomenological Inquiry of the Urban Identity*


## 12. Bibliography and Main Sources of Information


- David Watt & Belinda Colston, D.W. & B.C., 2003, Conservation of Historic Buildings and their Contents, Dorset, Donhead Publishing Ltd.
- Landes D A , 2012 *Maurice Merleau- Ponty Phenomenology of Perception USA*
- Local Authorities.
  Such as; Lefkosia Municipality (Greek& Turkish), the Leventis Museum, Evkaf.
- European and Worldwide Authorities.
  Such as; UNDP, UNOPS, ICOMOS, British Council Library.
- Libraries on both sides.
- Books, magazines, leaflets, published articles, archive websites, plans and drawings.
- Gathering information from Inhabitants and own observations.
- Site visits



**Acknowledgments**

I hereby would like to thank the Near East University who supported me financially to attend this conference. I would also like to thank esteem Assist. Prof. Dr. Kozan Uzunoglu, esteem Assist. Prof. Dr. Sema Uzunoglu and my dear colleague Ipek Yaralioglu for encouraging me to attend this conference.

IOP Conf. Series: Materials Science and Engineering **371** (2018) 012038 doi:10.1088/1757-899X/371/1/012038

# Plant Species Selection Based on Leaf Vibration Experiments

**Mengmeng Li [1] and Jian Kang [1, 2, *]**

1 Hei-Long-Jiang Cold Region Architectural Science Key Laboratory, School of Architecture, Harbin Institute of Technology, Harbin, China
2 UCL Institute for Environmental Design and Engineering, The Bartlett, University College London, London WC1H 0NN, UK
* Author to whom any correspondence should be addressed, E-mail: j.kang@ucl.ac.uk

**Abstract.** Noise pollution is exacerbated by the rapid urban development. Noise has a serious impact on both people's life and cities' ecological environment. Noise should be controlled immediately. Plants are one of the natural noise-reducing materials, which becomes increasingly important and popular. In urban landscape design, it is very important to select the plant species which have good noise reduction effect. The aim of this paper is to find out the characteristics of the plants with good noise reduction effect and apply it in urban landscape design. In this study, an experiment of sound attenuation by plants focusing on leaf vibration is carried out. This study investigates the vibration of leaves of 6 plant species in a sound field using a Keyence (IG- 1000 /CCD) Laser Micrometre. The results show that the vibration amplitudes of plant leaves increase significantly by about 4-12 μm after being stimulated by sound. In addition, driven by the same sound, the amplitudes of all leaves varied with the difference of leaf thickness, leaf area and leaf mass. The amplitudes of all leaves increase with the increase of leaf area and leaf mass, while decrease with the increase of leaf thickness.

## 1. Introduction

With the rapid development of urbanization, city gray infrastructure has increased significantly, and urban green space has become less and less. This triggers a series of rapid urbanization problems: the urban ecological environment is severely damaged, the various environmental elements necessary for people's well-being are also seriously polluted and destroyed, resulting in the deterioration of people's life quality [1]. At the same time, urban noise pollution is serious. An analysis of noise over the past few decades has found that traffic noise has become the major source of noise pollution [2]. In order to reduce the impact of traffic noise on residents living along the traffic lines, it is an effective way to set sound absorption barriers on both sides of roads. Urban green space can be used as a sound barrier to control traffic noise. By using plants to attenuate traffic noise, traffic noise can be effectively controlled and urban ecological environment can be improved. Renterhem et al. found that plant crowns had a better effect on noise attenuation at higher frequencies. Tree trunks did not attenuate low-frequency noise significantly, while the effect of intermediate-frequency noise attenuation was significant [3]. Yang et al. analysed the factors influencing plant noise reduction by studying the propagation path of sound through a tree [4]. Fan found that plant leaf types affected noise attenuation by plants, and the leaves with oval shape had more significant effects than long and narrow leaves [5]. Fang and Ling found that vegetation belts of different width, height and visibility had different effects on noise reduction [6]. Plants are one of the natural noise-reducing materials, which have become more and more important and popular. In urban landscape design, it is very important to select the plant species with good noise reduction effect for the sustainable development of urban ecology.

IOP Conf. Series: Materials Science and Engineering **371** (2018) 012038 doi:10.1088/1757-899X/371/1/012038

In the current study, an experiment of leaf vibration is carried out with leaves on the plants. 6 typical plant species are considered. In order to enhance the effect of traffic noise attenuation, this paper is based on the leaf vibration of plants, considering the leaf characteristics of tree species with better noise reduction effect, and provides the fundamental for the selection of plant species in urban landscape.

## 2. Methods and materials

### *2.1. Materials*

Six species of common plants are selected for experiment. The sound source is point sound source which is low-frequency pulse sound (93.5 dB), and the vibration tester is a Keyence laser micrometre (IG- 1000 / CCD). Table 1 shows some of the basic physical properties of the leaves of the 6 common plants tested.

**Table 1.** Physical characteristics of leaves.

| Plant | Leaf area ($mm^2$) | Leaf thickness (mm) | Petiole length (mm) | Petiole width (mm) | Petiole thickness (mm) | Mass (g) |
|---|---|---|---|---|---|---|
| Anthurium andraeanum Linden. | 7156.94 | 0.60 | 120.00 | 3.54 | 4.56 | 3.8392 |
| | 3115.17 | 0.74 | 95.50 | 2.76 | 3.00 | 1.5635 |
| | 3544.02 | 0.66 | 145.00 | 2.92 | 3.64 | 1.9649 |
| Gardenia jasminoides Ellis. | 931.91 | 0.80 | 4.00 | 2.00 | 1.86 | 0.2696 |
| | 619.77 | 0.58 | 2.00 | 1.50 | 1.58 | 0.1466 |
| | 625.23 | 0.56 | 3.00 | 3.50 | 1.54 | 0.1215 |
| Zamioculcas zamiifolia Engl. | 1434.65 | 1.00 | No | No | No | 0.7350 |
| | 1558.26 | 0.96 | No | No | No | 0.6414 |
| | 1014.72 | 0.58 | No | No | No | 0.4847 |
| Echeveria pulvinata. | 418.13 | 3.58 | No | No | No | 1.1896 |
| | 292.47 | 2.82 | No | No | No | 0.6339 |
| | 408.16 | 3.64 | No | No | No | 1.0146 |
| Plectranthus scutellarioides (L.) R.Br. | 3038.79 | 0.58 | 3.50 | 2.44 | 1.88 | 0.8834 |
| | 2173.72 | 0.64 | 3.35 | 2.12 | 1.64 | 0.5887 |
| | 2008.98 | 0.60 | 1.85 | 2.28 | 1.58 | 0.5638 |
| Hibiscus syriacus Linn. | 1294.21 | 0.74 | 22.50 | 1.98 | 2.00 | 0.5303 |
| | 2503.96 | 0.72 | 29.50 | 1.94 | 1.92 | 0.9828 |
| | 1649.32 | 0.60 | 30.50 | 1.70 | 1.62 | 0.4942 |

### *2.2. Development of a noise source*

Noise reduction by plant is mainly through two ways [5]-[14]. One is changing the sound energy into the kinetic energy by leaf vibration. The other is changing the sound energy into the thermal energy by leaf friction. The vibration of plant leaves mainly attenuates the low-frequency noise, so this experimental sound source chooses the low-frequency pulse sound. Four kinds of sound sources are selected during the experiment, as shown in Table 2, and the sound source 1 is finally selected.

IOP Conf. Series: Materials Science and Engineering **371** (2018) 012038 doi:10.1088/1757-899X/371/1/012038

**Table 2.** Sound source selection.

| | Sound source 1 | Sound source 2 | Sound source 3 | Sound source 4 |
|---|---|---|---|---|
| Spectrum | 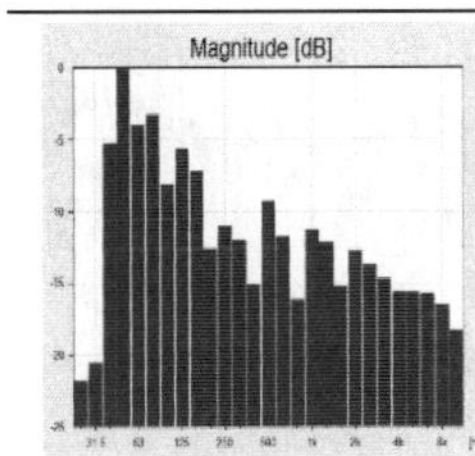 | 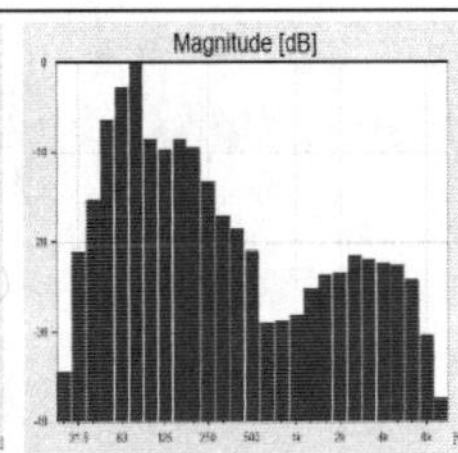 | 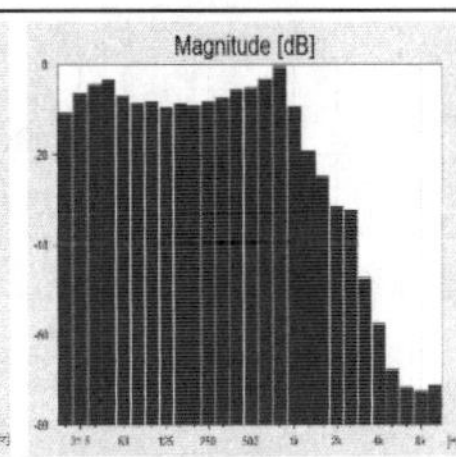 | 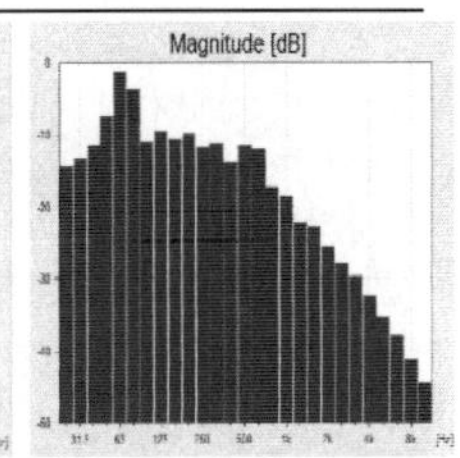 |

### 2.3. Experimental design

In order to prevent the impact of human activity on the experimental results, the experiment is carried out on a hermetically sealed wind-free vibration isolation table. The experimental process is shown in Figure 1 and Figure 2 (h is the vertical distance between the sound source and the plant leaf measuring point, h= 14-22 cm). In order to obtain the changes of amplitude of plant leaves after being stimulated by sound, the vibration of plant leaves is recorded for 24 s in each experiment, totalling 1.2 million measuring points. During the first 8-12 s, the self-vibration of leaves is recorded under no-sound excitation. Then the changes of the amplitude of plant leaves under the excitation of the sound are recorded. In this experiment, 3 leaves per plant are selected and tested, and each leaf is tested 5 times for analysis.

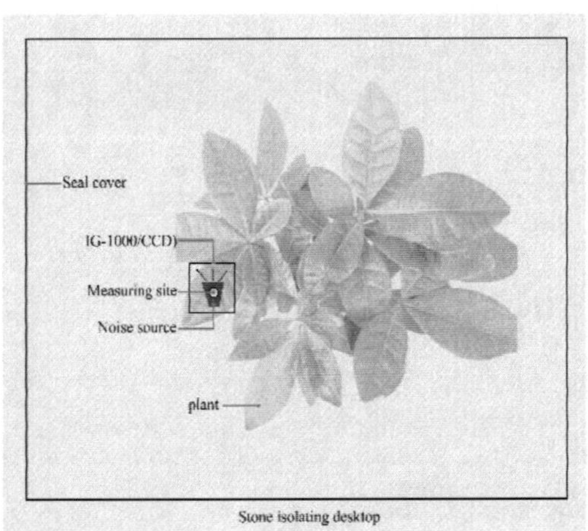


**Figure 1.** Top view of the plant leaf vibration.

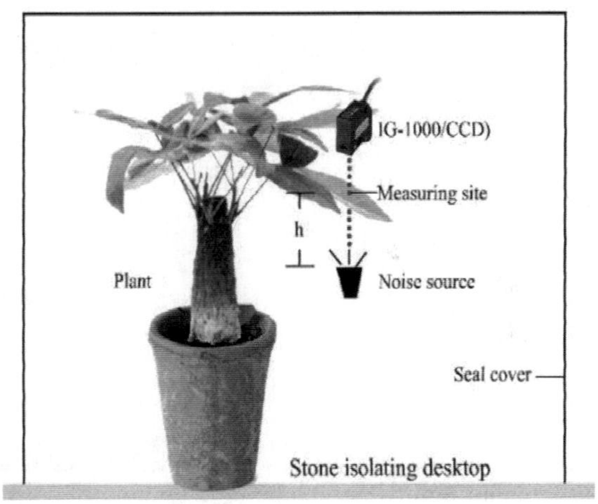


**Figure 2.** Side view of the plant leaf vibration.

## 3. Results

The results show that the self-vibration amplitude of plant leaves is 1 μm under no-sound. After stimulated by sound, the amplitude of plant leaf significantly increases about 4-12 μm, which is shown in Table 3. The amplitude of different leaves is significantly different under the same excitation. By analysing the influencing factors, it is found that the influence of physical factors such as mass, area and thickness of different plant leaves on amplitude is significantly different. Further analysis of these factors shows that plant leaf mass, area and thickness are the main factors that affect the increase of plant leaf amplitude.

The amplitudes of all leaves increase with the increase of leaf area and leaf mass, while decrease with the increase of leaf thickness.

IOP Conf. Series: Materials Science and Engineering **371** (2018) 012038 doi:10.1088/1757-899X/371/1/012038

**Table 3.** Vibration amplitude of plant leaves.

| Plant | Vibration amplitude of the leaf |
|---|---|
| Anthurium andraeanum Linden. (leaf 1) | 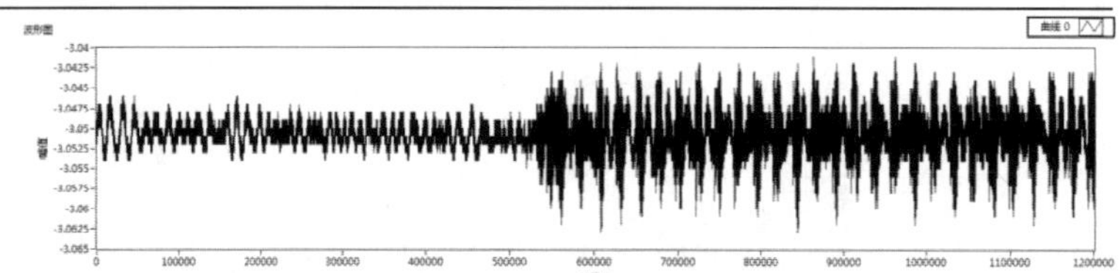 |
| Anthurium andraeanum Linden. (leaf 2) | 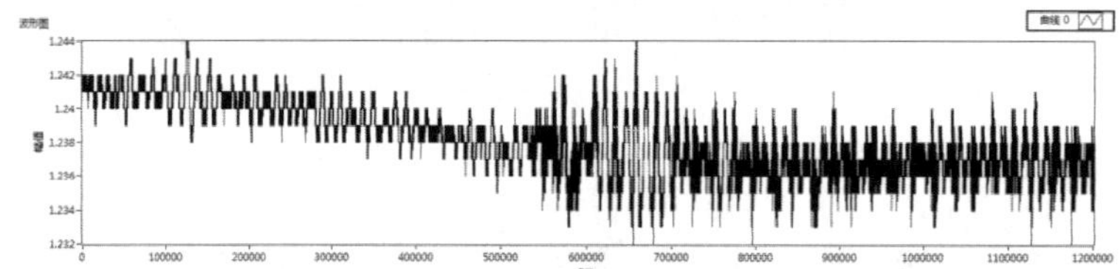 |
| Echeveria pulvinata. (leaf 2) | 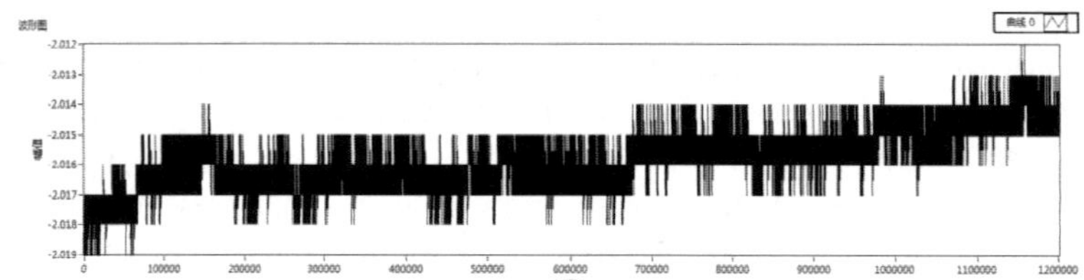 |
| Echeveria pulvinata. (leaf 3) | 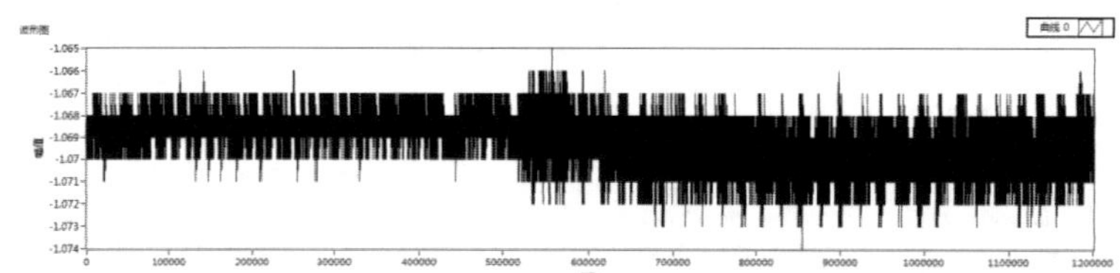 |
| Hibiscus syriacus Linn. (leaf 1) | 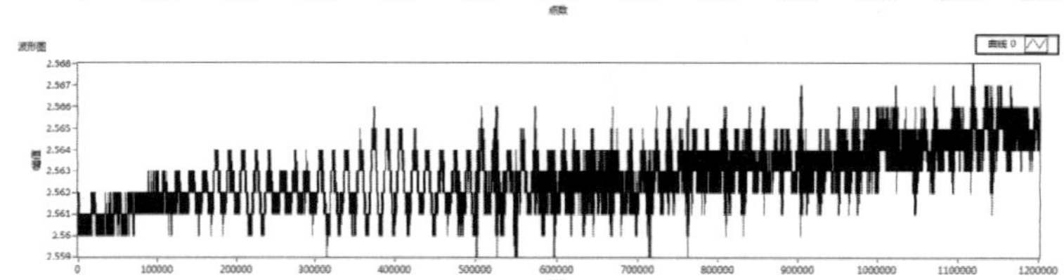 |
| Hibiscus syriacus Linn. (leaf 3) | 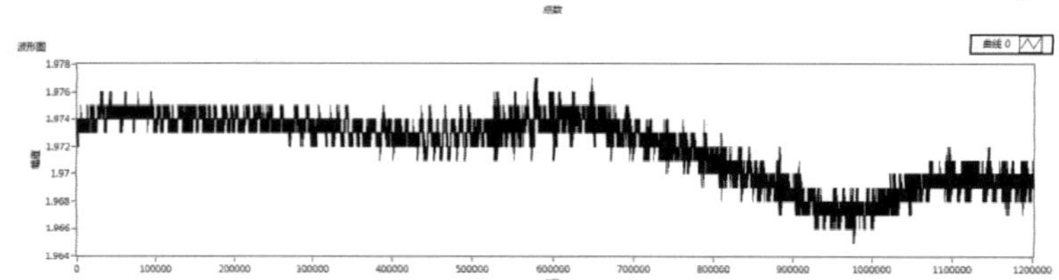 |

## 4. Discussion

Leaf mass and leaf area are the two most important factors affecting plant leaf vibration. It is also found that whether plant leaves contain petioles or not, and the length and thickness of petioles are the factors affecting the increase of plant leaf vibration.

The results show that there is a significant positive correlation between leaf mass, leaf area and leaf amplitude, while leaf thickness and leaf amplitude are negatively correlated. And the influence of leaf mass and area on amplitude is greater than the thickness. Therefore, in the future the plant species selection in landscape, in order to achieve better noise reduction effect, should focus on the selection of leaf weight, large area and thin plant species. In addition, in order to promote the sustainable development of urban ecology, other principles should also be followed in the selection of plant species: (1) Local species should be used. According to the criteria of suitable tree planting and local conditions, the local dominant tree species can better adapt to the local climatic conditions and soil conditions. (2) Ecologically-beneficial plants should be used to assist. Many plants have special ecological functions. Some plants can absorb the harmful gases which the traffic exhausts and enhance the overall ecological environment of the city.

## 5. Conclusion

In this study, 6 species of plants were tested on a sealed, wind-free platform to examine leaf amplitude, by considering various factors including leaf mass, leaf area and leaf thickness. When driven by the same sound, the amplitudes of all leaves varied with the difference of leaf thickness, leaf area and leaf mass. The vibration amplitudes of plant leaves increase significantly by about 4-12 μm after being stimulated by sound. The mass and area of plant leaves have a positive correlation with the amplitude, and the leaf thickness has a negative correlation with the amplitude. The influence of leaf mass and area on amplitude is greater than leaf thickness. In addition, many of physical properties of plant leaves have an impact on the amplitude of vibration. These findings can be used to guide urban landscape design and planting design. This is worthy of investigation which focuses on the other factors affecting leaf vibration. Further studies will be conducted on the other factors of plant leaf which can influence leaf vibration.

## 6. References

[1] V Tyagi, K Kumar and V K Jain 2006 A study of the spectral characteristics of traffic noise attenuation by vegetation belts in Delhi *Applied Acoustics* **67** pp 926-935

[2] T V Renterghem, D Botteldooren and K Verheyen 2012 Road traffic noise shielding by vegetation belts of limited depth *Journal of Sound & Vibration* **331** pp 2404-25

[3] T V Renterghem, D Botteldooren, M Hornikx and J Kang 2012 Road traffic noise reduction by vegetated low noise barriers in urban streets *Euronoise* (Prague, Czech Republic) pp 944-948

[4] H S Yang, J Kang, C Cheal, T V Renterghem and D Botteldooren 2013 Quantifying scattered sound energy from a single tree by means of reverberation time *Journal of the Acoustical Society of America* **134** pp 264

[5] Y Fan, Z Y Bao, Z J Zhu and Y N Lin 2010 The investigation of noise attenuation by plants and the corresponding noise-reducing spectrum *Journal of Environmental Health* **72** p 8

[6] C F Fang and D L Ling 2003 Investigation of the noise reduction provided by tree belts *Landscape & Urban Planning* **63** pp 187-195

[7] H S Yang, J Kang and C Cheal 2013 Random-Incidence Absorption and Scattering Coefficients of Vegetation *Acta Acustica United with Acustica* **99** pp 379-388

[8] K V Horoshenkov, A Khan, and H Benkreira 2013 Acoustic properties of low growing plants *Acoustical Society of America. J. Acoust. Soc. Am* **133** pp 2554-65

[9] H S Yang, J Kang, H J Kim and Y J Jeon 2015 Scale-model method for measuring noise reduction in residential buildings by vegetation *Building and Environment* **86** pp 81-88

[10] V Pathak, B D Tripathi and V K Mishra 2011 Evaluation of Anticipated Performance Index of some tree species for green belt development to mitigate traffic generated noise *Forestry & Urban Greening* **10** pp 61-66

[11] H Bediaf, R Sabre, L Journaux and F Cointault 2015 Comparison of leaf surface roughness analysis methods by sensitivity to noise analysis *Biosystems engineering* **136** pp 77-86

[12] T V Renterghem 2014 Guidelines for optimizing road traffic noise shielding by non-deep tree belts *Ecological Engineering* **69** pp 276-286.

[13] H Bédiaf, R Sabre, L Journaux and F Cointault 2015 Comparison of leaf surface roughness analysis methods by sensitivity to noise analysis *Biosystems Engineering* **136** pp 77-86.

[14] H S Yang, J Kang and M Choi 2012 Acoustic effects of green roof systems on a low-profiled structure at street level *Building and Environment* **50** pp 44-55.

IOP Conf. Series: Materials Science and Engineering **371** (2018) 012039 doi:10.1088/1757-899X/371/1/012039

# Agent-Based Simulation on Pedestrian Flow Capacity of Narrow Space

**Shi Sun[1,2], Cheng Sun[1,2,3] and Yunsong Han[1,2]**

1 School of Architecture, Harbin Institute of Technology, Harbin, China
2 Hei Longjiang Cold Region Architectural Science Key Laboratory
3 Author to whom any correspondence should be addressed
E-mail: suncheng@hit.edu.cn

**Abstract.** The narrow space in the urban pedestrian traffic system is the area which is liable to be congested. Recently, researches on the maximum pedestrian flow of space mainly focus on the width of exit, but there are not sufficient researches on the walking space with a certain length. The aim of this paper is to analyze the change of pedestrian flow capacity of the narrow space with a certain length under the influence of width. The simulation model of walking behavior is established by Net Logo software. By changing the space width and the number of pedestrians entering into the space per unit time, the quantitative relationship between the width of space and the space pedestrian flow capacity is simulated. In the simulation, the pedestrians' walking behavior is influenced by the target area, the space boundary and the visible surrounding pedestrian density. The results show that when the length of the walking space is 30 m and the space width is less than 10.33 m, the increase of the width will significantly improve the pedestrian flow capacity of the space. When the space width is larger than 10.33 m, the increase of the width will enhance the pedestrian flow capacity less significantly. The research results can be applied to the design of the width of urban long and narrow walking space.

## 1. Introduction

In recent years, with the sustained and stable economic development, the number of urban public places fulfilling material and cultural needs experienced a rapid development. The scale of gathering crowd grew in various public places. The design of urban walking areas suffering from pedestrian congestion had attracted great attention. Crowd congestion occurs mainly in pedestrian bottleneck areas with poor maximum pedestrian flow capacity, and therefore the importance of the study of pedestrian traffic bottleneck area is increasing [1]-[3]. Helbing et al. thought that every pedestrian felt the application of both "social" and "physical" forces. Social forces had no physical source, instead they reflected the pedestrian's intentions and did not collide with other people or walls in the room and moved in a particular direction at a specific speed [4]-[6]. Weng et al. divided pedestrian walking behavior into "move", "dodge" and "turn" 3 kinds of basic behaviors. The direction of walking was determined by walking weights, which were the sum of the basic behavior of each vector multiplied by the weights in the respective directions. The model could be used for path analysis and congestion simulation [7]. Based on the NetLogo platform, Lee *et al.* proposed a multi-agent path selection model based on field of view. With the consideration of the view angle, visual distance, visual disturbance, moving speed, attractiveness factors and other pedestrian factors, the problem of pedestrian path selection under the influence of building entity was analyzed [8]. Wang et al. proposed a new congestion evacuation model based on multi-agent to simulate the process of pedestrian evacuation in an obstructed or unobstructed stadium. The state of each agent was divided into normal condition, catching up and death. In this model, all pedestrians were divided into 4 types, each with 6 different competencies. The individual's course of action was influenced by its competencies, distance to exit, number of agents in field of vision, and crowd density. The simulation analyzed congestion and death phenomena [9]. Harada et al. studied that when evacuation numbers exceeded the capacity of the evacuation zone, those who could not enter the evacuation zone changed their destination to other zones. In order to simulate the change of destination, a switch action model was developed. The behavior of destination change during evacuation was considered as an important factor and the switch-effect algorithm was introduced into the traditional multi-agent model in order to be able to reproduce the destination change during the evacuation process [10]. Li *et al.* used the population dynamics theory to study the clogging of evacuation spaces during emergency evacuation, the consequent accidents and stampede. Research showed that the amount of the population flow was related to population and spatial factors. Based on the distribution model of population movement along evacuation routes, the influence of initial population density on congestion was analyzed in detail. The results showed that when the population density reached 2.17 persons·m$^{-2}$, the crowd would be hard to move [11]. Shi *et al.* conducted a series of controlled laboratory tests to examine the safety constraints that affected the mobility of people in the merger perspective and the direction of flow under slow operation and obstruction visual conditions. Studies showed that the direction of flow had a significant effect on the outflow of pedestrians, while visual disturbances had an impact on the interactions and mergers of pedestrian populations [12]. The previous studies provided a scientific basis for the establishment of walking behavior model and the setting of relevant parameters, but few researches have studied the pedestrian flow capacity of spaces with a certain length.

In this paper, aiming at the long and narrow walking space prone to traffic jam in walking environment, the simulation models with different widths are established and the traffic efficiency is analyzed, which provides reference for the design of walking space width in urban design.

## 2. Methods

### *2.1. Simulation environment settings*

In this paper, the model space length is 30 m, the grid size is 0.33 m, and the space width is increased gradually by 4 times grid width (1.33 m). The model space width is 3.67 m, 5.00 m, 6.33 m, 7.67 m, 9.00 m, 10.33 m, 11.67 m, 13.00 m and 14.33 m respectively. 9 different widths are simulated in total, as shown in Fig. 1.

Environment settings in space include target space, walls and pedestrians. The target space is located in the black region at the far right side of the space. All the pedestrians walk towards the target space gradually. The walls are the light grey regions located at the upper and lower edges of the space as the boundary of the simulation space. The pedestrians start from the dark grey region on the left side of the space, and leave on the dark grey region on the right. Taking Asian body size into account, the pedestrian diameter is set 0.33 m, and walking speed is set 1.3 m·s$^{-1}$. In the simulation, the number of pedestrians entering space per unit of time is gradually increased until the number of pedestrians entering space per unit time reaches the maximum flow rate in that space.

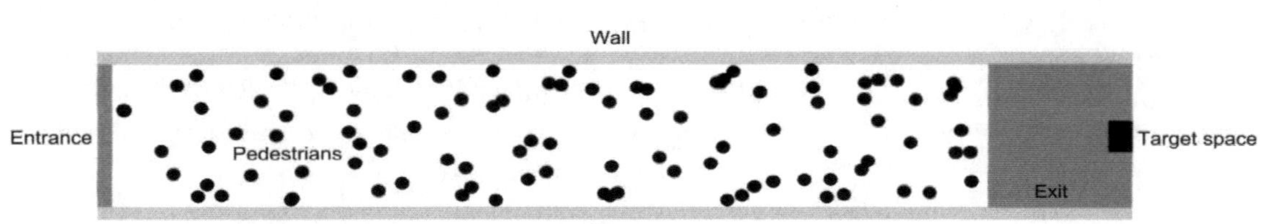


**Figure 1.** 30 m long and 7.67 m wide model simulation results.

### *2.2. Walking behavior setting*

The walking behavior of the population is mainly affected by the spatial environment and other pedestrians within the visible range. In this paper, aiming at the walking behavior of high-density crowd, the visible area is set to the area within 120° of the pedestrian's walking direction and within 2 m from the pedestrians' current location.

Pedestrians' response to the environment within the visible range is divided into 4 basic behaviors: moving forward, turning, bypassing and waiting. When there are no obstructions in front of pedestrians, the pedestrians move toward the target. When pedestrians are less than 2 steps away from the wall, pedestrians randomly rotate within 30° to the opposite direction of the wall in order to avoid of getting out the current region. When there are other pedestrians in the visible range of pedestrians, the degree of congestion in the visible area is evaluated by the number of pedestrians in different distance ranges of the target space. Pedestrians will bypass other pedestrians and select the routes with lower congestion when they go to the target area. According to the research of Li *et al*., when the pedestrian density is larger than 2.17 persons·m$^{-2}$, pedestrians stop walking [8]. According to the visible area condition set in this article, pedestrians will stop walking and wait for other pedestrians if the number of pedestrians exceeds 8 in the visible range. Then, pedestrians continue to walk after others' passing, as shown in Figure 2.

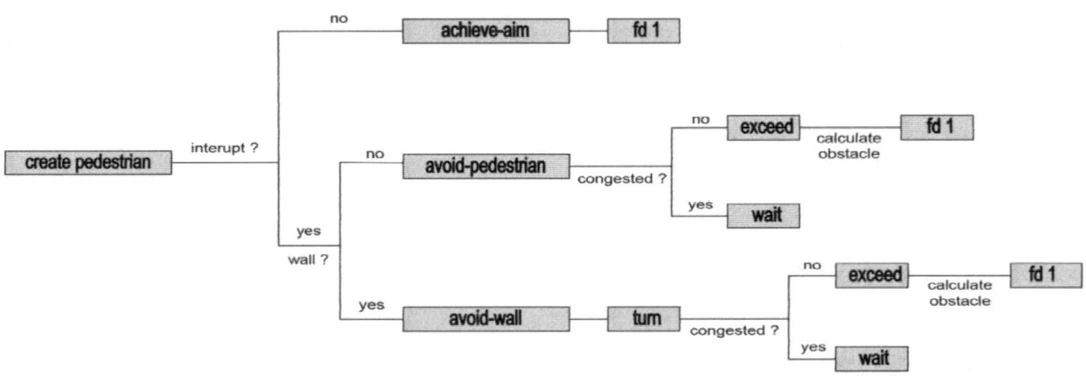


**Figure 2.** Walking behavior setting method.

## 3. Results

If the flow of people entering the space within a unit time does not reach the maximum flow rate of space, the number of people entering the space and the number of people leaving the space will reach a balance after a period of operation of the model, and the number of pedestrians in the space will no longer increase. When the number of passengers exceeds the maximum capacity of the space, congestion will occur, resulting in a continuous increase in the number of pedestrians in the space. When the space width is 6.33m, the maximum capacity of the space is 5 persons·s$^{-1}$. If the number of pedestrians entering the space exceeds the capacity, the number of pedestrians will continuously increase, as shown in Table 1.

**Table 1.** Simulation results of the change of population under the width of 6.33 m.

| 1 person·s$^{-1}$ | 2 persons·s$^{-1}$ | 3 persons·s$^{-1}$ | 4 persons·s$^{-1}$ |
|---|---|---|---|
| 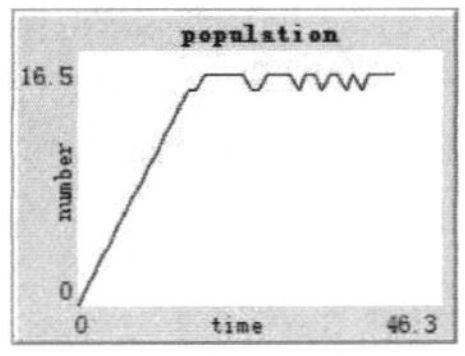 | 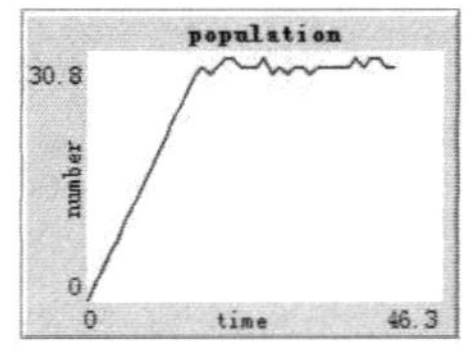 | 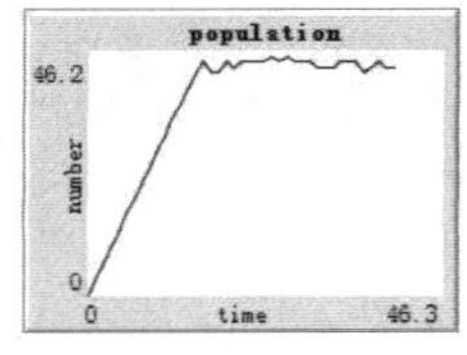 | 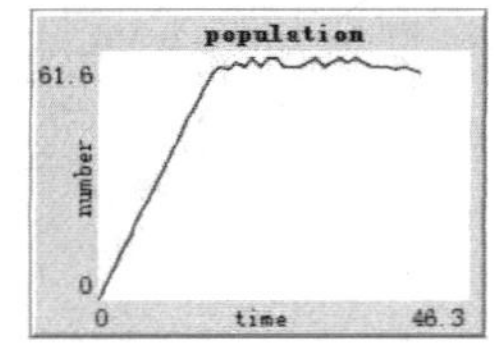 |
| 5 persons·s$^{-1}$ | 6 persons·s$^{-1}$ | 7 persons·s$^{-1}$ | 8 persons·s$^{-1}$ |

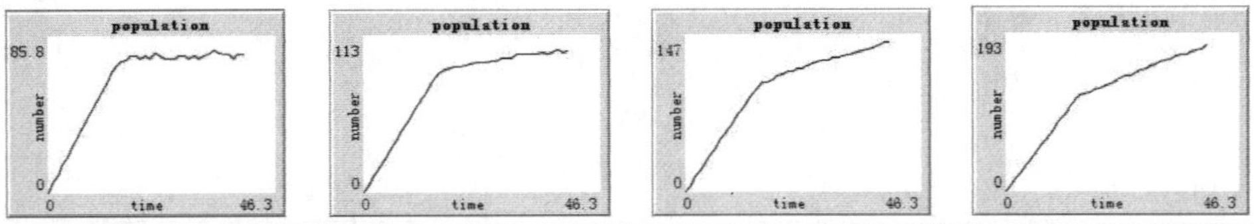


*3.1. The maximum flow rate*

For the walking space of 30m in length, the maximum flow has positive correlation with the space width. Yhe maximum flow rate of the space is 3 persons·s$^{-1}$ when the space width is 3.67 m, 4 persons·s$^{-1}$ when the space width is 5.00 m, 5 persons·s$^{-1}$ when the space width is 6.33 m, 6 persons·s$^{-1}$ when the space width is 7.67 m, 7 persons·s$^{-1}$ when the space width is 9.00 m, 8 persons·s$^{-1}$ when the space width is 10.33 m, 8 persons·s$^{-1}$ when the space width is 11.67 m, 9 persons·s$^{-1}$ when the space width is 13.00 m, and 9 persons·s$^{-1}$ when the space width is 14.33 m, as shown in Table 2.

**Table 2.** Simulation results of maximum flow in walking space of different width.

| 3.67 m | | 5.00 m | | 6.33 m | |
|---|---|---|---|---|---|
| 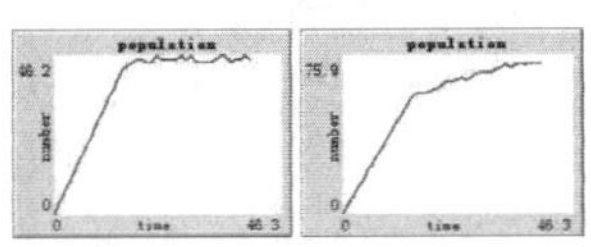 | | 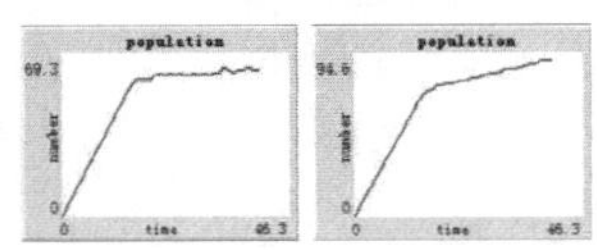 | | 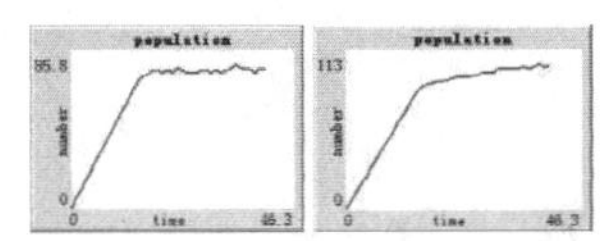 | |
| 3 persons·s$^{-1}$ | 4 persons·s$^{-1}$ | 4 persons·s$^{-1}$ | 5 persons·s$^{-1}$ | 5 persons·s$^{-1}$ | 6 persons·s$^{-1}$ |
| 7.67 m | | 9.00 m | | 10.33 m | |
| 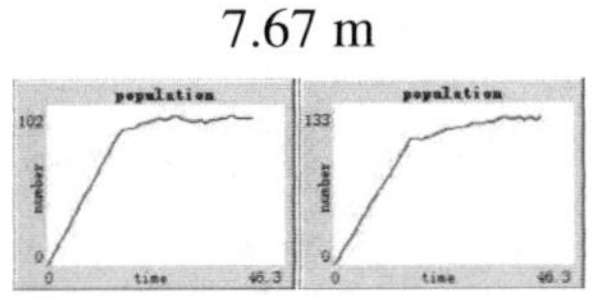 | | 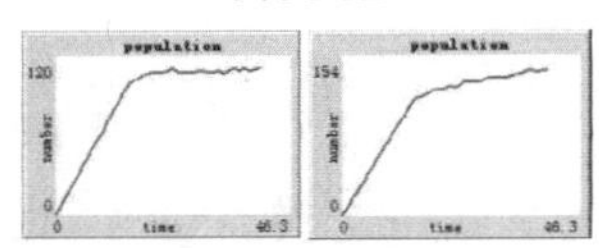 | | 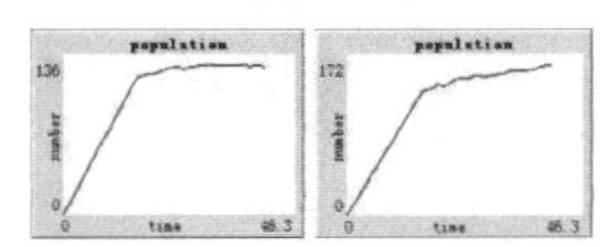 | |
| 6 persons·s$^{-1}$ | 7 persons·s$^{-1}$ | 7 persons·s$^{-1}$ | 8 persons·s$^{-1}$ | 8 persons·s$^{-1}$ | 9 persons·s$^{-1}$ |
| 11.67 m | | 13.00 m | | 14.33 m | |
| 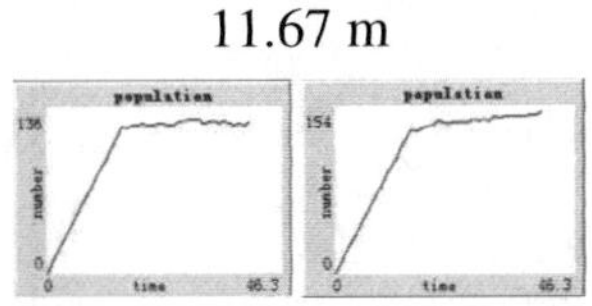 | | 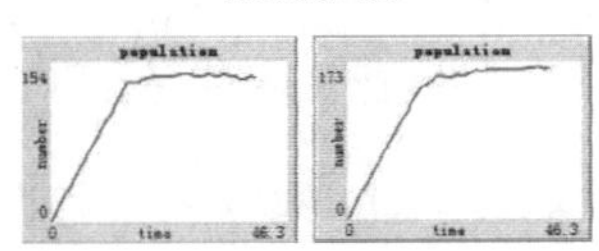 | | 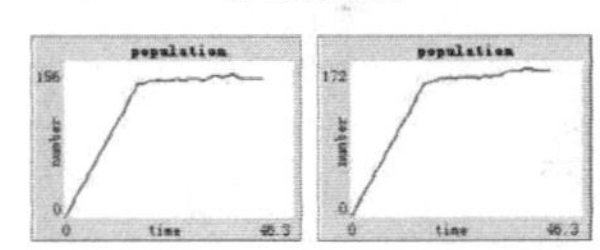 | |
| 8 persons·s$^{-1}$ | 9 persons·s$^{-1}$ | 9 persons·s$^{-1}$ | 10 persons·s$^{-1}$ | 9 persons·s$^{-1}$ | 10 persons·s$^{-1}$ |

*3.2. Pedestrian flow condition*

For simulation of the walking space of 30 m in length, the area of space, the maximum pedestrian flow, the maximum pedestrian flow per meter and pedestrian density according to the space width are shown in Table 3.

**Table 3.** The statistics when walking space reaches the maximum number of people.

| Width (m) | Area (m$^2$) | Maximum flow (persons·s$^{-1}$) | Capacity (persons·m$^{-1}$·s$^{-1}$) | Density (persons·m$^{-2}$) |
|---|---|---|---|---|
| 3.67 | 110 | 3 | 0.82 | 0.42 |
| 5.00 | 150 | 4 | 0.80 | 0.44 |
| 6.33 | 190 | 5 | 0.79 | 0.43 |
| 7.67 | 230 | 6 | 0.78 | 0.43 |
| 9.00 | 270 | 7 | 0.78 | 0.43 |

| | | | | |
|---|---|---|---|---|
| 10.33 | 310 | 8 | 0.77 | 0.43 |
| 11.67 | 350 | 8 | 0.69 | 0.38 |
| 13.00 | 390 | 9 | 0.69 | 0.38 |
| 14.33 | 430 | 9 | 0.63 | 0.35 |

## 4. Discussion

With the increase of walking space width, the maximum flow rate is on the rise. When the space width is between 3.67 m and 10.33 m, the maximum flow rate increases significantly with the increase of space width. When the space width increases by 1.33 m, the maximum space flow increases by about 1 person·$s^{-1}$. When the space width is between 10.33 m and 14.33 m, with the increase of width, the maximum flow rate in space has a decreasing tendency of increase. When the space width increases by 1.33 m, the maximum flow rate in space increases by about 0.5 persons·$s^{-1}$, as shown in Figure 3. Then, it is a more meaningful method to enlarge the space width when it is less than 10.33 m.

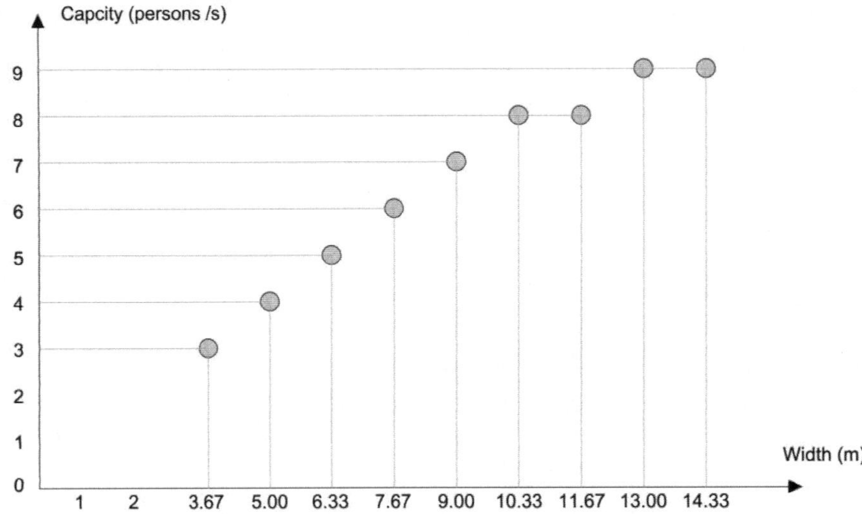


**Figure 3.** The trend of maximum flow rate of space with width.

With the increase of walking space width, the maximum pedestrian flow rate per meter of walking space shows a downward trend. When the space width is between 3.67 m and 10.33 m, the maximum flow rate per meter of walking space shows a slow downward trend when the walking space reaches the maximum and maximum flow rate. When the walking space reaches the maximum pedestrian flow rate, and the space width is between 10.33 m and 14.33 m, the drop of the maximum flow rate per meter with the increase of walking space width is significant. When the space width is between 3.67 m and 10.33 m, the pedestrian density keeps at about 0.43 persons·$m^{-2}$ when the pedestrian space reaches the maximum pedestrian flow rate. When the space width is between 10.33 m and 14.33 m, and the pedestrian space reaches the maximum pedestrian flow rate, the pedestrian density shows a downward trend.

With the analysis of the simulation results, it can be seen that when the width of the space is increased by 1.33 m and the number of pedestrians is increased by 1 person·$s^{-1}$, the pedestrian density in the spatial range almost remains stable. But when the space width is more than 10.33 m, the pedestrian density will reduce. The reason may be the increase of the complexity of mutual interference between pedestrians and the uneven distribution of pedestrians in space.

## 5. Conclusion

It can be seen from the simulation results that the width of walking space is directly related to the maximum traffic in space. When the width is between 3.67 m and 10.33 m, the effect of the increase of space width on the maximum flow rate is remarkable, and the maximum flow per unit width shows a slowly decreasing trend. The pedestrian density maintains 0.43 persons·$m^{-2}$. When the width is between 10.33 m and 14.33 m, the effect of the increase of space width on the maximum flow rate is subtle, and the maximum flow per unit width shows a significantly decreasing trend. The pedestrian density decreases. For the space with a certain length, if the space width is increased, the increasing trend of maximum flow will decrease when the space with reaches a certain threshold. In future, the space types can be more abundant, and different kinds of pedestrians, such as different genders and ages, can be taken into consideration to make the simulation applicable in some typical environments.

## 6. Acknowledge

This research was supported by the Heilongjiang Science Fund for Distinguished Young Scholars (2016JC008), Science and Technology Project of the Ministry of Housing and Urban-Rural Development of the People's Republic of China—Research and Development Project (2016-K1-001) and Key Technologies R & D Program of Heilongjiang (GZ15A502).

## 7. References

[1] M Hu, Z Fang, S Liu, W Chen and B Yang 2010 Study on evolution mechanism of fateful stampede accident based on graphical evaluation and review technique *IEEE Int. Conf. on Systems Man and Cybernetics* (Istanbul, Turkey) (IEEE) pp 2976-78

[2] Q Wang 2016 Subway Crowded and Stampede Neural Networks Safety Assessment Basing on SHEL Model *Man-Machine-Environment System Engineering: Pro. of the 16th Int. Conf. on MMESE.* (Springer Singapore) pp 455-468

[3] S R Jaime and O P Samuel 2017 Analysis of the 'News Divine' stampede disaster *Safety Science* **91** pp 11-23

[4] D Helbing and P Molnár 1995 Social force model for pedestrian dynamics *Physical Review E* **51** pp 4282-87

[5] D Helbing, I Farkas and T Vicsek 2000 Simulating dynamical features of escape panic *Nature* **407** pp 487-490

[6] T I Lakoba, D J Kaup and N M Finkelstein 2005 Modifications of the Helbing-Molnár-Farkas-Vicsek Social Force Model for Pedestrian Evolution *Simulation* **81** pp 339-352

[7] W G Weng, S F Shen, H Y Yuan and W C Fan 2007 A behavior-based model for pedestrian counter flow *Physica A Statistical Mechanics & Its Applications* **375** pp 668-678

[8] S J Lee, K H Lee and S J Kang 2013 Study on a Pedestrian Simulation Model of Natural Movement *Journal of Asian Architecture & Building Engineering* **12** pp 41-48

[9] J Wang, Q Shi, X Hu and P Yang 2014 Modeling and simulating for congestion evacuation based on multi-agent approach *Control Conference* (Nanjing, China) (IEEE) pp 6421-26

[10] E Harada, H Gotoh and N B A Rahman 2015 A switching action model for DEM-based multi-agent crowded behavior simulator *Safety Science* **79** pp 105-115

[11] J Li, B Zhang, S Tang and G Hu 2015 A minor modification in the model for describing congestion and recovery along evacuation paths *Land Use Policy* **42** pp 751-758

[12] X Shi, Z Ye, N Shiwakoti and W Wang 2016 Empirical investigation on safety constraints of merging pedestrian crowd through macroscopic and microscopic analysis *Accident analysis and prevention* **95** pp 405

IOP Conf. Series: Materials Science and Engineering **371** (2018) 012040 doi:10.1088/1757-899X/371/1/012040

# Thinking about the Contradictions of Space Use of Square Dance in Chinese Cold Cities through Newspaper Reports

**Liu Xiaobing[1, 2], Zhao Xiaolong[1, 2] and Zhang Bo[3, *]**
1 Department of Landscape Architecture, Harbin Institute of Technology, Harbin, CHN
2 Institute of Ecological Landscape, Room 109, 66 West-Dazhi Street, Harbin, CHN
3 Department of Landscape Architecture, Oklahoma State University, Stillwater, USA
e-mail: 65746174@qq.com

**Abstract.** The square dance is one of the most popular new physical activities in China in recent years, and has become the hotspot of Chinese garden research. This study attempts to interpret the space use contradictions of square dance from the angle of newspaper report, and then make up the deficiency of the planning designer's understanding of square dance space. We collected 749 news reports in 8 years of four authoritative newspapers in Harbin, and probed into the square use contradictions of usability, accessibility and climate environment of square dance. We hope to provide the content basis for future social science research in the field of planning and design, and provide the direction for technical research.

## 1. Introduction

In recent years, the public enthusiasm for health activities is unprecedentedly high, and people spontaneously formed many popular sports cultural achievements. The square dance is one of the typical representatives. Because of the spontaneity of its formative process, participants often explore their own use of the venues and tend to be more inclusive. In addition, the development of square dance activities is rapid, resulting in the space planning and design industry cannot create a profound understanding of the space use of square dance.

At present, the research of square dance activity space mainly concentrates on the related content of "square dance disturbing the public": on the one hand, they discuss the noise regulation or policy of square dance, such as Liang Qin-chao (2016) [1], Wang Qian-ni (2015) [2]; on the other hand, discusses how to solve this problem through reasonable planning and design methods, such as Yuan Ji-fang (2014) [3], Zhang Xin-si (2014) [4].

In fact, it was the mass media that initially brought the space contradictions of the square dance into the public eye. The advantage of traditional paper media, especially newspaper, lies in interview right, credibility, professional news content productivity and professional news editing team. According to Nielsen's survey of "global Advertising Trust," Consumers have a 58% trust in newspapers. This is much higher than in television, radio and the Internet. This shows that the newspaper has irreplaceable role in the credibility and authority. On the one hand, it is an important way for the contemporary people to understand the outside world, and then to influence the consciousness and behavior of the masses. On the other hand, it is a wind vane for the superstructure, and it can give the society information feedback to the government, so as to monitor public opinion and the convey people's willing. So in social phenomena and contradictions, newspapers are one of the most influential sources of data.so as to monitor public opinion and the convey people's willing.

Therefore, this study attempts to interpret the space use contradictions of square dance from the angle of newspaper report, and then make up the deficiency of the planning designer's understanding of square dance space. At the same time, in the winter, residents of the cold area differ from other places in the use of square dance space. So take "Harbin" this typical cold city as an example, with the reports of four kinds of most influential local newspapers-"New Evening News" "Daily Life" "Harbin Daily", "Heilongjiang Daily", and reflect on the use contradictions of square dance space. We hope to provide the content basis for future social science research in the field of planning and design, and provide the direction for technical research.

## 2. Data collection

### *2.1. Data sources*

All the reports we collected came from the "New Evening News" "Daily life" "Harbin Daily", "Heilongjiang Daily". According to the Meihua [5], these four kinds of newspaper are the best circulation in Harbin. Among them, consumers of "Daily Life" and "New Evening News" are mostly citizens. The consumers of "Harbin Daily" and "Heilongjiang Daily" are mostly government agencies. Through these four kinds of newspapers, we can look at our research problems comprehensively from both the public and the government perspectives.

### *2.2. Time filtering*

In fact, the coverage of this study is limited to 2000-2015. But the number of press reports about square dance for 2000-2007 was 0. The number of reports in 2008-2010 is very small, and the content of the article is not strong. So the real research data for this study are all the news reports from 2011 to 2015.

### *2.3. Location filtering*

"New Evening News" " Daily Life " and "Harbin Daily" are city-level newspapers. Its news reports are mainly concentrated in the city of Harbin, but there are still some reports on the surrounding towns of Harbin. "Heilongjiang Daily" is a province-level newspaper, covers a number of cities and towns. As this study takes the city of Harbin as a research object, so the news reports that focus on other cities and towns are not included in the study.

### *2.4. Content filtering*

As the core goal of this research is to explore and discuss the space use contradictions of square dance (a kind of fitness activities), and finally provide advice and guidance for the urban planners and venues designers. As a result, the report's screening principle is to look for news reports about three aspects of "square dance", " venues planning and design" and "health", not just the "square dance" itself.

## 3. Results

"Square Dance" or "fitness dance" as the key words for full-text search, we initially received 749 reports in 2008-2015. Through the further content reading, we finally obtains 350 reports: 143 from the life newspaper, 77 from " New Evening News", 55 from "Harbin Daily", 75 from "Heilongjiang Daily". From the time distribution of the number of reports (figure 1), in the period of 2011-2015, both the 4 newspapers ' respective and overall number of reports, generally show an upward trend. That is to say, since 2011, square dance has been getting more and more attention, and peaked in 2015. From the content of the report (figure 2), the number of reports involving usability contradictions is 194, that involving the accessibility contradictions is 82, and that involving climate and environment contradictions is 74.

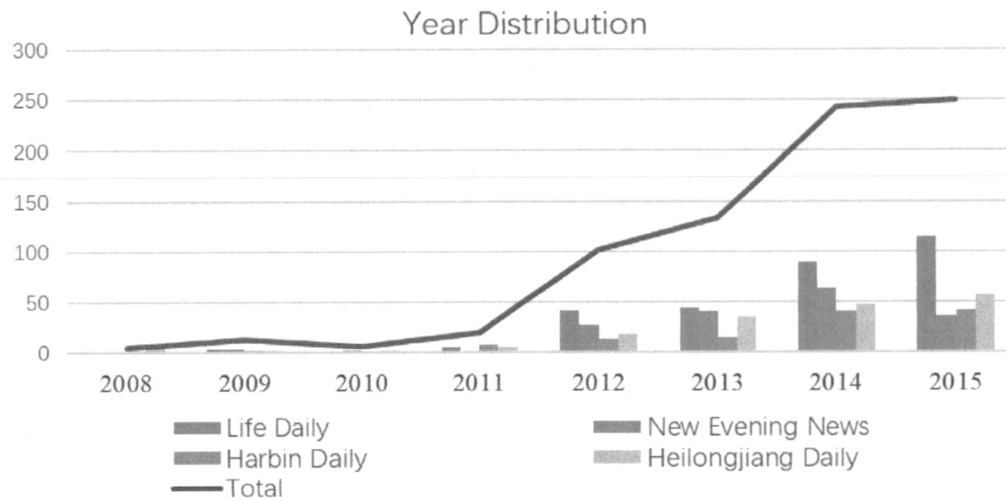


**Figure 1.** Reports' year distribution.

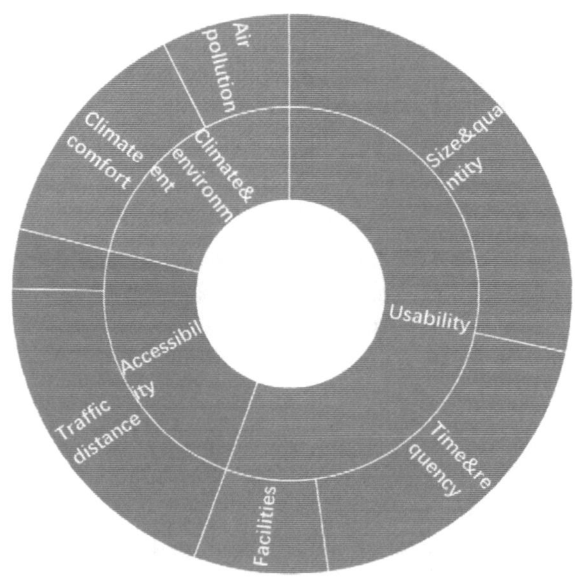


**Figure 2.** Reports' content distribution.

*3.1. Usability contradictions*

There are 194 reports covering the usability contradictions of the venues, 99 of which covered the size and quantity of the venues, 68 dealed with the time and frequency of space use, and 27 dealed with equipment and facilities in a venue.

*3.1.1. Size and quantity of venues.* In the reports on the size and quantity of the venues (Figure 3), 68 reflected the number of venues and 31 reflected the size of venues. The size and quantity of the square dance space are the main demands reflected in the reports. In a report, there were square dancers who danced in a community open space, feeling cramped and being reported by the surrounding residents because of the noise nuisance. The government later transformed a disused shanty town into a 2.1-hectare square, giving the square dancers a very good experience [6]. In addition, the study found that the study found that square dancers often occupied urban roads, car parks [7] and other venues like this. This not only made the square dancer's activity inconvenient, but also disturbed the normal

IOP Conf. Series: Materials Science and Engineering **371** (2018) 012040 doi:10.1088/1757-899X/371/1/012040

social order. This reflected the current number of square dance venuess was difficult to meet the needs of square dancers.

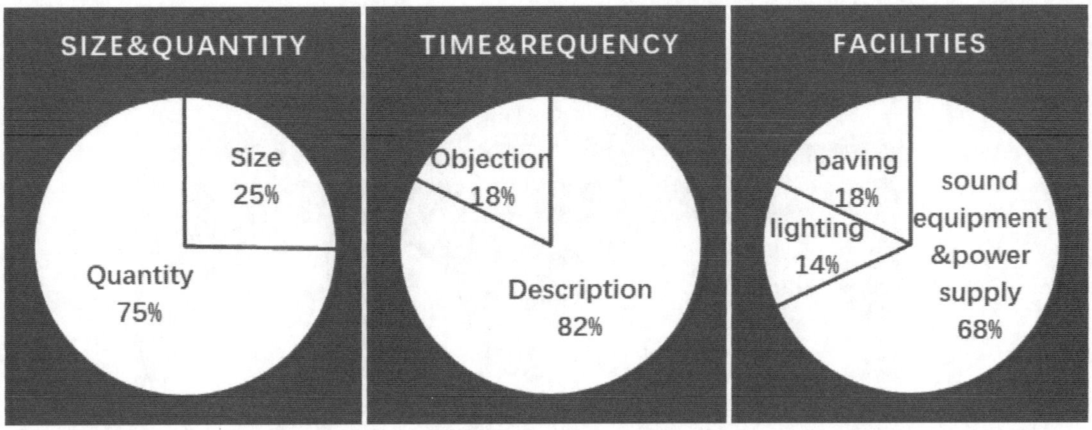


**Figure 3.** Quantitative distribution of reports on size and quantities of venues.

**Figure 4.** Quantitative distribution of reports on space use time and frequency.

**Figure 5.** Quantitative distribution of reports on facilities.

*3.1.2. Space use time and frequency.* In the reports on space use time and frequency (Figure 4), 56 objectively reflected the space use time and frequency of square dancers, and 12 objected to the rule of space use. Square dances are usually carried out regularly. As the square dancers are mainly middle-aged and elderly people, so the use of the site is usually in every morning and evening. However, there have been reports of objections to this practice, which suggested that the square dance would hinder the dancers from eating on time, thus negatively affecting the health of the square Dancers[8].

*3.1.3. Facilities.* Among the reports on facilities(Figure 5), 16 mentioned audio and power supplies, 6 mentioned site paving issues, and 5 mentioned site lighting issues. Audio equipment is a necessary soundtrack tool for square dances. For some organizers of square dance, the music equipment and power supply are important questions they considered at the beginning of their planning [9]. The venue facilities required by the square dancers include lighting and site paving. There were square dancers who thought that a good lighting environment will make them feel a lively atmosphere [10]. The pavement was closely related to the health of the square dancer, and the chilled concrete floor has a negative effect on the dancers ' legs and joints, especially in winter, when the cold and snow would make the situation worse [11], [12].

*3.2. Accessibility contradictions*

A total of 81 reports mentioned Accessibility contradictions, which involved the traffic distance 69, traffic safety 12.

*3.2.1. Traffic distance.* As a group activity, the square dance usually had a relatively fixed place of activity. Its group organization was divided by geographical space, so a square dance team was usually composed of residents living in the vicinity of active space. The vast majority of square dancers arrived at the arena by walking, with a small number of dancers arriving at the venues by bus [13]. Some square dancers would suspend the square dance or change the venue because the traffic was too far away. And the square dancers who traveled by bus would increase the public traffic pressure at rush hour [14]. But there was no accurate reference to the square Dancer's perception of the farthest distance of walking or bus.

*3.2.2. Traffic safety.* There were 12 reports mentioned square dancers had a traffic accident during the trip to and from the venue. This showed that the square dancers had traffic safety problems when commuting between home and venues, which to a certain extent, reduced the square dance venues of accessibility.

*3.3. Climate and environment contradictions*
There were 74 reports on climate and environment contradictions, including 48 reports on climate comfort, and 26 on air pollution.

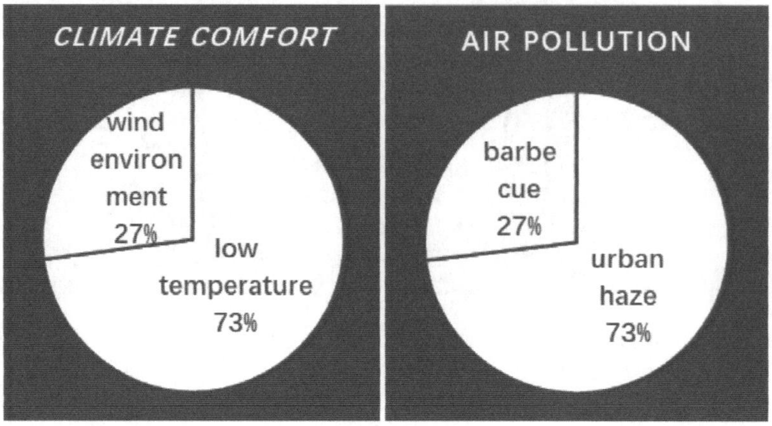


**Figure 6.** Quantitative distribution of reports on climate comfort.

**Figure 7.** Quantitative distribution of reports on Air pollution.

*3.3.1. Climate comfort.* The reports on climate comfort (Figure 6) mainly reflected two kinds of climate comfort, such as low temperature environment (35), wind environment (13). Winter mornings and evenings tend to have lower outdoor temperatures, which not only make people feel cold, but also increase the risk of joint injuries in square dancers. Therefore, it was reported that the winter square dance should be carried out as far as possible at noon, because the midday temperature is high and sunny enough to make up for the lack of sunshine caused by long time indoor activities. But the report did not verify the extent to which the square dancers were receptive to the proposal [8]. In the report, most of the problems surrounding the wind environment appeared in a riverside square in the early morning. Square dancers tended to be more inclined to the waterfront space activities, but t microclimate in the waterfront space is often not comfortable enough.But the report failed to arrive at the exact cause of the phenomenon.
*3.3.2. Air pollution.* The main sources of air pollution (Figure 7) affecting the square dance wrere barbecue (7) and urban haze (19). The former caused the participants to feel unwell during the square dance, while the latter directly affected the participants ' physical health seriously. There were square dancers had asthma attacks because of dancing in the haze. In the summer, some square dancers also complained about the serious air pollution in the venues near the open-air barbecue stall [15].

## 4. Discussion and conclusion

*4.1. urban planning*
For traffic distance, traffic safety, size and quantity of the site, adjusting urban planning and land use plan is an important method to solve these problems. Square dance usually has a relatively fixed place of activity and personnel composition. This makes it possible to determine the number of square dancers and the range of activities, so as to determine the location, size and quantity of the site, and to make rational allocation of resources through urban land use and planning methods. The objective of site selection is to control the traffic distance and to avoid factors that obstruct the square dance (such as the summer barbecue area) or reduce the accessibility of the site (the fast motorway on the way). In

addition, the urban land tension is also an important factor to limit the total space of the square dance. This requires the urban Planning department to find ways to increase the utilization of urban land, to make landscape regeneration (such as rebuilding obsolete shanty towns or factories) or vertical construction (such as building a rooftop plaza, etc.). And the regulation of the size and quantity of venues is based on the number of the square dance team in a certain area. And it is necessary to ensure satisfying the square dancers without causing the waste of urban land resources.

### *4.2. Space Activities Planning*

Square dance often occured in the city's multi-functional public space (such as city square, park, etc.). That is to say, the different activities in such spaces could have time and space conflicts. In addition, because the square dance needed music, it was difficult for different square dance teams to share space without interfering with each other. At the same time, in terms of health, there were still objections to the activity time (early morning and evening) of the square dancer (such as the low temperature of the morning in winter could damage the fitness of the square dancer). Therefore, the activity mechanism of square dancers still needs to be guided by more scientific norms. At present, some government departments have begun to register and manage the square dance teams in some areas of Harbin [16], [17]. This initiative can help planners to make reasonable site resource allocation. But this only alleviates the space use conflict between the different square dance teams, however the space use conflict between the other activities and the square dance has not received enough attention. This requires the investigation of potential activity types and their space use conflicts on the site, then to plan the space for the reasonable activity contents and time and put forward the management and guidance program.

### *4.3. Space design*

Space design is an important means to improve the utilization rate of space and enhance the user experience. Sound equipment, power supply, good lighting environment, suitable site paving, comfortable micro-climate can improve the square dancers' usage experience, reduce the negative impact of the site on the physical and mental health of square dancers. For the site designers, by means of design, add sound and power, improve the lighting environment and paving quality is relatively easy to complete, and the regulation of micro-climate environment, such as wind environment, is relatively complex, which need more research support.

## 5. References

[1] Liang Qin-chao, Li Yuan and Shi Zhen-guo 2016 The deep reason and the control of "the public square dance disturbing the public" *Journal of Wuhan Institute of Physical Education* **39** pp 36-41.

[2] Wang Qian-ni 2015 Stigma and conflict: square dance in the middle of times *Culture Crossings* 02(2014) pp 74-81

[3] Yuan Ji-fang and Chen Jian-guo 2014 The lack of public space in urban sports recreation from the square dance *Journal of Wuhan Institute of Physical Education* **48** pp 34-38.

[4] Zhang Xin-si, Liu Ming-hui and Zhao Li-na 2014 "Square Dance Contradiction" and the planning and management of urban public cultural space *Chinese Landscape Architecture* **30** pp112-5

[5] http://mediasearch.meihua.info/ListingHybrid.htm?q=--1-1-False-False-True-8,_,1,_,,0,0_,False,2,0,,-,_,1,_,0,_,False,,2,0-,_,1,,_,0_,False,1,0,,-,_,1,,_,0_,False,1,0,,-,_,1,,,_,_,0,0-,_,1,2,0,0-,_,1,_,0,0,0-,_,1,0,0_,0-1-

[6] Wang Yan-lin 2014 Generous inside landscape woodland and Harbin East Street garden has been put into use *Daily Life* (Harbin: Heilongjiang Daily) **4 August** p 2

[7] Duan Xiao-ning and Zhao Wei 2015 Pushing three cars for the square dance *Daily Life* (Harbin: Heilongjiang Daily) 2**4 July** p C15

[8] Qian Long 2015 Winter square dance with health tips *New Evening News* (Harbin: Harbin Daily) **29 January** p B06

IOP Conf. Series: Materials Science and Engineering **371** (2018) 012040 doi:10.1088/1757-899X/371/1/012040

[9] Jiao Bo 2012 Three Ms. Liu make the joyful dance popular *Daily Life* (Harbin: Heilongjiang Daily) **17 May** p 48
[10] He Xing-li and Li Wei-bing 2014 In August the central street area changed a lot *Daily Life* (Harbin: Heilongjiang Daily) **31 July** p 06
[11] Anon 2015 How to protect the joints *Daily Life* (Harbin: Heilongjiang Daily) **24 September** p B08
[12] Zhang Jie 2015 10 miles a day women two months practicing bad knees *Daily Life* (Harbin: Heilongjiang Daily) **7 July** p A07
[13] Lu Ming 2015 89-line bus turns rolling down aisle middle-aged woman *Daily Life* (Harbin: Heilongjiang Daily) **23 July** p A13
[14] Zhan Xin-yu 2014 Morning and evening traffic peak of the elderly accounted for 20% *Daily Life* (Harbin: Heilongjiang Daily) **26 November** p 08
[15] Tian Di and Yuan Ye 2015 Air pollution around outdoor barbecue stalls *New Evening News* (Harbin: Harbin Daily) **8 July** p A03
[16] Liu Ying 2015 Harbin Xiangfang District Square dance music does not exceed 80 db *Daily Life* (Harbin: Heilongjiang Daily) **4 March** p A04
[17] Yu Hai-xia 2014 160 fitness spots to keep you moving *Daily Life* (Harbin: Heilongjiang Daily) **29 August** p A21.

IOP Conf. Series: Materials Science and Engineering **371** (2018) 012041 doi:10.1088/1757-899X/371/1/012041

# Support Draft Calculation for a Ramp in the Form of Developable Helicoid

**M Rynkovskaya**[1]

1 Associate Professor, Peoples' Friendship University of Russia (RUDN University), Moscow, Russia
E-mail: rynkovskaya_mi@pfur.ru

**Abstract.** The article is about the analytical method of calculation a ramp in the form of developable helicoid on the support draft. The asymptotic method of small parameter is applied to solve the system of three differential equilibrium equations for developable helicoid stress-strain. The numerical results of displacements and bending moments are verified and coincide with engineering practice. The suggested approach can be extended for calculation of torso-helicoids with other boundary conditions.

## 1. Introduction

According to [1] the helical ruled surfaces may be divided into five types: oblique helicoids; right helicoids (a particular case of oblique helicoids, when inclination angle of a generatrix is equal to zero); developable helicoids; convolute helicoids; pseudo-developable helicoids (or right convolute helicoid that is a particular case of convolute helicoid). Developable helicoids can also be called evolvent helicoids, for example in [2], or torso-helicoids, for example in [3], or involute helicoids [4].
Developable helicoids are mostly used in mechanical engineering in metal elements of small sizes [5], [6], where plastic deformations are needed to be taken into account [7]; however, they can also be successfully used in architecture and civil engineering for designing ramps, helical elements, helical parts of car interchanges, geometric models of embankment slopes when the road is raised and rounded (as a surface of the equal slope) [8]; and even as parts of the whole buildings, as it was done in the building of the famous Guggenheim's museum of modern art in New York (USA) by an architect Frank Lloyd Wright [9].
In this paper, there is an attempt to solve a particular practical task by the means of analytical methods of calculation applied to developable helicoids.
A developable helicoid (figure 1) is a developable surface formed by tangents to the helical line of a constant step on a circular cylinder of radius a. The development of the developable helicoid on the plane is an annular region bounded by coaxial circles [10].
Therefore we can use the theory of thin elastic shells and calculate the ramp in the form of a long developable helicoid using the asymptotic method of a small parameter. This method is described in more details in the papers [11], [12]. The aim of this study is to show the analytical approach to the ramps in the form of developable helicoids calculation on a support draft.

## 2. Asymptotic Method of Small Parameter for Developable Helicoid

The most popular parametric equations for developable helicoids (figure 1) are the following [10]:

$$x = x(u,v) = a\cos v - \frac{au\sin v}{m},$$
$$y = y(u,v) = a\sin v + \frac{au\cos v}{m},$$
$$z = z(u,v) = bv + \frac{bu}{m}, \qquad (1)$$

where $m = \sqrt{a^2 + b^2}$,
$a$ is radius of the cylinder, tangents to which are the generators of a developable helicoids;
$b$ is a hitch of a curve $u = 0$ (return edge);
$v$ is an angle deflected from axis $Ox$.

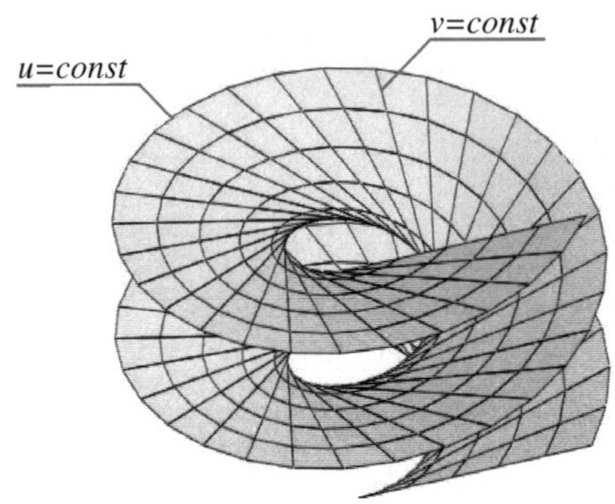


**Figure 1.** Developable helicoid

We determine the equations of developable helicoids as follows

$$x = x(u,s) = a_0\cos^2\varphi\left(\cos\frac{s}{m} - \frac{u}{m}\sin\frac{s}{m}\right),$$
$$y = y(u,s) = a_0\cos^2\varphi\left(\sin\frac{s}{m} + \frac{u}{m}\cos\frac{s}{m}\right),$$
$$z = z(u,s) = (s+u)\sin\varphi, \qquad (2)$$

Here $u$, $s$ are non-orthogonal conjugate coordinate lines (the coordinate lines $u$ coincide with the rectilinear generatrixes of the torso-helicoid, and the lines $s$ are equidistant helical lines of the surface);
$m = a_0\cos\varphi$;
$a = a_0\cos^2\varphi$;
$b = a_0\sin\varphi\cos\varphi$;
$a_0$ is a radius of a helical evolvent;
$\varphi$ – is inclination angle between the generatrix and the plane;
$tg\varphi = b/a$;
$s = mv$;
$s$ is an arc of a helical cuspidal edge.
The asymptotic method of small parameter in application to developable helicoids was introduced by Krivoshapko in [11] and developed by Rynkovskaya in [12], [13].
We can use here the system of three ordinary differential equations in dimensionless displacements $U$, $V$, $W$ obtained by Krivoshapko [11] for calculation of thin elastic shells in the form of long torso-helicoids subject to the dead load:

$$\frac{d}{d\alpha}\left[\alpha^3\frac{d}{d\alpha}\left(\frac{U}{\alpha}\right)\right]=-\frac{a_0\alpha^4}{CB^2}X+\mu\left[W-\frac{\alpha}{B^2}\left(1+v\alpha^2\right)\frac{dW}{d\alpha}\right]=-\frac{a_0\alpha^4}{CB^2}X+\mu E,$$

$$\frac{B^4}{\alpha}\frac{dV}{d\alpha}=-\frac{B^2}{\alpha}\frac{dU}{d\alpha}+2U+2\mu W-\frac{2a_0}{(1-v)C}\int\alpha(BY+X)d\alpha+A_1,$$

$$\frac{d}{d\alpha}\left\{\frac{t}{\alpha}\frac{d}{d\alpha}\left[\frac{B^4}{\alpha}\frac{d}{d\alpha}\left(\frac{1}{\alpha}\frac{dW}{d\alpha}\right)\right]+\mu\frac{1-v}{2\alpha}\left(\frac{dU}{d\alpha}+B^2\frac{dV}{d\alpha}\right)+\mu\alpha\frac{dU}{d\alpha}+\mu v(U+\mu W)\right\}+\frac{a_0\alpha}{C}\left(\mu X+\mu\frac{Y}{B}-Z\right)=0. \tag{3}$$

$U$, $V$, $W$ are dimensionless parameters of displacements;
$X$, $Y$, $Z$ are components of surface distributed load (analogue to dead load);

$$\alpha=\frac{u}{a_0};\; U=\frac{U_u}{a_0};\; V=\frac{U_s}{a_0B};\; W=\frac{U_z}{a_0};\; \mu=tg\varphi;\; t=\frac{h^2}{12a_0^{\,2}}; \tag{4}$$

where $U_u, U_s, U_z$ are vector components of elastic displacement of the shell middle surface (with the width $h$).
Dimensionless parameters of displacements $U$, $V$, $W$ may be expressed from Eq. 3 as follows:

$$U=\mu\frac{a_0Z}{C}\left(\frac{\alpha^2}{3}+\frac{1}{2}-\frac{B^2}{2\alpha}arctg\alpha\right)+\mu\alpha K_2+\frac{A_2}{\alpha}+\alpha A_3,$$

$$E=\left[W-\frac{\alpha}{B^2}\left(1+v\alpha^2\right)\frac{dW}{d\alpha}\right],$$

$$V=\mu\int\left[\left(\alpha^2-1\right)\frac{K_2}{B^4}-\frac{K_1}{\alpha^2B^2}+\frac{2\alpha}{B^4}W\right]d\alpha+\mu\frac{a_0}{2C}\left[\frac{1}{B^2(1-v)}-\frac{1}{3B^2}+\frac{\ln B^2}{1-v}+\right.$$
$$\left.+\frac{1}{\alpha}arctg\alpha\right]Z-\frac{A_1}{2B^2}-\frac{A_2}{\alpha B^2}-\frac{\alpha}{B^2}A_3+A_4,$$

$$tW=\frac{a_0Z}{64C}\alpha^4+\mu\left\{\frac{(1-v)}{32}A_1\left(\ln B^2\right)^2+\frac{(1-v)}{4}\left(A_2-A_3\right)\left[arctg\alpha\left(\ln B^2+\frac{7}{2}+\frac{\alpha^2}{2}\right)-\right.\right.$$
$$\left.\left.-\frac{7}{2}\alpha-\int\frac{\ln B^2}{B^2}d\alpha\right]+\frac{(1+v)}{3}A_3\left[\frac{arctg\alpha}{4}\left(3\alpha^2+5\right)-\frac{5}{4}\alpha-\frac{\alpha^3}{3}\right]\right\}+$$
$$+\mu^2\frac{a_0Z}{6C}\left\{(1+v)\left[\alpha arctg\alpha\left(\frac{\alpha^2}{3}+\frac{5}{4}\right)-\frac{1}{8}(arctg\alpha)^2\left(5+3\alpha^2\right)\right]-(2v+1)\frac{\alpha^4}{32}-\right. \tag{5}$$
$$\left.-\frac{5}{16}\left(\ln B^2\right)^2\right\}+\mu^2\int\left\{\alpha\int\left[\frac{\alpha}{B^4}\int\left(\frac{1-v}{B^2}\alpha^4K_2+\alpha^2K_2-\frac{1+v\alpha^2}{B^2}\alpha W\right)d\alpha-\right.\right.$$
$$\left.\left.-\frac{\alpha^4}{B^4}K_2\right]d\alpha\right\}d\alpha+C_1B^2\ln B^2+C_2\ln B^2+C_3\alpha^2+C_4$$

Since we use here the method of small parameter, the inclination angle between the generatrix and the plane must be $\varphi < 45^0$ ($\mu = tg\varphi < 1$), and solutions $U$, $V$, $W$ are found in the form of series in powers of the small parameter $\mu$:

$$U=U(\alpha,\mu)=\sum_{k=0}^{\infty}U_k(\alpha)\mu^k,$$

$$V=V(\alpha,\mu)=\sum_{k=0}^{\infty}V_k(\alpha)\mu^k,$$

$$W=W(\alpha,\mu)=\sum_{k=0}^{\infty}W_k(\alpha)\mu^k, \tag{6}$$

where $U_k, V_k, W_k$ are vector coefficients to be found.

IOP Conf. Series: Materials Science and Engineering **371** (2018) 012041 doi:10.1088/1757-899X/371/1/012041

The small parameter method is used to solve mechanical and physical problems with differential equations containing certain parameters in cases when it is possible to find a particular solution of the differential equation for certain fixed values of these parameters that satisfy the initial conditions [14]. Substituting expressions (Eq. 6) into equations (Eq. 5), and equating consecutively free members and coefficients for powers $\mu$, we obtain a system of $(3+3k)$ equations for computing $U_k, V_k, W_k$. For example, to determine the vector coefficients $U_0, V_0, W_0$, it is necessary to take in equations (Eq.5) that $\mu = 0$. As a result, we find solutions of the generating equations:

$$U_0 = \frac{A_{20}}{\alpha} + \alpha A_{30},$$

$$V_0 = -\frac{A_{10}}{2B^2} - \frac{A_{20}}{\alpha B^2} - \frac{\alpha}{B^2} A_{30} + A_{40},$$

$$W_0 = a_0{}^3 Z\alpha^4 / (64D) + C_{10}B^2 \ln B^2 + C_{20} \ln B^2 + C_{30}\alpha^2 + C_{40}. \quad (7)$$

## 3. Application for calculation of a support draft

We consider a long shallow developable helicoid with the following geometric characteristics (in the form of parametric equations Eq.1): $a = 3$ m; $b = 0{,}2$ m; $u_1 = 4$ m; $u_2 = 6$ m; $h$=0,01 m.

If we use the parametric equations Eq.2, these characteristics correspond to the slope angle of the rectilinear generatrix $\varphi = 0{,}06657\ rad = 3{,}81418\ ^0$; the inner radius $R1 = 4{,}993$ m, the external radius $R2 = 6{,}696$ m, the radius of the a radius of a helical evolvent of the torso-helicoid return to the plane $a_0 = 3.0133$ m.

Let us determine modulus of elasticity $E = 2 \cdot 10^8\ kPa$ and Poisson's ration $\nu = 0{,}3$.

We will assume that the surface load is $P_x = P_y = P_z = 0$, therefore, $X = Y = Z = 0$; and determine that the curvilinear edges $\alpha = \alpha_1$ and $\alpha = \alpha_2$ are rigidly clamped.

Let us calculate the stress-strain state of the developable helicoid when the internal curvilinear contour $\alpha = \alpha_1$ is displaced along the fixed $z$-axis by

$$\delta = \frac{u_z m^2}{a} = \frac{u_z \sqrt{a^2 + b^2}}{a}. \quad (8)$$

We will assume $u_z(\alpha = \alpha_1) = 0.005$ m.

For a developable helicoid determined by parametric equations Eq.1, when the return-edge equations of the developable helicoid are written in the form

$$x = a\cos v,$$
$$y = a\sin v,$$
$$z = bv, \quad (9)$$

angles between the fixed axis $z$ and the directions of displacements $u_u$, $u_v$, $u_z$ are

$$\cos(z_н \wedge u_u) = \frac{z'}{F} = \frac{b}{F},$$

$$\cos(z_н \wedge z_n) = \cos(z_н \wedge u_z) = \frac{a}{F}, \quad (10)$$

where $F = \sqrt{a^2 + b^2}$.

The angle between the tangent to the coordinate line $v$ and the fixed axis $z$ is found from the vector expression

$$e_v = \frac{r_v}{B} = \cos(x_н \wedge u_v)i + \cos(y_н \wedge u_v)j + \cos(z_н \wedge u_v)k,$$

where $e_v$ is the unit vector that is tangent to the coordinate line $v$;

$r_v = \frac{\partial r}{\partial v}$, where $r$ is a radius vector; and the vector equation of the surface of the developable helicoids is

$$r = r(u, v) = \rho(v) + ul(v),$$

where $\rho(v)$ is a current radius vector of a return edge:

$$\rho(v) = x(v)i + y(v)j + z(v)k,$$

$l(v)$ is the unit tangent vector given at each point of the return edge:

$$l(v) = \frac{x'(v)i + y'(v)j + z'(v)k}{\sqrt{x'^2(v) + y'^2(v) + z'^2(v)}}.$$

Then $\cos(z_н \wedge u_v) = \frac{b}{B}$, where $B = \sqrt{F^2 + u^2a^2 / F^2}$.

However, for the shape of the developable helicoid in the parametric form (Eq. 2), when the length of the arc $s$ of the return edge is taken as the parameter $v$, the formulas for the angles between the fixed $z$ axis and the displacement directions $u_u$, $u_s$, $u_z$ (Eq. 10) can be written in the following form:

$$\cos(z_н \wedge u_u) = \sin\varphi,$$

$$\cos(z_н \wedge z_n) = \cos(z_н \wedge u_z) = \cos\varphi. \qquad (11)$$

The angle between the tangent to the arc length of the return edge $s$ and the fixed axis $z$ is found from the vector expression

$$e_s = \frac{r_s}{B} = \cos(x_н \wedge u_s)i + \cos(y_н \wedge u_s)j + \cos(z_н \wedge u_s)k,$$

where $r = r(u, s) = \rho(s) + ul(s)$,

$\rho(s) = x(s)i + y(s)j + z(s)k$,

$l(s) = \rho'(s) = x'(s)i + y'(s)j + z'(s)k$.

We obtain here

$$\cos(z_н \wedge u_s) = \frac{a_0 \sin\varphi}{\sqrt{a_0^2 + u^2}},$$

$$\cos(z_н \wedge u_s) = \frac{\sin\varphi}{\sqrt{1+\alpha^2}} = \frac{\sin\varphi}{B}, \qquad (12)$$

$$B = \sqrt{1+\alpha^2}. \qquad (13)$$

The boundary conditions for this task will be written as:

$$W_i(\alpha = \alpha_2) = 0,\ W'_i(\alpha = \alpha_2) = 0,\ U_i(\alpha = \alpha_2) = 0,\ V_i(\alpha = \alpha_2) = 0, \qquad (14)$$

$$W_i(\alpha = \alpha_1) = \frac{u_z}{a_0} = \frac{\delta\cos(z_н \wedge u_z)}{a_0} = \frac{\delta\cos\varphi}{a_0},$$

$$W'_i(\alpha = \alpha_1) = 0,$$

$$U_i(\alpha = \alpha_1) = \frac{u_u}{a_0} = \frac{\delta\cos(z_н \wedge u_u)}{a_0} = \frac{\delta\sin\varphi}{a_0},$$

$$V_i(\alpha = \alpha_1) = \frac{u_s}{a_0 B} = \frac{\delta\cos(z_н \wedge u_s)}{a_0 B} = \frac{\delta\sin\varphi}{a_0 B}, \qquad (15)$$

where $B$ is determined in Eq. 13.

Equations (14) show that the curvilinear outer edge $\alpha = \alpha_2 = const$ of the developable helicoid is fixed and rigidly clamped. Equations (15) allow the rigidly constrained curvilinear contour $\alpha = \alpha_1 = const$ move along the fixed $z$ axis by a value $\delta$.

To solve the task, we can apply the method of a small parameter, for instance with the first terms of the series

$$U = U_0,\ V = V_0,\ W = W_0.$$

Dimensionless parameters of the first terms of the series $U_0, V_0, W_0$ are obtained from Eq. 7 as follows:

$$U_0 = \frac{A_{20}}{\alpha} + \alpha A_{30},$$

$$V_0 = -\frac{A_{10}}{2B^2} - \frac{A_{20}}{\alpha B^2} - \frac{\alpha}{B^2}A_{30} + A_{40},$$

$$W_0 = C_{10}B^2 \ln B^2 + C_{20} \ln B^2 + C_{30}\alpha^2 + C_{40}. \qquad (16)$$

To calculate the constants $C_{10}, C_{20}, C_{30}, C_{40}$, we apply the boundary conditions Eq. 14, 15 as follows:

IOP Conf. Series: Materials Science and Engineering **371** (2018) 012041 doi:10.1088/1757-899X/371/1/012041

$$W_0 = \frac{\partial W_0}{\partial u} = 0, \text{ when } \alpha = \alpha_2 = \frac{u_2}{a_0}, \text{ and } u_2 = 6 \text{ m};$$

$$W_0 = \frac{\delta \cos\varphi}{a_0}, \ \frac{\partial W_0}{\partial u} = 0, \text{ when } \alpha = \alpha_1 = \frac{u_1}{a_0}, \text{ and } u_1 = 4 \text{ m}.$$

To calculate the constants $A_{10}, A_{20}, A_{30}, A_{40}$, we use the following conditions

$$U_0(\alpha = \alpha_2) = 0,$$

$$V_0(\alpha = \alpha_2) = 0,$$

$$U_0(\alpha = \alpha_1) = \frac{\delta \sin\varphi}{a_0},$$

$$V_0(\alpha = \alpha_1) = \frac{\delta \sin\varphi}{a_0 B}.$$

Finally, for the first term in the series, the expressions for the bending moments in this method will be written in the following form:

$$M_u = -\frac{D}{a_0\alpha}\frac{d}{d\alpha}\left[\frac{B^2}{\alpha}\frac{dW_0}{d\alpha} - (1-\nu)W_0\right]. \tag{17}$$

## 4. Test numerical experiments and discussion

It is necessary to point out that this task is solved in the "pseudo-moments" $M_u$* (the asterisk ... * in Eq. 17 is not shown conditionally), because the equilibrium equations suggested by Goldenveiser A.G. [15] are used. There is an interrelation between "pseudo-moments" and traditional engineering moments, and it can be converted into traditional internal moments $M_u$, using the formula:

$$M_u = -M_u{}^* - M_{us}{}^* \cos\chi, \tag{18}$$

where $\cos\chi = \frac{F}{AB} = \frac{1}{B} = \frac{1}{\sqrt{1+\alpha^2}}$, and $\chi$ is an angle between coordinate lines $u$ and $v$.

As it is shown in Fig. 2, the normal displacements are distributed in accordance with engineering practice, and the boundary conditions are satisfied.

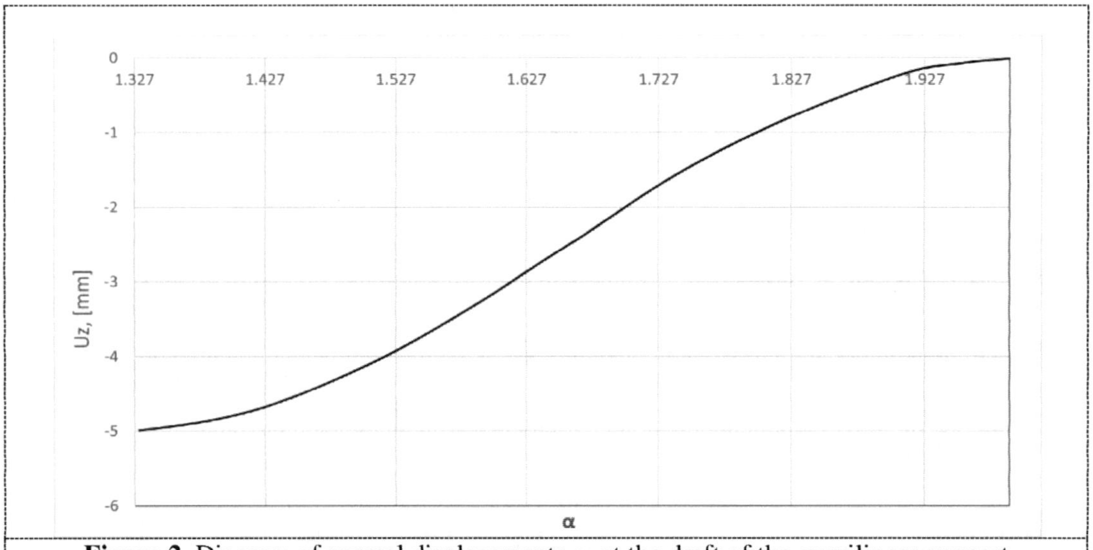


**Figure 2.** Diagram of normal displacements $u_z$ at the draft of the curvilinear support

Bending moments are shown in Fig. 3.

The results shown in Fig. 2, 3 correlate to results obtained by Krivoshapko S.N. [11] for the developable helicoid in the form of Eq.1.

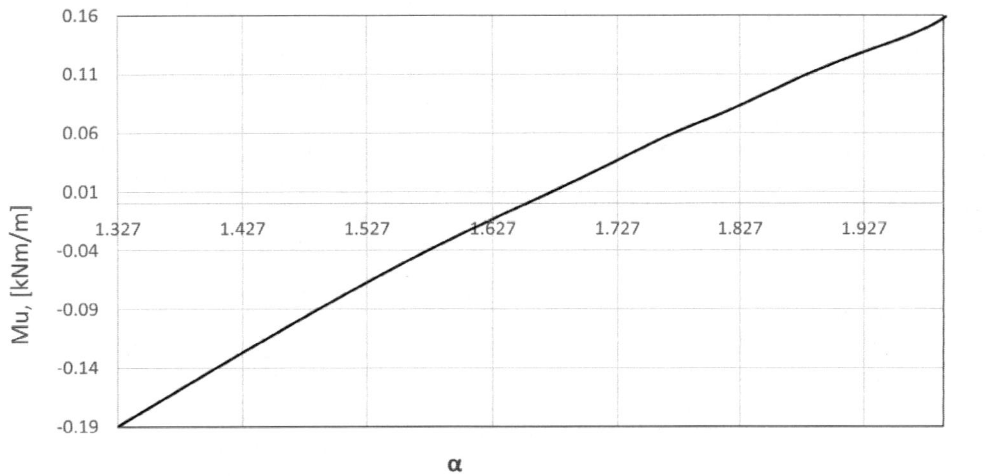


**Figure 3.** Diagram of bending moments $M_u$ at the draft of the curvilinear support

## 5. Conclusion

There is the investigation on the analytical approach to calculation of a developable helicoid curvilinear support draft which is based on the three ordinary differential equations in dimensionless displacements *U, V, W* solved by asymptotic method of small parameter for the developable helicoid determined through the axes $u$ and $s$, and a radius of a helical evolvent $a_0$ (Eq. 2).

Numerical experiments are conducted. The results for normal displacements and bending moments obtained by this method are shown and they correlate to the engineering practice and numerical results obtained by similar analytical method applied to the developable helicoid determined through the axes $u$ and $v$, and a radius of the cylinder $a$ (Eq. 1).

Despite the fact that the task is solved in "pseudo-moments", they can be easily transverse into traditional engineering bending moments by the compact formula.

It is shown that the analytical methods can also be successfully used for calculation of the complex form surfaces. As a future research, it would be interesting to solve the task using more members of series, as well as the task for developable helicoid with other boundary conditions.

## 6. References

[1] Krivoshapko S N and Rynkovskaya M 2017 Five Types of Ruled Helical Surfaces for Helical Conveyers, Support Anchors and Screws *MATEC Web of Conferences* **95**, 06002 (2017)

[2] Romanova V A and Thoma A 2017 Automatic modeling of the surfaces of the equal slope in AutoCAD system through language autolisp *Structural Mechanics of Engineering Constructions and Buildings* **5** p 5-11

[3] Krivoshapko S N 1998 Geometry and strength of general helicoidal shells *Appl Mech Rev* **52 5** p 161-175

[4] Dukkipati R V 2007 Mechanism and Machine Theory *New Age International Publisher* p 624

[5] Anisimov R and Tarapanov A 2017 Design of the tool for periodic not evolvent profiles *MATEC Web of Conferences* **129** 01039

[6] Lewis J R, Mccurdy A and Gaudet A D 1985 Laser printing system with developable helicoids reflector *US Patent 4494821*, 01.22.1985.

[7] Rynkovskaya M 2017 Analysis of displacements in beam structures and shells with middle developable surfaces *MATEC Web of Conferences* **108** 16001

[8] Kartashev A I 1954 Surface of the same slope: Diss. Cand. tech. sciences. - L .: LIIZhT, 1954.

[9] Letters to Architects 2nd Edition 1986 *Architectural Press* p 238

[10] Krivoshapko S N and Ivanov V N 2015 Encyclopedia of Analytical Surfaces *Springer* p 752

[11] Krivoshapko S N 2009 Geometry of ruled surfaces with an edge of a return and a linear theory of calculating torso shells *RUDN* p 357.
[12] Rynkovskaya M I 2009 On application of Bernoulli's numbers for thin torso-helicoids calculated by asymptotic method of small parameter *MOO Spatial Structures* **12** p 59-64
[13] Rynkovskaya M I 2015 Influence of the Poisson's Ratio on the Accuracy of the Results of Calculating the Torso-Helicoid Stress-Strain *Construction and Reconstruction* **4 (60)** p 51-56
[14] Berezin I S and Zhidkov N P 1959 Methods of calculation *Fimatgiz* p 620
[15] Goldenweiser A L 1961 Theory of Elastic Thin Shells *Pergamon Press* p 680

**Acknowledgments**
This paper was financially supported by the Ministry of Education and Science of the Russian Federation on the program to improve the competitiveness of Peoples' Friendship University of Russia (RUDN University) among the world's leading research and education centres in the 2016-2020. This publication was prepared with the support of the "RUDN University Program 5-100".

IOP Conf. Series: Materials Science and Engineering **371** (2018) 012042 doi:10.1088/1757-899X/371/1/012042

# Generic Vertical Frame Model for Estimation of Floor Vertical Vibration

**J B Sun[1, 2] and P Pan[2]**

1 Institute of Engineering Mechanics, China Earthquake Administration; Key Laboratory of Earthquake Engineering and Engineering Vibration of China Earthquake Administration, China
2 Key Laboratory of Civil Engineering Safety and Durability of the China Education Ministry, Tsinghua University, Beijing 100084, China
E-mail: panpeng@tsinghua.edu.cn

**Abstract.** Seismic-isolation is an appealing design alternative for enhancing both the seismic safety and functionality of building structures against large earthquake motions. However, large floor vertical accelerations was observed in the conventionally designed base-isolated buildings. Previous researches indicate that column and beam vibrations should be considered when evaluating the floor vertical accelerations. The column and beam vibrations lead the representation of the Single Degree of Freedom system and corresponding spectra analysis ineffective. To evaluate the vertical floor vibration of building structures, the generic vertical frame (GVF) model was developed in this study. Comparing with the conventionally frame analysis model, the GVF model significantly reduced the DOFs. The GVF model is capable of estimating the vertical acceleration at interior spans for multiple span structures with errors not more than twenty percent, and becomes more accurate for structural systems with strong columns.

## 1. Introduction

Seismic-isolation has been accepted as an appealing design alternative for enhancing both the seismic safety and functionality of building structures against large earthquake motions. Since base-isolation techniques essentially isolate only against horizontal motions, effects of vertical motions on the base-isolated structure becomes a concern. Previous researches indicate that conventionally designed base-isolated buildings have a large margin of safety against vertical ground motions, but maximum floor vertical accelerations can reach as large as more than five times of the peak ground acceleration. Floor vertical accelerations are closely related to structural functionality, thus quantification of acceleration amplifications is critical [2], [3]. In the dynamic response analysis of structures subjected to vertical ground motion, detailed finite element analysis is often required to take into account floor vibration. It is rather time consuming for both modeling and computation. To generalize the behavior of floor vertical vibration, simplified models that can preserve the prototype's vertical vibration characteristics are recommended [1]. Previous researches indicate that two types of vibrations should be considered when evaluating the floor vertical accelerations. One is the column vibration featured by column elongation and contraction, and the other is the beam vibration featured by beam deflection. The column and beam vibrations are coupled in many modes, leading the representation of the SDOF (Single Degree of Freedom) system and corresponding spectra analysis ineffective. To this end, a model, named the generic vertical frame (GVF) model, is developed. In this paper, development of the GVF model is detailed, and time history analyses are carried out to validate the effectiveness.

## 2. Development of Generic Vertical Frame Model

### *2.1. Vertical full structure, substructure, and generic frame models*

A full vertical frame model (designated as the FVF model hereafter) is given in Fig. 1(a), and the corresponding substructure vertical frame models (designated as the SVF model hereafter) [1] are constructed as in Fig. 1(b). The horizontal movements are restricted for both FVF and SVF models. In the SVF model, all middle columns and associated half beams are included, and the boundary condition at the end of the half-beams is taken to slide without rotation. This is justified because the maximum deflection is located approximately at the middle of the beams where the shear force is nearly zero. The SVF model is further simplified to a GVF model. Specifically, the floor vibration is simplified to an equivalent SDOF (single degree of freedom) system. A hanging spring and a hanging mass are used to represent the equivalent SDOF system. A schematic illustration of the GVF model is given in Fig. 1(c). In the GVF model, only vertical DOFs are considered. The hanging spring with a stiffness of $K_H$ and a mass of $M_H$ are connected at the beam-to-column joint. The column is also represented by a spring with a stiffness of $K_C$, and a mass of $M_C$ is lumped at the beam-to-column joint. Values of $K_H$, $M_H$, $K_C$, and $M_C$ are derived in the following discussions.

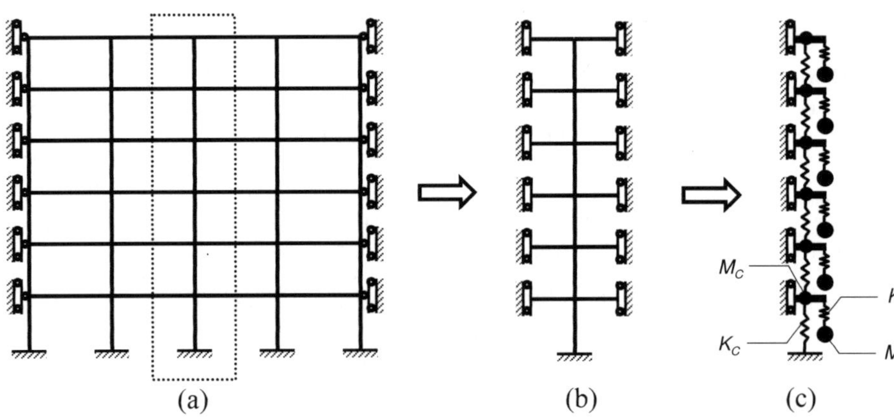


**Figure 1.** Full vertical frame (FVF), substructure vertical frame (SVF), and generic vertical frame (GVF) models: (a) FVF; (b) SVF; (c) GVF.

### *2.2. Assignment of mass and stiffness*

Rayleigh's method [4], [5] is adopted to determine the equivalent stiffness and equivalent mass of the hanging part. For explanation, one story of a SVF model is examined in Fig. 2. Since the middle column deflection is negligible, only the beam deflection is considered here.

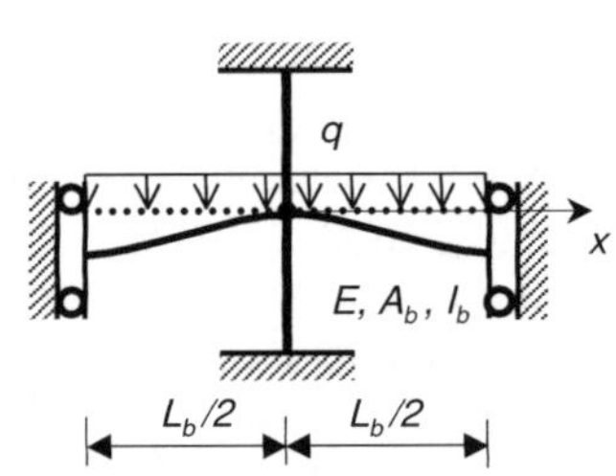


**Figure 2.** One story of SVF model for interior span.

The defected shape function $\psi_b(x)$ resulting from a uniformly distributed vertical load $q$ is given in Equation (1):

$$\psi_b(x) = \frac{q}{24EI_b}(x^4 - 2L_b x^3 + L_b^2 x^2) \tag{1}$$

where, $E$ is the modulus of elasticity, $I_b$ is the second moment of area of the beam, $L_b$ is the length of the beam, and $x$ is the coordinate along the beam (Fig. 2).
The generalized mass is given in Equation (2):

$$M_{in} = 2 \times \int_0^{\frac{L_b}{2}} m(x)\left(\psi_b(x)\right)^2 dx \tag{2}$$

Where, $m(x)$ is mass per unit length of the beam. Since the model is symmetrical about the column, the generalized mass of the left half-beam equals to that of the right half-beam.
The generalized stiffness is given in Equation (3):

$$K_{in} = 2 \times \int_0^{\frac{L_b}{2}} EI(x)\left(\psi_b''(x)\right)^2 dx \tag{3}$$

The circular frequency $\omega_{in}$ of the equivalent SDOF system is given in Equation (4):

$$\omega_{in} = \sqrt{\frac{K_{in}}{M_{in}}} \tag{4}$$

The participation factor $\beta_{in}$ is expressed in Equation (5):

$$\beta_{in} = \frac{L_{in}}{M_{in}} \tag{5}$$

where $L_{in}$ is defined in Equation (6):

$$L_{in} = 2 \times \int_0^{\frac{L_b}{2}} m(x)\psi_b(x)dx \tag{6}$$

The equivalent mass $M_{in}^*$ and stiffness $K_{in}^*$ are given Equation (7) and (8), respectively:

$$M_{in}^* = \beta_{in}{}^2 M_{in} \tag{7}$$

$$K_{in}^* = M_{in}^* \omega_{in}^2 \tag{8}$$

Note that the subscript "*in*" in Equations (2) to (8) refers to interior spans. The equivalent SDOF system that represents the floor consists of the hanging spring and mass in the GVF model, thus for the interior span:

$$M_H = M_{in}^* \tag{9}$$

$$K_H = K_{in}^* \tag{10}$$

Since only the vertical direction is considered, the stiffness $K_C$ of the spring that represents the column is as follows:

$$K_C = \frac{EA_C}{L_C} \tag{11}$$

where, $A_C$ is the cross-section area of the column, and $L_C$ is the length of the column.
To keep the total mass of the equivalent system equal to the original system, the mass $M_C$ lumped at the beam-to-column joint is assigned as:

$$M_C = 2\times\int_0^{\frac{L_b}{2}} m(x)dx - M_H \tag{12}$$

Note that in Equations (9) to (12), the subscripts "*H*" and "*C*" represent the parameters related to the hanging spring system and original column system, respectively.

### *2.3. Correlation on Floor Vertical Acceleration*

The following equations are used to correlate the floor vertical accelerations of the GVF model with those of the FVF and SVF models:

$$Acc(x) = Acc_C + \beta\times(Acc_H - Acc_C)\times\psi_b(x) \tag{13}$$

In Equation (13), $Acc_C$ is the acceleration at the beam-to-column joint in the GVF model, $Acc_H$ is the acceleration of the mass attached to the hanging spring, $\beta$ is the participation factor as defined in Equation (6), and $\psi_b(x)$ is the assumed shape function. If the assumed shape function $\psi_b(x)$ is normalized, i.e., $\psi_b(L_b/2)=1$, the floor acceleration at the middle span, is simplified in Equation (14):

$$Acc(L_b/2) = Acc_C + \beta\times(Acc_H - Acc_C) \tag{14}$$

## 3. Representative Structures

As shown in Fig. 3, two steel moment frames AO-04 and AO-12 were chosen as the representative structures. AO-04 is four-story and four-span, and AO-12 is twelve-story and four-span. They are treated as planar structures, and the gravity load is taken to be 60 kN/m.

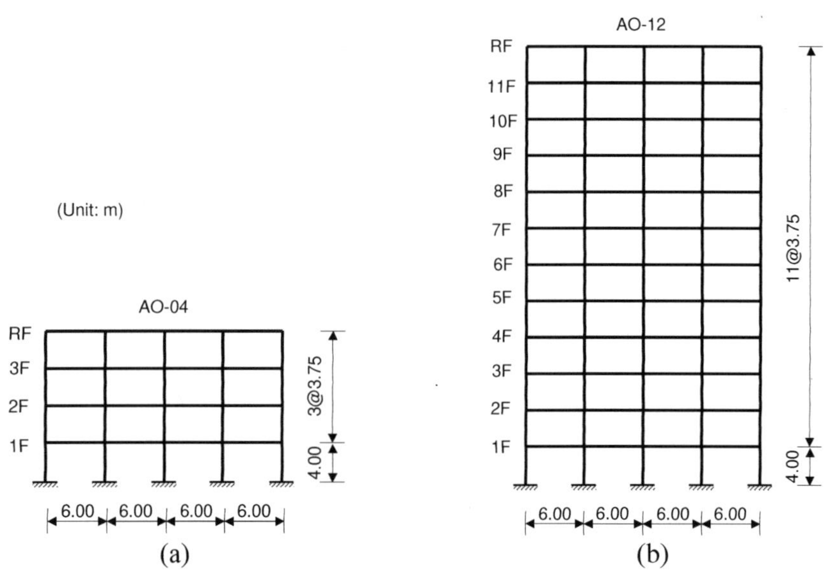


**Figure 3.** Sample structures: (a) AO-04; (b) AO-12.

Ten vertical ground motions recorded in the 1995 Hyogoken-Nanbu (Kobe) earthquake designated as the Kobe set [6], were chosen as representative ground motions. The pseudo acceleration spectra corresponding to 5% critical damping and normalized by the corresponding peak ground acceleration (by dashed line), and median of these values (by the solid line) are plotted in Fig. 4. The normalized spectra of the representative ground motions are anchored to unity and having peaks in the periods of 0.1 to 0.2 seconds. If the period is not greater than 0.05 seconds, the normalized spectrum is nearly unity, indicating that the amplification is very small [2].

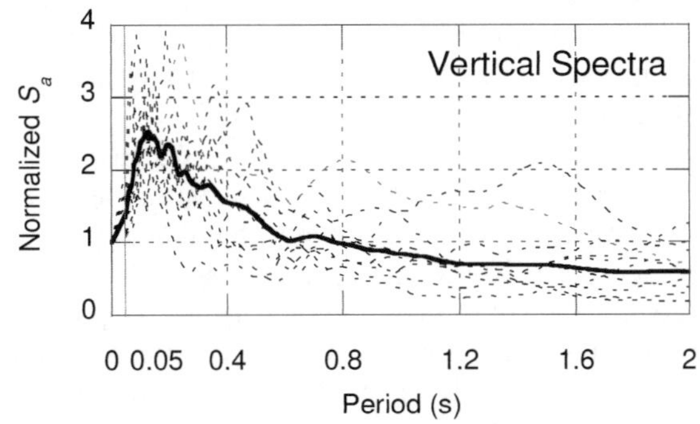


**Figure 4.** Pseudo acceleration spectra (5% critical damping) normalized by PGA.

## 4. Validation by Comparative Study

### *4.1. Analysis of FVF, SVF, GVF models*

A finite element program code named "CLAP" was used to analyse the FVF and SVF models. In order to take into account floor vertical vibration, each beam was divided into six elements in the FVF model, and each half beam was divided into 3 elements in the SFV model. In both FVF and SVF models, the associated lumped mass was assigned at each node.

In the GVF model, a deflection shape function should be specified. According to Equation (1), it is a four order polynomial function with the coefficients depending on the span length $L_b$. In the example frames, $L_b$. is 6 m for both AO-04 and AO-12, and the shape function is:

$$\psi_b(x) = x^4 - 12x^3 + 36x^2 \tag{15}$$

Substituting the shape function to the associated equations in the previous section, $K_H$, $M_H$, $K_C$, and $M_C$ of the GVF model are determined.

The number of DOFs of FVF, SVF, and GVF models are compared in Table 1 for AO-04 and AO-12. It is clear that the number of DOFs decreases significantly from the FVF to GVF models, with a ratio of 35 for these specific cases.

**Table 1.** Comparison of numbers of DOFs among FVF, SVF, and GVF models.

| | FVF | SVF | GVF |
|---|---|---|---|
| AO-04 | 280 | 72 | 8 |
| AO-12 | 840 | 216 | 24 |

### *4.2. Comparison between FVF, SVF, and GVF models*

In order to investigate whether the SVF model is capable of representing the FVF model, a series of time history analyses were carried out. Here the focused are the accelerations at the interior middle span of the FVF model and the beam end of the SVF model. To qualify the closeness of the responses obtained from the FVF and SVF models, statistical evaluation was conducted, and the ratio of the maximum acceleration obtained from SVF to that from FVF was adopted as an index. Figure 5 shows the medians and SD's [7] for the ratios. They are presented for all stories of each example structure. The figure also include the SD's of the maximum accelerations obtained from the FVF model, which serves as an index to examine the degree of dispersion associated with the ground motions. The medians of the ratios ranged from 0.8 to 1.0, and 0.9 to 1.0 for AO-04 and AO-12, respectively, and the SD's of the ratios are smaller than 0.15 and 0.10 for AO-04 and AO-12, respectively. Also, the SD's for the maximum acceleration obtained from FVF model are three to four times larger than the SD's of the ratios. These observations indicate that the responses of the SVF model are reasonably close to that of the FVF model. The SVF model gives better estimation for the twelve-story structure (AO-12) than the four-stories structure (A0-04). This is because the column vibrations, which are faithfully modeled in the SVF model, are more significant in taller frames.

Figure 6 shows detailed comparisons between the acceleration obtained from the FVF and SVF models. The responses subjected to JMA record, one of the most popular motions in the Kobe set, is shown in the figure. All the accelerations obtained from the time history analyses are included. For each time step, plotted is a pair of data, the floor maximum acceleration at top floor obtained from the FVF and SVF models. In the figure, all the data scatter near the 45-degree line. This indicates that the response obtained from the FVF and SVF models are similar not only in the absolute value but also in the vibration phase. This is a further confirmation of the conclusion drawn from Fig. 5.

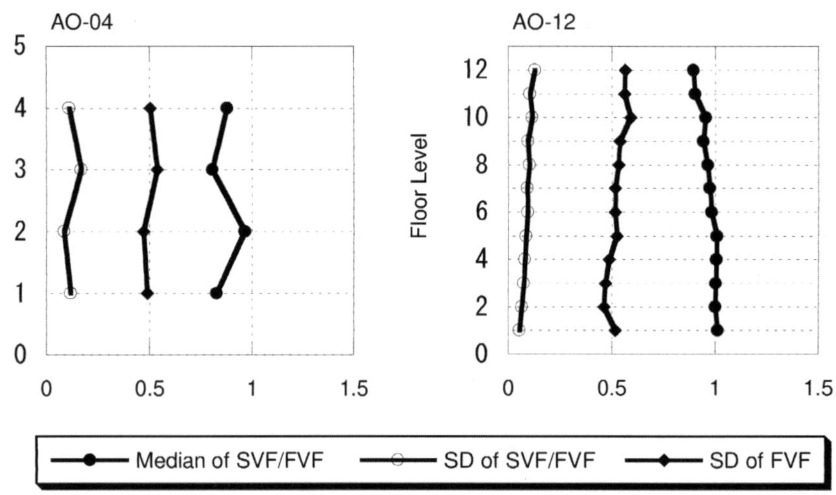


**Figure 5.** Comparison of vibration magnitude between SVF and FVF models.

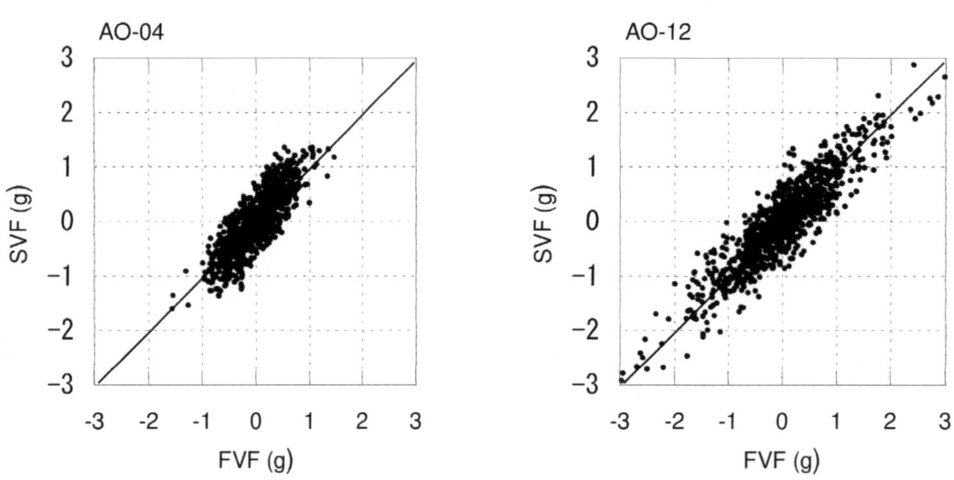


**Figure 6.** Comparison of vibration phase between SVF and FVF models.

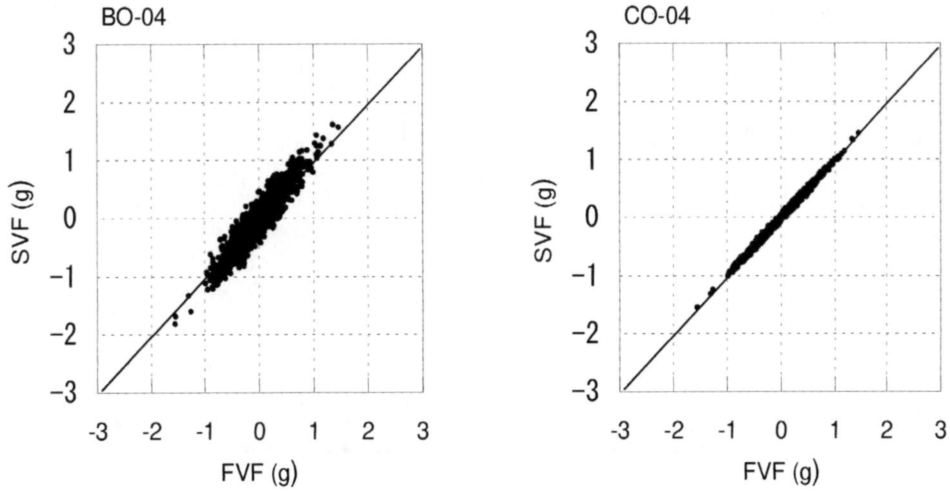


**Figure 7.** Comparison of vibration phase between SVF and FVF models for structures with strong columns.

It is also notable from Fig. 5 that although small, the SVF model tends to give a slightly smaller estimate. This is primarily because the constraints associated with the SVF model. The maximum deflection point is assumed to be located at the middle span when developing the SVF model. This is based on the assumption that all interior columns do not sustain any deflection. However, this is not entirely true since the columns next to the exterior columns sustain some deflection. The situation is less relevant if the columns are stiffer. Two further structures, designated as BO-04 and CO-04, were chosen as variation of AO-04. BO-04 and CO-04 are identical with AO-04 except that bending stiffness of columns are increased two and ten times, respectively. Following the same way in Fig. 6, the responses of the FVF and SVF models are compared for BO-04 and CO-04 in Fig. 7. It is apparent that compared with AO-04, the responses of the FVF and SVF models become closer for BO-04, and become almost identical for CO-04. This observation justified the interpretation of underestimation, and also indicated that the SVF model is more precise for structures with strong columns.

The responses are compared between the SVF and GVF models. Specifically, comparison is focused on the acceleration at the end of the half-beams for the SVF model and the acceleration of the hanging masses [corrected by Equation (14)] for the GVF model. Here the ratio of the maximum acceleration obtained from the GVF model to that obtained from the SVF model is adopted as an index to quantify the closeness of the responses. They are presented for all stories, and the ratios are estimated for each story. As shown in Fig. 8, the medians of the ratios are almost unity with a difference less than 0.02 for both AO-04 and AO-12, and the SD's of the ratios are almost zero. These observations indicate that the response obtained from the GVF model is almost identical with that obtained from the SVF model

Using the JMA ground motion, the accelerations at the top floor of AO-04 and AO-12 are investigated in detail to examine the closeness of responses between the GVF and SVF models. All the accelerations obtained from the time history analyses are included in Fig. 9. In the figure, all the data scatter very close to the 45-degree line. This indicates that responses obtained from the GVF and SVF model are very close in both vibration magnitude and phase, further confirming the conclusion drawn from Fig. 8.

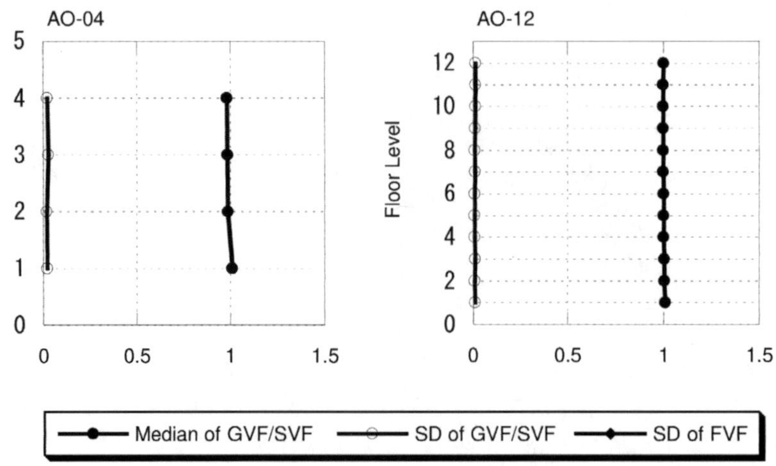


**Figure 8.** Comparison of vibration magnitude between GVF and SVF models.

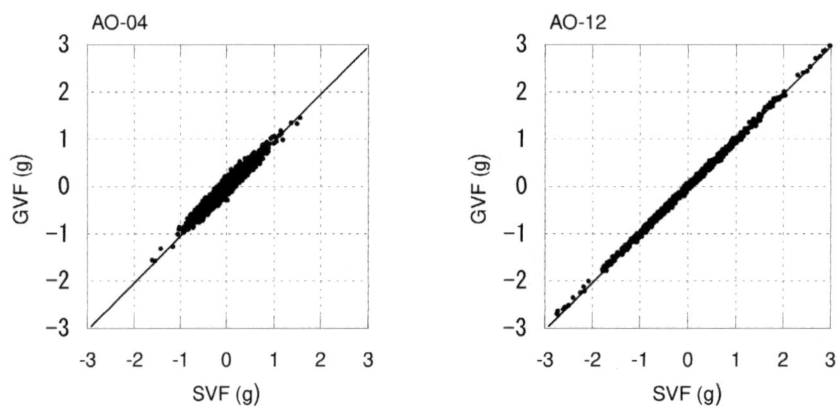


**Figure 9.** Comparison of vibration phase between GVF and SVF models.

## 5. Conclusions

To evaluate the vertical floor vibration of building structures, a generic vertical frame (GVF) model was developed in this study. Major conclusions obtained are summarized as follows:
(1) Compared with the FVF and SVF model, the GVF model significantly reduces the DOFs treated in the analysis.
(2) The GVF model is capable of estimating the vertical acceleration at interior spans for multiple span structures with errors not more than twenty percent.
(3) The GVF model becomes more accurate for structural systems with strong columns.

## 6. References

[1] Inoue K, Higashi K, Ogawa K, Tada M, and Hasegawa T 1995 Earthquake response of structural members of rigid frames with RHS columns (Part 1, design of model frames), *Summaries of Technical Papers of Annual Meeting C-1, AIJ*, pp 269-70.
[2] Elnashai AS 1997 Seismic design with vertical earthquake motion, *Seismic Design Methodologies for the next generation of codes*, Faifa & Krawinkler (eds), Balkema: Rotterdam
[3] Chopra A K 2001 *Dynamics of Structures: Theory and Applications to Earthquake Engineering*;

2nd edn, Prentice Hall: New Jersey
[4] Clough R W and Penzien J 1993 *Dynamics of Structures;* 2nd edn, McGraw Hill: New York
[5] Leung A Y T 1993 *Dynamic Stiffness and Substructures*; Springer-Verlag: London
[6] Nakashima M, Matsumiya T, and Asano K 2000 Comparison in Earthquake Responses of Steel Moment Frames Subjected to Near-Fault Strong Motions Recorded in Japan, Taiwan, and the U.S., *International Workshop on Annual Commemoration of Chi-Chi Earthquake*, Taiepi, Taiwan, pp 112-23.
[7] Luco N and Cornell C A 2000 Effect of connection fractures on SMRF seismic drift demands. *J Struct Eng-ASCE* **126(1)**:127-36

**Acknowledgments**
Financial support from the Scientific Research Fund of the Institute of Engineering Mechanics, China Earthquake Administration (Grant No. 2017D07); and that from the Opening Funds of the State Key Laboratory of Building Safety and Built Environment and the National Engineering Research Center of Building Technology (Grant No. BSBE2017-01), are gratefully acknowledged.

IOP Conf. Series: Materials Science and Engineering **371** (2018) 012044 doi:10.1088/1757-899X/371/1/012044

# A Study on the Importance of Water Conservancy Engineering Design

**Chuankai Yang**[1]

[1]Xinyang City Water Conservancy Survey and Design Institute, Xinyang, 464000, China

**Abstract.** Water conservancy project as a part of Chinese socialism infrastructure projects water plays a major role in agriculture, industry and the domestic residents. Water conservancy engineering design is the most basic and important work in the early stage of water conservancy construction. We must do the design work of water conservancy project, and the problems in the design should be dealt with a timely manner so that the construction of water conservancy project can be carried out better. This paper gives an overview of the water conservancy project, explains its importance, analyzes the design of the water conservancy project, and puts forward the improvement measures for the problem.

## 1. Introduction

With the development of the society, the increasing of the population and the prosperity of the economy, the importance of the water conservancy project has become more and more prominent. Most of China is still in shortage of water currently. How to make water conservancy projects well to ensure people's proper water use is the most difficult problems. The quality of the project requires that the construction is of the water conservancy is up to the standard and the demands and functions as well as the safety is refined. Therefore, the construction of water conservancy is profitable for the development of economy, society and the environment. It is quite necessary to have design in early stage of the construction of water conservancy which is also the promise about the smooth construction..

## 2. Introduction on water conservancy project

Water conservancy is the project which can control and allocate the surface water and ground water to eradicate harmful things and set up benefits for people. Water is a very important resource. Water is affluent for somewhere but some places are badly in shortage of water. Water conservancy is to improve the discordant situation. The construction of water conservancy is a way to guarantee the safety of life and properties, to meet the demands for water of people and to control floods and droughts. It is necessary and important to have the design of dams, dykes, spillways, water locks, canals, aqueducts, log chute and fishpass.

## 3. The imporatnce of water conservancy engineering design

### *3.1. Optimization Influences Project Costs*

Construction of engineering project includes three stages about project decision, project design and project implementation. The key on the investment controlling lies in the stages of decision and design. After the investment decision has been made, the key point is design. The project design, structural style and construction materials have great influence on economical efficiency of project, especially the scheme comparison, material selection, structural design and foundation type.

*3.2. The Quality of Design Products Influences Project Costs*
The geological environment and conditions of construction is quite different for the site selection of of water conservancy projects. The design have a direct impact on the project costs namely the the economic principles. Sometimes the design changes so frequently that the designers are confused which makes the design is in disorder and the details are ignored. Some projects even have problems of quality and potential safe hazard which brings a lot of loss for the country and investors and causes a waste of investment.

*3.3. The Influence On Operation Expense By Project Designs*
The quality of project design can not only influence the disposable investment of project construction, but also it can influence the operation expense during the usage stage. For example, the expense of maintenance and inspection should be increased. If some water supply project does not analyze the annual water consumption for design and the scale is too large, the expense will be so high that the project will be in deficit state. The disposable investment of projects may have inverse relations with the operation expense. But the best combination of the two can be achieved through the increase of project quality to decrease the operation cost of the project construction to the most.

*3.4. The Design Should Be Prospective*
Unreasonable function setting has bad impacts on the later function. The backward equipment shortens the lifetime of project so as to increase the repair work, which influences the project costs. Overlapping investment influences the economic principles of the project. Therefore when designed the projects should be prospective.

## 4. The problems in water conservancy engineering design

*4.1. Design Is Not Normative*
Some of the project designer is lack of economic concepts. Inconsideration can influence the project design when they are designing. Problems in the project design can lead to the change of the construction plans during the construction which means that it is difficult to control the cost of project. Some fundamental materials are copied without basic arguments on the details of the design. All of this can have bad effects on the normal construction of the projects which can cause the price of project to be too high or too low so that the budget of construction is in distortion to affect the accuracy of the project investment.

*4.2. The Design Is Poor*
Current technology is too low and cannot meet the development demands in the water conservancy design and construction. A variety of environmental and ecological problems will appear after the project comes into use or in use for the lack of some related technological arguments and theoretical arguments like ecological protection and the lack of the scientific ideas of acting according to the circumstances and without the combination of construction environment and geographical conditions, aquatic environment together.

*4.3. Water Conservancy Project Design Lacks Market Competition*
Many water conservancy project designs are similar to some inferior commodities and they are not competitive in the market. This is because the water conservancy projects lack innovation in technology and equipment and construction funds as well as information management. The careful, responsible and innovative designers are desperately needed. With the development of economy, the competition has become fierce. The water conservancy firms have to focus on innovation to enhance the economic and environmental benefits to get the high level with low costs to adjust to the market economy competition to win in the market competition.

### *4.4. The Design Unit Lacks The Service Consciousness Of The Subject*
The consciousness of service is quite weak which will lead to the arguments of the design with the proprietors. The designers will start without getting to know the exact requirements about the proprietors. The design faults will influence the clients' satisfaction. The firms should take the system of proprietors to make them give the same emphasize on the costs and the income of investment. Problems in the design will have impacts directly on the costs of water conservancy projects.

## 5. Improvement measures

### *5.1. Improving The Expertise Of Designers*
When we hire the water conservancy projects staff, the aptitudes have to be examined and the actual design level should be tested to ensure every staff is excellent in the expertise. The design institute should have training on the expertise and skill of the staff to provide the consciousness responsibility and quality management. The level should be ensured as the advanced so that the the project can be scientific and reliable.

### *5.2. Technology And Economy Should Be Prepared To Control The Project Costs*
The deficit of water conservancy projects lies in the technology. It is necessary to innovate the technology to put it into the water conservancy projects. The design institutes should be interacted with to provide the reliable data for the designers. The discussion about the projects should be involved to have analyzes of technology and economy. Informatization and high efficiency can be realized in the operation of the projects with the advanced technology and scientific innovation so that the projects can contribute to people's development.

### *5.3. Enhancing Competitive Strengths*
Project design should be made according to the geographical environment of selected area, hydrographic features and the requirements of the proprietors. The whole qualities of the staff should be improved and the scrutiny systems of construction schemes should be refined to avoid changing the design at will. When the project is under construction, the staff should save resources with the domestic water resource situation.

The considerations on land conservation, energy conservation, material conservation, water conservation can make the water conservation projects become the example of resources conservation. The water conservancy project should rely on the natural environment as much as possible but destroying the environment as little as possible and should get into the environment as much as possible. The project can be turned into water conservancy sightseeing zone to attract the social funds to promote the continuous development of water conservancy economy to realize the coordinated and sustainable development of engineering and environment.

### *5.4. The Design Concept Is People-oriented*
The concept of people-oriented should be taken into considerations as the most of the water conservancy projects are built in remote mountainous areas and the situations are complicated as well as the design principles and project costs are not changed. The best situation is let engineering design become formalized. The true design is the most practical with great flexibility. To make the design into the practice, the designers should have the return visit to realize the concept of people-oriented of the overall projects.

## 6. Conclusion
Water conservancy projects are to eliminate water disaster and develop and utilize water resources and the foundation of the smooth running of water projects to ensure that the water conservancy can have more economic benefits. Although the experts of water conservancy have done quite a few studies, the design cannot be refined. It is a long task to improve the level of water conservancy project to foreign level. Any construction design can be disturbed by a lot of factors. Only the design is controlled, the resources can be controlled well to refine the whole project.

IOP Conf. Series: Materials Science and Engineering **371** (2018) 012044 doi:10.1088/1757-899X/371/1/012044

## 7. References

[1] Xiaobo Jing 2010 Research On Problems And Measures In Water Conservancy Engineeringt Design. *J. Shanxi Water Conservancy* **6** 77
[2] Changjin Wang 2011 Discussion On Problems In Water Conservancy Engineering Design And Improvement. *J. Technology and Life* **2** 13-14
[3] Guoya Fu 2012 Problems and Measures In Water Conservancy Engineering Design. *J. Jiangxi Builiding Materials* **5** 121-122
[4] Liqiang Jin 2013 Problems In Water Conservancy Engineering Design And Improvement Measures. *J. Neimenggu Water Conservancy* **1** 171-172

IOP Conf. Series: Materials Science and Engineering **371** (2018) 012045 doi:10.1088/1757-899X/371/1/012045

# Issues of Specifications for Seismic Design of Highway Bridges and Suggestions

**Hongmei Cao**[1]

[1]Sichuan College of Architectural Technology,Sichuan 618000

**Abstract.** This article elaborates the development of the specifications for seismic design of highway bridges in China, analyzes the issues of Chinese seismic design of highway bridges from the vague construction index, unscientific material selection and low seismic precautionary intensity, and puts forward measures of specifications for seismic design of highway bridges based on the present status of seismic design of highway bridges in China. Hope that this article could provide some reference and recommendations for the relevant personnel in the field.

## 1. Introduction

In recent years, there are more and more earthquake disasters in China, especially the shocking earthquake in Wenchuan in 2008. Once the earthquake occurs, which not only damages the building seriously, but also makes the building collapse, and seriously threatens and damages the lives of the residents and social economy in our country. In order to ensure the safety of people's lives and properties, the normalization of the seismic design of highway bridges must be ensured, some seismic designs of highway bridges of advanced countries can be referred, and on this basis, combined with the Chinese own conditions of highway bridges to establish the Chinese seismic design concept of highway bridges, so as to achieve good anti-seismic effect and the integration with seismic design of highway bridge of advanced countries. However, according to the current situation, in the process of carrying out the seismic design specifications of highway bridges, the seismic index, seismic period, shear strength of bridges, etc. are not clear, which have brought difficulties to the seismic design of highway bridges. Therefore, it is necessary for us to take the corresponding improvement measures according to the issues of seismic design specifications of highway bridges, so as to improve the seismic capacity of highway bridges.

## 2. Elaboration of the development of the specifications for seismic design of highway bridges in China

With the steady development of bridge construction in our country, the original seismic design of bridge has been unable to meet the requirements of the current seismic design of highway bridges. Since China promulgated the relevant mechanisms of seismic design of highway bridges in 2008, existing seismic design concepts of highway bridges and the ways have been regularized. And they are also the basic factors to improve the seismic capacity of the bridge structure.

At present, in the seismic design of highway bridges, the principle of " highway bridges should not be damaged in small earthquakes, should be repaired in medium earthquakes and should not collapse in large earthquakes" should be upheld, although the related mechanisms of new seismic design of highway bridge has been perfected based on the original seismic standards, and solved the issues in the original seismic standards. However, there are still many issues compared with the seismic technologies of bridge construction in advanced countries.

## 3. Present status of seismic design of highway bridges in China

Most of the bridge structures in our country are simply supported bridges and bridges of simply-supported into continuous, and the top of bridge provides special shape supports to piers by laminated rubber bearings. When an earthquake occurs, the bearings will be damaged or moved, the bridge's way of force transmission is damaged, and the inertia force transmitted to the lower end of the bridge will be reduced gradually. Influenced by the earthquake, the structure of the lower end of the bridge will not be seriously damaged, but the appropriate measures of collapse-proof still need to be taken. The seismic failure mode of bridge structure is shown in figure 1:

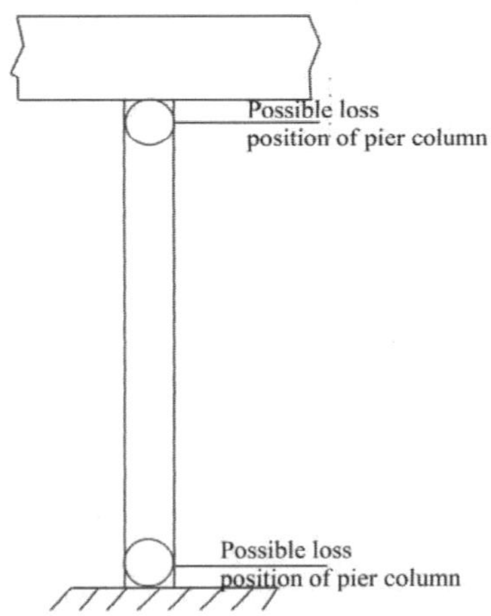


**Figure 1.** Seismic failure mode of bridge structure

When earthquake occurs, because different countries have different force transmission systems of bridge frame, so the characteristics of earthquake damage for bridge are also different, the bridges in USA have frame pier with the connection of pier and beam. Japan mainly uses steel support, which could effectively prevent the channel of force transmission of bridge form breaking. The current seismic design code of bridge in China mainly uses the content of bridges seismic design code of various countries, does not pay attention to differences in bridges' structure system, while strengthens force transmission without fracture when the force transmission system of bridge structure affected by the earthquake in China, namely, the bearing connecting the top structure of the bridge and the low structure of the bridge would be compromised, let the inertia force on top structure of the bridge be transmitted to the low structure of the bridge by bearing, at the same time the plastic hinge arranged in the pier dissipates force caused by earthquake, and thus achieves the anti- seismic effect.

But because the bridges in China mostly use open laminated rubber bearings without any measures of anchoring, so in the earthquake it is not possible that there are no damages in the bridge under the influence of earthquake. Therefore, in the process of making specifications of bridge's seismic design in China, we should take account of our bridges' own situation and structural force transmission system, in accordance with the relevant design standard of our country, compare the conditions of bridge damage by earthquake in different countries, we can learn that, laminated rubber bearings are widely used in bridges in China, in the earthquake disaster, they can play a good anti-seismic effect. When the earthquake comes, laminated rubber bearings will be damaged first, and the lower end of the bridge structure will be protected, which is the design concept with better effect.

## 4. Issues of Chinese seismic design of highway bridges

*4.1. Vague construction index*

With the rapid development of social economy in China, people pursue higher and higher quality of life, and also put forward new requirements for the seismic grade of the bridge. But in the graphic and vertical design of the bridge, there is always a degree of complexity caused by vague construction index. The gradually increasing irregular phenomenon of graphic and vertical design of the bridge, brings challenges to the seismic design of highway bridges, and also affects the seismic capacity of highway bridges [1].

*4.2. Unscientific material selection*

In recent years, there are more and more earthquake disasters in China, therefore, the scientific and reasonable seismic design of bridge structure must be ensured for the design of highway bridges in earthquake prone areas. In addition, in the selection of construction materials, we must choose high-quality, high standard building materials. However, some construction enterprises deliberately choose some cheap construction materials of substandard quality in order to save money, and the safety of the bridges cannot be guaranteed.

*4.3. Low seismic precautionary intensity*

In the past years, due to the low level of the overall economic development of our country, the requirements for seismic precautionary intensity are relatively low, and the general standard of seismic precautionary intensity was set at around 10%. In order to cut corners, some of the construction units do not design the grade of seismic precautionary intensity in accordance with the standard, leading to the generally low seismic performance of highway bridges. Once the earthquake occurs, it definitely will bring harm to the society and the people.

## 5. Measures of seismic design specifications of highway bridges

*5.1. Seismic concept design*

The occurrence time of most of the actual earthquakes is uncertainty, and the forms of actual earthquakes are also different, and when making specifications of seismic design of bridge, we should combine with the actual situation, at the same time take increasing the bridge's seismic performance as the design standard, and realize the full implementation of the specifications of bridge seismic design. In the process of bridge design, we should not only design the performance and the appearance of the bridge, but also fully consider the seismic performance of the bridge, and try best to use better seismic structure. In the design of seismic concept, the design of the weak links, such as the convergence point of upper end and the lower end of the bridge, should be strengthened, and the nature of the bridge pier and the rationality of the design of the hollow connection position should also be guaranteed [2]. According to the actual situation of the bridge construction, conduct a comprehensive analysis of the bridge seismic design and seismic response, check that whether the design meets the local seismic demand, and enhance the rigidity of bridge's weak link through the use of reinforcement and structural design, so as to enhance the seismic capacity of whole bridge. Finally, the seismic performance of the bridge should be comprehensively evaluated to ensure the rationality of the seismic structure system, so as to ensure the scientificity and efficiency of the bridge design scheme.

*5.2. Main points of regularizing seismic design of highway bridges*

First, reduce the weight of the bridge structure's itself, reduce the structure center of gravity, so as to reduce the impact and internal force of bridge structure brought by the earthquake; second, strengthen the stability of the bridge, and guarantee the bridge structure's hardness center of gravity and the center of the quality are together, in order to reduce the seismic stress enhanced by reverse [3]; third, in the process of designing bridge structure, ensure the rationality of length and height, as far as possible to avoid the damages caused by differences of various positions and natures; fourth, in the scope of the specification, the appropriately reduce the structural rigidity, use ductile materials to enhance the seismic performance of bridge structure, thereby reducing the impact of the earthquake;

fifth, pay high attention to the dressing and processing of bridge foundation, reduce the probability of deformation of the foundation, and avoid the failure of foundation.

Assume that the structure rigidity of bridge has symmetry, so its seismic performance is much higher than that of bridge whose structure rigidity does not have symmetry, and the bridge with different span is easier to be affected by the earthquake, especially the bridge with big difference of pier height, it can not only make pier appear higher horizontal seismic force, suppose span is different, it will make pier with a large span appear strong seismic force. Therefore, in the process making specifications of bridge seismic design, as far as possible to avoid building bridges with different spans in earthquake prone areas, combine with the actual situation, take appropriate bridge, and taking some measures of energy dissipation for the bridge with strong seismic horizontal force, so as to alleviate the concentrated stiffness on the top the bridge pier. At present, most of China's bridges adopt the seismic bearing, which resolutes the seismic capacity of bridges, a part of the bridge structures cannot be used in high seismic region, based on the original seismic performance of bridge structure, design a new type of bridge structure in accordance with the conditions of the area, and calculate the position of its displacement, standing in the perspective of design, carry out regulate design of the whole function of bridge structure for increasing the seismic performance of bridge structure [4].

### *5.3. Seismic design measures of new bridge*

The traditional seismic design code of bridge mainly enhances the seismic performance of the bridge in two aspects, the first is the strength of the bridge; the second is the ductility of the bridge. However, at present, we cannot accurately understand and explain the seismic performance of the bridge structure, and therefore the bridge structure will be damaged by the earthquake in varying degrees. In order to effectively prevent the emergence of this phenomenon, in the design of the bridge, we integrate steel reinforced concrete into the design, because of its strong bearing capacity and strong shear capacity, compared with reinforced concrete it has good ductility, can control the deformation of the bridge in a reasonable range, so as to improve the quality of highway bridges and increase the seismic performance of bridges.

### *5.4. New material for solving the issues*

For the original design of highway bridges, the reinforced concrete is usually used for the bridge building structure, and the main load construction of the whole bridge is reinforced concrete. Its defect of large gravity somehow influences the seismic capacity of long-span structures. At the same time, the reinforced concrete structure does not have strong cracking resistance, and once heavily influenced by strong force, the whole bridge structure will be changed. For this phenomenon, we can use steel reinforced concrete or prestressed concrete to replace the original reinforced concrete [5]. Compared with reinforced concrete, bearing capacity of steel reinforced concrete is much better, and steel reinforced concrete has good shear-bearing capacity and ductility, which can effectively reduce the probability of deformation of bridge under the earthquake. The bearing capacity of prestressed concrete structure is 30% higher than that of reinforced concrete. At the same time, in the process of bridge construction, because of their lighter weights, lower building heights and strong durability, so they can effectively enhance the ductility of the bridge structure, and the bridge will not crack because of earthquake.

## 6. Concluding remarks

In short, in the context of the increasing requirements of current seismic design of highway bridges, it is necessary to ensure the design staff to well analyze and evaluate the seismic capacity of highway bridges. Although the earthquake is uncertain, as long as we comprehensively study the influence of earthquake on the structure of highway bridge, comprehensively consider the influence factors when making the code of bridge's seismic design, continuously adjust and regulate the seismic program, take corresponding measures to enhance the seismic performance of bridge structure, strictly uphold the key points of the code of bridge's seismic design, so as to reduce the damage of bridge caused by the earthquake to minimum, and ensure the safety of bridge.

IOP Conf. Series: Materials Science and Engineering **371** (2018) 012045 doi:10.1088/1757-899X/371/1/012045

## 7. Reference

[1] Hailong J 2016 Problems and suggestions of code for seismic design of Chinese highway bridges. *J. Construction knowledge* **3** 165-166

[2] Ou L 2013 Thinking of the related problems of seismic design code for bridges in China and Europe. *J. Technology Innovation and Application* **4** 210-211

[3] Zhiqiang W, Yucai Y, Shichi J, Yongbo L, Xingfei Y and Zhuoming G 2013 Exploration of existing seismic reinforcement methods of bridge system. *J. Journal of Fuzhou University(Natural Science Edition)* **4** 556-561

[4] Xiaojin L, Lijun L and Xin W 2014 Problem analysis of seismic design of typical highway bridges. *J. Modern Transportation Technology* **1** 14-16

[5] Kehai W, Chong L, Qian L and Yue L 2014 Seismic design method for medium and small span bridges considering friction sliding of bearing. *J. Engineering Mechanics* **6** 85-92

IOP Conf. Series: Materials Science and Engineering **371** (2018) 012046 doi:10.1088/1757-899X/371/1/012046

# Study on Cross-regional Allocation of Construction Land in Beijing-Tianjin-Hebei Region based on Differentiation

**Shuqin Zhao[1,*], Guoyan Wang and Hongfei An[1]**

[1]Hebei GEO University, Huai'an East Road, Shijiazhuang, China
shuqinzhao999@163.com

**Abstract.** Shanghai and Beijing are two leading cities no matter in proportion of urban construction land, ability of attracting funds or in population concentration. In contrast, Hebei province is not comparable to the two cities in such three areas, but has higher planning potential of total city and rural construction land and higher potential of farmland transference. The cross-regional paid circulation of construction land is not only effective in overcoming utilization loss of construction land, land shortage of megacities and big cities, city economic competitiveness decline, but also an inevitable choice to attract capital and manpower from Beijing and Tianjing and enhance Beijing-Tianjin-Hebei Region cooperative development, providing reference for Xiong'an New Area development in construction land allocation.

## 1. Introduction

At present, supply-side structural reform has become a important strategy of the Communist Party of China and the State Council. The Fifth Plenary Session of the 18th CPC Central Committee clearly proposed to optimize allocation of land elements, create new supply and promote the development of new industries. Whether land supply side reform can provide driving force for overall supply-side structural reform has become a research focus. The city economic competitiveness decline caused by low supply rigidity of construction land, low supply ratio of residential land for megacity, low plot ratio and cross-regional circulation of construction-prohibited land. The voice for current land supply system reform, especially construction land allocation system reform, has always been there [1]. The attempt of cross-regional paid circulation of Beijing-Tianjin-Hebei construction land is of positive significance for Beijing-Tianjin-Hebei coordinated development and of reference signicance for Xiong'an New Area construction in land allocation.

## 2. Comparison of land element allocation among China three cities groups

### *2.1. Comparison of Construction Land Construction*

Table 1 shows Land Utilization Overall Planning (2006-2020) data on the administrative land area, agricultural land area, cultivated area, construction land area, land area for urban and rural construction, rural residential land area of Beijing, Tianjin, Hebei, Shanghai, Jiangsu, Zhejiang, Guangdong prescribed on web portal of Ministry of Land and Resources (Ministry of Land and Resources of the People's Republic of China). Urban construction land area date and Shanghai construction land area data are derived from Chinese City Statistical Yearbook (2016) (Urban Social Economic Bureau of State Statistics Bureau). The value of Beijing-Tianjin-Hebei Region is the sum of values of Beijing, Tianjing and Hebei; the value of Yangtze River Delta is the sum of values of Shanghai, Jiangsu and Zhejiang, and value of Pearl River Delta is equal to the value of Guangdong.

IOP Conf. Series: Materials Science and Engineering **371** (2018) 012046 doi:10.1088/1757-899X/371/1/012046

**Table 1.** Land use planning data of three major areas Unit: square kilometers

| To 2020 | Administrative land area | Agricultural land area | | Construction land area | | |
|---|---|---|---|---|---|---|
| | | | Cultivated area（ten thousand mu） | | Land area for urban and rural construction | |
| | | | | | | Rural residential land area |
| Shanghai | 8239 | 3875 | 374 | 2981 | 2600 | 400 |
| Jiangsu | 106743 | 67750 | 7127 | 20615 | 15601 | 9020 |
| Zhejiang | 105397 | 86363 | 3826 | 11326 | 7980 | 3520 |
| **Yangtze River Delta** | **220380** | 157988 | **11327** | **34922** | **26181** | **12940** |
| Guangdong | 179757 | 149971 | 4363 | 20060 | 15230 | 6100 |
| **Pearl River Delta** | **179757** | 149971 | **4363** | **20060** | **15230** | **6100** |
| Beijing | 16410.54 | 11960 | 322 | 3817 | 2700 | 730 |
| Tianjing | 11917.32 | 7073 | 656 | 4034 | 2500 | 750 |
| Hebei | 188434 | 142531 | 9454 | 19113 | 14980 | 9280 |
| **Beijing-Tianjin-Hebei Region** | **216762** | 161564 | **10432** | **26964** | **20180** | **10760** |

As can be seen in Figure 1, Yangtze River Delta has the largest land area for urban and rural construction, followed by Beijing-Tianjin-Hebei Region and then Pearl River Delta. However, in ratio between rural residential land area and total land area for urban and rural construction, Beijing-Tianjin-Hebei Region has the highest value, followed by Yangtze River Delta and then Pearl River Delta.

As shown in Figure 2, Jiangsu has the largest total land area for urban and rural construction, followed by Guangdong and then Hebei, and Tianjing ranks in last position. However, from the aspect of ratio between rural residential land area and total land area for urban and rural construction, Hebei is the highest, followed by Jiangsu, and Shanghai ranks the bottom.

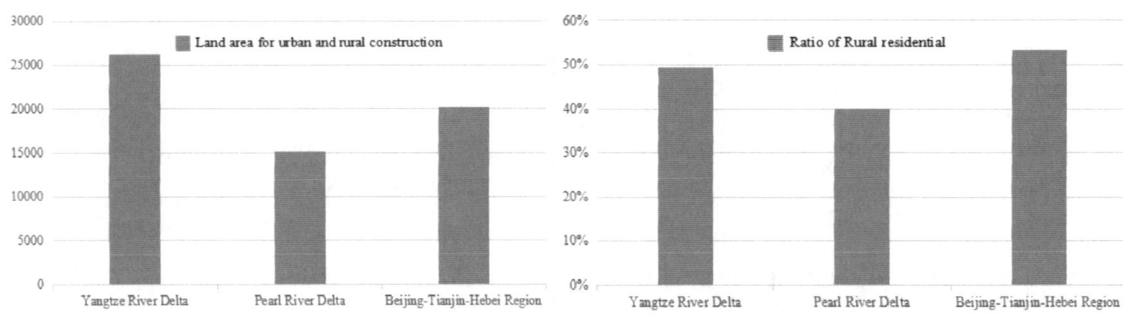


**Figure 1.** Comparison of construction land scale and structure among three major areas

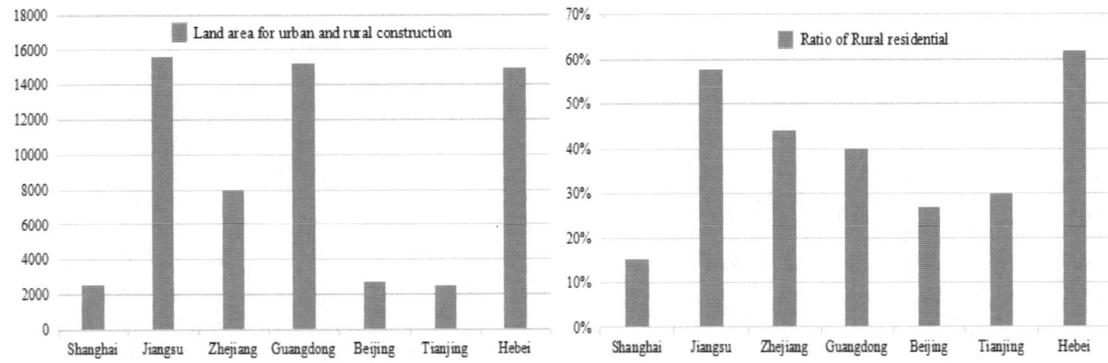


**Figure 2.** Comparison of construction land scale and structure among major cities

On one hand, the results indicate different areas have different construction land structure. For example, Yangtze River Delta has the largest land area for urban and rural construction, Beijing-Tianjin-Hebei Region has the largest proportion of rural residential land area, and Pearl River Delta has the lowest proportion of rural residential land in rural collective construction land. Among the three major areas, the construction land scales per household or per resident are different, i.e. the land wealth per household or per resident are different. This may be due to large construction land of Yangtze River Delta and Pearl River Delta in history or the farmland transference in recent years. On the other hand, the results also indicate Beijing-Tianjin-Hebei Region has larger planning potential of urban and rural construction land, and enjoys larger development space than Yangtze River Delta and Pearl River Delta.

### *2.2. Comparison of Potential for Farmland Transference Among Three Areas*

As shown in Figture 3, Pearl River Delta has the largest ratio of agricultural land area to administrative land area, while the lowest ratio of cultivated land to agricultural land area, showing a large potential of farmland transference. Compared with Yangtze River Delta, Beijing-Tianjin-Hebei Region has higher ratio of agricultural land area, but a lower ratio of cultivated land, which indicates that farmland transference potential of Beijing-Tianjin-Hebei Region is higher.

As shown in Figure 4, Guangdong has the highest ratio of agricultural land area, followed by Zhejiang and Hebei; from the aspect of the ratio of cultivated land, Jiangsu has ranks the first, Tianjing ranks the second, Hebei ranks the third, Guangdong is higher than Beijing; Guangdong has the highest potential of farmland transference, followed by Zhejiang and then Hebei, wherein Tianjian has the lowest potential. It should be noted that the larger the potential of farmland transference is, the larger the support of land can provide for future development.

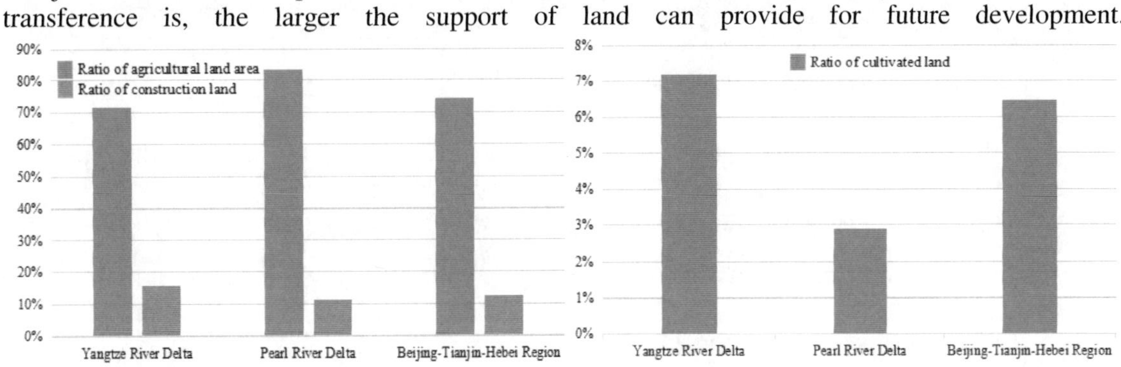

IOP Conf. Series: Materials Science and Engineering **371** (2018) 012046 doi:10.1088/1757-899X/371/1/012046

**Figure 3.** Land utilization structure of three major areas

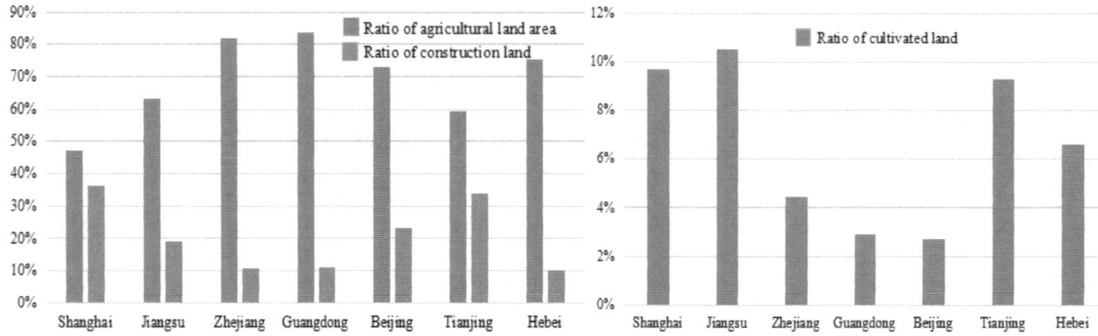


**Figure 4.** Land utilization structure among major cities

## 3. Comparison of capital and population gathering situation of China three major areas

### *3.1. Comparison of Capital and Capital Allocation Situation of China Three Major Areas*

Table 2 shows the 2016 year-end balance of deposits (domestic and foreign currency) in all financing institutions (including foreign capital), 2016 year-end balance of loans (domestic and foreign currency) in all financing institutions (including foreign capital) , 2016 household balance of deposits (RMB) over province (city), 2016 balance of deposits (RMB) of non-financial business over province (city), and deposit per capita (RMB). There data are collected from 2016 National Economic and Social Development Bulletin on web portal of National Bureau of Statistics of the People's Republic of China (National Bureau of Statistics of the People's).

**Table 2.** Total deposit balance (RMB) in all financing institutions Unit: 100 million rmb

| 2016 | Balance of deposit of all financing institutions | | | Balance of loan of all financing institutions | Deposit per capita（yuan） |
|---|---|---|---|---|---|
| | | Household deposit | Non-financial business deposit | | |
| Shanghai | 110510.96 | 25112.99 | 45105.14 | 59982.2 | 87720.02 |
| Jiangsu | 121106.6 | 439005 | 452779 | 91107.6 | 42915.60 |
| Zhejiang | 99530 | 38775 | 34562 | 81805 | 53402.11 |
| **Yangtze River Delta** | **331147.6** | **502892.99** | **532446.14** | **232894.8** | **-** |
| Guangdong | 179829.19 | 59768.75 | 63509.89 | 110928.41 | 47574.48 |
| **Pearl River Delta** | **179829.2** | **59768.75** | **63509.89** | **110928.4** | **-** |
| Beijing | 138408.9 | 28012 | 50998.3 | 63739.4 | 112292.42 |
| Tianjing | 30067.03 | - | - | 28754.04 | 52273.79 |
| Hebei | 55513.3 | 32710.9 | - | 37352.2 | 31941.56 |
| **Beijing-Tianjin-Hebei Region** | 223989.2 | - | - | 129845.6 | - |

As shown in Figure 5, Guangdong ranks the first in both deposit and loan of all financial institutions, Beijing ranks the second in deposit of all financial institution. Jiangsu and Zhejiang ranks the second in loan of all financial institution. Jinagsu Jiangsu ranks the first in both household deposit and non-financial business deposit. All these results indicate the areas mentioned above have strong money-making ability, while Tianjing and Hebei have lower money-making ability.

As shown in Figure 6, the three major areas have different money-making abilities, wherein Yangtze River Delta has higher money-making ability than Beijing-Tianjin-Hebei Region. There is also a large internal differentiation of money-making ability, for instance, Beijing has the largest per capita deposit, followed by Shanghai, and Hebei ranks the bottom. The internal differentiation of Beijing-Tianjin-Hebei Region is larger than that of Yangtze River Delta, and Beijing has a strong siphonage of capital from Tianjing and Hebei.

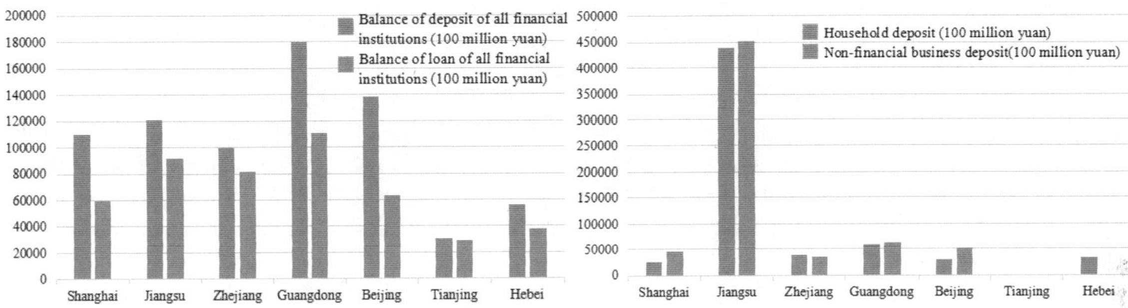


**Figure 5.** Comparison of major cities in loan and deposit

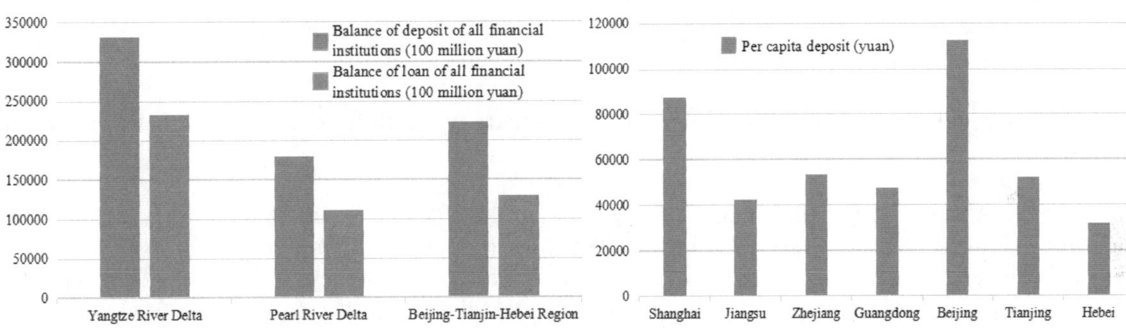


**Figure 6.** Comparison of three major areas and major cities in loan and deposit

### *3.2. Comparison of Population Aggregation Degree Among Three Major Areas*

Table 3 shows the population aggregation degree consisting of "proportion of urban population" and "population density" [2], which is calculated based on data in Chinese City Statistical Yearbook (2016). Population aggregation degree=a×proportion of urban population +ß×population density (wherein the weight of a and ß are both set to 0.5， a+ß=1). First, the proportion of urban population and population density are respectively subjected to N (0,1) statistical standardization processing: proportion of urban population=(proportion of urban population-mean value)/standard deviation, population density=(population density-mean value)/standard deviation, obtaining population aggregation degree. Secondly, the obtained population aggregation degree is subjected to differential treatment: population aggregation degree=(population aggregation degree-minimal value)/(maximum value-minimal value)×100. It can be seen from Table.3 that Shanghai has the largest population aggregation degree, followed by Beijing, and Hebei ranks the bottom.

**Table 3.** Regional population aggregation index and its location

| Province | Proportion of urban population (%) | Population density (people/km$^2$) | N(0,1) statistical standardization | | Population aggregation degree | | |
|---|---|---|---|---|---|---|---|
| | | | Urban population | Population density | Exponential value | Standardization (%) | Rank |
| Beijing | 86.51 | 812.50 | 1.18 | -0.08 | 0.55 | 56.7 | 2 |
| Tianjin | 82.64 | 853.12 | 0.82 | -0.02 | 0.40 | 51.8 | 3 |
| Hebei | 57.33 | 406.77 | -1.48 | -0.76 | -1.12 | 0.00 | 7 |
| Jiangsu | 66.52 | 728.97 | -0.64 | -0.22 | -0.43 | 23.30 | 4 |
| Shanghai | 87.60 | 2275.67 | 1.28 | 2.36 | 1.82 | 100.00 | 1 |
| Zhejiang | 65.80 | 466.49 | -0.71 | -0.66 | -0.69 | 14.80 | 6 |
| Guangdong | 68.71 | 500.36 | -0.44 | -0.61 | -0.53 | 20.20 | 5 |

It can be concluded that on one hand, Beijing and Shanghai are two leading cities no matter in urban construction land proportion, ability of attracting funds and population aggregation degree, while Hebei ranks lower in these three indexes. On the other hand, the land reserve for future development of Beijing and Shanghai are limited, while Hebei has great potential of land support for its future development, as long as the corresponding capital and population resources are matched, Hebei can gain a great future development.

Beijing ranks the top in ability of attracting fund and population aggregation degree, similarly Tianjing does a great job in these two aspect, while Hebei is less superior although it has the largest potential of land utilization. The cross-regional paid circulation of construction land is an inevitable choice to attract capital and human resource from Tianjing and Beijing and promote harmonic cooperative development of Beijing-Tianjin-Hebei Region, providing reference for millennium engineering project of Hebei Xiongan New Area.

## 4. The cross-regional paid land allocation path and strategy for Beijing-Tianjin-Hebei Region

### *4.1. China Land Circulation of Construction Land and Characteristics*

According to Article 10 of China's Constitution "Land in the cities is owned by the state. Land in rural areas and suburban areas of cities excluding those belonging to the state prescribed by law belongs to collective ownership; house sites and privately farmed plots of cropland and hilly land are also owned by collectives." According to current laws and regulations, the ownership of China construction land can be divided into state-ownership and collective ownership. Collective construction land can be further decomposed into house site, collective public construction land (street, school land, etc) and collective commercial construction land (village-run business and country-run business).

Since the release of "Temporary Regulations on the Assignment and Transfer of the Right to the Use of State-owned Land in the People's Republic of China" (State Council decree no.55) in May 1990, the circulation of urban construction land has been permitted. However, the circulation of rural collective construction land has successively experienced three stages: circulation of collective commercial construction land in "establishment and alteration of property right of township and village enterprises", circulation of rural collective construction land in "new rural construction and centralized residence of village people or rural land management project", and paid voluntary withdrawal or transfer of house site "within the collective economic organization".

As prescribed in Article 63 of "The Land Administration Law of the People's Republic of China", "the right of use of land owned by peasants shall not be transferred, transferred or leased for non-agricultural construction; however, enterprises that conform to the overall planning of land use and acquire land for construction in accordance with the law shall be exempted from the transfer of land use right according to law due to bankruptcy or merger". The collective commercial construction land in this stage can only be passively transferred.

Under the "new rural construction and villagers' centralized residence or rural land renovation project", the circulation of rural collective construction land has been permitted since "the Issue of a Pilot Management Method for the Increase and Reduction of Urban and Rural Construction Land", of which the purpose is to cover the gap of urban construction land index. The concrete operation is to transfer idle and inefficient collective commercial construction land into urban construction land via collective construction land saving in new rural construction renovation project, or via collective construction land saving in centralized residence renovation project for villagers, or via collective construction land by demolition and reconstruction of via house sites.

According to "The CPC Central Committee's Decision on Comprehensively Deepening Reform of Several Major Issues" agreed on the Third Plenary Session of the 18th CPC Central Committee and "The Decision of Standing Committee of the National People's Congress on Authorizing the State Council's to Temporarily Adjust and Implement Relative Legal Regulations for Beijing Daxing District and Other 33 Pilot Counties (Cities, Districts)" settled on the 13th meeting of the standing committee of the 12th National People's Congress in 2015 (The Decision of Standing Committee of the National People's Congress on Authorizing the State Council's to Temporarily Adjust and Implement Relative Legal Regulations for Beijing Daxing District and Other 33 Pilot Counties (Cities, Districts)), the proposal of "allowing the sale, leasing and demutualization of right to use of rural collective business construction land, and implementing equal share equal right equal price as with right to use of state-owned construction land" has been clearly approved. This indicates rural collective commercial construction land can not only be circulated, but also in equal price and equal share.

In 2014, "China New Urbanization Plan (2014-2020)" proposed to implement connecting policy of increasing urban construction land scale and attracting migration rural population to settle down. On the Fifth Plenary Session of the 18th CPC Central Committee, the ork conference on central urbanization proposed reform task – "Press Conference on Suggestion of Implementation on Establishing Hooking Mechanism between Increase Scale of Urban Construction Land and Settlement Number of Migration Rural Population". In 2016, the 13th five-year plan of the state lay further emphasis on accelerating the citizenization of migration rural population and on establishing connecting mechanism between increasing urban construction land scale and attracting migration rural population to settle down. In the end of 2016, ministry of land, housing, public security and environmental protection expressed that "allowing rural migrant residents to voluntarily have paid withdrawal or transfer of house sites within their own collective economic organizations based on requirements of rural residence land system pilot reform" in " The Opinions on The Implementation of Connecting Policy of Increasing Urban Construction Land Scale and Attracting Migration Rural Population to Settle Down" (Ministry of Land and Resources of the People's Republic of China). This basically opens the path of circulation for house sites.

Currently, a new pattern has formed on the circulation of China land usage right, i.e. "newly-added construction land is provided by annual plan of land utilization, urban and rural construction land is comprehensively planned as a whole, the use right of urban and collective construction land is circulated within province, and house site is circulated following migration rural population". System problems such as the city economic competitiveness decline caused by low supply rigidity of construction land, low supply ratio of residential land for megacity, low plot ratio and cross-regional circulation of construction-prohibited land have also been concerned.

### *4.2. Cross-regional Paid Land Allocation Strategy for Beijing-Tianjin-Hebei Region*

Under the premise of abiding by the land-use regulation system, provincial overall plan of land utilization and annual plan for land use index, firstly, the newly-added construction land of Beijing-Tianjin-Hebei Region is allowed to be sold on cross-regional market, and cross-regional requisition-compensation balance of cultivated land is allowed; secondly, realize market-oriented planning of cross-regional increase and reduction of inventory urban construction land; thirdly, realize cross-regional circulation of household collective construction land; fourthly, allow cross-regional ecological renovation of inventory ecological land without changing ownership. In all, realize manpower land-use through cross-regional paid planning of land use right.

According to "World Development Report 2009" announced by World Bank, nearly half of world's economic activities happen on 1.5% of land area. Beijing-Tianjin-Hebei Region, Yangtze River Delta and Pearl River Delta are three major economic area of China, wherein Yangtze River Delta has become one of 6 world's big city groups, but still lag behind New York City agglomeration and Tokyo urban agglomeration. The paid overall utilization of land resource, assets and capital can not only improve the economic aggregation and balanced development of Beijing-Tianjin-Hebei Region, but also contribute to the environmental protection and management of Beijing-Tianjin-Hebei Region.
Although the circulation of inventory urban construction land that conform to land use control has always been permitted within provincial domain, due to limited number of suppliers and demanders, the inventory land capital at "redundant area" of various cities are mainly digested by real estate development, which leads to a small trading volume of urban industry land and provides insufficient driving force industry transformation and upgrading of cities in "lack of fund". Allowing cross-regional paid circulation of construction land in Beijing-Tianjin-Hebei Region not only provides time-space matching channel for industrial land demand and land supply, but also provides new blood for new economic growth point of redundant cities, which is a win-win strategy.

## 5. Conclusions and suggestion

Currently, supply-side structural reform has become an important strategy of the Communist Party of China and the State Council. Land system, as one of the most basic economic systems, plays an vital role in supply-side structural reform, regarded as basic platform, core content and key point of whole supply-side structural reform. Giving play to the decisive role of market in resource allocation has long been a major policy view since the 18th CPC National Congress.
Allowing cross-regional trade and planning of newly added and inventory construction land and allowing cross-regional "requisition-compensation balance" of cultivated land are innovation points of land supply side reform. On one hand, peasants who have long been lived and worked in cities are allowed to transfer their house-site construction land at original country-side to the cities they work, serving as construction land for urban expansion, while the agricultural land inventory at their original countryside is increased via restoring cultivation of house-site. On the other hand, allowing cross-regional circulation of inventory construction land can activate the fiexed land capital, injecting blood into the new industry. It is of positive significance for promoting the coordinated development of Beijing-Tianjin-Hebei Region through the supply of land element supply, land asset price and land financial market transmission mechanism under the new normal, and promoting the linkage reform of capital and household registration through cross-regional allocation of construction land.

## 6. References

[1] Phoe N 2017 Annual Conference of Boao Forum for Asia. Shijin. *J. High Housing Prices May Lead to Industrial Decline* **4** 11-12
[2] Ming Lu 2017 The Power of Space (Geography, Politics and City Development). *J. Shanghai People's Publishing House* **2** 53-77

IOP Conf. Series: Materials Science and Engineering **371** (2018) 012047 doi:10.1088/1757-899X/371/1/012047

# Urban Building Change Detection Based on Improved Four-component Scattering Model

**Xingdong Wang**[1]

[1]College of Information Science and Engineering, Henan University of Technology, Zhengzhou, Henan 450001, China
zkywxd@163.com

**Abstract.** A four-component scattering model and the enhanced double-bounce scattering with the rotation of coherency matrix was previously studied, and on this basis, a method for improving Polarimetric Synthetic Aperture Radar (PolSAR) to precisely detect urban buildings is proposed. To this end, Helix scattering and enhanced double-bounce scattering components are combined to detect the urban building change. Application of the RADARSAT-2 polarimetric data of August 9, 2008 and May 24, 2009 to the reconstruction of Mianzhu City, Sichuan Province, China, and experimental results show that the improved model is feasible and effective for improving the accuracy of detecting changes in city buildings.

## 1. Introduction

SAR is independent of daytime and clouds cannot obstruct its view, ensuring continuous supply of information with every pass even under unfavorable weather conditions, and it is suitable for building detection in urban areas. PolSAR is used to obtain the complex scattering matrix, which is a complete description of the electromagnetic scattering phenomena. Therefore, this matrix can accurately detect target characteristics with the advantages of traditional SAR. PolSAR can acquire echo responses of ground objects to radar signals of different wavebands. Compared with the traditional single PolSAR, the outstanding advantages of PolSAR images lie in that the measurement data contain more abundant target information, enabling more accurate target detection and classification, and providing the possibility for information mining [1].

Many studies show that when flight direction of PolSAR is different, the same building has distinct characteristics in the radar images [2-4]. Because of the different directions, urban buildings show different backscattering; this phenomenon makes the results of urban building extraction problematic and unsatisfactory. In order to increase the accuracy of urban building detection and classification from PolSAR data, the Yamaguchi four-component decomposition scattering model was taken into account. When buildings are parallel to the flight direction of the radar, the signal intensity of the buildings is more than that of the surrounding objects [5], and the three-component and four-component scattering models can be used to detect this type of buildings by obtaining the double-bounce scattering signals between the wall and the ground. However, the strong double-bounce scattering angle reflection is less than 10° [6]. When there is a large angle between radar flight direction and buildings, the double-bounce scattering is not satisfactory. Considering the above mentioned factors, this paper presents a method that combines the spiral volume scattering component of the Yamaguchi four-component scattering model with the dihedral scattering component enhanced by the coherent rotation matrix to detect urban buildings.

## 2. Study area and datasets

In order to verify the effectiveness of the proposed method, Mianzhu City, Sichuan Province, China was selected as a test area. The city is located in the northwest of Sichuan Basin (east longitude 103° 54′ - 104° 20′, north latitude 30° 09′ - 31° 42′), and wavy terrain is very small. Urban area is 11.86 square kilometers including Jiannan Town, Southwest Town, Northeast Town, and Xiaode Town.

The main image sensors of RADARSAT-2 are C-band SAR with multiple imaging modes, which adopts full polarization (HH/HV/VH/VV) work mode. These sensors can significantly increase the number of detectable objects or target category, providing radar data with 3-100 m resolution and 10-500 km width. The main imaging mode can provide a strip image with a sample of 3 m each. The data used in this study is the SLC format data (ground range, unsigned 16-bit integer number, 8-m pixel spacing) from August 9, 2008 for the test site.

## 3. Urban Building Change Detection

### *3.1. Yamaguchi Decomposition Model*

The coherent matrix based on the Yamaguchi four-component scattering power decomposition model can be extended to [7]:

$$\langle [T]\rangle = f_s\langle [T]\rangle_{furface} + f_d\langle [T]\rangle_{double} + f_v\langle [T]\rangle_{vol} + f_c\langle [T]\rangle_{helix} \tag{1}$$

where $f_s$, $f_d$, $f_v$, and $f_s$ are surface scattering, double-bounce scattering, volume scattering, and helix scattering components, respectively. Thus, we can obtain $p_s$ (surface scattering), $p_d$ (double-bounce scattering), $p_v$ (volume scattering), and $p_c$ (helix scattering) components of the power by the following formula:

$$p_s = f_s(1+|\beta|^2),\ p_d = f_d(1+|\alpha|^2),\ p_v = f_v,\ p_c = f_c \tag{2}$$

Through this model, the double-bounce scattering component has been widely applied to the detection of non-natural objects or buildings [8], but when the angle between the flight direction of radar and buildings is relatively large, the detection rate is reduced. The helix scattering power is related to $S_{hh}S_{hv}^*$ or the imaginary part of $S_{vv}S_{hv}^*$, which is often used in complex urban areas. In a complex urban area, the reflection symmetry, $S_{hh}S_{hv}^* = 0$ or $S_{vv}S_{hv}^* = 0$, condition is often not satisfied. For example, when the angle between the flight direction of radar and buildings is relatively large, $S_{hh}S_{hv}^* \neq 0$ or $S_{vv}S_{hv}^* \neq 0$, the helix scattering component can be used to detect non-natural targets.

### *3.2. Enhanced Double-bounce Scattering*

As mentioned above, when there is a large angle between radar flight direction and buildings, the double-bounce scattering component does not work for non-natural targets. Therefore, the orientation angle needs to be standardized [9], that is, the coherent rotation matrix will effectively enhance the double-bounce scattering power [10].

Assume the following coherent matrix:

$$\langle [T]\rangle = \begin{bmatrix} T_{11} & T_{12} & T_{13} \\ T_{21} & T_{22} & T_{23} \\ T_{31} & T_{32} & T_{33} \end{bmatrix} \tag{3}$$

The coherent matrix can be obtained after rotating $\theta$ angle:

$$\langle [T(\theta)] \rangle = [R_P(\theta)] \langle [T] \rangle [R_P(\theta)]'$$
$$\begin{bmatrix} T_{11}(\theta) & T_{12}(\theta) & T_{13}(\theta) \\ T_{21}(\theta) & T_{22}(\theta) & T_{23}(\theta) \\ T_{31}(\theta) & T_{32}(\theta) & T_{33}(\theta) \end{bmatrix} \quad (4)$$

where the rotation matrix is:

$$[R_P(\theta)] = \begin{bmatrix} 1 & 0 & 0 \\ 0 & \cos 2\theta & \sin 2\theta \\ 0 & -\sin 2\theta & \cos 2\theta \end{bmatrix} \quad (5)$$

Therefore,

$$T_{33}(\theta) = T_{33}\cos^2 2\theta - \mathrm{Re}(T_{23})\sin 4\theta + T_{22}\sin^2 2\theta \quad (6)$$

Minimizing $T_{33}$, we get:

$$T'_{33}(\theta) = 2(T_{22} - T_{33})\sin 4\theta - 4\,\mathrm{Re}(T_{23})\cos 4\theta = 0 \quad (7)$$

Thus, we can obtain:

$$\tan 4\theta = \frac{2\,\mathrm{Re}(T_{23})}{T_{22} - T_{33}} = \frac{4\,\mathrm{Re}\langle S^*_{HV}(S_{HH} - S_{VV})\rangle}{\langle |S_{HH} - S_{VV}|^2 \rangle - 4\langle |S_{HV}|^2 \rangle} \quad (8)$$

Therefore, the rotation $\theta$ angle is:

$$\theta = \frac{1}{4}\left[\tan^{-1}\frac{4\,\mathrm{Re}\langle S^*_{HV}(S_{HH} - S_{VV})\rangle}{\langle |S_{HH} - S_{VV}|^2 \rangle - 4\langle |S_{HV}|^2 \rangle}\right] \quad (9)$$

In short, the basic steps for the enhancement of the double-bounce scattering component are as follows:
1) Estimate the offset of the direction angle.
2) Compensate the offset of the direction angle.

### *3.3. Urban Building Change Detection*

Therefore, the enhanced double-bounce scattering component is effective for small angles between the flight direction and urban buildings, and the helix scattering component is effective for the large angles between the flight direction and urban buildings. Therefore, the double-bounce and helix scattering components can be combined to effectively detect changes in urban buildings; the flow chart is shown in Figure1.

IOP Conf. Series: Materials Science and Engineering **371** (2018) 012047 doi:10.1088/1757-899X/371/1/012047

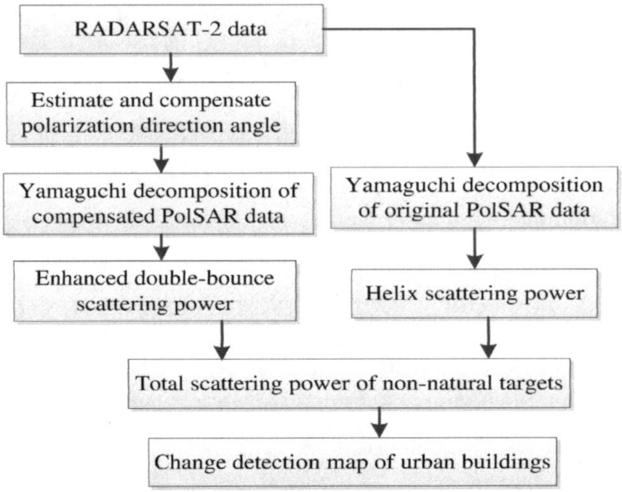


**Figure 1.** Flow chart of urban building detection

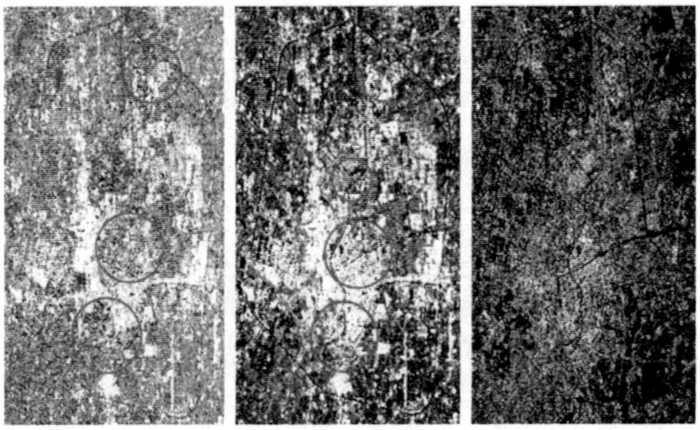

**Figure 2.** Scattering model decomposition in the study area on May 24, 2009 (a) Double-bounce scattering (b) Enhanced double-bounce scattering (c) Helix scattering

Figure 2 shows the double-bounce, enhanced double-bounce, and helix scattering components obtained by the improved Yamaguchi four-component scattering model for full polarimetric data of RADARSAT-2 on May 24, 2009 in Mianzhu City. White areas in Figure2 (a) show the strong double-bounce scattering produced by non-natural objects perpendicular to the flight direction. The directions of most buildings in most areas in the red circles are not perpendicular to the flight direction, so the double-bounce scattering component cannot effectively recognize non-natural objects. Comparing Figure2 (b) and Figure2 (a), it can be seen that most areas in the red circles, after the enhancement of the coherent rotation matrix, the angles between the flight direction and the buildings are less than or equal to 45° [11]. This implies that the double-bounce scattering component of non-natural objects was significantly enhanced. The angles between the non-natural objects and the flight direction in the areas in the red circles in Figure 2 (c) are large, and the helix scattering component is larger. Therefore, the enhanced double-bounce scattering and helix scattering are complementary, and the enhanced double-bounce scattering and helix scattering were combined to feasibly and effectively detect non-natural objects.

## 4. Results and Analysis

Figure 3 shows RGB composite images based on the improved Yamaguchi four-component scattering model of August 9, 2008 and May 24, 2009 for Mianzhu City from RADARSAT-2 data. Green areas represent strong double-bounce scattering caused by non-natural objects. The upper right corner in Figure 3 (a) shows the temporary shelters built after the Wenchuan earthquake. Surrounding areas of

Mianzhu City are agricultural areas, in May, the main crops were wheat and rape, and in August, the main crops were rice and soybeans. The surface scattering was mainly caused by the crops in the fields. It can be seen from the comparison of Figure 3 (a) and Figure 3 (b) : a large number of non-natural objects in the city have changed. In order to further verify the experimental results, on-site survey was carried out on November 13, 2010, and two major changes were observed in the city proper: 1) most of the temporary shelters were demolished; and 2) reconstruction was basically complete.

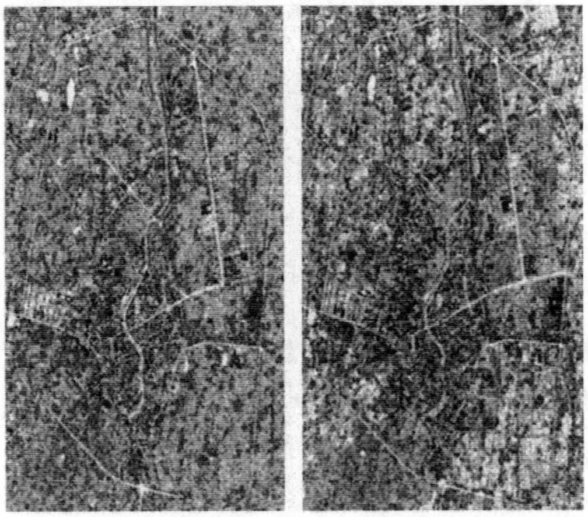

**Figure 3.** RGB composite images of the study area (a) August 9, 2008 (b) May 24, 2009

Change map of the topographic features of Mianzhu City for August 9, 2008 and May 24, 2009 based on the improved Yamaguchi four-component scattering decomposition model shown in Figure 4. Red areas show the presence of non-natural objects and green areas represent the absence of non-natural objects. Changes in the double-bounce scattering were mainly due to the post-disaster reconstruction, and some temporary houses were removed, and some new non-natural objects were reconstructed, and some vegetation was planted simultaneously. In farmland areas, changes in the green areas are due to the changes in the type of crops. The results of the experiments are consistent with our on-site survey and the high-resolution optical remote-sensing images. In order to further verify the accuracy of the proposed method in change detection, typical non-natural objects and crops were verified using high-resolution optical remote-sensing images and on-site survey. The statistical analysis showed that the correct rate of change detection of non-natural objects was about 78.8% and the correct rate of the detected unchanged non-natural objects was about 74.9%.

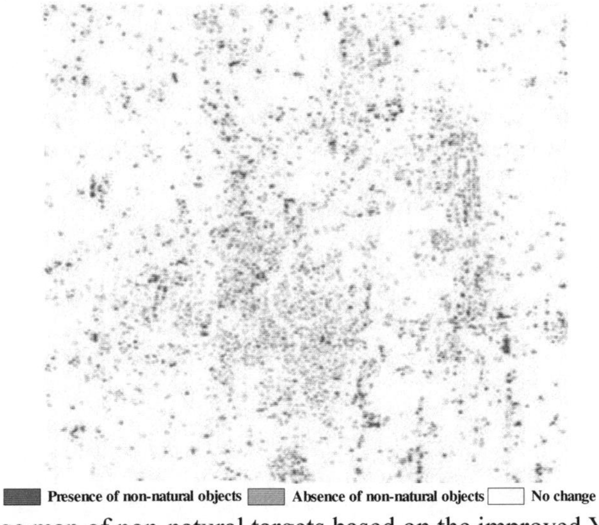


**Figure 4.** Change map of non-natural targets based on the improved Yamaguchi model

IOP Conf. Series: Materials Science and Engineering **371** (2018) 012047 doi:10.1088/1757-899X/371/1/012047

## 5. Conclusion

This paper proposes a new method for improving the accuracy of urban building detection based on the Yamaguchi decomposition model. The required rotation angle for enhancing the double-bounce scattering component could be determined by minimizing the cross-polarization component, and the double-bounce scattering component could be enhanced using the rotation method of the coherent matrix. Then, the helix scattering component and the enhanced double-bounce scattering component were combined for application to the detection of changes in urban buildings. The comparison of the detected changes using the full polarimetric RADARSAT-2 data and the corrected data for Mianzhu City, Sichuan Province, China showed the effectiveness of the proposed method in improving the accuracy of the detection of urban building changes. At the same time, it also provides a methodological support for detecting changes in urban buildings.

## 6. References


[1] Xu L.Y, Li S.Q and Deng Y.K 2014 Unsupervised classification of PolInSAR based on improved four-component decomposition. *J. Remote Sensing Letters* **3** 295-304

[2] Ainsworth T.L, Schuler D.L and Lee J.S 2008 Polarimetric SAR characterization of man-made structures in urban areas using normalized circular-pol correlation coefficients. *J. Remote Sensing of Environment* **6** 2876-2885

[3] Franceschetti G, Iodice A and Riccio D 2002 A canonical problem in electromagnetic backscattering from buildings. *J. IEEE Transactions on Geoscience and Remote Sensing* **8** 1787-1801

[4] Xia Z.G and Henderson F.M 1997 Understanding the relationships between radar response patterns and the bio- and geophysical parameters of urban areas. *J. IEEE Transactions on Geoscience and Remote Sensing* **1** 93-101

[5] Xiang D.L, Tang T and Ban Y.F 2016 Unsupervised polarimetric SAR urban area classification based on model-based decomposition with cross scattering. *J. ISPRS Journal of Photogrammetry and Remote Sensing* **11** 86-100

[6] Lee J.S, Grunes M.R and Pottier E 2004 Unsupervised Terrain Classification Preserving Polarimetric Scattering Characteristics. *J. IEEE Transactions on Geoscience and Remote Sensing* **4** 722-731

[7] Azmedroub B, Ouarzeddine M and Souissi B 2016 Extraction of Urban Areas From Polarimetric SAR Imagery. *J. IEEE Journal of Selected Topics in Applied Earth Observations and Remote Sensing* **6** 2583-2591

[8] Bhattacharya A, Singh G and Manickam S 2015. An Adaptive General Four-Component Scattering Power Decomposition With Unitary Transformation of Coherency Matrix (AG4U). *J. IEEE Geoscience and Remote Sensing Letters* **10** 2110-2114

[9] Zhang L, Li X.W and Guo H.D 2010 Effect of aspect angle normalized PolSAR data on urban building detection. *J. Canadian Journal of Remote Sensing* **3** 276-286

[10] Yamaguchi Y, Sato A and Boerner W.M 2011 Four-Component Scattering Power Decomposition With Rotation of Coherency Matrix. *J. IEEE Transactions on Geoscience and Remote Sensing* **6** 2251-2258

[11] Xiang D.L Ban Y.F and Su Y 2016 The cross-scattering component of polarimetric SAR in urban areas and its application to model-based scattering decomposition. *J. International Journal of Remote Sensing* **16** 3729-3752

IOP Conf. Series: Materials Science and Engineering **371** (2018) 012048 doi:10.1088/1757-899X/371/1/012048

# Assessment of Ecosystem Impacts by Engineering Measures Using the Concept of Building with Nature

**C Y Chen[1] and C R Ku[2]**

1 Department of Civil and Water Resources Engineering, National Chiayi University, Chiayi City 60004, Taiwan R.O.C.
2 Liu-Yi Development Co., Ltd, Tainan City, Taiwan R.O.C.

**Abstract.** A field investigation monitored a river restoration indicator, Kandelia, following the idea of Building with Nature (BwN). The main objective of the present study is to understand the relationship between Kandelia growth areas and river patterns; such an understanding can support river watershed management after engineering measures have altered rivers. The fitness-existence of Kandelia was evaluated through measurements and numerical simulations of river discharge, flow velocity, and depth of the river flow. The ecological survey results showed that although the quantity of Kandelia exhibited periods of growth and decline, the overall quantity did not change substantially. Even though the growing area shrank, the number of Kandelia plants did not decline and the biomass density increased. The flow rate was less than 0.4 m/s, equivalent to static. The numerical simulations indicated that the downstream sandbar area favored the growth of Kandelia, even under floodwater after typhoons had brought rain. When typhoons bring rain, the upstream has a high flow rate, which is not conducive to plant growth. A 5S model and a SWOT analysis were used for a comprehensive analysis. The indicator Kandelia is a local ecosystem index of environmental impacts after engineering measures have been conducted. The results may be used as a reference for river watershed engineering planning and to enhance the concept of BwN.

## 1. Introduction

Climate change has recently caused numerous natural disasters worldwide. We are forced to consider different measures for disaster prevention and mitigation and are obligated to alleviate stress on the natural ecosphere. Engineering measures for disaster prevention and mitigation require financial expenditures and have effects on the ecosystem. It would be desirable to reduce the ecological footprint of engineering measures by the restorative capacities of nature.

The program of Building with Nature (BwN) is sourced from the Netherlands [1]. The core element of the philosophy is to provide a solution that integrates both engineering design and protection of the natural environment. The philosophy entails learning the principles of ecosystem restoration and engineering design; the purpose is to develop more effective and sustainable hydraulic infrastructure. Engineering planning has rarely been concerned about the impacts on the natural environment; in 1970, in the Netherlands, scholars sought to promote both economic development and environmental protection simultaneously. Government provisions were made for planning engineering measures to minimize harm to the environment. The initial core concept was "Building in Nature," which asserted that when planning measures, engineers should consider their effects on nature. The concept was further enhanced into "Building of Nature." In the 1990s Environmental Impact Assessment laws of the Netherlands, the notion of "reshape nature" was introduced to mitigate the impacts of engineering

IOP Conf. Series: Materials Science and Engineering **371** (2018) 012048 doi:10.1088/1757-899X/371/1/012048

construction. A slogan, “Room for the River,” was proposed at this stage; this slogan was used in the plan of Black Stork [2]. The concept of the Room for the River Program was to allocate land area to the river to accommodate and manage higher water levels. The engineering planning for the program included lowering of floodplains, deepening of summer bed areas, storing water, relocating dikes, creating high water channels, lowering groynes, removing polders, removing obstacles, and strengthening dikes [3]. The habitat area of native aquatic plants in floodplain wetlands is a local ecosystem indicator of “Building of Nature.”

In recent years, BwN proposals have planned engineering projects for the benefit of the natural environment [4]. The idea of BwN is to promote natural development of the environment to reduce the ecological footprint of engineering measures; ultimately, it is hoped that this will benefit humankind. BwN combines diverse purposes, including safety, sustainability, nature, and socioeconomic development for the benefit of coastal communities and river delta communities [4-9]. The following five steps are adopted for the implementation of BwN design [4]:

- Understand the system (including ecosystem services, values, and interests).
- Identify realistic alternatives that use or provide ecosystem services.
- Evaluate the qualities of each alternative and preselect an integral solution.
- Fine tune the selected solution (in light of practical restrictions and the governance context).
- Prepare the solution for implementation in the next project phase.

Rivers in Taiwan are short in length and have steep slopes that cause soil deposition at river mouths; such rivers favour the growing of mangrove forests in the dry season. Nevertheless, these aquatic plants could be eroded and destroyed by flash floods in the rainy season. Climate change frequently causes typhoons, rainstorms, and abrupt changes in hydrological environments that influence ecological systems. It is necessary to combine engineering measures and environmental conservation for effective and economical natural disaster prevention engineering. The basis of BwN is the concept that humankind can live sustainably in the natural environment. This new design philosophy proposes to utilize the forces of nature, thereby strengthening the nature, economy, and society to prevent and mitigate natural disasters.

## 2. Study area and methodology

The study area is located at the mouth of Touchien River in Hsinchu County in northern Taiwan (Figure 1). The area at the confluence of a large river and its smaller branch has a sandbar called Jiugang Island. The branch has low discharge; masses of sand are deposited, which cause flood disasters when the area receives torrential rains. Engineering measures for flood disaster mitigation started in 2008 with a precast flood control dam and bypass flow engineering; river dredging works began in 2013. The field investigation area was located at the west side of Jiugang Island near the bypass flow and bank protection facility and the island’s flood detention pond. The study site was tidal land that formed after the formation of Jiugang Island; Kandelia grows in the study area because the bypass flow engineering facilities are upstream.

There are six types of relevant plants in the mangrove forest in Taiwan. The present study concerns the ecological system of Taiwan’s mangrove forest. The species Kandelia is selected as the indicator for the ecological restoration of the river after engineering measures. Kandelia is saltproof and submergence tolerant. It is commonly distributed at river mouths as a pioneer plant for newly formed beaches. Its growing status reflects the environmental tendencies of erosion or deposition.

Three specific field investigations were conducted in December 2014, April 2015 (before the flood season), and October 2015 (after the flood season). Items for the investigation included the number of Kandelia plants, their growth statuses, and distribution areas. Measuring tapes, counters, and GPS units were used for recording Kandelia plants’ heights, densities, and coordinates (Figure 2). Data were subjected to statistical analysis.

## 3. Results and discussion

### *3.1. Investigation in December 2014*

The first field investigation started in December 2014; the number and heights of Kandelia specimens were recorded, and the growing area was calculated using a geographical information system (GIS). The investigation was conducted in the winter season and thus avoided the effects of rainstorms; the weather was suitable for investigation of the Kandelia ecosystem. Results show that a growing area of 20,164 $m^2$ (Figure 3) had 99 plants with an average height of 17.57 cm (Table 1). Most of these Kandelia plants were just within two months of vivipation and were considered to be young plants. The Kandelia showed periodic succession that was affected by tides and floods. Some of the Kandelia plants were 30 cm in height and were speculated to be over 6 months of age. Biomass and biomass density are used to present the growing status of Kandelia in the environment. The biomass density is defined as the number of plants (Q) divided by their growing area ($m^2$), and the biomass is the height (cm) divided by area. High biomass density indicates that the Kandelia colony is healthy.

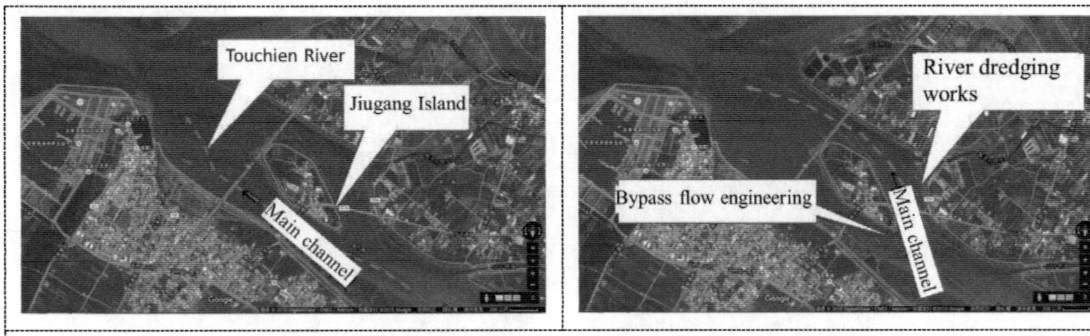


**Figure 1.** Study area at the mouth of Touchien River with (a) artificial dredging and (b) engineering measures in 2013.

**Figure 2.** Field ecology observation and investigation methodology.

*3.2. Investigation in April 2015*

The second field investigation started in April 2015. The growing area of Kandelia was separated into two parts (Figure 3), and the growing area decreased to 11,220 $m^2$. Between the first and second investigations, there were no typhoons and no torrential rains. It is speculated that the area was affected by tides, and the plants that were observed were adapted only for growth during nonflood seasons. The plant number and heights were increased during this survey, although the growing area was lower. The biomass density increased from 49 to 93 plants per $10^4$ $m^2$ (Table 1).

*3.3. Investigation in October 2015*

In October 2015, the third field investigation showed that the growing area of Kandelia was further decreased to 533 $m^2$ (Figure 3). In 2015, Typhoon Soudelor hit Taiwan on August 6 and Typhoon

Dujuan hit on September 29; they brought torrential rainfalls that caused high flow velocity and water levels that did not favour Kandelia growth.

The three investigations showed that the number of Kandelia remained stable at approximately 100 plants after one year of periodic changes (Table 1). The Kandelia undergoing vivipation, the aggregate plants, the biomass, and the density increased from 0.93% in April to 18.59% in October in 2015. Kandelia continued growing in the area, even when the area was separated into two zones (Figure 3). In the future, even if the Kandelia colony becomes a forest, plant aggregation can cause soil deposition, and the growing area may increase in area.

**Table 1.** Statistical results for three field investigations.

| Date | Plant number (Q) | Average height (cm) | Area ($m^2$) | Biomass density ($Q/m^2$) | Biomass ($cm/m^2$) |
|---|---|---|---|---|---|
| Dec. 2014 | 99 | 17.57 | 20164 | 0.49 % | 0.087 |
| April 2015 | 104 | 25.27 | 11220 | 0.93 % | 0.225 |
| Oct. 2015 | 99 | 28.42 | 533 | 18.59 % | 5.337 |

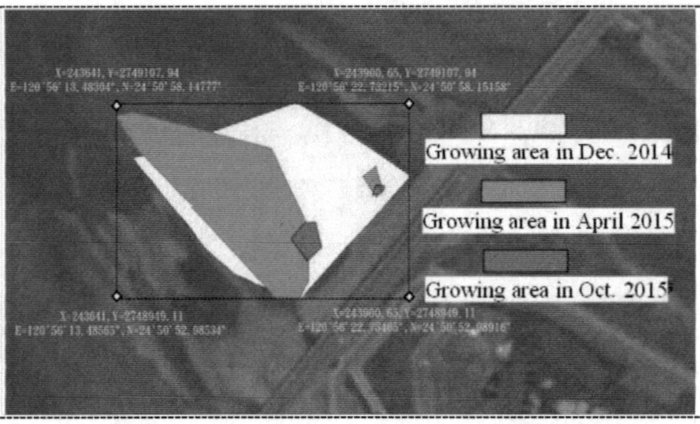


**Figure 3.** Field investigation results of Kandelia distribution.

## 4. Soil particle size distribution and growth of Kandelia

The riverbed soil particle size distribution within the Kandelia growing area can be examined using GIS spatial analysis. Some soil in the area is silt–fine sand soil in which the sizes of 90% of the particles ($d_{90}$) are smaller than 2 mm. Another soil is cobble–gravel, which has $d_{90}$ larger than 2 mm. In Taiwan, Kandelia grows on fluvial plains and tidal areas near river mouths or deltas; these areas have clayey muddy soil. The GIS cover layer analysis showed that Kandelia was growing only in areas with silt–fine sand soil area (Figure 4).

## 5. Flow velocity modeling and distribution of Kandelia

Typhoon Soudelor hit Taiwan in 2015 with a high sea level of EL 2.35 m. A numerical simulation of constant flow discharge (4.75 cm) at an average tidal sea level of EL 0.115 m by using CCHE2D modelling [10] for August 7–9 was analysed for hydrologic characteristics [11]. CCHE2D is a general surface water flow model that simulates the dynamic processes of water flow and sediment transport and water quality in rivers and along coasts. The peak discharge at high sea level was 1,180 cm at the river mouth according to the model.

The model showed that the peak flow velocity during Typhoon Soudelor was 1.6–2.4 m/s at the south channel, 1.2–1.6 m/s at the north channel, and 0.4 m/s downstream of the island (Figure 5). It showed that for low discharge (4.75 cm) and a sea level of EL 0.115 m, the study area was equivalent to static water and favoured the growing of Kandelia. However, high flow velocity caused by rainstorm-induced peak discharge (1,180 cm) could erode the soil of young Kandelia plants at the river channel.

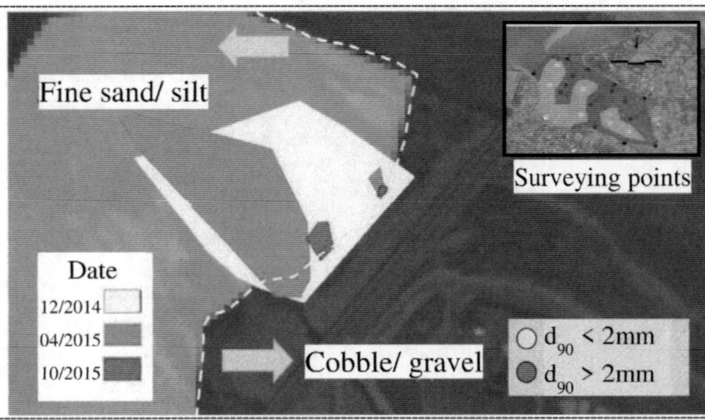


**Figure 4.** Kandelia distribution areas and corresponding soil types (the Second River Management Office of Water Resource Agency, http://www.wra02.gov.tw/).

Young Kandelia plants can survive beneath 0.5 m of water at static flow velocity. High flow velocity and long-term submergence (no more than 12 h) can threaten the young plants. Normal tidal conditions (12 h cycle of high and low tides) do not threaten the growth of Kandelia. However, if a typhoon brings a rainstorm, Kandelia plants can be submerged for periods longer than 12 h and could die even under low flow velocity.

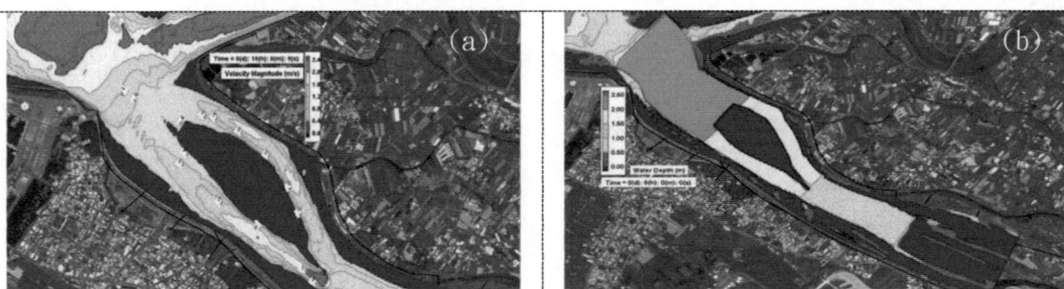


**Figure 5.** Modeled flow velocity during Typhoon Soudelor brought rainfall-induced (a) peak discharge, and (b) during average tidal of water depth [11].

## 6. 5S model analysis

A 5S model analysis featuring system conditions, stream hydrology, structures, substances, and species was used for the verification analysis of various parameters and the growing status of Kandelia (Table 2).

## 7. SWOT analysis

The study recorded the growing status of Kandelia in the specified river environment. For the assessment of the environmental impacts of engineering measures, a 5S model analysis formalized strengths and weaknesses (Table 3). A large Kandelia colony would have various strengths (S): it could incise the main channel of the river, purify water, and increase biological reproduction and biodiversity. Potential weaknesses (W) of a large Kandelia colony can include soil deposition at a river channel, attracting alien species that would be natural enemies, and imbalancing the ecosystem. A few engineering measures (opportunity, O) can control the threat (T) to optimize the ecosystem. For instance, reducing the number of Kandelia to avoid excessive proliferation can reduce the chance of flooding during rainy seasons.

**Table 2.** Comparisons between 5S model and Kandelia distribution areas.

| Remarks | Factors | Parameters | Kandelia | Field status | Fitness |
|---|---|---|---|---|---|
| Ecosystem status | Climate | average temperature per month (°C) | >20 | 22.7 | ++ |
| | Habitat type | wetland | ++ | ++ | ++ |
| | | intertidal | +++ | +++ | +++ |
| Hydrology | Hydrography | water depth (m) | <2.5 | 0.00–4.00 | + |
| | | daily soaking period (h) | <12 | +++ | +++ |
| | Hydraulic | flow velocity (m/s) | <0.5 | 0.00–4.983 | — |
| River type and structure | Riverbed material | mud | +++ | +++ | ++ |
| | | sand | +++ | ++ | ++ |
| | | cobble | ——— | ——— | ——— |
| | | gravel | —— | —— | —— |
| Water quality | Basic water quality | pH value | no data | 7.2–8.6 | |
| | | salinity | 5–15 | 3.0–22.2 | ++ |
| | Hydrated ions | conductivity (μs/cm) | no data | 299–563 | |
| RPI | Dissolved Oxygen (mg/L) | no data | no data | 6.6–13.7 | |
| | Biochemical oxygen demand (mg/L) | no data | no data | 1.2–3.7 | |
| Nutrients | Total phosphorus (mg/L) | no data | no data | 0.065–0.21 | |
| | Nitrate (mg/L) | no data | no data | 0.79–2.26 | |
| Species | Plant | degree of competition | Black Mangrove (extrusion the growing of young Kandelia) | none | +++ |
| | Animal | predators | Anoplophora maculata | none | +++ |

Note: The environmental fitness for the indicator Kandelia: + fair, + + favorable, + + + very favorable, - acceptable, - - unfavorable, - - - organism cannot survive.

IOP Conf. Series: Materials Science and Engineering **371** (2018) 012048 doi:10.1088/1757-899X/371/1/012048

The 5S model analysis shows that the temperature and environment are favorable to the growth of Kandelia without natural enemies in the study area. Both high sea levels and high flow velocities during rainstorms can adversely affect Kandelia growth. Rivers in Taiwan show high flow velocity upstream with mass sediments, which changes water flow and can adversely affect Kandelia growth downstream. The River Pollution Index of water quality was in the range of 2.25 to 3.75 and was classified as light to moderate pollution according to the criteria of the Environmental Protection Administration in Taiwan (http://web.epa.gov.tw/en/index.aspx). The field growth status of Kandelia shows that the effects of water quality do not pose a threat to survival; however, the analysis requires additional monitoring data.

**Table 3.** SWOT analysis of Kandelia distribution areas and ecosystem restoration strategy.

| Strategy | | Strengths (S) | Weaknesses (W) |
|---|---|---|---|
| | | 1. Incise the main river channel<br>2. Purify water quality<br>3. Biological diversity | 1.Could develop into forest and contribute to rising water table<br>2.Overdevelopment could affect other species |
| Opportunity (O) | Using engineering measures | 1. Environmental and engineering superiority can manage the growth of Kandelia.<br>2. Superiority of Kandelia can create extra benefits for the ecosystem. | Engineering measures can limit Kandelia's multiplication. |
| Threat (T) | Alien species are natural enemies | 1. Channel incision can limit Kandelia's growing areas to deep water areas<br>2. Biodiversity can change the numbers of natural enemies or can reduce the number of Kandelia by some alien species. | Engineering measures can limit Kandelia's growing area. |

## 8. Conclusions

This paper reports field investigations of the quantities of Kandelia near the Touchien River in northern Taiwan in the years 2014 to 2015. Results were analyzed to assess the environmental impacts of engineering measures. Kandelia was used as a local environmental indicator of the ecosystem after engineering measures. The study verified riverbed particle size distribution, flow velocity, and flow depth within the Kandelia growing area; the analysis explained the changes in hydrologic characteristics after engineering measures. Field investigations from October 2014 to October 2015 before and after two typhoon events showed that the biomass density of Kandelia was increasing. The water quality, flow velocity, and 5S model analysis showed that the study area was suitable for the growth of Kandelia after engineering measures. Some new Kandelia growing areas were vulnerable to floods and excessively fast river flows; these vulnerabilities limited the extent of the area suited to Kandelia. The study shows an example of a local indicator to be used when BwN for minimizing ecological harm caused by engineering measures. The species Kandelia can be used as an indicator for ecosystem impacts assessment after engineering measures have been conducted. Considerable time is required for observation, tracing, and collection of data to assess the ecological impacts of engineering measures for implementing the design of BwN.

IOP Conf. Series: Materials Science and Engineering **371** (2018) 012048 doi:10.1088/1757-899X/371/1/012048

## 9. References

[1] EcoShape 2012 *The Building with Nature Design Guideline* http://ecoshape.nl/en_GB/wiki-guideline.html (last visited January 11, 2017)

[2] De Bruin D, Hamhuis D, van Nieuwenhuijze L, Overmars W, Sijmons D and Vera F 1987 De toekomst van het rivierengebied (The future of the rivers region) Stichting Gelderse Milieufederatie, Arnhem, the Netherlands.

[3] UNEISO-IHE 2016 *Room for the River* (https://www.unesco-ihe.org/sites/default/files/13270-rvdr-brochure-governance-engels_def-pdf-a.pdf)

[4] De Vriend HJ, Van Koningsveld M, Aarninkhof SGJ., de Vries MB and Baptist MJ 2015 Sustainable hydraulic engineering through Building with Nature, *Journal of Hydro-environment Research* **9** pp 159–171

[5] De Vriend HJ and Wesselink A 2009 Building with Nature: ecodynamic design in practice *2nd German Environmental Sociology Summit Reshaping Nature*: Old Limits and New Possibilities Leipzig 6-7 November 2009 Keynote paper http://www.ufz.de/index.php?de=17540

[6] De Vriend HJ and Van Koningsveld M 2012 *Building with Nature: Thinking, acting and interacting differently*, EcoShape, Building with Nature, Dordrecht, Los Paises Bajos

[7] De Vriend HJ, Van Koningsveld M and Aarninkhof S 2014 Building with nature: The new Dutch approach to coastal and river works. *ICE Proceedings Civil Engineering* **167** pp 18–24

[8] De Vriend HJ, Van Koningsveld M, Aarninkhof SGJ, and Baptist MJ 2015 Sustainable hydraulic engineering through building with nature. *Journal of Hydro-environment Research* **9** pp 159-171

[9] Wilms T, Van der Goot F and Debrot AO 2017 Building with Nature - an integrated approach for coastal zone solutions using natural, socio-economic and institutional processes. Coasts & Ports 2017 Conference – Cairns, 21-23 June 2017

[10] Jia Y and Wang SSY 1999 Numerical model for channel flow and morphological change studies. *Journal of Hydraulic Engineering* **125** pp 924–933

[11] Yeh KC et al. 2015 *Study on overall planning of eco-hydraulics and benefit assessment of diversion dike in Touchien River's estuary* The Second River Management Office of Water Resource Agency (http://www.wra02.gov.tw/)

## Acknowledgments

Authors wishing to acknowledge financial support from Ministry of Science and Technology in Taiwan under contract No. MOST 106-2625-M-415-003.

IOP Conf. Series: Materials Science and Engineering **371** (2018) 012049 doi:10.1088/1757-899X/371/1/012049

# Disaster Resilience: A Sustainable Way for Niue

**Anthony Freddie[1,3] and Shenuka de Sylva[2,4]**

1 Candidate, MA in Architecture, SOA, Victoria University of Wellington, NZ
2 Lecturer, School of Architecture, Victoria University of Wellington, NZ
3 Email: aliuvaie@gmail.com
4 Email: shenuka.desylva@vuw.ac.nz

**Abstract.** The increasing frequency of natural hazards and disasters, due to climate change, severely compromises Niue's fragile economy and its future as a sovereign nation. Sustainable measures to building resilience and preventing natural hazards from turning into disasters are urgently needed to reverse Niue's population decline. Urban renewal and development projects formulated based on individual's experiences, outdated attitudes and approaches, or research conducted through pan-Pasifika and western lenses, and without the necessary policy controls, have severely compromised the safety of people and property. The dearth of research on sustainable and resilient planning, design and building practices, suited to the unique conditions of Niue and its people aggravate this situation. The paper presents traditional solutions and key findings from community-based studies conducted in Niue and from disaster preparedness and community resilience research in the Asia Pacific region. Using a case study approach the paper discusses new and sustainable solutions, methods and materials that are economically viable and practical, and resonate with local people, traditions, practices and the place. The aim of the paper is to demonstrate the opportunities that exist locally for devising and designing sustainable solutions to Niue's problems.

## 1. Introduction

'Savage Island' was the name given to Niue by early colonizers due to the fierceness they encountered from the local warriors fighting to protect their tiny island home from foreign invaders [1]. The resilience and tenacity of the Niuean people, which ensured the island's sovereignty from colonial dominance, had also enabled human inhabitation of a harsh island environment for several generations prior to the arrival of European explorers to the pacific. Known as the 'Rock of Polynesia' Niue is one of the largest raised coral atolls in the world with an area of 269sq.km [2]. Formed entirely of coral limestone and with no lakes or rivers, fresh water can only be harvested through deep bore wells; in addition its soil is considered unsuitable for large scale cultivation of most varieties of crops [2]. The steep limestone coastal cliffs that lift Niue 24m above sea level along with the coral reefs that surround it entirely with just a single break, makes fishing an arduous means of livelihood for the islanders [2]. Niue's geographical location in the pacific region does nothing to ease the difficulties of inhabiting it; lying in the path of numerous natural weather events cyclones and droughts are a regular occurrence, causing havoc on the already fragile sustenance sources and living environments of the island's population [2]. Recent studies on the impacts of climate change in the pacific region have shown an increase in the frequency of droughts and cyclones due to global warming [2, 3], and studies on climate change and food security in the region have identified the impacts of likely adverse weather conditions on food production and sourcing and the fragile future situation of small pacific island nations as Niue [2-4, 7].

Extreme weather events have had a direct and indirect devastating impact on Niue for several generations, and its continuity with increased frequency and intensity, will most certainly imply the future abandonment of this small island by its current youth and future generations who have a relatively safe haven with reliable educational and economic opportunities in New Zealand, which has granted Niuean's citizenship [5]. While natural disasters significantly undermined Niue's economy and its socio-cultural fabric [2, 3, 6], factors as climatically inappropriate and westernised building methods, materials and planning practices contribute to the widespread damage resulting from natural hazard situations as cyclones, and to the demoralising effect of ongoing damage to lives, property, public infrastructure, and livelihood sources [6] that drive Niuean's to find safety and security in New Zealand [3]. The devastating outcomes of Cyclone Heta in 2004 on Niue is a good example of this situation, it not only destroyed valuable infrastructure but also caused an exodus of Niuean's to New Zealand [7] upsetting the steadying of population numbers Niue was experiencing at the time, and desperately needed to sustain the future of the country.

The worst damage from Cyclone Heta, and the Tsunami that resulted, was experienced in Aliluki, the more modern built up south of the island's capital Alofi established on a highly exposed stretch of the island by early missionaries. The island's national infrastructure including its only hospital, national museum, public library, the Niue Hotel and the government-owned housing development Amanau Apartments were completely destroyed [2]. Other parts of the island, where local communities had established themselves several generations back in naturally protected areas and followed low built architecture traditions, experienced less damage. This situation sheds light on the seriousness of the issue of inappropriate planning and building methods and practices, and the importance of privileging traditional knowledge and resilient practices in fragile developing nations as Niue. New Zealand continues to support Niue's economy, as it has done for several decades, by providing nearly 50% of its GDP through various budget support programmes [2, 3] in addition to economic aid which was estimated at NZS$13.8 million for 2016-2017 [8].

## 2. Aim and Method

This paper presents the findings of a study carried out with the aim of (a). identifying a sustainable architectural solution to the effects of climate change on Niue, and (b). to increasing the resilience of the Niuean people to the damaging effects of ongoing natural events, and thereby (c). support the economic growth of Niue and reduce future population drain. The study focused on identifying a set of criteria which can be utiliesed as a guide for developing an ecological, resilient and sustainable architectural response to Niue's cyclone conditions, which can also contribute to building local capacity, skills and knowledge and thereby improves the economic condition and livelihood opportunities available to local people. The paper begins with a discussion of the key issues affecting the disaster resilience and recovery of communities in Niue. This is followed by key finding from community and site based studies and literature reviews, both historical accounts on Niue and recent studies on post disaster reconstruction in Asia and the Pacific. Finally the paper presents and discusses briefly a case study which applied the key findings to a hypothetical architecture design project and tested their potential to support design outcomes that meet the aims (a), (b) and (c) mentioned above.

## 3. Issues and Findings

### *3.1. Problem*

The dearth of research on sustainable practices in Niue has negatively affected its development for several decades. Modern development endeavors have been formulated based on individual's experiences, outdated attitudes and approaches, or the findings of research conducted through Pan-Pasifika or western lenses [3]. The lack of critical assessment of urban renewal and redevelopment projects and their sustainability and resilience to natural hazards, particularly since the 1990's and following the devastation caused by Cyclone Ofa [3], resulted in widespread damage to these development projects in 2004 following Cyclone Heta [2]. In addition, the unique contextual characteristics of Niue have made it a testing ground for a range of ad-hoc and experimental building projects [2]. However, the lack of evidence-based design and building practices has meant that these projects too have not been suitable to Niuean culture and context and as such has not benefited the people [2, 3].

The conditions that contributed to this situation is the ongoing and gradual westernisation that has taken place over several decades, through processes of christianisation, European education systems, modernization [3], and since of late globalisations. This has contributed to the undermining of the value of traditional knowledge and practices causing these to be ignored, discarded and eventually forgotten. This situation has also cultivated a preference for imported ideas, practices, materials, technologies and guidelines, most of which are not suited to Niue's unique local context or living practices, and as they are often applied without adaptation or modification to suit local environmental conditions and culture they fail [3].

Currently Niue imports most, if not all, of its building materials except for the aggregate used in the foundations of buildings, albeit the availability of some materials locally [2, 3]. The damage caused by collapsed buildings and debris in natural hazard conditions have demonstrated the disastrous effects of using heavy and solid materials as concrete or sharp flyaway materials as sheet iron roofing in volatile environments as Niue. What is not readily acknowledged in post disaster analysis is how these imported materials and poor adaptation of non-vernacular construction systems have exacerbated fragile economies, as in Niue, by increasing the damage to lives, property and the natural environment [6].

Sourcing resources locally by growing and manufacturing the materials needed for building projects, devising new vernacular planning and design solutions using local knowledge and skills offers huge advantages both for developing local capacity and improving livelihood opportunities and thereby the country's economy [2, 3, 6]. Such processes also ensure quicker recovery from disasters and greater resilience both in the short and long term [6]. Additionally, land in Niue is scarce and every effort must be made to explore sustainable and resilient methods for redeveloping sites destroyed by natural hazards and to reduce the damaging effects of natural hazards on living environments, livelihood sources and public infrastructure. Reviewing the issues Niue faces and identifying sustainable methods and practices to replace current practices is urgently needed and vital for improving the resilience of local communities and for ensuring a sustainable future for this small island nation currently under serious threat population drain to New Zealand.

### *3.2. Lessons*

Key findings from community-based research, empirical studies, and interviews conducted in Niue in 2017 and the analysis of post disaster accounts by locals demonstrated the negative consequences of local traditions and practices being replaced by westernised and imported methods and materials, and highlighted the lack of innovative solutions sensitive to the local context. A primary finding from these studies was that the replacing of traditions with modern knowledge and technologies have not necessarily ensured the safety of people and property in disaster situations [2, 3, 6]. The loss of lives in the situation of Cyclone Heta in 2004 was significantly reduced only because people had followed traditional safety practices and taken refuge in caves, amongst the large above ground root networks of well-established banyan trees, locally known as the 'Ovava' tree, and in similar natural shelters as their ancestors had done in the past. It is believed by some Niuean elders interviewed that traditional stone buildings with walls of thicknesses of over 1 meter, constructed by Niuean ancestors, and capable of withstanding the forces of natural hazards were demolished early in the 20th century due to westernised mindsets that feared their collapse in cyclones. The buildings that have replaced traditional thatch architectures of Niue and the early stone structures are made of imported concrete and sheet iron, which are neither suited to the tropical heat or the hazardous conditions of Niue [3]. Several such homes in local villages can be seen abandoned and damaged beyond repair after a cyclone.

Although there is no written record of Niuean building traditions and architecture prior to its settlement by missionaries, early European accounts hint at elongated oval forms of thatched structures arranged in community clusters and sitting sheltered amidst groves of tall coconut palms [1, 9]. Following settlement of Niue by Samoan and British missionaries from the London Missionary Society, buildings within the grounds of the mission were constructed in the same traditional

IOP Conf. Series: Materials Science and Engineering **371** (2018) 012049 doi:10.1088/1757-899X/371/1/012049

elongated oval form but using imported concrete instead of locally sourced thatch, and were whitewashed with locally sourced lime plaster [9]. The roofs of these cottages took on the high pitched thatched roof forms found in Samoa [9]. A description of this architecture is given by B.C Thompson who visited the island in 1901, he states that the natives preferring to live in the 'thatched hovels' of their ancestors had built traditional homes behind the stiffly arranged row of white washed concrete cottages [9]. Thompson also notes that the new houses stood on the landside of a grassy lawn that extended to the edge of the cliff [9]. From this we can gather that the new houses stood exposed to the natural elements and wind forces, unlike the tree sheltered pre-Christian villages.

From Thompson's accounts we know that in 1901, 11 traditional villages were positioned around the island, established most likely on the safest sites on the island, and the land granted to the missionaries was separated from these villages and in an area known as Alofi [9]. Alofi, is a coastal stretch to the south of the Tuapa, the village in which the king had resided and here was established a church and houses for the mission [9]. Subsequently Alofi developed into the capital of Niue. Following Cyclone Heta in 2004 Alofi sustained heavy damage and the more exposed south Alofi where most modern public infrastructure and housing had evolved from the foundations established by the missionaries was completely destroyed [10]. Villages located in and around traditional sites sustained only minor damage from the cyclone, perhaps due to the careful positioning of these based on the knowledge of ancestors, who most likely were competent sailors and navigators and understood the natural weather hazards that afflicted the island. Following its 2004 destruction and its history of vulnerability to cyclonic winds, storm surges and high waves Alofi as assessed as a high-risk area, and the Niuean government proposed relocating the capital to the upper plateau around Fonuakula [11].

From these site studies and literature reviews key actions were identified for increasing the resilience of local communities, living environments and recovery after disaster. These are listed below in Table 1.

**Table 1.** Key Actions informed by site studies and literature reviews.

| |
|---|
| 1. Identify, improve and build local capacity |
| 2. Improve and diversify livelihood sources |
| 3. Safeguard traditions and traditional knowledge |
| 4. Utilise traditional knowledge effectively to improve living environments |
| 4. Develop building technologies to suit local conditions, skills and capacity |
| 5. Adapt modern and proven technologies, methods and systems to local conditions |
| 6. Utilise traditional craft (eg. weaving, lashing) to improve design and construction |
| 7. Grow and source building materials locally |

*3.3. Lessons from Asia and the pacific*

To overcome the problems Niue faces it is essential to draws from lessons learnt from the wide range of studies conducted in the Asia Pacific region following recent natural disasters as the 2004 Asian Tsunami, 2008 Wenchuan Earthquake, China, 2008 Cyclone Nargis, Myanmar, 2010 and 2011 Christchurch Earthquakes, New Zealand, 2011 East Japan Earthquake, 2013 Typhoon Haiyan, Philippines, 2016 Cyclone Winston, Fiji, to name a few. The vast body of research that has emerged in recent years related to disaster preparedness, recovery and reconstruction from Asia and the Pacific including World Bank and UN publications provide valuable insight on reconstruction processes including post disaster governance structures and governance issues, funding and donor organisations involvement, disaster preparedness, and methods for building community resilience [3, 6, 11-17]. Research on reconstruction efforts following recent disasters also highlight the issues with top down and bottom up approaches to rebuilding, multiple actor involvement in rebuilding processes ranging from governments to local councils, local and international donors, religious organisations and insurance companies, and the consequences of this on the speed and success of recovery efforts [11-17]. They also highlight the impact of the lack of coordination across various organisations and at various levels, inequality in the distribution of funds and development benefits, unsuitability of development projects to culture and context and lack of accountability [6, 11-17]. In addition studies also demonstrate how poor and vulnerable communities are disempowered through these development endeavors, the value of being prepared for disaster and having in place well considered disaster risk reduction and recovery strategies as well as rebuilding strategies [6, 11-18]. The Sendai Framework for Disaster Risk Reduction 2015-2030, the outcome of the United Nations World Conference on Disaster Reduction in Sendai in 2015 together with those studies mentioned above reinforces the need to act 'with a renewed sense of urgency within the context of sustainable development and poverty eradication, and to integrate, as appropriate, both disaster risk reduction and the building of resilience into policies, plans, programs and budgets at all levels and to consider both within relevant frameworks'[18].

Table 2 lists some key lessons and recommends actions for consideration in Niue when planning for future disaster preparedness and recovery. These are some key points identified for the purpose of devising a sustainable solution for a resilient architecture that could reduce the risk of natural hazards from developing into disasters, prevent loss and injury to life and minimise damage to property and livelihood sources. The aim of the table is to clearly outline the current situation and the actions required to ensure development projects contribute to building the resilience of local people, local economies and local environments.

**Table 2.** Key Lessons and Actions for Disaster Resilience Planning in Niue.

| | |
|---|---|
| Lessons | 1. Funding: high dependence on NZ, international governments, donor organisations and private insurance for recovery after disaster due to the under-development of opportunities for economic growth and food production locally [2-8].<br>2. Relocation: no place to relocate to after a disaster other than to NZ but at severe negative consequence to Niue's economy and future [2-7,14].<br>3. Top down and westernised approaches: likely to be imposed in situations of external funding through bureaucratic processes, may not align with local aims [3,12,14-17],<br>4. Imported materials, methods and skills; does not build capacity and resilience at local levels [3,7,10,14-17]. |
| Actions | 1. Local engagement: communities to take ownership of all future development endeavor at various levels and capacities to ensure their suitability and sustainability.<br>2. Positive contribution: to the local economy, socio-culture and environment must be demonstrated by all proposed development projects.<br>3. Localised knowledge: clear understanding of local environments, capacity, traditions, local resilience structures, suitability of new technology and materials to Niuean culture and context to be demonstrated by all key actors involved in the planning and implementation of development projects.<br>4. Tried, tested and adapted: technologies, methods and materials tried and tested in similar environments conditions and adaptable to suit local conditions to be used. |

## 4. Case Study

One of the national development goals outlined in Niue's National Strategic Development Plan is to become self-sustainable and self-reliant [5]. Two main sources of income for Niuean's outside of government employment or welfare programmes, is from locally owned and operated small scale eco-tourism endeavors and agriculture [3, 7, 10]. Based on this information and the key lessons and actions listed in Table 1 and 2 above, the case study project focused on a speculative architectural proposal for facilitating the strategic economic and social development of local communities through a rural or village-based eco-tourism programme, managed and financially supported by a centrally based vocational training, tourism and cultural complex.

IOP Conf. Series: Materials Science and Engineering **371** (2018) 012049 doi:10.1088/1757-899X/371/1/012049

A site in South Alofi, the exposed coastal stretch on which the public center of Niue, destroyed by cyclone Heta in 2004 had once stood, was selected as the testing ground for this speculative resilient and sustainable architecture project. The diversity of the project programme and aims allowed for the design and construction solutions developed to be adapted and tested at various scales. The project intention was to satisfy the three aims of the study, **(a).** identifying a sustainable architectural solution to the effects of climate change on Niue, **(b).** to increasing the resilience of the Niuean people to the damaging effects of ongoing natural events, **(c).** support the economic growth of Niue and reduce future population drain.

### *4.1. Sustainable site intervention for safety from natural hazards*

Detailed site studies revealed the multiple hazards that needed to be considered and potential solutions [10]. From these it was established the safest location for the building complex would be further inland on the site and on higher ground. The gradual rise and depressions on the land would allow the building forms and foundations to wrap around and hug the hard limestone terrain and form protective cave like spaces. Studies on sites prone to similar natural hazards to Niue and precedent design projects from an international context demonstrated the benefits of planting heavily, with native coastal trees and shrubs the most exposed and dangerous parts of the site particularly the flat terrain close to the cliff edge and susceptible to water surges and tsunamis to minimize wave damage [10]. These solutions were found to be feasible alternatives to the building of massive seawalls and engaging in large engineering projects, and better suited to local capacity and context when combined with early warning systems.

### *4.2. Building form, structural systems and materials for resilience*

Achieving resilience to cyclones was the key aim of the building form and structure, inspiration for this was drawn from the structural principles of the native Ovava tree which has evolved an ingenious structural system that has sustained this tree in harsh weather conditions for several decades, and structurally supports its branches as they spread out far and wide. Through process of physical modelling, the capacity of an organically interconnected above and below ground structural system was established. From these studies emerged a built form and construction system suited to the site and context. Bamboo was identified, after study of its versatility, unique qualities and sustainable nature as a suitable material for the structure.

The structural capabilities of bamboo can be maximized by combining multiple reeds and by bending it to create curved forms and round shapes that deflect with cyclonic forces. Traditional tying and lashing techniques were tested for securing these forms and shapes in a manner that allows the building to flex and move within acceptable limits in cyclonic conditions. A large central and several secondary structural cores were designed to acts in a manner similar to the central trunk and supportive above ground root networks of the Ovava tree. This system supports the building laterally and vertically when cyclonic forces are applied. A system of raised concrete foundation piles, designed similar to the ground root network of the Ovava tree, protects the bamboo and anchors the sub structure to the ground. Formed as a circular maze these cave like foundations are also a protective vertical barrier to tsunami waves that may reach the building. In addition low entry points with step down entrance spaces, strategically located in response to the terrain and raised foundations, minimizes heavy and turbulent winds from forcing uplift in the curvilinear sub structure. The roof, a diaphragm system inspired by traditional basket weaving techniques, acts like a lattice brace that links and structurally reinforces the whole building to resist lateral loads. Local traditions of thatching and local palm material resources were considered for roof coverings. The proposed building design is adaptable to domestic and public scale buildings and is a structurally safe, sophisticated and aesthetically beautiful response suited to the environment, local culture and local capacity.

## 5. Discussion

Community resilience is a measure of how effectively a community bounces back after a natural disaster. As such the architectural response focused on not just structural resilience of the built form and reducing damage to lives, property and the environment in a disaster situation but also on how architecture can contribute to building the economic resilience of the community. With this aim, design concepts, construction methods and material that resonates with traditional knowledge and skills was identified. Bamboo a sustainable building material suited structurally for cyclonic conditions and architecturally for its environmental and aesthetic benefits to Niue was selected. Additionally bamboo is a resilient plant that can be locally cultivated, harvested within a short and treated at village levels cost effectively. Also current familiarity with the material and its wide range of uses beyond construction makes training attractive to local. This increases its viability as a sustainable livelihood source as pioneered by the Green School in Bali [19] and when cultivated locally a readily available environment friendly and versatile resource as evidenced in its use in contemporary architecture in southeast Asia [19, 20, 21].

Integrated into the building design and the landscaping of the complex are methods for renewable energy systems as solar, wind, biogas and biomass, and reliable food cropping systems such as hydroponics, aquaponics and greenhouses, and water collection and treatment. This was done with the intention of not only developing an efficient self-sufficient built complex but also as system to train and upgrade the skills and knowledge of local people to enable transfer and application of these systems at a village and domestic scales. In addition, urban planning and design guidelines for the complex and site were developed to be adapted and applied to similar exposed sites. Through this design the aim of having ecologically sensitive and locally accessible solutions that resonate with local knowledge and understandings and increase economic opportunity was achieved. Importantly architectural and planning solutions were developed around the cultural values, knowledges and beliefs of the people, and by utilising the natural environment as design inspiration and a key resource for a sustainable and resilient solution for Niue.

## 6. Conclusion

This project demonstrates the opportunities that exist locally for devising sustainable solutions to the key issues faced by Niue, primarily lack of resilience to the natural hazards that afflict this island nation. As the paper demonstrates in-depth research is essential for context and culture appropriate solutions and to prevent development projects that are potentially harmful to people and thereby to a fragile economy and resilience of a small island nation. In the last two decades a vast body of research has emerged from the Asia-Pacific region on disaster preparedness, disaster resilience and disaster recovery. These studies provide valuable lessons for Niue and must to be considered when planning for future disasters. In addition the building and material technologies, and methods, that have been developed in Asia and the Pacific in the last decade are proving to be highly advanced, more appropriate to local conditions and significantly different to the western modernaisation approaches and methods that have dominated development endeavors in emerging economies up to now [19, 21]. These offer a sustainable way forward for non-western small island nations as Niue. As research has shown the main aim of development agendas for Niue must aim prevent the transplanting of western and unsuitable methods and technologies and instead ensure projects that contribute to the sustainable development of the island and increase the resilience of its people.

## 7. References

[1] Tregear E 1893 *Niue: or Savage Island* The Journal of the Polynesian Society, Volume 2, p 11-16. Auckland, NZ.
[2] Pacific Adaptation to Climate Change 2006 *Niue Island: Report of In-country Consultations*. Available from: https://www.sprep.org/publications/pacc-report-of-in-country-consultations-niue (accessed on 15.05.2017).
[3] Barker C J 2000 *Hurricanes and socio-economic development on Niue Island* Asia Pacific Viewpoint, Vol 41, No. 2, August 2000, p 191-205.
[4] UN Report 2008 *Climate change and food security in Pacific Island countries: Issues and requirements* Food and Agriculture Organisastion of the United Nations, Rome.
[5] Government of Niue 2016 Niue *National Strategic Plan 2016-2018*, p 24-33.
[6] Oliver-Smith A 1996 *Anthropological Research on Hazards and Disasters*, Annual Review of Anthropology Vol 25, p 303-328.

IOP Conf. Series: Materials Science and Engineering **371** (2018) 012049 doi:10.1088/1757-899X/371/1/012049


[7] Barnett J 2012 *On the risk of engineering mobility to reduce vulnerability to climate change: insights from a small island nation* in eds Hastrup K & K Fog Olwig, Climate Change and Human Mobility: Challenges to the Social Sciences, Cambridge University Press, UK.

[8] NZ Ministry of Foreign Affairs and Trade 2017 Available from: https://www.mfat.govt.nz/en/aid-and-development/our-work-in-the-pacific/aid-partnership-with-niue/. (accessed on 10.12.2017).

[9] Thomson B C 1902 *Savage Island: An Account of a Sojourn in Niue and Tonga*. John Murray, Albemarle Street, London.

[10] Pickett S, Cadenasso M & McGrath B (eds) 2013 *Resilience in Ecology and Urban Design: Linking Theory and Practice for Sustainable Cities*, Springer Science & Business Media.

[11] ReliefWeb 2004 *Niue recovery at $23 million, Alofi to be relocated* Available from: https://reliefweb.int/report/niue-new-zealand/niue-recovery-23-million-alofi-be-relocated.

[12] Johnson L & Olshansky R 2016 *After Great Disasters: How Six Countries Managed Community Recovery* Lincoln Institute of Land Policy, MA, California.

[13] World Bank 2006 *Not If, But When: Adapting to Natural Hazards in the Pacific islands Region: A Policy Note* Washington, DC.

[14] Sanghi A, Ramachandran S, Fuente A, Tonizzo M, Sahin S & Adam B 2010 *Natural Hazards, UnNatural Disasters: The Economics of Prevention* International Bank for Reconstruction and Development (The World Bank) Washington DC.

[15] Lyons M, Shilderman T and Boano C (eds) 2010 *Building Back Better, Delivering People-centered Housing Reconstruction at Scale* Practical Action Publishing Ltd, UK.

[16] Anderson M, Brown D and Jean B 2012 *Time to Listen: Hearing People on the Receiving End of International Aid*. CDA Collaborative Learning Projects, Cambridge, MA.

[17] Daly P and R Feener (eds) 2016 *Rebuilding Asia Following Natural Disasters: Approaches to Reconstruction in the Asia-Pacific Region* Cambridge University Press, Cambridge, UK.

[18] UNISDR 2015 *Sendai Framework for Disaster Risk Reduction 2015-2030*, Switzerland.

[19] Hardy J 2010 *My Green school Dream* Technology Entertainment Design (TED) Global, available from: https://www.ted.com/talks/john_hardy_my_green_school_dream

[20] Minke G 2016 *Building with Bamboo: Design and Technology of a Sustainable Architecture*, Edition 2, Birkhaus, Basel.

[21] Newman P and Matan A 2013 *Green Urbanism in Asia: The Emerging Green Tigers,* World Scientific Publishing Company Pte Ltd, Singapore.

IOP Conf. Series: Materials Science and Engineering **371** (2018) 012050 doi:10.1088/1757-899X/371/1/012050

# The Effects of Windows and Reduced Night Temperature on the Heating Energy Demand in Different Regions of Russia

**Z Pásztory[1]*, G Gorbacheva[2], D Czimondor[1], P Rébék-Nagy[1], V Sanaev[2], S Rykunin[2], I Czupy[1], Z Börcsök[1]**

1 University of Sopron, 4 Bajcsy Zs. str. Sopron, Hungary, 9400
2 Federal State Budgetary Educational Institution of Higher Education «Bauman Moscow State Technical University» (Mytishchi Branch), 1st Institutskaya street, 1, Mytishchi, Moscow region, Russian Federation, 141005
* corresponding author
E-mail: pasztory.zoltan@uni-sopron.hu

**Abstract.** An energetic model of a typical log home was placed in ten different regions of Russia and examined the yearly heating energy demand. The building was supplied with three different types of windows to demonstrate the effect of windows on the heating energy demand. In a 108 $m^2$ building, changing the 3.5 $W/m^2K$ thermal insulation capacity windows to 1.4 $W/m^2K$ and 0.7 $W/m^2K$ windows caused an energy saving of 11.9% and 15.9% heating energy respectively. In a heated space reducing the night temperature by 2 degrees Celsius can result in only a 2.7% (1 865 kWh) saving in the colder Magadan region, while in the Mediterranean Krasnodar, there was a 4.48% (1 151 kWh) saving of the net heating energy amount. Calculations estimated that Russia possesses a very significant heating potential savings.

## 1. Introduction

Because of their large number and significant energy use, residential buildings are responsible for a high proportion of all utilized energy. Residential buildings consume 30-45% of all utilized energy. This value differs from country to country depending on climate conditions, and the thermal properties of buildings [1, 2]. Due to the high energy proportion and inefficient utilization, the energy saving potential is high mainly due to the thermal properties of buildings [3-6]. When building and retrofitting houses, with higher energetic requirements, the main goal is to improve thermal insulation and to increase energy efficiency, to reduce energy consumption. These modifications are important both for the environment and operation cost-efficiency, which is proved by several studies [7-10].

During the determination of the thermal characteristics of a building, special attention must be given to the windows, because their heat loss can significantly surpass that of other building parts. Several investigations emphasize the importance of the windows in the heat loss of buildings and the opportunity to harness solar gaining [11, 12]. Grynning et al. [13] determined that the possible heat loss ratio related to the windows can be over 40% according to the actual Norwegian regulation [14]. The optimal thermal parameters of windows are influenced by their size and orientation and annual heating-cooling energy demand which connects directly to the climatic conditions as Jber and Ajib determined [15].

Besides the thermal parameters of buildings, the geographical location and position play a very important role determining the heating-cooling energy. The same building requires different amounts of energy in different weather and orientation conditions to provide the same inner circumstances. The Russian Federation is the biggest country in the world. It stretches from Eastern Europe to the northern

Pacific Ocean. There are many types of climate in its 17 125 187 km$^2$ area. Mostly the continental features dominate, but near the Arctic circle the climate is subarctic, and subtropical climates can also be found near the Black and Caspian Seas.
The great majority of residential houses in rural areas of Russia are wooden houses [16, 17]. They have a centuries old tradition to build log homes, which are an advantageous solution in local conditions. Good quality wood as building material has always been available in great amount and so far the 18-25 cm thick log wall could satisfy the regulations [18]. The regulations set recommendations for "low-rise" buildings which includes the individual residential buildings, but does not determine the minimal thermal resistance of the wall structure.
The three main goals of the present research are: (1) determination of heat energy demand of the same building in ten Russian cities with different climates, (2) determination of the role of windows with three different insulation levels regarding the heat energy demand of an average Russian log house, (3) determining the available energy saving by reducing the building's inner temperature at night in theses ten regions.

## 2. Materials and methods

The heat energy model of a typical Russian log building was built with the help of Finite Element Model simulation. Then the building's virtual energetics model was placed into different regions of Russia and an energetics simulation was run to determine the heating energy demand. Although the solar gaining and wind have also significant effects on the heating demand, the necessary local data on sunshine and wind data was not available. That is why the orientation of the building was not considered in the calculations in other words it was the same in all cases. The heated area in two floors totals 108 m$^2$ with an almost square ground plan. The building stands on concrete point foundations. The layer order of the walls, roof, ceiling, floor is shown in Table 1.

**Table 1.** Layers of the building.

| No. | Floor | | Ceiling | | Roof | | Side wall | |
|---|---|---|---|---|---|---|---|---|
| | Thickness [mm] | Name | Thickness [mm] | Name | Thickness [mm] | Name | Thickness [mm] | Name |
| 1 | 40 | sawnwood | 40 | sawnwood | 0.7 | metal tile | 200 | log beams |
| 2 | 1 | water insulation | 200 x 50 | beam frame | 25 | purlin | | |
| 3 | 200x 50 | beam frame | 150 | rockwool thermal insulation | 25 | counter batten | | |
| | 200 | rockwool thermal insulation | 1 | vapor barrier | 2 | water insulation | | |
| 4 | 25 | lower floor boards | 20 | final decking | 200 | rafter | | |
| 5 | 25 | lath frame | | | 150 | thermal insulation | | |
| 6 | | air gap | | | 1 | vapor barrier | | |
| 7 | | soil | | | 25 | purlin | | |

The biggest proportion of the wall is made of a simple laminated 200 mm thick log, whose U value is 0.5 W/m$^2$K the thermal conductivity of coniferous wood material is 0.11 W/mK and the inner surface coefficient of heat transmission is 1/24 while the outer is 1/8. The virtual thermodynamic model of the building was created in WinWatt software (http://www.bausoft.hu) based on the data of Table 1. The characteristic values of each building part were chosen from the database of the software, which provides up-to-date data for big amounts of building material. The model was run with windows having three different thermal insulation capacity having unified thermal conductivity („U" value) of 3.5 W/m$^2$K, 1.4 W/m$^2$K and 0.7 W/m$^2$K. These values include the thermal parameters of the glass and frame weighting in proportion of the surface. Total surface of windows was 18 square meter including the frames. The thermal finite element model made it possible to calculate the current heat loss of the building for every temperature difference between indoors and outdoors.

The yearly heat energy demand was integrated from the hourly data of the inner and outer temperature difference with the help of a program made for this purpose (EnergiKalk software) [19]. The program allows us to determine the energy demand of the same building in different climatic conditions. and counts the heat energy consumption for a given period using the weather data for the given city and the inner temperature for the building. With the difference of the building's inner and the outer temperature, there is a specified heat energy demand. The energy demand of the house for a given period of the year is derived by integrating this hourly data. The yearly energy demand in the first case was determined with an inner temperature of 20 degree Celsius. In the second case, the temperature was lowered to 18 degrees between 10 pm and 6 am thus reducing the difference of the inner and outer temperature and consequently the heat loss. The calculated energy demand results the net energy needs of the building.

The chosen cities for modeling are Archangelsk, Chelyabinsk, Irkutsk, Krasnodar, Krasnoyarsk, Magadan, Moscow, Omsk, St. Petersburg and Vladivostok (Figure 1). The building had been investigated can be built legally any types of windows mentioned in this article which mean that all of them meet the official regulations. Near each of these cities there is a civil airport whose weather data between 01/01/2014 and 31/12/2014 were used during modeling [rp5.ru]. The exactness of the data was checked with the help of other weather stations and found that the data are identical.

**Figure 1.** The location of the cities studied.

Although in some of the cities the outdoor temperature hardly reaches the indoor temperature, in several other locations the temperature far exceeds the indoor temperature of 20 degrees. In this case cooling energy would be needed to keep the buildings inner temperature. The climate conditions of Russia require cooling for only a short period therefore the air conditioning systems generally are not used. Therefore during modelling was not calculate the demand for cooling energy. Table 2 presents the location and temperature data of cities.

The biggest yearly fluctuation of temperature is in Krasnoyarsk, the lowest yearly average temperature is in Magadan, the highest is in Krasnodar. The number of heating hours was the highest in Magadan and the lowest was in Krasnodar. Despite its northern location, Archangelsk is not the coldest due to the equalizing effect of the sea. In most cities the continental effect prevails on different scales. The number of heating hours is affected by several factors besides geographical location, such as sea currents, wind conditions and elevation and the terrain. The result is that the northernmost city has not got the coldest average temperature. In Russia's continental territories the number of heating hours is approximately the same although the minimum and maximum temperature values are different. It was determined the amount of the hourly temperature differences in the heating temperature of each region by the number of heating degree hours.

**Table 2.** Geographical data of the cities and extreme temperatures in 2014.

| Town | North latitude | East longitude | Mean-temperature (°C) | Maximum temperature (°C) | Minimum temperature (°C) | Number of heating hours* |
|---|---|---|---|---|---|---|
| Archangelsk | 64.55 | 40.56 | 2.4 | 32 | -30 | 8246 |
| Chelyabinsk | 55.16 | 61.44 | 2.9 | 32 | -35 | 7877 |
| Irkutsk | 52.29 | 104.31 | 1.0 | 31 | -34 | 7926 |
| Krasnodar | 45.03 | 38.97 | 12.6 | 37 | -17 | 6125 |
| Krasnoyarsk | 56.02 | 92.89 | 1.3 | 35 | -35 | 7927 |
| Magadan | 59.57 | 150.8 | -4.2 | 27 | -45 | 8627 |
| Moscow | 55.76 | 37.62 | 6.4 | 33 | -26 | 7592 |
| Omsk | 54.98 | 73.37 | 2.1 | 35 | -35 | 7707 |
| St. Petersburg | 59.93 | 30.34 | 6.6 | 33 | -22 | 7808 |
| Vladivostok | 43.13 | 131.9 | 5.4 | 32 | -29 | 7258 |

* Number of heating hours if the inner temperature is 20°C

## 3. Results and discussion

With the help of the EnergiKalk software, were calculated the yearly heating energy demand for the three types of windows and for two kinds of daily temperature rhythms in each of ten cities (Table 3). The table shows the yearly heating degree hours in both the 20 degrees and the lowered 18 degrees Celsius night temperature.

**Table 3.** Energy demands.

| City | Heating degree hours constant 20°C (h°C) | Heating degree hours 18°C at night and 20°C by day (h°C) | Energy demand constant 20°C (kWh/year) | | | Energy demand 18°C at night and 20°C by day (kWh/year) | | |
|---|---|---|---|---|---|---|---|---|
| | | | 3.5 $W/m^2K$ | 1.4 $W/m^2K$ | 0.7 $W/m^2K$ | 3.5 $W/m^2K$ | 1.4 $W/m^2K$ | 0.7 $W/m^2K$ |
| Archangelsk | 155 599 | 150 633 | 51 534 | 45 390 | 43 336 | 49 671 | 43 749 | 41 769 |
| Chelyabinsk | 152 980 | 148 560 | 50 396 | 44 386 | 42 379 | 48 724 | 42 914 | 40 973 |
| Irkutsk | 169 550 | 165 361 | 55 775 | 49 123 | 46 904 | 54 217 | 47 751 | 45 593 |
| Krasnodar | 77 357 | 74 183 | 25 710 | 22 645 | 21 618 | 24 559 | 21 632 | 20 651 |
| Krasnoyarsk | 167 009 | 162 086 | 55 077 | 48 509 | 46 317 | 53 205 | 46 861 | 44 743 |
| Magadan | 212 417 | 206 759 | 69 850 | 61 517 | 58 739 | 67 985 | 59 876 | 57 171 |
| Moscow | 124 180 | 119 457 | 41 141 | 36 236 | 34 597 | 39 400 | 34 703 | 33 132 |
| Omsk | 161 330 | 156 464 | 53 003 | 46 681 | 44 573 | 51 198 | 45 092 | 43 055 |
| St. Petersburg | 120 797 | 116 254 | 40 211 | 35 416 | 33 813 | 38 539 | 33 945 | 32 407 |
| Vladivostok | 132 475 | 129 008 | 43 510 | 38 322 | 36 590 | 42 219 | 37 186 | 35 504 |

(1) According to the calculations, the amount of heating energy is not related to the number of heating hours, but to the heating degree hours. Magadan has the highest energy demand and the number of heating hours; these values are the lowest in Krasnodar. In Archangelsk the heating degree hours are not the highest despite the high number of heating hours. Irkutsk, Krasnoyarsk and Omsk have a higher energy demand because of the greater number of heating degree hours. The reason is that although the temperature is permanently under the given value, the minimum and average temperatures are higher than in the middle part of the country, where the temperature could be much lower during a shorter heating period.

The energy consumption of a building in the Moscow region with 1.4 W/m$^2$K type windows is 335 kWh/m$^2$ per year (Table 3). This data is a good agreement with the calculation have been done in the GWD Engineering 2017 system (gwde.ru) which is about 298 kWh/m$^2$ per year [20].
(2) The different energy requirements from using different types of windows can be compared because the configuration and orientation of the house is constant. The window with the lowest insulation capacity and 3.5 W/m$^2$K thermal conductivity accepted as a basis. The 1.4 W/m$^2$K thermal insulation window saved 11.9%, and the 0.7 W/m$^2$K window saved an additional 4.0% compared to the initial value. In the coldest Magadan area, modern windows (U=0.7 W/m$^2$K) save 11 MWh energy annually compared to the 3.5 W/m$^2$K windows. In Krasnodar, where the energy demand is the smallest, the saving is still 4 MWh. The results show the building's energy demand by measuring heating degree hours with the different types of windows. Changing the windows from type 1.4 W/m$^2$K to type 0.7 W/m$^2$K produced much less (~4%) energy savings.
(3) If the temperature is reduced from 20°C to 18°C between 10 pm and 6 am, energy use can be reduced by as much as 4.5% depending on weather conditions characteristic of the area (Table 4). By reducing the inner temperature the heating degree hours are reduced as well, because the night temperature difference will be lower (Table 4).

**Table 4.** Heating energy saving by reduced night temperature.

| City | Energy saving 18°C at night and 20°C by day [kWh/year] | | | Energy saving [%] | Decrease of heating degree hours [%] |
|---|---|---|---|---|---|
| | 3.5 (W/m$^2$K) | 1.4 (W/m$^2$K) | 0.7 (W/m$^2$K) | | |
| Archangelsk | 1 863 | 1 641 | 1 567 | 3.62 | 3.19 |
| Chelyabinsk | 1 671 | 1 471 | 1 406 | 3.32 | 2.89 |
| Irkutsk | 1 558 | 1 372 | 1 311 | 2.79 | 2.47 |
| Krasnodar | 1 151 | 1 013 | 968 | 4.48 | 4.10 |
| Krasnoyarsk | 1 872 | 1 648 | 1 574 | 3.40 | 2.95 |
| Magadan | 1 865 | 1 641 | 1 567 | 2.67 | 2.66 |
| Moscow | 1 741 | 1 533 | 1 464 | 4.23 | 3.80 |
| Omsk | 1 805 | 1 589 | 1 518 | 3.41 | 3.02 |
| St. Petersburg | 1 673 | 1 472 | 1 406 | 4.16 | 3.76 |
| Vladivostok | 1 291 | 1 136 | 1 085 | 2.97 | 2.62 |

The saving in Magadan with the highest heating demand is 1 865 kWh, (2.67%). In Krasnodar it is 1 151 kWh, (4.48%) of the energy using the 3.5 W/m$^2$K windows. The energy saving in Magadan is almost one and a half times as much as in Krasnodar in absolute value. However, it indicates a smaller proportion. The amount of savings is mostly connected to the whole amount of energy needed for heating. Where the heating energy demand is bigger, more energy can be saved by lowering the inner temperature at night. The percentage of energy saved is the same with the better insulated windows, although its absolute quantity is lower.

## 4. Conclusions

Cities in the warmest areas require heating in 70% of the year while city in the coldest places need heating for more than 98% of the year. Due to the long heating period, a smaller improvement of insulation saves a large amount of energy. Therefore using windows with lower thermal conductivity and lowering the temperature at night reduces heating energy 20% in most areas of Russia. Windows play a significant role in the amount of heat building needs. Compared to the two times one layer glazing (3.5 W/m$^2$K), the windows with thermal insulation glazing and a low emission coating and argon gas filling (1.4 W/m$^2$K) can produce significant savings. The windows with three layers of glazing and excellent insulation (0.7 W/m$^2$K) can reduce a building's whole heating energy demand by almost one sixth. That is a really significant result if the low surface are of the windows is taken into account.
Reducing the night temperature 2°C can save 2.67% to 4.48% of the required energy depending on the heating demand and the region. This type of energy saving can be realized without investment.

At present the rural regions of the country mostly heat with wood, a fuel that is abundantly available everywhere in the country. Since Russia has plenty of fossil energy sources compared to the western countries, there is less pressure to use energy efficiently. This is reflected in the regulations for thermal characteristics of buildings. Nationwide Russia has a very significant potential to save energy and carbon dioxide. The key to this is to raise the standards for the energetics of buildings and to retrofit buildings.

## 5. References

[1] Swan L G and Ugursal V I 2009 Modeling of end-use energy consumption in the residential sector: A review of modeling techniques. *Renewable and Sustainable Energy Reviews* **13** 1819-35 DOI: 10.1016/j.rser.2008.09.033

[2] Estiri H 2015 The indirect role of households in shaping US residential energy demand patterns *Energy Policy* **86** 585-94 DOI: 10.1016/j.enpol.2015.08.008

[3] Balaras C, Droutsa K, Dascalaki E and Kontoyiannidis S 2005 Heating energy consumption and resulting environmental impact of European apartment buildings *Energy and Buildings* **37** 429-42 DOI: 10.1016/j.enbuild.2004.08.003

[4] Harvey L D D (2009): Reducing energy use in the buildings sector: measures, costs, and examples *Energy Efficiency* **2** 139-63 DOI 10.1007/s12053-009-9041-2.

[5] Lechtenböhmer S and Schüring A 2010 The potential for large-scale savings from insulating residential buildings in the EU *Energy Efficiency* **4**(2) 257-70 DOI: 10.1007/s12053-010-9090-6

[6] Nyers J, Tomić S and Nyers A 2014 Economic optimum of thermal insulating layer for external wall of brick *Acta Polytechnica Hungarica* **11**(7) 209-22

[7] Arumägi E and Kalamees T 2014 Analysis of energy economic renovation for historic wooden apartment buildings in cold climates *Applied Energy* **115** 540-8 DOI: 10.1016/j.apenergy.2013.10.041

[8] Basinska M, Koczyk H and Szczechowiak E 2015 Sensitivity analysis in determining the optimum energy for residential buildings in Polish conditions *Energy and Buildings* **107** 307-18 DOI:10.1016/j.enbuild.2015.08.029

[9] Jermyn D and Richman R 2016 A process for developing deep energy retrofit strategies for single-family housing typologies: Three Toronto case studies *Energy and Buildings* **116** 522-34 DOI: 10.1016/j.enbuild.2016.01.022

[10] Skarning G C J, Hviid C A and Svendsen S. 2016 Roadmap for improving roof and facade windows in nearly zero-energy houses in Europe *Energy and Buildings* **116** 602-13 DOI: 10.1016/j.enbuild.2016.01.038

[11] Grynning S, Gustavsen A, Time B and Jelle BP 2013 Windows in the buildings of tomorrow: Energy losers or energy gainers? *Energy and Buildings* **61** 185-92 DOI: 10.1016/j.enbuild.2013.02.029

[12] Arıcı M, Karabay H and Kan M 2015 Flow and heat transfer in double, triple and quadruple pane windows *Energy and Buildings* **86** 394-402 DOI: 10.1016/j.enbuild.2014.10.043

[13] Grynning S, Time B and Uvslřkk S 2011 An overview and some reflections on energy saving potentials by heat loss reduction through the building envelope. Project report to be published within the Research Centre on Zero Emission Buildings.

[14] TEK 2010 Technical regulations to the Norwegian building regulations, Forskrift om tekniske krav til byggverk (Byggteknisk forskrift), 2010.

[15] Jaber S and Ajib S 2011 Thermal and economic windows design for different climate zones *Energy and Buildings* **43** 3208–15 DOI: 10.1016/j.enbuild.2011.08.019

[16] Efimov E M 2011 Wooden housing construction in Russia: state, problems and prospects of development *Problems of economics and legal practice* **2** 239-41 ISSN: 2541-8025

[17] Petrova Z K 2012 The problem of development of low-rise life-supporting housing in Russia and abroad *Urban planning* **4** 59-66 ISSN: 2218-8762

[18] SNiP 2003 Building codes and regulations. SNiP 23-02-2003. Thermal performance of the buildings, 2003.

[19] Vados M 2013 Épületek hőveszteségének integrált energetikai modellezése (Integral energetic model of thermal loss of buildings) University of West Hungary, BSc. Thesis

[20] GWD Engineering. 2017 Guide to choosing fuel heating system. http://www.gwde.ru/articles/rukovodstvo-po-vyboru-topliva-sistemy-otopleniya/

**Acknowledgement**
The described work was carried out as part of the „Sustainable Raw Material Management Thematic Network – RING 2017", EFOP-3.6.2-16-2017-00010 project in the framework of the Széchenyi 2020 Program. The realization of this project is supported by the European Union, co-financed by the European Social Fund.

IOP Conf. Series: Materials Science and Engineering **371** (2018) 012051 doi:10.1088/1757-899X/371/1/012051

# BIM for Existing Buildings: Potential Opportunities and Barriers

**M A Hossain[1] and J K W Yeoh[2]**

1 Dept. of Civil Engineering, Nazarbayev University, Astana, Kazakhstan
2 Dept. Civil & Env. Eng., National University of Singapore, Singapore
E-mail: md.hossain@nu.edu.kz

**Abstract.** Building Information Modelling (BIM) can play a significant role for better operation and maintenance (O&M) and building facility management (FM) through the use of Information and Communication Technologies (ICT) tools. However, most of the existing buildings have no BIM. Research in creating BIM for existing buildings has received growing attention in recent years. It is evident from current literature that generating BIM for existing building is complex, tedious, time consuming and costly. The objective of this project is to explore potential opportunities and barriers in constructing BIM for existing buildings. Different techniques used in capturing building data such as imaging, 3D scanning, Ground Penetration Radar (GPR), 2D scanned plans and so on, have their pros and cons in terms of high modelling/conversion efforts necessary to construct a semantically rich BIM. Moreover, most of these techniques are complex; require expert knowledge; core steps are still manual; face challenges in handling occlusions/uncertain data; and BIM conversion becomes cumbersome with the BIM's increasing level of detail (LoD). A systematic analysis is done on various techniques used in data capturing and how they are converted into a semantically rich BIM, accuracy of the converted model, interoperability of the generated BIM in terms of functionality, potential barriers in the automation process and how they can be overcome.

## 1. Introduction

Building Information Modelling (BIM) is becoming the core of information management in Architecture, Engineering & Construction (AEC) industry as well as for operation and maintenance (O&M) and facility management (FM). It allows all stakeholders to exchange and manage information about building components throughout the lifecycle of a building [1]. Particularly, BIM can be used for energy usage analyses, defect detection, firefighting, renovation and demolition, safety in facility management and so on [2-5]. BIM adoption in AEC industry is well established and has received growing attention for O&M and FM [6-8]. Although, there is an increasing trend of BIM adoption in AECO (Architecture, Engineering, Construction and Operation) industries, it has only been used for some new projects in recent years [9]. Hence, most of the existing buildings do not have a BIM. In order to fully realize the benefit of BIM, it is necessary to create BIM for existing buildings.

Researchers have been working to create compatible BIM for existing buildings. Nevertheless, creation of a semantically rich BIM is challenging, complex and expensive considering the special equipment requirement for data capturing, high modeling/conversion effort, expert knowledge of skillful personnel, handling obstacle/uncertain data and so on [10-13]. Moreover, the complexity of BIM generation also depends on the BIM's level of detail (LoD), intended use, interoperability and functional issues [12, 14-16]. Hence, the potentials of BIM for existing building has yet to be fully realized.

This study aims to identify the potentials of BIM for existing buildings and major bottlenecks of BIM generation procedures considering the LoD and functionalities in O&M lifecycle of building facilities. A systematic analysis through extensive literature review has been done to perceive the knowledge gap between the necessity and existing theories and practice. Finally, the study explores the future opportunities to minimize the knowledge gap and overcome the potential barriers in implementing BIM for existing buildings.

## 2. Managing building's lifecycle information

The longest phase of the lifecycle of a building is the operation and maintenance and by far the most costly considering the monetary expense for labor and materials in upkeep, maintenance and refurbishment/replacement [17]. According to [18], 85% of the total project cost is spent in operation and maintenance from the owner's perspective. However, traditional practice of information management for O&M is fragmented, manual and tedious leading inefficient O&M and FM of buildings. Particularly, information management suffers following issues that hinder effective O&M.

- Current practice of asset/facility management is paper-based consisting manual inspection and proves to be time consuming, tedious, and prone to humane error.
- Building systems all over the world are becoming increasingly complex with utilization of sophisticated technologies for their operation, building automation, security issues and sustainability, making it challenging for the management and operation of the facility [19].
- According to the "Bureau of Labor Statistics" and [20], maintenance and repair requirements during the life cycle of O&M and FM pose high risk of injuries including falls, electric shocks, crushing, cuts and bruises to the workers as well as facility users.
- Cost effective retrofit for existing building is a great technical challenge considering complex interaction of energy audit system, building performance assessment, cost benefit analysis, risk assessment, and potential of energy savings [21].
- Operation of existing building contributes a large portion of total energy worldwide [22, 23]. Efficient energy use in building is a prime concern in mitigating the stress put on the energy system and bringing positive environmental impact and BIM can significantly improve energy analysis through simulation [24-27].
- Fragmented O&M has made it difficult to do energy simulation, lead error prone decision in retrofit/renovation, safety issues arise in operation, maintenance and repair [28, 29].

Hence, efficient information management for O&M phase of building's lifecycle is utmost important. BIM's dynamic, open access, digital environment and 3D visualization capabilities enable storage, sharing and integration of information for buildings' O&M phase [8].

BIM has been successfully implemented during design and construction as a means of digital database for information management and knowledge sharing between stakeholders, resolving construction schedule conflict and space congestion, enhanced safety through virtual reality and so on. However, use of the same BIM in O&M and FM are still limited considering the inconsistent naming conventions, absence of uncounted number of information necessary for O&M and FM, inadequate methodology to capture existing facilities and asset, and poor information synchronization [30, 31]. COBie (Construction Operations Building Information Exchange) and some other researchers [32-34] have been working on new data standards for BIM project to eliminate the interoperability issues between the model used for design and construction and the model to be used for FM. Nevertheless, all these efforts and BIM adoption have been done for new building construction projects. BIM integrated information management in O&M and FM for existing building is still its infancy.

## 3. Potentials of BIM for operation, maintenance and sustainability

As mentioned earlier, BIM has been well adopted in design and construction. BIM can be beneficial in O&M and FM phases by many means, such as, better facility management, efficient energy use, cost effective retrofit decision making, safe O&M and repair works, and organized demolition. Table 1 shows potentials of BIM for existing buildings.

BIM enables increased efficiencies in data consistency, single source of information storage and retrieval for FM, reduced effort (time and resource) necessary on finding equipment and material information [34, 41]. Accordingly, information is more easily shared and reused with BIM [3]. It is evident from current literature that the benefits of BIM in O&M and FM are well documented. Nevertheless, lack of BIM for existing buildings is one of the major bottlenecks of the benefits to be realized.

**Table 1.** BIM for operation, maintenance and sustainability.

| Item | Description | Source |
|---|---|---|
| Lifecycle analysis | Facility/asset management, monitoring, 3D visualization, real time data access, maintenance schedule, auditing, emergency management | [3, 5, 6, 8, 17, 35, 36] |
| Retrofit | Retrofit decision making, Repair and reconstruction | [21, 37] |
| Energy | Building Energy Modelling (BEM), Energy efficiencies and energy conservation | [25-27] |
| Safety | Safe facility management, hazard mitigation, fall prevention, fire prevention and disaster relief, Integrated Multidisciplinary Model for Security and Safety Management (IMMSSM) | [5, 7, 16, 17] |
| Sustainability | Sustainable facility management, Sustainable retrofit | [38, 39] |
| Decommissioning | Analysing material composition prior to demolition | [40] |

## 4. Challenges in creating BIM for existing buildings

Generation of BIM for existing buildings is complex and expensive considering the challenge in capturing building data and high modelling/processing effort required to create a semantically rich BIM. Moreover, the complexity increases with the BIM's level of detail (LoD), intended use, O&M and FM functionalities [10, 12, 14, 42]. As mentioned in [33], increasing effort is required to make a BIM with increasing benefit to be achieved by FM/O&M team which is depicted in Figure 1.

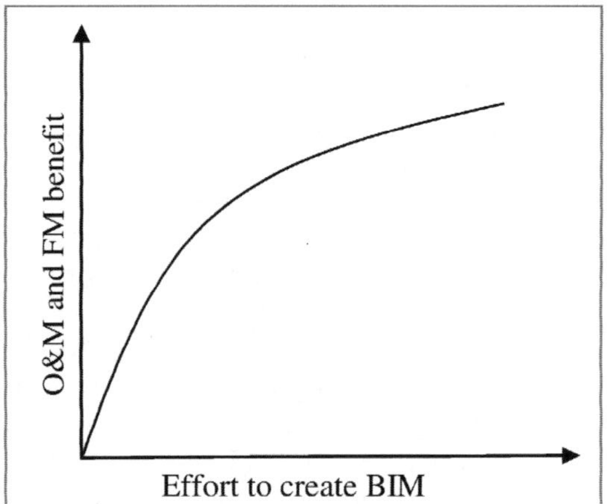


**Figure 1.** Effort vs Benefit curve (adapted from [33]).

### *4.1. Data capturing and processing challenges*

In this paper we review several different data acquisition technologies used in the industry to support the generation of as-built information for creating BIM: Digital Photogrammetry, Terrestrial Laser Scanner (TLS), and Ground Penetrating Radar (GPR). The typical function of these data acquisition technologies is to capture the geometric information of the object. Digital Photogrammetry works by capturing still images and turning these into 3D point clouds. Laser Scanners use multiple points of

IOP Conf. Series: Materials Science and Engineering **371** (2018) 012051 doi:10.1088/1757-899X/371/1/012051

amplified light to measure distances from the scanner to its target. Ground Penetrating Radar use a high frequency radio signal to infer the location of objects embedded.

Current techniques for as-built data acquisition involve laborious on-site surveys with manual measurement or visual assessment [43]. These current techniques suffer from being error-prone and tediously time consuming. The consequence is that such techniques are often unreliable, and create significant uncertainty for decision makers.

*4.1.1. Digital photogrammetry: application and challenges.* Digital Photogrammetry transforms the still images captured by a camera to derive 3D geometric information. For building applications, terrestrial or close-range photogrammetry is often utilized. At least two cameras with known coordinates are located *away* from an object to create a 3D representation. Mathematically, this creates at least two converging lines to identify a point in space; a technique commonly referred to as triangulation. Typically, this process requires several inputs: The X, Y, and Z coordinates of the location of each camera, as well as $\omega$, $\psi$, and $\kappa$ angles of rotation of the camera. Additionally, real-world (common) coordinates represented by control points should also be identified in each image captured, so that the pixel can be recognized.

With today's camera systems, photogrammetry is a relatively inexpensive technology [44]. Several commercially available softwares are available to process digital pictures for building construction applications (e.g. Bundler, Autodesk Remake, etc). The ease of use of this technology has reportedly increased the productivity of survey teams by three-fold [45].

Several problems and challenges exist with photogrammetry techniques. Firstly, digital photogrammetry is sensitive to the changing light conditions. Pictures taken at different times of the day of the same object may suffer from shadows. These shadows may affect the alignment of photos during post-processing. Secondly, image matching is often affected by noise in image sensor of the camera and the presence of occlusions [46]. Thirdly, fixing the absolute scale in digital photographs is a potential major source of error [47]. Lastly, adequate planning is required to obtain accurate models. Ideally, photographs should overlap significantly (some practice professionals use a 60% rule-of-thumb). Photos should also be captured 10 to 15 degrees horizontally or vertically around the object. Also, more photographs will need to be taken if overhangs, protrusions or holes in the building are present.

*4.1.2. Terrestrial laser scanning: application and challenges.* Terrestrial Laser Scanning is also known as Light Detection and Raging (LiDAR), and LADAR (Laser Distance and Ranging). Both terms refer to a similar technique, with different industrial applications. In general, the technique refers to the emission of pulses of light to the surface of the object of interest. The light is reflected, and a sensor detects the reflected light. The round trip time from the emission to detection is computed, and the distance derived [48].

Newer TLS have features allowing multiple returns of data [49]. At long ranges, the spot size of the TLS increases, because the light emitted from the TLS expands as the range increases, creating the effect of multiple returns. For example, the foliage of the trees will constitute the data of the first return, while the ground below the foliage is the second return of data. Such multiple returns can be important for dealing with partial occlusions.

TLS also has several challenges. Not all objects are suitable for TLS capture; particularly, if an object has a highly non-reflective surface, e.g. black surfaces or rough surfaces, TLS capture quality may be compromised. Secondly, some TLS allow for different selection of wavelengths. The choice of wavelength depends on application. Smaller wavelengths improve the detection rate of smaller objects, while longer wavelengths allow for better penetration through some occlusions. The choice depends on application and on the user's experience. Lastly, the planimetric accuracy of digital photogrammetry is higher, but terrestrial laser scanning can achieve higher accuracy with vertical elevations [50].

*4.1.3. Ground penetration radar: application and challenges.* Ground Penetration Radar has recently emerged as a complementary tool for acquiring as-built information from building structures. Two

applications of GPR are possible: One, is to use GPR to identify the location of buried utilities [51], and another is to use GPR to inspect concrete structures [52]. As part of this inspection, it is possible to use GPR to locate individual rebars within the concrete structures. Ground Penetration Radar works by emitting a high *frequency* radio signal. The object reflects the signals to the receiver, and the time taken for the pulse to travel to and from the target object is used to infer its depth and location. Depending on the frequency of the radio signal used, GPR can detect objects up to a depth of 30m.
Another operating characteristic is the use of either air-launched antennas or ground-coupled antennas [51]. Ground coupled antennas tend to be used for detection of deeper objects, but are typically slower. Air-launched antennas are typically used for shallow ground detection, but is able to cover a larger area.
The utilization of GPR is dependent upon many factors: the depth of the object to be detected, the material of the object, the presence of other sources of interference, and the material of the medium. For example, using GPR in marine clay is known to be limited to a few meters due to the high conductivity of marine clay. The presence of other objects distort the return signal, and increases the noise in the resultant waveform. Lastly, GPR has a relatively high energy consumption profile, limiting its use in extensive field surveys.

*4.1.4. Comparison of technologies.* Table 2 provides a short comparison of the different data acquisition techniques in the building operations context. It should be noted that each technique has its merits and disadvantages, and future work involving hybridizing or fusing the outputs of two or more techniques to gain a more comprehensive picture of the as-built condition should be explored.

**Table 2.** Relative comparison of the data acquisition techniques.

| | Digital Photogrammetry | Terrestrial Laser Scanning | Ground Penetration Radar |
|---|---|---|---|
| Relative Accuracy | Low (1cm to 1m) | High (1cm to 1mm) | NA |
| Use in Low-light conditions | No | Yes | Yes |
| Relative Weight | Low | High | Medium |
| Relative Cost | Low | High | High |
| Relative Degree of Expertise | Low | Medium | High |

### *4.2. Functionalities and interoperability issues*

Significant advancement has been made in capturing building data and processing technologies to generate BIM for existing buildings. Nevertheless, potential of BIM is yet to be fully realized due to functionalities and interoperability issues. For example, not all O&M related information is compatible for hosting in a BIM environment [8], dealing ambiguity due to incomplete building documentation [33], lack of user-friendly interfaces and incomplete as-is BIM to be used by staff workers in O&M [12] and so on. Moreover, there is hardly a standardized information transfer and storage system available to use BIM for building energy modelling (BEM) [25]. As mentioned by [53], building model and analysis model are separate and restricted by storage format. Furthermore, BIM's information storage capability restricts inclusion of large time series datasets generated by BEM simulations [25]. Hence, it is important to properly identify necessary information to improve the operational performance of building in creating a BIM in operations model [33].

### *4.3. Collaboration, responsibility and ownership issues*

Use of BIM in building's operation and maintenance is very limited till date. There is a clear lack of competent BIM experts in O&M and FM market. Staff workers in O&M phase have limited knowledge and experience of using and updating as-is BIM [12]. Hence, a framework for educating existing O&M and FM personnel is necessary about BIM process and technology [54]. Moreover, ownership of BIM and responsibilities in managing information must be clearly defined for efficient O&M to avoid any miscommunication.

## 5. Knowledge gap and future opportunities

The usefulness of BIM for O&M phase has been well acknowledged in literature. Nevertheless, the foremost challenge in realizing potential benefit of BIM for existing buildings is creation of economic and automatic BIM that is compatible for O&M and FM. Moreover, a certain level of BIM maturity should be attained among O&M and FM stakeholders. Quality, accuracy and completeness of a BIM will benefit facility owner with little technical knowledge [54]. It is also important to address organizational challenges that may arise due to the paradigm shift among stakeholders [37]. Furthermore, integration of BIM-GIS would enhance city/country wide energy performance and minimize global greenhouse gas emissions [28].

## 6. Conclusion

BIM has a great potential for operation and maintenance (O&M) and facility management (FM) of buildings. However, the benefit of BIM has not been fully realized in O&M and FM yet. In part, this is because most of the existing buildings do not have a BIM and creating a BIM for existing building is challenging. This paper has explored the current state of BIM for existing buildings in terms of potential opportunities and major bottlenecks in creating a semantically rich as-built BIM. As can be seen, BIM can be beneficial in O&M and FM phases by many means, such as, lifecycle analysis for better facility management, sustainable and efficient energy use, cost effective retrofit decision making, safe O&M and repair works, and organized demolition. To realize these benefits researchers have been working to create as built BIM for existing buildings. Current techniques used for as built data acquisition are mainly digital photogrammetry, terrestrial laser scanning, ground penetration radar. These techniques involve laborious on-site surveys with manual measurement or visual assessment, error-prone, tediously time consuming, often unreliable, and create significant uncertainty for decision makers. Moreover, creating BIM from the captured data requires high conversion/modelling effort from skillful personnel. The usefulness of BIM is further challenged by functionality and interoperability issues, lack of user-friendly interfaces and incomplete as-is BIM to be used by staff workers in O&M, and lack of competent BIM experts in O&M and FM market. Hence, there is a great potential of research opportunities in creating economic and automatic BIM for existing building and attain certain level of BIM maturity among O&M and FM stakeholders. Furthermore, a framework should be developed to educate existing O&M and FM personnel about BIM process and technology.

## 7. References

[1] I. Motawa and A. Almarshad, "A knowledge-based BIM system for building maintenance," *Automation in Construction,* vol. 29, pp. 173-182, 1// 2013.

[2] L. Gimenez, S. Robert, F. Suard, and K. Zreik, "Automatic reconstruction of 3D building models from scanned 2D floor plans," *Automation in Construction,* vol. 63, pp. 48-56, 3/1/March 2016 2016.

[3] R. A. Kivits and C. Furneaux, "BIM: Enabling sustainability and asset management through knowledge management," *The Scientific World Journal,* vol. 2013, 2013.

[4] P. E. D. Love, I. Simpson, A. Hill, and C. Standing, "From justification to evaluation: Building information modeling for asset owners," *Automation in Construction,* vol. 35, pp. 208-216, 2013/11/01/ 2013.

[5] E. M. Wetzel and W. Y. Thabet, "The use of a BIM-based framework to support safe facility management processes," *Automation in Construction,* vol. 60, pp. 12-24, 2015.

[6] P. E. D. Love, J. Matthews, I. Simpson, A. Hill, and O. A. Olatunji, "A benefits realization management building information modeling framework for asset owners," *Automation in Construction,* vol. 37, pp. 1-10, 2014/01/01/ 2014.

[7] M.-Y. Cheng, K.-C. Chiu, Y.-M. Hsieh, I. T. Yang, J.-S. Chou, and Y.-W. Wu, "BIM integrated smart monitoring technique for building fire prevention and disaster relief," *Automation in Construction,* vol. 84, pp. 14-30, 2017/12/01/ 2017.

[8] E. A. Pärn, D. J. Edwards, and M. C. P. Sing, "The building information modelling trajectory in facilities management: A review," *Automation in Construction,* vol. 75, pp. 45-55, 3// 2017.
[9] NBS, "NBS National BIM Report 2016," 2016.
[10] C. Wang, Y. K. Cho, and C. Kim, "Automatic BIM component extraction from point clouds of existing buildings for sustainability applications," *Automation in Construction,* vol. 56, pp. 1-13, 8/1/August 2015 2015.
[11] T. Gao, S. Ergan, B. Akinci, and J. Garrett, "Evaluation of Different Features for Matching Point Clouds to Building Information Models," *Journal of Computing in Civil Engineering,* vol. 30, p. 04014107, 2016.
[12] Q. Lu and S. Lee, "Image-Based Technologies for Constructing As-Is Building Information Models for Existing Buildings," *Journal of Computing in Civil Engineering,* vol. 31, p. 04017005, 2017.
[13] R. Volk, J. Stengel, and F. Schultmann, "Building Information Modeling (BIM) for existing buildings — Literature review and future needs," *Automation in Construction,* vol. 38, pp. 109-127, 3// 2014.
[14] D. F. Laefer and L. Truong-Hong, "Toward automatic generation of 3D steel structures for building information modelling," *Automation in Construction,* vol. 74, pp. 66-77, 2/1/February 2017 2017.
[15] C. Wang, Y. K. Cho, and M. Gai, "As-Is 3D Thermal Modeling for Existing Building Envelopes Using a Hybrid LIDAR System," *Journal of Computing in Civil Engineering,* vol. 27, pp. 645-656, 2013.
[16] M. G. Angelini, V. Baiocchi, D. Costantino, and F. Garzia, "Scan To BIM for 3D Reconstruction of the Papal Basilica of Saint Francis in Assisi in Italy," *Int. Arch. Photogramm. Remote Sens. Spatial Inf. Sci.,* vol. XLII-5/W1, pp. 47-54, 2017.
[17] J. D. Lucas, "Managing the Facility with Lifecycle Information," *Journal of Current Issues in Media & Telecommunications,* vol. 7, pp. 13-36, 2015.
[18] E. Teicholz, "Bridging the AEC technology gap," *IFMA Facility Management Journal,* vol. March-April, 2004.
[19] IFMA. (2007, Exploring the Current Trends and Future Outlook for Facility Management. Available: http://www.ifma.org/docs/fm-forecasts/2007.pdf
[20] N. Leveson, "A new accident model for engineering safer systems," *Safety Science,* vol. 42, pp. 237-270, 2004.
[21] Z. Ma, P. Cooper, D. Daly, and L. Ledo, "Existing building retrofits: Methodology and state-of-the-art," *Energy and Buildings,* vol. 55, pp. 889-902, 2012/12/01/ 2012.
[22] E. Asadi, M. G. da Silva, C. H. Antunes, and L. Dias, "Multi-objective optimization for building retrofit strategies: A model and an application," *Energy and Buildings,* vol. 44, pp. 81-87, 2012/01/01/ 2012.
[23] S. Roberts, "Altering existing buildings in the UK," *Energy Policy,* vol. 36, pp. 4482-4486, 1/1/2008 2008.
[24] G. Salvalai, M. M. Sesana, and G. Iannaccone, "Deep renovation of multi-storey multi-owner existing residential buildings: A pilot case study in Italy," *Energy and Buildings,* vol. 148, pp. 23-36, 2017/08/01/ 2017.
[25] T. Gerrish, K. Ruikar, M. Cook, M. Johnson, and M. Phillip, "Using BIM capabilities to improve existing building energy modelling practices," *Engineering Construction & Architectural Management (09699988),* vol. 24, pp. 190-208, 2017.
[26] T. Reeves, S. Olbina, and R. Issa, "Guidelines for Using Building Information Modeling for Energy Analysis of Buildings," *Buildings,* vol. 5, p. 1361, 2015.
[27] K. Lee, B. Koo, B. Park, and Y. H. Ahn, "The development of an energy-efficient remodeling framework in South Korea," *Habitat International,* vol. 53, pp. 430-441, 2016/04/01/ 2016.
[28] C. Koo and T. Hong, "Development of a dynamic operational rating system in energy performance certificates for existing buildings: Geostatistical approach and data-mining technique," *Applied Energy,* vol. 154, pp. 254-270, 2015/09/15/ 2015.

[29] S. Xiaonuan and S. L. A. U. SiuYu, "Existing buildings' operation and maintenance: renovation project of Chow Yei Ching Building at the University of Hong Kong," *International Journal of Low-Carbon Technologies,* vol. 10, pp. 393-404, 2015.

[30] M. Kassem, G. Kelly, N. Dawood, M. Serginson, and S. Lockley, "BIM in facilities management applications: A case study of a large university complex," *Built Environment Project and Asset Management,* vol. 5, pp. 261-277, 07 / 06 / 2015.

[31] P. M. Teicholz and I. Foundation, *BIM for Facility Managers*. Hoboken, New Jersey: Wiley, 2013.

[32] V. Thein. Industry Foundation Classes (IFC): BIM Interoperability Through a Vendor-Independent File Format [Online]. Available: http://consultaec.com.au/white-paper-ifc-bim-interoperability-through-a-vendor-independent-file-format/

[33] J. J. McArthur, "A Building Information Management (BIM) Framework and Supporting Case Study for Existing Building Operations, Maintenance and Sustainability," *Procedia Engineering,* vol. 118, pp. 1104-1111, 2015/01/01/ 2015.

[34] C. Eastman, P. Teicholz, R. Sacks, and K. Liston, *BIM Handbook: A Guide to Building Information Modeling for Owners, Managers, Designers, Engineers and Contractors*: Wiley, 2011.

[35] L. Barazzetti, F. Banfi, R. Brumana, M. Previtali, and F. Roncoroni, "BIM from Laser Scans… Not Just for Buildings: NURBS-Based Parametric Modeling of a Medieval Bridge," *ISPRS Ann. Photogramm. Remote Sens. Spatial Inf. Sci.,* vol. III-5, pp. 51-56, 2016.

[36] B. Becerik-Gerber, F. Jazizadeh, N. Li, and G. Calis, "Application areas and data requirements for BIM-enabled facilities management," *Journal of Construction Engineering and Management,* vol. 138, pp. 431-442, 2012.

[37] I. Deniz and E. Esin, "BIM for building refurbishment and maintenance: current status and research directions," *Structural Survey,* vol. 33, pp. 228-256, 2015/07/13 2015.

[38] K. R. Galamba and S. B. Nielsen, "Towards sustainable public FM: collective building of capabilities," *Facilities,* vol. 34, pp. 177-195, 2016.

[39] P. Appleby, *Sustainable Retrofit and Facilities Management*. Hoboken: Routledge, 2013.

[40] R. A. Kivits and C. Furneaux, "BIM: Enabling Sustainability and Asset Management through Knowledge Management," *The Scientific World Journal,* vol. 2013, p. 14, 2013.

[41] L. Ding, R. M. Drogemuller, P. Akhurst, R. Hough, S. Bull, and C. Linning, "Towards sustainable facilities management," in *Technology, Design and Process Innovation in the Built Environment*, ed: Spon Press, 2009, pp. 373-392.

[42] E. Orthubera and J. Avbelj, "3D BUILDING RECONSTRUCTION FROM LIDAR POINT CLOUDS BY ADAPTIVE DUAL CONTOURING," *ISPRS Annals of Photogrammetry, Remote Sensing & Spatial Information Sciences,* p. 157, 08/07/ 2014.

[43] H. Son, C. Kim, and Y. Turkan, "Scan-to-BIM-an overview of the current state of the art and a look ahead," 2015.

[44] M.-F. Siu, M. Lu, and S. AbouRizk, "Combining photogrammetry and robotic total stations to obtain dimensional measurements of temporary facilities in construction field," *Visualization in Engineering,* vol. 1, p. 1, 12/01/Number 1/December 2013 2013.

[45] J. Jarroush, "Regular Digital Camera as a Practical Geodetic Measurement Tool : Issues and Challenges," presented at the FIG Working Week 2013, Abuja, Nigeria, 2013.

[46] P. Fricker, R. S Andau, A. Walker, and S. Diego, *Digital Photogrammetric Cameras: Possibilities and Problems*, 2003.

[47] F. Dai, Y. Feng, and R. Hough, "Photogrammetric error sources and impacts on modeling and surveying in construction engineering applications," *Visualization in Engineering,* vol. 2, p. 2, April 28 2014.

[48] P. Tang, D. Huber, B. Akinci, R. Lipman, and A. Lytle, "Automatic reconstruction of as-built building information models from laser-scanned point clouds: A review of related techniques," *Automation in Construction,* p. 829, 2010.

[49] S. Hernandez-Marin, A. M. Wallace, and G. J. Gibson, "Bayesian Analysis of Lidar Signals with Multiple Returns," *IEEE Transactions on Pattern Analysis and Machine Intelligence,*

IOP Conf. Series: Materials Science and Engineering **371** (2018) 012051 doi:10.1088/1757-899X/371/1/012051

*Pattern Analysis and Machine Intelligence, IEEE Transactions on, IEEE Trans. Pattern Anal. Mach. Intell.,* p. 2170, 2007.

[50] E. P. Baltsavias, "A comparison between photogrammetry and laser scanning," *ISPRS Journal of Photogrammetry and Remote Sensing,* vol. 54, pp. 83-94, 1999/07/01/ 1999.

[51] K. R. Maser, "Condition Assessment of Transportation Infrastructure Using Ground-Penetrating Radar," *Journal of Infrastructure Systems,* vol. 2, pp. 94-101, 1996.

[52] C. Maierhofer, "Nondestructive Evaluation of Concrete Infrastructure with Ground Penetrating Radar," *Journal of Materials in Civil Engineering,* vol. 15, p. 287, 2003.

[53] P. Sanguinetti, S. Abdelmohsen, J. Lee, J. Lee, H. Sheward, and C. Eastman, "General system architecture for BIM: An integrated approach for design and analysis," *Advanced Engineering Informatics,* vol. 26, pp. 317-333, 2012/04/01/ 2012.

[54] B. Giel and R. R. A. Issa, "Framework for Evaluating the BIM Competencies of Facility Owners," *Journal of Management in Engineering,* vol. 32, p. 04015024, 2016.

**Acknowledgments**

The authors wish to acknowledge the funding provided by Nazarbayev University under “Social Policy Grant” for this research work.

IOP Conf. Series: Materials Science and Engineering **371** (2018) 012052 doi:10.1088/1757-899X/371/1/012052

# Why European Standardization is not Working as Predicted: Construction Sector

**N D Cortiços**

CIAUD - Research Centre for Architecture, Urbanism and Design, Faculty of Architecture, University of Lisbon, Department of Technology for Architecture, Urbanism and Design, Rua Sá Nogueira Pólo Universitário Alto da Ajuda, Lisbon, Portugal
E-mail: ncorticos@campus.ul.pt

**Abstract..** The experience of the European market began with the abolition of border limitations for certifications based on non-harmonized standards by creating the CE[1] mark - impossible to circumvent and recognized around the World. This followed the steps of its global counterpart, ISO, although on a different scale. In reading the results achieved in the EU, the proposal is to replicate the ISO experience: through the design of legal mechanisms (supranational) for the definition of applicable principles, easily understood, validated by associated members and accepted by consumers, on a global scale. However, 17 years later, much remains to be achieved for the desired hegemony. The results remain below the expected. These are the possible findings based on the interpretation of the less successful results, arising from the identification of unimplemented or thwarted measures, in some countries. The treatment for each country was divergent — in the definition of (general) principles and legal framework; in the issuing of technical and legal mechanisms; in determining (or tampering with) exception tools; in establishing deadlines for responding to intervention requests — causing difficulty in understanding and implementation for manufacturers, dealers or importers.

## 1. Introduction

Products, resulting from an emerging global economy, do not know destination. They are produced in a location somewhere in the world, and used in another, distant and often different in character: cultural, technical, practical, geographic, among others.

It is recognized the difficulty in obtaining safe and reliable certification, under the yoke of rules resulting from protective policies and verified by bodies, responsible for the validation of manufacturing processes, logistics operations, use and performance. Many construction products (CP) present, under specific conditions, a lifespan below the expected, is the fault of the manufacturing chain and logistics processes [1].

Markets demand for construction products and services an expedited manufacturing and competitive pricing.

The standards are the basis of innovation and technological challenges. In synchronous, ensure the safety and health of workers and consumers and deal with a picky demand for products and solutions of construction. Those are essential tools of engineering and fundamental to the social cohesion of a world increasingly globalized [2].

[1] The letters "CE" are the abbreviation of French phrase "Conformité Europèene" which literally means "European Conformity".

IOP Conf. Series: Materials Science and Engineering **371** (2018) 012052 doi:10.1088/1757-899X/371/1/012052

The certifications that accompany raw materials, materials, products and construction solutions should — in today's world, as idealized for the European Space — comply with only one standard. Including ratings for each country, adjusted to different idiosyncrasies.
Normalization has proved critical to the industry's growth, more efficient and globally, based on the satisfaction of a larger number of consumers. That gained weight and visibility, taking advantage of the technological exhibition — such as the world wide web — in particular for stronger and experienced economic blocks, e.g., Federal Communications Commission (FCC) in the United States of America (EUA) or the Declaration of Conformity (CE) from the Economic Union (EU).
Reported conditions force a new paradigm. Inevitably the approach will have to arise from a coordinated action with the "International Standarisation Organisation (ISO)", without the need for national standards in each country where it intends to operate.
This theme is essential for a globalized society that is to be based on the standards of manufacture and ensures, informed and safe, transactions of Goods and Services[2,3] (G&S). In the case in point: Construction Products (CP).
In the current market, the commercial development of G&S lives a unique momentum. A high-tech product tends to be considered obsolete, even, during the first year. Seven to ten years is the expectation for the low-tech. Some reasons, common to both situations, that determine the longevity of G&S: industrial improvements (development and production); renewal of the design; and competitive strategies [3][4].
Markets, previously, delimited by strategic principles (cultural, political, economic, geographic, for instance), accepted the fundamentals of free competition[4] [5]:

- increased competitiveness;
- compatibility and interoperability;
- control over the range and the efficient use of materials, energy and human resources;
- saving raw materials and production times, reducing waste;
- elimination of trade barriers;
- protection of consumers and community interests;
- expediting the entrance into new markets;
- reduction of market uncertainty level;
- reflect the research, development and innovation;
- safety, health, protection of life and the environment; and the
- simplification on the wide variety of products and procedures, in everyday life.

As a result, national markets see the offer of G&S grow beyond the capacity of the mechanisms of monitoring bodies[5]. By agreement, within the World Trade Organization (WTO), it was also decided to consider valid any standardization developed in origin / location of the applicant. This situation weakens the purposes of standardization, as it makes it impossible to compare methods and manufacturing characteristics between G&S.
At this time, the industry moves toward full globalization, supported by standard production and trade exchanges, between neighboring countries or antipodal points on the planet. However, on any imported product we can find the certification of origin, if implemented, according to the International Organization for Standardization (ISO), which defines rules, guidelines and features lines to both G&S themselves as to their expected performance, an optimized security for a given context[6].

[2] "The WTO's General Agreement on Trade in Services (GATS) (…) in January 1995 (…), negotiations on specific market access commitments to further liberalize trade in services were mandated to start five years later. Accordingly, in early 2000, member governments started these negotiations. (…)" in "PRESS PACK - briefing notes, World Trade Organisation, 5th Ministerial Conference, Cancún, Mexico, 10–14 September 2003".

[3] The "intellectual property commercial rights" because they are not directly involved, when comparing construction factsheets, will not be addressed in this article.

[4] The openness of markets has the controlled by standards-based regulation.

[5] The construction sector seeks, increasingly, less technology with traditional origins and based on the mastery of hand labor.

[6] Each standard is supported by the work of a group of experts, the Technical Committee, specialized in a subject area and formed with a specific purpose.

IOP Conf. Series: Materials Science and Engineering **371** (2018) 012052 doi:10.1088/1757-899X/371/1/012052

### *1.1.Importance of data*

"Our Studies in cost analysis show that between 15% to greater than 20% of companies operating revenue is spent on doing things to get around or fix data quality issues."[7].

The standard information, the basis for any contemporary and globalized society, serves the purpose of creating economic growth, sustained in the security of data reading, a fundamental principle for taking position/making decisions, concise and defensible. Like a summary of the information provided by the sender to the receiver, both equipped with the tools / rules for encoding and decoding of the 'message', respectively [6].

### *1.2.Standard data format*

Companies, mostly national scale produce targeting domestic markets based on formatted information (harmonized standards) measured by local regulators. Some of them understand, by choice or imposition of the markets in which they operate, to accumulate more stringent national standards that the congener EC[8] [7].

In order to create obstacles to the import and marketing of products, there have been attempts, by some countries, to override the EC with national standards, in the overall result they have been unsuccessful, but some managed to establish themselves, e.g., "VOC labelling", since Jan 1st 2012, construction products traded in France must be labelled with an emissions classification on the basis of a VOC emissions test (as stated in the Decree nr 2011-321).

The establishment of standards is entrusted of a member/(local) body associated with ISO, which only sets the 'bases'[9] which tends to adjust to the demands of the internal market. The set of rules specifies efficiency and quality requirements for G&S based on the performance requirements applied to the product, considering the purpose of it. From that it is expected to promote and develop standards, ensuring the interrelation and framework for operators; to update the institutional legal acquis; to foster the adoption of other classifiable laws for the G&S, e.g. EU Directives, legislation [8].

### *1.3.Technical evaluation*

Technicians, linked to the project and the construction work, appear at the end of the line, as decision makers relying on the certification documents. They understand the rules as the valences to the prerequisites of any construction. It is not surprising that a particular G&S features standards from various countries to evaluate different characteristics.

The non-normative alignment for G&S makes the work of the technical / makers (design and construction) harder, and require research and study by the proliferation of various standards, even if we exclude the aesthetic / artistic components, in parallel with the economy of use. This task requires a comparison methodology to realize the similarities or incompatibilities (local laws and regulations, technological framework and good practice), for different objectives and expected performance.

The theme is fruitful, target of several publications dedicated to the comparison of standards and documentation of their experience in the field.

## 2. Problem

### *2.1. Before standardization*

By definition, convention and practicality, the International System of Units (SI) opted for the "metric" unit. Established after the Second World War, it was a crucial decision for the development of commercial trade. Systems based on other units feel a loss of influence, despite the dominance in

---

[7] by Larry English, an internationally recognized speaker, consultant, and author on information quality improvement.

[8] PC, not innovative: "Ü-Zeichen" (Ü-mark), German DIBt (German Institute for Construction Technology), Germany; "BBA certificate," British Board of Agrément, United Kingdom; and, "AFAQ/AFNOR marque", Association Française de Normalisation, France.

P, innovative: "Ü-Zeichen" (Ü-mark), German DIBt, Germany; LPC (loss prevention certification board), United Kingdom; and, "AFAQ/NF marque", Association Française de Normalisation, France.

[9] Standards developed by convention between ISO and local bodies to define the characteristics of a common orientative body, based on information — accurate, complete, updated, relevant, consistent ('validated') and reliable — suitable and accessible.

IOP Conf. Series: Materials Science and Engineering **371** (2018) 012052 doi:10.1088/1757-899X/371/1/012052

the domestic market (e.g., *Imperial*[10] *System:* english, used in the Commonwealth [9], and the American[11], in USA [10]).

The American domestic market was severely affected by the decline of the construction industry, as a result of economic crises, which has conditioned the heavy industries, its entrepreneurs and investors, with the difficulty in trading their production. The necessity becomes an opportunity with the opening to international markets [11]. This option implied adopting the metric system, only to export, for the steel profiles which followed standardization ISO/IEC 80000[12,13]. The 'external' demand for goods sized in imperial units is non-existent. Most US industries do not have G&S accredited for other economic areas, Europe for example, and if the option was to submit the products to tests and trials they would have to assume the costs. To these reasons we can add the logistical implications for transcontinental export.

*2.2.Standardization is recommended not imposed*

The standards serve as guides to the industry, references are impartial: not seeking private income only increased market dynamism, in the interests of manufacturers, operators and consumers. The use of standards is common in legislation — only in this case, and this case alone, it is mandatory — accreditation, certification, metrology, technical information and trade processes.

Certain sectors — such as food, safety, hygiene and health, environmental protection and quality management of G&S — are required, by regulation, to comply with norms for security standards imposed by legislation of consumer protection. Outside the scope, above, operators are free to certify or not the G&S, conditioned by the opportunities identified in the market to justify the decision. In both cases, certification or approval costs are added, either mandatory or optional — to others like: normative measurement, control inspection processes and the inherent bureaucracy.

*2.3.Standardization and Business Revenue*

Standards, certification and approval of G&S, are controlled by local public entities under the guidance of international standardization, ISO, the costs being charged to applicants. What configures a revenues market for entities and governments. System that inhibits, conditioned by income, the use of standards by small and medium-sized operators (importers, exporters, only the regulated field), especially the venturers.

The lack of control over production standards — globally, even by the ISO, meaning the not exclusivity of a rule attributed to a specific origin — can make it impossible to be compared to another, similar but not exactly the same. Situation which, in certain circumstances, may not be free of intentionality. Avoiding competition, established, by favoring the introduction of CP in the market. In stricter standardization situations, operators choose to slightly change the manufacturing processes and/or the performance qualities of the product, or even formalize the certification obtained in a third country, 'compatible' with the desired market.

## 3. Efficacy Analisys

*3.1.European committee for standardization (CEN) and the European committee for electrotechnical standardization (CENELEC)*

[10] System adopted for market protection: "Eventually, the Weights and Measures Act of 1878 abolish the use of all measurement units, that is, units other than the Imperial Units or their multiples, by making their use in commercial transactions illegal. (41 & 42 Victoria C.49, 1878, Weights and Measures act.)"

[11] "Divergent standards peculiar to a nation or region, complex conformity assessment requirements, and a thicket of other standards-related barriers have been estimated to impede the sale of an additional $20 billion to $40 billion worth of U.S. goods and services."

[12] The standard introduces the International System of Quantities (ISQ). It is a style guide for the use of physical quantities and units of measurement, formulas involving them, and their corresponding units, in scientific and educational documents for worldwide use.

[13] "The change to metric steel sizes opens up a tremendous opportunity for USA industry to rationalize on fewer steel sheet, plate, wire, bar, tube and section sizes. Some companies have shown remarkably good results when rationalizing, steel plates, for example, from 19 inch sizes to 8 metric sizes" in K. Kverneland, METRIC STANDARDS for Worldwide Manufacturing, GO metric USA TM.org, Inc., Statesville, North Carolina, USA, 2012.

IOP Conf. Series: Materials Science and Engineering **371** (2018) 012052 doi:10.1088/1757-899X/371/1/012052

The European Space - a vast territory marked by diversity, transnational, multicultural and cosmopolitical is unified by memoranda of understanding, adjusted and updated in accordance with contemporary values of universality. Built on the idea of a single market, governed by a parliamentary system, for the regulation and development, with the power to set laws (directives) applicable in the various member countries, with national force of law: unmatched in international standardization systems. Has, today, the highest number of normalization policies. It takes some degree of knowledge and confidence to exchange G&S and share manufacturing methods. Within Europe there were many national standards for Construction Products (CP), derived from ISO, interpreted/replicated/repeated by each member state. In 1999, it was understood that those are at the heart of the commercial dysfunction between Member States. The confusion between the certifications that accompanied the CP forced a position by the EU through legal approvals, a measurement known as "Nando (New Approach Notified and Designated Organisations) Information System".National legislation, testing, certifications, approvals, et cetera, in short, the lack of specific and harmonized standards were the focus of the problem and prevented the cross-border market. Embryonic conditions for the design of the current legislative framework, that culminated in the founding of three standardization bodies supervised by the European Commission (EC) and European Free Trade Association (EFTA):

- 1961, European Committee for Standardization (CEN)[14];
- 1973, European Committee for Electrotechnical Standardization (CENELEC)[13]; and,
- 1988 a European Telecommunications Standards Institute (ETSI).[15]

The entities, mentioned above, worked to establish bases for the harmonized standards, applied to the European single market, stimulating the growth of trade in goods, technological innovation and environmental protection through the European Standard (ENh). Those cooperated with the ISO (CEN) and the International Electrotechnical Commission (IEC) (CENELEC), and allowed to aggregate, in accordance to common principles, thirty-three national agencies (NB), with the overall objective, 'one standard, one test, accepted everywhere's, by 2020.

There are two channels for the production of standards, distributed as follows:

- 70% are requested to the Notified Body (NB), by operators and stakeholders, with the automatism of becoming effective and binding on other countries; this percentage holds also voluntary assessments by Technical Assessment Bodies (TAB) issued by the European Technical Assessment (ETA) — advancing to the DP and, if approved, the CE mark — which were valid for five years, until July 1st, 2013; after that date, indefinitely or until the development of another harmonized standard with power to revoke the evaluation; and,
- 30% by the EC as a technical component for the european legal framework; also mandatory, in this case: singular context on a global scale, the production of harmonized standards prevents the coexistence of similar documents.

NB (local standard setters) jointly with stakeholders (associations and companies), constitute themselves in Standardization Technical Committee (TC), organized in Subcommittees (SC) and/or Working Groups (WG), when justified, to prepare harmonized standards, documents and opinions to the study specific situations.

### *3.2. 'One standard, one test, accepted everywhere'*

Any country in Europe has a NB, accredited laboratory to perform, for example, testing, inspections, certifications, approvals. This may, in accumulation, set harmonized standards according to international, european and national criteria. When a rule is executed it receives an identification[16]. Or it can choose to adopt other standards, external and previously recognized, to which an identity is

---

[14] Despite the separation, natural by the particularities of CEN and CENELEC, the (technological) the evolution of PC consolidated, in early 2010, and the reiterated collaboration, the host of both in the same premises, "CEN-CENELEC - Management Centre (CCMC)" in Brussels.

[15] The three entities, described, form the European Standardization Organizations (ESOs).

[16] In Portugal, e.g., there are: NP, NP ENh, NP ENh ISO, NP HD, NP ENV, NP ISO, NP IEC and NP ISO/IEC. Apart from these, were also assumed as portuguese standards all, ENh, ENh ISO, ENh ISO/IEC and ETS.

IOP Conf. Series: Materials Science and Engineering **371** (2018) 012052 doi:10.1088/1757-899X/371/1/012052

added, the intervention is limited to the translation of the information. In both situations, those shall be placed in the national legal collection.

Essential Requirements (ER) for construction, present in the Construction Products Directive (CPD)[17], 89/106/EEC (revogada), (repealed), were the basis of the Harmonised technical rules or Harmonised standards[18] (Euroclasses): applicable to the control of factory production based on best practices in manufacturing. Now established by methods and performance evaluation criteria (Declaration of Performance - DP) of the CP in the reading of the ER as Construction Products Regulation performance (CPR)[19], EU 305/2011:

- Mechanical Resistence and Stability;
- Safety in Case of Fire;
- Hygiene, Health and the Environment;
- Safety in Use;
- Protection against Noise;
- Energy, Economy and Heat Retention; and,
- Sustainable Use of Natural Resources.

The CEN-CENELEC provides a platform for the development of European Standards and other technical specifications to various sectors, ensuring:

- the protection of consumers;
- facilitate cross-border trade;
- the interoperability of products;
- encourage innovation and technological development;
- include environmental protection; and,
- enable businesses to grow.

NB, in EU, work as an entity accredited by a Member State to ensure that a product meets the expected standards before being placed on the market.

The CE marking, Directive 93/68/EEC, is a mandatory conformity mark on many products placed on the single market in the European Economic Area (EEA) since 1985. The CE mark is the declaration by the manufacturer on the conformity of the production process (for products) with the EU directives, to ensure three principles [12]:

- Existence of a European Policy — if not available, to promote its implementation and adoption, with force of law, in the country(ies) concerned — member states have taken CPR as a national law, like the previous one that established the concept; its importance lies in the definition of 'Basic requirements for construction works - Essential characteristics' which must be present to ensure the CE mark;
- The process with a specific description of how the CPR was applied; named 'European Assessment Document';
- The publication of the documents produced[20].

The CE mark and the Technical Document that accompanies it do not indicate, e.g., any safety rules, the load capacity, functionality or aesthetics of a CP, only attest, to the supervisory bodies of the common market, their suitability for transaction within Europe - nevertheless indicate the qualities and classification performance obtained during the manufacture of the product, for information and control by regulators.

---

[17] Council Directive 89/106/EECof 21 December 1988 on the approximation of laws, regulations and administrative provisions of the Member States relating to construction products, OJ N.º L 40 of 11 February 1989.

[18] Fundamental technical details for the manufacture of the product: evaluation systems and performance verification.

[19] Regulation (EU) No 305/2011of the European Parliament and of the Council of 9 March 2011 laying down harmonised conditions for the marketing of construction products and repealing Council Directive 89/106/EEC, OJ L 88 of 4 April 2011

The European Construction Products Regulation (CPR) describes the fundamental requirements on hygiene, health and environment protection.

[20] The CE mark is used to label a product or the accompanying documents: is not accompanied by any certificate or statement. There is, however, the obligation to support and expose.

IOP Conf. Series: Materials Science and Engineering **371** (2018) 012052 doi:10.1088/1757-899X/371/1/012052

DP[21] requires the involvement of two entities for CE marking: the manufacturer and the NB. The process goes trough articulated phases - with the progress of steps, the manufacturer's responsibility increases, gradually, in the same extent as it exonerates the NB, which will also decrease its level of control.

DP issued by NB:

- on-site verification of production, by the manufacturer; the NB may require additional assessment tests;
- the extraordinary inspection of production jobs;
- continuous monitoring; and,
- possible repetition of tests, without notice, to control the results, financially assumed NB.

DP issued by the manufacturer:

- initial inspections on products by NB; and,
- on-site verification of production, by the manufacturer.

Manufacturer, importer or distributor who places a product on the market, under his name or under a trademark, assumes responsibility for the technical document that holds the CE mark for the product, according to the DP or if an ETA (non-harmonized products) is issued by a TAB, or even another, concerning improved protection of users of the product, not covered by the Directive.

Stabilized products will be accepted, without testing ETA, for knowledge shown or if supported by technical data 'institutional', measured against empirical provision, as set out in the specifications, only to a certain level or class. Important measure for traditional and current products, which gain ground to maintain presence and usual performance.

If the manufacturer of a product, finds in the acquis the regulatory framework to certify the development of procedures can avoid the same overall costs, by simply requiring authorization for the us of standard(s) to third parties: this type of trading is valid. With this principle, there is the risk of creating a '(parallel) market' for standards.

If a producer intents to introduce a PC on the market, the CE mark is mandatory, which can be requested to the NB, according applicable standards. If non-existent, the opening of a procedure standardization is requested to NB, which forms a labor committee, local/national, and the CEN another, to follow the work of the first. When a standard is effective all the same, prior to 1989, are repealed even in other countries.

When a Innovator Product of Construction (CPI) or Special Product of Construction[22] (CPE) does not meet the harmonized standards or do not have pre-established and consolidated technical information on their performance, have to resort to a TAB to start a certification process. This allows the PCI / E can obtain the CE mark through a European Technical Assessment (ETA) or European Assessment Document (EAD), certification issued by TAB under European Organization for Technical Approvals (EOTA), parallel to NH, which is not obtained by procedure. The process focuses on the verification of the risks implicit in its use in construction. From risk analysis arises the technical criteria for assessing behavior and performance of the CPI / E, given the applicable national rules.

[21] Compliance with the specific harmonized standards through the production system control.

[22] A limited production for which does not compensate to set a standard.

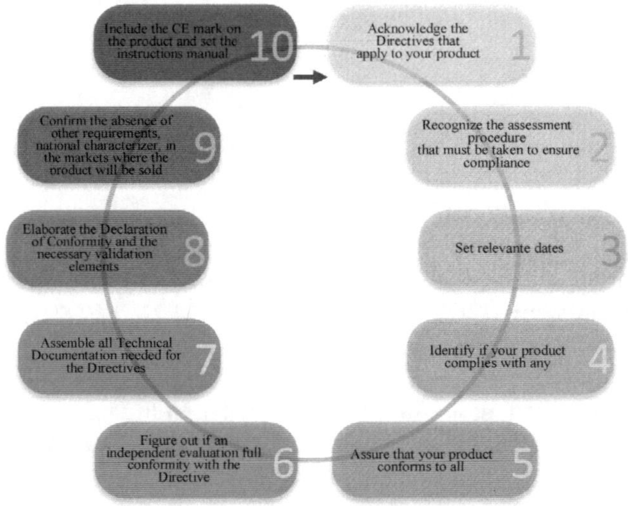


**Figure 1.** - European certification cycle.

ETA, requested by the manufacturer, is a documented technical evaluation that measures the performance ability of a construction product, according to a European Assessment Document organized by TAB. Applies to innovative and complex products (for which there is not, nor is expected, in the medium term, the existence of harmonized European standard).
Every country in the EU has, through the contact points, to parameterize the different climatic, geological, geographical and other perceived as relevant and prevalent conditions, to consider when establishing the performance classes of CP/I/E. The characteristics are defined by limit levels of performance for a given basic requirement. Detected restrictions must be included in the harmonized technical specifications. It is interesting to consider that the NB can select levels or classes of performance, to respond to local conditions and peculiarities, within the national territory.
Micro enterprises, if proven the condition may require simplified procedures for performance assessment, for products that do not involve additional concerns but ensure benefits similar to those counterparts.
PC similar in design and production or of 'variable' series can use the same DP[23], by request and authorization for simplified performance assessment procedures and demonstrated compliance with the requirements by the Commission after consulting the Standing Committee on Construction.
Importers, distributors and support chain[24], will have to ensure compliance of the CP with the DP specific characteristics and requirements: included in the technical document of the product that allows its transaction in the EU.
Exceptions are foreseen for the compulsory DP, e.g., in the case of a prototype product, developed only to a specific situation/condition; products manufactured on site with the monitoring of responsible for security; or a specific product, made using traditional methods/techniques, for a recovery or conservation, preserving the architectural and historical value.

*3.3.Before and after EN - model adjustment*

The interpretation and implementation of legal procedures differs from country to country. For some, the CE marking process is not only the first but also the only regulation for CP, in compliance with the

[23] The market for construction products in the EU is around: manufacturers, 178 million / year to 2 million / year in DP - does not include document management, cost of time spent and time of printing; distributors (about 76.5000 euros / year in the EU), EUR 476 million / year to 5800 million / year in DP - does not include document management, cost of time spent, printing time and delays in transport.

Examples: DD by each product costs 2,000 euros, but if the color is different, it needs another, always in multiples of 2000; if a manufacturing company purchases the same natural stone from different explorers it has to pay 2000 euros (cross value to EU countries), for each of them.

According to information available on the website of ANIET - Extractive Industry National Association and Manufacturing, May 9, 2016.

[24] Small and medium sized enterprises (SMEs), by directive, can see their access to legislation facilitated in the Member State where they intend to sell or make available their products.

steps in Figure 1. For other countries it is an agreement at the lowest level, because so far construction has been (and still is) organized by a multiplicity of standards and directives, including approvals and regulations by authorities, associations and safety bodies.

In accordance with the European Union politico-economic concept, more and more national regulations should be abolished and construction should be left to its own devices. For some that even applies to the safety of buildings, true to the motto: 'unsafe construction products (as those that collapse or fall down) will automatically disappear from the market if they are no good'.

This results in the state supervisory procedures and authorities being viewed with suspicion and their importance being reduced rather than being promoted. Voluntary checks on the basis of self-obligations e.g. at association level, are welcomed. Hence, in those countries in which construction had previously been state-controlled for reasons of safety, private quality organizations are being established that attempt to maintain the old high safety standards, preferably cross-border, on a voluntary basis.

The demand for (new) approvals in countries that had previously had no stipulated safety regulations (approvals or the like) is not in accord with the European goal of the free trade of goods, because new trade barriers are thereby erected. In particular, the demand for a European Technical Approval (ETA) when the CE mark, see Figure 2, is already borne due to a harmonized standard contradicts the construction product directive [13].

The safety of life and limb remains (for all products) a matter for the individual nations, because the CE mark does not give any indication about this. In order to meet these demands, obsolete normative or legal regulations exist in several countries (already for a long time in some cases), which are not (may not be) repealed by the CE mark, because this fundamental right of safety remains with the countries. Astoundingly, product approvals have nevertheless been withdrawn in at least one European country following the introduction of the CE marking obligation.

In order to maintain the safety of buildings, the compulsory approval remains in force, for example, in France, Germany, England and other countries too — as far as can be discerned up to now. Although the CE mark entitles a construction product to be freely traded in Germany, its use requires a Ü mark, which is only issued on the basis of a general building authority approval[25].

*3.4.Space for criticism*

The initial idea, which still stands, presupposed to boost the standards' market. Some of the factors contrary to the intended dynamics are: the time required to produce new standards; the revision of old standards (national and established); and, when this happens the NB resist to change, procrastinating processes, sometimes for years. A fateful bureaucratic process, the creation of technical committees (NB) and monitoring (CEN), without any failure to pass, until today; on the other hand it allows any of its representatives to deliver view/opinion even without experience or tradition in this type of PC [14].

Obtaining a CE, through an ETA/ETD can take several years, resulting in processes with hundreds of pages and burden on the requestor. It is being discussed how to simplify and overcome this process, based on a shared methodology with other NB for quick approval: Common Understanding of Assessment Procedure (CUAP).

The Construction Product Directive (CPD) is at present being revised and is to be replaced by a regulation (CPR) of the European Parliament and Council. The reason for this is that a regulation does not permit so many interpretations in terms of time and content, but must be implemented more strictly. The six essential requirements of the CPD will be supplemented by a seventh: the sustainable use of natural resources.The new regulation is intended to strengthen the credibility of the system and

[25] In Germany, the approval principles for health-evaluation of construction products by the German Institute for Structural Engineering (DIBt) specify these, especially concerning emissions. So in Germany, for reasons of health and environment, the Ü mark of conformity on the basis of a national technical approval (abZ) is required for: since 2005, resilient, textile and laminate floor coverings (EN 14041); since 2011, parquet and wood flooring (EN 14342); since 2011, parquet adhesives and coatings; since 2012, floor covering adhesives and flooring underlays; and, since 2014 on wall coverings.

IOP Conf. Series: Materials Science and Engineering **371** (2018) 012052 doi:10.1088/1757-899X/371/1/012052

the acceptance of the CE marking. The way in which these goals are to be achieved is clear in principle, but the detailed measures are still very much under discussion.

**4. Conflict of Norms**

The European standardization market tries, from the beginning, to prevent the formalization of constraints on, desired hegemony of, the CE mark (CPR's Article 8) through the DD. The fundamental idea seeks the elimination of technical barriers to free trade of PC, in accordance with the ENh and ETA, taking the essential characteristics as a response to basic requirements. As such, those are determined and 'static' to any geographical condition or safety issue, therefore, tend to be fitted by national/local regulations. There are, however, local manufacturers who can understand these differences in their interest, justified on technical issues in accordance with the CPR (Article 6): by means of an adjusted justification.

In 2008 and 2012, respectively, the Belgian and German governments were condemned by the European Court, for encouraging PC market operators to resort to national certification, in addition to the CE mark: BENOR, voluntary, and Ü, mandatory, in the same order. Many organizations are faced with such obstacles in countries with strong national values, standardization experience and pressure groups that influence decision-makers in favor of protection of the domestic economy, especially in markets that have industries with heavy production capacity and similar articles.

The German market is the most targeted by critics of the instigators of the PC free circulation. To these tests, in an institute recognized by DIBit, were required in order to be used in the construction sector. Thus, the applicant is subjected to administrative costs and, perhaps, in the production - investment in the repetition of tests and trials. In its defense the German government claimed concerns about the protection of health and the environment. In question was, in fact, the protection of three groups of domestic manufacturers dedicated to products: pipe joints seals made of thermoplastic elastomer, insulating materials made of mineral wool and gates, windows and exterior doors. Exposure was not accepted, October 16, 2014.

Sometimes, promoters impose and, even, require qualities greater than those indicated by the CPR, subordinating the CE to the national mark, for example, in UK, public roads are required to offer an asphalt with a minimum of two years durability, which can only be assured by AVCP System 2+, through the evaluation of the wear layer.

In some Member States, where the guarantee is assumed by the insurance companies, evaluations may be required to the constructed, six months after the entry into service of the building, with special interest in the parts considered 'sensitive'. These are, in the French case, the inspection targeted by independent entities, Centre Experimental de Recherches et d' Études du Bitiment et des Travaux Publics or int Rosenheim. Situation that induces the use of national standards, seen as more reliable by decision-makers and insurance companies, such as voluntary CEKAL certifications. This question, among others, is of the utmost importance to the insurance companies, which subject their actions to the existence of compliance with national standards, considered more demanding than similar EC. In particular, they tend to not consider or accept periods of more extensive warranty, as 10 years, and to inflate the value of the insurance premiums. To overcome the constraints created, stakeholders assume the resulting increase in costs for the necessary certifications and the delays in the commercialization of products, e.g.:

- in France, in particular, the insurance companies require, for more competitive conditions, the certificate Centre Scientifique et Technique du Bâtiment (CSTB) - 'third' party, regardless of the system AVCP (french NB) - based on Document Technique d'Application (DTA);
- for window manufacturers CEKAL standard is essential, because it facilitates the acceptance and marketing of products, especially those with high standards of insulation, tightness and durability;
- flooring manufacturers, in case of public use, are conditioned to NF UPEC and, in the same direction, Avis Technique (Atec);
- in Hungary and Poland, additional but volunteers requirements are suggested and, actually, carried out;

- in France and Spain in PC CE from other sources are required additional marks and testing, Asociación Española de Normalización y Certificación (AENOR) and Normalisation, Certification, Edition et Formation (AFNOR), respectively.

Sometimes, the co-existence with national rules leads to misunderstandings in favor of these. As a result, new processes are imposed, to stakeholders, using national standard taxes for alleged non-existence of equivalent ENh and outside the scope of ETA. An issue that limits the access to the single european market, e.g.:

- in the United Kingdom (UK), there is a predominance, in certain PC, of the British Board of Agrément (BBA) certificate, by the United Kingdom Accreditation Service (UKAS);
- also in the UK, it was decided to introduce three additional features to EN14351 in favor of a national rule, on Doors and Windows - in particular, air and water permeability and wind resistance —; an identical procedure was introduced in the Netherlands and Bulgaria for the same PC;
- in France, the NF228 in force for the tiles sector is considered an obstacle to market access, though that is at all similar to EN 12326 (adopted by the CE mark) is, however, regarded as inferior; in the UK, an early withdrawal of the British Standard BS 680 and the adoption of EN 12326 led the industry to hide behind the French standard NF228.

In parallel with the CE mark, the process of compliance with the National standard(s) tends to be, in such a way, so onerous and time consuming that many stakeholders end up abandoning their intent. The issues outlined in the previous section, are the example of the lack of effectiveness of European entities, despite the sanctions provided for in CPR, there is no record of such.

Sometimes, manufacturers and other interested parties, coming from a country inside or outside the EU, are conditioned to the use of additional markings (certifications). To access the fullness of the market — as double glazing and cements (as number 11 in the references) — it is required to some PC up to ten certifications, in as many Member States. Today, these situations tend to be accepted and even sought by the interested parties as if they were imposed obligations.

Ultimately, for the PC, there are national requirements that affect the decision of the designers and even consumers, in markets where national certification marks are, for many years, recognized by the performance of products. The products recently placed on the market with CE mark can not succeed, despite having qualities aligned with the others, there is a strengthening of the monopoly of local manufacturers, recognized by decision makers. There are consumers who devalue CE mark for the national equivalent. The imposed generalization, common to all kinds of products, from toys to complex hospital equipment, led to its trivialization and loss of importance in markets with local equivalents, with recognized and praised quality.

## 5. Some Technical Suggestions

Article 8, in its number 5, is vague and should be limited in a future revision of the CPR. In its writing, "(...) shall ensure that the use of construction products bearing the CE marking shall not be impeded by rules or conditions imposed by public bodies or private bodies acting as a public undertaking (...) when the declared performances correspond to the requirements". This last section leaves open other class requirements, in short, if so understood by the NB, it limits the marketing of some products or force manufacturers to request new tests to understand the local "user requirements".

In number 33 of the 'considerations' of the CPR we have: "The CE marking should be the only marking (...) However, other markings may be used, provided that they help to improve the protection of users of construction products and are not covered by existing Union harmonization legislation.". This is one of the anchors used by some Member States to impose other rules. Situations in some cases justified — e.g., the Dutch government has exposed the differences between the KOMO and EC, in the framework of the article mentioned — others are more difficult to accept, especially those which the implemented industrial capacity and high quality standards tend to protect the national economy.

In future CPR revisions the brand CE should be stratified — abandoning the idea of a 'static' quality brand — in levels of excellence: accurate and specific information for markets, of the PC efficiency, in order to replace or make redundant the national equivalents, understood as quality pillars. Like the efficiency rating for electrical household appliances, e.g., "D/C/B/A/A+/A++/A+++", wherein the above basic requirements of a particular class — provided for in the CPR, but with no clear distinction — would get a prominent ranking, as exemplified.

## 6. Conclusions

The EU aims to: protect the markets, protect operators and increase product safety, for interveners and users. Grounds for the creation of the CE mark[26]: transversal/common to all commercial activities. The scope of the CE mark, makes this an unfocused answer the unique problems of the construction sector, only adjusted to the particularities.

By force of situation, the EU has tools that enables it to compose supranational legal frameworks with precedence over the Member States: despite the divisions between peoples and differences in construction methods. Nevertheless, the CP are increasingly more widespread worldwide and, in particular, in Europe: proving that all generated legislation stimulated the dynamism of the market with products that meet the standards, although basic. However, countries like France and Germany managed to pass laws that increase the control of CP, for the protection of the environment, in the first, and safeguard the 'good construction' in the second.

The use of ISO standardization, as a starting point for ENL, was essential to aggregate participants around a European ideal. The path was integrator since it allowed to adjust the various NB around common legislation. Position which compromised the legislative framework, especially in a sector as 'influential' (powerful), with existing structures and vested interests to defend its 'natural/traditional' space: compromise which did not allow a strong position in face of some situations, e.g., the high number of ETA to somehow circumvent the costs and procedures involved in setting up new standards or assumption of similar.

Unlike the rest of the world, in Europe the production of standards documentation (Harmonized) is unique, resembling copyright: the registration of a rule prevents the publication of other similar (for established criteria). Also, in order to avoid duplication of testing, the manufacturer of a CP may request permission to use third parties test(s) results free of costs. Peculiarities instituted in order to reduce costs and time: to accelerate the introduction of the PC in the market. On the other hand, there is the growth of informal standardization markets, contrary to the spirit of European legislation.

CE certification was imposed on CP but the same does not apply to Construction Services. Some entities/operators, who voluntarily seek to establish a standard of rigor performance, have opted for Certification ISO 9001:2008[27,28], those are, however, a minority. While PCs have basic requirements and are controlled by different entities, those who manipulate them are not bound to any rigor or specific principle, having as tip of the balance the yoke of a technician responsible, trained and experienced. With the exception of the requirements made by the Member States as, e.g., carbon dioxide emissions or risk of health and safety working conditions.

The lack of tools for the assessment of buildings (finished product) constitutes an antagonistic case to other consumer goods, with CE marking. PC are handled, and sometimes tampered with, during the construction process, so its characteristics are altered making it different what it was when granted the CE mark. It can be assumed that, therefore, the safety of a citizen, living in the building or surrounding space, is not regulated by the basic requirements of CPR, since that is only dependent on national regulations.

As above and to complete, there are 'interests', both public and private, that act as NB, when it comes to regulating the market. What it is not and never was the spirit of the CPD or, later, CPR.

## 7. Acknowledgment

[26] Today, the CE mark is a source of pride for manufacturers. It is recognized within the Europe, by both operators and consumers (in general), and across borders (e.g., Canada signed a protocol with the EU to the free marketing of CP with CE marking), like the FCC Declaration of Conformity (US), the Canadian Standards Association (CSA), Underwriters Laboratories (UL) of the Electromagnetic Compatibility (EMC), to name a few.

[27] ISO 9001 is by far the world's most established quality framework, currently being used over 1.5 million organizations in 191 countries. Fonte: http://www.iso.org/iso/iso-survey.htm?certificate=ISO%209001&countrycode=AF

[28] In Europe, in 2005, extrapolating from the Italian example only 42% of the construction companies had ISO9001 certification. Source: P. Sampaio, P. Scott, A.G. Rodrigues, ISO 9001 Certification Forecasting Models, International Journal of Quality & Reliability Management Vol 28 Nr 1, 2011 pp. 5-26, DOI 10.1108 / 02656711111097526 2010.

IOP Conf. Series: Materials Science and Engineering **371** (2018) 012052 doi:10.1088/1757-899X/371/1/012052

The author would like to thank the Fundação para a Ciência e Tecnologia (FCT – Portugal) for the financial support to the research project UID/EAT/04008/2013, through Centro de Investigação em Arquitetura, Urbanismo e Design (CIAUD — Portugal).

**8. Appendix A World Trade Organization**

Was funded on five principles: 1. "Trade without discrimination"; 2. "Freer trade: gradually, through negotiation"; 3. "Predictability: through binding and transparency"; 4. "Promoting fair competition"; and, 5. "Encouraging development and economic reform"; to achieve the "Comparative advantage" trough available goods.

**Appendix B CE mark**

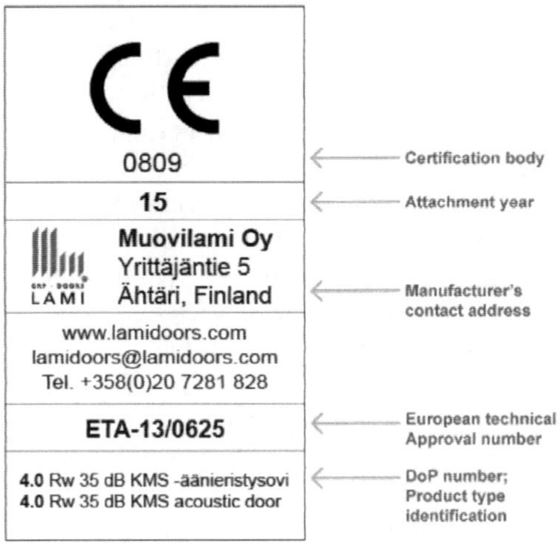


**Figure 2.** - CE mark (e.g.).

**Appendix C European Committee for Standardization (CEN)**

The CNE is composed of 30 national standards bodies (ONN): IQ - Instituto Português da Qualidade (Portugal); ON - Österreichisches Normungsinstitut (Austria); IBN - Institut Belge de Normalisation (Belgium); BDS - Bulgarian Institute for Standardisation (Bulgaria); CYS - Cyprus Organization for Standardisation (Cyprus); CNI - Cesky Normalizacni Institut (Czech Republic); DS - Danish Standards (Denmark); EVS - Estonian Centre for Standardisation (Estonia); SFS - Suomen Standardisoimisliitto r.y. (Finland); AFNOR - Association Française de Normalisation (France); DIN - Deutsches Institut für Normung (Germany); ELOT - Hellenic Organization for Standardization (Greece); MSZT - Magyar Szabványügyi Testület (Hungary); IST - Stadlard Islands (Iceland); NSAI - National Standards Authority of Ireland (Ireland); UNI - Ente Nazionale Italiano di Unificazione (Italy); LVS - Latvian Standards Ltd (Latvia); LST - Lithuanian Standards Board (Lithuania); SEE - Service de l'Energie de l'Etat (Luxembourg); MSA - Malta Standards Authority (Malta); NEN - Nederlands Normalisatie-instituut (Netherlands); SN - Standardiseringen i Norge (Norway); PKN - Polish Committee for Standardization (Poland); ASRO - Asociaţia de Standardizare din România (Romania); SUTN - Slovenský ústav technickej normalizácie (Slovakia); SIST - Slovenian Institute for Standardization (Slovenia); AENOR - Asociación Española de Normalización y Certificación (Spain); SIS - Swedish Standards Institute (Sweden); SNV - Schweizerische Normen-Vereinigung (Switzerland); and, BSI - British Standards Institution (United Kingdom).

**9. References**

[1] Analysis of implementation of the Construction Products Regulation, Risk & Policy Analysts Limited, London, Norfolk, UK, 2015.

[2] O. Ilvonen, "Assessing release of hazardous substances from construction products – Review of 10 years of experience with a horizontal approach in the European Union," *Build. Environ.*, vol. 69, pp. 194–205, 2013. https://doi.org/10.1016/j.buildenv.2013.08.010
[3] S. Mills, International Standards in the Emerging Global Economy, OpenStand publication, 2012.
[4] Economic benefits of standards, International Organization for Standardization, Genève, 2014.
[5] S. Vogel, Freer Markets, More Rules: Regulatory Reform in Advanced Industrial Countries, Cornell University Press, Cornell, 1996.
[6] Mallett, Robert L. "Why Standards Matter." Issues in Science and Technology 15, no. 2 (Winter 1999).
[7] R. Tricker, "3 - Structure of the CE conformity marking directive (93/465/EEC) BT - CE Conformity Marking," Oxford: Butterworth-Heinemann, 2000, pp. 55–96. https://doi.org/10.1016/B978-075064813-4/50015-1
[8] Regulation(EU) N.º 305/2011 of the European Parliament and of the Council, Official Journal of the European Union, Strasbourg, 2011.
[9] A. Velker, Markets and Measurements in Nineteenth-Century Britain, Cambridge University Press, Cambridge, 2012. https://doi.org/10.1017/CBO9781139151030
[10] A. Kylili and P. A. Fokaides, "Policy trends for the sustainability assessment of construction materials: A review," *Sustain. Cities Soc.*, vol. 35, pp. 280–288, 2017. https://doi.org/10.1016/B978-075064813-4/50015-1
[11] K. O. Kverneland, Metric Standards for Worldwide Manufacturing, 2012 Edition, "GO metric USA™.org", Inc., Statesville, North Carolina, 2012.
[12] CE mark now! - Mandatory CE marking of construction products, Swedish National Board of Housing, Building and Planning, Boverket inhouse, Boverket, 2013. ISBN 978-91-87131-78-3
[13] The CE mark and European Technical Approval, Kalzip GmbH · Part of Tata Steel Europe Ltd, Kobenz, 2010.
[14] T. Nwaogu, S. Marshall, Analysis of Implementation of the Construction Products, Risk & Policy Analysts Limited, London, United Kingdom, 2015.

IOP Conf. Series: Materials Science and Engineering **371** (2018) 012053 doi:10.1088/1757-899X/371/1/012053

# The Effect of Different Heights and Angles of Energy Director on Interface Temperature for Ultrasonic Welding of Thermoplastics

**NV Thang[1] and P Lenfeld[2]**

1 Ph.D. Student, Faculty of Mechanical Engineering, Department of Engineering Technology, Technical University of Liberec, Czech Republic
2 Professor, Faculty of Mechanical Engineering, Department of Engineering Technology, Technical University of Liberec, Czech Republic
E-mail: petrnguyen321@gmail.com

**Abstract.** Besides technological parameters in the ultrasonic welding process, height and angle of energy director (ED) also affect the quality of the joint strength owing to its energy transfer and melting. This paper deals with evaluating the effect of distinct ED heights and angles of butt-joint on temperature distribution at the weld interface. Experiments were simulated using transient thermal analysis of Ansys 18 for 3D-samples made of acrylonitrile butadiene styrene (ABS) and high-density polyethylene (HDPE). The results obtained from simulation of temperature distribution in various ED heights and angles and welding amplitudes are presented and compared with each other.

## 1. Introduction

For bonding plastics and composites often use welding technologies. One of the most modern welding methods is ultrasonic welding technology, which has various applications in the engineering industry, especially in the automotive and textile industry, in the production of electrical appliances, in the packaging technology, etc. The main advantages of ultrasonic welding are fast, clean, efficient and repeatable process, producing a strong bond with the consumption of a very little amount of energy. To create the welded joint by ultrasonic technology requires no solvent, adhesive or external heat and the resulting joint is undetachable.

Ultrasonic plastic bonding is the joining or reforming of thermoplastics by using heat generated from high-frequency mechanical motion. This can be achieved by converting electrical energy into high-frequency mechanical motion (vibrations), which generates frictional heat at the joint area. The vibrations applied to a part under pressure/force, generate frictional heat at the interface and cause melting of the material and creating a molecular bond between the welded plastic parts. The ultrasonic welding process is described in several phases shown in Figure. 1 [1].

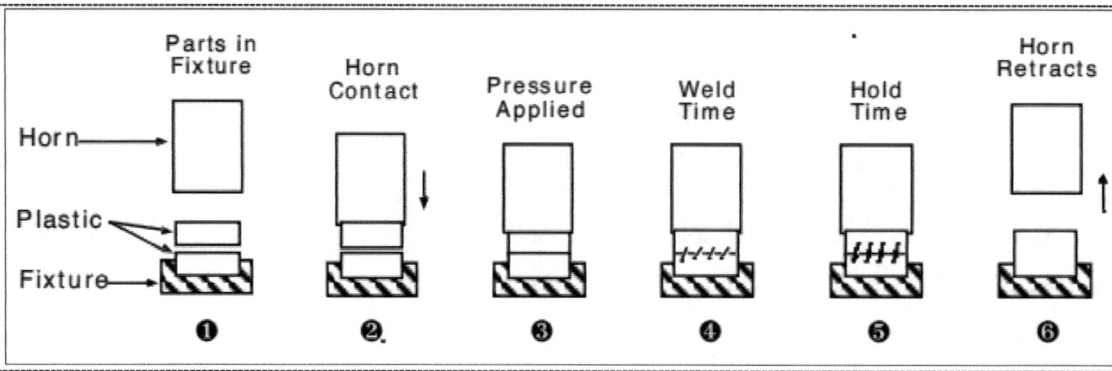


**Figure 1.** Phases of ultrasonic welding.

A whole range of welding parameters (frequence, amplitude, time, etc.) has an important effect on quality of the resulting welded joint. Besides the welding processing parameters, for proper evaluation, it was also necessary to determine the ED geometry, namely its height and angle.
The energy director (see Figure. 2) was developed to provide a specific volume of material for melting, in order to achieve a good bond strength without excessive flash. When ultrasonic energy is transmitted through the part under pressure and over time, the energy concentrates at the apex of the ED. This leads to a rapid generation of heat that causes the bead to melt. The molten material flows across the joint interface, forming a molecular bond between the two welded parts [1].

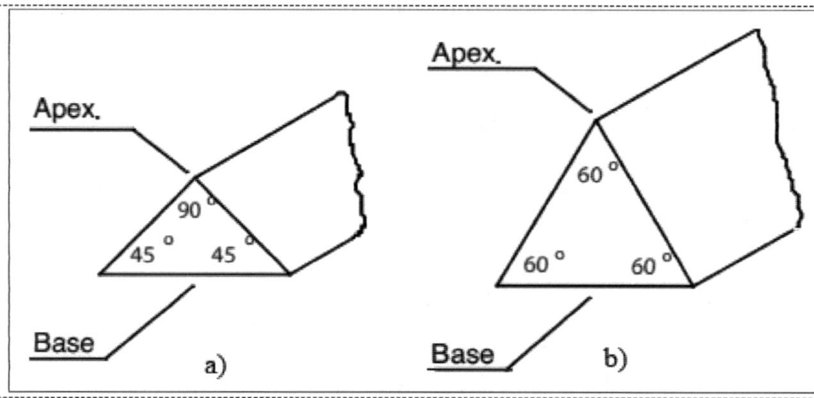


**Figure 2.** Energy director: a) for amorphous plastics, b) for semicrystalline plastics.

In this paper, the butt joint with 60°, 75° and 90° angle included ED was used to simulate temperature distribution in ultrasonic welding at a frequence of 20 (kHz) and at amplitudes of 40, 50 and 60 microns. In addition, for each individual angle of ED, its height ranged from 0.5 to 1.75 (mm).

## 2. Procedure for simulation analysis

### *2.1. 3D Model and thermal loads*

The two rectangular models having dimensions of 80*10*4 (mm) for the first one and of 80*11.2*4 (mm) for the second one are welded together. The first model with various EDs is on the top of the second one, as shown Figure. 3. The 3D model was meshed using the tetrahedral method. A transient thermal analysis was chosen to simulate for welding time of 0.8 [s] with a time increment of 0.05 [s]. The applied thermal loads were given as follows: the initial temperature of 25 °C, thermal convection to the environment with overall heat transfer coefficient U ($W/m^2K$), heat flux $\emptyset_q$ ($W/m^2$) for ED height applied as a thermal load at the interface. The properties of ABS and HDPE are presented in Table. 1.

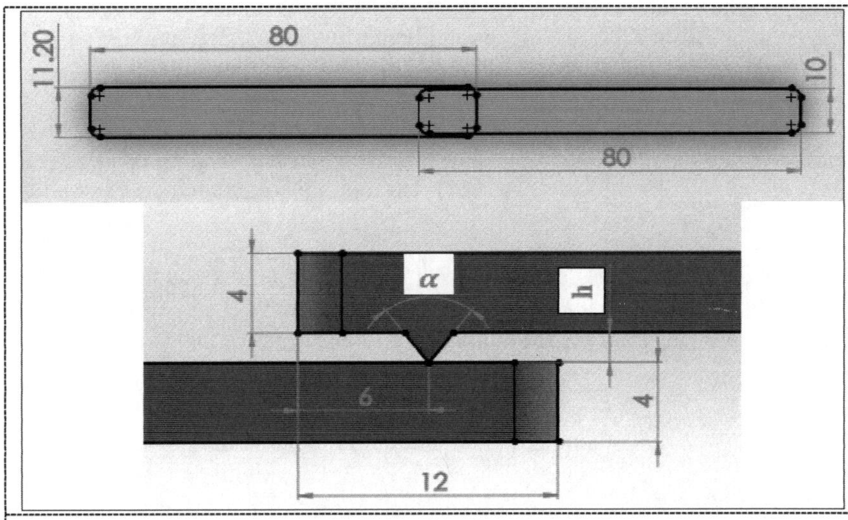


**Figure 3.** The position of workpiece models with various heights and angles of ED (all dimensions in mm).

**Table 1.** Material properties of ABS and HDPE.

| Material | Density $\rho$ (kg/m$^3$) | Specific heat capacity $C_p$ (J/kgK) | Coefficient of thermal conductivity k (W/mK) | Young's modulus (GPa) | Mechanical loss coefficient (-) |
|---|---|---|---|---|---|
| ABS | 1020 | 1386 | 0.23 | 2.0 | 0.0138 |
| HDPE | 952 | 1796 | 0.46 | 1.07 | 0.0437 |

### *2.2. Determination of values for required parameters*

*2.2.1. Calculation of overall transfer coefficients.*The overall heat transfer coefficient U per unit area can be expressed as [2]:

$$U = \frac{1}{\left(\frac{1}{h_A} + \frac{dx_w}{k} + \frac{1}{h_B}\right)} \tag{1}$$

Where:

U= overall heat transfer coefficient (W/m$^2$K)

k= thermal conductivity of the material (W/mK)

h= individual convection heat transfer coefficient for each fluid (W/m$^2$K)

$dx_w$= wall thickness (m)

A single plate exchanger with media A transfers heat to media B. Both of mediums are air with a convection heat transfer coefficient of $h_{air}$ = 50 W/m$^2$K. The wall thickness is 4 mm and the materials used are ABS and HDPE with coefficients of thermal conductivity listed in Table. 1. The calculated values of overall heat transfer coefficient are shown in Table. 2.

IOP Conf. Series: Materials Science and Engineering **371** (2018) 012053 doi:10.1088/1757-899X/371/1/012053

**Table 2.** Values of overall heat transfer coefficient.

| Material | Value of U ($W/m^2K$) |
|---|---|
| ABS | 17.32 |
| HDPE | 20.54 |

*2.2.2. Calculation of loss moduli.* Loss modulus is a measure of the viscous response of a material. It measures the energy dissipated as heat and is calculated based on Young's modulus and mechanical loss coefficient (tan delta). The Young's modulus is considered to be equal to [3]:

$$E = \sqrt{\left[(E')^2 + (E'')^2\right]} \quad (2)$$

Where

E=Young's modulus (GPa)

$E'$=storage modulus (GPa)

$E''$=loss modulus (GPa)

The ratio between the loss and storage modulus in a viscoelastic material is defined as the tan delta, which provides a measure of dampening in the material and can be calculated by equation (3) [3]:

$$\tan\delta = \frac{E''}{E'} \quad (3)$$

Loss moduli are given by using equations (2) and (3):

$$E'' = \sqrt{\frac{(E * \tan\delta)^2}{1 + (\tan\delta)^2}} \quad (4)$$

The calculated values of loss modulus for ABS and HDPE are shown in Table. 3.

**Table 3.** Values of loss modulus.

| Material | Value of $E''$ (GPa) |
|---|---|
| ABS | 0.028 |
| HDPE | 0.046 |

*2.2.3. Calculation of heat fluxes*

The heat generated at the joint is proportional to the square of the amplitude of the vibration. Hence, any change in the amplitude has a greater effect than a change of any other parameter [4]. In ultrasonic welding using ED, the average energy dissipated per unit time or the average internal heat generation rate can be calculated as [5]:

$$Q_{avg} = \frac{\omega * \varepsilon_o{}^2 * E''}{2} = \pi * f * \varepsilon_o{}^2 * E'' \quad (5)$$

Where

$Q_{avg}$=the average heating rate ($W/m^3$)

E''=loss modulus of the material (Pa)

f is welding frequency (Hz)

$\varepsilon_o$=the strain amplitude, which can be written as [6]:

$$\varepsilon_o = \frac{A_0}{2 * L} \tag{6}$$

Where

$A_0$ = vibration amplitude applied during welding, which is ranged from 40 to 60 microns

L = thickness of workpiece (m)

Heat flux is calculated according to the following equation [7]:

$$\phi_q = \frac{Q_{avg} * V_{ED}}{A_{ED}} = \frac{Q_{avg} * A_{ED} * h}{A_{ED}} = Q_{avg} * h \tag{7}$$

Where

$\phi_q$=heat flux ($W/m^2$)

$V_{ED}$=volume of ED ($m^3$)

$A_{ED}$=area of ED ($m^2$)

h=height of ED (m)

Chuah et al., 2000 [6] determined the energy absorption ratio '$r_{absorb}$' for the ABS and HDPE welded parts, which is defined as the ratio of the energy absorbed by the plastic material at the ED to that of the energy dissipated by the ultrasonic welder. The calculated values were 48.5 % for the ABS welded part and 21.1 % for that of HDPE. So, the heat flux is recalculated as (8) and the final calculated values of heat flux are listed in Table. 4.

$$\phi_q = Q_{avg} * h * r_{absorb} \tag{8}$$

Where

$r_{absorb}$=energy absorption ratio (%)

**Table 4.** Values of heat flux.

| Amplitude (μm) | Value ($W/m^2$) | Material | | | | | | | |
|---|---|---|---|---|---|---|---|---|---|
| | | ABS | | | | HDPE | | | |
| | | Height of ED | | | | | | | |
| | | h = 0.5 (mm) | h = 1 (mm) | h = 1.5 (mm) | h = 1.75 (mm) | h = 0.5 (mm) | h = 1 (mm) | h = 1.5 (mm) | h = 1.75 (mm) |
| 40 | $\phi_q$ | 10500.4 | 21000.8 | 31501.2 | 36751.4 | 7735.2 | 15470.3 | 23205.5 | 27073.0 |
| 50 | | 16406.9 | 32813.7 | 49220.6 | 57424.0 | 12086.2 | 24172.4 | 36258.5 | 42301.6 |
| 60 | | 23625.9 | 47251.8 | 70877.7 | 82690.6 | 17404.1 | 34808.2 | 52212.3 | 60914.3 |

## 3. Results and discussion

The simulation process was carried out on the 3D model with ED to determine temperature distribution, as well as maximum interface temperature (MIT) at the apex of ED, as shown in Figure. 4. Figure 5, 6 showed values of MIT for ABS and HDPE, respectively, which depended on welding amplitude, height and angle of ED. It could be seen in Figure. 5 that the highest value was 570.2 (°C) at welding amplitude of 60 (μm) with ED height and angle of 1.75 (mm) and 60 (°), respectively. Whereas the smallest value was 79.5 (°C) at welding amplitude of 40 (μm) with ED height and angle of 0.5 (mm) and 90 (°), respectively. Thus, it clearly showed that increasing welding amplitude caused a strong increase in the value of MIT. In terms of the percentage difference in MIT, the greatest value between welding amplitude of 40 (μm) and 60 (μm) was about 113 (%) at ED height and angle of 1.75 (mm) and 60 (°), respectively. The smallest value was approximately 86 (%) at ED height and angle of 0.5 (mm) and 90 (°), respectively. In contrast, the value of MIT slightly decreased with increasing ED angle: namely, the greatest percentage difference in MIT was almost 29.4 (%) between ED angle of 60° and 90° at welding amplitude of 60 (μm) and ED height of 1.75 (mm). The smallest value was 6 (%) between ED angle of 75° and 90° at welding amplitude of 40 (μm) and ED height of 0.5 (mm). On the other hand, value of ED height was moved from 0.5 (mm) to 1.75 (mm), causing a significant increase in MIT: namely, the greatest percentage difference in MIT was approximately 225 (%) between ED height of 0.5 (mm) and 1.75 (mm) at welding amplitude of 60 (μm) and ED angle of 60°. The smallest value was about 14 (%) between ED height of 1.5 (mm) and 1.75 (mm) at welding amplitude of 40 (μm) and ED angle of 90°.

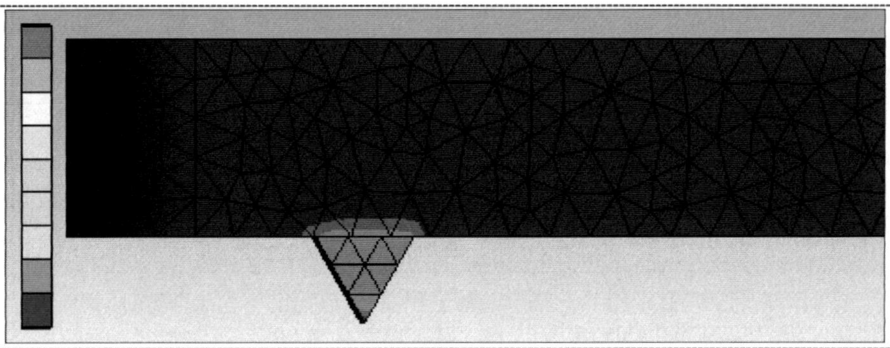

**Figure 4.** Temperature distribution at the weld interface in workpiece model with ED.

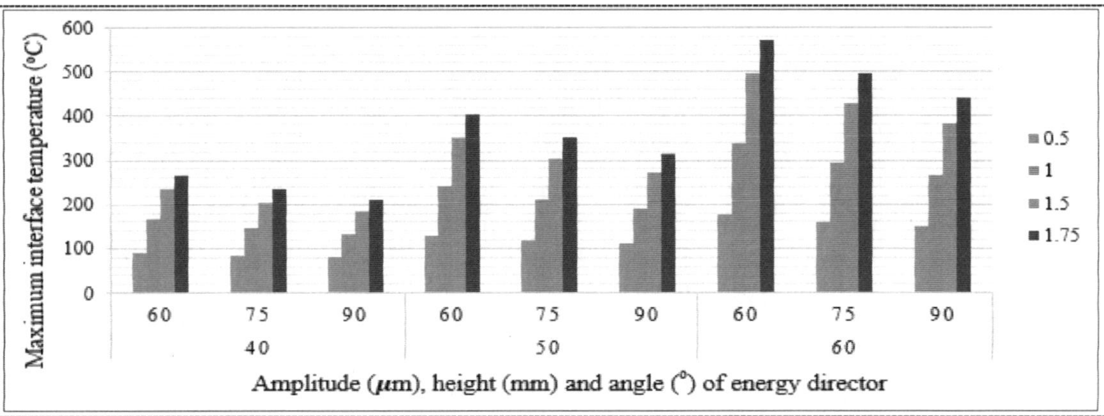


**Figure 5.** The maximum temperature at the weld interface for ABS.

Similarly, the data in Figure. 6 indicated that the highest value was 285.6 (°C) at welding amplitude of 60 (μm) with ED height and angle of 1.75 (mm) and 60 (°), respectively. Whereas the smallest value was 49.9 (°C) at welding amplitude of 40 (μm) with ED height and angle of 0.5 (mm) and 90 (°), respectively. Hence it clearly showed that when rising welding amplitude, it caused a strong rise in the value of MIT. In terms of the percentage difference in MIT, the greatest value between welding amplitude of 40 (μm) and 60 (μm) was almost 103 (%) at ED height and angle of 1.75 (mm) and 60 (°), respectively. The smallest value was about 63 (%) at ED height and angle of 0.5 (mm) and 90 (°), respectively. On the other hand, it can be observed that value of MIT slightly declined with increasing ED angle: namely, the greatest percentage difference in MIT was 28 (%) between ED angle of 60° and 90° at welding amplitude of 60 (μm) and ED height of 1.75 (mm). The smallest value was about 3.6 (%) between ED angle of 75° and 90° at welding amplitude of 40 (μm) and ED height of 0.5 (mm). It also revealed that rise of ED height from 0.5 (mm) to 1.75 (mm) caused a significant rise in MIT: namely, the greatest percentage difference in MIT was about 211 (%) between ED height of 0.5 (mm) and 1.75 (mm) at welding amplitude of 60 (μm) and ED angle of 60°. The smallest value was approximately 12 (%) between ED height of 1.5 (mm) and 1.75 (mm) at welding amplitude of 40 (μm) and ED angle of 90°.

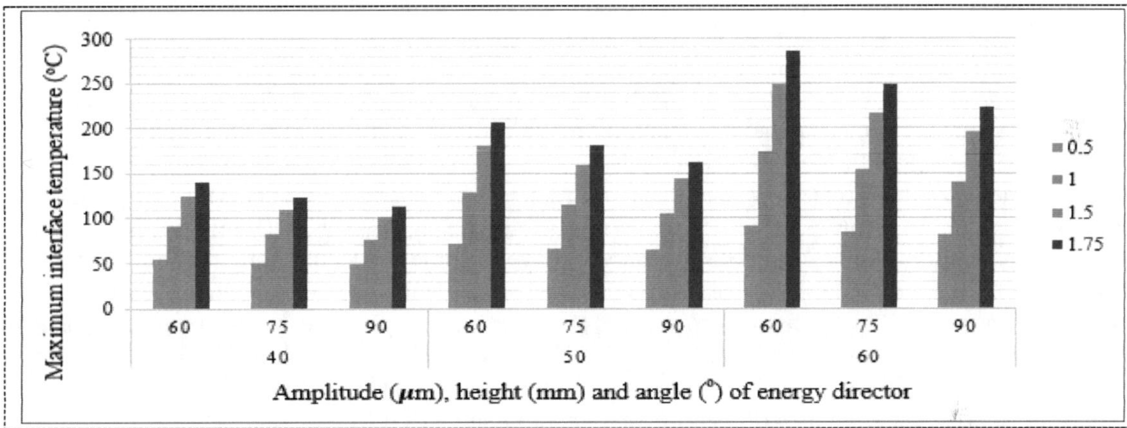


**Figure 6.** The maximum temperature at the weld interface for HDPE.

According to the following three reasons were attributed to the variation in values of MIT:

- The first and second reason – when increasing welding amplitude and value of ED height, caused an increase in MIT. This is explained as follows: according to equations (5)-(7), the value of heat flux was proportional to the value of welding amplitude and ED height. As a result, rising these values caused a strong rise in the rate of heat energy transfer through the surface of ED.
- The third reason – smaller ED angle concentrated more viscoelastic heat. This created higher temperature at the weld interface. Thus the value of MIT slightly dropped with rising ED angle.

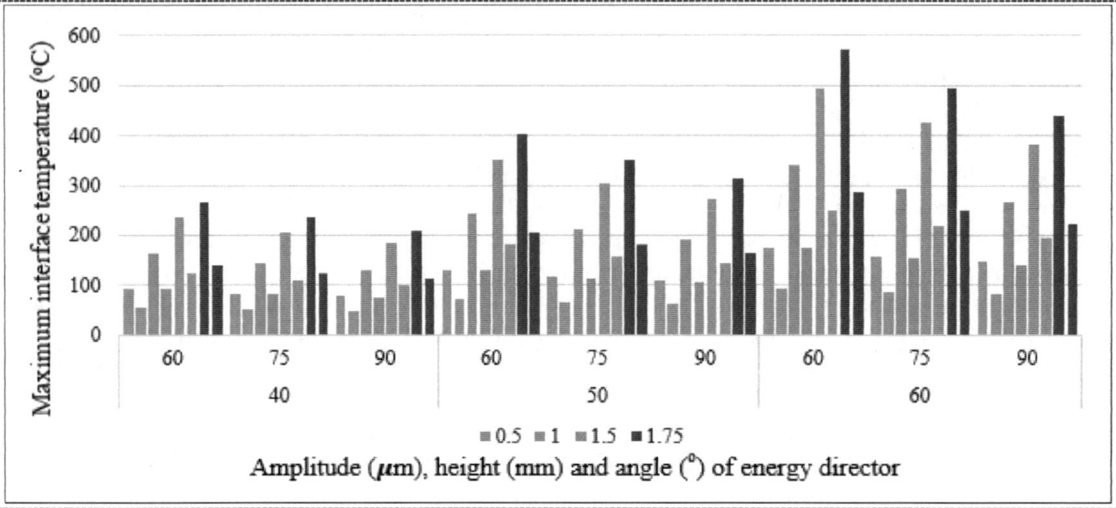


**Figure 7.** Comparison of maximum temperature at the weld interface between ABS (left) and HDPE (right).

Figure 7 clearly showed that values of MIT of ABS were always greater than that of HDPE. The greatest percentage difference in MIT was approximately 100 (%) between ABS and HDPE at welding amplitude of 60 (μm) with ED height and angle of 1.75 (mm) and 60 (°), respectively. In contrast, the smallest value was about 60 (%) between ABS and HDPE at welding amplitude of 40 (μm) with ED height and angle of 0.5 (mm) and 90 (°), respectively. As known, ABS is an amorphous polymer and HDPE is a semi-crystalline one. In amorphous polymers, the molecules are randomly oriented, allowing the vibrational energy to pass through them easily, with little attenuation. Therefore, they transmit ultrasonic vibrations more efficiently. Whereas semi-crystalline polymers tend to absorb vibrations due to their orderly molecular structure [1,5]. Furthermore, Chuah [8] determined value of energy absorption ratio for ABS being almost 2.5 times greater than that for HDPE. This resulted in increasing value of heat flux of ABS in comparison with that of HDPE.

## 4. Conclusion

This paper indicated that ED height, as well as welding amplitude strongly affected the interface temperature for ultrasonic welding of thermoplastics, viz increasing welding amplitude and ED height strongly increased MIT. Whereas the effect of ED angle on the interface temperature was less. With increasing ED angle, the value of MIT slightly decreased. The simulation figure of temperature distribution showed that the maximum interface temperature was found at the apex of ED, where the energy was concentrated. Ultrasonic welding of amorphous polymers is more suitable than that of semi-crystalline ones due to their ability to effectively transfer ultrasonic vibrations. In other words, energy absorption ratio for ABS was higher than that for HDPE. Hence values of MIT of ABS were always greater than that of HDPE.

The further research directions will be fabrication of the test samples with different energy directors by using injection moulding and then be welded under various welding conditions. Subsequently, the obtained results will be used to evaluate the effect of the main factors on the joint strength and the resulting dimension of workpiece after welding. In addition, the experimental temperature distributions during welding process will be measured by using thermocouples positioned at different points on the workpiece and the obtained results will be compared with the ANSYS simulation results.

## 5. References

[1] Guide to Ultrasonic Plastics Assembly, https://www.dukane.com/us/Documents/DesignGuides
[2] Overall Heat Transfer Coefficient, http://www.engineeringtoolbox.com/overall-heat-transfer-coefficient-d_434.html

[3] Menard K 2008 *Dynamic Mechanical Analysis: A Practical Introduction* (Florida: CRC Press) p 218
[4] Ultrasonic Welding, http://www.solvay.com/en/binaries/Ultrasonic-Welding_EN-199453.pdf
[5] Troughton M 2008 *Handbook of Plastics Joining: A Practical Guide* (New York: William Andrew Inc.) p 600
[6] Potente H 1984 Ultrasonic welding: principles and theory *Materials and Design* **5** 228–234
[7] Suresh K S, "Studies on temperature distribution in various joint designs and design of horns for ultrasonic welding of thermoplastics," Ph.D. dissertation, Dept. Mech. Eng., Anna Univ., Chennai, India, 2007
[8] Chuah Y, Chien L, Chang B et al. 2000 Effects of the shape of the energy director on far-field ultrasonic welding of thermoplastics *Polymer Engineering and Science* **40** 157–167

**Acknowledgments**
This publication was written at the Technical University of Liberec with the 117/2200 financial institutional support of the Department of Engineering Technology/Faculty of Mechanical Engineering.

IOP Conf. Series: Materials Science and Engineering **371** (2018) 012054 doi:10.1088/1757-899X/371/1/012054

# Plastic Deformations Occuring in Shells with Developable Middle Surfaces during Bending

**M Rynkovskaya**[1]

1 Associate Professor, Peoples' Friendship University of Russia (RUDN University), Moscow, Russia
E-mail: rynkovskaya_mi@pfur.ru

**Abstract.** A huge number of investigations on developable surfaces have been made, but the problem of analysis of plastic deformation during bending was usually out of the path. It is widespread in practice to make these surfaces with the means of bending. In previous papers, there has been found no note of the appearance of plastic deformations in a plate which is being bent into a cylinder. That is why analytical investigation of bending of developable shells is considered to be of great value. In the paper, analytical formulas which give the opportunity to determine the bending moments appearing due to parabolic bending of an elastic slab, as well as formulas for limitations to the slope angle of rectilinear generator of evolvent helicoids during the process of parabolic bending of an annulus into the given helicoids, are presented. The author also derived the formulas for determination of the minimum radius of a cylinder made by bending of a plate without the emergence of plastic deformation in it.

## 1. Introduction

Developable surfaces are widely used in industrial design [1] and manufacturing [2], shipbuilding [3], engineering [4] and architecture [5], computer graphics for texture mapping, some kinds of surface modeling [6], computer animation, etc. According to [5], a developable surface can be flattened onto a plane without stretching, tearing or creasing, and tangent lines of the edge of regression (cuspidal edge) of this surface generate it.

Planes, cylindrical surfaces, conical surfaces, surfaces of tangents to any curve, surfaces composed of patches of previous types (transition surfaces) are developable surfaces. A few articles [7] can be found about oloid, a rather amazing developable surface which is used for example in the oloid agitator, a device which consists of two oloids on two axes, functions like a paddle and can successfully imitate natural water movements. Considerable amount of information about geometry and stress-strain analysis of thin elastic developable shells can be found in a monograph [8]. An analytical stress-strain analysis of developable helicoids is presented in works [9], [10]. Developable helicoid is a well-known developable surface.

As far as modern architecture is concerned, free form shapes are becoming increasingly popular, but the relationship between shape and fabrication poses new challenges and requires greater sophistication from the underlying geometry. Most materials which are used for dry building enclosures are supplied as sheet goods, making developable surfaces – surfaces foldable from a flat sheet – the geometry of choice [11]. Non-developable curved surfaces are made primary by casting, stamping, or similar methods that need a die or mold, which lacks economies of scale if the individual components are different from each other. That is why a family of curved three-dimensional shapes that can be fabricated from two-dimensional sheet materials by way of bending or from curved creases is extremely desired either by industry design or by architecture.

IOP Conf. Series: Materials Science and Engineering **371** (2018) 012054 doi:10.1088/1757-899X/371/1/012054

Since developable surfaces are mostly used with materials that are not amenable to stretching, they may be of a benefit for manufacturing. In shipbuilding or automobile industry, surfaces are formed using only rollers or presses, and heat treatment is then used only to remove distortion induced by welding or other means. However, doubly curved surfaces must be heat treated after rolling to induce additional curvature. This process is very time-consuming and labour-intensive, because it is mostly done by hand and requires a skilled artisan with years of training to achieve the correct amount of bending [12]. Designing a ship hull or a vehicle entirely of singly curved, or developable, surfaces would reduce manufacturing costs.

Zhao and Wang [13] propose a new method for designing a developable surface by constructing a surface pencil passing through a given curve, which is quite in accord with the practice in industry design and manufacture. By representing the surface by a combination of the given curve, and the three vectors decomposed along the directions of Frenet trihedron frame, the authors derive the necessary and sufficient conditions for a surface to be developable.

In paper [14], the authors simplify the modelling process by introducing an intuitive sketch-based approach for modelling developables, and construct an algorithm that given an arbitrary, user specified 3D polyline boundary, constructed using a sketching interface, generates a smooth discrete developable surface that interpolates this boundary.

According to Williams and Skaggs [15] the process of bending metal to curved shapes by applying bending moments thereto is not new, and the use of heat to relieve stress in metals which have been formed by bending is a well-known, widely commercially used process. For example, each sheet can be formed to desired curvature by means of turnbuckles attached to rigid connecting brackets at the vertical edges of the sheets, thereby effectively producing a bending moment at the edge of the sheet. Despite the mention of providing the proper bending moment on the sheet metal, there is no analytical formula for its calculation.

It is shown in the paper [16] that multistage bending technique is an alternative way to produce cylindrical tubes. The authors use one pair of punch and die to experimentally investigate pure bending of a blank sheet into a cylindrical tube by multistage process, and numerical simulation was conducted on bending and spring back with LS-DYNA solver. The sequence is such that a blank sheet is bent into C-tube and then squeezed with welding into O-tube, while gap width and roundness of C-tube were the two dominant parameters to evaluate the bending performance. The method is considered to help reducing severe thinning and shape defects in the finished tubes.

Obviously, there are a lot of articles on application of developable surfaces in the fields of architecture, shipbuilding and design [17-21]. But the most of the papers are about geometry and modeling, some articles are about static strength analysis of shells with developable middle surfaces, very few works about dynamic analysis and almost all of the papers consider that the whole shell works in elastic stage. In general, a variety of investigations on developable surfaces has been made, [22-26] – the only few examples of the XXI century, but according to the investigation the problem of mathematical analysis of plastic deformation during bending is still avoided. Sometimes it is mentioned in articles, but most of the research relies on experimental data [16].

Folding elastically toward desired angels can cause plastic deformation near the folding line. The author has found out only a few works where plastic deformation is mentioned, furthermore, there has not been found any analytical methods for calculation or determination. At the same time some analytical issues have been developed at the Department of Strength of Materials and Structures in Peoples' Friendship University of Russia (Moscow, Russia). The paper written by Prof. S.N. Krivoshapko [27] was published in 2003, but it seems like it has not become widely known; in there the problem of determination of the shell stress-strain state in view of very large displacements $u_u$, $u_s$, $u_z$, but very small normal strains $\varepsilon_u$, $\varepsilon_s$ and shear strains $\varepsilon_{us}$ is investigated.

However, this issue cannot be ignored, because in most cases in industry design of ship hulls, automobile elements, building enclosures, etc., produced by bending of thin metal plates, developable surfaces are used. The most desired property of a developable surface for industry is that it can be obtained by bending from flat manufacture product. As far as the author's investigation is concerned,

IOP Conf. Series: Materials Science and Engineering **371** (2018) 012054 doi:10.1088/1757-899X/371/1/012054

it is extremely hard to find any useful practical or analytical methods on stress analysis of developable shells when plastic strain occurs.

Rather poor accuracy may develop in calculation of a cylindrical shell and obtaining strains from external loads in the case when the plastic stresses which arise in the process of manufacture of the product are not taken into account.

## 2. Materials and methods

The problem to discuss below mostly belongs to metal products with the yield point $\sigma_y$ which are produced by bending.

Analytical formulas giving the opportunity to determine the bending and twisting moments emerging in the process of parabolic bending of an elastic slab [27]

$$M_1 = D(\kappa_1 + \nu\kappa_2); \quad (1)$$

$$M_2 = D(\kappa_2 + \nu\kappa_1); \quad (2)$$

$$H = M_{su} = 0, \quad (3)$$

where $M_1, M_2$ are bending moments; $H, M_{su}$ are twisting moments; $D = Eh^3/[12(1-\nu^2)]$ is flexural rigidity of a shell; $\nu$ is a Poisson's ratio; $h$ is shell thickness; $\kappa_1$, $\kappa_2$ are parameters of curvature variations,

can be obtained from formulas of Hooke's law for shells in lines of principal curvatures by substituting the last expressions in them:

$$\kappa_1 = \frac{1}{R_1'} - \frac{1}{R_1} + \frac{\varepsilon_1}{R_1};$$

$$\kappa_2 = \frac{1}{R_2'} - \frac{1}{R_2} + \frac{\varepsilon_2}{R_2}, \quad (4)$$

where $R_1$ and $R_2$ are the principal radii of curvatures of a surface before bending; $R_1'$ and $R_2'$ are the principal radii of curvatures of a surface after bending.

Since the Lame coefficients of a developable surface under parabolic bending do not change, linear strains $\varepsilon_1$, $\varepsilon_2$ and angular displacement $\varepsilon_{12}$ are equal to zero, as well as the force resultants. Furthermore, it is predictable that $\frac{1}{R_1'} = \frac{1}{R_1} = 0$, because rectilinear generatrixes of the developable middle surface are considered to be unchanged when bending is parabolic. Besides, $\frac{1}{R_2} = 0$ if the bending process begins from a plane slab. Thereby, when parabolic bending from a plane slab is assumed, Equations (1-3) become

$\kappa_1 = 0$; $\kappa_2 = \frac{1}{R_2'}$ and the formulas for bending and twisting moments (Equations (4)) will be:

$$M_1 = D\nu\kappa_2 = \nu M_2 = M_u;$$

$$M_2 = D\kappa_2 = M_s;$$

$$H = M_{su} = 0. \quad (5)$$

It is also possible to find formulas for moments $M_u$, $M_s$, $M_{us}$ for curvilinear non-orthogonal coordinates u, s according to formulas of transition from internal moments in lines of principal curvatures:

$$M_u = M_2 \cos^2\chi + M_1 \sin^2\chi;$$

$$M_s = M_2;$$

$$M_{us} = (M_1 - M_2)\sin\chi\cos\chi, \quad (6)$$

IOP Conf. Series: Materials Science and Engineering **371** (2018) 012054 doi:10.1088/1757-899X/371/1/012054

where $\chi$ is an angle between two intersecting curvilinear coordinate lines *u* and *s*.
Let us illustrate the application of the formulas given above with an example.

## 3. Numerical experiments

A thin steel rectangular plate of dimensions $l \times b \times h$ with the following properties: Poison's ratio is $\nu = 0.3$; Young's modulus is $E = 2.06 \cdot 10^5 MPa$; thickness is $h = 0.001$ m; yield stress is $\sigma_y = 245 MPa$; is bent into an L-shaped one (figure 1). Let us calculate the minimum value of the radius ($R_{\min}$) of bending when the plate will remain in the elastic stage (may turn back to its original flat state). In other words, the plate is bent without plastic deformations ($\sigma_{s,\max} \le \sigma_y$).

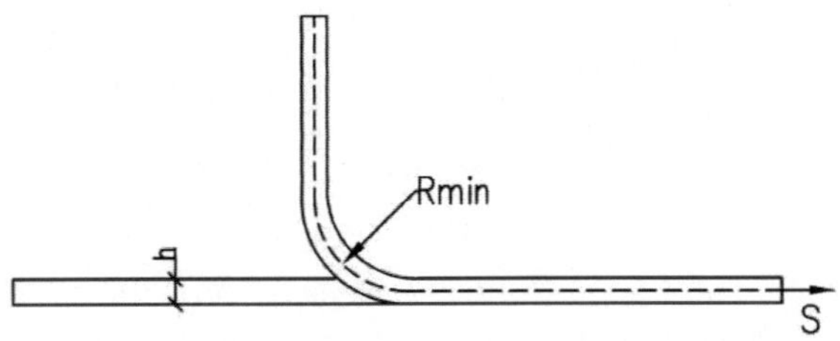


**Figure 1.** A parabolic bending of a thin plate into the L-shaped one

According to the Equations (1-4), the curvature valriations and bending moments:

$$\kappa_1 = 0;\ \kappa_2 = \frac{1}{R_{\min}};\ M_1 = \frac{D\nu}{R_{\min}} = M_u;\ M_2 = \frac{D}{R_{\min}} = M_s. \qquad (7)$$

Thus, the minimum radius of the developable surface to be manufactured without plastic deformations can be calculated by the formula $R_{\min} = \frac{D}{M_y}$, where $M_y$ is plastic moment.

The flexural rigidity of the shell is $D = 18.86 N \cdot m$, while the bending moment $M_y$, created by normal stresses, occurs when maximum normal stresses rise to the yield stresses ($\sigma_{s,\max} = \sigma_y$) is

$$M_y = \frac{\sigma_y h^2}{6} = 40.8 N \cdot m / m.$$

Bending moments $M_S$ and the radii correspondence is presented in figure 2. The point of intersection of the horizontal line $M_y = 40.8 N \cdot m / m$ and the $M_S$ – R graph shows the minimal bending radius for the plate without plastic deformations (*Rmin* = 0.462m in this example).

A similar calculation can be made for a radius of a cylinder (figure 3) or helicoids made by parabolic bending of a thin flat plate.

IOP Conf. Series: Materials Science and Engineering **371** (2018) 012054 doi:10.1088/1757-899X/371/1/012054

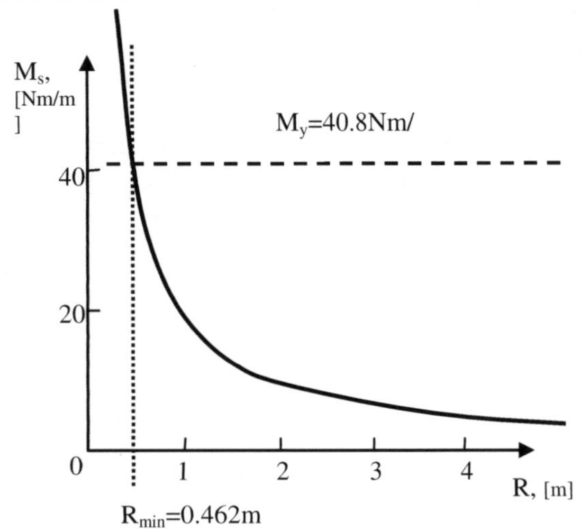


**Figure 2.** The Ms – Rmin diagram for the parabolic bending of the parabolic bending of the rectangular thin plate

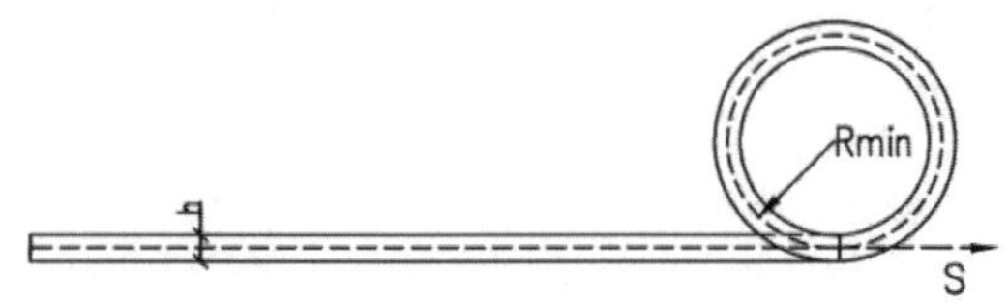


**Figure 3.** A parabolic bending of a thin plate into the O-shaped one (cylindrical product)

## 4. Results and Discussion

Generally, most engineers consider the task of bending of a thin plate as a task of bending of the middle surface, and do not take into account normal strains, because normal strains do not occur on the middle surface. In practice, since the factor of safety is usually very high for such products as steel tubes, such approach does not lead to any problems. However, it can be useful to consider plastic deformations when a plate is being bent into a product.

The main result of the paper is the formula for determination of the minimum radius which allows the plate to remain in elastic region. This formula is discussed in relation with the L-shaped plate, but can be applied to an O-shaped plate (cylindrical product).

The computer program based on the presented formulas was written with the tools of MathCAD for easy, rapid and accurate design of folded or unfolded surfaces. The similar approach was applied to a beam in [28], and numerical results obtained are satisfactory.

Besides, there are some other points for analytical investigation of bending of developable shells. One of them is a question about the range of the inclination angle that makes a thin helicoidal shell be under plastic stresses. The analytical formulas for limitations to the slope angle of rectilinear generator of evolvent helicoid during the process of parabolic bending of the annulus into the given helicoid may also be easily obtained. The author also gives an example for determination of the smallest allowable radius of L-shaped plate made by bending of a flat plate without the emergence of plastic deformation in it.

In the paper, the bending technology for manufacturing a product from a plate is not considered, as the issue does not relate to the topic of this study and can be selected as a subject for an another article. It may be also interesting to get experimental data that will confirm the suggested theoretical results.

## 5. Conclusion

When calculating, the designer should be aware that the material of the product is already under tension which was caused in the manufacturing process; otherwise it may lead to the destruction of products in the process of loading and/or operation.

In the paper, the analytical formulas for bending moments appearing due to parabolic bending of an elastic slab made of metals with yield plateau are presented. The same formulas may be used for limitations to the slope angle of rectilinear generator of evolvent helicoids during the process of parabolic bending of an annulus into the given helicoid; and determination of the minimum radius of cylinder made by bending of a flat plate without the emergence of plastic deformation in it.

## 6. References

[1] Liua Y J, Tangb K, Gongc W Y and Wuc T R 2011 Industrial design using interpolatory discrete developable surfaces *Computer-Aided Design* V 43 Iss **9** pp 1089–1098
[2] Krivoshapko S N and Gbaguidi G L A 2009 Developable shell products made by parabolic bending of thin metal slabs *Journal of the Ghana Institution of Engineers* V 6&7 **1** pp 51-56
[3] Oetter R, Barry C D, Duffty B and Welter J 2002 Block Construction of Small Ships and Boats Through Use of Developable Panels *Journal of Ship Production* V 18 **2** pp 65-72(8)
[4] Gotman A S 2005 Use of Developable Surfaces for Designing Well-Stream Shapes *Oceanic Engineering International* V 9 **2** pp 81-96
[5] Krivoshapko S N and Ivanov V N 2015 Encyclopedia of Analytical Surfaces *Springer* p 752
[6] Haeberli P 2002 Modeling and fabrication of objects represented as developable surfaces *US Pat 6493603 USA*
[7] Kuleshov A S, Hubbard M, Peterson D L and Gede G 2011 Motion of the oloid on the horizontal plane *Rus J Nonlin Dyn* V 7 **4** pp 825-835
[8] Krivoshapko S N 2009 Geometry of ruled surfaces with an edge of a return and a linear theory of calculating torso shells *RUDN* p 357.
[9] Rynkovskaya M I 2009 Analytical Methods of Analysis of Right and Open Helicoids *ASEE Conf Engineering in the New Global Economy*
[10] Rynkovskaya M I 2015 Influence of the Poisson's Ratio on the Accuracy of the Results of Calculating the Torso-Helicoid Stress-Strain *Construction and Reconstruction* **4 (60)** p 51-56
[11] Koschitz D, Demaine E D and Demaine M L 2008 Curved Crease Origami. *Abstracts from Advances in Architectural Geometry* pp 29–32
[12] Steel-Chalfant J 1997 Analysis and design of developable surfaces for shipbuilding *Master's thesis*
[13] Zhao H and Wang G 2008 A new method for designing a developable surface utilizing the surface pencil through a given curve *Progress in Natural Science* V 18 **1** pp 105-110
[14] Rose K, Sheffer A, Wither J, Cani M P and Thibert B 2007 Developable Surfaces from Arbitrary Sketched Boundaries *SGP '07 Proceedings of the fifth Eurographics symposium on Geometry processing* pp 63-172
[15] Williams G O and Skaggs R L 1980 Method of forming a parabolic trough *US Pat. 4236399*
[16] Xia M, Yan Q, Zuo D and Xie J 2011 An investigation on multistage bending of blank sheet into cylindrical tube by experiment and numerical simulation *Int J of Adv Manuf Tech* V 53 **1-4** pp 145-155
[17] Solomon J, Vouga E, Wardetzky M and Grinspun E 2012 Flexible Developable Surfaces *Computer Graphics Forum* V 31 **5** pp 1567-1576
[18] Dias M A, Dudte L H, Mahadevan L and Santangelo C D 2010 Geometric Mechanics of Curved Crease Origami *Physical Review Letters* **109** 114301
[19] Jung A et al 2015 Sketching Folds: Developable Surfaces from Non-Planar Silhouettes *ACM Transactions on Graphics* V 34 **5** p155
[20] Yoon J S, Ryu C and Lee J H 2008 Developable polynomial surface approximation to smooth surfaces for fabrication parameters of a large curved shell plate by Differential Evolution *Computer-Aided Design* V 40 **9** pp 905-915
[21] Kilian M et al 2008 Curved Folding *ACM Transactions on Graphics* (TOG) V 27 **3** 75
[22] Liu Y et al 2006 Geometric modeling with conical meshes and developable surfaces *ACM Transactions on Graphics* V 25 **3** pp 681-689
[23] Krivoshapko S N 1998 Static analysis of shells with developable middle surfaces *Appl Mech Rev* V 51 **12** pp 731-746
[24] Redont P 1989 Representation and deformation of developable surfaces *Computer Aided Geometric Design* V 21 (1) pp 13-20

[25] Aumann G 2003 A simple algorithm for designing developable Bezier surfaces *Computer Aided Geometric Design* V 24 (4) pp 601-609
[26] Rekach F V, Shambina S L and Sinichenko E K 2017 Influence of Pressure Stabilizer Perforation Area on Character of Unsteady Fluid Motion in Hydraulic Systems *International Journal of Mechanical Engineering and Robotics Research* V 6 **4** pp 269-271
[27] Krivoshapko S N 2003 Parabolic bending of thin elastic plane or developable half-finished products into developable shell *Struct Mech of Eng Constr and Buildings* **12** pp 12-19
[28] Rynkovskaya M 2017 Analysis of displacements in beam structures and shells with middle developable surfaces *MATEC Web of Conferences* **108** 16001

**Acknowledgments**

This paper was financially supported by the Ministry of Education and Science of the Russian Federation on the program to improve the competitiveness of Peoples' Friendship University of Russia (RUDN University) among the world's leading research and education centres in the 2016-2020. This publication was prepared with the support of the "RUDN University Program 5-100".

IOP Conf. Series: Materials Science and Engineering **371** (2018) 012055 doi:10.1088/1757-899X/371/1/012055

# Improve the Properties of Iraqi Bentonite Using Salts

**Prof. Dr. Kareem Hassan Alwan [1], Dr. Baydaa Hussain Maula1,[*]Qais Mikheal Bahnan[1]**

1Institute of Technology, Middle Technical University, Baghdad,Iraq
E-mail: dr.maula@itbaghdad.edu.iq

**Abstract.** After the success of experiments using the Iraqi Bentonite in the work of drilling oil wells as an alternative to the importer Bentonite which encouraged its use as an alternative for the work of civil excavation, such as diaphragm walls, drilling, bore pile, barriers, dams and tunnels. By changing the mixing and drilling method followed in the work of oil wells, which requires the conduct of many experimental mixtures to reach the optimal mix. Benefits were getting from the previous experiments conducted on such type of Bentonite were used by the General Company for Geological Survey and Mining and Ramadan Mubarak Company one of the formalization of the Ministry of Housing and Construction; in which the optimum mix was reached, the method of mixing was determined, the components and quantities of the materials included in the preparation of the optimal mix were determined in $1m^3$ of water for mixing time of 5 min. This study was supported by the results obtained by the Geological Survey and Mining Company, which relied on the use of (NaOH) as an alternative to the active calcium-centric (CMC) of Bentonite, which is considered to be one of the most expensive materials. This paper will be a good index for all concerned with implementation of civil excavations and field works.

## 1 Introduction

Bentonite is the name given to clay materials that have the ability to swell and bond when mixed or dissolved in water [1].

The name Bentonite originates from the discovery of a type of clay found near the Benton area in the United States in the nineteenth century. It was used in the drilling of oil wells in the state of Dakuna in the United States, the type of sodium [2]. As the time progressed, Bentonite began to work in the civil works in Iraq to its quality, its ability to swell and the formation of the thick, transparent wall that surrounds the drilling sides due to its chemical properties because it contains sodium; but because of its economic cost and the difficulty of transporting it from the United States to other parts of the world, the need of available locally Bentonite was raised in Iraq, the local Bentonite used to drill the oil wells has been directed in the countary. However, there is a problem in the Iraqi Bentonite due to its chemical composition as it is a calcium type which has a negative impact on the civil excavation [3]. The effect of calcium ion concentration on fluid viscosity which can be expressed with the minimum limits of 150-200 mg/liter from filtration loss. The negativity of these liquids is when the temperature greater than (100 ° C) or when drilling for deepest elevation; the oxide Calcium under a very alkaline medium leads to the formation of products similar to those formed at the hardening of the liquid. This means a sharp increase in viscosity and gel strength to reduce this phenomena is by reducing the concentration of the chloride, calcium oxide. It is worth noting this material in its natural state. It is a kind of soil characterized by high smoothness and free of calcareous materials which have a negative effect on the skin, and easily transform into very fine powder and form with water.

IOP Conf. Series: Materials Science and Engineering **371** (2018) 012055 doi:10.1088/1757-899X/371/1/012055

The gluten gel solution remains stuck and spread throughout the water content for a long time, in addition to the high purity of these soils with kaolin are almost ineffective and thus have no effect on the skin, especially when they are free of calcareous materials. All these features they play an important role in sustaining the skin's freshness and softness of hair.
Many people may be able to observe the material presence in different parts of the Kurdish land, especially in the taxonomic areas. However, the geological studies carried out by the General Company for Geological Survey indicate that the material is located in a mining facility in the Qarabah area of Kasr Kefri, called the Catapult [4].
The main task of the drilling fluid in civil works is to prevent the hole sides from collapsing. In addition, the liquid flowing through the ring cavity deposits a protective envelope porous formation which is called as a mud cake. As a consequence of drilling operation a part of water composed of the drilling fluid will enter through the permeable layers by the hydrostatic pressure of the liquid column, the part of the water that enters will be called as (Filtration loss), and the benefits of the mud crust is to install the hole walls and prevent it from collapsing. This paper deals with the effect of additions to drilling fluids in detail; the general and specific objectives of this research. In the following section deals with the properties of drilling fluids (Muds), in addition to the different types of drilling fluids such as the American and Iraqi Bentonite.
To illustrate the manner in which research was carried out and the various experiments undertaken for this study.At the end of this research, some recommendations and engineering proposals were presented based on laboratory results in the field of drilling fluids.

## 2. Statement of the Problem

Bentonite is commonly in interaction with water; therefore the properties of clay-water are of fundamental importance to the engineering works [5]. The chemical composition of Bentonite is vital due to changes in its physical properties that is combined to its chemical compositions. Generally there are three common kinds of Bentonite in the world, namely (natural sodium; natural calcium and sodium-activated) [6]. All Bentonite have a tendency to exchange cations more than other clays when it's mixed with water; this phenomenon is known as thixotrop see Figure 1.

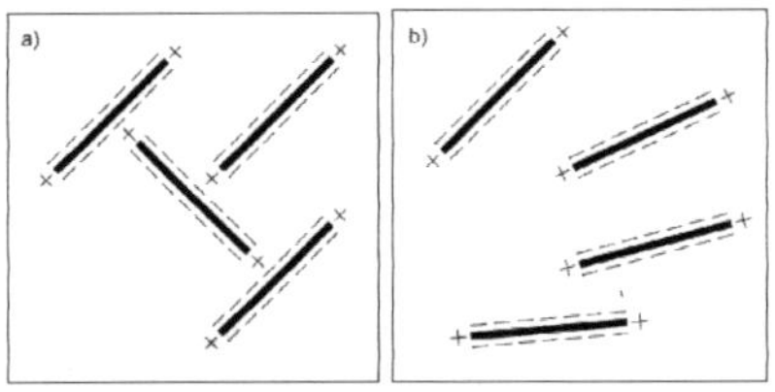


**Figure 1.** Thixotrop phenomenon: a) slurry to get; b) slurry becomes fluid.

Sodium bentonite is "characterized by very high swelling ability, high liquid limit and low filter loss and high cost. This Bentonite was used as the standard by which all other Bentonites were measured for many years. The predominant exchangeable cation in natural sodium bentonite is the sodium cation but there may also be significant amounts of other cations present.
Natural calcium Bentonite, where calcium is the predominant exchangeable cation, is mined worldwide. It has much lower swelling ability and liquid limit, and much higher filter or fluid loss than natural sodium Bentonite which is not suitable for the purpose of drilling"[6].
Sodium-activated Bentonite is created by the adding of $Na_2co_3$ to calcium Bentonite [7],[8]. This effects a base exchange on the surfaces of the clay particles, replacing calcium ions with those of sodium. The result is a Bentonite exhibiting many of the typical characteristics of a natural sodium Bentonite.Most Bentonites used in civil engineering to produce support fluids are sodium-activated. But there is a need to discover the differences between these additives ($Na_2co_3$ vs NaoH) on parameters (viscosity, filter loss and pH) under the same circumstances of testing.

## 3. Practical Side

The aim of the treatment is to improve the properties of the locally Bentonite liquid only, especially with regard to the mud cake properties Not to get special fluids. The properties of the liquid are secured by mixing (Bentonite + water) and the properties of the original liquid are improved by adding

IOP Conf. Series: Materials Science and Engineering **371** (2018) 012055 doi:10.1088/1757-899X/371/1/012055

some substances such as viscous dampers, lossless filtration loss and other substances that affect the properties of the original liquid.
To simulate the mixing device used in the formation of Bentonite solution which is located in the civil excavation sites, a similar device; mud mixer used in this survey.

### *3.1. Method of examination*
First, a mixture of non-activated local Bentonite with certain proportions shall be mixed in the mixing apparatus and shall be tested according to the international specifications after 24 hours of the mixing process, in which the solution shall be examined in terms of viscosity, density, permeability(as a filter loss) and PH. Second, taking a sample of Bentonite and activating it with sodium hydroxide in a certain percentage. It is tested according to the international specifications after 24, 48, 72, 96 hours of the mixing process, after which the solution is examined in terms of viscosity, density, permeability and pH. The same steps will be applied for Bentonite activated with soda ash (sodium carbonate).

### *3.2. The experimental procedure*
Figure 2, shows the mud mixer apparatus. The water cup was filled by the specified quantity of distilled water about (1000ml), the added, solids were weighed at a specific percent as shown in Table 1-2.
Turn on the machine to move the water inside it and then add the solids gradually according to the pre-determined quantity were done waiting, for 3 to 5 minutes until the mixture is homogenized. These steps are repeated for all tested samples, for all tested samples the density was calculated with the mud balance apparatus. While the apparent viscosity will be reported according to Marsh Funnel viscosity; which is express by the time required to fill a volume (964cm$^3$) during Marsh Funnel measured in size. Notice that the measured viscosity of water is 26 Sec. The filter loss was measured according to test presented in Figure 2(c); the amount of filtered water through filter paper (cm$^3$); the thickness of the filter cake is measured in millimeters.

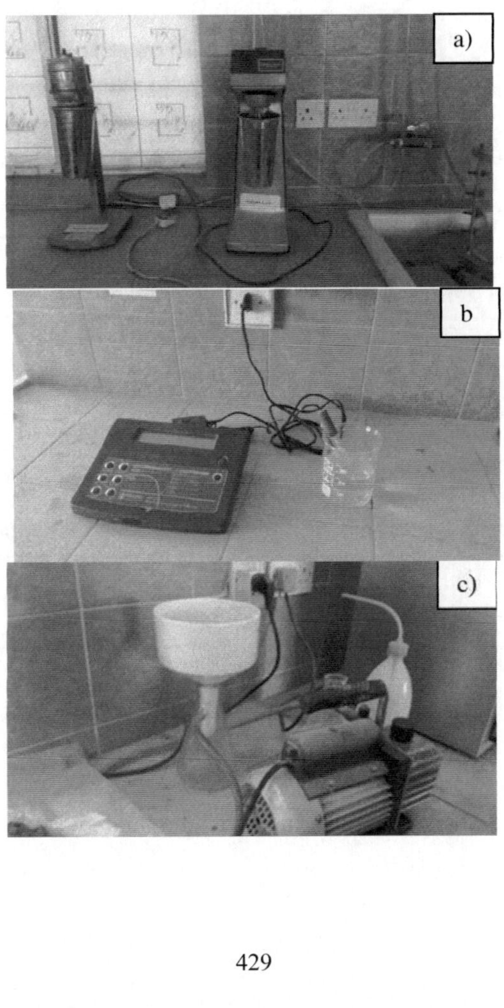

IOP Conf. Series: Materials Science and Engineering **371** (2018) 012055 doi:10.1088/1757-899X/371/1/012055

**Figure 2.** Testing apparatus: a) Mud mixer; b) pH-meter; c) Filter press.

## 4. Discussion of Results

As is known, one of the most important difficulties that hampered access to the best mix of several and most complex was the process of converting the local Bentonite from the type of calcium to the type of sodium by the process of activation by adding both sodium ash and caustic soda [5] during the mixing process.

Through this study, several key factors have been observed that have an effect on the functioning of the activation process such as (mixer type; the time period for mixing and the percentage of additives added).

These factors can be controlled on-site through experimental mixtures [6]. The results of this test showed that the viscosity of the solution was not matched to what is required, as well as the amount of separation when mixing the additives and increasing the mixing time more over than 48 hours compared to the results obtained for the same Bentonite with certain additives and for specific time periods. The results of the tests were significantly improved when using caustic soda (NaOH), which is an alternative to CMC, as shown in Table 1-2.

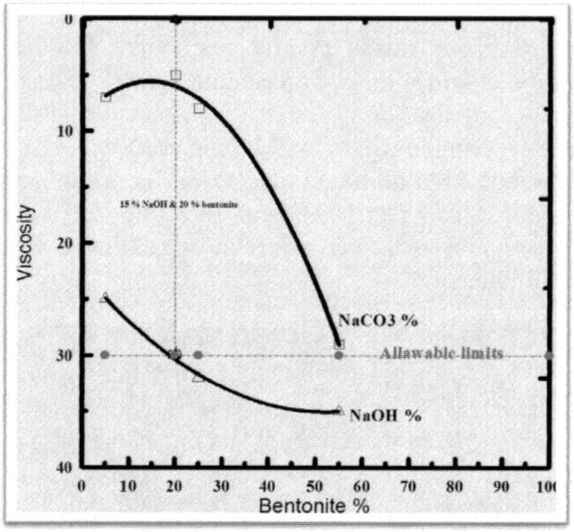


**Figure 3.** Effect of additives on Iraqi Bentonite.

**Table 1.** Effect of additives on Bentonite solution properties: a)$Na_2co_3$, b) NaOH, c) $Na_2co_3$+NaoH (After [8])

a)$Na_2co_3$

| (B-W) % | Time of Testing (h) | Additives % | Density of solution $g / cm^3$ | Viscosity | Loss of fluid | PH |
|---|---|---|---|---|---|---|
| 5 | 48 | 2 | 1.02 | 8 | - | 10 |
| 20 | = | 15 | 1.03 | 6 | - | 10 |
| 30 | = | 25 | 1.038 | 25 | - | 10 |
| 35 | = | 30 | 1.04 | 25 | - | 10 |
| 45 | = | 35 | 1.036 | 34 | - | 10 |
| 55 | = | 40 | 1.04 | 35 | 35 | 10 |
| (B-W) % | Time of Testing (h) | Additives % | Density of solution $g / cm^3$ | Viscosity | Loss of fluid | PH |
| 5 | 72 | 2 | 1.03 | 28 | - | 10 |
| 20 | = | 15 | 1.03 | 29 | - | 10 |
| 30 | = | 25 | 1.05 | 29 | - | 10 |
| 35 | = | 30 | 1.05 | 33 | - | 10 |

| | | | | | | |
|---|---|---|---|---|---|---|
| 45 | = | 35 | 1.05 | 34 | - | 10 |
| 55 | = | 40 | 1.05 | 34 | - | 10 |
| (B-W) % | Time of Testing (h) | Additives % | Density of solution g / cm 3 | Viscosity | Loss of fluid | PH |
| 5 | 96 | 2 | 1.05 | 30 | - | 10 |
| 20 | = | 15 | 1.06 | 33 | - | 11 |
| 30 | = | 25 | 1.06 | 35 | - | 11 |
| 35 | = | 30 | 1.06 | 35 | - | 11 |
| 45 | = | 35 | 1.06 | 35 | - | 11 |
| 55 | = | 40 | 1.06 | 35 | - | 11 |

b) NaOH

| (B-W) % | Time of Testing (h) | Additives % | Density of solution g / cm $^3$ | Viscosity | Loss of fluid | PH |
|---|---|---|---|---|---|---|
| 5 | 48 | 2 | - | 7 | 36 | - |
| 20 | = | 15 | 1.04 | 8 | 20 | 10 |
| 25 | = | 25 | 1.04 | 9 | 29 | 10 |
| 55 | = | 40 | 1.04 | 29 | - | 9 |
| (B-W) % | Time of Testing (h) | Additives % | Density of solution g / cm 3 | Viscosity | Loss of fluid | PH |
| 5 | 72 | 2 | - | 7 | 36 | - |
| 20 | = | 15 | 1.04 | 8 | 20 | 10 |
| 25 | = | 25 | 1.04 | 9 | 29 | 10 |
| 55 | = | 40 | 1.04 | 29 | - | 9 |
| (B-W) % | Time of Testing (h) | Additives % | Density of solution g / cm $^3$ | Viscosity | Loss of fluid | PH |
| 5 | 96 | 2 | - | 7 | 36 | - |
| 20 | = | 15 | 1.04 | 8 | 20 | 10 |
| 25 | = | 25 | 1.04 | 9 | 29 | 10 |
| 55 | = | 40 | 1.04 | 29 | - | 9 |

C) $Na_2co_3$+NaoH

| (B-W) % | Time of Testing (h) | Additives ($Na_2co_3$) % | Additives (NaoH) % | Density of solution g / cm $^3$ | Viscosity | Loss of fluid | PH |
|---|---|---|---|---|---|---|---|
| 5 | 48 | 2 | 1 | 1.02 | 29 | - | 8 |
| 5 | = | 3.5 | 1 | 1.06 | 30 | - | 9 |
| 6 | = | 2 | 1 | 1.05 | 28 | - | 8 |
| 6.5 | = | 5.5 | 1 | 1.06 | 30 | - | 10 |
| 6.5 | = | 5.5 | 1 | 1.05 | 28 | - | 10 |
| 5.5 | = | 5 | - | 1.04 | 28 | - | 10 |
| (B-W) % | Time of Testing (h) | Additives ($Na_2co_3$) % | Additives (NaoH) % | Density of solution g / cm $^3$ | Viscosity | Loss of fluid | PH |
| 5 | 72 | 2 | 1 | 1.04 | 28 | - | 9 |
| 5 | = | 3.5 | 1 | 1.05 | 28 | - | 9 |
| 6 | = | 2 | 1 | 1.05 | 30 | - | 9 |
| 6.5 | = | 5.5 | 1 | 1.05 | 28 | - | - |
| 5.5 | = | 5.5 | - | 1.05 | 28 | - | - |

After examining the results of the laboratory tests of the activated Iraqi Bentonite and compared them with the specifications of Bentonite solution adopted by the Union of experts of Piles (FPS) [6], [9] has been fixed limits of specifications required for the material activator Iraqi Bentonite and according to the table 3, noting that all the results were supported the results of the General Company for Geological Survey which was advised to use caustic soda. The percentage of addition of the best results for mixing according to this research study is 15% for sodium hydroxide for the mixing ratio of 20% Bentonite of water. Figure 3 below shows compassion between the optimum mixing percentages of both additive if used alone. The Figure illustrates that caustic soda has a great influence on viscosity more than $Na_2co_3$; while the same additives gave the same influence on Bentonite slurry if they mixed together under a specific quantity with less mixed percent for Bentonite. See Table 2.

**Table 2.** Optimum mixing percentage (After [8])

| (B-W) % | Time of Testing (h) | Additives ($Na_2co_3$) % | Additives (NaoH) % | Density of solution $g / cm^3$ | Viscosity | Loss of fluid | PH |
|---|---|---|---|---|---|---|---|
| 5 | 24 | 5 | 1 | 1.038 | 27 | - | 9 |
| 5 | 48 | 5 | 1 | 1.037 | 28 | - | 10 |
| 5 | 72 | 5 | 1 | 1.039 | 29 | - | 10 |
| 5 | 96 | 5 | 1 | 1.04 | 30 | - | 10 |
| (B-W) % | Time of Testing (h) | Additives ($Na_2co_3$) % | Additives (NaoH) % | Density of solution $g / cm^3$ | Viscosity | Loss of fluid | PH |
| 5 | 24 | 5 | 1.5 | 1.037 | 30 | - | 10 |
| 5 | 48 | 5 | 1.5 | 1.036 | 31 | - | 10 |
| 5 | 72 | 5 | 1.5 | 1.04 | 33 | - | 10 |
| (B-W) % | Time of Testing (h) | Additives ($Na_2co_3$) % | Additives (NaoH) % | Density of solution $g / cm^3$ | Viscosity | Loss of fluid | PH |
| 5 | 24 | 5 | 2 | 1.039 | 30 | - | 10 |
| 5 | 48 | 5 | 2 | 1.037 | 31 | - | 10 |
| 5 | 72 | 5 | 2 | 1.038 | 32 | - | 10 |
| 5 | 96 | 5 | 2 | 1.04 | 33 | - | 11.5 |

**Table 3.** The permissible limits for the properties of the suspended solution and according to the Union of experts of the pillars [6], [9].

| **According to international standards** | |
|---|---|
| Les 1.1 | Density |
| 30- 90 Sec | Viscosity |
| 28 | Loss of fluid |
| 9.5 -12 | PH |
| **Iraqi Bentonite Activated with NaOH** | |
| 1.03-1.04 | Density |
| 10-35 | Viscosity |
| 29-32 | Loss of fluid |
| 10-12 | PH |

## 5. Conclusion

Treatment of Iraqi Bentonite was used successfully, noted that there are some factors directly affected the efficiency and quality of this treatment, including the following:

1. Mixer quality, the speed of the rotation and the mixing time, which requires conducting experiments before the actual commencement of work to reach the required standard.
2. The quality of the Iraqi Bentonite that's mean the sourcess of the tested samples.
3. Quality and purity of the additive materials.
4. The Iraqi Bentonite gives an internal membrane filter more than the membrane rate of the foreign Bentonite. This requires the casting process and a high balance with the concrete slab to ensure that it rises inside the casting tube according to the required specifications.
5. Do not leave Bentonite solution more than 48 hours to ensure that the thickness of the membrane is not increased, which causes difficulty in the process of casting and this problem is required to avoid the site. Bentonite dried after the activation and examinations according to the specifications required and then use it to stabilize the chemical composition.
6. The results show the vital role of NoaH combined with $Na_2co_3$ as a success activator for Iraqi Bentonite besides to its role as Bactericides in dirling liquid.

IOP Conf. Series: Materials Science and Engineering **371** (2018) 012055 doi:10.1088/1757-899X/371/1/012055

7. Its recommended to do the testing steps on a large mixer device model like 1*1*1 m (L, W, H) and make a compassion with the obtaining results and study the time effect on filter cake thickness under the variety of additives
8. Its recommended to study more factors like the temperature; gel strength; effect of Polymer % on the yield point and PH; Corrosion Inhibitors; and Thinners & Dispersants.

## 6. References


[1] A. Asad., et al. Suitability of bentonite clay: an analytical approach. International Journal of Earth Science, 2.3:88-95 (2013)

[2] David J.DAppolonia and Christopher R.Ryan .Soil-Bentonite slurry trench cut –off walls. Geotechnical Exhibition and Technical Conference, Chicago-March 26(1979)

[3] Report of the General Contracting Company for Pillars and Foundations, (1980)

[4] Mustafa Ahmed Mohamed. Village of Jalabad and Al-Tayin Khawwah / Ghal Sahr. Journal of Kilgamesh, 2 (2006)

[5] G. Arthur. Industrial application of Bentonite

[6] Fedreration of piling specialists Bentonite support fluids in civil engineering (2006)

[7] Saleh, Shakhatreh K.; Mahasneh, Mehaysen A. Activation of Jordanian Ore Bentonite by Sodium Carbonates. Journal of Minerals and Materials Characterization and Engineering, 3.06: 477 (2015)

[8] Report submitted by the Ministry of Housing and Construction. Using the properties of Bentonite in Iraq (1993)

[9] Federation of Piling Specialists, modifications (1977) to Specification for Cast in place Concrete Diaphragm Wall (published 1973) and Specification for Cast in place piles Formed under Bentonite Suspension (published 1975).



**Acknowledgements**

The author gratefully acknowledges support for this research by the Laboratory of Water Resources department at the Baghdad Institute of Technology, Middle Technical University -Iraq.

IOP Conf. Series: Materials Science and Engineering **371** (2018) 012056 doi:10.1088/1757-899X/371/1/012056

# Empirical Stress Intensity Factor Equations for Cracked Steel Plates Repaired with Double-Sided FRP Patches

**B Do[a] and A Lenwari[b]**

Structural Engineering Division, Department of Civil Engineering, Faculty of Engineering, Chulalongkorn University, Bangkok THAILAND 10330
E-mail: [a]Bach.K@student.chula.ac.th, [b]Akhrawat.L@chula.ac.th

**Abstract.** This paper presents an approach that combines the finite element (FE) modeling and genetic programming (GP) to provide accurate empirical stress intensity factor (SIF) equations for center-cracked steel plates repaired with adhesive-bonded double-sided fiber-reinforced polymer (FRP) patches. Several past studies in recent years independently showed that the reduction on the SIF of cracked structures after the patch repair is dependent on many factors such as bonding techniques, material parameters, geometric parameters, environmental factors. In this study, the SIF of the repaired cracked steel plate was considered to be a function of seven parameters including the crack length, elastic modulus of FRP material, shear modulus of adhesive material, dimensions (width, length, and thickness) of rectangular FRP patches, and thickness of adhesive layers. Empirical SIF equations were created by the data mining process of genetic programming analyses performed on a database created from the FE results. The SIF values obtained from these equations were also compared with an analytical equation to assess whether their ability to perform well on a certain design or not. It was found that the proposed SIF equations fitted well with the FE results as the squared Pearson correlation coefficients $R^2$ are higher than 0.9. In addition, the correlations between proposed equations and the analytical equation are approximately 0.8.

## 1. Introduction

Fiber reinforced polymer (FRP) materials that consist of high strength fibers and tough resin matrix possess a lot of prominent and beneficial features such as high modulus and strength, low density, excellent fatigue resistance, great resistance to corrosion, and great flexibility [1]. In recent years, these features have become more attractive in structural engineering for strengthening structural members as well as for repairing cracked or defective ones.

Numerous previous studies showed that bonding of carbon fiber reinforced polymer (CFRP) patches strongly influences flexural strength of steel beams [2, 3], slightly increases lateral-torsional buckling capacity of steel girders [4], successfully opposes local buckling in steel hollow section members [5], and enhances the strength and the ductility of concrete-filled steel tubes [6]. The FRP bonding also increases flexural stiffness of repaired members [7] and reduces stresses in these members as transferred stresses to external components through adhesive shear stress [8]. In addition, it has been demonstrated that the crack-bridging effect from the adhesively CFRP patching [9] and the redistribution of stresses in a repaired structure cause a significant reduction on both the crack opening displacement and effective stress range at the crack tip [10]. The crack repair technique also shows the beneficial effects of increasing the fatigue performance and reducing the mode I of stress intensity factor (SIF) of cracked structures. Previous investigations showed that the repair technique can significantly increase the fatigue life of cracked steel beams from 50% to 4 times [4] and more than

20% for cracked steel plates [11]. A significant reduction of the SIF due to the FRP patches was also reported for both cracked steel girders [12] and cracked steel plates [13].

The effectiveness of the repair design for a cracked structure is assessed based on the reduction of SIFs at the crack tip while adhesive shear stress and stresses in the FRP patch are within acceptable limits [10]. To facilitate the repair design, the closed-form SIF equation should be developed. However, the development of an accurate empirical closed-form SIF equation that incorporates the combined effects of several design parameters simultaneously is still a challenging work.

This paper presents an approach that combines the FE modeling and genetic programming (GP) to provide accurate empirical equations of the SIF in terms of the crack length, and six design parameters for a center-cracked steel plate repaired with adhesive-bonded double-sided FRP patches.

## 2. Rose's approach for the SIF solution

An important concept for a preliminary estimate of the SIF of a crack patching problem is the two-stage process proposed by Rose [14], as depicted in Figure 1. In the first stage, the un-cracked steel plate is assumed. The normal stress in the un-cracked steel plate underneath the patch, $\sigma_s$, is determined based on one-dimensional linear-elastic technique [15] as specified in equation (1). In the second stage, the steel plate is cut along a line segment as the central crack of length 2a. The stress $\sigma_0=\sigma_s(y=0)$ is applied internally as the pressure to the crack faces. The SIF for the crack configuration in stage II is determined by the well-known formula in equation (2).

$$\sigma_s(y)=2\left[\frac{\dfrac{\sigma_y t_s}{E_s t_s}}{t_s\left(\dfrac{1}{E_p t_p}+\dfrac{2}{E_s t_s}\right)}\right]\frac{\cosh(\lambda y)}{\cosh(0.5\lambda L_p)}+\frac{\dfrac{\sigma_y t_s}{E_p t_p}}{t_s\left(\dfrac{1}{E_p t_p}+\dfrac{2}{E_s t_s}\right)} \tag{1}$$

$$K=\sigma_s(y=0)(\pi a)^{1/2}\sqrt{\sec\frac{\pi a}{\mathrm{W}_s}} \tag{2}$$

where

$$\lambda^2=\frac{G_a}{t_a}\left(\frac{1}{E_p t_p}+\frac{2}{E_s t_s}\right) \tag{3}$$

$E_s$, $E_p$, and $G_a$ are the modulus of steel, the longitudinal modulus of FRP, and the in-plane shear modulus of adhesive, respectively; $t_s$, $t_p$, and $t_a$ represent the thicknesses of the steel plate, FRP patch, and adhesive layer, respectively; $\sigma_y$ is the uniform tensile stress at the steel plate ends; $L_p$ is the length of the rectangular FRP patch.

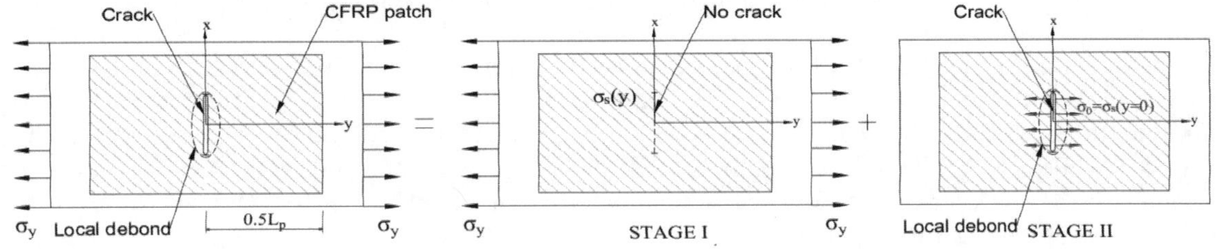


**Figure 1.** Two-stage analysis process for crack patching problem.

## 3. Finite element analyses

### *3.1. Geometric and material properties*

To evaluate the average of the SIF value along the crack front of the repaired cracked steel plate, the three-dimensional FE analyses were performed using the general FE program ABAQUS/CAE. The subscript letters s, a, and p denote the steel plate, the adhesive layer and the FRP patch, respectively.
A total 175 FE models were generated for the steel plate with the length, $L_s$=500 mm, the width, $W_s$=90 mm, and the thickness, $t_s$=10 mm. As shown in Figure 2, the central crack of length, 2a, with a circular hole R = 5 mm was included. Different crack lengths of 20%, 30%, 40%, 50%, 60%, 70%, and 80% of the steel plate width were chosen.
The steel plate was subjected to a uniform pressure of $\sigma_y$=150 MPa on two ends. Due to symmetry about the two axes (i.e. the x-axis and the y-axis), a quarter model with symmetric constraints applied on the boundaries was used, as shown in Figure 3.

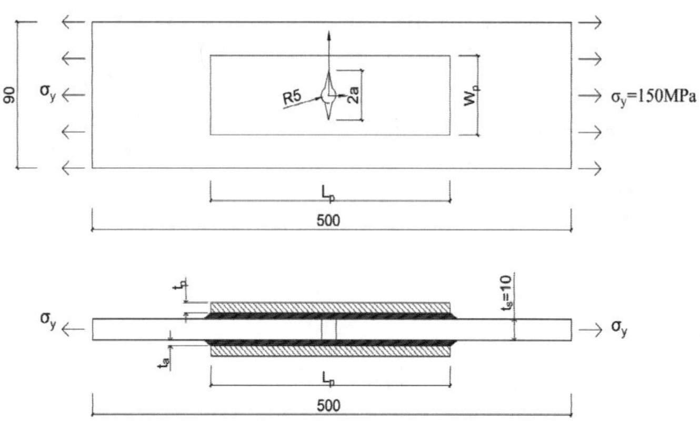


**Figure 2.** Geometry of cracked steel plate repaired with double-side FRP patches.

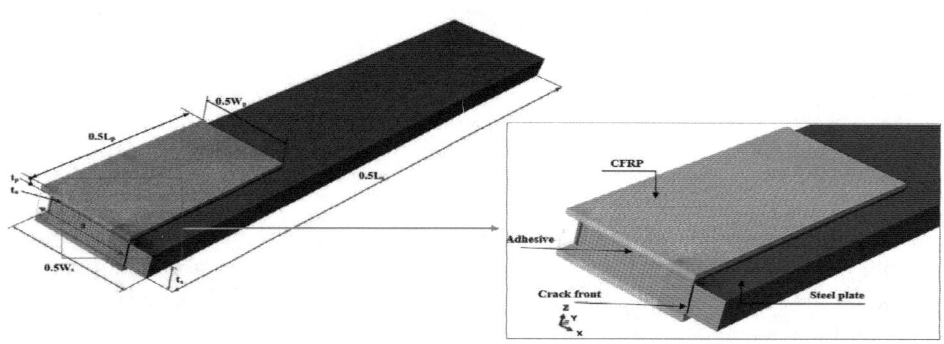


**Figure 3.** Quarter finite element model in ABAQUS.

The double-sided rectangular FRP patches and adhesive layers have the same dimensions: the length, $L_p$=$L_a$, and the width, $W_p$=$W_a$. The FE analyses were carried out for different geometrical configurations of $L_p$ and $W_p$. The FRP patch width, $W_p$, was subdivided into five levels in the interval [2a, $W_s$] while five levels for the length of FRP patch, $L_p$, were 50, 100, 150, 200, 250 mm. For example, if the crack length was 2a=20%$W_s$=18 mm, the interval of $W_p$ is [18, 90] and five levels for $W_p$ were 18, 36, 54, 72, 90 mm.
The thickness of FRP plate, $t_p$, changed in the range from 1 to 1.8 mm with five levels, $t_p$={1, 1.2, 1.4, 1.6, 1.8}. The thickness of adhesive layers, $t_a$, varied from 0.25 to 1.25 mm with five levels, $t_a$={0.25, 0.50, 0.75, 1.00, 1.25}.
The material properties of the steel plate, five different FRP types and adhesive materials were adapted from a previous work [10], as given in Table 1. The mechanical behavior of all materials was linear elastic. Steel and adhesive were analyzed as linear isotropic elastic materials; the behavior of FRP material was orthotropic elastic in plane stress [16].

**Table 1.** Material properties of cracked steel plate, patches, and adhesives [10].

| Materials | Material properties | | | |
|---|---|---|---|---|
| | Longitudinal modulus $E_1$ (GPa) | Transverse in-plane modulus $E_2$ (GPa) | In-plane shear modulus $G_{12}$ (GPa) | Major in-plane Poison's ratio $\upsilon_{12}$ |
| **Steel plate** | 200.00 | n/a | 76.92 | 0.30 |
| **FRP composite** | | | | |
| Boron/Epoxy 5505 | 219.90 | 21.40 | 6.89 | 0.21 |
| Boron/Epoxy 5521 | 195.10 | 19.40 | 5.52 | 0.21 |
| AS4/3501-6 | 124.10 | 11.00 | 5.52 | 0.34 |
| IM6/SC1081 | 177.20 | 10.80 | 3.93 | 0.27 |
| T300/F934 | 148.10 | 9.70 | 4.55 | 0.30 |
| **Adhesive** | | | | |
| FM-73 | 0.96 | n/a | 0.36 | 0.35 |
| FM-300 | 1.29 | n/a | 0.46 | 0.40 |
| FM-300K | 1.37 | n/a | 0.52 | 0.32 |
| A-9321 | 1.50 | n/a | 0.55 | 0.36 |
| FM-36 | 1.81 | n/a | 0.67 | 0.35 |

### *3.2. Finite element models*

A twenty-node quadratic brick element, C3D20, was used to represent steel and adhesive while a continuum shell element, SC8R, which allows for thick and thin shell applications, was used for FRP. To capture the singular stress and strain fields near the crack front, a collapsed three-dimensional element (collapsed C3D20) was assigned as steel elements near the crack front. All element types and their applications are clearly introduced in ABAQUS analysis user's manual [16]. The global size of steel plate elements was approximately 1 mm; the element size of 0.75 mm was used for both FRP and adhesive elements.

Furthermore, to construct an analysis matrix for performing FE analyses as well as to create a database for GP analyses, a total seven L-25 Taguchi arrays corresponding to 175 ABAQUS models were created. Each L-25 array with 25 observations was constructed based on the Taguchi methodology [17] for a particular crack length combined with the rest six design parameters; each parameter has five levels.

## 4. Genetic programming methodology

### *4.1. Genetic algorithm*

Genetic algorithms (GAs) are iterative numerical algorithms for solving optimization problems motivated by natural selection and natural genetics [18]. Each GA operates on a population of candidate solutions represented by binary strings in a computer program; an example of numeric numbers encoded by binary strings is shown in Figure 4a.

| Numeric values | | Length-10 binary strings | | | | | | | | | |
|---|---|---|---|---|---|---|---|---|---|---|---|
| **30** | encode | **0** | **0** | **0** | **0** | **0** | **1** | **1** | **1** | **1** | **0** |
| **60** | ⟶ | **0** | **0** | **0** | **0** | **1** | **1** | **1** | **1** | **0** | **0** |

a) GA individuals

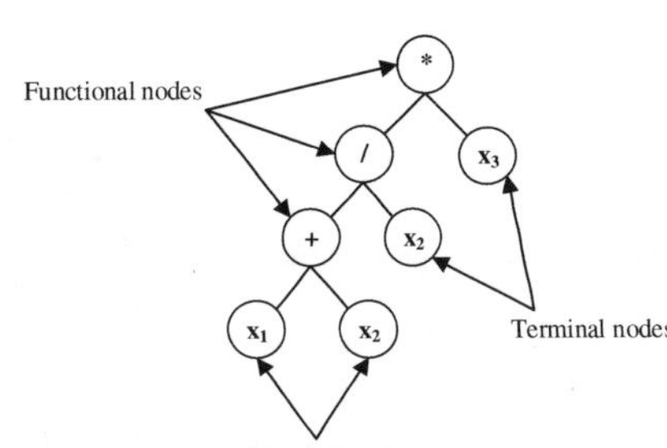


b) A GP individual $\left[(x_1+x_2)/x_2\right]x_3$

**Figure 4.** Example of candidate solutions in GA and GP.

At the start of each GA, randomly numeric values of independent variables in the solution space are encoded to corresponding computer strings in the computer program with respect to the 1-to-1 mapping property in which each binary string in coding space represents exactly one point in the numeric solution space and vice versa. GA then determines directly the fitness of each string, which is the value of a given objective function at each corresponding numeric point. After that, the algorithm scores all existing strings in the current computer program to determine which strings are good and which strings are weak. Based on this determination, good strings are combined randomly in pairs (crossover) while weak strings may change their own structure (mutation) to produce a better population of new strings for the next generation. This loop will be repeated until reaching stopping criterion as a given number of generations or iterations of the algorithm.

### *4.2. Genetic programming*

Genetic programming (GP) is one important application of GA. The major difference between GP and GA is the representation of candidate solutions, which are binary strings in GA, as shown in Figure 4a and hierarchical computer programs (tree structures) in GP, as shown Figure 4b.

In the present study, GP maximized the objective function as the squared Pearson correlation coefficient $R^2$, which effectively measures the strength of a linear relationship between two sources obtained from FE models and obtained from GP models to provide the best fit models for the SIF. The main reason of this choice is that the SIF is affected by many factors; there is no evidence to decide what type of functional form to be defined first as the property of traditional regression techniques. In this case, GP property of simultaneously finding the model that most appropriately fits the data and corresponding numerical coefficients of the model, namely symbolic regression, is reasonably required.

The process of performing the GP includes three steps. In the first step, the initial population of a certain number of the tree structures is created by using materials from members of function (F) and terminal (T) sets. The F set includes arithmetic operations, mathematical functions, Boolean operations, conditional operators, or any user-defined functions [19]. Within a tree structure, each member of the F can occupy nodes that have two arguments, namely functional nodes. Meanwhile, the T set contains independent variable atoms or constant atoms [19]. These atoms may locate at terminal nodes, which have no explicit argument. Examples of the functional nodes and terminal nodes are given in Figure 4b. In the second step, GP computes $R^2$ values that represent the correlation between the FE results and those obtained from all tree structures in the current generation. Based on these $R^2$ values, GP stores all strings in a column vector and arranges them in a descending order of fitness value. After the arrangement, GP performs three genetic operators in a sequence: elite transfer (Figure 5a), crossover (Figure 5b), and mutation (Figure 5c). Finally, the last step defines the solution of the GP analysis that is the best-so-far tree structure. The GP's solution merely depends on the stopping condition of the GP analysis where the algorithm states the current population of tree structures and reports the best tree structure stored in the computer cache at the same time.

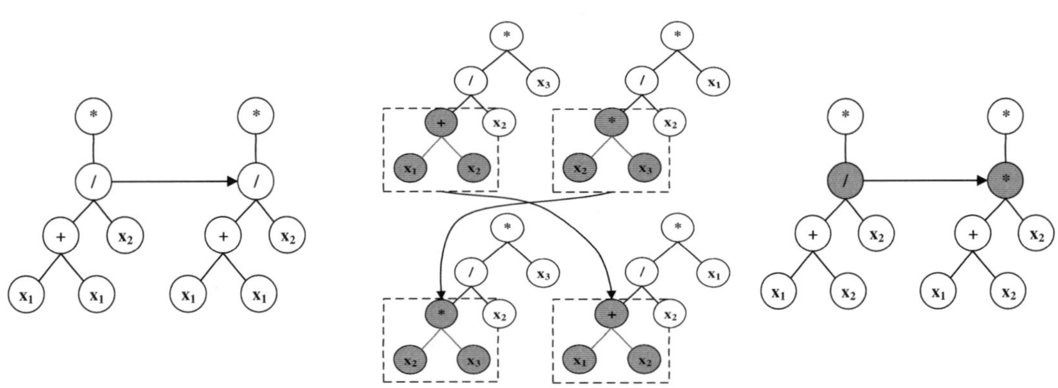

a) GP elite transfer b) GP crossover c) GP mutation

**Figure 5**. Example of three genetic operators of GP.

IOP Conf. Series: Materials Science and Engineering **371** (2018) 012056 doi:10.1088/1757-899X/371/1/012056

This study used HeuristicLab [20] as the GP tool for the SIF equations.

## 5. Results and discussions

### *5.1. GP Analyses*

A database of 175 samples taken from FE results was randomly shuffled into seven different groups with 25 samples for each group, i.e., group A, B, C, D, E, F, and G to avoid the bias of the GP analysis on a certain group of samples. Corresponding to these groups, seven different models, i.e., M1, M2, M3, M4, M5, M6, and M7, were assigned. For example, in model M1, as given in Table 2, data group A was used for testing, and the rest six groups (B, C, D, E, F, and G) were used for training. This is consistent with the principle of separating the dataset in data mining where most of the data are assigned as a training set and a smaller portion is assigned as a testing set. The use of both the training set to produce an initial fitting model and testing this model by making predictions oppose the testing set provides a considerable significance to avoid overfitting of regression solution.

**Table 2.** Models used for GP analyses.

| Model | Training set | | Testing set | |
|---|---|---|---|---|
| | Data group | Number of data | Data group | Number of data |
| M1 | B, C, D, E, F, G | 150 | A | 25 |
| M2 | A, C, D, E, F, G | 150 | B | 25 |
| M3 | A, B, D, E, F, G | 150 | C | 25 |
| M4 | A, B, C, E, F, G | 150 | D | 25 |
| M5 | A, B, C, D, F, G | 150 | E | 25 |
| M6 | A, B, C, D, E, G | 150 | F | 25 |
| M7 | A, B, C, D, E, F | 150 | G | 25 |

All empirical equations were also developed with the same GP input of both function set F= {+, -, *, square, power} and terminal set T= {$x_1$, $x_2$..., $x_7$, [-5,5]}. Here, the constant atoms were in the interval [-5,5]. The stopping condition of each GP analysis was equal to the maximum number of generations of the algorithm loop, gen=500, 1000, 1500, and 2000. Additional control parameters defined at the beginning of each GP analysis are also provided in Table 3.

**Table 3.** Control parameters used in GP analyses.

| Parameter | Value |
|---|---|
| Number of tree structures | 10 000 |
| Probability of mutation | 25% |
| Elite count (reproduction option) | 2 |
| Maximum number of tree depth | 10 |
| Maximum number of tree length | 50 |

### *5.2. Empirical stress intensity factor equations*

The relationship between the SIF and seven parameters including the crack length, FRP modulus, FRP patch width, FRP patch length, FRP patch thickness, adhesive modulus, and adhesive layer thickness is characterized by the following mathematical model:

$$K = f\left(x_1, x_2, x_3, x_4, x_5, x_6, x_7\right)\sigma_y\left(\pi a\right)^{1/2} \quad (4)$$

where

$$x_1 = \frac{2a}{W_s}; x_2 = \frac{E_p}{E_s}; x_3 = \frac{W_p}{2a}; x_4 = \frac{L_p}{2a}; x_5 = \frac{t_p}{t_s}; x_6 = \frac{G_a}{E_s}; x_7 = \frac{t_a}{t_s} \quad (5)$$

Let, the GP analyses produced acceptable closed-form equations for the *f* factor when the corresponding coefficient $R^2$ is higher than 0.9.

Table *4* shows the coefficients $R^2$ of the three best models (M2, M3, and M7) obtained from a total of seven GP analyses for both training and testing sets. The scatter plots of Pearson correlation analyses of these models are also given in Figure 6a, b, and c. It can be clearly observed that after reaching the maximum number of generations as stopping criterion of the GP analyses, the $R^2$ values for training data of models M2, M3, and M7 are 0.941, 0.925, and 0.935, respectively. In
Table *5*, the coefficients $R^2$ of the rest four models (M1, M4, M5, and M6) at gen=2000 are also almost 0.92. These results indicate a significantly high correlation between all GP models and FE results.

**Table 4.** Results of the best three GP analyses M2, M3, and M7 for f factor model.

| | Model M2 | | Model M3 | | Model M7 | |
|---|---|---|---|---|---|---|
| gen | Training set $R^2$ | Testing set $R^2$ | Training set $R^2$ | Testing set $R^2$ | Training set $R^2$ | Testing set $R^2$ |
| 500 | 0.870 | 0.847 | 0.868 | 0.878 | 0.908 | 0.903 |
| 1000 | 0.890 | 0.900 | 0.904 | 0.940 | 0.928 | 0.892 |
| 1500 | 0.917 | 0.928 | 0.914 | 0.889 | 0.931 | 0.908 |
| 2000 | 0.941 | 0.901 | 0.925 | 0.947 | 0.935 | 0.900 |

**Table 5.** Comparison of seven f factor equation models.

| Model | Gen | Training set, $R^2$ | Testing set, $R^2$ |
|---|---|---|---|
| M1 | 2000 | 0.921 | 0.857 |
| M2 | 2000 | 0.941 | 0.901 |
| M3 | 2000 | 0.925 | 0.947 |
| M4 | 2000 | 0.916 | 0.907 |
| M5 | 2000 | 0.916 | 0.934 |
| M6 | 2000 | 0.916 | 0.912 |
| M7 | 2000 | 0.935 | 0.900 |

In Figure 6d, the predicted *f* factor values of all seven proposed models are compared with the *f* factor values evaluated from FE results for the training samples and testing samples. It is recognized that proposed models are very similar even though they were built based on different data groups. Hence, it can be concluded that these models can have the same ability to provide a powerful tool for the prediction of the SIF for a cracked steel plate repaired by bonded double-sided FRP patches.
The closed-form SIF equation of model M2 with the highest coefficient $R^2$ for the training data ($R^2$=0.941) is expressed by:

$$K=\left[\left(f_1+f_2\right)f_3+0.9598\right]\sigma_y\left(\pi a\right)^{1/2} \tag{6}$$

where

$$f_1=c_0x_5\left[\left(c_1x_7\right)^{-1}+c_2c_3x_3+c_4x_4+\left(c_5x_1\right)^{-2}+c_6x_1-1.2218\right] \tag{7}$$

$$f_2=\left\{c_7c_8x_3\left[\left(c_9x_7\right)^{-1}-\left(c_{10}x_1\right)^{5}-\left(c_{11}x_4\right)^{-2}\right]-\left(c_{12}x_1\right)^{-2}\right\}c_{13}x_7 \tag{8}$$

$$f_3=c_{14}x_1c_{15}x_2c_{16}x_3c_{17} \tag{9}$$

where $c_0$, $c_1$,…,$c_{17}$ are constants, as given in Table 6.

**Table 6.** Coefficient values for SIF empirical equation provided by the GP model M2.

| $c_0$ | $c_1$ | $c_2$ | $c_3$ | $c_4$ | $c_5$ | $c_6$ | $c_7$ | $c_8$ |
|---|---|---|---|---|---|---|---|---|
| 3.7443 | 2.4381 | 4.7712 | 1.3121 | -0.5499 | 1.2800 | 1.7375 | 5.8520 | 1.2888 |
| $c_9$ | $c_{10}$ | $c_{11}$ | $c_{12}$ | $c_{13}$ | $c_{14}$ | $c_{15}$ | $c_{16}$ | $c_{17}$ |
| 2.4381 | 1.7375 | -0.5374 | 0.6311 | 1.1647 | 1.3527 | 0.3121 | 0.3193 | -87.0620 |

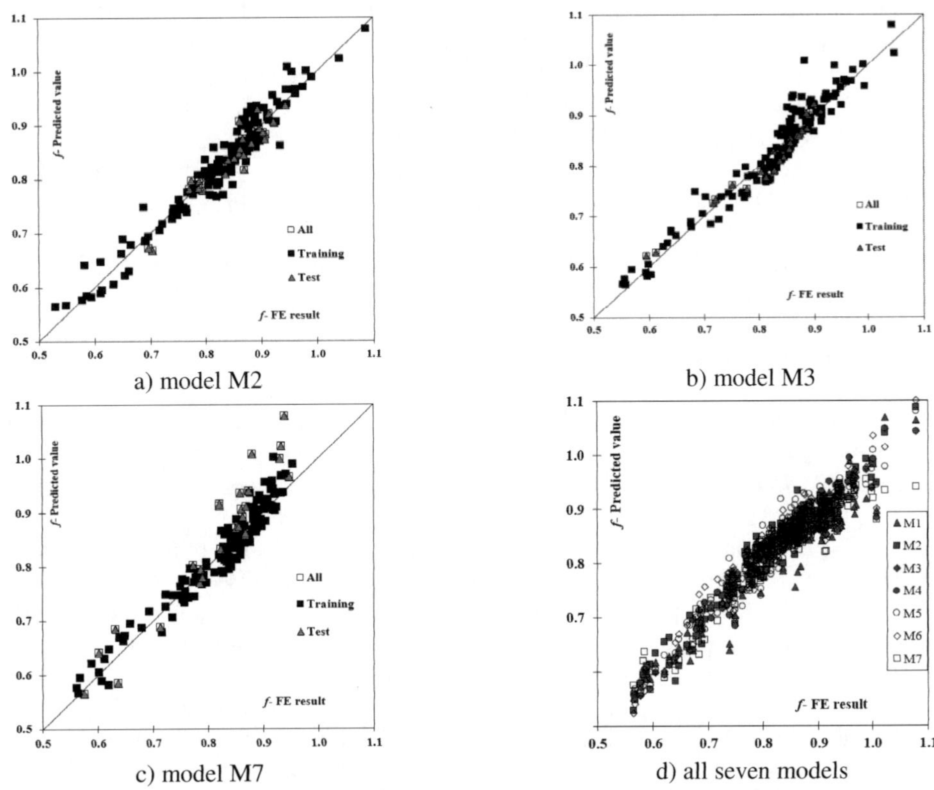


**Figure 6.** Association for f factor from FE results and predicted values from a) model M2; b) model M3; c) model M7; d) all seven proposed models.

### *5.3. Comparison of the SIFs obtained from Empirical Equations and Rose's Approach*

Recently, based largely on analyses derived from Rose's approach, some software programs were developed for the crack repair in the field of aerospace engineering [10]. This fact promotes the need for an evaluation of how well the analytical equation is represented by the proposed empirical equations. A comparison of the SIFs obtained from proposed equations and those obtained from Rose's approach in section 2 was made. It could be a good choice for checking the rationality of proposed models since so far, there have not been any equations that fully express the relation between the SIF and the stated seven design parameters. As the result, the SIFs obtained from proposed empirical equations are relatively smaller than the ones obtained from the analytical equation.

Furthermore, the correlation analyses used to assess the strength of the relationship between proposed equations and the analytical one were also performed. It was found that the correlations between the three best-proposed equations corresponding to the GP models M2, M3, and M7 and the analytical equation are 0.766, 0.761, and 0.770, respectively.

## 6. Conclusions

This paper is aimed at indicating the possibility of adapting the FE method combined with genetic programming to provide accurate empirical SIF equations for a cracked steel plate repaired with adhesive-bonded double-sided FRP patches. The results obtained from the numerical work led to the following conclusions:

- The SIF empirical equations obtained from GP analyses remarkable correlate with the FE models. The empirical equation that has the best correlation with FE results was also given.
- The SIFs predicted by the empirical equations are relatively smaller than those derived from Rose's approach. In other words, from a structural designer standpoint, each proposed equation is more economical than the analytical one.
- Relatively strong associations between proposed equations and the analytical equation are drawn as the correlations between them approach 0.8.

## 7. References

[1] A. Baker, "Bonded composite repair of fatigue-cracked primary aircraft structure," *Composite Structures,* vol. 47, no. 1, pp. 431-443, 1999/12/01/ 1999.

[2] A. Lenwari, T. Thepchatri, and P. Albrecht, "Flexural response of steel beams strengthened with partial-length CFRP plates," *Journal of Composites for Construction,* vol. 9, no. 4, pp. 296-303, 2005.

[3] A. Lenwari, T. Thepchatri, and P. Albrecht, "Debonding strength of steel beams strengthened with CFRP plates," *Journal of Composites for Construction,* vol. 10, no. 1, pp. 69-78, 2006.

[4] X.-L. Zhao, *FRP-strengthened metallic structures*. CRC Press, 2013.

[5] X.-L. Zhao and L. Zhang, "State-of-the-art review on FRP strengthened steel structures," *Engineering Structures,* vol. 29, no. 8, pp. 1808-1823, 2007.

[6] J. Teng, T. Yu, and D. Fernando, "Strengthening of steel structures with fiber-reinforced polymer composites," *Journal of Constructional Steel Research,* vol. 78, pp. 131-143, 2012.

[7] T. C. Miller, M. J. Chajes, D. R. Mertz, and J. N. Hastings, "Strengthening of a steel bridge girder using CFRP plates," *Journal of Bridge Engineering,* vol. 6, no. 6, pp. 514-522, 2001.

[8] J. Deng, M. M. Lee, and S. S. Moy, "Stress analysis of steel beams reinforced with a bonded CFRP plate," *Composite structures,* vol. 65, no. 2, pp. 205-215, 2004.

[9] C. Lin and P. Kao, "Effect of fiber bridging on the fatigue crack propagation in carbon fiber-reinforced aluminum laminates," *Materials Science and Engineering: A,* vol. 190, no. 1-2, pp. 65-73, 1995.

[10] C. N. Duong and C. H. Wang, *Composite repair: theory and design*. Elsevier, 2010.

[11] H. Liu, R. Al-Mahaidi, and X.-L. Zhao, "Experimental study of fatigue crack growth behaviour in adhesively reinforced steel structures," *Composite Structures,* vol. 90, no. 1, pp. 12-20, 2009/09/01/ 2009.

[12] A. Hmidan, Y. J. Kim, and S. Yazdani, *Stress Intensity Factors for Cracked Steel Girders Strengthened with CFRP Sheets*. 2014, p. 04014085.

[13] L. Gu, A. R. M. Kasavajhala, and S. Zhao, "Finite element analysis of cracks in aging aircraft structures with bonded composite-patch repairs," *Composites Part B: Engineering,* vol. 42, no. 3, pp. 505-510, 2011/04/01/ 2011.

[14] L. Rose, "Theoretical analysis of crack patching," *Bonded Repair of Aircraft Structures,* pp. 77-105, 1988.

[15] A. Albat and D. Romilly, "A direct linear-elastic analysis of double symmetric bonded joints and reinforcements," *Composites Science and Technology,* vol. 59, no. 7, pp. 1127-1137, 1999.

[16] U. s. M. Abaqus and E. U. s. Manuals, "Version 6.3, Hibbitt, Karlsson & Sorensen," *Inc. Rhode Island,* 2002.

[17] P. J. P. J. Ross, *Taguchi techniques for quality engineering: loss function, orthogonal experiments, parameter and tolerance design*. 1996.

[18] D. E. Goldberg, "Genetic algorithms in search, optimization, and machine learning, 1989," *Reading: Addison-Wesley,* 1989.

[19] J. R. Koza, *Genetic programming: on the programming of computers by means of natural selection*. MIT press, 1992.
[20] S. Wagner *et al.*, "Architecture and Design of the HeuristicLab Optimization Environment," in *Advanced Methods and Applications in Computational Intelligence*, R. Klempous, J. Nikodem, W. Jacak, and Z. Chaczko, Eds. Heidelberg %@ 978-3-319-01436-4: Springer International Publishing, 2014, pp. 197-261.

**Acknowledgments**
The authors gratefully acknowledge financial support for this study provided by the ASEAN University Network/Southeast Asia Engineering Education Development Network (AUN/SEED-Net).

IOP Conf. Series: Materials Science and Engineering **371** (2018) 012057 doi:10.1088/1757-899X/371/1/012057

# Fabrication of Activated Charcoal Adsorption $Bi_2O_3/TiO_2$ Composite Powders and the Study of Their Photocatalytic Activity

**Z F Wu[1] and L Y Hu[2]**

1School of Materials Science and Engineering, Baise University, PR China 533000
2 Library of Baise University, PR China 533000
E-mail: zhifuwu2013@163.com

**Abstract.** A visible light-driven activated charcoal adsorption $Bi_2O_3/TiO_2$ composite photocatalyst was prepared by an sol–gel method in which activated charcoal acted as a adsorption to capture metal ions by reacting with bismuth and titanium sources via a complex polycondensation pathway. The phase and particle size of the photocatalyst were characterized by X-ray diffraction (XRD), scanning electron microscopy (SEM).The results revealed that adsorption of $Bi_2O_3$ / $TiO_2$ composite powder with activated carbon is not easy to agglomerate and sinter during the preparation and calcination process, and it has the advantages of good dispersibility, less agglomeration and uniform particles. The photocatalytic activity of the as-prepared catalyst was evaluated by degradation of methylene blue under visible light irradiation. The results showed that the decomposition rate of methylene blue was 93% at 50 min catalyzed by composite powder.

## 1. Introduction

Sewage treatment, indoor air quality and other environmental problems has an important link with human health and has a very important impact on peoples living environment. The semiconductor titanium dioxide has a broad application prospects based on dioxide visible light photocatalytic degradation of organic pollutants and the provision of many potential environmental solutions. $TiO_2$ has many advantages such as: high stability, high sensitivity, non-toxic and so on. However its light quantum efficiency is only 4%, in the practical application of a greater limitation. So the modification of $TiO_2$ has aroused widespread concerning the chemical industry. The modification methods of $TiO_2$ are as follows: precious metal deposition [1] , metal ionization [2] , non-metallic element doping [3], composite semiconductor doping [4] and other methods, which the most effective improve the photocatalytic performance of $TiO_2$is the way to be combined with other semiconductors. And the composite oxide photocatalyst will generally have better catalytic performance than pure $TiO_2$. The difference in energy levels between two kinds of semiconductor can make effective separation of charge, so that its excitation wavelength extends to the visible light region [5-6]. The band gap of $Bi_2O_3$ is 2.8 eV, there are many reports on the modification of $TiO_2$ with $Bi_2O_3$[7-10]. Known methods are sol-gel method [11], hydrolysis method [12] , etc. But the methods have many the shortcomings, such as: poor dispersion, particle radius is not easy to control, preparation, drying, and roasting particles easy to agglomerate and sintering and so on [13].

In the reaction solution to prepare activated charcoal adsorption $Bi_2O_3$/ $TiO_2$ composite powders, the nanoparticles are adsorbed to the surface of the activated carbon in the reaction solution. Thereby the agglomeration of the granules reduced during the preparation. The photocatalytic activity of activated charcoal adsorption $Bi_2O_3/TiO_2$ composite powders was investigated by using the composite powders as the photocatalyst and methylene blue solution as the target degradation product under the UV irradiation of xenon lamp.

## 2. Results and Discussion

### *2.1 Activated carbon adsorption and sol–gel method*

It is not uncommon to study the preparation of $Bi_2O_3/TiO_2$ nanoparticles by ordinary methods. However, this method has many disadvantages, such as particle size is difficult to control and it's easy to easy agglomeration and sintering during sample drying and roasting. The sol-gel method in which activated charcoal is used as adsorbent can effectively avoid the shortcomings of the traditional method. Adding the right amount of activated carbon with strong adsorption capacity in the reaction solution, precursor once formed in the solution will be adsorbed on activated carbon surface. And the activated carbon has a good dispersion, so the product obtained in the drying and roasting process is not easy to reunite and sintering of the situation. $Bi(NO_3)_3$ will produce bismuth nitrate when it soluble in water, bismuth nitrate will generates Bi $(OH)_3$ with low-solubility in aqueous solution. The calcination process was heated by a high-temperature furnace (temperature ramp rate of 20℃ / min). when the temperature increased from 200 ℃ to 400 ℃, Ti $(OH)_4$ begins to lose crystal water generating $TiO_2$ and particals of Bi $(OH)_3$ is decomposed into $Bi_2O_3$.When the temperature reaches 500 ℃, the activated carbon begins to be oxidized. As the temperature increases, the activated carbon will gradually decrease. When the temperature reaches 700 ~ 800℃, the activated carbon tends to be completely oxidized, the composite powder will appears sintering situation after a long time. Therefore, the addition of activated carbon can effectively control the particle size of $Bi_2O_3/TiO_2$ composite powder adsorbed by activated carbon at the nanoscale level. And the composite powder has a larger specific surface area and higher photocatalytic performance.

### *2.2. Phase analysis*

The XRD pattern of activated charcoal adsorption $Bi_2O_3/TiO_2$ composite photocatalyst was prepared by an sol–gel method under different calcination temperature are shown in figure1. Where in the calcination temperatures of the figure1a, figure1b, figure1 c and figure1 d are 400 °C, 600 °C, 700 °C and 800 ° C, respectively. Due to the low content of $Bi_2O_3$ with high dispensability, The peaks in figure1 is essentially diffraction peaks of $TiO_2$. According to the diffraction peak, it can be seen that the crystallization rate is not high and the diffraction peak is weak with calcination temperature 400 ℃ composite photocatalysts are crystalline under the 600 ℃, 700 ℃ and 800 ℃.

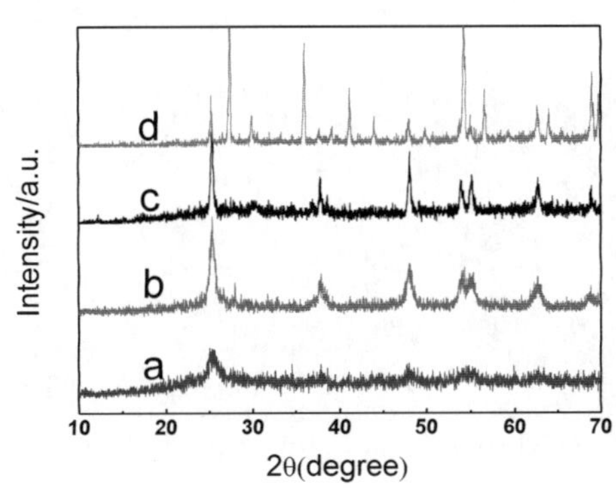


**Figure 1.** XRD pattern of and $Bi_2O_3/TiO_2$ composite synthesized at different calcination temperature (a:400 ℃, b:600 ℃, c:700 ℃, d:800 ℃).

The pattern of activated charcoal adsorption $B_{i2}O_3$/ $TiO_2$ composite powders was prepared by a sol–gel method shows that the peaks both are anatase diffraction peaks when the calcination temperature is 600 ℃ and 700 ℃. The composite powder is still anatase phase, and no rutile phase conversion under 800 ℃ of the calcination temperature. It is known that the powder prepared by ordinary sol-gel

IOP Conf. Series: Materials Science and Engineering **371** (2018) 012057 doi:10.1088/1757-899X/371/1/012057

method is converted from anatase phase to rutile phase when the calcination temperature reaches 800 ℃.

*2.3 SEM analysis*

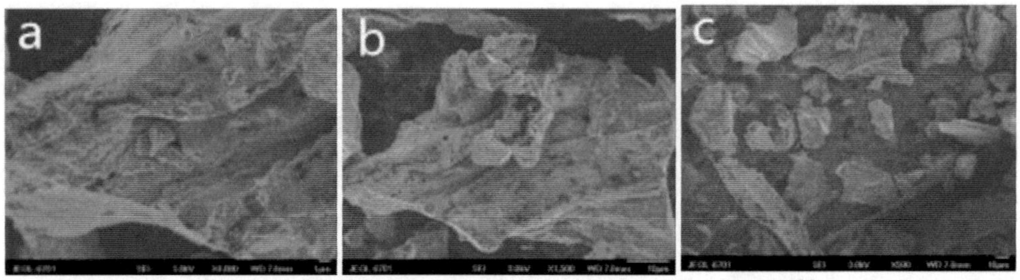

Figure2 SEM images of composite powders at 400 ℃

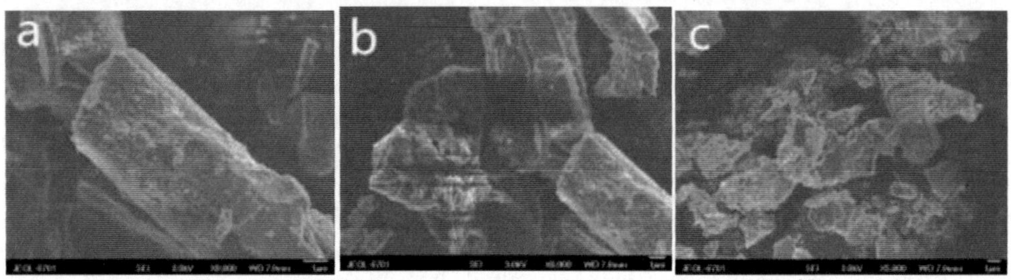

**Figure 3.** SEM images of composite powders at 600 ℃.

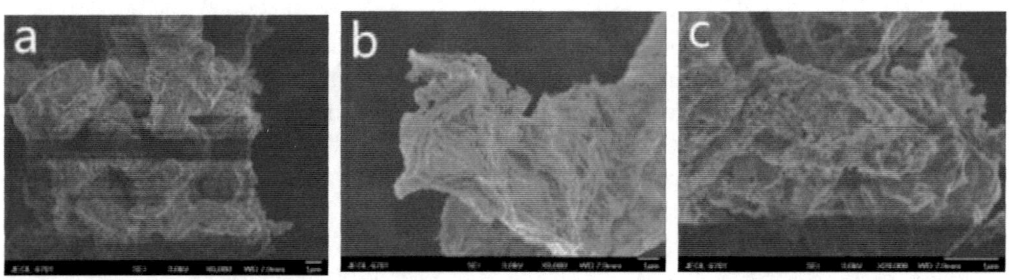

**Figure 4.** SEM images of composite powders at700 ℃.

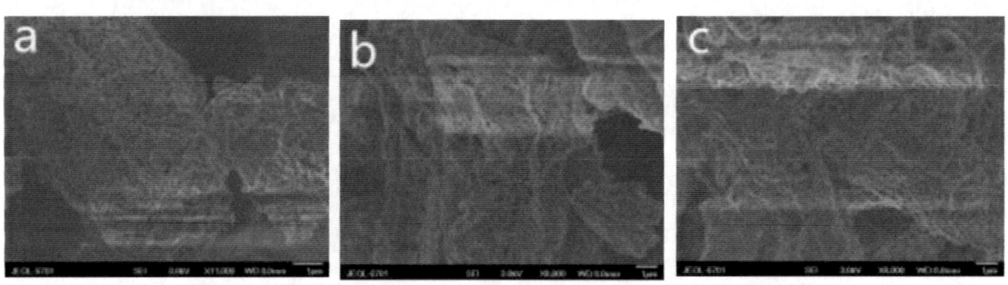

**Figure 5.** SEM images of composite powders at 800 ℃.

In figure 2, the length of figure 2a unit is 1 μm, and the length of figure 2b and figure 2c is 10 μm. It can be seen from the three figures that the composite powder is well dispersed and the particles are uniform with no obvious agglomeration. The three graphs unit length is 1μm in figure 3. The crystal is very regular, while uniform dispersion, no agglomeration phenomenon at calcination temperature 600 ℃. The three graphs unit length is 1μm in figure 4. From the three figures to see when the calcination temperature is 700 ℃, there is a slight reunion phenomenon, but the overall dispersion is more uniform. When the calcination temperature reaches 800 ℃, there is a clear phenomenon of agglomeration are shown in figure 5. This result is similar to the previous XRD analysis results. Those show that activated carbon adsorption $Bi_2O_3$ / $TiO_2$ composite powder at calcination temperature of 600 ℃ light catalytic effect is best.

### *2.4. Photocatalytic performance*

Figure 6 shows the different degradation rates of methylene blue under different catalysts. The degradation rate of methylene blue is only 1.5% without catalyst, which shows that the stability of methylene blue is very good in the absence of catalyst. The degradation rate of methylene blue increased obviously with the increase of time after adding different photo catalysts. Degradation rate is maximum when the calcination temperature was 600 ℃.And the degradation rate is the smallest at calcination temperature of 800 ℃.The decomposition rate of methylene blue was 86% at 40 minutes With activated carbon adsorption of composite powder as catalyst prepared by activated carbon adsorption sol-gel method and was calcined at 600 ℃， and the decomposition rate of methylene blue was 93% at 50 minutes. The decomposition rate of methylene blue was only 24% at 40 min. With $Bi_2O_3/TiO_2$ as catalyst prepared by ordinary sol-gel method and was calcined at 600 ℃.When adding activated carbon to the reaction solution, the formation of nanoparticles in the initial reaction solution will immediately activated by carbon adsorption. This effectively prevents the superfine particles agglomerating and sintering in the subsequent preparation and drying process and roasting stage. The nano-composite $Bi_2O_3/TiO_2$ powder with small particle size and large specific surface area can be obtained when the temperature is raised to 600 ℃. The temperature continues to rise, the composite powder will have sintering phenomenon, so the catalytic effect will be worse.

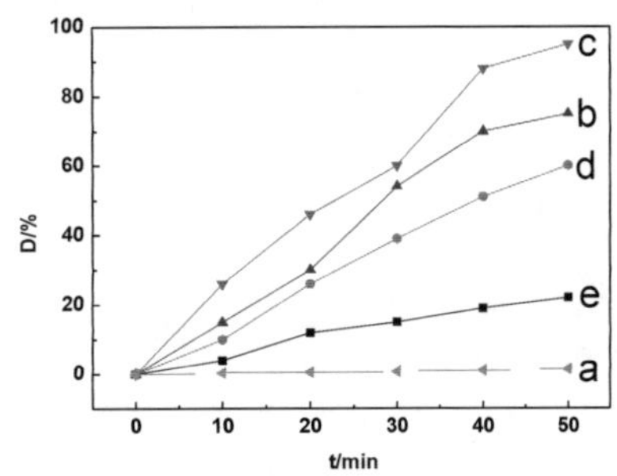


**Figure 6.** The chart of photo-degradation rate changing with the illumination time(calcination temperature: a: control experiment without catalyst b: 400 ℃, c:600 ℃, d:700 ℃, e:800 ℃,).

## 3. Conclusion

Activated carbon adsorption $Bi_2O_3/TiO_2$ composite powders average particle diameter is of about 19.9 nm at calcination temperature of 600 ℃. The adsorption of activated carbon can effectively reduce the agglomeration and sintering of $Bi_2O_3/TiO_2$ composite powder during the preparation, drying and high temperature calcination, thus inhibiting the growth of the particles and suppressing the composite powder from Anastasia phase to futile phase. The photocatalytic activity of the as-prepared catalyst was evaluated by degradation of methylene blue under visible light irradiation. The results showed that the decomposition rate of methylene blue was 93% at 50 min catalyzed by composite powder.

## 4. References

[1] Sai W L, Ken C, Tuti M L 2007 *Appl. Catal. B: Environ,* **72**: 363
[2] Wu S X, Ma Z, Qin Y N 2005 *Photographic Sci Photochem*, 23: 94
[3] Yu J C, Yu J G, Ho W K 2001 *Chem. Mater*, **14**: 3808
[4] Xiao Y F, Liu S 2007 *Inorg Chem Industy*, **39**: 5
[5] Tian J T, Chen L J, Yin Y S 2009 *Sur & Coatings Techno*, 204: 205
[6] Do Y R, Lee W, Dwight K 1994 *J Solid State Chem*, **108**: 198
[7] Yang Y, Li J T, Miao J 2011 *Chinese J Inorg Chem*, **27**: 547
[8] Xu J J, Ao Y H, Fu D G 2008 *Appl Surf Sci Sunlight Irradi*, **255**: 2365
[9] Zhao X, Liu H J, Qu J H 2011 *Appl Surf Sci*, **257**:4621

[10] Xu J J, Chen M D, Fu D G 2011 *Transactions Nonferrous Metals Soc China*, , **21**: 340
[11] Liu Y D, Feng X 2010 J *Alloys and Comp*, **498**: 179
[12] Lei J, Chen H J, Shi W S 2010 J *Xinyang Normal Univ*, **23**: 562
[13] Yuan W J, Zhou Y M 2003 *Materials Rev,* **22:** 59

**5. Acknowledgements**
This project is supported by Guangxi colleges and universities key subject of material physics and chemistry, Baise University, Baise, 533000, China.
And the novel aluminum matrix composites characteristic research team in Baise University. Baise, 533000, China.

IOP Conf. Series: Materials Science and Engineering **371** (2018) 012058 doi:10.1088/1757-899X/371/1/012058

# Research on Polymer Modifiers of Bio-asphalt Synthesized by Controlled/Living Radical Polymerization

**Jia Hui[1],Tong Wang[1], Ruinan Jiang[1], Baolin Zhu[1], and Xuji Wang[2]**

1 China Academy of Transportation Sciences, Beijing, China
2 Bayannaoer City traffic construction project quality supervision station, Bayannaoer, China
E-mail: huijiabuct@126.com

**Abstract.** Bio-asphalt is a new material compared with petroleum asphalt. The bio-asphalt is prepared by bio-oil, which is produced by fast pyrolysis of biomass. Disadvantages of bio-oil are instability and poor high and low temperature performance. Considering these shortcomings, a polymer modifier with controlled number-average molar mass and well-defined microstructure was prepared by controlled/living radical polymerization. In this research, the modifier was added in bio-asphalt to prove its performances which were evaluated by three major indexes tests of asphalt. Test result shows that the modified bio-asphalt has good stability and good high and low temperature performance. In conclusion, the polymer modifiers, synthesized by controlled/living radical polymerization, have excellent modification effect on bio-asphalt.

## 1. Introduction

In recent years, the unstable situation of oil resources has been forcing road researchers to keep searching for alternative materials of petroleum asphalt binders [1]. Therefore, it has become a hot spot in the field of pavement engineering to research and develop new material to replace petroleum asphalt [2]. As a new type of asphalt material, bio-asphalt has been paid more and more attention by researchers. It is extremely urgent to study the production, performance improvement and pavement use of bio-asphalt [3].

In this research, alkenes polymer modifiers were prepared by controlled/living radical polymerization. These modifiers have active groups that can be react with active groups in bio-asphalt. In addition to this, the number-average molar mass $M_n$ of the active groups in the modifiers is in an appropriate range (15000~100000), which is similar with the saturated main chain in conventional asphalt, and this improves compatibility between the bio-oil and the asphalt. Mix the bio-asphalt with different kinds of polymer modifiers to prepare modified bio-asphalt. Conduct consistency test, softening point test and ductility test on the modified bio-asphalt, and then compare the performance of the modified bio-asphalt and conventional asphalt. At last, it was analyzed how the different polymer modifiers affected the bio-asphalt.

## 2. Experimental Sections

### *2.1. Design discipline*

(1) There are polar groups like cyan groups, ester groups and carbonyl groups in bio-oil. Therefore, polymer modifiers with similar polar groups on the molecular chains are needed to improve stability of bio-asphalt.

IOP Conf. Series: Materials Science and Engineering **371** (2018) 012058 doi:10.1088/1757-899X/371/1/012058

(2) Make the number-average molar mass of the polymer modifiers synthesized by controlled/living radical polymerization in an appropriate range (15000~100000). Meanwhile, choose styrene monomers, acrylate monomers and butadiene monomers that all contain active groups to polymerize polymeric substances with active groups.

### *2.2. Raw material*

*2.2.1. Bio-oil.* Bio-oil contains almost all kinds of Oxygen-Bearing Organic Matters like ethers, esters, aldehydes, ketones, organic acids, alcohols and so on. By organic elemental analysis, it is known that the components of the bio-oil are 74.42% of carbon, 13.39% of oxygen, 10.095% of hydrogen, 1.92% of nitrogen and 0.17% of sulphur.

*2.2.2. Basis bitumen*
In this research, 90# basis bitumen was selected. Table 1 shows its technical indexes.

**Table 1.** Performance parameters of the 90# basis bitumen

| Entry | Units | Criterion | Testing results |
|---|---|---|---|
| Penetration(25℃, 5s, 100g) | [0.1mm] | 60～80 | 62.7 |
| Ductility(5℃, 5cm/min) | [cm] | ≥30 | 43 |
| Softening point | [℃] | ≥42 | 54 |
| Kinematic viscosity(135℃) | [Pa•s] | ≤3.0 | 1.136 |
| Flash point | [℃] | ≥230 | 258 |
| Relative density | [—] | — | 1.015 |
| Quality change | [%] | ≤±0.8 | -0.523 |
| Residual penetration ratio(25℃) | [%] | ≥65 | 78 |
| Residual ductility(10℃,5cm/min) | [cm] | ≥15 | 20 |

*2.2.3. Synthesis of polymer modifiers.* The controlled/living radical polymerization was used to synthesize the polymer modifiers. Normally, there are two methods of the controlled/living radical polymerization, Reversible addition-fragmentation chain transfer (RAFT) polymerization and Degenerative transfer radical polymerization (DTRP) [4-5].
①Synthesis of homopolymer:
The typical RAFT homogeneous polymerization of vinyl monomers is as following: At first, put vinyl monomer, RAFT reagent, initiator (azo) and solvent (benzene) in dry round bottomed flask equipped with a magnetic bar. In this polymerization, the molar concentration ratio of the monomer and the RAFT reagent is 150 – 1000, and the molar concentration ratio of the RAFT reagent and the initiator is 3 – 5. Then remove oxygen in this hierarchy by placing the hierarchy in frozen brine and argon bubbling for 10 minutes. Afterwards, the flask was degassed by five freeze-thaw-pump cycles, kept under argon and placed in an oil bath thermostated at 60℃. At the designed time, add mineral ether in the flask to stop the reaction. Use the polymer obtained for follow-up tests. In the RAFT polymerization, the RAFT reagent has selective regulation on vinyl monomers. For example, ethyl 2-(ethoxycarbonyl)prop-2-yl dithiobenzoate (EPDTB) is an effective regulatory agent of styrene(St) and

chloroprene(CP) as well as the 4-cyano-4-(phenylcarbonothioylthio) pentanoic acid (CPDTB) is an effective regulatory agent of methyl methacrylate (MMA).
The typical DTRP homogeneous polymerization of vinyl monomers is similar with RAFT homogeneous polymerization. In DTRP, it is also needed to select appropriate DTRP to regulate a particular monomer. For example, methenyl iodide is an effective regulatory agent of St and St, and the cyanopropyl iodide (CP-I)is an effective regulatory agent of n-Butylacrylate (BA)and MMA.
The polymer modifiers with appropriate structure and formula weight could be synthesized in heterogeneous polymerization by the controlled/living radical polymerization. Chloroprene monomer is a kind of vinyl monomers with high activity. Take the controlled/living radical polymerization of chloroprene monomer in emulsion as an example to demonstrate how to conduct synthesis of polymer modifiers.
Oil phase: Put 34.2g disproportionated rosin, 2.97-5.85g of the RAFT reagent (DBTTC) or degenerative transfer agent ($CHI_3$) into 900g of chloroprene monomer and dissolve fully. Water phase: Put 6.57g of NaOH, 1.80g of lauryl sodium sulfate, 3.60g of Lomar PW, 1.80g of $NaCO_3$ into 990g of distilled water and dissolve fully.
Emulsion polymerization: At first, put the prepare water phase into the low temperature emulsion polymerization hierarchy and then stir it. Meanwhile, put the prepared water phase into the hierarchy and purge nitrogen to isolate the hierarchy from the air. Secondly, pre-emulsificate the hierarchy at 20℃ for 35 minutes and then cool it down to 10℃ and add aqueous solution of -$K_2S_2O_8$ as an oxidant. Next, use peristaltic pump to add prepared deoxidizer, aqueous solution of $Na_2SO_3$ and thiourea dioxide sodium salt. Keep the temperature of polymerization at 9℃. During polymerization, take a little polymer emulsion and use weighing method to monitor the conversion rate of the chloroprene monomer. After the polymerization finished, put the emulsion into a large amount of methanol to gel precipitated. After desiccation, use the product to conduct follow up tests like GPC、$^1$H NMR.
②Synthesis of block copolymer:
The idea of using RAFT to synthesize polymeric modifiers is shown in figure 1 as following. Firstly, use the method of synthesis of homopolymer mentioned above to synthesize macromolecule RAFT reagent. Then, add in monomer B, conduct polymerization and obtain block copolymer modifiers.

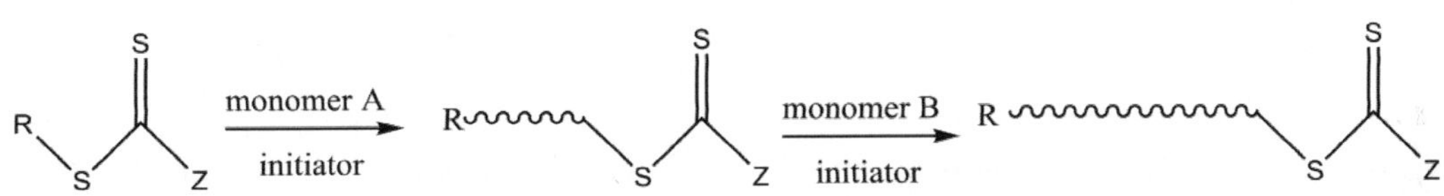


**Figure 1.** Add monomers in series to synthesize AB diblock copolymers.

Take PSt-*b*-PCP and PMMA-*b*-PCP as examples to demonstrate the detailed process of the synthesis.
Synthesis of PSt-*b*-PCP: According to synthesis of homopolymer, synthesize the RAFT reagent, Pst. Wash the methanol gel precipitation for three times and obtain PSt ($M_n$ = 2900 g·mol$^{-1}$, $M_w/M_n$ = 1.23), and then use PSt as the RAFT reagent to conduct the block polymerization of chloroprene. The prescription is PSt (2g), chloroprene monomer (20.21g, 138mmol), initiator AIBN (0.0283g, 0.17mmol) solvent benzene (30.3g). Use the product to conduct follow up tests.
Synthesis of PMMA-*b*-PCP: Similar as above, synthesize the RAFT reagent, PMMA ($M_n$ = 1800 g·mol$^{-1}$, $M_w/M_n$ = 1.33). Then use PMMA as the RAFT reagent to conduct the block polymerization of chloroprene. The prescription is PMMA (2g, 1.11mmol), chloroprene monomer (25g, 113mmol), initiator AIBN (0.0096g, 0.588mmol) and solvent benzene (30g). Use the product to conduct follow up tests.
The idea of using DTRP to synthesize polymeric modifiers is as following. Firstly, synthesize polymer DTRP agent. Then, add in monomer B, conduct polymerization and obtain block copolymers modifiers. This process is similar with the synthesis process of PSt-*b*-PCP and PMMA-*b*-PCP as mentioned above.
③Synthesis of bio-asphalt

IOP Conf. Series: Materials Science and Engineering **371** (2018) 012058 doi:10.1088/1757-899X/371/1/012058

First, add the polymer modifiers into bio-oil as designed dosage and keep the temperature at 165℃. Melt the basis bitumen and add in the modified bio-oil and use high speed shearing emulsifying machine shear the mixture at 3000r/min for 10 minutes, then the modified bio-asphalt is obtained.

④Three indicators of asphalt

Penetration is the viscosity index of asphalt, which reflects the softness of asphalt under certain conditions and temperatures. The smaller the penetration, the harder the asphalt. The softening point is the temperature when the asphalt meets the required viscosity. The higher the softening point, the higher the stiffness modulus of the same condition. The emulsification point can be used as the evaluation index of high temperature stability. Low temperature ductility is an important index for evaluating the cracking resistance of asphalt. When the temperature drops, due to the deformation ability of the asphalt chemical composition, the dislocations and sliding ability of the molecules, the decrease of asphalt ductility is very significant.

## 3. Testing results analysis

### *3.1. Synthesis of polymer modifiers*

It is obtained that the number-average molecular weight $M_n$ of the polymer modifiers produced by the controlled/living radical polymerization is 30900. Its polydispersity index of molecular weight $M_w/M_n$ is narrower(2.22) than the polymer produced by normal radical polymerization. This indicates that its microstructure is clear, see in figure 2.

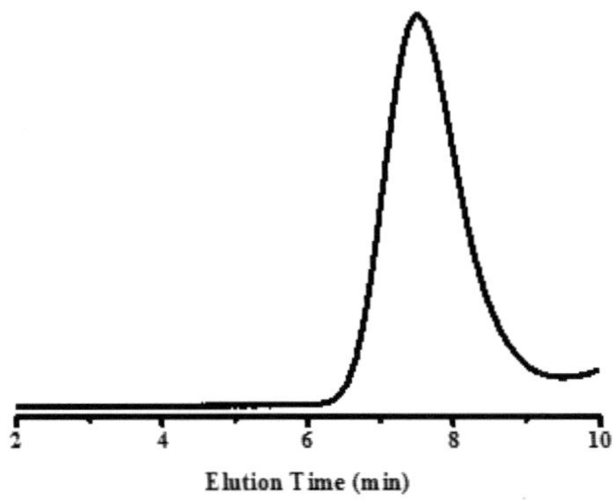


**Figure 2.** Typical GPC curve graph of polymer modifier synthesized by controlled/living radical polymerization ($M_n$ = 30900, $M_w/M_n$= 2.22).

Figures 3, 4 and 5 are nuclear magnetic hydrogen spectra of photopolymers synthesized by controlled/living radical polymerization. From these figures, it can be observed distinct characteristic peaks of the monomers and molecular weight regulators. This indicates that the reactions have been conducted according to the mechanism of controlled/living radical polymerization.

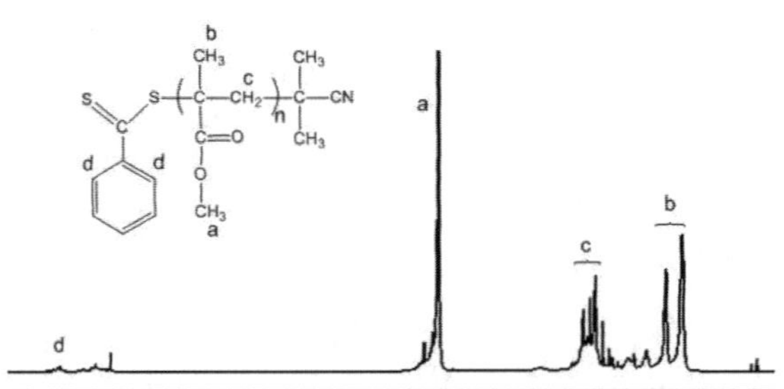


**Figure 3.** Nuclear magnetic hydrogen spectrum of PMMA synthesized under regulation with CPDB as RAFT reagent.

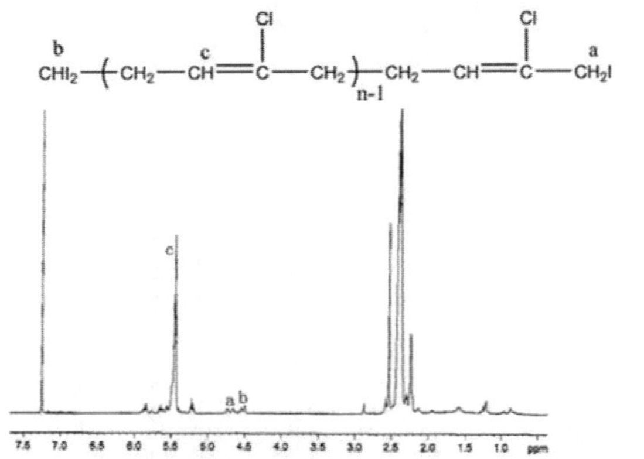

**Figure 4.** Nuclear magnetic hydrogen spectrum of PCP synthesized with methenyl iodide as DTRP agent.

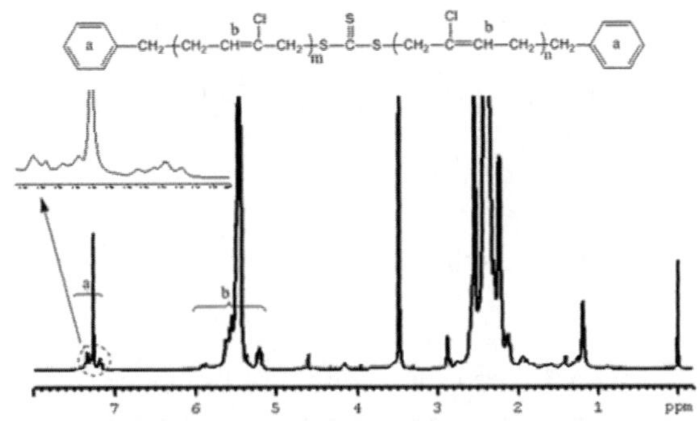

**Figure 5.** Nuclear magnetic hydrogen spectrum of PCP synthesized by chloroprene emulsion polymerization with DBTTC as RAFT reangent.

Conduct $^1$H-NMR on RAFT synthesized block copolymer. As shown in figure 6, the $^1$H NMR spectrum indicates that the characteristic peaks of chemical shift at 6.3-7.2pmm and 5.2-5.9pmm respectively belongs to polystyrene and the methylidyne protons of RAFT end group. This proves that the structure of PSt-*b*-PCP is diblock polymers. Considering the controlled radical polymerization, the synthesized block copolymer is PSt-*b*-PCP.

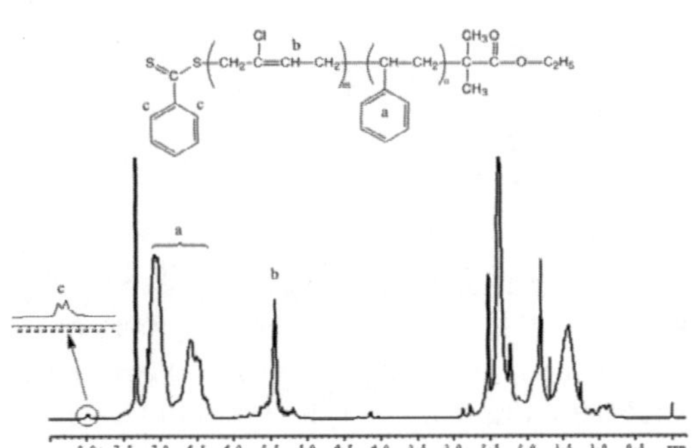

**Figure 6.** Nuclear magnetic hydrogen spectrum of PSt-*b*-PCP.

Conduct $^1$H-NMR on DTRP synthesized block copolymer. As shown in figure 7, the $^1$H NMR spectrum indicates that the a peak at 5.1-5.9 ppm chemical shift belongs to the characteristic peak of the methylidyne on main chain of polychloroprene. The b peak at 3.6 ppm is the characteristic peak of the methyl protons of PMMA. The c peak at 2.9 ppm is the characteristic peak of the β-end methylene

IOP Conf. Series: Materials Science and Engineering **371** (2018) 012058 doi:10.1088/1757-899X/371/1/012058

protons adjacent to iodine of chain end. This testing result further proved PCP-*b*-PMMA was successfully synthesized [7].

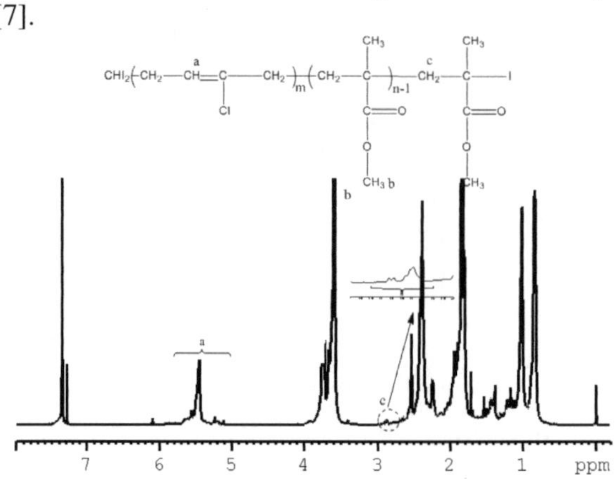

**Figure 7.** Nuclear magnetic hydrogen spectrum of PCP-*b*-PMMA.

Based on the analysis above, the designed polymer modifiers are successfully synthesized by controlled/living radical polymerization.

*3.2. The three indexes of bio-asphalt*

Conduct the tests of the three index of asphalt on the modified bio-asphalt, and the testing results is shown in table 2.

**Table 2.** The three indexes of bio-asphalts.

| Three indexes / Modifiers | Penetration [25℃,dmm] | Softening point [℃] | Ductility [10℃, cm] |
|---|---|---|---|
| PMMA(CPDTB) | 64 | 50 | 47.6 |
| PCP($CHI_3$) | 71 | 43 | 51 . 2 |
| PCP(DBTTC) | 68 | 44 | 50.5 |
| PMMA-*b*-PCP(CPDB) | 67 | 46 | 49.2 |
| PCP-*b*-PSt($CHI_3$) | 65 | 49 | 49 |

From table 2, it can be told that the three index of the modified bio-asphalt are excellent so the bio-asphalt has good high and low temperature performance. The reason why the modified bio-asphalt has good performance is as following. Firstly, the synthesized polymers have appropriate number-average molecular weight (15000-100000) which is similar with the basis bitumen. Secondly the monomers chosen to be synthesized are styrene monomers, acrylate monomers and butadienes monomers so that the synthesized modifiers have active groups that can react with the active groups of the bio-asphalt. This increases the stability of the bio-asphalt. Meanwhile, the modifiers and the basis bitumen have similar saturated main chains that makes they are compatible to each other. On the other hand, traditional modifiers of asphalt always have huge number-average molecular weight which is way higher than 100,000. In addition to this, their main chains are saturated alkanes. Therefore, the way they modifying the bio-asphalt is just physical blending but on chemical reaction that leads to poor compatibility of polymer modifiers and bio-asphalt.

From table 2, there are two laws for the penetration and the ductility of the modified bio-asphalts. For the penetrations, PCP($CHI_3$) > PCP(DBTTC) > PMMA-*b*-PCP(CPDB) > PCP-*b*-PSt($CHI_3$) > PMMA(CPDTB); for the ductilities, PMMA(CPDTB) > PCP-*b*-PSt($CHI_3$) > PMMA-*b*-PCP(CPDB) > PCP(DBTTC) > PCP($CHI_3$). The reason for this phenomenon is that PMMA has the highest rigidity of main chain; PSt takes second place and PCP is less than both of the formers.

## 4. Summary

(1) Use controlled/living radical polymerization method to polymerize the alkene monomers (styrene monomers, acrylate monomers and butadienes monomers) to synthesize modifiers for bio-asphalt. The testing results of the gel permeation chromatography indicate the number-average molecular weights of the modifiers are in an appropriate range (15,000-100,000). The nuclear magnetic hydrogen spectra indicate the modifiers are homopolymers or two block copolymers with clear structure. All these mean that it is successfully synthesized designed modifiers.

(2) Use the synthesized modifiers to modify the bio-asphalt. Testing results show that the modified bio-asphalt is stable and has good performances.

## 5. References

[1] J. A. Shen and F. P. Li 1997 *The Highway Research Institute of the Ministry of Communications* 11 12

[2] F. Ma, X. Ren, and Z Fu 2015 *Highway Engineering* 40 63

[3] M. He, D. W. Cao, H. Y. Zhang, X. W Wu, and T. G. 2015 Li *Journal of Highway & Transportation Research & Development* 32 8

[4] J. Chiefari, Y. K. B. Chong, F. Ercole, J. Krstina, J. Jeffery, T. P. T. Le, R. T. A. Mayadunne, G. F. Meijs, C. L. Moad and S. H. Thang 1998 *Macromolecules* 31 5559

[5] H. Sawada, T. Tashima, Y. Nishiyama, M. Kikuchi, Y. Goto, G. Kostov, and B. Ameduri, 2011 *Macromolecules* 44 1114

[6] M. Tanishima, A. Goto, L. Lei, A. Ohtsuki, H. Kaji, A. Nomura, Y. Tsujii, Y. Yamaguchi, H. Komatsu, and M. Miyamoto 2014 *Polymers* 6 311

[7] J. Hui, Y. Shi, T. Li, J. Wu, and Z. F. Fu 2015 *RSC Adv.* 5 44326

## Acknowledgments

The authors acknowledge the financial support from The central level, scientific research institutes for basic research and development special fund business (no. 20177503).

IOP Conf. Series: Materials Science and Engineering **371** (2018) 012059 doi:10.1088/1757-899X/371/1/012059

# An Efficient Method of Finding Stress Solutions in Porous Material under Axial Symmetry

**S Alexandrov[1,2], E Lyamina[3] and T Nguyen-Thoi[4,5]**
1 Visiting Professor, Division of Computational Mathematics and Engineering, Institute for Computational Science, Ton Duc Thang University, Ho Chi Minh City, Vietnam
2 Visiting Professor, Faculty of Civil Engineering, Ton Duc Thang University, Ho Chi Minh City, Vietnam
3 Senior Researcher, Institute for problems in mechanics RAS, Moscow, Russia
4 Professor, Division of Computational Mathematics and Engineering, Institute for Computational Science, Ton Duc Thang University, Ho Chi Minh City, Vietnam
5 Professor, Faculty of Civil Engineering, Ton Duc Thang University, Ho Chi Minh City, Vietnam
E-mail: sergeialexandrov@tdt.edu.vn

**Abstract.** In the mechanics of porous and powder materials the system of equations comprising the pyramid yield criterion together with the stress equilibrium equations under conditions of axial symmetry forms a statically determinate system at edge regimes. The results presented here for this system are consequently independent of any flow rule that may be chosen to calculate the deformation and also independent of whether elastic strains are included. The stress equilibrium equations are written relative to a coordinate system in which the coordinate curves coincide with the trajectories of the principal stress directions. Then, a method of finding the trajectories of the principal stresses is developed.

## 1. Introduction

In the case of several yield criteria widely used in applications, the system of equations comprising the yield criterion together with the stress equilibrium equations forms a statically determinate system under certain conditions. This system can be studied without using any flow rule that may be chosen to calculate the deformation and independently of whether elastic strains are included. In particular, much work has been carried out for plane strain deformation. The geometric properties of the principal line coordinate system (i.e. the coordinate curves of this coordinate system coincide with trajectories of the principal stress directions) for the Tresca yield criterion have been derived in [1]. This result has been extended to the Mohr-Coulomb yield criterion in [2]. Several methods are available for calculating the field of stress for the Tresca yield criterion. In particular, the R – S method has been developed in [3]. Another widely used method is based on Mikhlin's variables [4, 5]. The R – S method has been extended to the Mohr-Coulomb yield criterion in [6]. The pyramid yield criterion for porous and powder materials has been proposed in [7]. The R – S method and the method based on Mikhlin's variables have been developed for this yield criterion in [8] and [9], respectively. In the present paper, a new method for determining the field of stress under axial symmetry is proposed for the pyramid yield criterion. The development of this method requires the generalization of the results reported in [1, 2] on axisymmetric deformation of materials obeying the pyramid yield criterion.

## 2. Material Model

The piece-wise linear yield criterion in terms of the principal stresses $\sigma_1$, $\sigma_2$ and $\sigma_3$ proposed in [7] reads

$$\frac{|\sigma_i - \sigma_j|}{2\tau_s} + \frac{|\sigma|}{p_s} = 1 \tag{1}$$

where $i, j = 1, 2, 3$, $i \neq j$ and

$$\sigma = \frac{\sigma_1 + \sigma_2 + \sigma_3}{3} \tag{2}$$

is the hydrostatic stress. Also, $\tau_s$ is the shear yield stress and $p_s$ is the yield stress in hydrostatic compression. Both $\tau_s$ and $p_s$ depend of on the porosity, $\upsilon$. In particular, $p_s \to \infty$ as $\upsilon \to 0$. In this case, the yield criterion (1) approaches the Tresca yield criterion. In what follows, it is assumed that the porosity is uniformly distributed. Therefore, $\tau_s$ and $p_s$ are independent of space coordinates. In the case of axisymmetric deformation the circumferential stress $\sigma_\theta$ is one of the principal stresses. Without loss of generality, it is possible to assume that

$$\sigma_\theta \equiv \sigma_3 \quad \text{and} \quad \sigma_1 > \sigma_2 \tag{3}$$

In general, under conditions of axial symmetry it is necessary to consider several edge and face regimes of the yield criterion (1). For practical applications, the edge regimes at which $\sigma_1 \geq \sigma_\theta \geq \sigma_2$ are most important. There are four regimes of this kind. Using (1) these regimes can be represented as

$$\frac{\sigma_1 - \sigma_2}{2\tau_s} - \frac{\sigma}{p_s} = 1, \quad \frac{\sigma_1 - \sigma_\theta}{2\tau_s} - \frac{\sigma}{p_s} = 1 \tag{4}$$

in the case of regime 1,

$$\frac{\sigma_1 - \sigma_2}{2\tau_s} - \frac{\sigma}{p_s} = 1, \quad \frac{\sigma_\theta - \sigma_2}{2\tau_s} - \frac{\sigma}{p_s} = 1 \tag{5}$$

in the case of regime 2,

$$\frac{\sigma_1 - \sigma_2}{2\tau_s} + \frac{\sigma}{p_s} = 1, \quad \frac{\sigma_1 - \sigma_\theta}{2\tau_s} + \frac{\sigma}{p_s} = 1 \tag{6}$$

in the case of regime 3, and

$$\frac{\sigma_1 - \sigma_2}{2\tau_s} + \frac{\sigma}{p_s} = 1, \quad \frac{\sigma_\theta - \sigma_2}{2\tau_s} + \frac{\sigma}{p_s} = 1 \tag{7}$$

in the case of regime 4. In the case of regimes 1 and 2 the hydrostatic stress can vary in the range $0 \geq \sigma \geq -p_s$ and in the case of regimes 3 and 4 in the range $0 \leq \sigma \leq p_s$. In the case of regimes 1 and 3

$$\sigma_2 = \sigma_\theta \tag{8}$$

IOP Conf. Series: Materials Science and Engineering **371** (2018) 012059 doi:10.1088/1757-899X/371/1/012059

and in the case of regimes 2 and 4

$$\sigma_1 = \sigma_\theta \tag{9}$$

The velocity equations follow from the associated flow rule. These equations are given in [7]. However, they are not involved in the subsequent analysis.

## 3. A Geometric Property of Principal Stress Trajectories

The yield criterion formulated in the previous section should be complemented with the equilibrium equations. It is convenient to introduce two coordinate systems, a cylindrical coordinate system $(r, \theta, z)$ whose $z$- axis coincides with the axis of symmetry of the boundary value problem such that its solution is independent of $\theta$ and a principal line coordinate system $(\xi, \theta, \eta)$. The coordinate curves of the principal line coordinate system coincide with the trajectories of the principal stresses. Without loss of generality it is assumed that the $\xi-$curves coincide with the trajectories of the stress $\sigma_1$ and the $\eta-$curves with the trajectories of the stress $\sigma_2$. Since the shear stresses vanish in the principal line coordinate system, the equilibrium equations referred to this system take the form [10]

$$\frac{\partial(rh_\eta\sigma_1)}{\partial\xi} - \sigma_\theta h_\eta \frac{\partial r}{\partial\xi} - \sigma_2 r\frac{\partial h_\eta}{\partial\xi} = 0, \quad \frac{\partial(rh_\xi\sigma_2)}{\partial\eta} - \sigma_1 r\frac{\partial h_\xi}{\partial\eta} - \sigma_\theta h_\xi \frac{\partial r}{\partial\eta} = 0. \tag{10}$$

In the case of regimes 1 and 3 equations (8) and (10) combine to give

$$(\sigma_1 - \sigma_2)\left(h_\eta \frac{\partial r}{\partial\xi} + r\frac{\partial h_\eta}{\partial\xi}\right) + rh_\eta \frac{\partial\sigma_1}{\partial\xi} = 0, \quad -(\sigma_1 - \sigma_2)\frac{\partial h_\xi}{\partial\eta} + h_\xi \frac{\partial\sigma_2}{\partial\eta} = 0. \tag{11}$$

In the case of regimes 2 and 4 equations (9) and (10) combine to give

$$-(\sigma_1 - \sigma_2)\left(h_\xi \frac{\partial r}{\partial\eta} + r\frac{\partial h_\xi}{\partial\eta}\right) + rh_\xi \frac{\partial\sigma_2}{\partial\eta} = 0, \quad (\sigma_1 - \sigma_2)\frac{\partial h_\eta}{\partial\xi} + h_\eta \frac{\partial\sigma_1}{\partial\xi} = 0. \tag{12}$$

Eliminating the difference $\sigma_1 - \sigma_2$ in (11) by means of (4) and (6) yields

$$2\tau_s\left(1 \pm \frac{\sigma}{p_s}\right)\left(h_\eta \frac{\partial r}{\partial\xi} + r\frac{\partial h_\eta}{\partial\xi}\right) + rh_\eta \frac{\partial\sigma_1}{\partial\xi} = 0, \quad -2\tau_s\left(1 \pm \frac{\sigma}{p_s}\right)\frac{\partial h_\xi}{\partial\eta} + h_\xi \frac{\partial\sigma_2}{\partial\eta} = 0 \tag{13}$$

where the upper sign corresponds to regime 1 and the lower sign to regime 3. Eliminating the difference $\sigma_1 - \sigma_2$ in (12) by means of (5) and (7) yields

$$-2\tau_s\left(1 \pm \frac{\sigma}{p_s}\right)\left(h_\xi \frac{\partial r}{\partial\eta} + r\frac{\partial h_\xi}{\partial\eta}\right) + rh_\xi \frac{\partial\sigma_2}{\partial\eta} = 0, \quad 2\tau_s\left(1 \pm \frac{\sigma}{p_s}\right)\frac{\partial h_\eta}{\partial\xi} + h_\eta \frac{\partial\sigma_1}{\partial\xi} = 0. \tag{14}$$

where the upper sign corresponds to regime 2 and the lower sign to regime 4. It follows from (2), (8) and (9) that $3\sigma = \sigma_1 + 2\sigma_2$ in the case of regimes 1 and 3 and $3\sigma = 2\sigma_1 + \sigma_2$ in the case of regimes 2 and 4. Using these equations together with equations (4) to (7) it is possible to express $\sigma_1$ and $\sigma_2$ in terms of $\sigma$. As a result,

$$\sigma_1 = (1+2k_s)\sigma + 4\tau_s/3, \quad \sigma_2 = (1-k_s)\sigma - 2\tau_s/3 \tag{15}$$

in the case of regime 1,

$$\sigma_1 = (1+k_s)\sigma + 2\tau_s/3, \quad \sigma_2 = (1-2k_s)\sigma - 4\tau_s/3 \tag{16}$$

in the case of regime 2,

$$\sigma_1 = (1-2k_s)\sigma + 4\tau_s/3, \quad \sigma_2 = (1+k_s)\sigma - 2\tau_s/3 \tag{17}$$

in the case of regime 3, and

$$\sigma_1 = (1-k_s)\sigma + 2\tau_s/3, \quad \sigma_2 = (1+2k_s)\sigma - 4\tau_s/3 \tag{18}$$

in the case of regime 4. In equations (15) to (18), $k_s = 2\tau_s/(3p_s)$.
In the case of regime 1, eliminating $\sigma_1$ and $\sigma_2$ in (13) by means of (15) yields

$$\frac{\partial \ln(rh_\eta)}{\partial \xi} + \frac{(1+2k_s)}{2\tau_s(1+\sigma/p_s)}\frac{\partial \sigma}{\partial \xi} = 0, \quad \frac{\partial \ln h_\xi}{\partial \eta} - \frac{(1-k_s)}{2\tau_s(1+\sigma/p_s)}\frac{\partial \sigma}{\partial \eta} = 0 \tag{19}$$

Each of these equations can be immediately integrated to give

$$\frac{3k_s}{(1+2k_s)}\ln(rh_\eta) + \ln\left(1+\frac{\sigma}{p_s}\right) = \ln[C_1(\eta)], \quad \frac{3k_s}{(1-k_s)}\ln h_\xi - \ln\left(1+\frac{\sigma}{p_s}\right) = \ln[C_2(\xi)]. \tag{20}$$

Here $C_1(\eta)$ is an arbitrary function of only $\eta$ and $C_2(\xi)$ is an arbitrary function of only $\xi$. Eliminating $\sigma$ between the equations in (20) yields

$$\frac{3k_s}{(1+2k_s)}\ln(rh_\eta) + \frac{3k_s}{(1-k_s)}\ln h_\xi = \ln[C_1(\eta)C_2(\xi)] \tag{21}$$

It is seen from this equation that different choices of the functions $C_1(\eta)$ and $C_2(\xi)$ merely change the scale of the $\eta-$ and $\xi-$ curves, respectively. Therefore, it is always possible to choose $C_1(\eta) = C_2(\xi) = 1$. Then, equation (21) becomes

$$rh_\eta h_\xi^{t_1} = 1, \quad t_1 = \frac{1+2k_s}{1-k_s}. \tag{22}$$

In this case equation (20) supplies

$$\frac{\sigma}{p_s} = h_\xi^{t_0} - 1, \quad t_0 = \frac{3k_s}{1-k_s}. \tag{23}$$

Using equations (13) to (18) regimes 2, 3 and 4 can be treated in a similar manner.

IOP Conf. Series: Materials Science and Engineering **371** (2018) 012059 doi:10.1088/1757-899X/371/1/012059

## 4. Mapping between the Cylindrical and Principal Line Coordinate Systems

The cylindrical and principal line coordinate systems are shown in Fig. 1 where $\psi$ is the orientation of the $\xi-$curves relative to the $r-$axis.

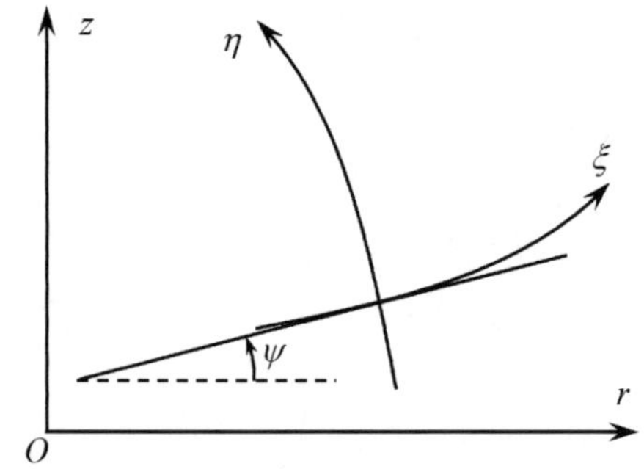


**Figure 1.** Cylindrical ($r$, $z$) and principal line ($\xi$, $\eta$) coordinate systems.

It follows from the geometry of this figure that

$$\frac{\partial r}{\partial \xi} = h_\xi \cos\psi, \quad \frac{\partial r}{\partial \eta} = -h_\eta \sin\psi, \quad \frac{\partial z}{\partial \xi} = h_\xi \sin\psi, \quad \frac{\partial z}{\partial \eta} = h_\eta \cos\psi. \tag{24}$$

If regime1 is operative then this equation and (22) result in

$$\frac{\partial r}{\partial \xi} = h\cos\psi, \quad \frac{\partial r}{\partial \eta} = -\frac{\sin\psi}{rh^{t_1}}, \quad \frac{\partial z}{\partial \xi} = h\sin\psi, \quad \frac{\partial z}{\partial \eta} = \frac{\cos\psi}{rh^{t_1}} \tag{25}$$

where $h \equiv h_\xi$. The compatibility equations are

$$\frac{\partial^2 r}{\partial \xi \partial \eta} = \frac{\partial^2 r}{\partial \eta \partial \xi}, \quad \frac{\partial^2 z}{\partial \xi \partial \eta} = \frac{\partial^2 z}{\partial \eta \partial \xi}. \tag{26}$$

Substituting (25) into (26) gives

$$\cos\psi \frac{\partial h}{\partial \eta} - h\sin\psi \frac{\partial \psi}{\partial \eta} + \frac{\cos\psi}{rh^{t_1}} \frac{\partial \psi}{\partial \xi} - \frac{t_1 \sin\psi}{rh^{t_1+1}} \frac{\partial h}{\partial \xi} = \frac{\sin\psi \cos\psi}{r^2 h^{t_1-1}},$$
$$\sin\psi \frac{\partial h}{\partial \eta} + h\cos\psi \frac{\partial \psi}{\partial \eta} + \frac{\sin\psi}{rh^{t_1}} \frac{\partial \psi}{\partial \xi} + \frac{t_1 \cos\psi}{rh^{t_1+1}} \frac{\partial h}{\partial \xi} = -\frac{\cos^2\psi}{r^2 h^{t_1-1}}. \tag{27}$$

Multiplying the first of (27) by $\sin\psi$ and subtracting from the result the second of (27) multiplied by $\cos\psi$ yields

$$\frac{\partial \psi}{\partial \eta} + \frac{t_1}{rh^{t_1+2}} \frac{\partial h}{\partial \xi} = -\frac{\cos\psi}{r^2 h^{t_1}}. \tag{28}$$

In a similar manner, if the first and second of (27) are multiplied by $\cos\psi$ and $\sin\psi$ respectively, and the sum of the resulting equations is taken, it is found that

$$rh^{t_1}\frac{\partial h}{\partial \eta}+\frac{\partial \psi}{\partial \xi}=0. \tag{29}$$

Using a standard procedure it is possible to show that (28) and (29) is a hyperbolic system of equations for $h$ and $\psi$ with characteristic curves (termed $\alpha-$ and $\beta-$ lines) given by

$$\frac{d\xi}{d\eta}=\pm\frac{\sqrt{t_1}}{rh^{t_1+1}}. \tag{30}$$

where the lower and upper signs refer to the $\alpha-$lines and $\beta-$lines respectively. The characteristic relations are

$$\pm d\psi+\frac{\sqrt{t_1}}{h}dh=-\frac{h\cos\psi}{r\sqrt{t_1}}d\xi. \tag{31}$$

As before, the lower and upper signs correspond to the $\alpha-$lines and $\beta-$lines respectively. Equations (30) and (31) should be solved together with equation (25) for $r$. This system of equations is similar to that that describes axisymmetric flow of Tresca's solids [11]. Numerical techniques for solving the latter are well documented [11]. These techniques can be adopted for solving the system of equations derived. Once this system has been solved, equation (23) supplies the distribution of $\sigma$ and equation (15) the distributions of $\sigma_1$ and $\sigma_2$. Since the angle $\psi$ has been found, the distribution of the stress components in the cylindrical coordinate system follows from the standard transformation equations for stress components.

Regimes 2, 3 and 4 can be treated in a similar manner.

## 5. Conclusions

On the assumption of axial symmetry conditions the system of equations comprising the pyramid yield criterion and the equilibrium equations has been studied in a curvilinear orthogonal coordinate system in which the coordinate curves coincide with trajectories of the principal stress directions. It has been shown that the scale factors of the coordinate curves should satisfy equation (22) in the case of regime 1 or similar equations in the case of the other edge regimes of the pyramid yield criterion. Using this equation a method of finding stress solutions has been developed. In particular, one of the scale factors of the principal line coordinate system and the orientation of the trajectory of the major principal stress should satisfy equations (28) and (29). These equations form a hyperbolic system with the characteristic curves given by (30) and the characteristic relations given by (31). Numerical techniques for solving this type of equations are available in the literature [11].

## 6. References

[1] Sadowsky M A 1941 *ASME J. Appl. Mech.* 63 A74
[2] Alexandrov S and Harris D 2017 *ZAMM* 97 473
[3] Hill R, Lee E H and Tupper S J 1951 *ASME J. Appl. Mech.* 18 46
[4] Hill R 1950 *The Mathematical Theory of Plasticity* (Oxford: Clarendon Press)
[5] Chakrabarty J 1987 *Theory of Plasticity* (Singapore: McGraw-Hill)
[6] Alexandrov S 2015 *ZAMM* 95 1296
[7] Druyanov B 1993 *Technological Mechanics of Porous Bodies* (New-York: Clarendon Press)
[8] Aleksandrov S E and Lyamina E A 2015 *Mech. Solids* 50 171
[9] Alexandrov S 2017 *Proc. CIRP* 63 511

IOP Conf. Series: Materials Science and Engineering **371** (2018) 012059 doi:10.1088/1757-899X/371/1/012059

[10] Malvern L E 1969 *Introduction to the Mechanics of a Continuous Medium* (Englewood Cliffs: Prentice-Hall)
[11] Druyanov B. A and Nepershin R I 1994 *Problems of Technological Plasticity*, (Amsterdam: Elsevier)

**Acknowledgment**
This research was made possible by a grant from the Russian Foundation for Basic Research (Project 17-58-540007).

IOP Conf. Series: Materials Science and Engineering **371** (2018) 012060 doi:10.1088/1757-899X/371/1/012060

# A Semi-analytic Stress Solution for Elastic/Plastic FGM Discs Subject to External Pressure

**S Alexandrov[1,2], S Usov[3] and T Nguyen-Thoi[4,5]**

[1] Visiting Professor, Division of Computational Mathematics and Engineering, Institute for Computational Science, Ton Duc Thang University, Ho Chi Minh City, Vietnam
[2] Visiting Professor, Faculty of Civil Engineering, Ton Duc Thang University, Ho Chi Minh City, Vietnam
[3] Professor, Moscow Aviation University, Moscow, Russia
[4] Professor, Division of Computational Mathematics and Engineering, Institute for Computational Science, Ton Duc Thang University, Ho Chi Minh City, Vietnam
[5] Professor, Faculty of Civil Engineering, Ton Duc Thang University, Ho Chi Minh City, Vietnam
E-mail: sergeialexandrov@tdt.edu.vn

**Abstract.** A new semi-analytic plane stress solution for the elastic/plastic distribution of stress in a thin annular disc subject to pressure over its outer radius is presented. The yield stress varies along the radius of the disc and is a monotonically decreasing function of the radius. It is shown that the general solution consists of several stages and the general structure of the solution depends on material and geometric parameters. An example illustrating the general solution is provided.

## 1. Introduction

The expansion/contraction of a disc is one of the classical problems of plasticity. Therefore, there is a vast amount of literature on this topic. However, most studies for elastic/plastic models for stationary and rotating discs have only focused on deformation theories of plasticity, unless Tresca's yield criterion is adopted [1-6]. An overview of solutions for the flow theory of plasticity based on several smooth yield criteria has been provided in [7]. All these solutions are for discs of uniform properties. Purely elastic solutions for functionally graded discs are given, for example, in [8, 9]. Modelling of creep in functionally graded discs is presented in [10]. The objective of the present paper is to extend the method for finding stress and strain distributions in thin discs developed in [7] to functionally graded discs subject to external pressure. The solution for a chosen distribution of the yield stress along the radius is semi-analytic. A numerical technique is only necessary to solve transcendental equations.

## 2. Statement of the problem

It is assumed that a thin hollow disc of Poisson's ratio $\nu$, Young's modulus $E$, outer radius $b_0$, and inner radius $a_0$ is subject to uniform pressure $p_0$ over its outer radius. It is assumed that both Poisson's ratio and Young's modulus are constant. The disc has no stress at the initial instant. Strains are supposed to be infinitesimal. It is convenient to choose a cylindrical coordinate system $(r, \theta, z)$ in which the solution is independent of $\theta$. The stress components referred to this coordinate system are denoted as $\sigma_r$, $\sigma_\theta$ and $\sigma_z$. These stresses are the principal stresses. Moreover, $\sigma_z = 0$. The circumferential displacement vanishes everywhere. The stress boundary conditions are

$$\sigma_r = 0 \tag{1}$$

for $r = a_0$ and

$$\sigma_r = -p_0 \tag{2}$$

for $r = b_0$. It is assumed that the magnitude of $p_0$ is large enough to initiate plastic yielding. Therefore, there are elastic and plastic regions in the disc. The constitutive equations in the elastic region are

$$\varepsilon_r^e = \frac{\sigma_r - \nu\sigma_\theta}{E}, \quad \varepsilon_\theta^e = \frac{\sigma_\theta - \nu\sigma_r}{E}, \quad \varepsilon_z^e = -\frac{\nu(\sigma_r + \sigma_\theta)}{E}. \tag{3}$$

The superscript *e* denotes the elastic part of the strain. In the elastic region the whole strain is elastic. Equation (3) is also valid in plastic regions. However, this equation is not required in such regions for finding stress solutions. The von Mises plane stress yield criterion for functionally graded materials reads

$$\sigma_r^2 + \sigma_\theta^2 - \sigma_\theta\sigma_r = \sigma_0^2\Phi^2(r). \tag{4}$$

Here $\sigma_0$ is a reference stress and $\Phi(r)$ is some given function of *r*. In the present paper, it is assumed that

$$d\Phi(r)/dr \le 0. \tag{5}$$

The flow rule associated with the von Mises yield criterion is

$$\dot{\varepsilon}_r^p = \lambda(\sigma_r - \sigma), \quad \dot{\varepsilon}_\theta^p = \lambda(\sigma_\theta - \sigma), \quad \dot{\varepsilon}_z^p = \lambda(\sigma_z - \sigma) \tag{6}$$

where $\dot{\varepsilon}_r^p$, $\dot{\varepsilon}_\theta^p$ and $\dot{\varepsilon}_z^p$ are the plastic strain rates, $\lambda$ is a non-negative multiplier and $\sigma = (\sigma_r + \sigma_\theta + \sigma_z)/3$. Since $\sigma_z = 0$ in the case under consideration, equation (6) becomes

$$\dot{\varepsilon}_r^p = \frac{\lambda}{3}(2\sigma_r - \sigma_\theta), \quad \dot{\varepsilon}_\theta^p = \frac{\lambda}{3}(2\sigma_\theta - \sigma_r), \quad \dot{\varepsilon}_z^p = -\frac{\lambda}{3}(\sigma_r + \sigma_\theta). \tag{7}$$

Since the boundary value problem is statically determinate, the flow rule is not required for the stress solution. The only nontrivial equilibrium equation reduces to

$$\frac{\partial\sigma_r}{\partial r} + \frac{\sigma_r - \sigma_\theta}{r} = 0. \tag{8}$$

It is convenient to introduce the following dimensionless quantities

$$a = \frac{a_0}{b_0}, \quad \rho = \frac{r}{b_0}, \quad p = \frac{p_0}{\sigma_0}. \tag{9}$$

## 3. Purely elastic solution

The general elastic solution is well known (see, for example, [11]). In our nomenclature, this solution can be written as

$$\frac{\sigma_r}{\sigma_0} = \frac{A}{\rho^2} + B, \quad \frac{\sigma_\theta}{\sigma_0} = -\frac{A}{\rho^2} + B. \tag{10}$$

Here $A$ and $B$ are constants of integration. These constants are found from the boundary conditions (1) and (2). In particular, using (9) it is possible to get $A = A_e = pa^2(1-a^2)^{-1}$ and $B = B_e = -p(1-a^2)^{-1}$. Let $p_e$ be the value of $p$ at which the initiation of plastic yielding occurs. Replacing $A$ and $B$ in (10) with $A_e$ and $B_e$, respectively, and substituting the resulting expressions into (4) shows that if the initiation of plastic yielding occurs at $\rho = a$ then the corresponding value of $p$ is $p = p_{e1} = \Phi(a_0)(1-a^2)/2$. Analogously, if the initiation of plastic yielding occurs at $\rho = 1$ then the corresponding value of $p$ is $p = p_{e2} = \Phi(b_0)(1-a^2)/\sqrt{3a^4+1}$. Thus $p_e = \min\{p_{e1}, p_{e2}\}$. It follows from this equation that there are three different general structures of the solution. In particular, (*i*) plastic yielding initiates at $\rho = a$ if $\Phi(a_0)\sqrt{3a^4+1} < 2\Phi(b_0)$, (*ii*) at $\rho = 1$ if $\Phi(a_0)\sqrt{3a^4+1} > 2\Phi(b_0)$, and (*iii*) at $\rho = a$ and $\rho = 1$ simultaneously if $\Phi(a_0)\sqrt{3a^4+1} = 2\Phi(b_0)$. In what follows, it is assumed that

$$\Phi(a_0)\sqrt{3a^4+1} < 2\Phi(b_0). \tag{11}$$

## 4. Elastic/plastic solution with one plastic region

The following solution is for the function $\Phi(\rho)$ in the form

$$\Phi(r) = (r/b_0)^n = \rho^n. \tag{12}$$

The condition (5) requires that $n \le 0$ and the condition (11) that $a^n\sqrt{3a^4+1} < 2$. The yield criterion (4) is valid in the plastic region. This criterion is satisfied by the following substitution:

$$\frac{\sigma_r}{\sigma_0} = -\frac{2\rho^n \sin\psi}{\sqrt{3}}, \quad \frac{\sigma_\theta}{\sigma_0} = -\rho^n\left(\frac{\sin\psi}{\sqrt{3}} + \cos\psi\right). \tag{13}$$

Here equation (12) has been taken into account. Substituting (13) into (8) yields

$$\frac{2\cos\psi d\psi}{\left[\sqrt{3}\cos\psi - (1+2n)\sin\psi\right]} = \frac{d\rho}{\rho}. \tag{14}$$

It follows from (1), (9) and (13) that $\psi = 0$ for $\rho = a$. The solution of equation (14) satisfying this boundary condition is

$$\ln\frac{\rho}{a} = \frac{\sqrt{3}\psi}{2(1+n+n^2)} - \frac{(1+2n)}{2(1+n+n^2)}\ln\left[\cos\psi - \frac{(1+2n)}{\sqrt{3}}\sin\psi\right]. \tag{15}$$

The distribution of the stresses in the plastic region is determined from (13) and (15) in parametric form. The distribution of the stresses in the elastic region is given by (10) but $A \ne A_e$ and $B \ne B_e$. This distribution must satisfy the boundary condition (2). Then, using (9) and (10) it is possible to find that

$$A + B = -p. \tag{16}$$

and

$$\frac{\sigma_r}{\sigma_0} = -\frac{p}{\rho^2} + B\left(1 - \frac{1}{\rho^2}\right), \quad \frac{\sigma_\theta}{\sigma_0} = \frac{p}{\rho^2} + B\left(1 + \frac{1}{\rho^2}\right) \tag{17}$$

in the elastic region. Let $\rho_c$ be the dimensionless radius of the elastic/plastic boundary and $\psi_c$ be the corresponding value of $\psi$. It follows from (15) that

$$\ln\frac{\rho_c}{a}=\frac{\sqrt{3}\psi_c}{2\left(1+n+n^2\right)}-\frac{(1+2n)}{2\left(1+n+n^2\right)}\ln\left[\cos\psi_c-\frac{(1+2n)}{\sqrt{3}}\sin\psi_c\right]. \tag{18}$$

Moreover, both the radial and circumferential stresses should be continuous across the elastic/plastic boundary. Then, it follows from (13) and (17) that

$$\frac{2\rho_c^n\sin\psi_c}{\sqrt{3}}=\frac{p}{\rho_c^2}-B\left(1-\frac{1}{\rho_c^2}\right),\qquad -\rho_c^n\left(\frac{\sin\psi_c}{\sqrt{3}}+\cos\psi_c\right)=\frac{p}{\rho_c^2}+B\left(1+\frac{1}{\rho_c^2}\right). \tag{19}$$

Solving these equations for $p$ and $B$ yields

$$2\sqrt{3}p=\rho_c^n\left[2\left(1+\rho_c^2\right)\sin\psi_c+\left(1-\rho_c^2\right)\left(\sin\psi_c+\sqrt{3}\cos\psi_c\right)\right],\quad B=-\frac{\rho_c^n}{2}\left(\sqrt{3}\sin\psi_c+\cos\psi_c\right). \tag{20}$$

Eliminating here $\rho_c$ by means of (18) supplies the dependencies of $p$ and $B$ on $\psi_c$. Using (16) it is now possible to find the dependence of $A$ on $\psi_c$. Then, the variation of the stresses with $\rho$ at a given value of $\psi_c$ in the elastic region is determined from (10). The solution with one plastic region is valid if the yield criterion is not violated in the elastic region. Using (4), (10) and (12) this condition can be written as

$$F(\rho)\equiv\frac{3A^2}{\rho^4}+B^2-\rho^{2n}\le 0 \tag{21}$$

in the range $\rho_c\le\rho\le 1$. The only critical number of the function $F$ in the range $0<\rho<\infty$ is

$$\rho_0=\left[-\frac{6A^2}{n}\right]^{n_1},\quad n_1=\frac{1}{2(n+2)}. \tag{22}$$

The second derivative of the function $F$ with respect to ρ is

$$\frac{d^2F}{d\rho^2}=\frac{2}{\rho^6}\left[10A^2-n(2n-1)\rho^{2(n+2)}\right]. \tag{23}$$

Since $n\le 0$, it is evident from (23) that $d^2F/d\rho^2>0$ at $\rho=\rho_0$ and that the function $F$ attains a minimum at this number. Therefore, another plastic region may start to develop from the surface $\rho=1$ if $\rho_c<\rho_0<1$. The corresponding value of $\psi_c$ is denoted as $\psi_{c1}$ and is determined from the equation $F(1)=0$. Using (21) this equation can be rewritten as

$$3A^2+B^2-1=0. \tag{24}$$

This equation should be solved numerically. The value of $p$ at which the second plastic region starts to develop is determined from (20) where $\psi_c$ should be replaced with $\psi_{c1}$. This value of $p$ is denoted as $p_1$. There are two plastic regions and one elastic region at $\psi_c>\psi_{c1}$ (or $p>p_1$).

## 5. Elastic/Plastic Solution with Two Plastic Regions

Let $\rho_s$ be the dimensionless radius of the second elastic/plastic boundary and $\psi_s$ be the corresponding value of $\psi$. In the elastic region, $\rho_c\le\rho\le\rho_s$, the solution (10) is valid. The solution (13) and (15) is valid in the plastic region $a\le\rho\le\rho_c$. Equations (13) and (14) are valid in the plastic region $\rho_s\le\rho\le 1$. It follows from (2), (9) and (13) that the boundary condition to equation (14) is

$$\psi=\psi_f=\arcsin\left(\sqrt{3}p/2\right) \tag{25}$$

for $\rho = 1$. The solution of equation (14) satisfying the boundary condition (25) is

$$\ln\rho = \frac{\sqrt{3}(\psi - \psi_f)}{2(1+n+n^2)} - \frac{(1+2n)}{2(1+n+n^2)}\ln\left[\frac{\cos\psi - (1+2n)\sin\psi/\sqrt{3}}{\sqrt{1-3p^2/4} - (1+2n)p/2}\right]. \tag{26}$$

It follows from this equation that

$$\ln\rho_s = \frac{\sqrt{3}(\psi_s - \psi_f)}{2(1+n+n^2)} - \frac{(1+2n)}{2(1+n+n^2)}\ln\left[\frac{\cos\psi_s - (1+2n)\sin\psi_s/\sqrt{3}}{\sqrt{1-3p^2/4} - (1+2n)p/2}\right]. \tag{27}$$

Both the radial and circumferential stresses should be continuous across the surfaces $\rho = \rho_c$ and $\rho = \rho_s$. Using these conditions together with equations (10) and (13) gives

$$\begin{aligned} &\frac{A}{\rho_c^2} + B = -\frac{2\rho_c^n \sin\psi_c}{\sqrt{3}}, \quad \frac{A}{\rho_c^2} - B = \rho_c^n\left(\frac{\sin\psi_c}{\sqrt{3}} + \cos\psi_c\right), \\ &\frac{A}{\rho_s^2} + B = -\frac{2\rho_s^n \sin\psi_s}{\sqrt{3}}, \quad \frac{A}{\rho_s^2} - B = \rho_s^n\left(\frac{\sin\psi_s}{\sqrt{3}} + \cos\psi_s\right). \end{aligned} \tag{28}$$

It is possible to find from these equations that

$$\begin{aligned} A &= \frac{\left[\rho_c^n\left(\sin\psi_c + \sqrt{3}\cos\psi_c\right) - \rho_s^n\left(\sin\psi_s + \sqrt{3}\cos\psi_s\right)\right]\rho_s^2\rho_c^2}{\sqrt{3}\left(\rho_s^2 - \rho_c^2\right)}, \\ B &= -\frac{\left[\rho_c^n\left(\sin\psi_c + \sqrt{3}\cos\psi_c\right) - \rho_s^n\left(\sin\psi_s + \sqrt{3}\cos\psi_s\right)\right]\rho_s^2}{\sqrt{3}\left(\rho_s^2 - \rho_c^2\right)} - \frac{2\rho_c^2\sin\psi_c}{\sqrt{3}}. \end{aligned} \tag{29}$$

Since $\rho_c$ is a function of $\psi_c$ and $\rho_s$ is a function of $\psi_s$ due to (18) and (27), eliminating *A* and *B* in the third and fourth equations in (28) by means of (29) supplies a system of two equations for determining $\psi_c$ and $\psi_s$ as functions of *p*. This system should be solved numerically. This solution allows $\rho_c$, $\rho_s$, *A* and *B* to be found from (18), (27) and (29). Then, the distribution of stresses follows from (10) and (13). The disc becomes fully plastic when $\rho_c = \rho_s$.

## 6. Illustrative example

Equations (24), (28) and (29) have been solved numerically at $a = 0.3$ and $n = -0.5$. It has been found that $\psi_{c1} \approx 0.078$ and $p_1 \approx 0.898$. The distribution of the radial and circumferential stresses is depicted in Figs. 1 and 2, respectively, at $p = 0.91$, $p = 0.93$, $p = 0.95$, $p = 0.97$, and $p = 0.99$. It worthy of note that the size of the elastic region is very sensitive to the value of *p* at $p > 0.99$.

## 7. Conclusions

A new semi-analytic solution for stress has been derived for functionally graded discs subject to external pressure under plane stress conditions. A numerical technique is only necessary to solve transcendental equations such as (24), (28) and (29). It has been shown that there are several stages of the process. The disc is purely elastic if $p \le p_e$. In the case treated in the present paper, a plastic region starts to develop from the surface $\rho = a$ at $p = p_e$. The solution with one plastic region exists in the range $p_e \le p \le p_1$. Another plastic region starts to develop from the surface $\rho = 1$ at $p = p_1$. The solution with two plastic regions exists until the disc becomes fully plastic at $\rho_c = \rho_s$.

IOP Conf. Series: Materials Science and Engineering **371** (2018) 012060 doi:10.1088/1757-899X/371/1/012060

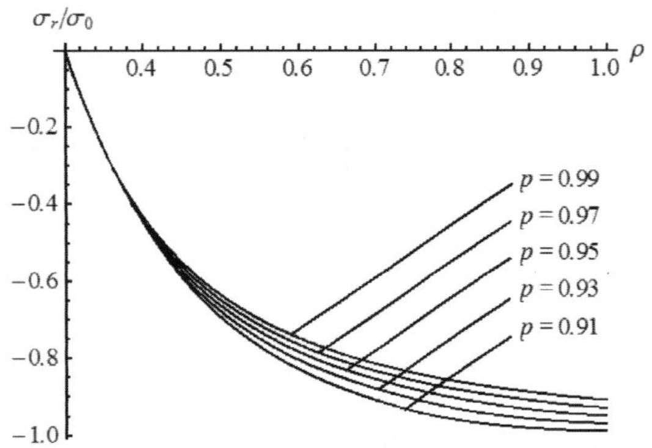


**Figure 1.** Variation of the radial stress with $\rho$ at several values of $p$.

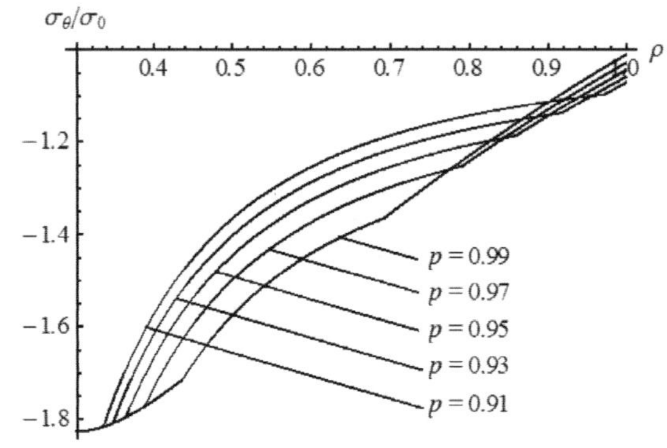


**Figure 2.** Variation of the radial stress with $\rho$ at several values of $p$.

## 8. References


[1] Alujevic A, Legat J and Zupec J 1993 *ZAMM* **73** T283
[2] Mack W and Bengeri M 1994 *Int. J. Mech. Sci.* **36** 699
[3] Parmaksizoglu C and Guven U 1998 *Mech. Struct. Mach.* **26** 9-20
[4] Mack W and Plochl M 2000 *Int. J. Eng. Sci.* **38** 921
[5] Eraslan A N and Akis T 2003 *Mech. Based Des. Struct. Mach.* **31** 529
[6] Arslan E, Mack W and Eraslan A N 2008 *Acta Mech.* **195** 129
[7] Alexandrov S 2015 *Elastic/Plastic Discs Under Plane Stress Conditions* (Springer)
[8] Callioglu H, Sayer M and Demair E 2011 *Indian J. Eng. Mater. Sci.* **18** 111
[9] Eray A and Ahmet N 2015 *J. Multidiscip. Eng. Sci. Technol.* **2** 2843
[10] Vandana Gupta and Singh S B 2016 *Regen. Eng. Transl. Med.* **2** 126
[11] Hill R 1950 *The Mathematical Theory of Plasticity* (Oxford: Clarendon Press)

IOP Conf. Series: Materials Science and Engineering **371** (2018) 012061 doi:10.1088/1757-899X/371/1/012061

# Study on the Influencing Factors of Urban Subway Engineering Construction Safety

**Shixia Duan[1], Rui Sun[1]**

[1]Management Engineering College,Zhengzhou University,Zhengzhou,China
duanshixia@zzu.edu.cn, 1506187239@qq.com

**Abstract.** The application of subway in the construction of urban modernized transportation facilities is increasing day by day. At the same time, the safety of urban subway construction has been paid more and more attention by the society. Based on the AHP-entropy method, this paper constructs the AHP-entropy 's safety analysis model of the subway construction. Take Zijingshan Station as an example, the weight of the index is determined from the subjective and objective aspects, finding out the key factors influencing the construction safety of the subway and putting forward reasonable proposals of subway construction safety control measures, hoping to reduce the occurrence of subway safety accidents.

## 1. Introduction

From the end of 2016, a total of 30 cities in mainland China opened up urban rail operations, with a total length of 4152.8 km, of which 3168.7 km is subway accounting for 76.3% according to the 2016 Annual Urban Rail Transit Statistical and Analysis Report. Moreover, the scale of construction will plan to further expand. With the rapid development of urban rail transit, subway safety accidents occur frequently, resulting in huge economic losses and harsh social impact. Subway construction safety accidents have occurred in China, especially in economically developed cities, such as Beijing, Shanghai, Shenzhen, etc., causing great economic losses and adverse effects to the society.

The analysis of subway construction safety factors is helpful for the prevention strategy and countermeasures in the process of subway construction. The reduction of safety accidents is of vital importance to the state, society and people. Domestic scholars have never stopped practice and research on subway construction safety. Hui Zhang applied the hierarchy fuzzy comprehensive evaluation method to evaluate the construction risk of Shanghai rail transit line 12, and put forward the corresponding measures and suggestions. Guixiang Chen studied the major risk factors, probability and loss distribution of the four stages of planning, design, construction and operation of the Metro [2]. Fan Chen constructed the subway construction safety risk warning model to identify potential safety risk factors in subway construction, through the combination of BP neural network technology and factor analysis method [3]. Chengfang Zhang built a information security management model for the time and space of subway construction based on BIM technology, which can be used for the safety management of the construction site [4]. Honglin Wang summed up 132 domestic urban rail transit construction safety accidents and the analysis results showed that collapse, object strike, falling height is the most common type of safety accident [5]. Qunfang Hu analyzed the safety risk accidents occurred in the construction of urban rail transit, and obtained the frequency of construction safety risk accidents. The central tunnel was 2 times as much as that of the station foundation pit [6]. Dongsheng made statistical analysis of the subway construction safety accident and the statistical results showed that the lack of information communication, geological exploration, shortage of construction defects,

design mistakes are the main cause of an accident [7]. Zhipeng Zhou suggested the control of construction safety risk should be considered from the aspects of personnel factors, equipment factors, material factors and environmental factors with the help of fault tree analysis of Hangzhou Metro accident [8].
This paper establishes the evaluation index system of construction safety and the analysis model of subway construction safety based on AHP-entropy method. Taking Zhengzhou line 2 Zijingshan hill station as an example, to find out the key factors affecting the safety of subway construction, put forward reasonable suggestions on construction safety.

## 2. Establishing the subway construction safety evaluation index system

In the construction of evaluation index system, with the spirit of science, authenticity and validity principle, on the basis of previous studies, questionnaire survey and expert judgment, concludes 14 an index of metro construction safety evaluation system from the unsafe behavior of human, mechanical equipment, construction environment, and construction management, see Table 1.

**Table 1.** The safety evaluation index system of subway construction

| First grade evaluation index | Second grade evaluation index |
|---|---|
| Unsafe behavior of human being $U_1$ | Education level $U_{11}$ |
| | Professional skills and abilities $U_{12}$ |
| | Physiological status $U_{13}$ |
| | Psychological quality $U_{14}$ |
| Mechanical equipment $U_2$ | Mechanical structure $U_{21}$ |
| | Operation method $U_{22}$ |
| | Maintenance and repair $U_{23}$ |
| Construction environment $U_3$ | Geological and hydrological conditions $U_{31}$ |
| | Pipeline distribution $U_{32}$ |
| | The density of adjacent buildings $U_{33}$ |
| | Operating environment(ventilation, power supply, etc.) $U_{34}$ |
| construction management $U_4$ | Management organization and staff $U_{41}$ |
| | Management system $U_{42}$ |
| | The level of enforcement $U_{43}$ |

On the basis of the analysis of index synthesis, we modeled the risk grade. Risk grade is: Grade I - unacceptable; Grade II - unwilling to accept; grade III - acceptable; Grade IV - negligible. Thus, the city subway construction safety can be divided into 5 levels as follows: Grade I - serious unsafety; Grade II - no serious safety; Grade III - less safe; Grade IV - safety; Grade V - very safe. See table 2 for details.

**Table 2.** Safety grade division of Metro Construction.

| grade | Security | Scaling |
|---|---|---|
| Grade I | serious unsafety | (0,0.2) |
| Grade II | no serious safety | (0.2,0.5) |
| Grade III | less safe | (0.5,0.7) |
| Grade IV | safety | (0.7,0.9) |
| Grade V | very safe | (0.9,1) |

## 3. Safety analysis model of subway construction based on ahp- entropy method

According to the actual situation of the construction of subway construction project, select the appropriate analysis method for the evaluation index, determine the key parameters affecting the safety of subway construction, and puts forward effective measures. In this paper, AHP- entropy method is used to determine the weight value of subway construction safety evaluation index.

### *3.1. Using AHP method to calculate weight*

AHP method belongs to the system analysis method, using the principle of combining qualitative and quantitative, calculates the corresponding index weight [9]. First of all, we need to analyse the problems in detail, compare the indicators one by one, get the judgment matrix, calculate the feature vector, and then unify the processing to get the weight of each evaluation index.

### *3.2. Using entropy method to calculate weight*

The Entropy method is a method to determine the weight by evaluating the index value matrix. According to the magnitude of index information, entropy is obtained through a series of calculations, and the entropy is inversely related to the amount of information. This method can effectively eliminate the interference of subjective factors, and more objectively reflect the index weight. We can rely on the security level node and the evaluation index value to constitute the original data, and then calculate the weight through the entropy method, then get the weight of each index.

1) The acquisition of raw data

A number of experts and personnel participating in the subway construction are selected to assign the indexes and take the average value to form the evaluation matrix $A_{1\times n}$, remember as $A=[x_{01} \quad x_{02} \quad x_{03} \quad \Lambda \quad x_{0n}]$. According to a security grade node, the benchmark evaluation matrix is constituted, remember as $B_{(m-1)\times n}$:

IOP Conf. Series: Materials Science and Engineering **371** (2018) 012061 doi:10.1088/1757-899X/371/1/012061

$$B = \begin{vmatrix} x_{11} & x_{12} & x_{13} & \Lambda & x_{1n} \\ x_{21} & x_{22} & x_{23} & \Lambda & x_{2n} \\ x_{31} & x_{32} & x_{33} & \Lambda & x_{3n} \\ \mathrm{M} & \mathrm{M} & \mathrm{M} & \mathrm{M} & \mathrm{M} \\ x_{m-1,1} & x_{m-1,2} & x_{m-1,3} & \Lambda & x_{m-1,n} \end{vmatrix} \tag{1}$$

2) Constructing evaluation matrix

The evaluation matrix is composed of the evaluation index data and the reference matrix made up of safe and reliable nodes. Remember as $X_{m\times n}$:

$$X = \begin{pmatrix} A \\ B \end{pmatrix} = \begin{vmatrix} x_{01} & x_{02} & x_{03} & \Lambda & x_{0n} \\ x_{11} & x_{12} & x_{13} & \Lambda & x_{1n} \\ x_{21} & x_{22} & x_{23} & \Lambda & x_{2n} \\ \mathrm{M} & \mathrm{M} & \mathrm{M} & \mathrm{M} & \mathrm{M} \\ x_{m-1,1} & x_{m-1,2} & x_{m-1,3} & \Lambda & x_{m-1,n} \end{vmatrix} \tag{2}$$

3) Making results dimensionless:

$$y_{ij} = \frac{x_{ij}}{\sum_{i=1}^{m} x_{ij}} \tag{3}$$

In the formula (3): $m$ is the number of objects to be evaluated.

4) The entropy of the index of J is:

$$e_j = -k\sum_{i=1}^{m} y_{ij} \ln y_{ij}, K = \frac{1}{\ln m} \tag{4}$$

5) The difference coefficient of the index of J is:

$$d_j = 1 - e_j \tag{5}$$

6) Calculating the weight of the J index:

$$w_j = \frac{d_j}{\sum_{i=1}^{n} d_j} \tag{6}$$

### *3.3. Using AHP-entropy method to calculate the weight of the model*

Firstly, the subjective weight of the evaluation index is obtained by AHP method, then the corresponding objective weight is obtained by the entropy method. Finally, the AHP-entropy method is used to modify the subjective weight. The new method reduces the impact of human subjective reasons, so that the results are more reasonable.

$$\beta_j = \frac{v_j w_j}{\sum_{j=1}^{n} v_j w_j} \tag{7}$$

In the formula (3): the subjective weight of the index is obtained by AHP method, remember as $v_j$, $n$ is the number of indicators.

## 4. Engineering example

### *4.1. Project Overview*

Zhengzhou Metro Line 2 is basically a north-south direction of the line, starting from the Huiji District, only to Zhengzhou Xinzheng International Airport T2 Terminal 6 kilometers east of Zhengzhou South

IOP Conf. Series: Materials Science and Engineering **371** (2018) 012061 doi:10.1088/1757-899X/371/1/012061

Station, a total length of 71.25 kilometers. Zijingshan subway station is Zhengzhou Metro Line 1 and 2 transfer station and is the first underground four-store building. Zijingshan subway station is surrounded by the main commercial buildings, shopping plaza, municipal traffic roads, and traffic flow has been great.

### *4.2. Analysis on Construction Safety Evaluation Index of Zijingshan Station on Zhengzhou No.2 Line*

(1) Using AHP method to calculate the weight

According to the evaluation index to prepare the corresponding questionnaire, the site staff and experts are invited to score the construction process of the indicators with the nine scale method, and calculate the weight of the evaluation index system in Table 3.

**Table 3.** AHP method for weight calculation in evaluation index system

| First grade evaluation index | Weight | Second grade evaluation index | Weight | Combination weight |
|---|---|---|---|---|
| $U_1$ | 0.27 | $U_{11}$ | 0.245 | 0.066 |
| | | $U_{12}$ | 0.265 | 0.072 |
| | | $U_{13}$ | 0.260 | 0.070 |
| | | $U_{14}$ | 0.230 | 0.062 |
| $U_2$ | 0.21 | $U_{21}$ | 0.349 | 0.073 |
| | | $U_{22}$ | 0.336 | 0.071 |
| | | $U_{23}$ | 0.315 | 0.066 |
| $U_3$ | 0.28 | $U_{31}$ | 0.285 | 0.080 |
| | | $U_{32}$ | 0.251 | 0.070 |
| | | $U_{33}$ | 0.264 | 0.074 |
| | | $U_{34}$ | 0.200 | 0.056 |
| $U_4$ | 0.24 | $U_{42}$ | 0.318 | 0.076 |
| | | $U_{42}$ | 0.335 | 0.080 |
| | | $U_{43}$ | 0.347 | 0.083 |

(2) Using the entropy method to calculate the weight

According to the index system established in Table 1, the relevant experts and construction personnel of the construction site were invited to complete the professional assignment of 14 indicators according to the situation of construction safety management. The assignment data are from 1 to 10, indicating that the greater the value, the greater the impact of the index. And then form an evaluation matrix together with a reference matrix composed of a security level table, remember as $X$ .

$$X = \begin{pmatrix} A \\ B \end{pmatrix} = \begin{vmatrix} 5.45 & 6.55 & 6.25 & 5.75 & 6.80 & 7.55 & 5.70 & 7.80 & 7.65 & 6.60 & 5.80 & 7.50 & 7.75 & 8.00 \\ 0.20 & 0.20 & 0.20 & 0.20 & 0.20 & 0.20 & 0.20 & 0.20 & 0.20 & 0.20 & 0.20 & 0.20 & 0.20 & 0.20 \\ 0.50 & 0.50 & 0.50 & 0.50 & 0.50 & 0.50 & 0.50 & 0.50 & 0.50 & 0.50 & 0.50 & 0.50 & 0.50 & 0.50 \\ 0.70 & 0.70 & 0.70 & 0.70 & 0.70 & 0.70 & 0.70 & 0.70 & 0.70 & 0.70 & 0.70 & 0.70 & 0.70 & 0.70 \\ 0.90 & 0.90 & 0.90 & 0.90 & 0.90 & 0.90 & 0.90 & 0.90 & 0.90 & 0.90 & 0.90 & 0.90 & 0.90 & 0.90 \end{vmatrix}$$

According to the entropy method, the information entropy, the difference coefficient and the combined weight of each index are calculated. As the calculation amount is large, it cannot be presented in the limited layout. The weight data of each index are shown in Table 4.

IOP Conf. Series: Materials Science and Engineering **371** (2018) 012061 doi:10.1088/1757-899X/371/1/012061

**Table 4.** Evaluation Index System Entropy Method Weight Calculation

| First grade evaluation index | Combination weight | Second grade evaluation index | Combination weight |
|---|---|---|---|
| | | $U_{11}$ | 0.0712 |
| $U_1$ | 0.2854 | $U_{12}$ | 0.0715 |
| | | $U_{13}$ | 0.0714 |
| | | $U_{14}$ | 0.0713 |
| | | $U_{21}$ | 0.0716 |
| $U_2$ | 0.2146 | $U_{22}$ | 0.0718 |
| | | $U_{23}$ | 0.0713 |
| | | $U_{31}$ | 0.0711 |
| $U_3$ | 0.2857 | $U_{32}$ | 0.0718 |
| | | $U_{33}$ | 0.0715 |
| | | $U_{34}$ | 0.0713 |
| | | $U_{41}$ | 0.0718 |
| $U_4$ | 0.2154 | $U_{42}$ | 0.0718 |
| | | $U_{43}$ | 0.0719 |

Using AHP-entropy method to calculate the weight
The weight of the index calculated by the AHP method and the weight obtained by the entropy method are determined by the formula (7), and the final comprehensive weight of the index is determined. See table 5.

**Table 5.** AHP-entropy method to determine the weight

| Second grade evaluation index | Weight of AHP | Weight of entropy | comprehensive weight | First grade evaluation index |
|---|---|---|---|---|
| $U_{11}$ | 0.066 | 0.0712 | 0.0671 | |
| $U_{12}$ | 0.072 | 0.0715 | 0.0736 | $U_1 = 0.2752$ |
| $U_{13}$ | 0.070 | 0.0714 | 0.0714 | |
| $U_{14}$ | 0.062 | 0.0713 | 0.0631 | |
| $U_{21}$ | 0.073 | 0.0716 | 0.0747 | |
| $U_{22}$ | 0.071 | 0.0718 | 0.0728 | $U_2 = 0.2146$ |
| $U_{23}$ | 0.066 | 0.0713 | 0.0672 | |
| $U_{31}$ | 0.080 | 0.0711 | 0.0812 | |
| $U_{32}$ | 0.070 | 0.0718 | 0.0718 | $U_3 = 0.2857$ |
| $U_{33}$ | 0.074 | 0.0715 | 0.0756 | |
| $U_{34}$ | 0.056 | 0.0713 | 0.0570 | |
| $U_{41}$ | 0.076 | 0.0718 | 0.0779 | |
| $U_{42}$ | 0.080 | 0.0718 | 0.0820 | $U_4 = 0.2452$ |
| $U_{43}$ | 0.083 | 0.0719 | 0.0851 | |

From Table 6 we can draw in the first-level indicators, that is, $U_3 > U_1 > U_4 > U_2$. In the subway construction process, the construction environment is the highest degree of insecurity for the safety of accidents.

According to incomplete statistics, more than 20 security incidents occurred in 2010 to 2016, and the construction environment of the project collapse of the accident up to 10. In the subway safety accident classification, the proportion of collapse accidents is the highest frequency as high as 55%; In addition, water damage, mechanical damage, object strikes and other accidents occur frequently. The collapse of the accident is because of subway project construction characteristics and construction process. Construction of the subway, the first is to dig tunnels. There is a high degree of difference in the subway mouth if the material and soil are improper accumulation, this prone to collapse accident. And the occurrence of water damage accidents due to underground construction in the process of underground pipe rupture, groundwater penetration in the tunnel and so on.

Located in the top four are respectively: $U_{43}$、$U_{42}$、$U_{31}$、$U_{41}$,is the level of enforcement, management system, geological and hydrological conditions and management organization and staff. In the 14 secondary indicators, the construction management of the three indicators are located in the forefront, indicating that in the construction process from the bottom of the management is particularly important. Perfect construction management can effectively reduce the safety accidents, to avoid casualties and economic losses.

### *4.3. Measures to Improve the Construction Safety of Zijoushan Station*

The safety status of subway construction depends on the coordination effect of people, machine, environment and management, and the people is subjective and dynamic. Therefore, the quality of the operator will directly affect the security of the operating system. To reduce the casualties and property losses during the construction of the subway, the safety risk of the subway construction stage is to be well controlled. It is necessary to strengthen the construction safety operation education and understand the safety situation of the construction interface. So that every site operator spontaneously comply with the rules and regulations of the construction site, the formation of the whole process of construction safety risk production concept.

There are many machines and equipment in Zijingshan subway station construction site. Mechanical injury is one of the reasons for the safety accident and some prominent equipment to be controlled for the safety of subway construction is very vital. In the subway construction site, the performance of machinery and equipment need to be normal, the operator to certificates, with the ability to temporarily solve the problem.

Zijingshan station with open excavation construction, the construction environment is complex, affected by multiple factors, such as foundation excavation, groundwater, etc.. It may cause the settlement of the surrounding facilities deformation. Therefore, it is necessary to monitor settlement deformation during the construction of Zijingshan station and to protect the surrounding residents and buildings safety. According to the geological characteristics of Zijingshan and underground pipeline distribution, the surrounding buildings sinking, envelope displacement, underground pipelines, and steel pipe support the internal are forced to monitor.

In the management of the organizational structure, the responsibility of reciprocity is clearly. During the construction of the subway project, the safety responsibilities and corresponding rights of each department shall be clarified first. Second, the security management system will be developed. At the same time, strengthen the simultaneous construction of multi-type coordination and subway construction safety management.

## 5. Conclusion

The analysis of safety factors is of great significance to the subway construction of modern city in the construction of urban subway project. AHP-entropy method is used to construct the model, and the weight of the index is determined synthetically from the two aspects of subjective and objective, and the example of Zijingshan station on Zhengzhou No.2 is analyzed. The results of the study are important to the indicators, and the key factors influencing the construction safety of the subway in

IOP Conf. Series: Materials Science and Engineering **371** (2018) 012061 doi:10.1088/1757-899X/371/1/012061

Zijingshan Railway are found out. The results of the existence of bias, we need to collect more data to conduct empirical research so that the results more scientific and reasonable.

## 6. References


[1] Hui Zhang, Wenyong Li and Zhifang Ni 2010 Study on Risk Assessment and Control Measures of Shanghai Rail Transit Line 12 Project Construction *J. Urban Rail Transit Research.* **11** 46-51

[2] Guixiang Chen 2004 Study on Risk Management of Subway Project *J. Shanghai: Tongji University.* **5** 89-90

[3] Fan Chen and Hongtao Xie 2012 Study on Safety Early Warning of Subway Construction Based on Factor Analysis and BP Neural Network *J. Journal of Chinese Society for Safety Science.* **8** 1-4

[4] Chengfang Zhang and Chao Li 2013 Application of BIM technology in subway construction safety *J. Henan Science and Technology.* **5** 130-131

[5] Honglin Wang 2013 Study of risk management of the urban rail transit construction project. *J. Xuzhou:China University of Mining and Technology.* **9** 20-22

[6] Qunfang Hu and Jiabao Qin 2013 Statistical analysis of accidents of subway tunnel construction from 2003 to 2011 in China *J. Chinese Journal of Underground Space and Engineering.* **3** 51-55

[7] Dongsheng Xie, Qihu Qian and Xiaoli Wan 2012 Risk management rail transit construction *J. Journal of Civil Engineering and Management.* **29** 61-67

[8] Zhipeng Zhou, Qiming Li and Xiaopeng Deng 2009 Analysis of Metro collapse accidents based on accident mechanism and management factors: With the collapse of Hangzhou Metro as an example *J. China Safety Science Journal.* **9** 139-145

[9] Yang Xu, Yan Zhou and Xin Sun 2009 Comprehensive evaluation of mine safety based on fuzzy analytic hierarchy process *J. Journal of Chinese Society for Safety Science.* **15** 147-153

IOP Conf. Series: Materials Science and Engineering **371** (2018) 012062 doi:10.1088/1757-899X/371/1/012062

# Oroqen Decorative Pattern Design and Research on the Application of Creation Thought

**Jian Huang[1], Xiaojie Zhou[1]**

[1]Art and Design Department, Heihe College, Heilongjiang, China

**Abstract.** The dress had distinctive regional and national costume pattern design theme forms simple and delicate rough and chic and unique style dress patterns through the investigation analysis of the patterns of dress of the Oroqen nationality clothing patterns of the decorative techniques and design theme from the theme selection, design, use of color, form 4 aspects of research the patterns of the characteristics of the Rukai clothing patterns hierarchy, life custom, ancestor worship, spirit, carry forward the traditional cultural connotation shows the Rukai people for a better life of love and longing for the Rukai clothing patterns research may promote the costume cultural heritage and protection for the modern design reference the essence of national culture and the creation inspiration. This article mainly from the Oroqen nationality clothing patterns modelling techniques and related materials and other aspects of the analysis, in order to have a more in-depth understanding of the design and creation of the decorative patterns of other ideas.

## 1. Preface

The main distribution in Inner Mongolia and Heilongjiang Heihe City, a population of about eleven thousand people, mainly in the Oroqen nationality is a patriarchal society, has a very strict internal rating system, can be divided into nobles, warriors, civilian leaders, four class, class as the hereditary system, to a noble family as a system center. The more the clothing category, style simple, strong color contrast decorative techniques Yuyishenke complex patterns, the dress patterns are wide. There are hundreds of snakes, lilies, pottery pots, the sun, stoves, ethnic creation theme. The theme of memorial ceremony with swing and stab ball. The theme of life scenes with hunting and dancing [1]. And the butterfly, diamond, cloud shaped totem worship genre, Oroqen nationality clothing patterns rich regional features rich folk culture has high research value. With the "guidance" issued by the state on promoting the development of cultural industry has more and more people realize the history of costume culture and artistic value, multicultural development will become a new direction of fashion creation.

## 2. The dress patterns of the decorative techniques

The dress is very rich with techniques of beadwork, sticking cloth, embroidery, weaving and other clamp techniques have glass beads and shells, and silver patterns show fine patterns techniques decoration materials.

### *2.1. Beads*

The main process of bead material for glass beads can also use glass beads and shells with line string together different colored beads, according to the idea of designers with embroidery stitch in costumes.

IOP Conf. Series: Materials Science and Engineering **371** (2018) 012062 doi:10.1088/1757-899X/371/1/012062

By hanging beads combined with a variety of embroidery techniques to form a variety of patterns, the beads were thought to be able to bless evil Oroqen, aristocratic dress patterns of this material to design alternative.

### *2.2. Clip Weaving*

Clip is the most traditional of the fabric weaving skills, weaving material mainly used in Ramie weaving process, with mobile level strap loom warp weft weave fixed, different colors form a decorative pattern, woven fabric thick, making many men after open pants or used as clothing edge decoration edge.

### *2.3. Satin Embroidery*

The surface is embroidered Satin decorative clothing used in minority nationalities, the embroidery method is very difficult, not the direction of embroidery The binding lines and lines must be arranged neatly together. There is no space or overlap. The patterns of the front and back sides of the embroidery are the same, and the surface is smooth and smooth Used for decorative collar and cuffs, slashed parts.

### *2.4. Sticking Cloth Embroidery*

The embroidered fabrics, with larger and fabric color contrast, fabric pattern outline and cut symmetrical shape or radial pattern, the pattern edge suture in the decorative parts of clothing with lock stitch form, the use of decorative decorative techniques of large area, simple patterns color, contrasting texture texture clear form the more overall. The dress patterns through traditional techniques to decorative patterns on clothing, the formation of planar and three-dimensional decorative effect, Satin Embroidery, weaving yarn clamping techniques by the color change of the formation of plane decorative pattern, visual effects, beadwork, embroidered fabrics by adding other decorative material to form a certain thickness of the three-dimensional decorative pattern. The performance of both the visual effect and increase tactile texture effect.

## 3. The decorative pattern design

The decorative pattern design theme - from the closely related to the living things or scene inspired by myth, history, religion and other factors, Yuyishenke, meaning rich, using the form of abstract or figurative expression of Oroqen people emotional sustenance, the use of decorative cultural symbols convey Oroqen people looking for a better life.

### *3.1. The Creation of Mythological Themes*

#### *3.1.1 Hundred Step Snake Pattern*

According to legend, in ancient times a clay pot two eggs were hatched two snake snake, later become the ancestors, the ancestors miss disintegin decorative patterns in the decoration that people respect for our ancestors, the snake pattern, design to combine with the jug snake patterns of symmetrical form, snake climbing patterns show strong effect, patterns in the circles represent the ethnic relations of solidarity, triangle representing mountains living environment, on behalf of the transverse rivers, patterns with bead decoration techniques, in front of the dress, trousers or knee after open middle legs, shoulders, sleeves in the back, and other parts of the decoration disintegin pattern is a symbol of nobility of power for noble special patterns, civilians have no right to use.

#### *3.1.2 Sun pattern*

According to legend, the sun has produced a white and red two eggs in the mountains, a snake hatched a pair of men and women to become the leader of the ancestors.At the same time, a green snake eggs hatch the civilians, civilian leader and has formed the social class level

The sun pattern, a symbol of nobility as noble special patterns, the sun pattern design techniques using personification, into people's facial features to the sun patterns, patterns around the extension lines, the formation of the radiant visual extension of a sense of the whole pattern will point line and surface elements of the composition and the perfect combination of style variety, the patterns in site is relatively wide, clothing can be inside and outside decoration, decoration techniques mainly compose bead based.

*3.1.3 Stove Pattern*

The stove pattern is inheriting the tribal ancestors origin legend formulated a race emblem logo, reflects the primitive totem worship, the concrete design patterns into a fire burning stove, stove through patterns, colors and shapes of different tribes will differentiate to form a social and cultural symbol, enhance the tribe's cohesion and pride, for ancestor worship derived blessing children happy and healthy range of decorative pattern, this pattern in the bead decoration techniques to form a continuous pattern, decorative decoration in the edge position.

*3.2. Life Scene Theme*

*3.2.1 Hunting Patterns*

Hunting is an important social man Oroqen traditional livelihood activities, although the hunting can not at any time, to the tribal shaman blessing before hunting, the hunting patterns symbolize smoothly, hunting patterns is described with the hunter spear and knife, headband leaves camouflage articles hunting scene to the hunting techniques of realistic design the scene perfectly, the patterns for men only decoration, with independent patterns in the form of decorative use of embroidery, weaving, bead clamping performance in various parts of the decorative techniques on decoration.

*3.2.2 Dance Patterns*

The Oroqen people sing and dance, dance source in production and life, farming, harvesting, festivals, adult ceremony various activities are held every event is an integral part of its life, reflecting the Hubei Lun passion for work, love of life, dance pattern composition to people hand in hand dancing posture for design, image to show the Oroqen tribe, Qingfeng, memorial ceremony, the dance of the pace of patterns in continuous patterns used in clothing can make the edges of the decorative parts to extend the visual effect, full screen, dynamic strong, commonly used decorative techniques of embroidery, bead.

*3.3. Totem Worship Theme*

*3.3.1 Butterfly Pattern*

Butterfly elegant, colorful, dance, as people will add beauty of nature, it is connected with beauty, rich cultural symbolism, other variety, with the dancing style design patterns showing a strong color contrast effect of rhythm, the pattern shows contrast and balance effect of visual, decorative pattern to the people of life bring happiness, reposing people yearning and pursuit of spiritual emotion, with the butterfly pattern embroidery decoration technique, commonly used in women's decoration, decoration in the front of the dress, back and other parts, shape and color of the pattern has many elements in modern clothing design.

*3.3.2 Diamond Pattern*

Diamond pattern is one of the most widely used in the decorative patterns. The pattern is full of infinite beauty and charm is the important spiritual sustenance and ethnic identity, highlighting the rich characteristics of this nation, reflects the ethnic strong cohesion

The diamond pattern is inspired by the snake's back pattern. Because one of the ancestors of the ethnic group is a hundred step snake. Therefore, the pattern was valued and respected by the people. The diamond patterns are characterized by many decorative techniques such as weaving, embroidery and so on. They have concise design patterns and rich forms of change. In the form of continuous patterns used in the edge of clothing, has a strong sense of rhythm.

*3.3.3 Cloud Shaped Pattern*

Cloud pattern is a beautiful symbol of good luck pattern form beautiful, profound meanings and flowing style. Oroqen

Believe everything has a spirit worship of everything in the world, people believed that heaven is God the gift of wisdom, cloud shaped patterns in the decoration is in line with the psychological theoroqen ethnic habits and aesthetic characteristics is a natural and simple beauty and contains the cultural connotation of the eternal symbol of a successful, with cloud shaped patterns of multi in the form of continuous pattern decoration in decorative patterns with edge length and density of regularly arranged, forming a composition in the form of rhythm. The decorative pattern of the pattern is most common with cloth embroidery.

## 4. Decorative patterns of cultural meaning

### *4.1. Grade System*

The society has obvious level, divided into specific leader, nobles, warriors and civilians in four grades. Each level of Patrimonialism, provisions of the Oroqen society aristocrats wearing gorgeous decoration, reflects the nobility rank and power with certain exclusive patterns. Snake pattern, lily flower pattern, swing pattern, pattern belongs to the noble leader and thorn ball logo, these patterns become the symbol constraints of civilians, civilians can not use noble patterns, can only use simple / geometric patterns

### *4.2. Life Customs*

The activities of decorative patterns in many scenes, into the nation's local customs and practices reflects the people, love of life, through the swing pattern, thorn ball patterns, hunting patterns, dance patterns show the folk activities, these customs through patterns recorded down, become a theoroqen ethnic spirit, through the custom theme will the inherent spiritual consciousness and humanistic value behaved most incisive, through the historical culture and customs of the precipitation, showing a profound historical accumulation of the ancient Oroqen nationality and cultural connotation show the innovation of national wisdom, reflects the development of living standard, records the local customs and practices changes, the formation of the characteristics of cultural symbols.

### *4.3. Ancestor Worship*

The ancestor worship is a kind of religious belief, people believed that dead ancestors of the soul still exists, has an impact on the world, Hubei will disintegin Lun, lily, clay pots, sun, kitchen decoration on clothing creation, artistic creation through ancestral creation theme, shows both the decorative beauty, and memorial their ancestors believed that the ancestors will continue to bless the people, descendants of blessing. The people with patterns in memory of ancestors, put the emotion into the decorative patterns, the cultural heritage of its aesthetic identity.

### *4.4. Spiritual Sustenance*

The spiritual sustenance from the totem worship is ancient and profound nature worship, totem patterns or the figurative realism into decoration, reflects the formation of Hubei Lun cultural beliefs form in the process of life close to nature, and through the spirit totem worship, totem worship and ancestor worship of totem worship is different, pray the benefits of ancestor worship is affection expression, the butterfly pattern and cloud shaped patterns are reflected in the theoroqen ethnic yearning for life and emotional sustenance. The decorative patterns from the myths and legends of many people reflect to the ancestors of the totem worship, admiration and longing for life.

### *4.5. Carry Forward Tradition*

The decorative patterns and decorative techniques retained with a long cultural and historical background, these methods must be after a long time study to master, making beadwork techniques as a leader and noble decorative patterns will use the techniques of since 1600 has been a history bright color glass beads to ignition, ornamentation, cross section, grinding and dyeing five completion of the process, reflects the Hubei Lun people have deep knowledge of traditional crafts, satin stitch is the

IOP Conf. Series: Materials Science and Engineering **371** (2018) 012062 doi:10.1088/1757-899X/371/1/012062

most characteristic of textile, weaving exquisite level Seiko xizuo clip weaving technique is the most traditional manufacturing technology of the complex process, superb skill level.

## 5. Conclusions

The use of a variety of decorative patterns forms of the folk beliefs and cultural consciousness, into the vision will be better to decorative patterns, with patterns of the national sustenance of the sentiment, decoration exquisite technique, display and develop theoroqen ethnic confirm the traditional techniques, patterns of various topics, myths and legends the memorial activities, daily, sight, totem worship can be used as a design theme, reflects the historical and cultural landscape deposition, pattern composition is clever, bright color, strong visual impact, performance of the Oroqen people's aesthetic consciousness and unique patterns of the profound cultural connotation, the moral of the Oroqen people the vision of a better life, to study the decorative patterns form and intrinsic cultural implication, provide unique imagination of modern fashion creation And creativity, so that the decorative art presents a unique national cultural charm.

## 6. Acknowledgements

This thesis is a research project of Heilongjiang Province Department of art and science project" "Inheritance and application of Oroqen decorative's the design and creation of." (project number: 2017D085)"

## 7. References

[1] Abel J 2011 From image to image: Butterfly art form in the view of folk custom culture on the study of national art **12** 91-96

IOP Conf. Series: Materials Science and Engineering **371** (2018) 012063 doi:10.1088/1757-899X/371/1/012063

# Research on 3D Architectural Performance and Urban Landscape Design

**Bingjiang Wang [1], Weibao Wang[1], Yongzhi Li[2], Yulong Liu[3], Jiaan Zhang[4], Yang Wu[5], Yan Wang[1] and BoYan[2]**

[1]Institute Wanjiang College of Anhui Normal University, Anhui, Wuhu,China
[2] Donghua University, Shanghai,China
[3] Suzhou University, Jiangsu, Suzhou,China
[4] Greenland Group , Shanghai,China
[5] Shanghai Yixun Cultures and Arts College, Shanghai,China

**Abstract.** The abstract should summarize the contents of the paper and should contain at least 70 and at most 200 words. It should be set in 9-point font size, justified and should have a hanging indent of 2-centimenter. There should be a space before of 12-point and after of 30-point.

## 1. Introduction
With the rapid development of computer technology, architectural renderings have been difficult to meet customer needs, architectural animation gradually replaced the architectural renderings, into the architectural performance to come. Three-dimensional architectural performance technology to today, in the production of new things at the same time, but also for the social transformation made a prominent contribution to the three-dimensional architectural performance art is particularly important[1]. In recent years the rapid development of the construction industry, led to the progress of three-dimensional architectural performance art[2]. In recent years, with the development of computer hardware and software technology, three-dimensional animation this emerging technology came into being, followed by a building animation. The so-called architectural animation, is the use of digital technology means combined with the film's performance techniques, the architectural appearance, interior design, and outdoor landscape for all-round, multi-angle display and interpretation[3]. It is a combination of real estate, interior, garden and other planning and design drawings, the interior structure, real estate appearance, community environment, life support, property management and other future life scenes completed in advance interpretation and display , So that people can easily and clearly understand the quality of life in the future home. Architectural animation of the lens free unlimited, can not show the real shot of the lens deduction out of the real estate as a whole to show the overall image of the real estate designer's thinking, accurate and perfect expression, so that people feel like immersive The future of home is true and beautiful.

## 2. Landscape Design Content Classification
A comfortable city living environment is inseparable from the city planning, that is, a city land plan and road arrangements, that is, the city's landscape design[4]. So, in a city in the city planning, we must first consider the urban landscape related issues. Urban landscape planning is a comprehensive reference to the landuse plan, and according to the relationship between the overall and local planning. First of all, to investigate the city's landscape resources, through the collection of access to information

IOP Conf. Series: Materials Science and Engineering 371 (2018) 012063 doi:10.1088/1757-899X/371/1/012063

and photographs and field visits to determine the city's social value, human values, economic value and a variety of potential factors, and ultimately its planning and design. Most of the city can be divided into living area, traditional characteristics area, business district, industrial area, natural scenic area, etc., for different landscape areas, to develop consistent with the improvement plan, such as the city skyline and the city outline Control, spatial visual induction and the overall visual induction control, architectural form and architectural form and building height control, building volume ratio and building coverage control, the use of landscape materials and landscape color ratio, etc. are convenient to reflect The details of the landscape design. A good urban design is based on the overall urban season as a guide to make the masses as the main body with a variety of material elements of the integrated environment design, making it a link between architectural design and urban planning and design of the bridge. The speed of urban construction development is complementary with its urban planning, they exist both differences and interdependence, the difference between the urban architecture design mainly to function, space design, and focus on urban construction research methods, While the urban landscape planning and design not only focus on the artificial environment, but also focus on the natural environment, the most important thing is to pay attention to people's life landscape design. The two are the same is for the development of the city made a great contribution. With the wide application of computer 3D technology, people through different views, with a dynamic way to observe the city's space shape design, so that designers do design entities before the design of the finished product and a good grasp, so Both grasp the overall relationship and achieve their own design purposes. Urban landscape planning and design of its space and its body design has a higher demand, both from the landscape color, landscape form, location address on the potential impact of the landscape environment have done a detailed and careful consideration. Urban landscape planning and design should consider the relationship between the whole and the division, taking into account people's feelings and environmental factors, and thus began to design the overall planning and design of urban landscape, in order to speed up the formation of the city's overall landscape design.

Building signs, flower beds, city road signs, city streets, center parks, artwork, bustling shops, Street corner, historical relics, lighting, avenue, advertising and other topics are the elements of landscape architecture design. The overall urban landscape is composed of unique single building, single building architectural form, color, scale and texture of the city's overall landscape plays an important role can not be ignored. The overall architectural landscape can not be separated from the single architectural landscape, single architectural landscape must consider the overall requirements, in general, the single architectural landscape must adapt to its surrounding environmental factors and people in different perspectives, and then the overall design, the final Become able to integrate into the overall urban landscape of the single landscape architecture. Landscape architecture design is the combination of all the factors related to urban construction and design, in the urban landscape design plays a vital role, will largely affect the development and formation of environmental architecture landscape. The formation of a successful urban landscape requires not only a good geographical and natural environment, but also a good artificial environment, that is, strict management system. In China, there are many successful urban construction, but the coastal city in the country's urban landscape construction in the lead. Here we take the city of Dalian as an example, the city of Dalian, according to its own unique geographical environment, waters, climate factors, choose a different regional planning, develop a strict planning and design, different geographical areas are shared with different urban landscape construction Dalian, the business district, also play a part of the construction of urban landscape, the commercial area of the landscape architecture in terms of color, the color of the landscape, Texture, content has a strong business atmosphere. Second, the management system is also very strict, for each building has strict design requirements, such as the height of the building, shape, color and so have strict checks, and its design style also has certain standards. Therefore, it is precisely because the management of the entire city of every building details of the strict rules, making the city of Dalian city landscape planning and construction of the true meaning of people to bring the soul and life of comfort, to achieve a higher level of urban landscape quality. In this, also hope that other cities can follow the strict management of Dalian city system, to create people in the eyes of the beautiful home.

**Figure 1.** City square

## 3. Landscape Design Application

Society in the development of the times in progress, urban landscape construction is also vigorously promoted by the government under the mushroomed development. In recent years, the city's renovation project and the construction of new technology park development, a large number of new house construction, but also accelerated the pace of development of urban landscape. But the speed of urban development, but greatly overlooked the city's landscape design, urban landscape design must be based on the culture of the city and residents of the understanding, due to the understanding of the various urban culture and folk, unique to the city and Did not appear with the increase in the building, did not completely study the reasons, making the overall urban architectural landscape style is no substantial differences, and ultimately caused the landscape between the various cities in the same style, no features , The formation of the city landscape no characteristic phenomenon. The meaning of the city is not just for people to provide survival of the place, the greater significance is to focus on the spirit of the people to bring the feelings. The physical environment of a city and the external manifestation of social life is the city style, a city-specific landscape here in the performance of the most incisive, depicting the city's temperament and character, the public's mental outlook in this also show Especially prominent. The city to provide people with living space environment at the same time, but also affect the spirit of each person. When people are satisfied with the material life, people's requirements for the quality of space environment has become increasingly strict, many cities have begun to attach importance to urban landscape problems, urban landscape has also been a qualitative change. From the city's environmental planning, from the strict management of the city, from the choice of the city's vegetation, from the city square and the establishment of the business district, the urban landscape is facing a new challenge with a new look. Urban landscape is not a simple programming, but a vast project, it not only requires the designer's creative ideas, but also the government's strict planning and management. The rapid development of urban development today, we bear the brunt of the city in recent years to sum up the experience and lessons learned, and learn from foreign advanced technology and advanced ideas to promote the development of China's urban planning and design.

## 4. 3D Architecture And Landscape Design

In recent years, computer hardware and software technology, especially worth mentioning is the three-dimensional technology and high-tech difficult virtual technology, because of the rapid development of these technologies, so that the visual development of the landscape with a more solid scientific and technological strength. In particular, the 1999 "Our Visual Landscape" international symposium held in Switzerland once again, it is set off a landscape of the visual development of a new wave of research. Many foreign researchers focus on technical research in computer technology and video mapping, such as the output strength of landscape images based on GIS spatial data, and the vegetation landscape modeling aspects of visualization. Compared with the first phase, the rapid development of software technology, microcomputer computing speed of more than one billion times, the computer graphics graphics card technology greatly accelerated the speed of computing and processing of landscape data, landscape visualization of the development of the system to achieve rapid interaction The new level. Specialized visualization of the equipment has also been a lot of development, a lot of production is also slowly. Software development has also sprung up, developed a series of three-dimensional landscape visualization software.

The main development of the stage is characterized by: by the two-dimensional landscape entity show that the three-dimensional landscape entity show forward, to the three-dimensional landscape of its own unique physical display model to show the landscape model, landscape connotation and the meaning of its performance, the establishment of three-dimensional landscape model , Use the technical software VRGIStool to draw the regional landscape map. In short, the visual development of this period from the two-dimensional plan to the three-dimensional entity display, from the static display method to the simulation of the display method, from the landscape two-dimensional performance to the landscape three-dimensional simulation and other advanced level. Although the visualization of China's landscape development research later, but still still made a lot of achievements, such as 3D technology urban landscape visual research and development, three-dimensional scene model interactive operation, three-dimensional dynamic interactive visualization model, visualization technology and its geological The application of exploration and so on.

Artificial modification as a means of making the traditional renderings, there are factors that mislead others. But the real estate to three-dimensional animation in the form of show to the developers, which can be through the personal experience of the program to give the characteristics and quality of the correct evaluation, so as to choose the best strategy to make the land utilization and construction project development rate To maximize the size of the increase in the cost of greatly increased. through the vivid three-dimensional technology to show larger projects, you can build lifelike three-dimensional model, multi-angle, multi-level, multi-directional to meet the requirements of the masses, the building itself for the various points and channels for a reasonable allocation, and People's design ideas resonate, which also makes the relationship between designers and people more harmonious. The development of computer digital technology is not only the designer to get rid of the heavy work such as hand-painted, but also for the architectural design provides a new rich information age characteristics of the performance techniques. To three-dimensional production software (mainly 3ds max) and other multimedia development software, for example, the computer can be a variety of three-dimensional building model according to the requirements of the planning program layout, according to the actual terrain contours generated analog terrain, and In the three-dimensional animation software with the help of creating an animation tour path, and eventually complete a three-dimensional software with the realization of the building roaming three-dimensional animation files. Building roaming animation can be multi-angle, comprehensive will be able to show the reality of the city scene to convey to the people, and can achieve the real sand table model can not reach the simulation of the architectural space effect, in addition, the original sand table model is not conducive to modification, And now a variety of virtual models is very easy to modify it. Finally, after the completion of the production here, the latter through the software to achieve its text description, music background, graphic information, etc., some simple query function can be achieved. In summary, the three-dimensional building performance has been far more than the original traditional method.

The traditional design is accomplished by means of pen and paper and designer's ideas, and the emergence of 3D building performance technology undoubtedly reduces the traditional work of large workload mode. For example, if we now need to build a building, according to the original traditional practice, we first need to do a lot of renderings, but know and can understand the construction drawings and renderings are not many people, but if we exercise three-dimensional performance Technology, you can show in detail the outside and inside the building, including the surrounding environment, can be like an immersive feel, so through the three-dimensional technology to do out of the building, first in the workload has been reduced, followed by Is that people can more easily feel the information people need, and finally, due to the upgrading of three-dimensional technology, people living in a virtual environment, any place where they are not satisfied with any place to modify, greatly improving the time and quality. people on the concept of sales more and more clear, a good product packaging sales, will bring endless benefits for this commodity, in the moment, the sale of long land projects is highlighted the superiority of three-dimensional technology, first of all, The traditional sand table model does not disappear, but the buildings in the sand table because of the proportion of the problem, so that people are not very satisfied, people can not feel through their own perspective of its ratio. But the emergence of three-dimensional building table technology, people only need to feel through the big screen can be the real scene, completely immersed in a virtual environment, three-dimensional building performance technology to our perspective, to lead people in a virtual environment, Increased people's awareness of the goods, which undoubtedly accelerated the proportion of sales of goods.

## 5. Three-Dimensional Construction of the Technical Means

With the development of science and technology, technology improvement, virtual reality environment projects have repeatedly seen. Roaming technology 360-degree all-dimensional three-dimensional real can be achieved at home on the network to see housing units, real estate architectural appearance three-dimensional animation video. Convenient and efficient electronic building books, various industries to show, studio, performance stage, space scene, office interior design, business hall environment construction, commercial space exhibition hall, hotel interior decoration and many other projects are used in life. More realistic immersive experience to interact, change the past passive viewing mode, the user can interact with the virtual environment for human-computer interaction is one of the most convenient features. Life in the clothes, food, live, are inseparable from this technology, such as shopping malls, luxury hotels, leisure hotels, luxury apartments, leisure and entertainment exhibitions, landscape display, office buildings, tourism, Display, exhibition hall, subway, station, airport, terminal, and other industry projects are widely used.

Three-dimensional virtual demonstration, also known as three-dimensional animation, through the three-dimensional animation software in the computer to establish a simulation model or scene, according to the requirements of the model, the camera movement and animation parameters, and finally render the material on the model, and give its light, the output three-dimensional animation file. It is also widely used in many areas such as housing construction, military, medical, educational, industrial, leisure, entertainment and so on because of its precise real and infinite maneuverability to simulate the real scenes of the unit building performance animation technology is a particularly fast Favorable tool. 3D virtual presentations fall into two categories: virtual reality (real-time interactive 3D animation) and conventional 3D animation (fixed mode playback). Virtual reality referred to as VR, the English name Virtual Realit. The viewer can take the initiative to watch, watch the time is not restricted, and can be observed for a long time; observation angle is more free from the restraint, you can replace multiple viewpoints, you can also like the conventional three-dimensional animation to develop the same route tour. Has a strong interaction and strong expression. Its main application areas for the virtual real estate, industrial simulation, urban planning, digital teaching, game entertainment.

Image editing software, which is relatively powerful application of a wide range of Photo Shop software, the software has many kinds of special filter function, improve and modify the artistic effect until the desired effect, Print the perspective effect graph by color printer. The scope of the landscape is great, not just in sightseeing. The construction of the premises, as well as the role of improving the climate of cities and towns, fine-tuning the surrounding areas of environmental humidity and airflow, and reduce sand, environmental protection can use these clever green design to protect, reduce air pollution, reduce noise , Greening the environment, controlling water pollution and soil pollution, white pollution, thus completing the highest level of design. Design drawings are determined by the designer to determine the effect of the map at the same time, with reference to the design requirements, and then with the computer data and hardware facilities, landscape projects in addition to the ornamental art value, as well as beautify the value of the environment.

## 6. Three-Dimensional Digital City Landscape Materialization

The technical means of carrying out the three-dimensional landscape of the commercial landscape around us, so that people use this means to enjoy the way in which they feel the mystery and the pursuit of their own needs. Urban commercial landscape animation is an important part of the performance of the three-dimensional building, showing the relationship between the city and the business; urban commercial landscape, the landscape atmosphere is particularly important, that is, the artistic atmosphere of the commercial landscape and technical means to be vividly, through three-dimensional software Of the intervention, to reflect the image on the commercial landscape animation and emotional beauty. So that it has a unique creative artistic feeling. It is not just business, from the cultural level of analysis, it is a unique cultural products; from the design level, it is a design aesthetic and gives the artist's art of art; from the economic level To analyze the words, it is the real estate developers on the landscape building products sales promotion of a means. So, the city commercial landscape animation is the city landscape commercial activity space image carrier.

Real estate business three-dimensional performance is a new experience, the main form of expression is three-dimensional dynamic image browsing, through the three-dimensional animation technology, from all angles to show the real estate real estate effect. Three-dimensional dynamic images are artistic, aesthetic, creative and emotional works of art, and for architects it is the architectural exhibits of the form of performance and packaging, for real estate developers for the real estate Sales promotion and promotion played a very unquestionable very important role. It is mainly through the virtual interactive way to show the real estate real estate space, with a very good interactive entertainment. Usually this image has the auditory and visual art recognition effect, while the particles into the special effects of the expression of the season and the weather under the real estate, the use of light effects difference between day and night performance of different effects. The use of audio-visual technology can be better performance of the real estate image of the perfect and authenticity, the perfect picture, smooth lens movement, as well as the perfect sound highlights the real atmosphere of the environment, to real estate customers immersive feeling. Compared with the traditional monochrome boring show, real estate three-dimensional dynamic browsing more attractive to customers, enabling users to understand the full range of real estate information to help users make positive decisions. This real estate dynamic images can even be published through the Internet to facilitate customers to view real estate information, so that customers can fully understand the various real estate information.

The development of a city, inseparable from the cultural heritage of the city itself, vertical analysis of Chinese culture up and down five thousand years of history, have to mention is China's Yuanmingyuan landscape architecture, the location of the Yuanmingyuan in the western suburbs of Beijing, the eastern part of Haidian District The Early for the Qing Dynasty Royal Garden, Yuanmingyuan is divided into Yuanming, Changchun, Qichun three parks, with a total area of 350 hectares. Known as the "Garden of the Garden" of the Yuanmingyuan is a very precious landscape, which has jewelry. Calligraphy, treasure and other precious cultural relics, the ground area and the size of the existing Palace, the lake area is equivalent to the current area of the Summer Palace, Yuanmingyuan garden landscape brings together the ancient garden technology, learn the essence of ancient garden art, Take the art of the garden for the garden. Unfortunately, the 1860 Yuanmingyuan was looted by the British and French forces, the buildings in the park were burned, a large number of cultural relics were looted, and the fabulous Yuanmingyuan became a ruin, and only a few remaining buildings could be seen. April 2010, Yuanmingyuan Management Office and the Tsinghua University Planning and Design Institute decided to computer panoramic three-dimensional virtual technology "reconstruction" Yuanmingyuan, co-sponsored the "Digital Yuanmingyuan" research projects, again through the Yuanmingyuan research data of digital finishing, complete The establishment of the Yuanmingyuan historical data database. Has completed the Yuanmingyuan 22 scenic restoration design works.

**Figure 2.** Yuanmingyuan recovery figure

## 7. Conclusion

People through the virtual three-dimensional building performance technology, exposure to a near real space, breaking the original single design pattern, personal experience a more real and more vivid and realistic level. In the rapid development of computer today, people need fast-paced lifestyle, but not only the need for fast speed, people also need a full sense of art to wrap our own eye. Through the elaborate design of the designers, both from the color, sound, visual experience will give people slowly bring a non-general artistic experience, but in view of the virtual reality technology in China's development time is shorter, in technology and application is not mature, There is an urgent need to solve the problem. Such as high demand for hardware and software equipment, hardware and software equipment, expensive and architectural design technology and virtual reality technology integration issues. However, I believe that as long as the architectural designers do not rigidly adhere to the traditional design and design concepts, virtual reality technology (VR) in the architectural design practice will be popular, and ultimately one day, we can through the use of Digital technology, to create the ideal of the immortal city.

China's three-dimensional landscape architecture performance of the development of mixed. Happiness is that people are slowly aware of the advantages of landscape architecture visualization, and slowly accept, due to the shortage of talent, the university three-dimensional landscape design professional also slowly more up, cultivate professionals to contribute to China's economic and cultural construction. It is worry that the professional level of the thousands of graduates is mostly unsatisfactory. The reason is because our three-dimensional landscape has just started, still try to explore the forward, people also need a lot of time to digest it. Since several large real estate landscape companies began to intervene in the industry, and achieved better results than expected. We have greatly improved in the exploration of our own quality, and find our own unique production style, speed up the development of three-dimensional landscape architecture, especially to take the Chinese characteristics of three-dimensional landscape performance path, has been the common voice of people from all walks of life.

## 8. Acknowledgements

Foundation for Distinguished Young Talents in Higher Education of anhui, China.2016(Project number: gxyqZD2016403)
Key projects of Humanities and social sciences of Anhui(Project number: SK2015A655)

## 9. References

[1] Yingying, L 2009 Virtual reality salvation *J. Internet Weekly* 1 182-189
[2] Changda L 2010 Analysis of Virtual Reality Technology on Internet *J. Knowledge Economy* **11** 378-389
[3] Zhou Jian S and Xiaotian Lu 2010 Three - dimensional digital city animation art research *J. Journal of Nanjing Art Institute* **3** 239-247
[4] Shengde Z, Shizhong Z and Jihua M 2000 Visualization Technology and Its Application in Geological Exploration *J. Geological Exploration Safety* **4** 23-29

IOP Conf. Series: Materials Science and Engineering **371** (2018) 012064 doi:10.1088/1757-899X/371/1/012064

# Research on Thematic Landscape Design

**Bing Gao**[1]

[1]Institute of Environmental Design ,Art and Design Department, SuiHua College, Heilongjiang, China

**Abstract.** Theme is the soul of landscape design, guiding and throughout the whole landscape design, the establishment of the theme is conducive to the formation of landscape structure and the combination of the optimization of the landscape elements. Theme is the soul of landscape design, and the concept is the basis of landscape design. The diversity of theme determines the diversity of landscape design, and different landscape has different theme characteristics. The establishment of the theme of landscape design is helpful to the layout of the landscape structure and the optimal combination of the various elements of the garden. Therefore, the theme should be clearly defined in the landscape design. At the same time, we must follow the design concept of garden landscape, which can play the real role of landscape architecture. In this paper, the theme of landscape design concepts and design themes were carefully studied, for reference only.

## 1. Preface

With the development of society and economy, people's living standard is gradually improved. After the material life has been satisfied, people have put forward higher requirements for the living environment. As an important part of urban planning, landscape architecture has played an important role in the improvement of urban environment. The landscape design should follow the corresponding design concept, clear the theme of the design, to play the true value of the landscape. It can be said that the theme of landscape design is the core of the landscape, the design concept is an important basis for landscape design, only on the study of landscape design concept and design theme, can play the role of landscape in city planning. People oriented, sustainable development as the criteria of ecological planning and design has become the development trend of landscape design, and ecological garden will be the only way for sustainable development of the city. Symonds said: "the landscape architect's lifelong goal and work is to help mankind, make people, buildings, communities, cities and their lives live in harmony with the earth[1]"." After living in high-rise buildings, as the car sound noisy, air pollution in the city, people yearning for nature and return to nature, and recover the original simplicity into fashion. With the global tide of ecology, people have to re-examine the landscape industry from a scientific point of view, and designers are beginning to link their mission with the whole earth ecosystem. The design no longer stops in the garden design narrow world or the picture idle talk, but starts to intervene in the more widespread landscape design domain. Respect for the ecological development of the site, the recycling of material and energy, the promotion of site self maintenance and sustainable treatment technology, specific to each design, embodies a strong ecological concept. In the design, the pursuit of ecology has been as important as the pursuit of function and form, and sometimes even beyond the latter two, occupying the primary position. Eco - ism has become the intrinsic and essential consideration of landscape architects, and it creates a landscape of sustainable development.

## 2. Landscape Design Concept

### *2.1. People Oriented*
In the design should be based on the "people-oriented" principle, in space design to take full account of people's multidimensional sense. The landscape in the leisure and entertainment facilities, such as kiosks, sofa, water platform and flower etc. are human centered design, both functional and beautiful, reflecting the requirements of green ecological modernization. Taking into account the climate characteristics of the region, and evaluating the environmental characteristics of the surrounding areas, the harmonious coexistence of human and natural environment can be realized.

### *2.2. Adjust Measures to Local Conditions*
Respecting the site, suiting measures to local conditions, seeking the design concept closely linked with the surrounding environment and forming the whole, has become the basic principle of modern landscape design. The role of landscape architects is not to deliberately innovate, but more to discover, to observe and understand the original features of the site with professional eyes, and to find out its positive aspects and guide them.

### *2.3. Pay Attention to Space*
The landscape is composed of two parts, one is the entity composed of some landscape elements, the other is the space made up of the entity. Entities are more likely to be noticed, and spaces tend to be ignored. Especially the current design method, often only pay attention to the scene of soft hard solid, solid scenery is relatively ignored, shape and extension of space, and the neighborhood space contact attention enough, the design method of various pile of scenery. Therefore, it is very important for us to pay attention to the design of spatial structure and landscape pattern, emphasize the design concept of space than entity, and design method for the visual space field.

### *2.4. Depth Mining and Honest Performance*
The first is the simplicity of the design method, which requires careful study and analysis of the design object, so as to seize the key factors, in order to avoid detours, with minimal change to achieve the maximum results. Two is the simplicity of expression, concise and general, with the least elements, scenery, the most important features of the landscape. The three is the simplicity of the design objectives, requires full understanding and compliance with the site context, texture, characteristics, minimize the man-made interference to the original landscape.

### *2.5. Pay Attention to Time*
The biggest difference between landscape design and architectural design is that the landscape changes with season and time, and it is alive, and it is in constant growth, movement and change. Therefore, the designer proposes to make the garden in motion a natural and lasting work. Landscape architects must carefully study the timeliness and timeliness factors, pay attention to landscape changes with time, in order to shape with time and can be updated, stable landscape.

## 3. Theme Of Landscape Design

### *3.1. Pay Attention to the Global Design*
The theme is induced in the planning, architecture and landscape design in 3 aspects, will play out the biggest value of project land potential, arouse people's emotional resonance, while the use of a new design concept to guide the overall maturity, Gangjumuzhang, given the central theme for all aspects of the subsequent, creating a series of connotation and value of landscape. The representative feature of art works on the whole is called style. Under the same theme, there may be many styles, and different themes may also be expressed in the same style. Based on the theme, the use of various methods of landscape art and technology, to make the theme of the artistic image is vivid, it is through the process of meaning by the operation, Italy Yu Wen, is also the financial situation of the entry, the process designer sustenance thoughts and feelings. Looking for the exact and suitable garden form for

the theme is called "Fu". The layout of garden art, the choice of the form of water, mountains, and the location of the outline is enough, with different functional division and connection, and quiet scenic landscape layout, main position are in the subject under the guidance of unified coordination arrangements. Garden landscape elements, landscape plants, buildings, roads, etc. the title is an important material theme in the form attached, is based on the framework of different types of patch. Sequence can continue, landscape cultural context and the original subject, theme landscape, undertake markers to strengthen the theme atmosphere, landscape node functions simple, volume small, unique style, artistic conception with small buildings or small art form is the theme of the punchline. Within the scope of the site, the use of the transformation elements to form the form related to the theme is one of the most intuitive expression of the theme on the planning level.

### *3.2. Design Theory Should be Innovative*

The theme is to carry out the design in 3 aspects of planning, architecture and landscape to bring the greatest value of the project into play and arouse people's sympathy. Under the innovative design concept, we can create a series of valuable landscapes. A theme can show different garden styles, and a style can also be expressed through different themes, which is the charm of theme creation. Once the theme is determined, the technical methods used to highlight the theme should be created according to the actual situation to achieve the image of the central image of the theme. In this process, designers integrate their feelings into a certain situation to support ideas, and find the appropriate form for the realization of the theme, which is called "Fu". The choice of garden layout and form should be based on landscape and contour selection. The arrangement of different artistic forms in the garden should be arranged in the theme.

The landscape elements of the landscape include mountains, water, buildings, roads, plants and so on. These themes are the form of subject attachment. The theme is expressed through these elements. In the group scene, the theme atmosphere can be strengthened by linking the culture with the original theme. In the landscape, some smaller buildings or artistic modeling can also fully reflect the theme of the main body. In the field of river to express the theme through the transformation of the combination and form. In order to form a kind of mountain and water fusion, the artistic form of the winding and winding mountain streams can be expressed by the ups and downs of the land, which not only makes people feel from the visual angle, but also causes psychological feelings. To determine the representative theme of landscape architecture is the main direction of the theme creation of landscape architecture. The designer should design landscape with humanistic characteristics and beautiful ecology according to local conditions.

## 4. Theme and culture in landscape planning and design

The Birthplace of Oriental Gardens

The traditional garden in China can be said to be the birthplace of Oriental Garden, more than five thousand years of historical and cultural heritage in the garden art has been greatly reflected. In the planning and design of the traditional landscape, influenced by the cultural background that history, "nature and nature" is the most basic principle. Under the influence of such thinking in the traditional garden in landscape planning and design of the landscape, vegetation, building organically, make full use of natural conditions, in the limited space within the natural landscape simulation, thus creating a "harmony" as the theme of the landscape, such as the Summer Palace, Beijing Hangzhou West Lake ten, Suzhou garden, Taiyuan Jinci etc.. At the same time, in the planning and design of the traditional garden, emphasizing the "garden at the beginning, meaning before", pay attention to traffic at the scene and the feeling and meaning in the matter, to Peter, to show the moral and spiritual beauty of people through the natural landscape. Not only that, the traditional garden on the creation of artistic conception, proposition into a quality suggestive of poetry or painting in the landscape garden, the tendency of Yu Qingxin elegant style, the artistic expression of the unique cultural connotation, which is the essence of Chinese traditional garden. For example, Songjukmae, collectively referred to as "pine, bamboo and plum blossom". The ancient "pine, peach apricot, spring pine, bamboo and plum blossom; a". The three plants are named after the strong vitality in the cold winter season. They are the symbol of noble personality in Chinese traditional culture, as well as the faithful friendship. The ancients often "Songjukmae symbol" as a metaphor for the beautiful things, open-minded, section, defying frost character and spirit. In landscape design, designers often put it into the landscape, making the landscape more far-reaching.

### *4.1. Humanization of Modern Landscape Architecture*

The current society emphasizes the "people-oriented", pay attention to the harmonious development of man and society, man and nature. Diversification: under the impetus of economic globalization, the political culture of various countries is also constantly blending, all kinds of new materials and technology for the modern landscape planning has injected new vitality. Nature: from the analysis of the current landscape construction, landscape planning should not be applied mechanically to the theme, but should be under the premise of respect for nature, the natural landscape is simulated and reproduced by means of abstract art, the landscape quality at the same time, emphasis on ecology and sustainable development, the expression of modern landscape the theme of landscape planning and culture closer to nature.

## 5. Modern theme Jingdezhen view design suggestions

First, to achieve the overall consideration in landscape planning and design, the need to always adhere to the principles of overall consideration, on the one hand, grasp the overall landscape design, the theme effectively contrast, on the other hand, the reasonable layout of the details, and to express the theme through the landscape culture. Therefore, the designers should fully consider various factors, into Chinese characteristics in the modern landscape design, at the same time to strengthen the in-depth study on landscape geographical and cultural background differences, etc., in satisfy the public aesthetic at the same time, enhance the garden landscape.

Second, the landscape design in the modern landscape planning and design for the theme, design personnel should be from the city's architectural style and the cultural history of the landscape design, determine the theme, to guide landscape planning and design. Before the design, we should do a good job of on-site inspection, analysis of the original architectural style of the garden, combined with the development history and cultural heritage of the city, lay a good foundation for the planning and design of the landscape. Not only that, the surrounding climate conditions, vegetation morphology, growth habits also need to be included in the analysis of the scope of consideration, determine the theme of landscape architecture, around the theme of landscape planning and design. Third, integrating into the ecological protection thought, the modern urban landscape is not only an important part of the urban landscape, but also the key link of the development of green ecological city. The influence of the concept of sustainable development, landscape designers not only need to pay attention to the theme of the garden and concise expression of culture, should also be prepared for influence on the ecological environment of city landscape system, integrating ecological protection thought in the design, as far as possible to reduce the field of plant graft, promote the recovery and development of city ecological environment.

## 6. Conclusions

Today, people pay close attention to the environment, the design of landscape architecture has become a hot social professional, and at the same time put forward higher requirements to designers. Designers of landscape architecture should not only have rich professional knowledge, but also have extensive knowledge, development ability and innovation ability. In the future landscape design should comply with the design concept of garden landscape, clear the embodiment of the theme of the landscape, and create an ecological, beautiful and practical landscape. In a word, it is the direction of the future landscape planning to determine the representative landscape theme and build a unique rich landscape on this basis. Landscape architects should design a humanized, novel and not impetuous, beautiful and practical, simple and meet the ecological requirements of the garden landscape.

## 7. Acknowledgements

Fund Project: Heilongjiang philosophy and social science research project (stage results); project name: red culture theme of ice and snow landscape design research; project number: 17YSE394

## 8. References

[1] Lin J 2012 On the theme and cultural analysis and research of landscape planning on science and technology innovation and application **4** 451-456

# Research on Ice and Snow Landscape Design

**Lei Huang**[1]

[1] Institute of Environmental Design, Art and design Department, Suihua College, Heilongjiang, China

**Abstract.** Snow and ice as the gift of nature, culture of the northern black culture at the same time, cultivate the northerners unique simplicity and fortitude character, ice sculpture is to create and the wisdom of the people in the north of the snow sublimation material. The design concept is the main landscape of ice and snow as the core, to highlight the strengthening of ice characteristics as the direction, the design method of the system as the main line, the landscape design of the function, combined with other studies, aesthetics, psychology, many factors of plants, demonstration and development about the existing design. Based on ten, from the design point of view to solve the problem of ice and snow landscape design. In this paper, starting from the ice and snow landscape design in Harbin, analysis of the ice snow landscape of interest, origin, and finally analyzed several ice and snow theme landscape design theory.

## 1. Introduction

The snow is a precious gift of nature to give Harbin of Northland ice city, also has unique northern snow landscape and soil culture, cultivate more north people honest and strong character. The North contact ice in the activity, the formation and development of the ice sledge, snowball, snowman in rich and colorful folk activities; the use of ice and snow in life, frozen food, goods and other ways to improve and enhance the people's daily life quality; watch the snow carving, crystal clear ice and snow are perfectly plastic represents the highest pursuit of northern art; snow bearing development, mining snow infinite value, form a set of ice and snow culture, art, education, tourism, ice snow and ice snow and a series of economic development strategy. Harbin International Ice and snow festival has become one of the four international ice and snow festivals, which is tied with the ice and snow Carnival in Quebec, Sapporo Snow Festival and Norway Olympic road skiing festival[1].

## 2. The rise of ice and snow landscape

Ice art originated in the northern folk activities, people, ice and ice cold warm snow making and decorated with festive atmosphere. In 1963, Harbin held the first fair will first use of natural ice as a landscape element for the landscape design for public viewing. Based in Harbin city in January 5, 1985 in the traditional winter tourism. the Ice Festival on the founder of the Harbin International Ice and Snow Festival, is the first snow activities for the content of the International Festival in the history of our country. The 2014 Harbin International Ice and Snow Festival after 30 years of rapid development of the planning and construction of a total of 13 snow theme square, 9 street, 17 points, a total of 39 snow landscape, foreign tourists more than 2000 people, has become an important carrier of cultural landscape in Harbin. The rise of ice and snow art is also the best reward for nature[2]. The origin of China North snow landscape of the city of North China belongs to ten "winter city in geographical conditions" (the cold city is a more general concept, according to the characteristics of city winter climate region defined by the long winter, harsh climate to adversely affect the city of city life. The climate is temperate climate Shihan. This area has four distinct seasons, seasonal changes of

IOP Conf. Series: Materials Science and Engineering **371** (2018) 012065 doi:10.1088/1757-899X/371/1/012065

the Ming Dynasty _ conspicuous long winter. The average winter temperatures below zero degrees Celsius, in the form of precipitation, snow water freezes at a short time. _ solar altitude angle is smaller. Such natural conditions make the ice of the landscape are richly endowed by nature development in city landscape, but affected by temperature, the single plant species, people go out the activities of the inconvenience also makes snow landscape ten more isolated position in the history of development, lack of coordination with other landscape Wu Su, and other seasons in the city landscape in the landscape is relatively disjointed. To create a better landscape of ice and snow, we must attach importance to the local geographical and climatic conditions, starting from the basic conditions, to adapt to the cold climate, unfavorable conditions for the favorable factors, pay attention to and mobilize other related factors, to better serve the city landscape image.

## 3. Origin analysis of ice and snow landscape in northern China

Chinese ancient society, the earth will multiply the Northeast Han, Sushen, Fuyu, Goguryeo and later Wu Ji, Khitan, Mongolia, Manchu and other ethnic groups, these ethnic groups have created the northeast culture and the characteristics of myths and legends, such as Shaman culture, Mongolian Maider female creation myths, "from these northeast first literary and historical records, we can deeply feel the ancestors in the power of nature under the unyielding spirit and tenacious survival consciousness, Yong ten pioneering, hard pioneer spirit and collective heroism". The harsh conditions in which the north had survived since ancient times also varied from the cold of the ten winter and the depression brought about by it. The northern culture influenced by the heroic struggle too much. In the long winter. depression in the environment, the north also formed a warm, cheerful personality, and Chinese southern culture is different. Sing praises of nature and scenery also do not have a flavour. After Fuyu has created the splendid culture of Koguryo period, the poem monk set on the creation of the "Yong master" Gushi: "back to the stone Zhisheng empty flat, more red. Only pull peaks, isolated white clouds." The boulder based scenery description, not only the performance of the poet with popular feeling - so, more give a person with cool open the vicissitudes of life beauty.

### *3.1. The vast natural beauty*

The northern terrain is plain, the terrain is vast, including the frontier area, including the desert culture and northeast culture. Wide area to the vast expanse, the boundless plain feeling, since ancient times, people in the struggle with nature, enjoying the vast natural landscape. Therefore, the northern aesthetic appreciation of large-scale, flat, broad beauty. In the north, winter is long, snow often remains, and the earth remains ten. The plants in the field retain branches in winter, which does not hinder people's sight, and white also expands the visual sense of human beings.

### *3.2. The majestic momentum of the United States*

The vast and cold north can always cast a heroic personality and strong. The vast world, the single scene in winter, the cold and dry climate make people feel more pure. Under this unique geographical scale, it always causes a sense of grandeur. Winter ice and snow have strong coverage, bring tremendous shock.

### *3.3. The hale and hearty lines of beauty*

Flat, broad sense of space and large scale space, coupled with the characteristics of northern people tough, more contact with the winter snow and ice making modern concise style is more popular, highlighting the tough line of beauty. People's clean white color has become a simple color tendency.

## 4. The design of ice and snow landscape

### *4.1. Ice*

The self luminous sculpture specially designed for the night uses the solidified carrier to contain the light of the changing light and creates the colorful and mysterious meaning. Using a combination of ice groups, to create a Crystal Palace like environment, so that people get a fantastic visual experience.

IOP Conf. Series: Materials Science and Engineering **371** (2018) 012065 doi:10.1088/1757-899X/371/1/012065

As a temporary artificial landscape in winter, it suddenly appears in people's accustomed living space, and brings fresh and refreshing spiritual experience to people.

### *4.2. Snow*

Snow is the unique gift of nature to cold city, for thousands of years, has been as a symbol of purity, showing romantic feelings. This beautiful image is a deep cultural implication. Snow is snow pure color and delicate texture of the condensation in the creation, with meticulous carving techniques, shaping the artistic image of every kind of change unpredictably. Because of the particularity of the material, the carving process costs less power, while the content of the molding is arbitrary. The central axis of the park landscape, "a scene of bustling activity long 180m" high 15m, with the amount of snow 17000m3, is the world's largest existing snow. The landscape with two head up the dragon, with a tiger, the other three majestic momentum majestic, true to life, is a rare snow plastic treasures. Of course, they will eventually melt and cease to exist, so they are more treasured in the limited period of existence. As a kind of temporal sculpture, it records the exquisite artistic temperament of the northerners.

### *4.3. Winter plants*

Because of the heavy snow, the details of the plants and woods are covered, but the overall layout of the garden is highlighted. At this time, the garden seems to carry out "map bottom conversion", the green outline is more clearly discernible, to create a concise and rich form of landscape architecture. The sudden drop in temperature, the water vapor condensation moment on the branches, so the tree became part of the winter. The far view hazy and illusory, almost did not see its crystal prescription exaggerated form, this particular, add a static aura for winter garden. Designers also clever use of the tree in winter, it is covered with colorful lighting, create an EVERYBRIGHT entrance square art effect day.

### *4.4. Architecture*

Different from the pure art, it should not only in the garden in the snow and ice landscape brilliance, but also need to assume some functional requirements such as diet, rest, indoor entertainment venues, to provide services for outdoor activities and visitors. Of course, the design can make use of the group combination of buildings to form small blocks, such as the northern snack street, folk custom village, etc., to add more strong northern culture to the ice and snow garden. The temporary landscape created by human beings aims to add a new and significant temporary theme to the environment by superimposing a new element with the ice and snow environment. Big world designers use a few rows of ice wall enclosure space, hanging celebrity painting and calligraphy, each straight to the theme, enhance the ice and snow culture charm. The park organizers also ordered the world's largest Korean drum with a diameter of nearly 2m, to reflect the rich exotic style. The amusement park in Harbin, the staff will have numerous red umbrella in a world of ice and snow on the trees in countless red ribbon winding walks, bring vitality to the quiet aura and ice environment. The design of these temporary landscape to offset the icy environment to a great extent bring cold monotonous feeling, become white body in the environment design of dotting pen sublimation.

### *4.5. Humantistic Landscape*

Static artificial landscape frozen once after the completion of a certain stability, the subsequent plastic space is limited, and the combination of dynamic cultural landscape can be in a fixed condition, create brilliant artistic effect, to expand the theme, prolong the life cycle for the park. Of course, the formation of these elements of ice and snow garden, need to be through different combinations, it may produce or cordial simplicity, or elegant, dignified, or light, elegant, or cheerful local space effect. The spatial effect of the whole, on the contrary, endows the functional features and figurative meanings of the elements. Combined with the specific layout method of the ice and snow world, several combinations of the summary of these landscape elements: axis combination of ice and snow.

*4.5.1 Snow world layout*

There is a whole garden axis obviously, it consists of the entrance door, the avenue of stars, like the Forbidden City Hall three buildings, two ice carving and Snow Dragon overall the whole park Friendship Square. This axis runs through the entire garden, clear and rigorous, magnificent.

*4.5.2 Contrast combination*

The use of icy creation classic architectural image is the tradition of previous world ice and snow. This year, it adopts the contrast of Chinese and Western architecture, which not only promotes national culture, but also highlights the park theme of "China Korea Friendship ice world". The use of light color contrast, sometimes can produce strong artistic effect.

*4.5.3 Progressive combination*

The climax of the theme environment is generated by the progressive and progressive environmental relations, which produce the contrast and gradual change. In the music square, this approach was adopted to lead people into the perfect combination of music and light.

*4.5.4 Freestyle combination*

All kinds of amusement items are various, and the area is different, so this combination is especially suitable for entertainment area. Through the formation of a free tour line, can make the tourists in different recreation experience with many twists and turns, avoid boring and monotonous feeling. Designers are not only reasonable in spatial organization, but also more skillful in dealing with many details. In order to emphasize the function of different partitions, the use of music style, prompting people to identify the characteristics of the region. For example, the lively and lively dance music is played in the recreation area, and the melodious folk music is changed in the Korean amorous feelings area. The use of the limited Dingbian structures.

## 5. Conclusions

Today, with the rapid development of modern society and economy, the concept of green design is the basic quality that designers should possess. The ice and snow landscape design mainly aims at the northern region, the long winter climate, uses the ice and snow material as the main element, integrates the ornamental, entertainment, sports, leisure as one of the integrated landscape design activities. Natural ice from frozen river ice, through artificial cutting to form a unified "ice cream"; natural snow quality soft, not as a long time to watch. Therefore, the snow material is mostly man-made snow, and the material itself is also fine particles atomized by water atomization. Ice and snow landscape materials are natural materials without any pollution, and are renewable resources, and less energy consumption is collected and transported. Snow and ice landscape itself is curing state of water, as a landscape display will have gasification sublimation phenomenon in the spring snow landscape will condense into water. The formation and melting process of the entire ice and snow landscape are all the three forms of water transformation and cycle development process. Green design needs to fully consider the energy and resource consumption in the design process, and strive to minimize consumption, but also an important part of urban planning development and landscape environment design.

## 6. Acknowledgements

Foundation project: Heilongjiang art science planning project (research results): Project name: Based on red culture theme snow landscape design research: Project number: 2017D012

## 7. References

[1] Zhu BW 2013 Experience and theme park landscape design on Harbin Institute of Technology **8** 451-456
[2] Zhang BX 2015 Study on waterscape design of northeast frigid University on Jilin University of architecture **14** 341-344

# A Safe Sharing Authentication Scheme for Marine Images on Cloud

**Yong-jun Li[1], Liang Zhou[2], Zhuo-ka Li[2]**

[1]Personnel department, Shanghai Ocean University, Shanghai 201306, China
[2]Network & Information Center， Shanghai Municipal Educational Examinations Authority, Shanghai 200433, China

**Abstract.** Nowadays, Cloud computing has become a very hot technology. with the emergence of cloud storage model, more and more users choose cloud to store the image data, while it brings challenges for image data security and usability due to the openness of cloud. An image authentication scheme is proposed to protect the ocean remote sensing image confidentiality on cloud, combining the Hash function and the threshold scheme, detecting the changes of the image data in sensitive regions and verifying the consistency for Encryption and Recovery images, The sensitive-region image can't be restored when the case that more than n-k out of n sub-secrets get lost or tampered happened. To avoid this, this paper divided the sensitive region image into blocks, and implemented secret sharing for each sub block, guarantee the lossless recovery of partial image and enhance the availability of image data. Through experiment contrast analysis, the algorithm can effectively prevent fraud during the process of secret image recovery, and it is more safe than the traditional methods.

## 1. Introduction

Cloud storage environment requires higher standard for data security. With the development of remote sensing technology, the data of marine remote sensing image is increased by TB per day[1]. The emergence of cloud storage, more and more users choose to immigrate remote sensing image data and its application to the cloud. In the military, marine remote sensing image data is useful, has an important strategic significance for the war, the enemy intelligence surveillance and target locking . Marine remote sensing images with large scale, sensitive, and other characteristics, the practical application of marine remote sensing image in the port, bay and islands of information is the core of image data, image data and sensitive area change. Therefore, the safety and accuracy of the remote sensing image in the sensitive area is of guiding significance to the marine environment monitoring, marine resources management, marine disaster early warning and rescue. However, the openness of the cloud storage environment makes remote sensing image data managers lose controlling over the security of the data. Therefore, how to ensure the confidentiality of sensitive area in the cloud storage environment is the key issue.

In cloud storage environment, data is usually stored in plaintext, lack of data privacy protection, therefore, the encryption of data is the key to protect the confidentiality of data. For the effective protection of the security sensitive area of marine remote sensing image, secret sharing method Hash verify the threshold based on the secret sharing scheme, first by Shamir and Blakley independently proposed threshold secret sharing scheme is the shared secret information generating sub secret, if and only if or more participants can recover the joint secret. With the explosive growth of the marine remote sensing data, the core area of the image encryption is becoming more and more important, and in the process of image restoration in the presence of false secret will result in poor image quality after data recovery, which greatly reduces the accuracy of the image data after recovery. Therefore, this paper uses hash function to detect whether the digital image tampering, a hash function is very sensitive to the data of each bit changes, small changes will make the generated hash values are changed, which can effectively prevent the recovery process, because the participants provide false secret and lead to the occurrence of recovery the secret of distortion, to ensure that the sensitive area of ocean remote sensing images before and after the restoration of data consistency and accuracy.

## 2. Related works

At present, the cloud environment often uses homomorphic encryption technology for data encryption, which uses vector and matrix of various calculation for encryption and decryption, so it is with low efficiency and high cost. KAmara and his fellows at the Microsoft research institute put forward a encryption storage framework[2] for the public cloud, which is less secure. Feng C.S[3] presents a decentralized cloud storage security framework that does not guarantee data integrity.

For image encryption, many studies has carried out at home and abroad, the early by statistical properties characterization of image hash of image is encrypted, though the method of LAN has stronger robustness, but for tiny disturbance is not sensitive to illegal, so it is lack of security[4]. Lou D.C[5] presents a method based on image block luminance histogram to extract the hash, the method adopts the brightness of each image block histogram feature, a malicious attacker can be generated with the same brightness histogram but image semantic content completely different image block image hash forged, so the method to extract the hash value of the security of the poor also[6]. NiuXia priests and others with gray levels image block as image characteristics to generate the hash, first of all, the image block, get the middle hash, and through the compression technology eventually to the hash of the image sequence. Colour histogram is used to characterize the image features, by capturing the image color histogram, the image region partition, and with the color to represent the image segments of each region. The two images were measured by measuring the overlap between the two images. Because in the process of getting color histogram, need to comprehend the image of the most representative of color, there is strong subjectivity, so the robustness is not strong. Swaminathan and his fellows have proposed an image Hashing algorithm based on the Fourier transform. Fourier transform, the algorithm first will be the image and under polar coordinates will be multiplied by the Fourier coefficient of every Angle amplitude random weight coefficient, produced by key to linear accumulative sum, fixed-length Hash sequence is obtained by quantitative compressed again. However, the attacker can fake the image by manipulating the phase information of the Fourier coefficient and keeping the image Hash value unchanged. Monga and his fellows introduced non - negative matrix factorization (NMF) into image Hashing, with non-negative constraints, NMF is different from such as singular value decomposition (SVD) traditional matrix dimension reduction techniques, such as in the literature, NMF is first used to decompose the randomly selected from the input image sub-block, NMF is on the second image decomposition and projection onto the random vector approximation matrix, obtained by series approximation matrix elements in the final Hash value, experiments show that this method has good robustness, but higher computational complexity[7].

In 1994, Naor and Shamir put forward a new image encryption technology - image secret sharing, the basic idea of this method is to put the secret image segmentation into sub image, any image and above the shadow image, can restore the original image. The flaw in the scheme is that the size of the shadow image is larger than the original secret image, and the image is distorted. The existing study of image encryption methods mainly solved the security problem of centralized storage environment, under the open cloud storage environment, according to the safety of the Marine remote sensing image problem, the above image encryption method for certain deficiencies. Considering marine remote sensing images with large quantities, large scale and high sensitive features, we put forward a cloud environment marine remote sensing image authentication scheme based on secret sharing, effectively protect the sensitive areas of marine remote sensing image data in cloud environment .

## 3. Marine remote sensing image authentication scheme Based on secret sharing

Marine remote sensing image is a kind of big data, and it is very important to protect the sensitive area of remote sensing image in the cloud storage environment. In this paper, the image of the sensitive region of the remote sensing image is segmented by the method of minimum external matrix. At the same time, the image is divided into blocks, and each sub image block is processed by secret sharing scheme, which effectively avoids the problem that the secret is lost or tampered with.

### *3.1. Issue description*

In order to ensure the security of remote sensing data in cloud environment, this paper proposes an image authentication scheme to improve the accuracy of remote sensing image data in the cloud environment. For the original remote sensing image set of the ocean, the image of the remote sensing image is restored by using the high security cloud environment, which is shown in figure1:

**Figure 1.** Storage and acquisition of Ocean remote sensing image on cloud

The method of hash function and threshold secret sharing is used to verify the consistency of the image before and after restoration. The related concepts of the cloud environment under the marine remote sensing image secret sharing scheme in the description and definition, data including the sensitive area of ocean remote sensing image sensitive sub image data sensitive area image data of marine remote sensing data set, marine remote sensing image sets, marine remote sensing image collection and recovery set.

In the authentication scheme, through the combination of hash function and threshold secret sharing method, the consistency of the image before and after restoration is verified to ensure the accuracy of the recovered secret image data. The gray value of the image is used as the image feature to calculate the hash value of the image in the sensitive area of the ocean remote sensing image. For sensitive area images (k, n) threshold secret sharing, using the method of projection matrix based on L.Bai image secret sharing, the value as a shared secret gray matrix, through the matrix transformation secret matrix generates a plurality of low dimension matrix, as a sub secret. The specific implementation scheme of the secret sharing scheme of the image of the sensitive area of the remote sensing image in the cloud environment is shown in Figure 2.

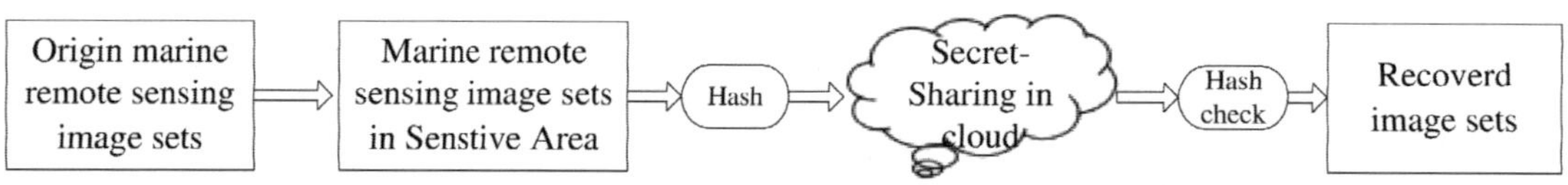


**Figure 2.** Ocean remote sensing image authentication scheme on cloud

In Figure 2, the encryption of the original ocean remote sensing image can be used to encrypt the remote sensing image data of the sensitive area after segmentation. The sensitive area of the marine remote sensing image secret sharing, first calculate the hash sequence image encryption, and the use of secret generating projection matrix method based on L.Bai, and the generated sub secret stored in the cloud environment. In the recovery phase image, gray image restoration is calculated again value hash sequence matrix, compared with hash sequence image encryption before if consistent, that image can be recovered, on the other hand, indicates that the restored image distortion. In order to restore the image, we need to choose the sub secret to restore the image.

Sensitive area image for ocean remote sensing image secret sharing, first on sensitive areas of images are divided into blocks, and every block is the sensitive area of sub images to calculate the hash sequence, then use the generated sub secret projection matrix method based on L.Bai, and the generated sub secret stored in the cloud environment. In the recovery phase of the sub image, the hash sequence of the gray value matrix of the restored image is calculated again.

### *3.2. Secret sharing algorithm for ocean remote sensing image*

The projection matrix of secret sharing scheme based on L.Bai is not validated on image quality after recovery, leading to the recovery of the image after the data accuracy is not high, this paper use hash function to verify the restored image based on, in order to ensure the accuracy of data recovery after image information. Image sub block after the sensitive area of cloud under the environment of the secret sharing of marine remote sensing image includes image sensitive area on marine remote sensing image secret sharing and the marine remote sensing image secret sharing, image sensitive area on marine remote sensing image secret sharing algorithm in 1, and the secret sharing scheme in sensitive area of image blocks after the marine remote sensing image the image of the implementation of algorithm . The Algorithm is as follows:

Input: ocean remote sensing images in sensitive area $T(id, S, q, A, H)$ 。

Output: a sensitive remote sensing image data set after restoration $C_{Z_1'}^{Z_\partial'}$ 。

Step 1：Extract Grey scale matrix $A$ of sensitive area images from ocean remote sensing images T；

Step 2：A hash sequence of sensitive region images of ocean remote sensing images is calculated, named $H(A)$；

Step 3：The sensitive areas are divided into molecular images；

Step 4: For $Z^{\partial}$ L.Bai threshold secret sharing, obtained n secret;

Step 5: Recover images base on sub secret;

Step 6: Merge sub images.

## 4. Experiment and analysis

In order to verify the validity of the proposed scheme for remote sensing image authentication based on secret sharing in cloud environment, this paper verifies the method in 2 aspects: correctness and security.

### *4.1. Accuracy analysis*

Experiments were conducted using Matlab 2010 to capture the sensitive region images of the Shanghai images obtained in July 2016. The image format is TIFF format, the size is 1024 * 1024, the image resolution of 30m. and using hash sequence sensitive area image of the original remote sensing image of the secret sharing scheme to restore the hash sequence matrix are compared, whether the secret image verification recovered correctly. In the process of image restoration in secret, the first choice to restore the image, after the restoration, by comparing the original image and the restore image found the secret image after the restoration of the damaged, which indicates that there exists at least one false sub secret. the restored image is not consistent with the original image, indicating that there is also false secret. Therefore, the existence of false secret leads to the poor quality of image restoration and greatly reduces the accuracy of image data recovery. In order to obtain the secret image lossless, we re-select the sub secret image, gray generation recovery value matrix, the hash sequence calculated by gray image before and after the restoration of the value of the hash sequence consistent matrix, verify the correctness of the proposed method.

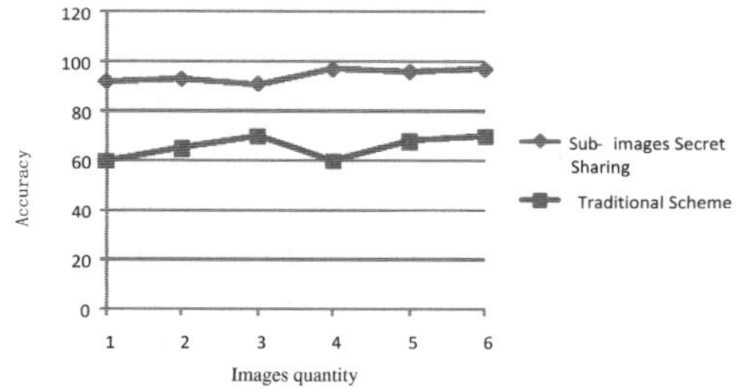


**Figure 3.** Comparison of data accuracy

### *4.2. Security analysis*

The experiment carried out 10 secret sharing of each scene image, and the vertical coordinate in Figure 3 was the ratio of the restoration of image accuracy, and the abscissa was the label of image. The experimental results show that the image data sharing the highest safety sensitive areas for marine remote sensing image block after sub images using this method, and the sensitive area of the traditional image secret sharing method to restore the image of the lowest safety data. Therefore, in the cloud environment, the secret sharing of the sub images of the sensitive images of the remote sensing images in the cloud environment has obvious advantages in ensuring the security of the data.

## 5. Conclusion

In the cloud storage environment, based on Shamir secret sharing system, combined with the sensitive property of the hash function of data changes, a threshold secret sharing authentication method is studied, which can prevent the fraud, through the sensitive area of Marine remote sensing image data is encrypted, greatly reduce the time needed for image encryption. Through analysis of experimental results, this method has high security, effectively detected the deceiver, avoid the restoration of image distortion, to protect the image data confidentiality. In addition, in order to avoid the threshold secret sharing scheme, due to some secret lost cause the whole image unrecoverable, we divide sensitive areas of the image into small pieces, then process each image block to ensure that each part of the image lossless recovered. Finally, we found the data security of the sub-images method is the most secure.

## 6. Reference

[1] Takabi H, Joshi J B D and Ahn G J 2010 Security and privacy challenges in cloud computing environments *J. IEEE Security & Privacy.* **6** 24-31
[2] Xue M, Xue W and Shu J 2015 A Secure Storage System Over Cloud Storage Environment *J. Chinese Journal of Computers.* **8** 31-32
[3] Feng C S, Qin Z G and Yuan D 2015 Techniques of Secure Storage for Cloud Data *J. Chinese Journal of Computers.* **15** 109-112
[4] Pang LJ, Pei QQ, Jiao LC and Wang YM 2008 An identity(ID)-based threshold multi-secret sharing scheme *J. Journal of Software.* **10** 2739-2745
[5] Lou D C and Liu J L 2000 Fault resilient and compression tolerant digital signature for image authentication *J. Consumer Electronics.* **46** 31-39
[6] Wang GL and Qing S 2005 Security Notes on Two Cheat-Proof Secret Sharing Schemes *J. Journal of Computer Research and Development.* **11** 1924-1927
[7] Monga V and Mihçak M K 2007 Robust and Secure Image Hashing via Non-Negative Matrix Factorizations *J. Information Forensics & Security IEEE Transactions on.* **3** 376-390.

IOP Conf. Series: Materials Science and Engineering **371** (2018) 012067 doi:10.1088/1757-899X/371/1/012067

# Study on the Practical Exploration and Realistic Dilemma of Sponge City Construction from the Perspective of Public Management Based on the Analysis of Research Data from Hubei Province

**Huayan Cui[1], Jie Gu[1]**

[1]School of Literature, Law and Economics, Wuhan University of Science and Technology, QingLing Street, Wuhan, China
943669662@qq.com,gujie0101@126.com

**Abstract.** In rainy season, urban flood frequently occurs in many cities of China. The sponge city construction is still filled with many problems that need resolving. In order to steadily push the sponge city construction forward. The research team carried out an research on the sponge city construction in Hubei Province, collected the relevant data and had an analysis of the research data of sponge city construction in the province. This paper summarizes the practical exploration and realistic dilemma of sponge city construction in Hubei Province, and the paper has a conclusion of the suggestions for the construction of sponge city in Hubei Province. Hope this research can provide reference for later sponge city construction.

## 1. Introduction

As an important part of the construction of ecological civilization in our country, sponge city construction is an important project of people's livelihood, which is a huge challenge in the process of public administration. "In many cities not only fails to make full use of rainwater resources, but also caused a very serious flood disaster [1]. As a result of the city's expansion, the proportion of impervious areas on the earth's surface continues to increase. Runoff coefficient is also growing"[2].The Province is known as the "province of thousands of Lake", the water is the core competitiveness of the future development. Promoting the construction of sponge city, solving the city problems including rain flood, dry disaster , "heat island effect" and other issues as soon as possible is the province strategic initiatives to seize the green development opportunities. "Sponge city is the specific expression of ecological civilization in urban rain water management and the Chinese wisdom that is in line with international advanced rainwater management concept" [1]. Therefore, it is of great significance to study the present situation of sponge city construction in the province.

## 2. Practical Exploration of Sponge City

### *2.1. Adhere to the Planning Lead*

The sponge city construction of the province adheres to the basic principle of planning to lead, first the planning, after the construction, which reflects the controlling and leading effect of planning. In September 2016, the Office of the provincial Government issued "The notice of accelerating the province's underground integrated corridors and sponge city construction". The notice stressed: make scientific planning and effectively enhance the Science, system and implementation of the planning. In

IOP Conf. Series: Materials Science and Engineering **371** (2018) 012067 doi:10.1088/1757-899X/371/1/012067

May2017, "the General Office of Hubei Provincial Government implementation views on the promotion of sponge city construction " issued.

### *2.2. Pay Attention to Ecological Priority*

"The sponge city construction pays attention to the basic principle of ecological priority" [2]. In the urban planning, the province makes the scientific delineation of the blue line and green line; in urban development and construction, changes the traditional ideas that only focus on the construction of drainage facilities, takes the systematic management model, makes the Urban topography, vegetation soil, wetland and water have an effect to the rainwater accumulation, penetration and purification, focuses on the protection to rivers, lakes, wetlands, ponds, ditches and other water-sensitive areas, strengthens the source control, gives priority to the use of natural drainage systems and low impact development facilities, makes the initial realization of the natural accumulation of rain, natural infiltration, natural purification and sustainable water cycle, and gradually improves the natural restoration of water ecosystem capacity, basically sustains the ecological function of the city.

### *2.3. Emphasize the Safety Importance*

The sponge city construction of the province emphasizes the basic principles of safety,regards the protection to people lives and property safety and social economy security as the starting point of all, takes the comprehensive use of engineering and non-engineering measures to improve the quality of low-impact development facilities and management level, eliminates security risks, enhance the capacity of disaster prevention and mitigation so the province is able to protect urban water security. In the construction procession of sponge city in the province, great importance is attached to drainage and flood work. According to the urban drainage and flood planning, make the construction plans. At the same time, on the basis of the previous drainage and flood prevention work, the province carries out inspection work including the drainage network clearance, drainage facilities conservation, flood point remediation and emergency response mechanism before the flood disaster, especially is strict with the implementation of flood Disaster reporting system.

### *2.4. Focus on the Local Conditions*

According to the local natural and geographical conditions, water characteristics, water resource status, rainfall law, water environment protection and flood control requirements, the Sponge city construction of the province adheres to the basic principles of local conditions, reasonably determines the objectives and indicators of development control by the way of low impact , scientifically plans and selects the sunken green space, planting grass ditch, rain wetlands, permeable pavement, multifunctional storage and other low-impact development facilities and its combined system.

### *2.5. Promote the Overall Construction*

The sponge city of the province promotes the overall construction, the local government carries out the overall planning and construction in consideration of both Urban and rural areas. In all types of construction projects, the development control objectives, indicators and technical requirements of low impact are Strictly implemented, having a whole construction. In January 2017, work conference system of underground integrated corridor, sponge city construction and urban black smelly water remediation was formally established in the province, marking the construction of sponge city in the province towards the overall construction track. In practice, the Province has also gradually accelerated the pace of co-ordination in the construction of sponge city, not only the co-ordination of new and old urban sponge city construction, but also the sponge-type community and public building sponge construction.

## 3. Realistic Dilemma of Sponge City

### *3.1. The Construction Concept is Relatively Backward*

"Unlike the conventional, engineering, inelastic gray infrastructure, the ecological infrastructure is a living system"[3].At present," high-speed construction on the ground, low-speed construction

IOP Conf. Series: Materials Science and Engineering **371** (2018) 012067 doi:10.1088/1757-899X/371/1/012067

underground" the phenomenon of urban construction is very common, the traditional thinking has not yet effectively changed. Sponge city construction management is a cross-system engineering involving many disciplines and fields, therefore, the relevant departments, industries, majors also have to change the construction concept, namely from a simple urban drainage to the economic, environmental, social awareness transition.

*3.2. System Design Still Needs Optimization*

Firstly, the top-level system design is not perfect. Sponge city construction involves planning, design, construction, operation, maintenance and other aspects, the success of the implementation depends on the top of the system design. However, some pilot cities did not develop a perfect top-level design. Secondly the relevant laws and regulations is not perfect. Sponge city construction is still in the new stage, the need for strong regulatory policy package is urgent, for example: planning laws and regulations, investment, financing and price policies of major construction projects, land development policies, economic levers policy and so on.

*3.3. Management Mechanism Needs to Be Improved*

At first, the experience of construction management is lacking. There is no mature domestic sponge city construction management system. Therefore, the construction of sponge city needs to accumulate management experience in exploration. Sponge city management needs the cooperation of lots of departments including planning, construction, municipal administration, housing management, garden, water, urban management, land, environmental protection. There is still a problem of convergence and integration between the existing administrative departments, and it is necessary to get rid of the shackles of the management mechanism. Secondly, Sponge city construction and management is dialectical unity. Due to the information lack of transfer process, the construction of the main body and the management of the main body is not uniform, the lack of effective information management platform and other reasons, the construction of sponge in the province pay more attention to the construction than management.

*3.4. Lack of Technical Support*

"Sponge philosophy embodies in the following three strategies: Consumption, deceleration and adaptation "[4].On one hand, the technical support system needs to be strengthened. In the construction of sponge city, we must establish a set of all-round and full-cycle technical system involving planning, construction, acceptance, operation, etc .However, due to the tight time, lack of technical reserves, the technical system of the province has not yet formed. On the other hand, there is a lack of technical strength. Sponge city construction requires a number of innovative talents with strong talents. However, in this area, professional and technical personnel reserves are relatively weak.

*3.5. Return on Investment and Financing is Difficult to Guarantee*

Sponge city construction has the typical characteristic, such as so many projects, scattered investment, long cycle, and sponge city construction projects are mostly public projects so there is no long-term sustainable income, return rate is difficult to guarantee, resulting in most investment units discouraged. "Sponge city will bring about a wide range of social welfare, but it requires huge investment" [5]. At present, most of the first batch of pilot civic construction funds still come from the state financial subsidies and local governmental direct investment, social capital injection is significantly insufficient.

## 4. Suggestions On the Implementation of Sponge City

*4.1. Cultivate Advanced Construction Concept*

"The essence of Sponge city is changing the concept of traditional urban construction, achieving the coordinated development of resources and the environment"[6].Ideas are the forerunner of action. All levels of governmental departments have to fully understand the importance and urgency of building a sponge city, and further enhance the sense of responsibility and mission, firmly establish the concept of new development and scientific achievements, actively create a social atmosphere of sponge city.

IOP Conf. Series: Materials Science and Engineering **371** (2018) 012067 doi:10.1088/1757-899X/371/1/012067

To this end, the construction of sponge city must Cultivate advanced construction concept, namely from the "rapid emission, the end governance" into "the source emission reduction, process control, system management, co-ordination", from "engineering water" into "ecological water control". Realize the organic combination among Gray and green ,ground and underground ,source and end, storage and row, return to urban natural circulation.

### *4.2. Improve the Laws and Regulations*

The construction management of sponge city needs to be combined with the policy and law support, actively carries out research on water ecological laws and policies, conscientiously sums up the practical experience, builds a more complete sponge city law and policy system. The government should issue guidance, formulate policies and regulations, establish a sponge city planning and implementation legal system based on the whole life cycle. The legislative department should clean up all kinds of wading laws and regulations, establish a unified wading legal system, provide legal Guarantee for sponge city construction. At the same time, the government should also develop supporting policies and improve the policy standard system.

### *4.3. Optimize the Overall Management Mechanism*

The city people's governments should set up special leading organizations and working groups, implement the working responsibilities, optimize the working mechanism, have a clear sponge city planning and construction of the leading sector and the relevant departments and units, comprehensively carry out management work. On the basis of a clear division of responsibilities, the relevant departments should form a coordinated linkage mechanism, according to work objectives from the leadership team , promote the phased work, further optimize the approval process of the sponge city construction , strengthen the department linkage, explore online approval, and effectively improve the efficiency of approval.

### *4.4. Consolidate Technology Support*

Relying on the Wuhan University sponge city construction research center, establish sponge city technology demonstration base. Improve the sponge city planning and technical system of implementation. Establish the intelligent sponge information management platform, construct the intelligent sponge city. To build a planning and construction management integrated information platform of sponge city, The government should pay attention to the Internet + technology in rain flood monitoring applications. And, The government should use large data analysis and cloud computing technology to achieve intelligent control of urban water systems and fine management, so that the city have fast, intelligent, flexible response to the water problem. More importantly, strengthen technical personnel protection, build expert service platform. Sponge city construction involves a wide range. The government should reserve technical personnel from investment, design, construction to operation, management and other aspects.

### *4.5. Innovative Investment and Financing Mechanism*

Governmental economic support for the future large-scale sponge city construction is more like an investment-oriented, which can only play a leverage. To really ease the funding bottleneck, innovation of investment and financing is the only way. Therefore, the construction of sponge cit should adhere to the principle of governmental guidance and social participation. Play the decisive role of market allocation of resources and the guiding role of governmental regulation, increase political support, create a favorable environment for development. And actively promote the governmental and social cooperation in capital (PPP), franchise and other models to attract social capital to participate extensively in sponge city construction.

### *4.6. Carry Out Publicity and Training*

Sponge city concept is deeply rooted is the social foundation of sponge city construction. The sponge city construction has formed a number of demonstration projects which can be promoted and replicated in the construction of future. When the experience matures, the government should have a

IOP Conf. Series: Materials Science and Engineering **371** (2018) 012067 doi:10.1088/1757-899X/371/1/012067

timely publicity and effective promotion, which requires governmental leading, social participation, comprehensive publicity and training, so that the concept of sponge city can become the consensus of the whole society.

## 5. Conclusions

As a major measure of governmental public management, sponge city construction is still in the exploratory stage. There are some difficulties in concept, system, management, technology, investment and financing. The sponge city construction process in the future, the society, government and academic and research institutions must continue to explore and summarize the practical experience so as to ensure the implementation of sponge city construction can orderly advance.

## 6. References

[1] Yang Y F 2017 Reflection on the sponge city construction of Xiamen *J. China Water & Wastewater.* **2** 27-30
[2] Zhang L W 2015 The overview of sponge city construction *J. Water & Wastewater Engineering.* **6** 1-7
[3] Yu K J 2017 Discussion about "sponge city" and "sponge land" *J. Architectural Design Management.* **2** 5-6
[4] Yu K J 2015 Three Key Strategies to Achieve A Sponge City:Retention, Slow Down and Adaptation *J. South Architecture.* **3** 4-7
[5] Xia S Z 2016 Sponge City *J. Chinese Public Administration.* **2** 160
[6] Qiu B X 2015 The Connotation, Way and Prospect of Sponge city(LID) *J. Water Supply and Sewerage.* **3** 1-7